AF566999

DRYLAND AGRICULTURE

AGRONOMY

A Series of Monographs

The American Society of Agronomy and Academic Press published the first six books in this series. The General Editor of Monographs 1 to 6 was A. G. Norman. They are available through Academic Press, Inc., 111 Fifth Avenue, New York, NY 10003.

1. C. EDMUND MARSHALL: The Colloid Chemical of the Silicate Minerals, 1949
2. BYRON T. SHAW, *Editor:* Soil Physical Conditions and Plant Growth, 1952
3. K. D. JACOB: Fertilizer Technology and Resources in the United States, 1953
4. W. H. PIERRE and A. G. NORMAN, *Editors:* Soil and Fertilizer Phosphate in Crop Nutrition, 1953
5. GEORGE F. SPRAGUE, *Editor:* Corn and Corn Improvement, 1955
6. J. LEVITT: The Hardiness of Plants, 1956

The Monographs published since 1957 are available from the American Society of Agronomy, 677 S. Segoe Road, Madison, WI 53711.

7. JAMES N. LUTHIN, *Editor:* Drainage of Agricultural Lands, 1957 — *General Editor,* D. E. Gregg
8. FRANKLIN A. COFFMAN, *Editor:* Oats and Oat Improvement — *Managing Editor,* H. L. Hamilton
9. C. A. BLACK, *Editor-in-Chief,* and D. D. EVANS, J. L. WHITE, L. E. ENSMINGER, and F. E. CLARK, *Associate Editors:* Methods of Soil Analysis, 1965.
   Part 1—Physical and Mineralogical Properties, Including Statistics of Measurement and Sampling
   A. L. PAGE, *Editor:* Methods of Soil Analysis, 1982
   Part 2—Chemical and Microbiological Properties, Second Edition — *Managing Editor,* R. C. Dinauer
10. W. V. BARTHOLOMEW and F. E. CLARK, *Editors:* Soil Nitrogen, 1965
    (Out of print; replaced by no. 22) — *Managing Editor,* H. L. Hamilton
11. R. M. HAGAN, H. R. HAISE, and T. W. EDMINSTER, *Editors:* Irrigation of Agricultural Lands, 1967 — *Managing Editor,* R. C. Dinauer
12. R. W. PEARSON and FRED ADAMS, *Editors:* Soil Acidity and Liming, 1967 — *Managing Editor,* R. C. Dinauer
13. K. S. QUISENBERRY and L. P. REITZ, *Editors:* Wheat and Wheat Improvement, 1967 — *Managing Editor,* H. L. Hamilton
14. A. A. HANSON and F. V. JUSKA, *Editors:* Turfgrass Science, 1969 — *Managing Editor,* H. L. Hamilton
15. CLARENCE H. HANSON, *Editor:* Alfalfa Science and Technology, 1972 — *Managing Editor,* H. L. Hamilton
16. B. E. CALDWELL, *Editor:* Soybeans: Improvement, Production, and Use, 1973 — *Managing Editor,* H. L. Hamilton
17. JAN VAN SCHILFGAARDE, *Editor:* Drainage for Agriculture, 1974 — *Managing Editor,* R. C. Dinauer
18. GEORGE F. SPRAGUE, *Editor:* Corn and Corn Improvement, 1977 — *Managing Editor,* D. A. Fuccillo
19. JACK F. CARTER, *Editor:* Sunflower Science and Technology, 1978 — *Managing Editor,* D. A. Fuccillo
20. ROBERT C. BUCKNER and L. P. BUSH, *Editors:* Tall Fescue, 1979 — *Managing Editor,* D. A. Fuccillo
21. M. T. BEATTY, G. W. PETERSEN, and L. D. SWINDALE, *Editors:* Planning the Uses and Management of Land, 1979 — *Managing Editor,* R. C. Dinauer
22. F. J. STEVENSON, *Editor:* Nitrogen in Agricultural Soils, 1982 — *Managing Editor,* R. C. Dinauer
23. H. E. DREGNE and W. O. WILLIS, *Editors:* Dryland Agriculture, 1983 — *Managing Editor,* D. A. Fuccillo

DRYLAND AGRICULTURE

Editors

H. E. Dregne and W. O. Willis

Managing Editor
DOMENIC A. FUCCILLO

Assistant Editor
KRISTINE E. GATES

Editor-in-Chief ASA Publications
DWAYNE R. BUXTON

Number 23 in the series
AGRONOMY

American Society of Agronomy, Inc., Crop Science Society of America, Inc.,
Soil Science Society of America, Inc., Publishers
Madison, Wisconsin, USA
1983

American Society of Agronomy, Inc.
Crop Science Society of America, Inc.
Soil Science Society of America, Inc.
677 South Segoe Road, Madison, Wisconsin 53711 USA

Library of Congress Cataloging in Publication Data

Dryland agriculture

(Agronomy; no. 23)
Bibliography
Includes index.
Contents: Dryland agriculture

IV. Series.
83-71807
ISBN 0-89118-075-3

Printed in the United States of America

CONTENTS

FOREWORD

A significant portion of the world's production of basic crops such as wheat, grain sorghum, and millet as well as lesser production of a number of other important crops occurs on lands, commonly known as drylands, which use only rainfall or stored rain from a previous fallow year. In North America the major dryland farming is found on prairie land located in the provinces of Manitoba, Saskatchewan, and Alberta in Canada and on drylands of the central and western USA located west of the Mississippi River. Success of dryland agriculture depends primarily on the weather, which can be highly variable from year to year. But, improved tillage practices, better adapted cultivars, better management techniques, and other advances in soil and water conservation methods that have been achieved through research have added considerable stability to dryland production.

This monograph, one in a series first published by Academic Press beginning in 1949 and by the American Society of Agronomy from 1957 through the 18th monograph in 1977 when the Crop Science and Soil Science societies became copublishers, is the first in the series dealing with the subject of dryland agriculture. The monographs represent an important and continuing effort by the officers and 11 700 members of the societies, who are located in more than 100 countries, to provide needed worldwide information on advances in research on crops and soils.

The authors and editors are among the outstanding scientists engaged in dryland research, and the information is presented to persons interested in dryland farming everywhere. On behalf of the associated societies we congratulate and extend thanks to the authors, editors, reviewers, and the Headquarters staff for the production of this monograph.

C. F. Eno, *President*
American Society of Agronomy

W. L. Colville, *President*
Crop Science Society of America

W. H. Gardner, *President*
Soil Science Society of America

PREFACE

Dryland agriculture is a dynamic and highly complex system in which the major limitation on food and fiber production is a deficiency of water. At best, dryland farming is a risky enterprise, with the possibility of drought, hail, high winds, and destructive storms providing an ever-present threat. What has been called the first great defeat of the American farmer came about in the late 1800's when a protracted drought forced abandonment of countless farms in the Western Great Plains. Since then, farmer experience and scientific research have reduced the risk, but the vagaries of weather still count heavily among the factors that control crop yields.

Crop production in the semiarid and subhumid drylands of the world is dominated by three crops: wheat, grain sorghum, and millet. Corn, cotton, peanut, pea, and bean are other important crops in some regions. In Canada and the USA, wheat is by all odds the major crop, followed by grain sorghum and cotton. Over 60% of the wheat involved in international trade comes from the drylands of North America, as does much of the world's grain sorghum exports. Climate and soil differences within and between dryland agriculture regions of North America make it necessary to employ different crop management practices in different regions. Temperature usually varies from north to south and precipitation from east to west. Soil differences follow climatic differences with many local exceptions. For those reasons, the drylands of North America are divided into crop production regions in this monograph for discussions of cropping, soil conservation, and water conservation practices. The principles that underlie the practices provide the common thread that unites the science and art of dryland farming. Success depends on adapting the principles to the particular environment.

The monograph is intended to set forth the principles and the best practices currently known of dryland agriculture in North America. It represents a distillation of the knowledge of the leading authorities in the field from the federal governments, universities, and private industry. Extrapolation of North American practices to other dryland farming regions of the world cannot be done without modifications to fit local conditions, but the practices can serve as useful guides to what may be successful in similar environments elsewhere. Certainly, the principles set forth are applicable anywhere.

Special thanks are due the other members of the Editorial Committee who guided preparation of the monograph: Drs. Glen H. Cannell, Jerry D. Eastin, and T. L. Jackson. We also express our deep appreciation to the authors for the forbearance they displayed in responding to requests and in tolerating delays and to the editorial staff of the American Society of Agronomy for their help in editing the manuscript. Common names of most plants are used throughout the monograph, with scientific names provided in a glossary later in the book.

W. O. Willis, *Editor*
Fort Collins, Colorado

H. E. Dregne, *Editor*
Lubbock, Texas

CONTRIBUTORS

Raymond R. Allmaras	Soil Scientist, Columbia Plateau Conservation Research Center, Agricultural Research Service, U.S. Department of Agriculture, Pendleton, Oregon
Armand Bauer	Soil Scientist, Northern Great Plains Research Laboratory, Agricultural Research Service, U.S. Department of Agriculture, Mandan, North Dakota
Alfred L. Black	Soil Scientist, Northern Great Plains Research Laboratory, Agricultural Research Service, U.S. Department of Agriculture, Mandan, North Dakota
Floyd E. Bolton	Associate Professor, Department of Crop Science, Oregon State University, Corvallis, Oregon
Earl Burnett	Soil Scientist, Grassland, Soil, and Water Research Laboratory, Agricultural Research Service, U.S. Department of Agriculture, Temple, Texas
Glen H. Cannell	Soil Physicist, Department of Soil and Environmental Sciences, University of California, Riverside, California
Norris E. Daniels	Associate Professor, Texas Agricultural Experiment Station, Bushland, Texas; now retired
Eeltje de Jong	Professor, Department of Soil Science, University of Saskatchewan, Saskatoon, Saskatchewan, Canada
T. E. Dickson	Associate Agronomist, DeKalb-Pfizer Genetics, Inc., DeKalb, Illinois
Ben Doupnik, Jr.	Professor, University of Nebraska, South Central Station, Clay Center, Nebraska
H. E. Dregne	Horn Distinguished Professor of Soil Science, Department of Plant and Soil Science, Texas Tech University, Lubbock, Texas
Jerry D. Eastin	Crop Physiologist, Department of Agronomy, University of Nebraska, Lincoln, Nebraska
Charles R. Fenster	Professor, University of Nebraska, Panhandle Experiment Station, Scottsbluff, Nebraska; now Professor Emeritus
George R. Foster	Research Hydraulic Engineer, U.S. Soil Erosion Laboratory, Agricultural Research Service, U.S. Department of Agriculture, West Lafayette, Indiana
Richard A. Frederiksen	Professor, Department of Plant Science, Texas A&M University, College Station, Texas
C. Kerry Gee	Agricultural Economist, National Economics Division, U.S. Department of Agriculture, Fort Collins, Colorado

Bentley W. Greb	Soil Scientist, Central Great Plains Research Station, Agricultural Research Service, U.S. Department of Agriculture, Akron, Ohio
Lawrence J. Hagen	Agricultural Engineer, Wind Erosion Research Unit, Agricultural Research Service, U.S. Department of Agriculture, Manhattan, Kansas
Ardell D. Halvorson	Supervisory Soil Scientist, Northern Plains Soil and Water Research Center, Agricultural Research Service, U.S. Department of Agriculture, Sidney, Montana
R. J. Hanks	Professor, Department of Soils and Biometeorology, Utah State University, Logan, Utah
Jerry L. Hatfield	Biometeorologist, Department of Land, Air, and Water Resources, University of California, Davis, California
G. O. Hinze	Professor, Department of Agronomy, Colorado State University, Akron, Colorado; now Agronomist, Kelley Bean Company, Inc., Scottsbluff, Nebraska
Robert F. Holt	Soil Scientist, North Central Soil Conservation Research Laboratory, Agricultural Research Service, U.S. Department of Agriculture, Morris, Minnesota; now retired
Thomas L. Jackson	Professor, Department of Soil Science, Oregon State University, Corvallis, Oregon
David R. Johnson	Professor, Department of Sociology, University of Nebraska, Lincoln, Nebraska
Wendell C. Johnson	Soil Scientist, Conservation and Production Research Laboratory, Agricultural Research Service, U.S. Department of Agriculture, Bushland, Texas; now retired
W. Earl Johnson	Sessional Lecturer, University of Saskatchewan, Saskatoon, and Agrology Consultant, Regina, Saskatchewan, Canada
Ordie R. Jones	Soil Scientist, Conservation and Production Research Laboratory, Agricultural Research Service, U.S. Department of Agriculture, Bushland, Texas
Rex J. Kartchner	Research Animal Scientist, Livestock and Range Research Station, Agricultural Research Service, U.S. Department of Agriculture, Miles City, Montana
Hans A. Krauss	Agronomist, Soil Conservation Service, U.S. Department of Agriculture, Spokane, Washington
Ronald D. Krenz	Agricultural Economist, Economic Research Service, U.S. Department of Agriculture, Stillwater, Oklahoma
D. R. Krieg	Professor, Department of Plant and Soil Science, Texas Tech University, Lubbock, Texas
M. J. Lindstrom	Soil Scientist, North Central Soil Conservation Research Laboratory, Agricultural Research Service, U.S. Department of Agriculture, Morris, Minnesota

Ralph E. Luebs Soil Scientist, Agricultural Research Service, U.S. Department of Agriculture, University of California, Riverside, California; now Consulting Agronomist, Lakewood, Colorado

Leon Lyles Research Leader, Wind Erosion Research Unit, Agricultural Research Service, U.S. Department of Agriculture, Manhattan, Kansas

A. Bruce Maunder Vice President, Agronomic Research, DeKalb-Pfizer Genetics, Inc., Lubbock, Texas

Donald K. McCool Agricultural Engineer, Agricultural Research Service, U.S. Department of Agriculture, Washington State University, Pullman, Washington

Robert I. Papendick Soil Scientist, Agricultural Research Service, U.S. Department of Agriculture, Washington State University, Pullman, Washington

Robert E. Ramig Soil Scientist, Columbia Plateau Conservation Research Center, Agricultural Research Service, U.S. Department of Agriculture, Pendleton, Oregon

Kenneth G. Renard Research Hydraulic Engineer, Southwest Rangeland Watershed Research Center, Agricultural Research Service, U.S. Department of Agriculture, Tucson, Arizona

Clarence W. Richardson Agricultural Engineer, Grassland, Soil, and Water Research Laboratory, Agricultural Research Service, U.S. Department of Agriculture, Temple, Texas

Francis H. Siddoway Supervisory Soil Scientist, Agricultural Research Service, U.S. Department of Agriculture, Sidney, Montana; now Consultant

Edward L. Skidmore Soil Scientist, Wind Erosion Research Unit, Agricultural Research Service, U.S. Department of Agriculture, Manhattan, Kansas

Darryl E. Smika Soil Scientist, Central Great Plains Research Station, Agricultural Research Service, U.S. Department of Agriculture, Akron, Colorado

Harold Steppuhn Research Scientist, Division of Hydrology, University of Saskatchewan, Saskatoon; now Research Scientist, Research Branch, Canada Agriculture, Canada Agriculture Research Station, Swift Current, Saskatchewan, Canada

Billy B. Tucker Extension Agronomist, Cooperative Extension Service, Department of Agronomy, Oklahoma State University, Stillwater, Oklahoma

Paul W. Unger Soil Scientist, Conservation and Production Research Laboratory, Agricultural Research Service, U.S. Department of Agriculture, Bushland, Texas

C. H. M. van Bavel Professor, Department of Soil and Crop Sciences, Texas A&M University, College Station, Texas

Odell L. Walker Professor, Department of Agricultural Economics, Oklahoma State University, Stillwater, Oklahoma

Lynn K. White — Associate Professor, Department of Sociology, University of Nebraska, Lincoln, Nebraska

Allen F. Wiese — Weed Scientist, Texas Agricultural Experiment Station, Bushland, Texas

J. Ross Wight — Range Scientist, Northwest Watershed Research Center, Agricultural Research Service, U.S. Department of Agriculture, Boise, Idaho

J. Allen Williams, Jr. — Professor, Department of Sociology, University of Nebraska, Lincoln, Nebraska

Wayne O. Willis — Soil Scientist, Agricultural Engineering Research Center, Agricultural Research Service, U.S. Department of Agriculture, Fort Collins, Colorado

Robert A. Young — Agricultural Engineer, North Central Soil Conservation Research Laboratory, Agricultural Research Service, U.S. Department of Agriculture, Morris, Minnesota.

SECTION A

INTRODUCTION

1 Regional Setting

1

G. H. Cannell
University of California
Riverside, California

H. E. Dregne
Texas Tech University
Lubbock, Texas

Regional Setting

Dryland agriculture is practiced in various parts of the world, but the North American region covering the Canadian Prairies and the Great Plains of the USA, the Pacific Northwest and Southwest of the USA, and parts of the intermountain areas is a classic example of a great food producing area. Gras (1946) suggests that historically six general types of farming developed in the USA. The Gras description generally fits the movement of settlers from east to west across the continent. Gras lists these farm types in terms of chronological development as woodland, prairie, ranch, irrigated, dryland, and leftover farming. Five of the six farming types are found west of the 95th meridian. In Canada, a similar east-west progression occurred.

Woodland farming began with the European colonization of the forested region along the eastern seaboard of North America. It spread to the west until it encountered the tall grass prairies of the northern Mississippi Valley and Manitoba, from Illinois to the mid-grass and short grass prairies typical in the western reaches of the Dakotas, Nebraska, Kansas, and Saskatchewan. Farther west were the ranch lands in the drier regions, among which irrigated farming was begun first by the settlers in Utah in about 1849. It was not generally perceived by farmers until about 1880 that grain crops would be successfully grown—when conditions were favorable—in some of the areas previously considered suitable only for ranching. The delay in recognizing cropping possibilities in the Great Plains was due partly to the belief that the "great American desert," as the plains were shown on early maps, was akin to the Sahara, a concept that persisted in school geographies and atlases as late as the 1850's (Hargreaves, 1957).

The conception of the Great Plains as a desert delayed the advance of farmers into the region. Nevertheless missionaries, fur traders, and Army post personnel established vegetable gardens in the settlements. Around

Dryland Agriculture—Agronomy Monograph no. 23.

such centers small farming units emerged, and oat, wheat, corn, barley, rye, and vegetables were successfully grown. Hargreaves (1957) stated that in the 1860's farmers in the Red River district of North Dakota appeared to have developed an early maturing wheat and that the enterprising farmers recognized the value of summer fallow at least every fifth or sixth year. The development of new agricultural practices, railroad transportation, decreases in Indian forays, use of new farming implements, land policy changes such as the Homestead Act of 1862, and subsequent land policy legislation to change the acreage to fit the land's capabilities led to a rapid increase of dryland farming. The Morrill Land Grant College Act of 1862 and establishment of the U.S. Department of Agriculture (USDA) in the same year provided a basis for scientific study of farming. These measures were greatly enhanced by the enactment of the Hatch Experiment Station Act of 1887, whereby each state was provided with one agricultural experiment station. By 1900, much of the dry farm practice was well established, and by 1910 the term "dry farming" became a common term.

One of the persons most responsible for widely publicizing the dry farming concepts was Hardy Webster Campbell. A detailed description of his background and personal leadership in the dry farming propaganda is given by Hargreaves (1957). Campbell (1907) published *Campbell's 1907 Soil Culture Manual* (excerpts are found in Hargreaves, 1957, and Rasmussen, 1975b), which expressed his recommended practices for the Northern Great Plains. This system was eventually associated with his name. Campbell drew on the conclusions of King (1895) and cited the work of Hilgard (1892), Whitney (1895), and other scientists of the period for authoritative support in discussions of water movement in the soil and the efficacy of mulch in water conservation.

Various definitions have been used to delineate the dry farming areas. Hargreaves (1957) mentioned that dry farming may be generally defined as agriculture without irrigation in regions of scanty precipitation. Higbee (1958) states that dry farming is an agriculture dependent on rain and snow; it generally implies farming carried on under the handicap of low rainfall. Duley and Coyle (1955) state that the annual precipitation is generally low, about 200 to 500 mm. The use of total rainfall as a general definition is misleading, because the evaporative index varies considerably within the North American region. A given amount of precipitation in the northern states of the semiarid region may be sufficient for crop production, but in the southwestern states where the average temperature is much higher and the growing season is longer, the same amount would be inadequate.

Dry farming areas of North America are clustered within six geographical regions lying in the 17 western states of the USA and the three prairie provinces of Canada (Fig. 1–1). The USA part of the Great Plains has three subdivisions (Southern, Central, and Northern), and the Canadian part of the North American Great Plains is called the Canadian Prairies.

For purposes of discussion in this monograph, a second type of regionalization has been used in the chapters on soil conservation (Chapt. 10 to 19) based on land resource regions for the USA and agroclimatic regions for the Canadian Prairies (Fig. 1–2; Tables 1–1 and 1–2). The region defini-

Fig. 1-1. Dryland agriculture regions, USA and Canada.

tions are different for the USA and Canada, and consequently, the regional boundaries across the international border do not always match.

The land resource region system of classification in the USA (Austin, 1965) was designed to organize information about land primarily as a resource for agriculture. The regions of Fig. 1-2 are made up of major land resource areas, which are in turn made up of land resource units. Land resource units are "geographic areas of land, usually several thousand acres (1 acre = 0.4 ha) in extent, that are characterized by particular patterns of soil (including slope and erosion), climate, water resources, land use, and type of farming" (Austin, 1965). This level of characterization bears close similarity to that of general soil maps. A soil mapping unit is "a portion of the landscape that has similar characteristics and qualities whose limits are fixed by precise definitions" (Klingebiel and Montgomery, 1973). This definition indicates that the level and detail of description are similar for the soil mapping unit and the land resource unit. The land resource unit scale, included in the discussion of the water or wind erosion control planning guides in a later chapter, provides detailed supplemental information about what practices may be necessary and successful for soil erosion control.

Major land resource areas are made up of geographically contiguous and similar land resource units; land resource regions are made of geographically contiguous and similar major land resource areas. The regional grouping generalizes information at the land resource unit level but preserves grouping of climatic, topographic, and major crop interactions for

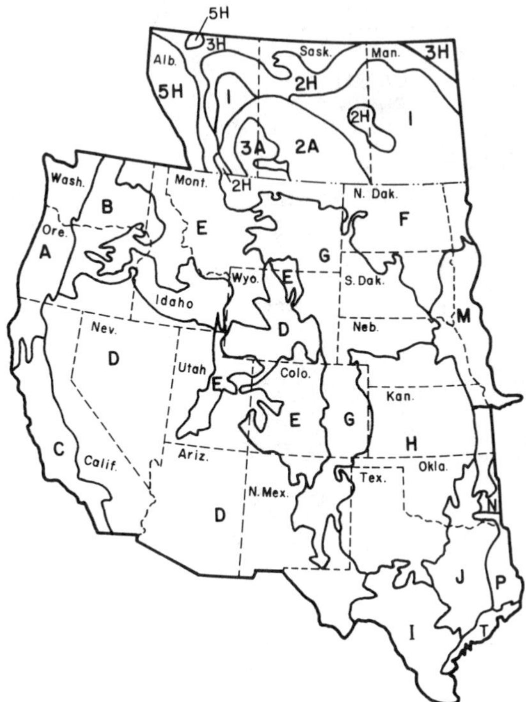

Fig. 1–2. Major land resource regions (USA) and agroclimatic subregions (Canada).

areas as large as the USA portion of the North American dryland region. In Canada, agroclimatic subregions approximate, to some degree, U.S. land resource regions. The climatic criteria for Canadian subregions are given in Table 18–1.

For this monograph, the major dryland farming areas are those shown in Fig. 1–3. Many different opinions exist about the eastern limit of the dryland region. Various persons contend that the north-south boundary should follow the 98th meridian, the 100th meridian, the eastern boundary of the mid-grass prairies, the line marking the Thornthwaite (1931) precipitation-effectiveness (P-E) index of 16, the separation of pedocal soils from pedalfer soils, or some other criterion. For our purpose, dryland agriculture refers to the type of agriculture requiring cropping practices that include adaptations to soil water deficiency during the growing season. These areas are in the semiarid and subhumid climatic zones shown on the UNESCO map of the arid regions of the world (UNESCO, 1979).

1–1 REGIONAL CLIMATES

Average annual precipitation for the 17 states and three provinces is shown in Fig. 1–4. Precipitation distribution patterns during the year may vary from the continental type in the Great Plains on the east to the mediterranean type along the west coast. Variations on those two patterns are

Table 1-1. Major land resource regions of USA.

Designation	Full name of region	Abbreviated name of region
A	Northwestern Forest, Forage, and Specialty Crop	Northwestern Forest
B	Northwestern Wheat and Range	Northwestern Wheat
C	California Subtropical Fruit, Truck, and Speciality Crop	California Subtropical Fruit
D	Western Range and Irrigated	Western Range
E	Rocky Mountain Range and Forest	Rocky Mountain Range
F	Northern Great Plains Spring Wheat	Northern Spring Wheat
G	Western Great Plains Range and Irrigated	Western Great Plains Range
H	Central Great Plains Winter Wheat and Range	Central Winter Wheat
I	Southwestern Plateaus and Plains Range and Cotton	Southwestern Plateaus
J	Southwestern Prairies Cotton and Forage	Southwestern Prairies
M	Central Feed Grains and Livestock	Central Feed Grains
N	East and Central General Farming and Forest	East and Central Farming
P	South Atlantic and Gulf Slope Cash Crop, Forest, and Livestock	South Atlantic and Gulf Slope Cash Crop
T	Atlantic and Gulf Coast Lowland Forest and Truck Crop	Atlantic and Gulf Coast Lowland Forest

Table 1-2. Agroclimatic subregions of Canada.

Designation	Description
1	Agroclimatic conditions adequate for most dryland crops
2A	Precipitation usually limiting in 50% of years
2H	Frost damage to wheat 30% of years
3A	Precipitation usually severely limiting for most dryland crops
3H	Severe frosts render wheat production impossible
5H	Frost-free period inadequate for cereals

found in the areas in between. Figure 1-5 shows precipitation distribution patterns at four representative stations.

The frost-free period differs greatly across the dryland regions (Fig. 1-6) and influences not only the choice of crops to be grown but also the magnitude of the water deficit. Wind direction and velocity also vary markedly across the dryland regions. The prevailing wind erosion direction in the Northern Great Plains is from the northwest, whereas in the Southern Great Plains that direction is from the south and southwest (Chepil et al., 1964).

1-1.1 Southern Great Plains

The rainfall in this region ranges from about 875 mm in the main wheat-growing areas to about 400 mm in the drier areas of eastern New Mexico. The rainfall is characterized by a summer maximum of the general Great Plains type, but the amount received depends on the position relative to the main atmospheric moisture source, the Gulf of Mexico. Temperature decreases markedly from the Texas Panhandle to eastern New Mexico, where the elevations change from about 300 to 1400 m. This temperature change is reflected by the Thornthwaite P-E index, which ranges from 20 to

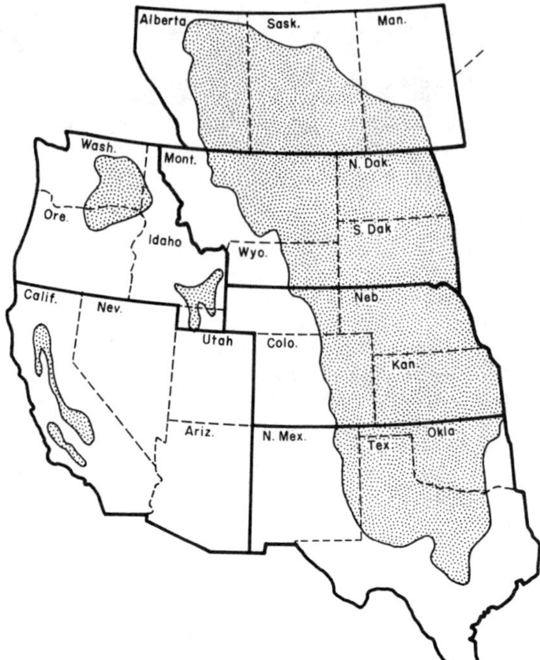

Fig. 1–3. Major dryland farming areas of the USA and Canada.

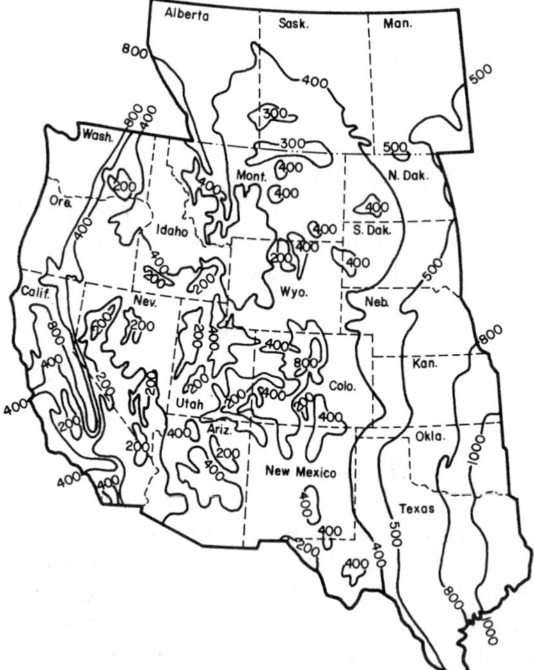

Fig. 1–4. Isohyets for western USA and prairie provinces of Canada. Precipitation given in millimeters.

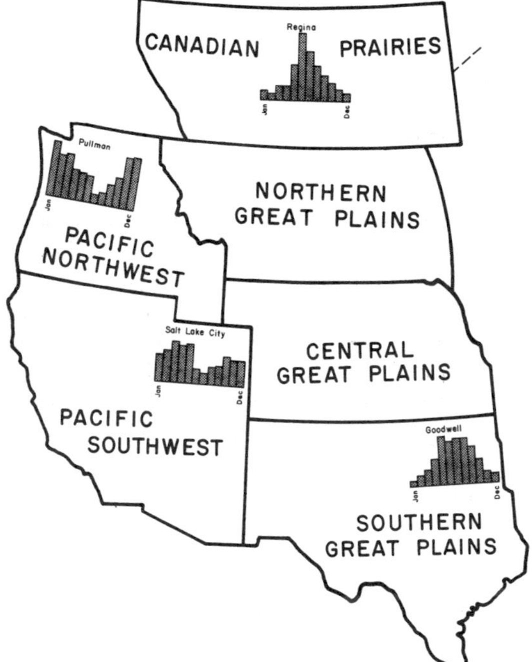

Fig. 1-5. Annual precipitation distribution patterns at four stations.

Fig. 1-6. Mean annual frost-free period, days.

32 in the dry region and progressively increases to over 64 in the southeast (Thornthwaite, 1931).

1-1.2 Central Great Plains

The region between the 96th and 98th meridians falls within the climatic continental subhumid zone, and west of the 98th meridian to the Rocky Mountains the climate is primarily semiarid (Thornthwaite, 1931; USDA, 1941). Thornthwaite (1931) expressed this climate as a precipitation-effectiveness (P-E) index across the Central Great Plains, which showed about 75% of the region as semiarid, 5% as arid, and 20% as subhumid. Precipitation is erratic and in some years may be half or double the long-term rainfall averages. The annual precipitation varies from less than 300 mm in the drier regions to more than 800 mm in the extreme subhumid part of the region. About 75 to 80% of the precipitation occurs during the growing season from April to October. Annual snowfall depth ranges from 500 to 875 mm from the southeast part to the northwest part of the region.

1-1.3 Northern Great Plains

The climate of this region is continental in character. The average mean temperature from April to September is about 15 °C, and the mean annual temperature is 5 °C. Summer temperatures often reach 38 °C, and winter temperatures of more than 45 °C below freezing have been recorded.

Precipitation varies considerably from year to year at any given location. The average annual rainfall is slightly above 500 mm along the eastern border and about 250 mm along the western border. About two thirds of the annual precipitation occurs through the spring and summer months, and snow accounts for about 20% of the total precipitation.

1-1.4 Canadian Prairies

The frost-free period in the Canadian Prairies is the shortest of any of the major dryland areas in North America, ranging from about 80 days in the north to about 120 days in the south. Average annual precipitation varies from about 300 to 400 mm. On the average, two thirds of the precipitation is in the form of rain and one third as snow. Most of it falls in the months of April, May, and June. The climate is continental, as with the remainder of the Great Plains.

Maximum summer temperatures reach as high as 43 °C, and winter temperatures may fall below a bitterly cold −48 °C. Winter wheat is produced in the western part of the dryland area but primarily spring-seeded grain is grown in the eastern part because of winter kill.

1-1.5 Pacific Northwest

The climate of the northwest region, including the Idaho panhandle, is marine and is determined primarily by the prevailing westerly winds spawned by the cyclonic storms that move in from the Pacific Ocean. The winters are cool and moist, and 70% of the annual precipitation occurs from November to April. Annual precipitation ranges from 200 to 750 mm in the grain-producing areas.

The climate in southeastern Idaho and northern Utah is in a transitional zone between the maritime climate of the Pacific Coast and that of the Great Plains. Precipitation in the high plateau, 1400 to 2000 m above sea level, ranges between 300 and 500 mm. Distribution is about equally divided for all months except May and June, when rainfall increases from about 25 mm/month to 40 to 50 mm/month. At higher elevations, snow prevails during the winter months.

1-1.6 Pacific Southwest

California's climate is described as mediterranean, with hot dry summers extending from southern California to the central interior regions. Rainfall occurs largely during December through March and accounts for 50% of the annual amount in the northeast and 75% in the south. Rainfall in dry farming areas of California amounts to an average of 200 to 500 mm annually. In the Central Valley area, the average is about 375 to 500 mm. Snowfall may be a significant source of water only in the northeast.

Nevada and Arizona are mostly arid and account for only about 120 ha each of dryland farming. Utah is the principal area of dryland farming. The precipitation there ranges from 250 to 350 mm.

1-2 NATURAL VEGETATION IN DRYLAND FARMING REGIONS

The natural vegetation of the western regions is shown in Fig. 1-7. The vegetation was described by Shantz and Zon (1936) and mapped in the *National Atlas of the United States of America* (Gerlach, 1970). The natural grassland vegetation can be divided into four categories:

1. Prairie: Tall grass, mostly east of the 100th meridian.
2. Plains: Short grass lying west of the 100th meridian.
3. Desert: Curly mesquite, which lies south of the short grass and occupies much of the higher land in west Texas and the more southern parts of New Mexico and Arizona; the mesquite and desert grass savanna borders on Mexico from south central Arizona to the Gulf of Mexico.

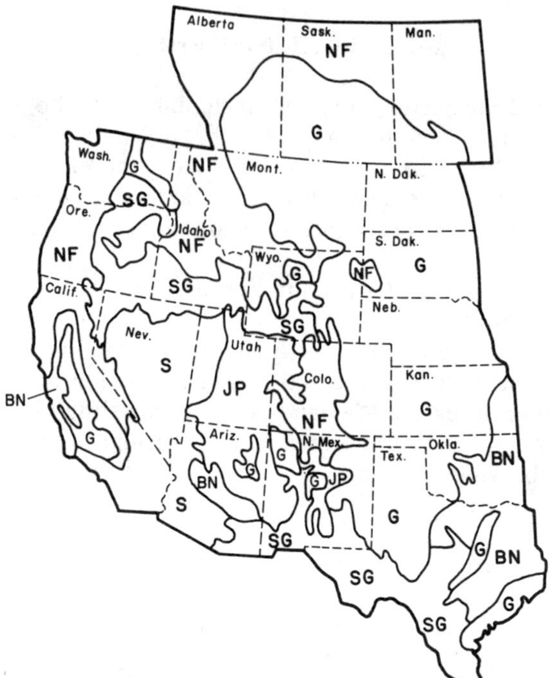

Fig. 1–7. Natural vegetation of western USA and prairie provinces of Canada. *G* = grassland; *SG* = shrub-grassland; *S* = shrub; *JP* = juniper-pinyon woodland; *BN* = broadleaf-needleleaf forest; *NF* = needleleaf forest.

4. Pacific: Bunchgrass, which occurs in high plateaus and foothills of eastern Washington, Oregon, and western Idaho, as well as the foothills and mountains of California and Utah.

The short grass extends on the south into the highlands of New Mexico and Arizona, and isolated areas of tall grass occur in the central and coastal regions of Texas.

Tall grasses grow in a region where the soil water is distributed to a depth of 60 cm or more below the soil surface, and growth is not usually limited by water. Tall grasses are mentioned here only because of the gradual change from tall to short grasses in the change of climate from subhumid to semiarid where there is a continual interrupting of dry and wet years.

The short grass characteristic of the Western Great Plains develops in a region of early spring and summer rainfall. The average annual precipitation ranges from 300 to 560 mm. Water does not on the average penetrate more than 60 cm below the soil surface. These grasses grow for only a short period of time and then pass into a drought rest period that is long in the south and short in the north. Often, the grasses produce little seed in the summer due to drought. Grama grasses take about 100 days to produce seed, which is characteristic of the north, whereas in Colorado the buffalograss requires 40 days. The short grass vegetation region in which dry farming has been undertaken may be divided into several associations:

1. Blue grama grass: The area of mixed grama and western needle grass on the west is separated from the grama grass–buffalograss areas to the east by a line running near the North Dakota–Montana boundary, cutting west of the Black Hills and extending down close to the mountains in Colorado. This area also includes the higher valleys of Colorado, New Mexico, Arizona, and Utah. The vegetation is dominated largely by grama grass, which often constitutes pure sod.
2. Galleta grass: This grass dominates vast areas of New Mexico and Arizona and extends deeply into the Utah desert. It is used widely as grazing land.
3. Grama grass–buffalograss mix: This mixture extends over Nebraska, New Mexico, and the Texas Panhandle and is dominated by almost equal quantities of grama grass and buffalograss. The grasses occur as a uniform sod and are the most typical short grass of the Great Plains grassland. Although many other plants occur in this association, the typical areas appear as pure sod except during wet years. Within this area, especially in the north, the small grains do fairly well in medium or wet years but fail in dry years. Winter wheat is grown in the northern and eastern portions. Farther east, where the rainfall is greater or on coarser-textured soils where water penetrates rapidly, tall plants are more abundant.
4. Wire grass: Wire grass forms a belt between grama grass–buffalograss of the Great Plains grassland and the bunchgrass of the prairie grassland and is most extensive in Nebraska, Kansas, and Texas. This area is more suited for crop production than is the pure grama grass–buffalograss land.
5. Western wheatgrass: This area is characterized by an even sod of grama grass and buffalograss and scattered growth of western wheatgrass. This area produces a heavier crop than grama grass–buffalograss alone and is found in South Dakota, Nebraska, and Colorado on heavy gumbo soils (fine-textured soils). Agriculturally, the area is about as productive as the grama grass–buffalograss area.
6. Curly mesquite: Curly mesquite is characteristic of the desert grassland. It occurs in Texas, New Mexico, and Arizona and reaches far into Mexico, not conspicuously different from the short grasses of the Great Plains. The growth period begins in July and August with summer rains. The area is characterized by a high evaporative index, with average rainfall of 450 to 500 mm. On most of this land, the conditions for dry farming are not as good as on the short grassland. The dry hot summers with the growth period limited to the hotter months of July, August, and September restricts dryland production to sorghums and similar warm weather crops and assures almost certain failure of small grains. Areas of pure grass are not great, and much of the area is covered with desert shrubs. Typical grasses are (i), black grama (ii), rothrock grama, and (iii), curly mesquite. Black grama is most extensive in New Mexico, Texas, and

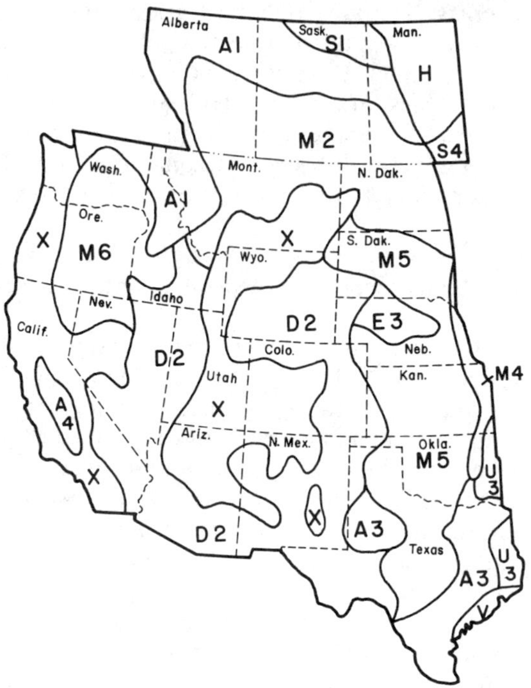

Fig. 1-8. Suborders of soils in western USA and prairie provinces of Canada. *A1* = Boralfs; *A3* = Ustalfs; *D2* = Argids; *E3* = Psamments; *H* = Histosols; *M2* = Borolls; *M4* = Udolls; *M5* = Ustolls; *M6* = Xerolls; *S1* = Spodosols (varied); *S4* = Orthods; *U3* = Udults; *V* = Vertisols; *X* = mountain soils.

Arizona, and rothrock grama and curly mesquite are found mainly in southeastern Arizona.

1-3 SOILS OF PRESCRIBED REGIONS

Figure 1-8 shows the dominant soil orders and suborders of soils covering the 17 western states and Canadian Prairies. The soils are classified according to the U.S. comprehensive soil classification system (Soil Survey Staff, 1975). The soils map in Fig. 1-8 (SCS, 1971) is general, and many inclusions of soils within each group are not shown. More detailed soil maps of the regions can be found in the *National Atlas of the United States of America* (Gerlach, 1970) and in the publication *Soils of Canada* (Agric. Canada, 1977). Soils of the various dry farming U.S. regions as outlined in Fig. 1-1 are treated with more detail in the USDA publication on summer fallow (USDA-ARS, 1974).

A general perception of soils in the various U.S. regions can be obtained from Fig. 1-8 by noting the orders and suborders. In the North American Great Plains, the soils fall predominantly into the order of Mollisols and suborders of Aquolls, Borolls, Udolls, and Ustolls. In the northwest region, the order of Mollisols also dominates, but the suborders are Ustolls and Zerolls. In the southwest region the predominant order is Aridisols, and the suborders are Argids and Orthids. In the Mollisols order in southern Utah and Arizona, the suborders Ustolls and Xerolls predominate.

1-4 DROUGHT

In the dry farming regions, normal rainfall is always accompanied by variability of rainfall during and between the seasons. Along with these normal rainfall occurrences often comes drought, one of man's worst natural enemies. Drought and agriculture are discussed by the World Meteorological Organization (1975). A comprehensive listing of drought definitions are given in Appendix I of that publication (WMO, 1975, g. 109–119). The following definitions that apply to North America are examples taken from that appendix.

Henry (1906) defined drought in terms of rainfall as 21 days or more when rainfall is 30% or less than average for the time and place, and extreme drought is when rainfall fails to reach 10% of normal for 21 days or more. Hoyt (1936) states the definition as any amount of rainfall less than 5% of normal. Palmer (1968) gives the definition in terms of soil water crop parameters. The severity of agricultural drought is defined in terms of magnitude of the computed abnormal evapotranspiration deficit and expressed as a crop moisture index. The inputs for this latter definition are based on weekly values of temperatures and rainfall. Thornthwaite (1931) related drought to the concepts of a precipitation effectiveness index based on mean temperature to give water requirement for optimum growth. The water balance model gives estimate of soil water status. Farmers consider drought as a shortage of water for their crops; economists define drought as a water shortage affecting the established economy of a region. Prolonged drought affects all segments of society within the given region, and often its effects extend to areas far from the drought situation.

Droughts have been recorded in the USA from the days of George Washington. Letter to William Strickland, 15 July 1797 (Rasmussen, 1975a) stated:

> From the cultivation of the common black-eye peas, I have more hope, and am trying them this year, both as a crop, and for ploughing in as a green manure; but the severe drought, under which we labor at present, may render the experiment inconclusive.

Drought is also mentioned by Taylor in his discussion of gypsum experiments in Virginia in 1809 (Rasmussen, 1975a). Merril (1910) states that 1910 will probably pass into history as the year of the drought not only in Utah and the intermountain country but throughout a large portion of the Midwest. Hargreaves (1957) mentions a prolonged drought of 1917 through 1919 in the northern Great Plains. The drought from 1934 to 1937 is well documented by many writers and was so severe in 1934 and 1936 that the government took emergency steps to save dying cattle and relieve destitute families to save human life (Rasmussen, 1975c). Droughts are still common in these regions. California and the Rocky Mountain areas suffered through a major drought in 1976 and 1977.

1-5 CONCLUSIONS

Dryland agriculture has been practiced for millenia and has been and is the major supplier of the world's wheat, grain sorghum, and millet. Research on dryland crops has been carried out in North America since the early 1900's. Research, along with the practical experience of perceptive farmers, has turned the "great American desert" into the breakbasket of the world. Risk of crop failures during the inevitable fluctuations in annual and seasonal precipitation and in the length of the growing season has been lessened. Improved management techniques, better adapted cultivars, crop insurance and disaster assistance programs, more suitable tillage practices, advances in soil and water conservation methods, and other changes have increased stability in dryland agriculture, especially during the past 40 years. How long that relative stability will continue is an unanswerable question, given our imperfect knowledge of atmospheric processes.

Despite the advances that have been made, climatic variations present a hazard that has been reduced but by no means eliminated. In addition, differences in annual and seasonal average precipitation and temperature, soils, wind direction and velocity, slope and exposure of land, and insect and disease infestations make management recommendations location specific to a large degree. Plant growth models integrating plant, soil, and climatic factors for dryland crops are either nonexistent or only rudimentary. Until the data needed for such models have been collected and refined models constructed, we must rely on local adaptive testing.

1-6 LITERATURE CITED

Agriculture Canada. 1977. Soils of Canada. Agriculture Canada, Ottawa.

Austin, M. E. 1965. Land resource regions and major land resource areas of the United States. USDA Agric. Handbook no. 296.

Campbell, H. W. 1907. Campbell's 1907 Soil culture manual. H. W. Campbell, Lincoln, Neb.

Chepil, W. S., F. H. Siddoway, and D. V. Armbrust. 1964. Prevailing wind erosion direction. J. Soil and Water Conserv. 19:67–70.

Duley, F. L., and J. J. Coyle. 1955. Farming where rainfall is 8–20 inches a year. p. 407–415. *In* Yearbook of agriculture. U.S. Dep. of Agric., U.S. Government Printing Office, Washington, D.C.

Gerlach, A. C. (ed.) 1970. National atlas of the United States of America. U.S. Dep. of Interior, U.S. Geological Survey, U.S. Government Printing Office, Washington, D.C.

Gras, N. S. B. 1946. A history of agriculture in Europe and America. F. S. Crofts and Co., New York. p. 341–343.

Hargreaves, M. W. M. 1957. Dry farming in the northern Great Plains, 1900–1925. Harvard University Press, Cambridge, Mass., p. 26–27, 87–90.

Henry, A. J. 1906. Drought in climatology of the United States. U.S. Weather Bureau Bull. Q. 361:51–58.

Higbee, E. 1958. American agriculture: Geography, resources, conservation. John Wiley & Sons, Inc., New York.

Hilgard, E. W. 1892. A report on the relations of soil to climate. U.S. Weather Bureau Bull. no. 3.

Hoyt, J. C. 1936. Droughts on 1930–34. U.S. Dep. of Interior, U.S. Geological Survey Water Supply Paper no. 680.

King, F. H. 1895. The soil, its nature, relations, and fundamental principles of management. Rural Sciences Series, Macmillan, Inc., New York.

Klingebiel, A. A., and P. H. Montgomery. 1973. Land capability classification. USDA Agric. Handbook no. 210.

Merril, L. A. 1910. The drought of 1910. Desert Farmer 7:1–12.

Palmer, W. C. 1968. Keeping track of crop moisture conditions, nationwide: The new crop moisture index. Weatherwise 21:156–161.

Rasmussen, W. D. (ed.) 1975a. Agriculture in the United States: A documentary history, Vol. 1. Random House, Inc., New York, p. 409–410.

Rasmussen, W. D. (ed.) 1975b. Agriculture in the United States: A documentary history, Vol. 2. Random House, Inc., New York, p. 1774.

Rasmussen, W. D. (ed.) 1975c. Agriculture in the United States: A documentary history, Vol. 4. Random House, Inc., New York.

Shantz, H. L., and R. Zon. 1924. The physical basis of agriculture, Section E: Natural vegetation. p. 15–36. *In* USDA atlas of American agriculture. U.S. Dep. of Agric., U.S. Government Printing Office, Washington, D.C.

Soil Conservation Service. 1971. Soils of the world. USDA-SCS, Hyattsville, Md.

Soil Survey Staff. 1975. Soil taxonomy. USDA Agric. Handbook no. 436.

Thornthwaite, C. W. 1931. Climate of North America according to a new classification. Geog. Rev. 21:633–655.

U.S. Department of Agriculture. 1941. Climate and man. p. 177–187. *In* Yearbook of agriculture. USDA, U.S. Government Printing Office, Washington, D.C.

U.S. Department of Agriculture, Agricultural Research Service. 1974. Summer fallow in the western United States. USDA-ARS Conserv. Res. Report no. 17.

United Nations Educational, Scientific, and Cultural Organization. 1979. Map of the world distribution of arid regions. Explanatory note. MAB Technical Notes no. 7, UNESCO, Paris.

Whitney, M. 1895. Reasons for cultivating the soil. p. 123–130. *In* Yearbook of agriculture. U.S. Dep. of Agric. U.S. Government Printing Office, Washington, D.C.

World Meteorological Organization. 1975. Drought and agriculture. WMO Technology Note no. 138, Geneva.

SECTION B

WATER CONSERVATION

2

W. O. Willis
Agricultural Research Service
U.S. Department of Agriculture
Fort Collins, Colorado

Water Conservation: Introduction

Water conservation is of paramount importance for plant and animal production in dryland agriculture. Seldom is sufficient, adequately distributed rainfall received during a growing season such that a crop can produce to its potential. Therefore, it is imperative to have stored soil water available to supplement the inadequate growing season rainfall.

Although inadequate amounts and unfavorable distribution of precipitation are cited as primary limitations to good crop yields, our inability to properly manage the limited precipitation may be as much of the problem as is the limited resource. Estimates show that production could be at least doubled, in most cases, if all the precipitation received were used only by cropped plants. Such estimates assume that no soil water is lost by evaporation (Wendt et al., 1970). At present, no economic means of controlling evaporation on large fields or rangelands are available.

Production of forage species increases as soil water increases, and the production response is dramatic if soil fertility is adequate. However, with wheat and similar crops that are common in the semiarid dryland region, about 18 to 25 cm of soil water is required before any grain is produced (Fig. 2-1). Daily water use by wheat is increased by higher soil fertility; an example for winter wheat is shown in Fig. 2-2.

Economical production depends on efficient use of various factors. Almost anything that increases yield also increases water use efficiency (Viets, 1962). Again with wheat as an example, Table 2-1 gives values of water use efficiency (yield per unit of water) for nonfertilized wheat in various dryland regions of the western USA. Similar data for wheat on fallow from the southern Canadian Prairies (Staple and Lehane, 1954) show water use efficiencies from about 3 kg ha^{-1} mm^{-1} to about 7.5 kg ha^{-1} mm^{-1}, depending on available soil water. Adding fertilizer such as N or P if the soil supply

Contribution from the U.S. Department of Agriculture, Agricultural Research Service, Fort Collins, Colorado.

Dryland Agriculture—Agronomy Monograph no. 23.

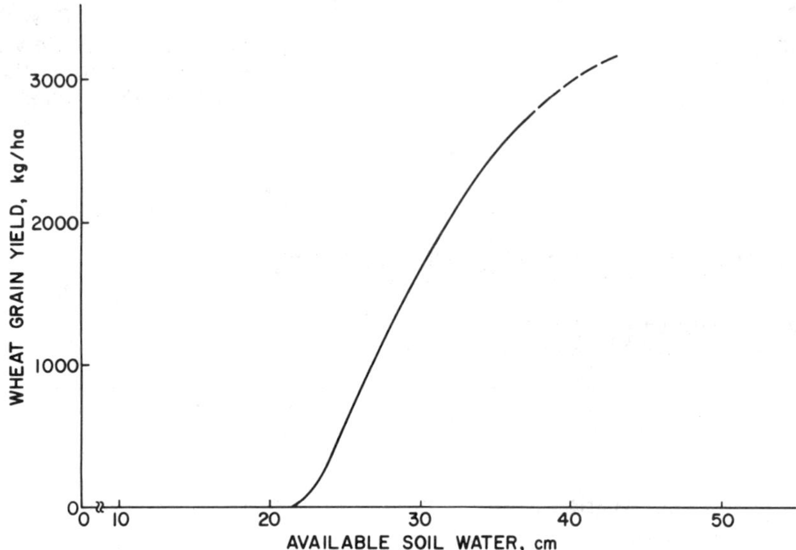

Fig. 2-1. Yield (kilograms per hectare) of winter wheat as a function of available soil water (centimeters) (Good and Smika, 1978).

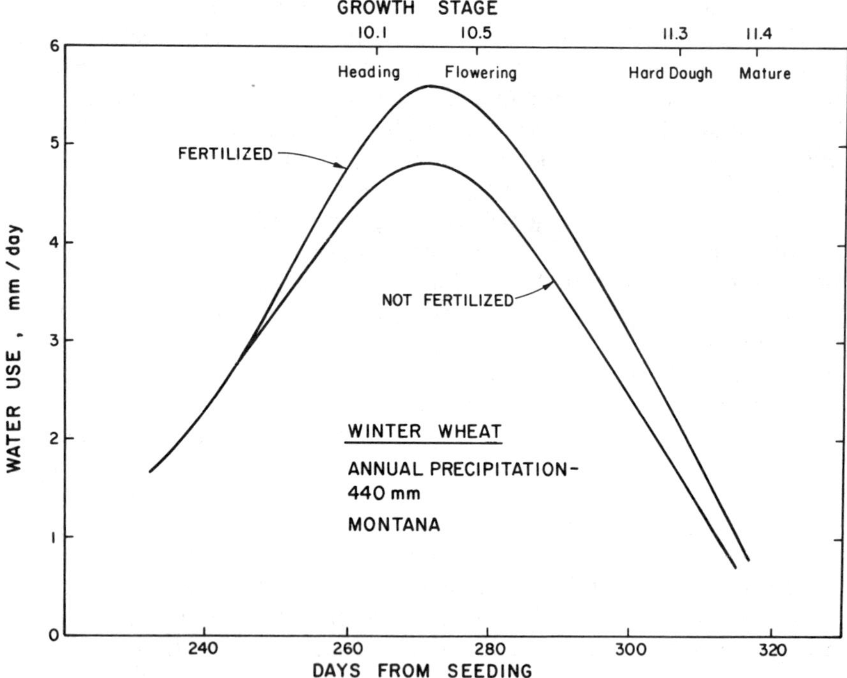

Fig. 2-2. Daily water use (millimeters per day) as a function of time during the spring and summer for nonfertilized and fertilized (67 kg of N/ha) winter wheat in Montana (Brown, 1971).

Table 2-1. Average values of annual precipitation, grain yields of nonfertilized spring or winter wheat, and water use efficiency for annually cropped and alternate wheat-fallow over five selected semiarid regions in the western USA (Willis, 1982).

Region	Years of crop record	Type of wheat	Annual precipi-tation†	Yield		Water use efficiency‡	
				Fallow	Annual	Fallow	Annual
			mm	kg/ha		kg ha^{-1} mm^{-1}	
Northern Great Plains	359	Spring	394	1427	962	1.81	2.44
Central Great Plains	165	Winter	462	1629	653	1.76	1.41
Northwest	76	Winter	368	2403	--	3.26	--
	60	Winter	421	--	1010	--	2.40
Northwest	70	Spring	349	1958	895	2.81	2.56
Southern Great Plains	28	Winter	500	1010	646	1.01	1.29
Southwest	4	Spring (fall planted)	417	2376	767	2.85	1.84

† Harvest to harvest for annually cropped wheat.

‡ Water use efficiency is the weight of marketable grain per unit area divided by the total precipitation from harvest to harvest. Thus, precipitation for 2 years is used for wheat-fallow, and precipitation for 1 year is used for annual wheat. This definition does not imply that all water received is transpired.

is limited will possibly double the values of water use efficiency for annual wheat given in Table 2-1. However, N fertilizer will increase the water use efficiency values of wheat on fallow by only 20 to 50%, because fallowing typically increases the level of soil NO_3-N (Haas et al., 1957).

Good water conservation results in higher crop yield, increased stability of production, and higher water use efficiency. However, increasing the level of management to gain greater water conservation but failing to use adequate fertilizer or the appropriate cultivar is not rational. Therefore, best management for water conservation should include cultural practices that provide the best water use efficiency. Little can be done to change regional climate, geology, or topography, but use of research-developed cultural practices has significantly mitigated the impacts of drought. Over the past 50 years, average yields have increased, and from 40 to 50% of the increase has been attributed to development and application of better soil water conservation practices. For example, fallowing typically has shown only 20 to 30% efficiency in conserving precipitation, but more recent practices such as no-tillage have shown up to 50% efficiency.

The following six chapters deal with practical application of principles of soil water flow, evaporation, and transpiration. They also discuss water conservation practices that are developed in the semiarid regions of the Southern Great Plains, Central Great Plains, Northern Great Plains, Canadian Prairies, Pacific Northwest, and Pacific Southwest. Such practices as summer fallow are common in all regions, but other distinct differences exist in regional concepts and practices. In northern regions, snow can be managed to significantly increase soil water storage and plant growth. In all regions, evaporation is the largest nonbeneficial use of soil water.

2-1 LITERATURE CITED

Brown, P. L. 1971. Water use and soil water depletion by dryland winter wheat as affected by nitrogen fertilization. Agron. J. 63:43–46.

Good, L. G., and D. E. Smika. 1978. Chemical fallow for soil and water conservation in the Great Plains. J. Soil Water Conserv. 33:89–90.

Haas, H. J., C. E. Evans, and E. F. Miles. 1957. Nitrogen and carbon changes in Great Plains soils as influenced by cropping and soil treatments. USDA Tech. Bull. no. 1164.

Staple, W. J., and J. J. Lehane. 1954. Wheat yield and use of moisture on substations in southern Saskatchewan. Can. J. Agric. Sci. 34:460–468.

Viets, F. G., Jr. 1962. Fertilizers and efficient use of water. Adv. Agron. 14:223–264.

Wendt, C. W., T. C. Olson, H. J. Haas, and W. O. Willis. 1970. Soil water evaporation. p. 207–227. *In* Evapotranspiration in the Great Plains. Great Plains Agric. Coun. Publ. no. 50, Univ. of Nebraska, Lincoln.

Willis, W. O. 1982. Soil water management practices for wheat production in semiarid climates of the United States. *In* J. Burgos (ed.) Proc. U.S.–Argentina Bilateral Workshop on Drought, Mar del Plata, Argentina. 4–9 Dec. 1978. Instit. Nacion di Tech. Agropec., Buenos Aires. (In press.)

3

C. H. M. van Bavel
Texas A&M University
College Station, Texas

R. J. Hanks
Utah State University
Logan, Utah

Water Conservation: Principles of Soil Water Flow, Evaporation, and Evapotranspiration

The purpose of this chapter is to discuss theoretical concepts that appear directly applicable and useful in the planning, interpretation, and application to management of field experiments that relate to water conservation in dryland agriculture.

One must recognize at the outset that the processes involved in the conservation of water, within the framework of a given soil and crop management scheme, are but partially understood. Thus, from first principles alone we cannot discriminate between alternative practices as to which is more desirable and by what measure. Furthermore, soils are variable in space and weather is variable in time. This fact additionally limits what theory alone can accomplish.

Rather, it will continue to be necessary to conduct field tests and accept them as always time consuming and expensive and sometimes not conclusive. Also, we must live with the realities of a changing economic environment and unpredictable government actions that cause shifts in the type of crops farmers plant and in the price ratios between commodities and the inputs such as land, fertilizer, machinery, fuels, and water. For this reason, field experiments, however well conducted and interpreted, may become irrelevant, and new information must be generated continually. For-example, in North Dakota and the High Plains of Texas, an explosive ex-

Dryland Agriculture—Agronomy Monograph no. 23.

pansion of the area planted to sunflower has occurred, but only a negligible amount of information exists on water conservation in sunflower production compared with wheat and grain sorghum production.

Theoretical concepts are needed for at least three reasons. First, theory is helpful in designing experiments. It suggests possibilities, or it predicts effects that are worthwhile to explore. Perhaps even more significant may be that theoretical concepts can be used to diagnose and prevent trivial or absurd experiments or, at least, may cause one to take a hard look at a superficially interesting proposal.

Second, theory assists in generalizing field data. Because it is patently impossible to evaluate a management practice for all significant combinations of soils and weather patterns, it is necessary to speculate on the results under different conditions by theorizing from the available data.

Theoretical concepts, finally, are readily integrated. Whereas in past research, both theory and experiment often concentrated on single elements of a complex problem, this approach has not been capable of providing a solution. Today, theory enables us to draw up a comprehensive structure of the problem.

There is, of course, a choice between the use of a statistical theory and one based on an admittedly incomplete concept of the processes involved. A good example is the prediction of evapotranspiration rates based on weather. Attempts have been made to predict evapotranspiration by multivariate regression and by mechanistic models based on our knowledge of energy transformation, radiative processes, and the aerodynamics of the atmosphere. The mechanistic approach has, in this instance, proved to be more satisfactory than regression. Just the same, many mechanistic theories must include statistical or empirical components, and in practice, we generally combine the two approaches.

In the following we will consider theoretical concepts dealing with (i) water transport in soils and (ii) water uptake by plant root systems.

3-1 WATER TRANSPORT IN SOIL

Soil water transport has received wide attention since about 1950, and a vast literature deals with water transport in the liquid phase.

In liquid or vapor phase flow may occur in an isothermal system as the result of differences in the water potential, or as the direct result of differences in temperature, in an anisothermal system. In practice systems are always anisothermal; therefore, both effects are present at the same time. The question is whether practical significance should be attached to all four modes of transport: isothermal liquid and vapor, and anisothermal liquid and vapor.

In either form, the practical impact of vapor flow appears to be questionable if the data obtained by Jackson et al. (1974) can be generalized. For wet, bare soil at 10-mm depth, they found that the daily net upward flow of about 3 mm was essentially liquid. When the water content at 10 mm was less and corresponded to a water potential of about −1.5 MPa, the vapor components tended to oppose and cancel each other, whereas a steady upward liquid component prevailed to give a daily net flux close to 1 mm. When the soil had become air-dry at the 10-mm depth, they found essentially the same result, except for a daily net flux of about 0.5 mm.

Inasmuch as the findings of Jackson et al. (1974) appeared to support the general theory of Philip and de Vries (1957), we may conclude that from an applied vantage point and on a 24-h basis, vapor flow can be considered negligible as a transport process, even in relatively dry soil and close to the soil surface. To this conclusion we should add that the measurement of vapor diffusivities as a function of temperature and water content is a tedious and highly specialized task.

Hanks et al. (1967) drew a similar conclusion. They concluded that computation of evaporation, if isothermal conditions for soils initially wetted to near saturation were assumed, would probably approach actual evaporation within 10% and probably be sufficiently accurate for most purposes.

For liquid flow, we note that the measurements of Jackson et al. (1974) imply an even less significant role for thermally driven liquid movement. Thus, we are left with the application of principles dealing with the isothermal behavior of liquid water.

The oldest and simplest is the principle of conservation of mass. Applied to a root zone it states that

$$\text{Change in soil water content} = \text{Precipitation} - \text{Runoff} - \text{Drainage} + \text{Upward flow} - \text{Evapotranspiration.} \quad [1]$$

Equation [1] is the *water balance equation.* In many instances it has been applied by assuming that runoff, drainage, and upward flow are essentially zero, and it has provided useful conclusions about the efficiency of storage and the significance of the amount of soil water stored at planting time.

However, Eq. [1] may also give erroneous estimates of water availability, particularly because drainage and upward flow are ignored. These components can be found by applying the classical equation of Darcy:

$$\text{Liquid flux} - \text{Hydraulic conductivity} \times \text{Hydraulic head gradient.} \quad [2]$$

The negative sign indicates that water flows along a path of decreasing potential. The applicability of Eq. [2] to saturated and unsaturated flow is not in question. Complications arise from salinity gradients or electrostatic fields, under which conditions Eq. [2] is insufficient. However, the occurrence of such exceptions with dryland water conservation problems remains to be demonstrated.

Darcy's equation, combined with the law of mass conservation, gives rise to the Richards equations, which can be stated as

$$\text{Change of water content/Unit time} = \text{Change in liquid flux/Unit depth.} \quad [3]$$

Equation [3] contains the key to many questions important in soil water management: infiltration of rainfall, redistribution of soil water during and following infiltration, rise of water to the surface in consequence of evaporation, redistribution following root uptake, deep drainage and recharge of aquifers, and upward flow from the unsaturated zone into the root zone.

Literature of the last 20 years reports the valid and practical solution of the Richards equation on the one hand and the determination of the relevant soil properties on the other. The second subject, though of crucial importance, will not be dealt with here. For a thought-provoking review, see Gardner (1974).

The general problem faced by practitioners of water conservation in dryland agriculture is to correctly anticipate, starting with a known water content profile, how this profile will change with time. The prediction must account for the disposition of rainfall, the loss by surface evaporation, the uptake of water by roots, and the ensuing transpiration loss from vegetation. Practitioners must also estimate the loss of water by drainage and the upward rise from subsurface layers not reached by roots or from a water table.

The problems of evapotranspiration will be discussed later. We shall discuss here the classical problem of agricultural hydrology applied to a special set of circumstances, for which it would seem that a solution is readily available.

Preliminary attempts to solve the problem with a simplified water balance (Eq. [1]) were reported by Ritchie (1972) and Hanks (1974). However, no general solution has been reported for a profile composed of layers of differing properties and in which infiltration, ponding, redistribution, and rise from a water table can occur simultaneously or in an arbitrary order of succession as it occurs in nature.

Nielsen et al. (1972) give a partial review of solutions for simplified cases, but the only feasible approach to a comprehensive treatment is by numerical approximation, as exemplified by Hanks et al. (1969), Hillel (1977), De Wit and Van Keulen (1972), Item (1974), Van Der Ploeg and Benecke (1974), and many others.

This method consists of dividing the soil profile into a number of layers, or other finite elements, and assigning to each layer the properties that were found in field investigations. This procedure does not require the assumption of specific mathematical expressions for the relation between pressure potential and water content or between hydraulic conductivity and water content. The condition and description of the surface layer is critical for the infiltration process, as is the nature of the deepest layer at its lower boundary for drainage or upward rise from water table. All other layers are gaining or losing water in the manner described by the Richards equation (Eq. [3]).

The numerical solution is repeated over small time steps to maintain a realistic approximation of a continuous process. A variable time step provision that responds to the relative rate of change of the system is a desirable feature. Hysteresis can be taken into account (Hanks et al., 1969).

Talsma (1978) stated that numerical solutions can give only specific solutions. This is a true statement, but analytical solutions need specific boundary conditions, Furthermore, all practical problems are specific, and the cost of repetitive solutions of the problem is no longer excessive. The computer programs that produce the estimates of soil water dynamics are inherently simple. By and large they consist of interpretation of tables, plus simple arithmetic operations, carried out separately for each designated soil layer. The prodigious speed and memory capabilities of modern computers have made the successive calculation of profiles, even when done thousands of times to simulate the events of a single month, quite feasible.

For a general discussion of philosophy and techniques, see the introduction of *Soil Water* (Nielsen et al., 1972) and the examples by Hillel (1977). These references are specific enough to serve as instruction manuals for a design engineer or an experimental agronomist.

Experience with numerical methods soon leads to two general conclusions. One is that it is critical to measure soil or profile characteristics in the field rather than in the laboratory on samples. The most important aspects are the measurement of the hydraulic conductivity at the maximum water content and the nature of the water characteristics at pressure potentials close to zero, e.g., from 0 down to − 10 kPa.

The second conclusion is that because of the conservative character of the basic equations and the opposing interplay between the changes over time in hyraulic conductivity and hydraulic gradient, the profile system as a whole has a large time constant. Using analytical and numerical studies, several reports show that the diurnal variations in surface evaporative flux and the irregular occurrence of rainstorms is usually unnoticeable at depths of 0.5 m or more; a quasi-steady flow exists that may show only seasonal variation.

Even though it is hazardous to generalize and much field experimental work needs yet to be done, we can draw a few conclusions pertinent to dryland water conservation from theoretical considerations. First, the magnitude and persistence of downward drainage is significantly more than generally assumed. Second, given the necessary elapsed time, an important reversal of flow is predicted and has been confirmed experimentally, in which agronomically significant amounts of water rise from deeper layers into the root zone. Both conclusions can be stated as a single one if we consider that a zone of zero flow never exists, only a plane of zero flow, a plane that moves up and down following wetting and drying episodes originating at the surface.

Third, in the absence of root extraction and transpiration, the zero flow plane moves slowly beyond depths of 0.1 m or so after an initial rapid descent from the surface after infiltration. This "self-mulching" of bare soil has been observed for a long time, and its enhancement by tillage or use of natural and artificial mulching materials has been investigated often. Only recently has the problem been studied in a general fashion (Hillel et al., 1975).

3-2 WATER UPTAKE BY PLANT ROOT SYSTEMS

Changes in the water content of the profile are not solely determined by its boundary conditions and the ensuing internal redistribution. Root water uptake resulting from transpiration losses from vegetation must be accounted for by adding a sink term to the equation of continuity. The integral of this term over the entire root zone equals the transpiration, which in turn can be estimated from meterorological, soil, and crop data.

Water losses from the surface layer by evaporation can be included as a similar sink term, so that the sum of all equals the evapotranspiration. When the relative humidity of the surface is close to 100% and the soil water potential does not result in partial stomatal closure, estimation of evapotranspiration can be done accurately enough for the purpose at hand. For extensive discussion, see *Irrigation of Agricultural Lands* (Hagan et al., (1967) and Jensen (1973).

However, two difficult problems remain. One is how to partition the total evapotranspiration by depth. It has been directly or indirectly proposed to partition it simply in proportion to the density distribution of the root system. Rarely will it be justified to give this distribution a mathematical form. Moreover, this function of depth obviously depends on the stage of crop development.

Thus, in addition to an empirical "crop coefficient" to modulate the maximum evapotranspiration as a function of time, we need an equally empirical root distribution function to estimate from which soil layer and at which rate the water will be taken. For practical purposes, such information can be handled only by a numerical method. A computer does the complex job of tracking of water in each soil compartment. Hanks (1974) uses a water balance equation with different zones. A growing root is considered. If roots are present, water, on any one day, is withdrawn from the layer with the highest relative water content.

The above procedure obviously simplifies the real situation but has been shown to give good predictions for seasonal water use. Several authors have proposed methods to reflect interaction of soil water status, depth, root distribution, and climatic "demand" on root water uptake (Nimah and Hanks, 1973; Feddes and Rijtema, 1972; Van Bavel and Ahmed, 1976). Nimah and Hanks (1973) allow for an integrated extraction of water from different depths depending on the conditions. These proposals give reasonable predictions but have not been validated in many situations; moreover, they make the computation much more complex.

The second difficult problem is to obtain an adequate representation of the overall reduction in evapotranspiration rates as soil water content in the root zone declines. This question is not negligible in the context to dryland problems; rather it is probably more important than any other.

Many methods have been devised to handle the declining phase of evaporation from the soil surface. Ritchie (1972) used an empirical inverse square root of time relation, which required that two empirical numbers be found for the surface soil in question. Other more fundamental methods have been proposed by Van Bavel and Hillel (1976) and Nimah and Hanks (1973). The procedure of Nimah and Hanks (1973) provides for a "switching" surface boundary condition. Evaporation is found to be equal to the climatic demand if water can be transmitted fast enough to the surface without causing the surface soil water content to decrease to the air-dry value. As the soil dries, the above condition cannot be met, and the soil surface water content reaches the air-dry value, after which the evaporation rate declines continuously below the climatic demand rate. Thus, when the soil is wet, the surface boundary flux is equal to the climatic demand, but as the soil dries, the boundary condition switches to a constant air-dry water content. Conditions in which the soil surface is totally or partially bare represents the dryland situation for much of the time, hence the significance of estimating the effect of drying on evaporative losses.

We do not have a completely adequate description of the role of soil water status as a modifier of transpiration, but we have no shortage of proposed solutions, and they all have some limitations (van Bavel and Ahmed, 1976; Feddes, 1980).

Transpiration is controlled not only by leaf stomata but also by leaf attitude and rolling and can be accelerated by senescence and declining expansion rate of new leaves. In the dryland environment, all these phenomena are commonplace, but to recognize each mechanism explicitly and adequately seems a formidable undertaking.

Nimah and Hanks (1973) use a concept of root water potential (which could be related to leaf stomata resistance and other plant behavior) to account for water limiting transpiration. In their method the integrated transmission of water, as influenced by soil water, gravitational, salinity, and root density factors, from the soil into the roots is simulated and is equal to the climatic demand as long as the root water potential is above a lower value simulating wilting. The driving force for root extraction is the water potential gradient from the soil to the root. As the soil dries, at some point the root extraction cannot meet the climatic demand if the root water potential stays above the limiting value. Thereafter, transpiration decreases because the potential gradient causes a decreased flow. This approach, as shown by Childs and Hanks (1975) and Childs et al. (1977), gives reasonable results.

All of the methods proposed depart from reality to some extent. However, the conservative nature of the evapotranspiration process makes reasonable approximations possible, especially over fairly long periods of time. The simpler methods have the advantage of less complexity and lower computer time requirements, whereas the more fundamental models will better predict unusual situations. For example, the simple model of Hanks (1974) takes about one-tenth the computer time of Nimah and Hanks (1973) and works equally well if salinity is not a problem and if water flow is downward only. However, to account for salinity conditions or water flow up from a water table, a fundamental method such as Nimah and Hanks (1973) is necessary.

3-3 FROZEN SOILS

Frozen soils commonly occur in cooler dryland systems such as those of the Northern Great Plains. When soils are frozen, infiltration of water into the soil decreases drastically. Consequently, the depth of the frost line is important to runoff, flooding, soil erosion, and seasonal water supplies. Cary et al. (1979) have developed equations and verified them in the field to account for water and salt movement in frozen soils. In another paper, Cary et al. (1978) describe simple equations to predict the depth of freezing from climatic records.

The equations consider flow of water in frozen soils as analogous to the liquid flow of water in unsaturated, unfrozen soils. Water flows from unfrozen to frozen soil, which is analogous to flow from wet to dry soil. Similarily, the hydraulic conductivity in frozen soil decreases rapidly as temperature decreases, which is analogous to the decrease of hydraulic conductivity with water content.

Thus, water in frozen soils flows from high to low temperatures. This liquid flow is so closely coupled with temperature that heat flow must be considered simultaneously with liquid flow in unfrozen soil. Cary et al. (1978) also found that soil solute concentration had a large influence on water flow in frozen soils, with water flowing from high to low salt concentrations.

3-4 CONCLUSIONS

The applicability of existing hydrologic theory to water conservation problems under dryland conditions is severely limited in spite of a great amount of research. At least, we better appreciate the size of the errors associated with the experimental procedures used to characterize soils. We have also more recently discovered that within a field that is managed and observed as a unit, these physical properties exhibit a large variability within short distances.

Because, in contrast, the rainfall and the conversion rate of solar radiation to sensible and latent heat are relatively uniform processes, we can calculate with adequate precision the recharge, drying, and deep drainage of bare soil profiles, because the principles of conservation of mass and energy make up for the uncertainty in the transfer and retention characteristics of the soil.

One difficulty is easily overlooked: these processes are driven by weather that is extremely variable from year to year. To know, for example, what the carryover of water is from an irrigated crop to a dryland crop in a given location would require that the process be simulated with actual series of weather data as found over a 25- to 50-year period. The average carryover cannot be computed as the carryover based on average meteorological data. Theory, however simplified and approximate, is the only approach to such an elementary question, because the required experimental time would probably exceed the time even the most patient decision maker would want to allot for an answer.

Theory, likewise, may be helpful in outlining the potential of surface modifications, such as residue management, tillage, artificial mulching, and the timing of surface treatments in relation to weather events.

We seem to be far from an adequate method to estimate the impact of alternative crop management methods, such as double as opposed to single rows, planting density in general, time of planting, and selection of cultivars of different growth habits with respect to either the aerial or the underground parts. At this time we hardly know what the transpiration and evaporation rates are under alternative cropping systems and known meteorological and water supply conditions. Thus, we are in a poor position to evaluate existing theory and even worse off when it comes to making predictions, or, as the current phrase goes, to make a technology assessment.

But as agronomists continue to examine the value of water conserving ideas through field experiments, theory dictates that they should at least attempt to measure soil and weather parameters in some detail. They must study also crop leaf area index development, stomatal behavior, root system development, stomatal behavior, root system development and activity, and dry matter accumulation rates.

3-5 LITERATURE CITED

Cary, J. W., G. Campbell, and R. I. Papendick. 1978. Is the soil frozen or not? An algorithm using weather records. Water Resour. Res. 14:1117–1122.

Cary, J. W., R. I. Papendick, and G. S. Campbell. 197. Water and salt movement in unsaturated frozen soil: Principles and field observations. Soil Sci. Soc. Am. J. 43:3–8.

Childs, S. W., J. R. Gilley, and W. E. Splinter. 1977. A simplified model of corn growth under moisture stress. Trans. ASAE 20:858–865.

Childs, S. W., and R. J. Hanks. 1975. Model of soil salinity effects on crop growth. Soil Sci. Soc. Am. Proc. 39:617–622.

De Wit, C. T., and H. van Keulen. 1972. Simulation of transport processes in soils. PUDOC, Wageningen.

Feddes, R. A. 1981. Water use models for assessing modifications. p. 347–390. *In* G. S. Arkin and H. M. Taylor (ed.) Modifying the root environment to reduce crop stress. ASAE Monogr. 17. Am. Soc. of Agric. Engineers, St. Joseph, Mich.

Feddes, R. A., and P. E. Rijtema. 1972. Water withdrawal by plant roots. J. Hydrol. (Amsterdam) 17:33–59.

Gardner, W. R. 1974. The permeability problem. Soil Sci. 117:243–249.

Hagan, R. M., H. R. Haise, and T. W. Edminster (ed.) 1967. Irrigation of agricultural lands. Am. Soc. of Agron., Madison, Wis.

Hanks, R. J. 1974. Model for predicting plant yield as influenced by water use. Agron. J. 66: 660–664.

Hanks, R. J., H. R. Gardner, and M. L. Fairbourne. 1967. Evaporation of water from soils as influenced by drying with wind and radiation. Soil Sci. Soc. Am. Proc. 31:593–598.

Hanks, R. J., A. Klute, and E. Bresler. 1969. A numeric method for estimating infiltration, redistribution, drainage, and evaporation of water from soil. Water Res. 5:1064–1069.

Hillel, D. 1977. Computer simulation of soil-water dynamics. Int. Development Res. Centre, Ottawa.

Hillel, D. I., C. H. M. van Bavel, and H. Talpaz. 1975. Dynamic simulation of water storage in fallow soil as affected by mulch of hydrophobic aggregates. Soil Sci. Soc. Am. Proc. 39:826–833.

Item, H. 1974. Model for the water regime of a deciduous forest. J. Hydrol. (Amsterdam) 21: 201–210.

Jackson, R. D., R. J. Reginato, B. A. Kimball, and F. S. Nakyama. 1974. Diurnal soil-water evaporation: Comparison of measured and calculated soil-water fluxes. Soil Sci. Soc. Am. Proc. 38:861–866.

Jensen, M. E. 1973. Consumptive use of water and irrigation water requirements. Am. Soc. of Civil Engineers, New York.

Nielsen, D. R., R. D. Jackson, J. W. Cary, and D. D. Evans (ed.) 1972. Soil water. Am. Soc. of Agron., Inc., Madison, Wis.

Nimah, M. N., and R. J. Hanks. 1973. Model for estimating soil water and atmospheric interrelation: I. Description and sensitivity: II. Field tests of the model. Soil Sci. Soc. Am. Proc. 37:522–532.

Philip, J. R., and D. A. de Vries. 1957. Moisture movement in porous materials under temperature gradients. Trans. Am. Geophys. Union 38:222–232.

Ritchie, J. T. 1972. Model for prediction evaporation from a row crop with incomplete cover. Water Resour. Res. 8:1204–1213.

Talsma, T. 1978. Soil water interchange processes. p. 45–53. *In* Studies of the Australian arid zone, Part 3. Commonwealth Scientific Industrial Research Organization, Melbourne.

Van Bavel, C. H. M., and J. Ahmed. 1976. Dynamic simulation of water depletion in the root zone. Ecol. Modell. 2:189–212.

Van Bavel, C. H. M., and D. I. Hillel. 1976. Calculating potential and actual evaporation from a bare soil surface by simulation of concurrent flow of water and heat. Agric. Meteorol. 17:453–476.

Van Der Ploeg, R. R., and R. Benecke. 1974. Unsteady, unsaturated, n-dimensional moisture flow in soil: A computer simulation program. Soil Sci. Soc. Am. Proc. 38:881–885.

Van Keulen, H. 1975. Simulation of water use and herbage growth in arid regions. PUDOC, Wageningen.

4

Paul W. Unger
Agriculture Research Service
U.S. Department of Agriculture
Bushland, Texas

Water Conservation: Southern Great Plains

The parts of New Mexico, Oklahoma, and Texas that are within the Southern Great Plains lie between about long. 97 and 105°W and lat. 32 and 37° N. Within this region, soils range from deep sands to clays. Water storage in soil is influenced by limited and erratic precipitation, high solar radiation, high summer temperatures, low relative humidity, and high winds. Soil factors that limit water storage are low infiltration rates, which cause runoff during intense storms, and, in some soils, limited storage capacity. With clean tillage and stubble mulch tillage, precipitation storage efficiencies during fallow range from 15 to 25% in the Great Plains (Mathews and Army, 1960). Storage efficiency values generally are lowest in the southern part of the region (Johnson and Davis, 1972; Mathews and Army, 1960; Unger, 1972). Because of the low efficiencies, soils often are not filled to capacity at crop planting, and subsequent yields are low unless precipitation is timely and adequate.

Water storage with clean tillage and stubble mulch tillage tends to be low; hence, yields generally are low (Johnson and Davis, 1972). Irrigation increases crop yields, but water for irrigation is limited, and irrigation is an energy-intensive practice (Allen et al., 1977). For more reliable crop production without irrigation, a larger percentage of precipitation between crops must be stored in soil, and the stored water and growing season precipitation must be used more efficiently for grain and forage production.

4-1 WATER CONSERVATION PRACTICES

Although precipitation is limited, average annual precipitation in dryland farming areas would be adequate for favorable yields if all of it were effectively used for crop production. For example, grain sorghum in Okla-

Contribution from U.S. Department of Agriculture, Agricultural Research Service, in cooperation with the Texas Agricultural Experiment Station, Texas A&M University, Bushland.

Dryland Agriculture—Agronomy Monograph no. 23.

homa yielded 6270 kg/ha of grain using only 17.8 cm of water from soil (Griffin et al., 1966). The soil was covered with plastic to prevent evaporation and rainfall infiltration during the growing season. In the Northern Great Plains, corn on ridged and 90% plastic-covered soil yielded 4330 kg/ha of grain using 37.8 cm of water in 1960 and 3930 kg/ha of grain using 19.8 cm of water in 1961 (Willis et al., 1963). Thus, when evaporation was prevented, crop water use was well below that of average annual precipitation in the Southern Great Plains and, in some cases, below the lowest recorded annual precipitation in the region. This indicates that if water conservation and crop management were improved, yields could be greatly increased.

Improving water storage involves increasing infiltration; reducing runoff, evaporation, and transpiration; and eliminating undesirable plants. Relationships between these factors are complex. The following is a discussion of these relationships and of several principles that are generally involved in the practices for improving water conservation.

4-1.1 Summer Fallow

Two major cropping systems involving fallow are alternate wheat and fallow and wheat-sorghum-fallow. The practice of fallowing becomes more widespread as one moves westward, because precipitation decreases and more water must be stored. Fallowing, however, is not influenced by precipitation alone. Of the total dryland wheat planted from 1968 to 1970, that planted after fallow made up from 4 to 16% in the eastern part of the region and from 29 to 41% in the western part. For 1974 and 1975, these amounts ranged from 0 to 6% and from 15 to 26%, respectively. The shift to planting less wheat on fallowed land in 1974 and 1975 was attributed to improved grain prices and to the policy for increased crop production (Johnson and Unger, 1976). At a given time, grain supplies and land set-aside programs may affect the trend.

Much research with fallow and other water conservation practices has been conducted at the Research Laboratory near Bushland, Texas, on Pullman clay loam (fine, mixed, thermic Torrertic Paleustoll). Taylor et al. (1963) described this soil in detail. The Pullman and closely related soils cover about 4.86 million ha in the Oklahoma and Texas panhandles and eastern New Mexico.

Pullman clay loam at Bushland has a water storage capacity, based on −0.033- and −1.5-MPa matric potentials, of about 23 cm to the 1.8-m depth, the depth to which winter wheat often extracts water. Filling the soil to capacity is difficult, however, because of a very slowly permeable clay horizon at the 23- to 71-cm depth. The infiltration rate decreases to about 1.3 mm/hour after shrinkage cracks and other temporary storage volumes are filled (Taylor et al., 1963). To fill the profile, long wet periods or frequent additions of water are needed.

A study involving wheat-fallow and continuous wheat was conducted at Bushland, Texas, from 1941 to 1977. Tillage methods were one-way and

Table 4–1. Effect of cropping system and tillage method for winter wheat on soil water content at planting, water storage efficiency, and grain yield (Johnson and Davis, 1972).

Cropping system and tillage method	Available soil water at planting†	Water storage efficiency‡	Grain yield†
	cm	%	kg/ha
Continuous wheat			
One-way	9.1	20	585
Stubble mulch	10.3	22	685
Wheat-fallow			
One-way	12.8	10	927
Stubble mulch	15.4	15	1060
Delayed stubble mulch	14.4	13	1040

† Average for 1942 to 1969, determined to a 1.8-m depth.
‡ Average for 1958 to 1969.

stubble mulch for continuous wheat, and one-way, stubble mulch, and delayed stubble mulch for wheat-fallow. For delayed stubble mulch, tillage after harvest was delayed until weed growth began the next spring. Although wheat-fallow resulted in higher water contents at wheat planting than continuous wheat, the trend was opposite for water storage efficiency (water storage divided by precipitation for the period and multiplied by 100) (Johnson and Davis, 1972) (Table 4–1). Differences in storage efficiencies were related to the time intervals between crops, which were about 3 and 15 months for continuous wheat and wheat-fallow, respectively. Since storage is highest in dry soil, as at the time of wheat harvest, precipitation during the first summer is stored rather effectively. Thereafter, storage is lower because water moves more slowly into the soil and evaporation is high, even though the water storage reservoir is not filled to capacity.

Wheat grain yields at Bushland averaged 58 and 54% higher for one-way and stubble mulch tillage, respectively, with the wheat-fallow than with the continuous wheat system. Since yields were not doubled with wheat-fallow, the average total grain production was higher with continuous wheat. Grain yields after conventional and delayed stubble mulch tillage were similar, even though weeds grew and used water the first summer and fall after wheat harvest on the delayed tillage plots. By wheat planting time, soil water contents for the two systems were similar (Table 4–1). Delayed stubble mulch was more economical because it required about 30% less tillage than stubble mulch during the entire fallow, but delayed stubble mulch is seldom used because weed seed production is high (Johnson and Davis, 1972).

In a 13-year study involving stubble mulch tillage in wheat-fallow, continuous wheat, wheat-sorghum-fallow, and continuous sorghum systems on Pullman clay loam, water contents at planting and storage efficiencies for comparable treatments were lower (Unger, 1972) than those reported by Johnson and Davis (1972), apparently because of the precipitation variations. Grain yields were similar in both studies. In the wheat-sorghum-fallow system, water storage efficiencies during fallow before wheat and sorghum were almost identical (Table 4–2), but continuous sorghum resulted in

Table 4–2. Effect of cropping systems for winter wheat and grain sorghum on soil water content at planting, water storage efficiency, and grain yield (Unger, 1972).

Cropping system	Available soil water at planting†	Water storage efficiency	Grain yield
	cm	%	kg/ha
Continuous wheat	5.8	14.8	700
Wheat-fallow	9.3	8.3	970
Wheat-sorghum-fallow			
Wheat	8.6	13.9	850
Grain sorghum	9.4	14.0	1740
Continuous sorghum	7.6	20.1	1270

† Determined to a 1.2-m depth.

higher efficiency than continuous wheat because most of the precipitation between crops occurred within about 45 days of sorghum planting, thus the relatively high water contents at planting. For continuous wheat, precipitation between crops occurs during summer when temperatures and evaporation potentials are high, thus the low water storage. In wheat-sorghum-fallow and continuous sorghum systems, sorghum grain yields were higher than wheat grain yields (Table 4–2) because growing season precipitation was more timely for sorghum than for wheat.

In contrast to results at Bushland, where fallowing resulted in about 40 to 50% higher yields of wheat and grain sorghum than did continuous cropping, fallowing at Woodward, Oklahoma, increased yields only about 5 to 20% for several *Sorghum* species, (kafir, milo, sorgo, and broomcorn) (Locke and Mathews, 1955). The lower yield was attributed to climatic and soil differences. Annual precipitation at Woodward, which was about 120 mm higher than at Bushland, favored continuous cropping. In addition, the soils at Woodward were silt loams, sandy loams, or loamy fine sands, which filled with water more readily than the clay loam at Bushland. Because the sandier soils at Woodward also had lower storage capacities, precipitation between continuous crops was adequate to fill soils to near capacity.

Mathews and Brown (1938) compared fallowing on sandy loam soils at Dalhart, Texas, and Woodward, Oklahoma. Annual precipitation at Dalhart was about 125 mm lower than at Woodward. At Dalhart, both row crops and wheat responded to fallow. At Woodward, wheat responded some, but row crops responded very little to fallow. Row crops responded poorly at Woodward because the soil generally was filled to capacity at planting, even on annually cropped land. Wheat at Woodward responded to fallowing because time between harvest and planting with continuous cropping was too short to permit water storage equal to that stored in fallowed land. When water storage between crops approached that on fallowed land, fallowing did not increase yields. The strong response to fallowing at Dalhart resulted from the lower precipitation, which was too low to fill the soil with continuous cropping. Similar results were obtained on sandy soils at Big Spring, Texas, where most crops yielded more after fallow than with continuous cropping (Keating and Mathews, 1957). Soils and precipitation were similar to those at Dalhart.

Fallowing generally increased yields on harvested areas, especially in the drier western part, but not enough so that total production with fallow equaled that with continuous cropping. However, where large amounts of land are farmed by one individual and production costs are considered, fallowing may be more economical; in addition, fallowing stabilizes crop production.

Although average water storage during fallow is low, storage during individual fallow periods can vary greatly. In general, storage is highest when the soil is dry at crop harvest and precipitation is adequate, and lowest when the soil is wet at harvest and precipitation is low. Water may be lost during fallow when water contents are high at crop harvest (Unger, 1972). Because water contents at harvest vary greatly due to precipitation near harvest, flexible cropping systems are needed. When soil water is adequate at harvest, continuous cropping, or possibly even double cropping with a forage or other short-season crop, would result in more efficient use of precipitation than with cropping systems involving fallow.

4-1.2 Mulching

Mulching is perhaps as old as agriculture itself (Jacks et al., 1955). Ancient Romans placed stones and the Chinese placed pebbles from stream beds on soil to conserve water. These and similar practices were practical when hand labor was plentiful, but they are impractical for modern, large-scale, mechanized agriculture. The current trend is to use crop residues on farmland and artificial mulches for some high-value crops (Unger, 1975).

Mulches conserve water by controlling storm water runoff, increasing infiltration, decreasing evaporation, and aiding in weed control. If they control evaporation, mulches increase yields with less soil water use than the average annual precipitation in dryland crop areas (Griffin et al., 1966; Willis et al., 1963). Many studies have been conducted involving mulches for water conservation. Plastic, paper, crude oil, and gravel mulches generally increased soil water contents and crop yields (Unger, 1971a, b; Wendt, 1973a, b; Wendt and Runkles, 1969). These mulches have potential for use on relatively small areas for high-value crops, but they are neither practical nor economical for large areas. Consequently, crop residue mulches have received much attention for water conservation in practical farming situations. The effectiveness of grown-in-place mulches for conserving water is limited in the Southern Great Plains, however, because of the meager residue generally produced by dryland crops. Even when these residues were left on the soil by stubble mulch tillage or chemical fallow, water storage and crop yields often were low (Johnson and Davis, 1972; Mathews and Army, 1960; McCalla and Army, 1961; Unger, 1972; Wiese et al., 1967; Zingg and Whitfield, 1957) because the residue amounts were too low to enhance infiltration and decrease evaporation.

Greb et al. (1967, 1970) showed that increasing amounts of wheat straw mulch on the soil during fallow increased water storage in the Central and Northern Great Plains. With 0 to 6720 kg/ha of mulch, precipitation stored

Table 4-3. Straw mulch effects on water storage efficiency and grain sorghum yield (Unger, 1978).

Mulch rate	Water storage efficiency†	Grain yield
t/ha	%	kg/ha
0	22.6c‡	1780c‡
1	31.1b	2410b
2	31.4b	2600b
4	36.5b	2980b
8	43.7a	3680a
12	46.2a	3990a

† Water storage determined to a 1.8-m depth. Precipitation averaged 318 mm.
‡ Column values followed by the same letter are not significantly different at the 5% level (Duncan multiple range test).

ranged from 16 to 37%, respectively. Unger (1978) obtained similar increases when he placed 0 to 12 t/ha of wheat straw on Pullman clay loam. Subsequent sorghum grain yields with 8 and 12 t/ha of mulch were about double those with no mulch (Table 4-3).

Because irrigated crops such as winter wheat produce more residue than dryland crops, recent studies on Pullman clay loam have involved the management of irrigated wheat residue during fallow. Unger et al. (1971) used combinations of disk, sweep, and herbicide treatments for weed and volunteer wheat control during fallow from wheat harvest in July until the following June. The irrigated wheat produced 11 000 kg/ha of residue. Water storage with herbicide treatments was about double that of the average for tillage-only treatments.

In 1970, winter wheat was planted after irrigated corn without tillage or after rotary tillage and on a fallowed area to obtain high-, medium-, and low-residue levels, respectively. Within these treatments, different amounts of residue production resulted from irrigating the wheat one to five times. Low-residue treatment plots were disked once after wheat harvest to incorporate some residues with soil. All plots were treated with atrazine [2-chloro-4-(ethylamino)-6-(isopropylamino)-*s*-triazine] and 2,4-D [(2,4-dichlorophenoxy)acetic acid] at 3.4 and 1.1 kg/ha, respectively, to control weeds and volunteer wheat. Surface residues, water storage during fallow, and sorghum grain yields were significantly affected by the treatments. Precipitation storage ranged from 11 to 45% and generally increased as residue levels increased. The highest residue level, however, did not result in the most storage, possibly because initial soil water contents were higher or precipitation was intercepted by residues. The sorghum, planted without tillage and not irrigated, yielded from 2970 to 6010 kg/ha of grain. The high yield was about 2240 kg/ha greater than sorghum yielded on an adjacent fallowed area where conventional production practices were used (Unger and Parker, 1975).

No-tillage, sweep tillage, and disk tillage methods were used during fallow for wheat residue management and weed control in an irrigated winter wheat–dryland grain sorghum–fallow cropping system at Bushland. Precipitation storage averaged 35.2, 22.7, and 15.2%; plant available water to a 1.8-m depth at sorghum planting averaged 21.7, 17.0, and 15.2 cm; sorghum grain yields averaged 3140; 2500; and 1930 kg/ha; and water use efficiencies for grain averaged 88.6, 77.3, and 66.3 kg ha^{-1} cm^{-1} for the respective treatments (Unger and Wiese, 1979).

In the Southern Great Plains, cotton bur or gin trash mulches often are used to control wind erosion on sandy soils. Besides controlling erosion, these mulches increase precipitation storage. On Amarillo sandy clay loam (fine-loamy, mixed, thermic Aridic Paleustalf) at Big Spring, Texas, the gain in soil water was nearly 30% as surface coverage by mulch increased from 0 to 100% (Fryrear and Koshi, 1971). About 11 t/ha of gin trash completely covered the surface. For the 0, 11.2, and 22.4 t/ha gin trash treatments, storage efficiencies averaged 41, 58, and 73% with 337 mm of average precipitation in 1968 and 1969. Soil water content was increased to a 3-m depth, and cotton lint yields averaged 197, 260, and 282 kg/ha with the respective treatments.

Davidson and Santelmann (1973) evaluated effects of stubble mulch tillage, clean tillage, chemicals, and no weed control on soil conditions and wheat yields on Grant silt loam (fine-silty, mixed, thermic Udic Argiustoll) at Cherokee, Oklahoma. Although minimum or no-tillage treatments (stubble mulch or chemical) resulted in equal or better soil physical conditions and higher water contents, average grain yields were higher with clean tillage (moldboard plowing). The effect on yield seemed inversely proportional to the amount of residue on the soil in May. At residue sampling in May, wheat seedlings on high-residue plots were chlorotic, similar to that caused by a N deficiency. Nitrogen applications, however, did not change the results, and the reason for plant chlorosis under high-residue conditions has not yet been determined.

4–1.3 Weed Control

Uncontrolled weed growth decreases water storage, and weeds and volunteer crop plants compete strongly with planted crops for stored water. Weeds are strong competitors because of their vigorous growth habits and because their root systems are often more extensive than those of planted crops (Davis et al., 1965, 1967). Competition between weeds and crops is influenced also by the soil water conditions under which they grow (Wiese and Vandiver, 1970). For example, corn, barnyard grass, cocklebur, sorghum, and crabgrass grew best in wet soil and less in relatively dry soil. In contrast, kochia, Russian thistle, buffalobur, and tumblegrass grew much less under wet conditions than the more competitive species, but relatively dry soil did not decrease their growth. In this test, species were grown in competition with each other.

To obtain maximum water storage for a subsequent crop, land generally must be kept weed free. Provided weed control was similar, the control method (disk, sweep, or chemical) had minor effects on soil water content throughout fallow and at planting of the next crop when initial surface resi-

Table 4–4. Effect of tillage frequency and timing during 11-month fallow on average number of tillage operations, soil water, and grain yields in a wheat-sorghum-fallow system (Lavake and Wiese, 1979).

Tillage treatment	For wheat crop			For sorghum crop		
	Tillage operations	Soil water at planting†	Grain yield	Tillage operations	Soil water at planting†	Grain yield
	Number	cm	kg/ha	Number	cm	kg/ha
Every 2 weeks	10.3	11.8a‡	567ab‡	10.6	9.0a‡	2410ab‡
Days after weed emergence						
4	5.3	11.4a	629a	6.1	9.0a	2600a
10	4.3	10.7ab	583ab	5.1	8.9a	2530a
17	3.6	9.7b	564ab	4.0	8.4a	2100bc
24	2.7	9.1b	500b	4.0	7.9a	1900c

† Plant available water determined to a 1.2-m depth.
‡ Column values followed by the same letter or letters are not significantly different at the 5% level (Duncan multiple range test).

due levels were low (Johnson and Davis, 1972; Wiese and Army, 1958). The differences noted (Table 4–1; Wiese and Army, 1958) may have resulted from the different amounts of residue that the various tillage methods maintained on soil. With high-residue levels, however, the weed control method had a major effect on water storage (Unger et al., 1971; Unger and Wiese, 1979).

Timeliness of weed control has a major effect on water storage. Controlling weeds with sweep tillage at 4, 10, 17, or 24 days after emergence or by repeated sweep tillage at 2-week intervals during the growing season affected water storage during fallow in a wheat-sorghum-fallow system. For both crops, controlling weeds at 4 or 10 days after emergence or at 2-week intervals did not significantly affect soil water at planting or grain yields (Table 4–4), but delaying weed control until 17 or 24 days after weed emergence decreased soil water content and yield (Lavake and Wiese, 1979). Because repeated tillage did not increase water content or yield, tillage can be delayed until weeds use more water than that lost by evaporation, thereby affording major savings in production costs and energy. Similar findings were reported by Wiese (1960).

As with tillage, timeliness of weed and volunteer plant control with herbicides is important for water conservation. Treating 15-cm tall pigweed plants with 2,4-D at 0.28 kg/ha stopped growth temporarily but did not kill the plants. Transpiration rate decreased within 2 days and remained depressed throughout the study. Smaller plants undoubtedly would have been killed. Sorghum plants treated with a toxic oil collapsed within 1 day and stopped transpiring within 6 days. For sorghum treated with dalapon (2,2-dichloropropionic acid) at 11.2 kg/ha, evapotranspiration decreased to the same or lower rate than evaporation from bare soil within 3 days. Treating 15-, 17-, or 34-cm tall soybean plants with 2,4-D at 1.12 kg/ha reduced evapotranspiration more than when it was applied at 0.56 kg/ha, regardless of plant height. The 0.56-kg/ha rate reduced evapotranspiration but did not kill soybean plants. Soybean plants 15 cm tall were killed within 2 days by 1.12 kg/ha of 2,4-D, and evapotranspiration by these plants and evaporation from bare soil were similar. The 17-cm tall plants treated with 1.12 kg/ha of 2,4-D died slowly. At 16 days, evapotranspiration was the same as evaporation from bare soil. The 2,4-D did not kill the largest soybean plants, and evapotranspiration remained higher than evaporation from bare soil throughout the study (Wiese et al., 1966). Under field conditions, delayed or incomplete weed kill by herbicides, therefore, would be detrimental to water conservation between crops and while crops are growing. Because weeds continue to use water for several days after herbicide application, whereas tillage kills weeds almost immediately, herbicides should be used earlier than tillage for effective water conservation.

4–1.4 Vertical Mulching

When precipitation rates exceed infiltration rates during intense rainstorms, the excess water normally flows laterally across the surface and collects in depressions or flows into streams. To prevent runoff water from leaving the field, vertical mulching was developed to quickly channel the water into soil. Wendt (1973c) found no benefit from vertically mulching Olton loam (fine, mixed, thermic Aridic Paleustoll) in 1970 because maximum daily rainfall was 31 mm, apparently not enough to cause runoff. In 1971, water content differences among treatments were slight from March until early August because the highest monthly rainfall (in May) was only 56 mm. Rainfall in late August, September, and October (159, 129, and 51 mm, respectively) caused higher water contents at the 30- to 90-cm depth in vertical-mulched plots and the water contents remained higher until the sorghum was killed by frost. Sorghum grain yields were 2090, 2490, and 3110 kg/ha on check, vertical mulch, and vertical mulch plots with oil (sprayed on the soil), respectively. The differences were significant.

Seemingly, vertical mulching would be most beneficial for water conservation on slowly permeable soils, such as Pullman clay loam. However, no vertical mulching studies have been conducted on this soil under dryland conditions. Vertical mulching is not practical for large-scale farming.

4–1.5 Terracing

Runoff from cropland may occur during any month in the Southern Great Plains. Runoff is influenced by crops grown and soil surface conditions as well as other factors. At Bushland, average runoff was low from areas cropped to winter wheat or in fallow after wheat. It was much higher from adjacent areas cropped to grain sorghum or in fallow after grain sorghum (Fig. 4–1) (Jones and Hauser, 1975). To control erosion and conserve runoff water, various types of terraces were studied.

Graded and level terraces have been evaluated for water storage and crop yields at several locations. Early results showed that runoff was lower

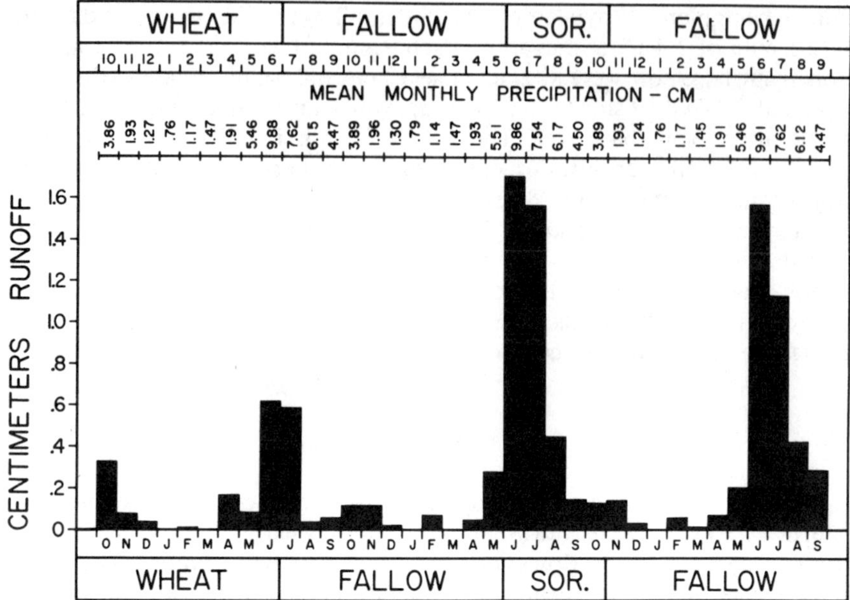

Fig. 4–1. Fourteen-year (1959–1972) mean monthly precipitation and runoff from three graded terraces cropped in a 3-year wheat-sorghum-fallow sequence. One terrace was in each phase of the sequence every year (Jones and Hauser, 1975).

and water contents and crop yields were higher on graded terrace fields than on nonterraced fields (Dickson et al., 1940; Finnell, 1944). Soil water content and yields generally were even higher with level terraces, especially with ends closed and contour furrows between terraces, or when areas between terraces were leveled (Burnett and Fisher, 1956; Dickson et al., 1940; Fisher and Burnett, 1953). When Foard silt loam (fine, montmorillonitic, thermic Typic Natrustoll) in Oklahoma was kept relatively smooth for wheat, runoff water was stored mostly on the upslope side of closed-end level terraces. During dry seasons, this area produced higher wheat yields, but during wet seasons, wheat yield in terrace channels was often less than that on intervals between terraces. During wet periods, water had to be drained from terrace channels to prevent damage to wheat (Harper, 1941).

On Pullman clay loam, wheat and grain sorghum yields from 1949 to 1960 were similar with graded and closed-end level terraces in a wheat-sorghum-fallow system. Since yields were similar, Hauser et al. (1962) suggested using open-end level terraces, which would avoid both the need for high terrace ridges to store large volumes of runoff and the need to drain water from terrace channels during wet periods.

To spread potential runoff water over larger areas and to decrease the need for high terrace ridges and draining of terrace channels during wet periods, bench terraces, with and without a contributing watershed, were evaluated at Bushland and Big Spring, Texas. At Bushland, bench terraces with a contributing watershed (conservation bench terraces) were constructed in 1955 on land with 1.0 to 1.8% slope. The watershed to bench ratio was 2:1. The benches were continuously cropped to grain sorghum, and the watersheds were in a wheat-sorghum-fallow sequence. The conservation bench terraces controlled runoff water, prevented water erosion, and uniformly distributed runoff water for storage in soil on the leveled area (Hauser, 1968; Zingg and Hauser, 1959). Water storage was greater on level benches of the conservation bench terraces than on level benches without a watershed (also under annual cropping) and about the same as on level-terraced fields that were fallowed for 11 months. Crop production was about 1.5 times more with the conservation bench terraces than with level terraces because all of the level-terraced area was cropped in a wheat-sorghum-fallow system, whereas the benches of the conservation bench terraces were continuously cropped to grain sorghum. Total production was highest on bench terraces without a watershed because these terraces were annually cropped to sorghum. Further evaluation of the systems for 1959 to 1972 substantiated earlier results (Jones and Hauser, 1975). Again, overall production was highest on level benches without a watershed, but with this system, the probability of poor yields increased due to low water storage in dry years. The major advantage of the conservation bench terraces over level benches without a watershed was that only one third of the area was leveled for the conservation bench terraces. The higher yields on benches without a watershed were not adequate to offset the additional construction costs compared with yields and construction costs for the conservation bench terraces (Jones and Shipley, 1975). To further decrease construction costs, Jones (1981) developed a system of conservation bench terraces that employed narrow benches.

Armbrust and Welch (1966) evaluated the conservation bench terraces at Big Spring on Amarillo fine sandy loam with a 1.3 to 1.9% slope. Watershed to level bench ratios were 2:1, 4:1, and 6:1. A level bench without a watershed and a check treatment were also evaluated. Using the conservation bench terraces did not increase cotton or grain sorghum yields. The sandy loam had a high infiltration rate and an available water storage capacity of only 10.2 cm to a 1.2-m depth. Because runoff occurred only from large, high-intensity rains or rains that followed previous rains within 1 or 2 days, the soil water reservoir was filled before impounding the runoff by level benches. The impounded runoff water, therefore, was lost through deep percolation.

4–1.6 Contouring

Contouring is normally practiced in conjunction with terracing. The extra water stored with contouring and terracing rather than with terracing alone results from potential runoff water being held on a major portion of a field. When listers are used, each ridge serves as a miniature terrace. At Spur, Texas, runoff averaged 3.6, 2.2, and 0.0 cm from areas with sloping rows, contoured rows, and closed-end level terraces, respectively. Cotton lint yields were 114, 138, and 168 kg/ha on the respective areas (Dickson et al., 1940). For the period from 1927 to 1952, runoff from the three areas averaged 7.0, 5.0, and 0.0 cm, and cotton lint yields averaged 131, 164, and 211 kg/ha (Fisher and Burnett, 1953).

Finnell (1944) attributed only 56 kg/ha of a total dryland wheat yield of 557 kg/ha in 11 areas to contouring. He attributed the relatively poor response of wheat to contouring compared with that of row crops to the flat tillage methods (sweep or one-way plowing) used for wheat. For row crops, ridges were normally formed and maintained during tillage, planting, and cultivation.

4–1.7 Basin Listing and Furrow Blocking

Basin listing was a practice introduced into the Southern Great Plains in the 1930's to more uniformly hold and distribute potential runoff water over the entire field. By 1950, however, the practice was little used because of the slowness of the operation, difficulties of weed control and seedbed preparation, furrow planting, subsequent tillage, and greater erosion when dams washed out during high rainfall. Stubble mulching, terracing, and other conservation practices that were easier to manage became more popular (Clark and Hudspeth, 1976), and yield increases with basin listing were slight (Daniel, 1950; Locke and Mathews, 1953).

With better background information and modern technology, Clark and Hudspeth (1976) felt that most of the problems that caused early failures with basin listing could be overcome. Long-term data for Bushland (Fig. 4–1) showed that runoff was highest just before or soon after the time that grain sorghum and cotton are normally planted. By blocking furrows at this time rather than during fallow after wheat when runoff is low, as was done previously, the extra water could be used almost immediately, and evaporation losses would be minimized.

In early basin listing studies, weeds were controlled by cultivation. Now herbicides are used, and many fields are not cultivated after planting. When cultivation is necessary, furrow blocks can be removed by a device ahead of the tractor, the crop is cultivated, and the blocks are reestablished, all in one operation.

Clark and Hudspeth (1976) reduced dam overtopping and subsequent erosion by using large furrows to increase furrow storage capacity. Also, the crop used the stored water between rains, thus increasing subsequent storage capacity and efficiency. Furrow blocking on Pullman clay loam resulted in higher water contents throughout the growing season for grain sorghum than those resulting from open furrows and flat planting. In 1975, grain yields were 2920, 2580, and 2470 kg/ha on blocked-furrow, open-furrow, and flat-planted areas, respectively. On Amarillo fine sandy loam with a 0.2% slope at Lubbock, runoff occurred from open furrows but not from blocked furrows. Cotton lint yields were 224 and 279 kg/ha with open and blocked furrows, respectively.

4-1.8 Planting Dates

Within limits, dryland crops are planted when soil water conditions are favorable. Wheat is normally planted between 20 August and 15 October, but most is planted by 1 October if soil water conditions are favorable. If the wheat is to be grazed, planting may be earlier, but early-planted wheat may deplete most available water during fall and winter, making grain production largely dependent on spring rainfall. Early planting also increases the potential for damage by insects and diseases. According to Porter et al. (1952), the most desirable time to plant wheat is from 20 September to 10 October. Delayed planting decreases water use by plants, thus conserving the water for later use (Hanway, 1976). However, wheat planted after 15 October may not provide enough ground cover for erosion control. Another consideration is that not all water conserved by late planting is available for plant use because of evaporation.

Planting dates strongly influence growing season length for winter wheat but have less influence on grain sorghum. The growing season for sorghum is largely determined by its maturity class. Although rainfall distribution may influence growing season length, and thereby water use, little water can be conserved by varying sorghum planting dates. Sorghum should be planted so that critical growth stages (booting to grain filling) occur when the probability for rainfall is highest (July to August), yet within limits imposed by soil and climate factors early in the season.

Grain sorghum planting dates are affected not only by soil water conditions but also by location and elevation. In the Texas Rolling Plains, grain yields are similar for April, May, or June plantings, provided soil water and rainfall conditions are similar. At higher elevations, as in the Texas High Plains, temperatures become favorable about 15 May, but grain yields are generally higher when sorghum is planted between 10 and 25 June (Quinby et al., 1958). Sorghum planted after 1 July in the Texas High Plains may be killed by early frost before grain maturity. However, if frost occurs at near-normal dates, yields comparable to those with late-May plantings are possible (Unger and Parker, 1975).

With adequate soil water, time from sunflower emergence to full flowering was 80 and 51 days for 4 April and 27 June plantings, respectively (Unger et al., 1976), suggesting a potential for conserving water by delayed planting. However, total water use may have been similar because of different evapotranspiration at different times in the year. Under limited soil water conditions, water use by sunflower was strongly influenced by growing season rainfall distribution, which also strongly influenced yields (Jones and Unger, 1977). Because rainfall is erratic, sunflower should be planted so that critical growth stages (budding to late flowering) occur when rainfall probability is highest (June, July, or August), with the same early season limitations as for sorghum.

4-1.9 Antitranspirants

Over 99% of the water absorbed by plant roots is lost by transpiration from leaves; therefore, a potential means of conserving water in crop production is reducing transpiration by stomatal control. Factors controlling stomatal movement have been studied extensively and reviewed by Stone (1978) and Zelitch (1963, 1965), but stomatal control is not fully understood (Zelitch, 1965). Consequently, stomatal control for water conservation is not common. Research regarding stomatal control has been limited in the Southern Great Plains (Stone, 1978; Wendt, 1973d).

4-2 EQUIPMENT

The selection of tillage and planting equipment for dryland crop production was discussed in a report by the Food and Agricultural Organization of the United Nations (1971). The equipment used is dictated largely by the management system used. Equipment, especially tillage equipment, differs when clean or conservation tillage systems are used.

4-2.1 Clean Tillage

In a clean-tillage system, most crop residues are incorporated into soil, and weeds are controlled by implements that partially or completely invert the surface layer. One-way disk plows are widely used, but tandem and offset disks have become popular in recent years. Generally, each operation with any of these plows incorporates about 50% of the surface residues. After three or four operations, the surface is essentially bare. In some cases, weed growth after initial plowing is controlled with sweep plows, rod weeders, or other implements that do not invert the soil.

For row crops, such as cotton and grain sorghum, listers may be used once or twice before planting to prepare the seedbed. Weeds on lister ridges are controlled with sweep cultivators, rolling cultivators, or sweep rod weeders.

In eastern areas of the Southern Great Plains, soil is frequently tilled initially with a moldboard plow, which inverts the surface layer and loosens compact soil to the tillage depth. Then tandem disks, sweep plows, rod weeders, or listers are used.

4-2.2 Conservation Tillage

One goal of conservation tillage is to keep crop residues on the surface, which improves erosion control and water storage. Common implements that till soil beneath the residues are sweep plows and rod weeders. Unlike the large (up to 213 cm) sweep or blade plows used in the Central and Northern Great Plains, plows commonly used in the Southern Great Plains have sweeps 50 to 107 cm wide. Sweep plows reduce surface residues about 10% with each operation.

Repeated sweep tillage may be used for weed control between crops. Sometimes rod weeders are used after the soil has been initially loosened by sweep tillage. When residue amounts are excessive or weed growth is lush after prolonged wet weather, one-way or tandem disk plows may be used once to partially reduce the residues or to control weeds.

For reduced tillage, limited tillage, or no-tillage systems, weeds are controlled primarily with herbicides. Occasionally, when herbicides do not control weeds, they are controlled by one or more sweep plow operations. Sweep plowing also loosens surface soil that has been compacted by equipment traffic, livestock, or natural settling.

4-2.3 Planting Equipment

Small grains are generally planted with shovel, hoe, or disk opener drills. Shovel opener drills work well for placing seeds in moist soil that is overlain by dry surface soil, and they form ridges that help control wind erosion. Shovel opener drills perform well in both clean tillage and some conservation tillage systems because of their high clearance and their staggered shanks that support the openers and seed spouts. They do not perform satisfactorily when large amounts of residue are on the soil surface.

Disk openers do not ridge the soil as much as shovel openers do; hence, they are less satisfactory for planting through dry surface soil and less effective for controlling erosion. Disk opener drills also tend to destroy surface clods remaining from previous tillage, which further decreases their effectiveness for controlling wind erosion. If disks are large and spaced wider (25 to 35 cm) than on conventional drills (18 to 25 cm), disk opener drills perform satisfactorily when there are large amounts of residue on the soil. Disk opener drills with close spacing are well suited to planting after clean tillage, especially when the soil is moist at or near the surface.

Lister planters are used widely for row crops, such as sorghum and cotton. The listers open furrows into moist soil and are followed by planting units that have disk, shoe, or shovel openers. Lister ridges help control erosion. Drills with some of the seed spouts closed are sometimes used to plant row crops on lister ridges.

Planters without listers are frequently used to plant row crops on the ridges of lister-plowed land, on flat-plowed land, and on no-tillage land. Where residues do not interfere, double-disk, sweep, shoe, or shovel openers work satisfactorily. Disk openers can operate in more residues than the other types, but with high residues, coulters may be needed to cut residues ahead of the openers.

Table 4–5. Advantages and disadvantages of various water conservation practices.

Practice	Advantages	Disadvantages
Summer fallow	Increases soil water content at planting Increases production reliability (fewer crop failures) Improves control of troublesome weeds Increases yields on harvested-area basis Is possibly more economical than continuous cropping	Results in low water storage efficiency Results in low yields on total area basis Produces less than one crop annually Lowers organic matter content
Mulching	Increases soil water storage Reduces storm water runoff Reduces evaporation Moderates soil temperatures Increases yields Reduces erosion	May require special tillage and planting equipment May require repeated tillage if weeds continue to grow from rainfall occurring soon after tillage May require tillage if herbicides do not control weeds Carries ''trash farmer'' stigma Requires above-average management May harbor insects, disease pathogens, and small animals May not be economical except for high-value crops
Weed control	Conserves water for crop use Reduces tillage and planting problems Reduces competition with crops for water, nutrients, space, and light Increases yields	May leave surface free of residues, thus permitting erosion
Vertical mulch	Provides path for rapid water entry into soil Reduces runoff and erosion Conserves water for crop use Increases yields	May be costly May not have enough residues grown in place to fill trench Is difficult to maintain slots open to surface Requires intensive management
Terracing	Controls runoff water Reduces erosion hazard Conserves water for crop use Increases yields	May have high initial costs Difficulty in farming with large equipment May need to drain channels during long wet periods Severe rainstorms may cause washouts from overtopping
Contouring	Controls runoff water Reduces erosion hazard Conserves water for crop use Increases yields	Causes difficulty in farming with large equipment Severe rainstorms may cause washouts from overtopping
Basin listing–furrow blocking	Controls runoff water Reduces erosion hazard Conserves water uniformly on area for crop use Increases yields	Requires special equipment for building and removing blocks Severe rainstorms may cause washouts from overtopping

(continued on next page)

Table 4–5. Continued.

Practice	Advantages	Disadvantages
Planting dates	Reduces total evapotranspiration Increases water supply at critical growth stages	May not have favorable soil water conditions at best time for planting Correlates poorly with yields because of low probability of rainfall at critical periods
Use of antitranspirants	Reduces water use Increases yields	Mechanisms of stomatal control not fully established Must apply materials frequently May cause variable results

4–2.4 Specialized Equipment

Most dryland crop production operations are performed with readily available equipment. However, where furrow blocking is used, equipment is needed that builds the blocks and then removes them before cultivation or, if necessary, before harvest. Such special equipment has been developed by Lyle and Dixon (1977) and is commercially available.

4–3 ADVANTAGES AND DISADVANTAGES OF PRACTICES

The advantages and disadvantages listed in Table 4–5 for various water-conserving practices are based on the previous discussions, comments in the literature and by co-workers, and personal experiences. Although the advantages and disadvantages listed are not applicable under all conditions, they are presented without discussion. For example, stubble mulching certainly has advantages different from those of plastic mulches. Where comparative statements are made, the practice listed is compared with no use of the practices, e.g., summer fallow vs. continuous cropping or mulching vs. no mulching.

4–4 PROMISING RESEARCH DEVELOPMENTS

Some water conservation practices that are currently used in the Southern Great Plains are based on research that was conducted more than 30 years ago. Even though the research is still applicable in many cases, recent research has led to improved water conservation. Among the most promising developments are improved herbicides for weed control, improved management practices, and improved and larger implements. Herbicides control weeds and decrease the need for frequent tillage. With less tillage, crop residues can be kept on soils, which enhances water infiltration, decreases evaporation, and helps to control erosion.

Probably the foremost crop management practice developed in recent years is the no-tillage cropping system, which decreases labor, energy, and equipment needs. Although results with no-tillage are limited and may not be as spectacular in the Southern Great Plains as in some more humid locations, these systems have potential for becoming the prime production method wherever crops are grown. Equipment is available for performing no-tillage operations, such as spraying, planting, and harvesting, but careful management is necessary for successful no-tillage crop production.

Other machinery has been developed that permits rapid and more timely weed control in conventional tillage systems; therefore, water use by weeds has decreased. New machinery, along with improved herbicides, has also created renewed interest in furrow blocking for conserving water.

4-5 RESEARCH NEEDS

More tillage than a scientifically determined minimum is undesirable because it exposes the soil to erosion by decreasing surface residues, and it causes water loss by evaporation (Hanway, 1976). Therefore, our ultimate goal should be crop production and water conservation on dryland without tillage. If systems such as no-tillage are to be widely accepted throughout the Southern Great Plains for dryland crop production, research is needed on the following subjects.

More Effective Herbicides. Highly effective and specific herbicides are needed to control weeds and volunteer crop plants between crops and within growing crops. Herbicides must not unduly affect subsequent crops.

Improved Equipment. Although planters are available for no-tillage crop production, residues sometimes clog them, which makes planting slow and causes uneven plant populations. Planters capable of trouble-free operation and of producing uniform populations under high-residue conditions are needed.

Better-adapted Cultivars. Present cultivars were developed for clean-tillage conditions. Cultivars need to be adapted for the higher residue conditions, higher soil water contents, lower soil temperatures, and different fertilizer status than those under older conventional dryland conditions.

Increased Applicability of Research Results. Most conservation tillage research has been conducted on a limited number of soils at a few locations. For general applicability, research must be conducted on more soils at different locations.

More Comprehensive Research. Most tillage research involves two to four controlled variables that are studied by one or two researchers. Research and development teams are needed to simultaneously study more variables and to develop widely applicable, practical, and functional integrated cropping systems.

Better Education and Extension. Even though improved water conservation practices are available for most crops, ineffective practices are still widely used. To promote adoption of improved practices, producers must be informed of these practices through education and extension activities. Tradition has no place in crop production when water, soil, and other resources are not effectively conserved.

4-6 LITERATURE CITED

Allen, R. R., B. A. Stewart, and P. W. Unger. 1977. Conservation tillage and energy. J. Soil Water Conserv. 32:84–87.

Armbrust, D. V., and N. H. Welch. 1966. Evaluation of Zingg conservation bench terraces on Amarillo fine sandy loam soil. J. Soil Water Conserv. 21:224–226.

Burnett, E., and C. E. Fisher. 1956. Land leveling increases dryland cotton production. Texas Agric. Exp. Stn. Progress Report no. 1914.

Clark, R. N., and E. B. Hudspeth. 1976. Runoff control for summer crop production in the Southern Plains. Am. Soc. of Agric. Engineers, Paper no. 76-2008.

Daniel, H. A. 1950. Water conservation for wheat production in Oklahoma. Soil Sci. Soc. Am. Proc. 15:408–412.

Davidson, J. M., and P. W. Santelmann. 1973. An evaluation of various tillage systems for wheat. Oklahoma Agric. Exp. Stn. Bull. no. B-711.

Davis, R. G., W. C. Johnson, and F. O. Wood. 1967. Weed root profiles. Agron. J. 59:555–556.

Davis, R. G., A. F. Wiese, and J. L. Pafford. 1965. Root moisture extraction profiles of various weeds. Weeds 13:98–100.

Dickson, R. E., B. C. Langley, and C. E. Fisher. 1940. Water and soil conservation experiments at Spur, Texas. Texas Agric. Exp. Stn. Bull. no. 587.

Finnell, H. H. 1944. Water conservation in Great Plains wheat production. Texas Agric. Exp. Stn. Bull. no. 655.

Fisher, C. E., and E. Burnett. 1953. Conservation and utilization of soil moisture. Texas Agric. Exp. Stn. Bull. no. 767.

Food and Agricultural Organization of the United Nations. 1971. Tillage and seeding practices and machines for crop production in semiarid areas. FAO Agric. Development Paper no. 92.

Fryrear, D. W., and P. T. Koshi. 1971. Conservation of sandy soils with a surface mulch. Trans. ASAE 14:492–495, 499.

Greb, B. W., D. E. Smika, and A. L. Black. 1967. Effect of straw-mulch rates on soil water storage during summer fallow in the Great Plains. Soil Sci. Soc. Am. Proc. 31:556–559.

Greb, B. W., D. E. Smika, and A. L. Black. 1970. Water conservation with stubble mulch fallow. J. Soil Water Conserv. 25:58–62.

Griffin, R. H., II, B. J. Ott, and J. F. Stone. 1966. Effect of water management and surface applied barriers on yields and moisture utilization of grain sorghum in the southern Great Plains. Agron. J. 58:449–452.

Hanway, D. G. 1976. Cultural advances in integrated dryland production systems. p. 210–227. *In* Conservation tillage. Great Plains Agric. Coun. Pub. no. 77, Univ. of Nebraska, Lincoln.

Harper, H. J. 1941. The effect of terrace ridges on the production of winter wheat. Soil Sci. Soc. Am. Proc. 6:474–479.

Hauser, V. L. 1968. Conservation bench terraces in Texas. Trans. ASAE 11:385–386, 392.

Hauser, V. L., C. E. Van Doren, and J. S. Robins. 1962. A comparison of level and graded terraces in the southern High Plains. Trans. ASAE 5:75–77.

Jacks, G. V., W. D. Brind, and R. Smith. 1955. Mulching. Commonwealth Bureau of Soil Sci. (England) Technical Commun. no. 49.

Johnson, W. C., and R. G. Davis. 1972. Research on stubble-mulch farming of winter wheat. USDA Conserv. Res. Report no. 16.

Johnson, W. C., and P. W. Unger. 1976. Mechanical fallow in southern Great Plains. p. 94–123. *In* Conservation tillage, Great Plains Agric. Coun. Pub. no. 77, Univ. of Nebraska, Lincoln.

Jones, O. R. 1981. Land forming effects on dryland sorghum production in the Southern Great Plains. Soil Sci. Soc. Am. J. 45:606–611.

Jones, O. R., and V. L. Hauser. 1975. Runoff utilization for grain production. *In* G. W. Frasier (ed.) Water Harvest Symp. Proc., Phoenix, Ariz. 26–28 March 1974. USDA Report no. ARS-W-22.

Jones, O. R., and J. L. Shipley. 1975. Economics of land leveling for dryland grain production. J. Soil Water Conserv. 30:177–181.

Jones, O. R., and P. W. Unger. 1977. Soil water effects on sunflowers in the southern High Plains. Sunflower Forum Proc. 2:12–16.

Keating, F. E., and O. R. Mathews. 1957. Soil and crop studies at the Big Spring (Texas) Field Station. USDA Production Res. Report no. 1.

Lavake, D. E., and A. F. Wiese. 1979. Influence of weed growth and tillage interval during fallow on water storage, soil nitrates, and yield. Soil Sci. Soc. Am. J. 43:565–569.

Locke, L. F., and O. R. Mathews. 1953. Relation of cultural practices to winter wheat production, Southern Great Plains Field Station, Woodward, Oklahoma. USDA Circ. no. 917.

Locke, L. F., and O. R. Mathews. 1955. Cultural practices for sorghums and miscellaneous field crops. USDA Circ. no. 959.

Lyle, W. M., and D. R. Dixon. 1977. Basin tillage for rainfall retention. Trans. ASAE 20: 1013–1017.

Mathews, O. R., and T. J. Army. 1960. Moisture storage on fallowed wheatland in the Great Plains. Soil Sci. Soc. Am. Proc. 24:414–418.

Mathews, O. R., and L. A. Brown. 1938. Winter wheat and sorghum production in the Southern Great Plains under limited rainfall. USDA Circ. no. 477.

McCalla, T. M., and J. T. Army. 1961. Stubble mulch farming. Adv. Agron. 13:125–196.

Porter, K. B., I. M. Atkins, and C. J. Whitfield. 1952. Wheat production in the panhandle of Texas. Texas Agric. Exp. Stn. Bull. no. 750.

Quinby, J. R., N. W. Kramer, J. C. Stephens, K. A. Sahr, and R. E. Karper. 1958. Grain sorghum production in Texas. Texas Agric. Exp. Stn. Bull. no. 912.

Stone, J. F. 1978. Evapotranspiration control on agricultural lands. Soil Crop Sci. Soc. Fla. Proc. 37:1–11.

Taylor, H. M., C. E. Van Doren, C. L. Godfrey, and J. R. Coover. 1963. Soils of the southwestern Great Plains field station. Texas Agric. Exp. Stn. Miscellaneous pub. no. 669.

Unger, P. W. 1971a. Soil profile gravel layers: I. Effect on water storage, distribution, and evaporation. Soil Sci. Soc. Am. Proc. 35:631–634.

Unger, P. W. 1971b. Soil profile gravel layers: II. Effect on growth and water use by a hybrid forage sorghum. Soil Sci. Soc. Am. Proc. 35:980–983.

Unger, P. W. 1972. Dryland winter wheat and grain sorghum cropping systems—northern High Plains of Texas. Texas Agric. Exp. Stn. Bull. no. 1126.

Unger, P. W. 1975. Role of mulches in dryland agriculture. p. 237–258. *In* U. S. Gupta (ed.) Physiological aspects of dryland farming. Oxford & IBH Publishing Co., New Delhi.

Unger, P. W. 1978. Straw-mulch rate effect on soil water storage and sorghum yield. Soil Sci. Soc. Am. J. 42:486–491.

Unger, P. W., R. R. Allen, O. R. Jones, A. C. Mathers, and B. A. Stewart. 1976. Sunflower research in the southern High Plains—a progress report. p. 24–29. *In* Proc. First Sunflower Forum, Fargo, N.D. 7–8 January. Sunflower Assoc. of Am., Fargo, N.D.

Unger, P. W., R. R. Allen, and A. F. Wiese. 1971. Tillage and herbicides for surface residue maintenance, weed control, and water conservation. J. Soil Water Conserv. 26:147–150.

Unger, P. W., and J. J. Parker. 1975. No-tillage dryland grain sorghum after irrigated wheat with intervening fallow. Texas Agric. Exp. Stn. Progress Report no. 3330.

Unger, P. W., and A. F. Wiese. 1979. Managing irrigated winter wheat residues for water storage and subsequent dryland grain sorghum production. Soil Sci. Soc. Am. J. 43:582–588.

Wendt, C. W. 1973a. Influence of plastics on soil moisture, soil temperature, and yield of sweet corn at Lubbock, Texas, 1969. Texas Agric. Exp. Stn. Progress Report no. 3151.

Wendt, C. W. 1973b. Applying degradable plastic to cotton, onions, and carrots in the High Plains, 1971. Texas Agric. Exp. Stn. Progress Report no. 3170.

Wendt, C. W. 1973c. Effects of minimum tillage–vertical mulch concept on soil moisture and yield of grain sorghum at Lubbock, Texas, 1970–1971. Texas Agric. Exp. Stn. Progress Report no. 3152.

Wendt, C. W. 1973d. Effect of an antitranspirant on the yield of crops in High Plains of Texas during 1969 and 1970. Texas Agric. Exp. Stn. Progress Report no. 3165.

Wendt, C. W., and J. R. Runkles. 1969. Influence of soil mulches on the environment, vegetative development and certain fiber properties of cotton. p. 96–99. *In* The 23rd Cotton Defoliation and Physiology Conference Proc., New Orleans. 7–9 January.

Wiese, A. F. 1960. Effect of tansy mustard (*Descurainia intermedia*) on moisture storage during fallow. Weeds 8:683–685.

Wiese, A. F., and T. J. Army. 1958. Effect of tillage and chemical weed control practices on soil moisture storage and losses. Agron. J. 50:465–468.

Wiese, A. F., T. J. Army, and J. D. Thomas. 1966. Moisture utilization by plants after herbicide treatment. Weeds 14:205–207.

Wiese, A. F., E. Burnett, and J. E. Box, Jr. 1967. Chemical fallow in dryland cropping sequences. Agron. J. 59:175–177.

Wiese, A. F., and C. W. Vandiver. 1970. Soil moisture effects on competitive ability of weeds. Weed Sci. 18:518–519.

Willis, W. O., H. J. Haas, and J. S. Robins. 1963. Moisture conservation by surface or subsurface barriers and soil configuration under semiarid conditions. Soil Sci. Soc. Am. Proc. 27:577–580.

Zelitch, I. (ed.) 1963. Stomata and water relations in plants. Connecticut Agric. Exp. Stn. Bull. no. 664.

Zelitch, I. 1965. Environmental and biochemical control of stomatal movement in leaves. Biol. Rev. 40:463–482.

Zingg, A. W., and V. L. Hauser. 1959. Terrace benching to save potential runoff for semiarid land. Agron. J. 51:289–292.

Zingg, A. W., and C. J. Whitfield. 1957. Stubble-mulch farming. USDA Tech. Bull. no. 1166.

5

B. W. Greb
Agricultural Research Service
U.S. Department of Agriculture
Akron, Colorado

Water Conservation: Central Great Plains

5-1 AREA

Areas of the Central Great Plains that receive less than 560 mm average annual precipitation are southeastern Wyoming, eastern Colorado, and west of the 100th meridian in Kansas and Nebraska. Drought probability for the region increases southward because of increasing temperatures and, to a lesser extent, westward because of decreasing average annual precipitation. Elevation increases from about 670 m along the eastern boundary of this semiarid area to over 1890 m along the western boundary at the base of the Rocky Mountains. About 70 to 75% of the annual precipitation occurs during April to October. Snowfall ranges from about 10% of the total precipitation in southwestern Kansas to over 30% in extreme western Nebraska.

The cropped areas are level to gently rolling and occupy the uplands between drainage systems that include five soil orders: Alfisols (Red-Brown), Aridisols (Sierozen to Light Brown), Entisols (Brown, Sandy, or shallow), Inceptisols, and Mollisols (Brown to Chestnut). The Mollisols include the largest area of cultivated dryland. Some of the more important great groups within the above orders include Argiustolls, Haplargids, Paleustolls, Ustorthents, and Torriorthents.

Before World War II, annual cropping prevailed, with little specialized wheat-fallow in the dryland area. About 9.30 million ha of land is now under cultivation. Of this, 3.64 million ha is usually in winter wheat, 3.52 million ha in fallow, and 2.14 million ha divided among plantings of grain and forage sorghum, barley, foxtail and proso millet, corn, oat, rye, and sunflower.

Contribution from the U.S. Department of Agriculture, Agricultural Research Service, Akron, Colorado.

Dryland Agriculture—Agronomy Monograph no. 23.

5-2 WATER LOSSES

Water losses from precipitation in the Central Great Plains are large. Of the water received, 50 to 70% is lost by evaporation (Greb et al., 1974; Good and Smika, 1978). The second largest loss is to weeds and volunteer crop plants (Greb, 1974; Wicks and Smika, 1973). Losses occur from runoff (Mickelson et al., 1965; Mickelson, 1966, 1968) and snow blowoff from cultivated fields (Black and Siddoway, 1971; Greb and Black, 1971; Greb, 1975). In addition, water is wasted by flooding of crops in drainless basins (Mickelson and Greb, 1970).

Researchers since early in the century have been evaluating the magnitude of these water losses and then devising various strategies to minimize the losses so that more water can be used for crop production.

5-3 REVIEW

Early water conservation studies and practices involved long-term rotations of adapted and nonadapted crops, date and rates of seeding, fall-spring plowing, and some fallow trials comparing various tillage implements (Brandon and Mathews, 1944; Kuska and Mathews, 1956; Zook and Weakley, 1944). Most early implements were not well designed for the existing crop, weed, soil, and weather conditions. Nevertheless, these investigators recognized the need for stubble mulching, better weed control, and short-term fallow-crop rotation systems.

Water conservation research has been greatly expanded in the last 25 years, and the findings have been more rapidly applied in commercial production. Most studies now consider all water aspects with a given cropping method or conservation system:

1. Precipitation: income and outgo during entire year.
2. Improved water use efficiency practices.
   a. Infiltration and storage.
   b. Use by plants.

Recent innovations and progress on water conservation research practices for the semiarid Central Great Plains are described below. Because fallow is dominant in this region, most other water conservation practices involve a fallow system directly or are supplementary to fallow.

5-4 WATER INFILTRATION AND STORAGE

5-4.1 Summer Fallow

Fallow is necessary in the semiarid Central Great Plains because the water requirement for initial grain development of winter wheat begins at about 22 cm of evapotranspiration in relation to the 300 to 560 mm/y of precipitation expectancy in this region (Greb et al., 1974; Good and Smika,

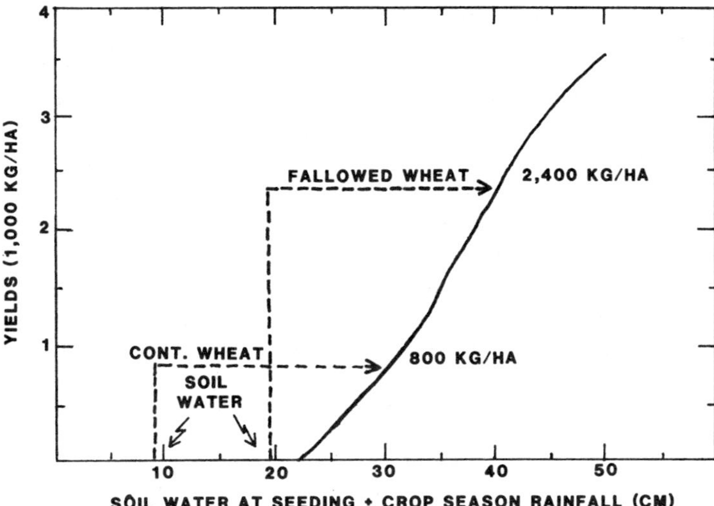

Fig. 5-1. Wheat yield expectancy at North Platte, Nebraska, from 1921 to 1967 (Greb et al., 1974; Smika, 1970).

1978; Smika, 1970). Grain yields then increase about 105 to 180 kg ha^{-1} cm^{-1} of water after the minimum water requirements are satisfied (Fig. 5-1). In a winter wheat-fallow sequence, the fallow period is roughly from harvest early in July to planting early in September of the next season. A 3-year system of fallow-winter wheat-sorghum is commonly used in the 500- to 620-mm average precipitation zone.

Water storage during fallow is subject to many climate, soil, and management factors. Years of research and experience suggest the following requirements for improving water conservation and erosion control in the fallow-crop cycle (Fenster et al., 1969; Fenster and McCalla, 1970; Greb et al., 1967, 1970; Greb, 1974; Harris, 1963; Smika and Whitfield, 1966; Smika and Wicks, 1968; Smika, 1976a, b):

1. Weed control for entire fallow period.
2. Stubble left standing over winter.
3. Straw mulches kept on the surface during the warm season.
4. Hard clods 1 to 8 cm in diameter kept on the soil surface.
5. Management for favorable water content in the seedbed to germinate seeds.

Meeting these requirements maximized available soil water storage and soil NO_3-N, reduces soil erosion potential to near zero, and minimizes the energy and other monetary inputs.

5-4.1.1 Weed Control

For good reason, much of the research with fallow has been devoted to weed control systems. Measurements have shown that uncontrolled weeds may consume soil water at about 0.5 cm/day (Wicks and Smika, 1973).

Certain broadleaf weeds, such as kochia, Russian thistle, and cocklebur, have a feeding depth and radius of root extension equal to or larger than that of most domestic crops. After harvest growth of weeds in undisturbed wheat stubble in the Central Great Plains has produced from 800 to 2700 kg/ha of dry matter (Greb, 1974). Dry matter production of 1000 kg/ha by weeds consumes 8 to 10 cm/ha of water and about 30 kg/ha of available N (Greb, 1974; Wicks and Smika, 1973). Thus, for each 100 kg/ha reduction of potential fall weed growth, wheat yields of grain and straw were increased by about 60 and 120 kg/ha, respectively (Greb, 1974).

Mechanical implements control weeds with varying success. Tillage, however, causes 0.5 to 0.8 cm of evaporative water loss from each disturbance of moist soil (Good and Smika, 1978). Herbicidal weed control, on the other hand, does not disturb the soil, but to be effective, herbicides must rapidly and completely kill all unwanted vegetation. A stunted live weed continues to use water.

5–4.1.2 Standing Stubble Over Winter

In the Central Great Plains north of lat. 39° where snowfall exceeds 600 mm with an average water content of 72 mm/season, from 30 to 55% of the soil water recharge during fallow is from snowmelt (Greb, 1975). Snowfall in standing stubble during the first winter of fallow plus snowfall on wheat in the second winter accounts for an estimated 35 to 45% of this area's wheat production. Snowmelt storage efficiency averaged 53% in undisturbed stubble over a 16-year period (Greb, 1975). Consequently, destroying surface stubble by disking weeds in the fall has proved detrimental. Water saved by controlling weeds with disking was more than offset by the snow blow-off caused by the lack of stubble (Greb, 1974).

5–4.1.3 Straw Mulch

The recent development of large high clearance V-blade implements and rod weeders with attached semichisels (tongs) in combination with increased power has made stubble mulching a widely adaptable type of tillage. By absorbing raindrop impact, mulches help prevent soil puddling and facilitate water infiltration. Mulches also decrease evaporation by cooling the soil by insulation, reflecting the sun's energy, and decreasing wind speed near the soil surface. Table 5–1 compares data for bare fallow and mulched fallow at seven Central Great Plains locations. Water storage during fallow is also influenced by the quantity of mulch available at the beginning of fallow (Table 5–2). Straw available for mulching ranges from 3400 to 4700 kg/ha for most of the Central Great Plains.

Table 5–1. Net soil water gain at end of fallow for bare vs. straw mulches at seven Central Great Plains locations (Smika, 1976a).

Location	Years reported	Soil water gain with	
		No mulch	Mulch
	Number	cm	
Akron, Colo.	6	14.2	17.3
Alliance, Neb.†	8	2.9	3.1
Archer, Wyo.†	2	2.8	4.2
Colby, Kan.	4	11.5	14.1
Garden City, Kan.	6	8.6	9.0
North Platte, Neb.	8	14.6	20.3
Oakley, Kan.	4	8.2	13.1
	Average	9.0	11.6

† Sampled spring to fall (4 months); other locations from harvest to planting (14 months).

Table 5–2. Net soil water gain at end of fallow as influenced by straw mulch rates at four Great Plains locations (Data from Greb et al., 1967, 1970; Smika, 1976a; Unger, 1978).

Location	Years reported	Soil water gain (t/ha of mulch)			
		0	2.2	4.4	6.6
	Number	cm			
Bushland, Tex.	3	7.1	9.9	9.9	10.7
Akron, Colo.	6	13.4	15.0	16.5	18.5
North Platte, Neb.	7	16.5	19.3	21.6	23.4
Sidney, Mont.	4	5.3	6.9	9.4	10.2
	Average	10.7	12.7	14.5	15.7
	Gain by mulching		2.0	3.8	5.0

5-4.1.4 Soil Clods

Only a small amount of moist, fine soil is needed around a seed for germination. Thus, most of the surface soil should be in hard clods. Surface soil clods resist wind erosion, help anchor mulches, slow runoff water, and provide shade and physical protection for young plants, but they do not provide a good medium for weed seed germination (Wittmuss et al., 1973).

5-4.1.5 Seedbed Water Availability

Modern, deep furrow planters can penetrate 12 cm of dry soil to place seeds in moist soil. Mulches maintain soil water closer to the surface and thus permit shallower planting, which is often critical for assuring germination and emergence in dry climate areas (Smika, 1976b). Disk drills are not very good for penetrating dry seedbeds; thus, drilling often results in a low percentage of wheat seed germination. During the drought years of the 1930's and 1950's, poor germination because of lack of surface soil water

Table 5-3. Progress in fallow systems and winter wheat yield at Akron, Colorado (Greb, 1970; Witmuss et al., 1973).

Years	Management system	Average precipi- tation	Precipitation during fallow received (14 months)	Soil water stored	Fallow efficiency†	Wheat yield
		mm/y	mm	cm	%	kg/ha
1916–1930	Maximum tillage: plow, harrow (dust-mulch)	439	531	10.1	19	1070
1931–1945	Conventional tillage: shallow disk, bare rod weeder	401	467	11.2	24	1105
1946–1960	Improved conventional tillage: began stubble mulch 1957	416	507	13.7	27	1725
1961–1975	Stubble mulch: began minimum tillage with herbicides	389	476	15.7	33	2165

† Based on percentage of soil water storage of precipitation received from wheat harvest in July to end of fallow in September (14-month period).

was a major factor resulting in enormous crop losses. Improving seed-zone water for wheat has increased yields by an estimated 8% in the Central Great Plains (Greb, 1979).

5-4.1.6 Improvements in Fallow Systems

Fallow efficiency, as used here, is the percentage of precipitation stored in the soil profile, measured to the 180-cm depth, from wheat harvest to wheat planting. Data obtained at the U.S. Central Great Plains Research Station (Table 5-3) show that development of improved equipment and new knowledge have improved fallow systems and fallow efficiency. Similar results were obtained at the Colby Branch Station, Kansas, and the North Platte Experiment Station, Nebraska (Greb et al., 1974). All systems mentioned in Table 5-3 are being used today in the Central Great Plains.

Data in Table 5-3 show that improving weed control and straw retention increases fallow efficiency. Wheat yields have doubled from 1916–1930 to 1961–1975, even though annual precipitation from 1961 to 1975 was 51 mm less. Some of the increase in wheat yield was also attributed to better varieties and improved planting and harvesting equipment, but increased fallow efficiency accounted for the largest share (Greb, 1979).

5-4.1.7 Types of Fallow Systems

Maximum tillage (dust-mulch system) involves 7 to 10 operations per season beginning in the spring with plowing and followed by shallow harrowing. Although the system kills weeds, it also destroys all stubble and pulverizes the surface soil. Fallow efficiencies range from 16 to 22%.

Conventional bare fallow begins in the spring and includes several shallow diskings and one rod weed operation for seedbed preparation and requires five to seven operations per season. Surface straw and soil clods are broken down but at a slower rate than with maximum tillage. Fallow efficiencies range from 20 to 24%.

Modified conventional fallow begins in the spring with single disking followed by chiseling with small sweep blades and rod-weeding with tongs attached for the last two seedbed operations. This system saves some straw, results in a moderately cloddy soil surface at planting, and requires four to six operations per season. Fallow efficiencies range from 24 to 27%.

Stubble mulch requires four or five subsurface (V-blades and rod weeder with tongs attached) operations per season when initiated in the spring. During wet springs, however, undercut weeds often reestablish roots. Undercutting of weeds in wheat stubble shortly after harvesting can save an additional 1 to 4 cm of water (Table 5-4) and thereby increase subsequent wheatyields 130 to 560 kg/ha compared with spring tillage (Greb, 1974; Wicks and Smika, 1973). A secondary fall-sweep operation is sometimes

Table 5-4. Effect of fall weed control treatments in new wheat stubble on soil water storage in dryland rotations.

Rotation and fall weed control treatments	Fall weed growth	Soil water storage at: Fall dormancy†	Soil water storage at: Wheat or millet planting‡	Grain yield
	kg/ha	cm	cm	kg/ha
Wheat-fallow			Winter wheat	
North Platte, Neb. 1963-1969 (Wicks and Smika, 1973)				
Check, spring disk	1800	−4.4	14.6	2690
Fall sweep, single	1480	0.5	20.3	2880
Fall sweep + atrazine	400	1.2	21.5	2910
Contact herbicide + atrazine + fall sweep	350	3.4	23.7	3040
Wheat-fallow			Winter wheat	
Akron, Colo. 1969-1972 (Greb, 1974)				
Check, spring disk	1140	2.8	8.9	2420
Fall sweep, single	650	5.3	11.2	2690
Fall sweep, double	370	6.4	12.7	2940
Contact herbicide + atrazine + fall sweep	325	6.4	12.7	2940
Fall-wheat-millet			Millet	
Akron, Colo. 1973-1977 (Greb, 1978a)				
Check, spring disk	1175	0.1	11.4	1990
Fall sweep, double	505	1.9	13.3	2260
Contact herbicides + fall sweep	560	1.3	12.7	2210
Fall sweep + atrazine	390	1.7	13.5	2410

† From 25 October to 10 November.
‡ From 1 to 10 September for wheat and 1 to 10 June for millet.

necessary if volunteer wheat germinates after the first fall-sweep operation. Fallow efficiencies range from 27 to 30% for initial tillage in the spring to 30 to 33% for initial tillage after wheat harvest in the early fall.

Minimum tillage substitutes herbicides for mechanical and weed control during part of the fallow season. The present system implies application of contact or preemergence herbicides shortly after wheat harvesting to rapidly kill all weeds and to inhibit any weeds or volunteer wheat from germinating until late in the following spring. One sweep operation and one or two rod weed (tongs attached) operations are then required until planting in early September. Data in Table 5–4 show the importance of fall weed control in increasing soil water storage and crop yields. When well managed, fallow efficiencies range from 33 to 38%.

No-till relies completely on herbicides, both contact and preemergence, for weed control throughout the entire fallow season. No-till is being tested at Akron, Colorado; Sidney and North Platte, Nebraska; and Tribune, Kansas. Minimum disturbance of stubble and maximum control of weeds should maximize water storage efficiency within today's practical limits of commercial feasibility. Fallow efficiencies have ranged from 40 to 55% (Good and Smika, 1978). Problems with no-till include high cost of certain contact herbicides, possible residual carryover of preemergence herbicides in sandy and high-lime surface soils, and difficulty of planting wheat in heavy stubble.

5–4.1.8 Commerical Wheat Yield Trends

Improved water conservation practices during fallow periods, along with other technical inputs, are being rapidly adopted commercially. Winter wheat yields in the semiarid Central Great Plains have doubled in recent decades (Greb, 1979). Gross yields for western Kansas, western Nebraska, and eastern Colorado averaged 755, 930, 1215, and 1580 kg/ha, respectively, for 1936 to 1945, 1946 to 1955, 1956 to 1965, and 1966 to 1975. These yields were calculated on the basis of planted area, not harvested area, because of the risk factor involved in some districts within the region. Similarly, wheat hectares abandoned decreased from 28% during 1936 to 1945 to 22, 20, and 16% for the next three decades. Wheat yields during 1976 to 1980 averaged 1800 kg/ha with only 12% of the planted hectares abandoned. By 1990, wide spread adaptation of present technology of mulching, minimum tillage, and no-till could increase average wheat yields to 2000 kg/ha, if present climatic conditions prevail.

5–4.2 Runoff Water Control

Annual runoff from dry cropland in the Central Great Plains varies from 0.6 to 5.0 cm on given sites (Mickelson et al., 1965; Zingg and Hauser, 1959). Routine agronomic practices for controlling runoff include (i) using straw mulches, (ii) chiseling to increase surface clods, (iii) planting on the

Table 5-5. Earth-moving practices for capturing water runoff, Akron, Colorado (Mickelson et al., 1965; Mickelson, 1966, 1968; Mickelson and Greb, 1970).

Engineering system	Years reported	Season gain over check runoff water†	Yield increase and crop†	
	Number	cm	kg/ha	
Land leveling	2	6.1	1315	Sorghum grain
	3	4.7	200	Winter wheat grain
Waterway flood pans	3	10.9	2685	Sorghum grain
	3	15.7	5130	Sorghum forage
Conservation terraces	4	3.8	450	Sorghum grain
			280	Winter wheat grain
			540	Winter wheat straw
Lagoon leveling	3	12.7	2275	Corn forage
	2	6.7	1485	Sorghum forage‡

† Based on comparison with check plots composed of contributing area or leveled area receiving no runoff.
‡ Years 1964 and 1966; year 1965 not included because of poor stand.

contour, (iv) stripcropping, and (v) reseeding steep croplands to permanent grass. Using engineering practices to control runoff water, however, involves earth moving. Land leveling, enlarging lagoons, and using waterway flood pans and flat channel terraces have all shown promise on limited site areas (Mickelson et al., 1965; Mickelson, 1966, 1968; Mickelson and Greb, 1970). Earth-moving costs and exposure of less fertile subsoil are weighed against better water utilization and erosion control. Table 5-5 is a summary of engineering practices used to control runoff water. All combined, these systems may be used on 15 to 20% of cultivated dryland.

5-4.3 Snow Control

Average snowfall amounts range from 40-mm water equivalent per annum in southwestern Kansas to 120-mm water equivalent per annum in extreme western Nebraska. About 40% of all snowstorms, which contribute 60% of the snowfall precipitation, are wind transport types that result in the snow being blown off planted wheat land (Greb, 1975). Experimental systems to control blowing snow on croplands include vegetative barriers and wood-slat fences of various porosities (Black and Siddoway, 1971; Greb and Black, 1971; Greb, 1975, 1978b, 1979). Single rows of trees and bushes are not acceptable as other snow barriers in this region because of excessive root extension into adjacent cropped areas (Greb and Black, 1961). Soil water gain per winter from snowmelt, over conventional systems, has averaged 4 to 6 cm across 10-m wide strips of cropland leeward of sudangrass (Greb and Black, 1971) and tall wheatgrass (Greb, 1978b) barriers. This increased water has increased wheat grain yields by 270 kg/ha and forage yields (wheat, rye, hay millet, and sudangrass) by 1200 kg/ha, including the space occupied by the barriers. Snowmelt storage efficiency varied from 45 to 72% and averaged 65%, depending on quantity of snow and exposure time to evaporation. Runoff from frozen soil snowmelt is rare in this region.

5-4.4 Deep Plowing

Deep plowing is a treatment that increases soil water intake by breaking up and diluting thin layers of impervious clay and hardpans that are shallow in the soil profile. An example is the Weld silt loam (Aridic Paleustoll), a benchmark soil of the semiarid Central Great Plains that contains a 10- to 12-cm thick Argillic horizon at the 10-cm depth. Native sod of this soil was plowed to a 45-cm depth as a dilution technique in the spring of 1967 at Akron, Colorado (Greb, 1970, 1979). Soil water storage during five fallow seasons was improved an average 1.8 cm by deep plowing compared with shallow-disk plowing. Straw mulches applied at the rate of 2.2 to 4.4 t/ha as a secondary treatment saved an additional 2.0 cm, thus showing that water conservation practices can give additive benefits. Total dry matter yields of five small grain crops (millet 1 year, barley 1 year, winter wheat 3 years) were increased an average of 765 kg/ha with deep plowing alone and 1550 kg/ha with deep plowing plus straw mulching. Likewise, barley and wheat grain yields were increased an average of 300 kg/ha with deep plowing alone and 490 kg/ha for deep plowing and straw mulch combined.

The benefits of deep plowing tended to diminish with time, suggesting that a second deep plowing would be necessary after about 6 years.

5-5 WATER USE EFFICIENCY BY PLANTS

Systems to improve water use efficiency by plants is another broad area of water conservation. High soil fertility, crop selection, improved crop varieties, evaporation suppression, and varous cultural manipulations increase crop production from a given water supply. These items are discussed below.

5-5.1 Soil Fertility

Many experiments have shown much higher water use efficiency by plants in high fertility soils compared with lower fertility soils. Plants are either not growing or growing slowly because of a lack of nutrients to promote new growth (Viets, 1962). For example, from 1966 to 1969, average water use efficiency of three grass species at Akron, Colorado, was increased from 81 kg ha^{-1} cm^{-1} of water without supplemental N to 110 and 135 kg ha^{-1} cm^{-1} of water with application rates of 28 and 56 kg/ha of available N, respectively (Greb and Black, 1971).

In the semiarid Central Great Plains, most loams as well as silt loams and clay loams are fertile with a moderate-to-good reservoir of organically bound N, plus mineral P and K, concentrated mostly in the surface 30 cm of soil. However, continued erosion and crop consumption of nutrients can reduce these reserves to deficiency levels in relation to water supply. This condition has already occurred on about 35% of the cultivated land area involving coarse-textured soil and on eroded hillsides of finer-textured land. Thus, soil fertility relationships are expected to play a larger future role in maintaining good water use efficiency.

5-5.2 Crop Selection

Crops do not vary much in daily water requirements, but peak demand can be double the average daily use (Viets, 1971). However, crops do vary in length of growing season. For example, the sugar beet has a long growing season, requires more water, and therefore yields poorly on semiarid dryland. Various crops also have peak growth periods at different times of the year. For example, fall-planted winter wheat has a peak water demand from mid-May to mid-June. Soil water accumulated during fallow maintains wheat through a low water demand winter, and wheat can then use peak spring rainfall for peak growth demands. In contrast, dryland corn and grain sorghum have peak water demands in August when rainfall is lower and more erratic and evaporation rates are high. Today, corn and sorghum are seldom grown without irrigation where annual precipitation averages less than 450 mm.

Short-season crops, like the millets, grain, and hay types, are quite water efficient. Maximum total water demand for grain millet is only 30 to 35 cm/season compared with 40 to 50 cm/season for winter wheat (Greb, 1978a). Second, most grain millet has a low straw-grain ratio (1.1) compared with modern wheat varieties (1.7) and dryland corn and sorghum (over 2). Field results at Akron, Colorado, showed that grain production for millet begins at 14.6-cm water consumption compared with 22.2-cm water consumption for winter wheat. Millet has yielded 2800 kg/ha of grain at Akron with a total water use of 33 cm, whereas, wheat yielded only 1000 to 1200 kg/ha of grain with the same water use.

5-5.3 Crop Varieties (Cultivars)

The advent of shorter straw and stalks for small grains and other crops was a logical outgrowth of the water use efficiency concept. Tall and thick plant stems consume too much water in relation to the grain produced, if grain is the desired product. Breeding of winter wheat in the Central Great Plains has followed this trend historically, and new varieties at Akron, Colorado, now have a 30% grain yield advantage over the original 'Turkey' and 'Kharkof' varieties introduced early in the century (Greb, 1979). Breeding for disease and pest resistance has also improved wheat varieties, and, there is evidence of varietal differences in water use efficiency.

5–5.4 Cultural Adaptations

A listing of some of the most widely used cultural practices adaptable to limited water supply are given below.

1. Reducing row-crop populations to match water supply is still used on limited areas of dryland corn, grain and forage sorghums, and sunflower (Brown and Shrader, 1959; Carlson et al., 1959).
2. Wide row spacing promoted root extension into untapped stored soil water and delays drought stress until more rainfall arrives (Greb, 1962).
3. Preemergence herbicides are used for summer crops such as millet. Each kilogram of weed tissue produced during the noncrop and crop season reduces the yield of desired dryland crop by 1.8 to 2.0 kg (Greb, 1974, 1978a).
4. Cropping on the contour is used to control runoff water and soil erosion potential.
5. Using deep-furrow planters for winter wheat provides a better microclimate for snow catchment and for protecting plant crowns from high velocity wind and soil impact.

5–5.5 Exotic Water Conservation Schemes

Several water conservation schemes have been tried for suppressing evaporation or improving water utilization by plants. These approaches to date are severely limited in dryland cropping because the effects are often short lived, expensive, labor intensive, and interfere with conventional farming operations. None of these schemes has proved successful. Details of data can be obtained from the literature cited for each water conservation practice listed below:

1. Applying chemicals to reduce transpiration by stomatal closure (Van Bavel and Ehrler, 1968).
2. Applying long chain organic monomolecular films (hexadeconal) to soils to suppress evaporation (Olson et al., 1964).
3. Applying polyethylene films between crop rows to suppress evaporation and increase soil temperature (Black and Greb, 1962; Harrold et al., 1959; Willis, 1962; Willis et al., 1963).
4. Using metal, cement, butyl rubber, plywood, etc. for water concentration on microslopes and microwatersheds (Aase and Kemper, 1968; Fairbourn and Gardner, 1974).
6. Using plastic films and pellets to prolong soil-clod stability and straw mulches needed for improving erosion control, water infiltration, and evaporation suppression (Fairbourn and Gardner, 1975).
7. Using reflectance shields to change heat radiation and thereby change plant development and timing of water utilization (Aase and Kemper, 1968).

Presently, these approaches seem to be more applicable to small-scale gardens and highly intensive truck farming situations than for large-scale dryland agriculture.

5-6 RESEARCH NEEDS

5-6.1 Source Value of Water

The semiarid Central Great Plains receives precipitation as rainfall, fog, hail, snow, and dew in variable quantities throughout the crop and noncrop seasons where temperatures range from −40 to 40°C. What percentage of these forms of water contributes to plant growth under all these different conditions? Little detailed information is available. Some preliminary estimates on the source value of water for Akron, Colorado, were recently tabulated, as shown in Table 5-6 (Greb, 1979). Although these data do not include a uniform set of crops and years, they do show that water availability for plants is related to evaporation potential. The overwhelmingly greater value of stored soil water compared with randomly occurring rainfall is conclusive evidence that crop yields, dryland or irrigated, are greater when the soil profile is wetted to the expected rooting depth at seeding. As shown, stored soil water per unit is worth four to five times as much as warm season rainfall. The data also show nearly identical benefits from snowmelt water, sprinkler irrigation, and captured runoff. Greater

Table 5-6. Estimated source value of water by descending order for Akron, Colorado (Greb, 1970).

Water source	Crop (data-base years)	Water use efficiency†	Comparative index
		kg ha^{-1} cm^{-1}	%
Soil water at seeding	Winter wheat (21), winter rye (4) grain millet (9), hay millet (4) sudangrass (4), grain sorghum (5)	377	100
Sprinkler irrigation	Corn (1), grain sorghum (1)	234	62
Captured snowmelt	Winter wheat (4), winter rye (4) hay millet (4), sudangrass (4)	216	57
Captured runoff	Winter wheat (5), grain millet (6) corn (7), grain sorghum (9)	216	57
Crop season rainfall, cool season	Winter wheat (4), winter rye (4)	95	25
Crop season rainfall, warm season	Hay millet (4), sudangrass (4)	81	21

† Based on total dry matter for combined group of crops and data years for each water source, planting to harvest.

knowledge in this concept of source value of water may provide clues for filling water loss voids using simple cultural changes.

5-6.2 Chemical Fallow

The technology for complete no-till chemical fallow is undergoing experimentation and development. The implications of reduced tillage in fallow made possible by herbicides are impressive: (i) near elimination of wind erosion threats, (ii) doubling of soil water storage compared with conventional bare fallow, and (iii) much higher wheat yields obtained with a significant decrease in energy expenditure per unit of production.

A substantial decrease in the cost of certain contact herbicides would make this system commercially feasible in the foreseeable future. Experimentation should continue to test both broader spectrum and more highly specific herbicides than those available today. As a supplement to experimentation with chemical fallow, water consumption by various weed types during specific noncrop periods should also be accurately determined.

5-7 CONCLUSIONS

Excellent progress in water conservation concepts and practices has been made in the last 25 years in the Central Great Plains. Much of this work has been concentrated on summer fallow. The easier means of filling water loss voids have been largely accomplished on experimental plots and to some degree in commercial dryland fields. Additional refinements in water management on all land sites in combination with continued improvement of plant varieties are needed to maximize water utilization.

5-8 LITERATURE CITED

Aase, J. K., and W. D. Kemper. 1968. Effect of ground color and microwatersheds on corn growth. J. Soil Water Conserv. 23:60–62.

Black, A. L., and B. W. Greb. 1962. Nitrate accumulation in soils covered with plastic mulch. Agron. J. 54:336.

Black, A. L., and F. H. Siddoway. 1971. Tall wheatgrass barriers for soil erosion control and water conservation. J. Soil Water Conserv. 26:107–111.

Brandon, J. F., and O. R. Mathews. 1944. Dryland rotation and tillage experiments at the Akron (Colorado) Field Station. USDA Circ. no. 700.

Brown, P. L., and W. D. Shrader. 1959. Grain yields, evapotranspiration and water use efficiency of grain sorghum under different cultural practices. Agron. J. 51:339–343.

Carlson, C. W., J. Alessi, and R. H. Mickelson. 1959. Evapotranspiration and yield of corn as influenced by moisture level, nitrogen fertilization, and plant density. Soil Sci. Soc. Am. Proc. 23:242–245.

Fairbourn, M. L., and H. R. Gardner. 1974. Field use of microwatersheds with vertical mulch. Agron. J. 66:740–744.

Fairbourn, M. L., and H. R. Gardner. 1975. Water-repellent soil clods and pellet as mulch. Agron. J. 67:377–380.

Fenster, C. R., C. E. Domingo, and O. R. Burnside. 1969. Weed control and plant residue maintenance with various tillage treatments in a wheat-fallow rotation. Agron. J. 61: 256–259.

Fenster, C. R., and T. M. McCalla. 1970. Tillage practices in western Nebraska with a wheat fallow rotation. Nebraska Agric. Exp. Stn. Bull. no. 507.

Good, L. G., and D. E. Smika. 1978. Chemical fallow for soil and water conservation in the Great Plains. J. Soil Water Conserv. 33:89–90.

Greb, B. W. 1962. Extra wide row spacing of grain sorghum. Colorado State Univ. Exp. Stn. Pub. no. PR-62-39, p. 1.

Greb, B. W. 1970. Deep plowing a shallow clay layer to increase soil water storage and crop yield. Colorado State Univ. Exp. Stn. Pub. no. PR-70-23, p. 1–2.

Greb, B. W. 1972. Snowfall characteristics and snowmelt storage at Akron, Colorado. p. 45–64. *In* Snow management on the Great Plains. Great Plains Agric. Coun. Pub. no. 73, Univ. of Nebraska, Lincoln.

Greb, B. W. 1974. Yield response to fall weed control in new wheat stubble in a fallow-wheat rotation. p. 33–45. *In* B. L. Bohmont (ed.) Fourth Annual Colorado Crop Protection Inst. Proc., Fort Collins, Colo. 13–14 November. Colorado State Univ., Fort Collins.

Greb, B. W. 1978a. Millet production with limited water. Colorado State Univ. Exp. Stn. Pub. no. PR-78-15, p. 1–3.

Greb, B. W. 1978b. Tall wheatgrass snow barriers boost continuously cropped forage yields. Colorado State Univ. Exp. Stn. Pub. no. PR-78-16, p. 1–3.

Greb, B. W. 1978c. Reducing drought effects on cropland in the West Central Great Plains. USDA Agric. Information Bull. no. 420.

Greb, B. W., and A. L. Black. 1961. Effect of windbreak plantings on adjacent crops. J. Soil Water Conserv. 16:223–227.

Greb, B. W., and A. L. Black. 1971. Vegetative barriers and artificial fences for managing snow in the Central and Northern Plains. p. 96–111. *In* A. O. Haugen (ed.) Snow and Ice in Relation to Wildlife and Recreation Proc., Ames, Iowa. 11–12 February. Iowa Cooperative Wildlife Research Unit and Iowa State Univ., Ames.

Greb, B. W., D. E. Smika, and A. L. Black. 1967. Effect of straw mulch rates on soil water storage during summer fallow in the Great Plains. Soil Sci. Soc. Am. Proc. 31:556–559.

Greb, R. W., D. E. Smika, and A. L. Black. 1970. Water conservation with stubble mulch fallow. J. Soil Water Conserv. 25:58–62.

Greb, B. W., D. E. Smika, N. P. Woodruff, and C. J. Whitfield. 1974. Summer fallow in the Central Great Plains. p. 51–85. *In* Summer fallow in the western United States. USDA-ARS Conserv. Res. Report no. 17.

Harris, W. W. 1963. Effects of residue management, rotations, and nitrogen fertilizer on small grain production. Agron. J. 55:281–284.

Harrold, L. L., D. B. Peters, F. R. Dreibelbis, and J. L. McGuinness. 1959. Transpiration evaluation of corn grown on a plastic-covered lysimeter. Soil Sci. Soc. Am. Proc. 23: 174–178.

Kuska, J. B., and O. R. Mathews. 1956. Dryland crop-rotation and tillage experiments at the Colby Kansas Branch Experiment Station. USDA Circ. no. 979.

Mickelson, R. H. 1966. Level pan systems for spreading and storing watershed runoff. Soil Sci. Soc. Am. Proc. 30:388–392.

Mickelson, R. H. 1968. Conservation bench terraces in eastern Colorado. Water Resour. Res. 4:95–101.

Mickelson, R. H. 1975. Performance and durability of sheet-metal, butyl rubber, asphalt roofing, and bentonite for harvesting precipitation. p. 93–102. *In* G. W. Frasier (ed.) Water Harvest Symp. Proc., Phoenix, Ariz. 26–28 March. USDA Report no. ARS-W-22.

Mickelson, R. H., M. B. Cox, and J. Musick. 1965. Runoff water spreading on leveled cropland. J. Soil Water Conserv. 20:57–60.

Mickelson, R. H., and B. W. Greb. 1970. Lagoon leveling permits annual cropping in semiarid areas. J. Soil Water Conserv. 35:13-16.

Olson, S. R., F. W. Watanabe, F. E. Clark, and W. D. Kemper. 1964. Effect of hexadeconal on evaporation of water from soil. Soil Sci. 97:13-18.

Smika, D. E. 1970. Summer fallow for dryland winter wheat in the semiarid Great Plains. Agron. J. 62:15-17.

Smika, D. E. 1976a. Mechanical tillage for conservation fallow in the semiarid Central Great Plains. p. 78-92. *In* Conservation tillage. Great Plains Agric. Coun. Pub. no. 77, Univ. of Nebraska, Lincoln.

Smika, D. E. 1976b. Seed zone soil water conditions with mechanical tillage in the semiarid Central Great Plains. p. 1-6. *In* The Seventh Int. Soil Tillage Res. Organization Proc., Section 37, Uppsala, Sweden. College of Agriculture of Sweden, Uppsala.

Smika, D. E., and C. J. Whitfield. 1966. Effect of standing wheat stubble on storage of winter precipitation. J. Soil Water Conserv. 21:138-141.

Smika, D. E., and G. O. Wicks. 1968. Soil water storage during fallow in the Central Great Plains as influenced by tillage and herbicide treatments. Soil Sci. Soc. Am. Proc. 32:591-595.

Unger, P. W. 1978. Straw mulch rate effect on soil water storage and sorghum yield. Soil Sci. Soc. Am. Proc. 42:486-491.

Van Bavel, C. H. M., and W. L. Ehrler. 1968. Water loss from a sorghum field and stomatal control. Agron. J. 60:84-86.

Viets, F. G., Jr. 1962. Fertilizers and the efficient use of water. Adv. Agron. 14:223-264.

Viets, F. G., Jr. 1971. Effect drought control for successful dryland agriculture. p. 57-76. *In* K. L. Larson and J. D. Eastin (ed.) Drought injury and resistance in crops. Crop Sci. Soc. of Am. Pub. no. 2, Madison, Wis.

Wicks, G. A., and D. E. Smika. 1973. Chemical fallow in a winter wheat-fallow rotation. Weed Sci. 21:97-102.

Willis, W. O. 1962. Effect of surface covers on evaporation from soil. Soil Sci. Soc. Am. Proc. 26:598-601.

Willis, W. O., H. J. Haas, and J. S. Robins. 1963. Moisture conservation by surface and subsurface barriers and soil configuration under semiarid conditions. Soil Sci. Soc. Am. Proc. 27:577-580.

Wittmuss, H. D., G. B. Triplett, and D. W. Greb. 1973. Concepts of conservation tillage systems using surface mulches. p. 5-12. *In* Conservation tillage. Proc. Natl. Conserv. Tillage Conference, Des Moines, Iowa. 28-30 March. Soil Conserv. Soc. of Am., Ankeny, Iowa.

Zingg, A. W., and V. L. Hauser. 1959. Terrace benching to save potential runoff from semiarid land. Agron. J. 51:289-292.

Zook, L. L., and H. E. Weakley. 1944. Summer fallow in Nebraska. Nebraska Agric. Exp. Stn. Bull. no. 362.

6

W. O. Willis
Agricultural Research Service
U.S. Department of Agriculture
Fort Collins, Colorado

A. Bauer and A. L. Black
Agricultural Research Service
U.S. Department of Agriculture
Mandan, North Dakota

Water Conservation: Northern Great Plains

The Northern Great Plains encompasses North and South Dakota, Montana, and most of the northern half of Wyoming. The western edge of the region is generally delineated at an elevation of about 1520 m. Most of the cultivated land is on nearly level to gently rolling topography. The rolling and rougher topographic positions are usually grasslands used for grazing livestock.

6-1 CLIMATE, CROPS, AND SOILS

The climate of the easternmost part has been referred to as dry subhumid, and that part west of the 98th or 100th meridian has been referred to as semiarid. Annual precipitation ranges from about 250 mm in the west to about 620 mm in southeastern South Dakota. Of this precipitation, about 70% falls during April through September, with the peak about mid-June coinciding with temperatures favorable for plant growth. Differences in monthly precipitation between the subhumid and semiarid areas are greatest during July through October and largely account for the differences in average annual amounts received in those two climatic areas.

Average annual pan evaporation ranges from about 900 mm in the northeastern part of the region to about 1500 mm in the southwestern part.

Contribution from the USDA, Agricultural Research Service.

Dryland Agriculture—Agronomy Monograph no. 23.

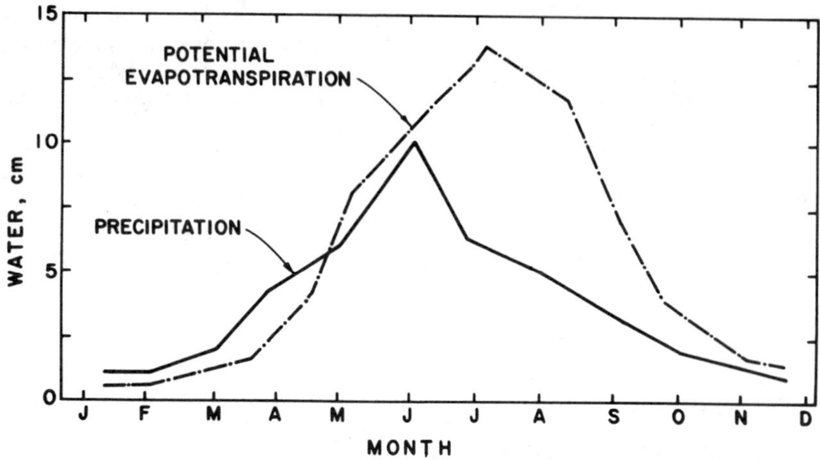

Fig. 6-1. Average annual precipitation and calculated potential evapotranspiration for Dickinson, North Dakota.

Figure 6-1 depicts the average monthly precipitation and calculated potential evapotranspiration for Dickinson, North Dakota. This figure, used as an example, identifies two characteristics that affect dryland agriculture in the Northern Great Plains. One is that average precipitation and average potential evapotranspiration are nearly equal until mid-June; thus, the climate is suitable for raising small grains and cool season grasses. The other is that with the high probability of potential evapotranspiration considerably exceeding precipitation after June, a reserve of available soil water is needed to supply crops in the latter part of the growing season.

The potential for soil water recharge before the next crop season is greater in the subhumid than in the semiarid region because of the greater average precipitation during July through October. More precipitation during this period permits a greater diversity of crops to be grown in the subhumid region. However, some crops have a growing season extending to and beyond the frost-free period, thereby reducing the time available for building of soil water reserves for the next crop.

As precipitation increases from west to east in the region, the variety of annual crops increases and the area devoted to perennial grass decreases. In the western part of the region, agriculture diversifies to include livestock and rangelands. Cultivated crops of the region are hard red spring wheat, hard red winter wheat, durum wheat, barley, flax, and sunflower, and to a lesser extent oat, corn, safflower, sorghum, alfalfa, dry beans, sugar beet, potato, and soybean. The latter four crops are usually grown only in the higher rainfall areas. The region produces almost all of the durum wheat and about 80% of the spring wheat grown in the USA. Of the approximately 7.79 million ha in dryland wheat production for the region in 1977, about 4.98 million ha, or 64%, was seeded on land that had been summerfallowed (USDA-ESCS. 1977. Crop statistic).

Briefly following is a list of soils of the region according to the comprehensive system of classification (USDA, 1970) with former great soil group class names given parenthetically and the associated land resource areas (Austin, 1965):

1. Udic Borolls (Chernozem) occur primarily on the Black Glaciated Plains in eastern North and South Dakota.
2. Typic Borolls (Chestnut and Brown) are on the Dark Brown Glaciated and Rolling Soft Shale Plains in western North Dakota, northern South Dakota and central Montana.
3. Aridic Borolls (Chestnut, Brown, Regosol, and Lithosol) occur on the Northern Rolling High Plains and Brown Glaciated Plains in northeastern Wyoming and across eastern and north central Montana.
4. Udic Ustolls (Chernozem) in the Central Feed Grains and Livestock Region land resource area occur in the southeast corner of South Dakota.
5. Typic Ustolls (Chernozem, Chestnut, and Brown) are found in the Dark Brown and Black Glaciated Plains, Rolling Pierre Shale Plains, and Dakota-Nebraska Eroded Tableland in south central South Dakota.
6. Aridic Ustolls (Chestnut, Brown, Regosol, and Lithosol) occur in the Pierre Shale Plains and Badlands, Black Hills Footslopes, and Mixed Sandy and Silty Tableland in western South Dakota and extreme northeastern Wyoming.
7. Ustollic Aridisols (Regosol, Lithosol, and Brown) occur in the Northern Rolling Hill Plains in south central Montana and northeastern Wyoming.

6-2 CONCEPTS

Principles of water conservation for dryland agriculture in the Northern Great Plains are well established. Essentially these are (i) to eliminate or reduce evaporation and (ii) to retain precipitation where it falls. The latter principle is directed especially to snow. Water conservation is best considered as a two-part system, one associated with the frost-free season and the other with managing snow and water from snowmelt. Snow as a water source is not presently being used to best advantage (Wight et al., 1975; Willis and Haas, 1965; Willis et al., 1969; Willis and Haas, 1969; Schneider et al., 1978). The two-part system is separable but must be considered interactive to gain best management.

An exception to the application of these principles for best management is associated with the fine-textured soils developed on the bed of glacial Lake Agassiz of subhumid eastern North Dakota (Bauer and Kucera, 1978). These soils are affected by poor drainage. Management to encourage winter evaporation and prevent snow accumulation will facilitate spring seeding and planting operations. Yield gains otherwise associated with increased water supply are offset and even reduced by delay in seeding.

A focal point of water conservation is residue management. Residues simultaneously reduce water evaporation rate and retain precipitation (especially snow) in place. Intuitively, we know that effective water conservation should be realized where residues are retained in place, including the period when the next crop is seeded and established. However, seeding spring crops directly into stubble may result in low seedling population and, hence, crop failure or low grain yield. Prevention of conventional drill disk penetration to depths ensuring seed coverage (for maximum seed–soil contact to enhance germination) has contributed to these "failures." Soil strength (penetrability) apparently is great enough to prevent penetration on moderately well-drained, medium-textured soils with surface bulk density of greater than about 1.2 g/cm^3 and wetter than 1.5 MPa of soil water tension (Bauer and Kucera, 1978).

6–3 WATER CONSERVATION PRACTICES

6–3.1 Cultivated Land

Summer fallow is included in virtually all cropping sequences. Alternate crop-fallow is practiced in much of the semiarid region, primarily as a means to increase water storage. Fallow is practiced less frequently in the subhumid region. Fallowing was used before the turn of the century (Howard, 1959), shortly after extensive cultivation began, and the practice steadily increased until about 1973 (USDA, 1974). Use of fallow is considered to contribute to stability of income because of the increased soil water storage, accumulation of available N, and better weed control. The probability of a given yield is higher for wheat on fallow than on annually cropped land in the semiarid region (USDA, 1974), but not in the eastern, or subhumid, part of the region.

Both spring wheat (principally in North Dakota) and winter wheat (mainly in Montana, western North Dakota, Wyoming, and South Dakota) are grown in a crop-fallow sequence, but the systems are different for the two crops as shown in Fig. 6–2. Soil is fallowed about 21 months (over two winters) with spring wheat, and about 14 months (over one winter) with winter wheat. In either case, 2 years are required to produce one crop. Yields of spring wheat after fallow are about 40% higher than with annual

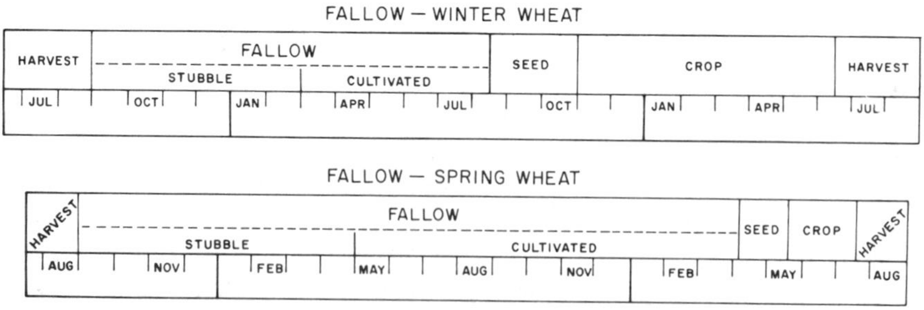

FALLOW — WINTER WHEAT
HARVEST | FALLOW (STUBBLE, CULTIVATED) | SEED | CROP | HARVEST
JUL OCT JAN APR JUL OCT JAN APR JUL

FALLOW — SPRING WHEAT
HARVEST | FALLOW (STUBBLE, CULTIVATED) | SEED | CROP | HARVEST
AUG NOV FEB MAY AUG NOV FEB MAY AUG

Fig. 6–2. Schematic time base for fallow–winter wheat and fallow–spring wheat cultural systems in the Northern Great Plains (Willis, 1976).

cropping, and winter wheat yields after fallow are about 50% higher than with annual cropping. Percentage yield differences favoring fallow (Young, 1962) are greater with no fertilizer added than with fertilizer applied (Table 6-1).

From a study at 14 locations in the Northern Great Plains, spring wheat yields were higher when wheat followed corn than when wheat followed wheat. Differences of yields after fallow compared with yields after corn were small, indicating that corn can be a good substitute for summer fallow (Haas et al., 1974). However, if the corn is harvested for silage rather than for grain, the soil is as vulnerable to wind and water erosion as if fallowed. Thus, the erosion susceptibility should be considered along with other factors if corn is used in place of summer fallow.

In the spring wheat region, usually 20 to 50% of the precipitation received from harvest to seeding time is stored with annual cropping. The extremes in storage efficiency have ranged from none to over 70% (Bauer and Conlon, 1974; Bauer and Kucera, 1978; Thysell, 1938). Distribution and quantity of precipitation are determinants of storage efficiency, as well as the time between when precipitation is received and when soil water is measured. Storage efficiency usually is greater where residues remain on the surface during the nongrowing season (Bauer and Kucera, 1978), because residues increase snow trapping and reduce runoff and evaporation rate.

Compared with annual cropping in the spring wheat region, storage efficiency of fallow is low; only about 14 to 25% of the total precipitation during the 21-month fallow period was stored (Haas et al., 1974; Mathews and Army, 1960). The average soil water storage was 32% (6.35 cm) with annual cropping and 19% (19.67 cm) with crop-fallow. Another study (Haas and Willis, 1962) showed the average difference in water storage between annual and alternate crop-fallow for spring wheat was about 4.47 cm and that of the water stored during the entire fallow period, 84% was stored

Table 6-1. Fertilization effect on wheat yield advantage of fallow over annual cropping in North Dakota, 1958-1961 (Young, 1962).

Area	Trials	Fertilization			
		None	Adequate†	None	Adequate†
		Yield advantage			
	Number	kg/ha		%	
West	24	276	289	22	18
West central	14	652	538	52	32
East central	21	719	450	49	21
Red River Valley	10	578	-40	33	-2
Statewide	69	531	343	38	18

† Average N and P rate for adequate fertilization was:

Area	Fallow	Annual cropping
	kg/ha N + P	
West	0 + 15	12 + 10
West central	2 + 15	17 + 15
East central	2 + 15	35 + 15
Red River Valley	0 + 17	45 + 19

Table 6-2. Average precipitation received and water stored in the 180-cm depth over the fallow period 1915-1954 at Mandan, North Dakota (Haas and Willis, 1962).

Subperiod and average data	Precipitation	Stored	Stored
	cm		%
Harvest to seeding (3 Aug.-20 Apr.)	18.16	5.99	33.0
Seeding to harvest (20 Apr.-3 Aug.)	21.41	3.71	17.3
Harvest to seeding (3 Aug.-20 Apr.)	18.16	1.38	7.6
Total (21 months)	57.73	11.08	19.2

between harvest and July (or about 11 months of the first part of the fallow period). These same data for the total fallow period, divided into three subperiods and presented in Table 6-2, show storage efficiency was greatest (33%) during the subperiod corresponding to the nongrowing portion of an annual cropping sequence and least (about 8%) during the third subperiod. On clean-tilled fallow, storage efficiency during the second subperiod of fallow (spring to autumn) was less than 3% when available water content exceeded 11 cm in the upper 122 cm in a medium-textured soil (Bauer and Conlon, 1978). Available soil water content in the upper 30 cm was lower in the autumn of the second subperiod than in the spring of the same subperiod; the difference, attributed to evaporation loss, was as much as 2.5 cm.

In 69 trials on farm fields in North Dakota, the average available soil water content to the 150-cm depth was about 4.8 cm greater in summer fallow than in comparable soil cropped the previous growing season (Young, 1962). The least difference (3.3 cm) was in soils developed on deposits of glacial Lake Agassiz (Table 6-3). These data, supportive of all others, show the relatively small average difference in water storage between similar time periods of summer fallow and the nongrowing season of annually cropped soils. Thus, if methods could be devised to conserve about 55% (rather than 32%) of the precipitation from harvest to seeding in an annual cropping system, as much water would be stored as during an entire summer fallow period. With increased water conservation, yields should increase.

With winter wheat, average water storage was about 28% (9.4 cm) over the 14-month fallow period (Black et al., 1974). Of the water stored, about 71% (7.87 cm) was stored over winter, and 5.9% (1.52) was stored during the summer of fallow. Efficiency of water conservation for a winter wheat crop-fallow system generally ranges from 20 to 30%.

Table 6-3. Increase in available water (150-cm depth) and NO_3-N (60-cm depth) of soils by fallowing compared with annual cropping in North Dakota, 1958-1961 (Young, 1962).

Area	Trials	Increase in available water	Increase in NO_3-N
	Number	cm	kg of N/ha
West	24	5.1	75
West central	14	5.1	81
East central	21	5.3	122
Red River Valley	10	3.3	115
Statewide	69	4.8	96

Long-term data from four geographic areas in North Dakota showed that only in the western one third of the state could a spring wheat–fallow system be more profitable than annual spring wheat, and then only with adequate fertilization (based on specific assumptions of wheat prices and the then current production costs). Fallow should increase spring wheat yield by 50% or more to be economical, and this will occur, on the average, when annual precipitation is about 300 to 380 mm (Bauer, 1968, 1971).

Historically, the practice of summer fallow has been controversial (Emory, 1894). Some advantages of the crop-fallow system are greater yield per planted hectare, more stable production because of higher soil water content and higher levels of available N, better weed control, reduced insect and disease problems, and more desirable distribution of the farmer's work load. Some disadvantages of the system are increases in both wind and water erosion, a greater decline in soil fertility, low soil water storage efficiency, and low precipitation use efficiency (Black et al., 1974; Haas et al., 1974).

The widespread use of a crop-fallow system in the semiarid region has apparently adversely affected some agricultural lands, namely, incidence of saline seeps has increased on hillsides, slopes, and drainageways. Geology, climate, and cultural practices are all parts of the problem. These seeps develop where the downward flow of water is intercepted and laterally transmitted by a geologic layer (different than the overburden) to a point on the landscape where the transmitting layer has been truncated by geologic erosion. The recharge areas are local (Doering and Sandoval, 1976a; Halvorson and Black, 1974), and therefore, the quantity of water contributed to downward flow is dependent on management of the recharge area. The increase in number and size of seeps was partly due to a higher annual precipitation during the 1962–1972 period. The other cause was identified as the practice of summer fallowing, including almost complete weed control (Doering and Sandoval, 1976b). The problem can be corrected by including alfalfa, permanent grass cover, or a flexible cropping system (Black et al., 1981) on the recharge area. A dilemma is possible in that returning residues on the soil surface to increase soil water storage needed for annual cropping may increase the supply of water for leaching in some years, particularly if the recharge area is fallowed and has predominately sandy soils. However, with perennial crops such as grass and alfalfa or the use of more intensive flexible cropping systems, the plants normally use all available soil water during a growing season. Therefore, it is important that good water conservation practices be utilized to improve the opportunity for annual or flexible cropping practices.

With continuously grown small grains, weeds often become a problem within 3 to 5 years. Most broadleaf weeds can be controlled by herbicides. Grassy weeds such as green foxtail, yellow foxtail, and wild oat are more of a problem than broadleaf weeds. However, a combination of undercutting small grain stubble within 2 weeks after harvest and simultaneously applying a residual herbicide is a recently developed treatment that appears promising. This practice leaves erect stubble to trap snow over winter, provides erosion control, and controls weed growth during the summer fallow

(Wicks and Smika, 1973). Including an oilseed crop (sunflower or safflower) periodically in the cropping system with adequate herbicidal weed control provides a means of controlling annual grassy weeds (Black and Siddoway, 1976). Another recently developed management system is a spring wheat–winter wheat–fallow sequence that appears promising for the winter wheat areas of the Northern Great Plains (Black and Siddoway, 1976). Winter wheat is seeded directly into erect (standing) spring wheat stubble, which provides a means of trapping snow. Snow trapped by the stubble provides insulation and, hence, protects the winter wheat against winterkilling. In addition, snow trapped in 28- to 38-cm upright stubble provided 2.0 to 2.5 cm of additional soil water compared with winter wheat seeded in a tandem-disked seedbed without standing stubble (Black and Siddoway, 1977). With adequate N fertilization, yields of winter wheat seeded in this manner were comparable with winter wheat grown on summer fallow (Black and Siddoway, 1977; Halvorson et al., 1976).

Standing stubble of small grain is very effective in trapping snow, and alternate relatively narrow strips of tall (greater than 30 cm) and short (no) stubble area even more effective. The latter practice increases total water content of the snowpack compared with a uniform stubble height across a field (Willis, 1978).

Snow can be managed to improve water conservation over the winter. Two objectives must be considered: (i) to trap and hold the snow where it is wanted and (ii) to hold the snowmelt in place until the soil thaws enough to allow infiltration. The first is accomplished by altering wind velocity with some type of wind barrier such as erect stubble taller than 25 cm (Black and Siddoway, 1977; Halvorson et al., 1976), tall wheatgrass barriers (Aase and Siddoway, 1974; Black and Siddoway, 1976), or field shelterbelts (Frank and George, 1975; Frank et al., 1976). Both objectives can be accomplished by using a land-forming practice such as bench terraces with unharvested smooth bromegrass on the downslope dike (Haas and Willis, 1968; Rauzi, 1975). Infiltration and storage of snowmelt are largely controlled by the water content of the upper 15 to 30 cm of soil at the time the soil freezes (Willis et al., 1961). Infiltration of snowmelt and increase in efficiency of soil water storage are enhanced by a dry soil profile. With the barrier system of tall wheatgrass (Black and Siddoway, 1976), overwinter soil water storage the first 9 months after harvest was increased twofold compared with land not protected (no stubble or barriers). Continuous cropping sequences within the barrier system stored 10 cm of water in 9 months, including the first winter, compared with 9 cm of water stored in 21 months of fallow outside the grass barriers.

On either cultivated or rangelands, the largest loss of soil water is by evaporation. Calculations of expected wheat yield, if evaporation were controlled, show that even in the drier years yields would exceed present yields on summer fallow (Wendt et al., 1970). Generally, wheat yields are increased by 60 kg/ha or more for each centimeter of soil water used (Bauer et al., 1966; Black, 1973). Forage from native range is increased by about 150 kg/ha for each additional centimeter of available soil water with nominal adequacy of available N (30 to 40 kg/ha). Thus, preventing evaporation is a worthwhile objective.

There is no apparent way to reduce evaporation from rangelands, but evaporation from cultivated land can be reduced with a mulch. Many materials, such as gravel, soil clods, plastic film, asphaltic emulsions, and plant residue (straw), have been studied. Of these, plant residues are the most economical and widely used. Shallow tillage (3 to 5 cm), accomplished as soon as possible after the soil has been wetted, apparently can be effective (Willis and Bond, 1971), but care must be taken not to till when the soil is too wet.

With straw mulch, the reduction in evaporation is a function of the quantity of surface residue present up to about 6 t/ha (Bond and Willis, 1970, 1971). Over an 8-year period, average available soil water stored in the 1.5-m depth increased significantly after 14 to 21 months of fallow as mulch levels increased. Each metric ton of wheat straw residue on the soil surface increased available soil water supplies an average of 0.6 cm (Black, 1973). Furthermore, spring wheat grain yields were positively correlated with stored soil water at seeding only with N–P fertilization.

Extraction of available soil water during the crop year improved the effectiveness and the efficiency of summer fallow the succeeding year. Fertilized winter wheat rooted deeper and extracted nearly twice as much soil water as the nonfertilized wheat (Brown, 1970). Because soil profiles of both the nonfertilized and fertilized treatments were completely recharged by fallowing, the precipitation storage efficiency was greater for the fertilized soil. Soil management and cropping practices that result in greater depths of soil water removal will increase subsequent soil water storage efficiency because of the greater volume of water needed to recharge the soil profile.

Surface residue also reduces rainfall runoff and thus potentially increases soil water content. However, mulch can cause lower soil temperature, which may adversely affect plant growth (Willis and Power, 1975), particularly in the northernmost part of the region where the frost-free growing season averages fewer than 135 days.

Stubble mulching has not been as widely adopted in the Northern Great Plains as in the Central Great Plains, because wheat yields in a wheat-fallow sequence are often less when stubble mulch is employed unless proper fertilization practices are used (Black, 1973). One study clearly showed reduced yield resulted from inadequate control of grassy weeds. With weed control, soil water content and subsequent yields were higher with stubble mulch than with bare soil or conventional stubble mulch tillage (Bond et al., 1971).

6–3.2 Research Needs on Cultivated Land

Research has developed better ways of conserving and utilizing a limited water resource to improve crop production under dryland conditions. Concurrently this research has recognized the need for conserving the soil resource itself by integrating better methods for controlling wind and water erosion.

Snow is important in soil water conservation in the North Great Plains (Black and Siddoway, 1976; Frank and George, 1975; Nichols, 1969; Rauzi, 1975; Wight et al., 1975; Willis et al., 1969; Willis and Haas, 1969; Schneider et al., 1978). However, many of the factors controlling infiltration and storage of water from snowmelt are not well understood. Some factors influencing snow trapping, infiltration, and water storage include soil type, vegetation, cropping systems, climate, topography, soil temperature, antecedent soil water content, crop residue management, and vegetative barriers. Soil and crop management systems that take these factors into consideration to provide best management of the snowmelt resource and a more flexible cropping sequence should be developed.

No-till and minimum-till concepts for dryland farming in the Northern Great Plains need to be developed to include residue management provisions for trapping snow (Schneider et al., 1978). Better information is needed on how to manage crop residue to increase the water use efficiency of different crops. Straw may be a host for small grain diseases, and quantitative data are needed. Concurrently, new and improved practices need to be developed for weed control, planting, and fertilization within a no-till cropping system.

6–3.3 Grasslands

Water is the major factor controlling plant growth on semiarid rangelands. Any water conservation measure that increases water supply to plants is likely to increase forage production. Water controls the annual variation in forage production and, with soil type, the vegetation types or species that occupy a specific range site.

6–3.3.1 Water Management Techniques

Mechanically manipulating the soil can improve rangeland forage production by improving the soil water regime, improving species composition, and modestly increasing soil fertility (Wight, 1973). Improving the soil water regime is the primary objective of most mechanical treatments. The effects of several types of mechanical treatments on forage production in the Northern Great Plains are summarized in Table 6–4 (Wight, 1973). The wide range in response is due to differences in intensity of treatment, climate, soil, and vegetation complex.

Bench terracing is the most intensive and costly of the mechanical methods of water management, but it has produced the greatest increase in forage yield (Haas and Willis, 1968). Water-spreading is a less intensive practice but has exhibited a relatively wide range of adaptability (Branson, 1956; Houston, 1960). Success of a water-spreading system is dependent on

Table 6–4. Mechanical treatments and their effects on forage production in the Northern Great Plains.

Treatment	Location of study	Study period	Soil texture	Dominant species	Yield of check	Increase due to treatment
					kg/ha	%
Bench Terracing						
Haas and Willis (1967)	Mandan, ND	1963–1966	Silty clay loam	Blue grama grass Western wheatgrass	678	536
Water spreading						
Branson (1956)	Alzada, MT	1952–1955	Clay	Western and streambank wheatgrass	419	164
Houston (1960)	Miles City, MT	1955–1959	Clay loam	Western wheatgrass	728	164
Houston (1960)	Miles City, MT	1955–1959	Sandy loam	Blue grama grass	954	353
Pitting						
Wight and White (1974)	Sidney, MT	1967–1972	Sandy loam	Western wheatgrass Sedges	1052	12
Ryerson et al. (1970)	Havre, MT	1962–1968	Loam	Clubmoss Needle-and-thread grass	605	73
Ryerson et al. (1970)	Glasgow, MT	1966–1968	Loam	Clubmoss Needle-and-thread grass	629	50
Rauzi (1968)	Archer, WY	1961–1965	Sandy loam	Blue grama grass Western wheatgrass	804	49
Rauzi (1974)	Archer, WY	1967–1971	Fine sandy loam	Blue grama grass Western wheatgrass	947	33
Nichols (1969)	Newell, SD	1963–1965	Clay	Western wheatgrass	986	-13
Contour furrowing						
Wight (1973)	Ekalaka, MT	1968–1972	Clay	Thickspike wheatgrass Sandberg blugrass	273	140
Wight (1973)†	Plevna, MT	1971–1972	Clay	Western wheatgrass Blue grama grass	1842	110
Lister interseeding						
Wight and White (1974)‡	Sidney, MT	1967–1972	Sandy loam	Western wheatgrass Sedges	1052	30

† Contour furrows were seeded with alfalfa.
‡ A mixture of native species was seeded into the scalped area.

a reliable source of runoff water, good soil productivity potential, adaptable grass species, and how well the spreader system prevents prolonged flooding.

Contour furrowing has been used on several million hectares of western rangeland, primarily on fine-textured soils where infiltration is very slow. Several types of contour furrowers have been used, but the ones used most commonly create furrows about 50 cm wide and 15 to 25 cm deep on 1.5-m centers. Longevity of the furrows depends on the effectiveness of the interfurrow dams (Neff and Wight, 1977). Contour furrowing will probably be of greatest benefit where an introduced species, such as alfalfa, is planted in the furrows. On the Plevna site (Table 6–4), forage yields were increased from 1800 kg/ha to 3800 kg/ha when alfalfa was seeded in the furrows. Alfalfa is deep rooted and can extract soil water from deeper in the profile than can grasses on Northern Great Plains rangelands. On an irrigated pasture with 35 to 54% alfalfa (by weight), grazing cows and calves "did require the feeding of poloxalene. . .to control bloat" (Dunn and Olson, 1978).

Lister interseeders have been designed to prepare a seedbed for interseeding species, usually climax species, into native range (Wight and White, 1974). The practice is similar to contour furrowing but without interfurrow dams.

Pitting is a relatively mild mechanical manipulation usually performed with a modified one-way disk. Most range pitters leave small pits 1.2 to 1.8 m long, 10 to 15 cm wide, and 5 to 8 cm deep. Pitting was designed primarily to reduce runoff, but it showed no benefit on clay soils, because weathering caused rapid filling of the pits (Nichols, 1969), or on a coarse-textured site (Rauzi, 1968; Rauzi, 1974; Wight and White, 1974) where infiltration rates were already high (Table 6–4). Pitting has shown the most success on sites severely infested with clubmoss (Ryerson et al., 1970).

Snow management research on rangelands in the Northern Great Plains has been minimal and usually incidental to other objectives. While studying the effect of pitting (Nichols, 1969; Rauzi, 1968; Ryerson et al., 1970; Wight and White, 1974), contour furrowing (Neff and Wight, 1977), and level bench terracing (Haas and Willis, 1968), researchers observed increases in snow depth relative to untreated rangeland. On furrowed and nonfurrowed panspot and saline upland range sites, an average of 6.4 and 4.1 cm of snowmelt were trapped per year, respectively, over a 5-year period (Neff and Wight, 1977). Corresponding infiltration of the snowmelt was 50% on furrowed and virtually zero on nonfurrowed rangeland. With level benches, nearly all the increase in soil water compared with untreated rangeland came from snow accumulation on the benches and not from runon from upslope contributing areas (Haas and Willis, 1968).

Others reported the importance of vegetative roughness (height) to snow trapping (Rauzi, 1968; Van Haveren and Striffler, 1976; Wight et al., 1975). Crested wheatgrass stubble trapped snow more effectively than adjacent native range grass stubble (Rauzi, 1968). In northeast Montana (Wight et al., 1975), stubble heights, representing (i) smooth, short stubble of 5 cm; (ii) rough, intermediate stubble of 5 to 25 cm; and (iii) rough, tall stubble of 30 to 40 cm, trapped 2.5, 7.6, and 20.3 cm snow, respectively, representing about 0.7, 2.3, and 6.1 cm of water equivalent in the snow, respectively. Snow drifts induced by vegetative (sorghum-sudangrass and tall wheatgrass) and artificial barriers (wood slats) have significantly increased soil water supplies (Greb and Black, 1971). Introduced grasses, e.g., Russian wildrye, responded better than native grasses to the increase in soil water supplies. In another study, Russian wildrye and intermediate wheatgrass produced 4000 kg ha^{-1} y^{-1} over a 5-year period when adequately fertilized and grown in 76- and 106-cm contour rows. This yield represents a sixfold increase in forage production over typical nonfertilized rangeland (Black and Reitz, 1969).

6-3.3.2 Water Infiltration and Utilization

Infiltration of precipitation can be improved by grazing management as well as by mechanical methods. Water intake is directly related to quantity of old and new vegetative cover (Rauzi and Fly, 1968); therefore, degree of grazing intensity has a direct bearing on the quantity of water infiltrating a given soil.

The productivity and subsequent water use efficiency of grassland ecosystems can be greatly increased by fertilization. Nitrogen fertilization has increased both root growth (Black, 1968; Lorenz and Rogler, 1967) and magnitude of soil water extraction. On a mixed prairie, upland range site, N fertilization increased soil water extraction, overwinter soil water recharge, and subsequent plant water use and precipitation use efficiencies (Smika et al., 1965; Wight and Black, 1978). Overwinter water recharge was inversely related to fall soil water content.

Because grazing intensity and fertilization alter the quantity of vegetative cover available, they affect rangeland water use efficiency through their indirect effect on reducing evaporation. Infiltration increases proportionately to the quantity of vegetative cover, and evaporation decreases proportionately to the quantity of crop residue on the soil surface (Rauzi and Fly, 1968). Because fertilization increases soil water extraction, the size of the water storage reservoir is increased relative to that of nonfertilized grassland soil profiles (Wight and Black, 1978).

Water losses to evaporation in grassland ecosystems can be reduced by soil and crop-grazing management techniques that provide enough crop residue to effectively influence water movement in and out of the soil (Van Haveren and Striffler, 1976). Although the production potential of some portions of the rangeland in the Northern Great Plains is so low that little residue is available, even very low quantities of vegetative cover may affect the soil water regime and influence subsequent forage production (Van Haveren and Stiffler, 1976). Any small change in production becomes highly significant when applied to over 70 million ha (Lodge, 1970) of grassland in the Northern Great Plains.

6-3.4 Research Needs on Grasslands

More research is needed to determine the effect of small changes in water management, water infiltration, and evaporation losses on productivity of grassland ecosystems.

Although the amount of water added to the soil through snow management techniques may be relatively small in some years, the increase in amount at the beginning of the growing season may be more valuable than the same amount randomly distributed throughout the summer. In some years, the amount of water available from properly managed snow can mean the difference between an adequate and an inadequate supply of forage for livestock. Additional research is urgently needed to develop better ways of effectively using the snowmelt resource of the Northern Great Plains.

6-4 LITERATURE CITED

Aase, J. K., and F. H. Siddoway. 1974. Tall wheatgrass barrier and winter wheat response. Agric. Meteorol. 13:321–338.

Austin, M. E. 1965. Land resource regions and major land resource areas of the United States. USDA Agric. Handbook no. 296.

Bauer, A. 1968. Evaluation of fallow to increase water storage for dryland wheat production. N.D. Farm Res. 25(5):6–9.

Bauer, A. 1971. Effects of water supply and seasonal distribution on spring wheat yields. North Dakota Agric. Exp. Stn. Bull. no. 490.

Bauer, A., and T. J. Conlon. 1974. Effect of tillage interval of fallow on available soil nitrogen: Some soil physical properties and wheat yields. North Dakota Agric. Exp. Stn. Res. Report no. 51.

Bauer, A., and T. J. Conlon. 1978. Influence of tillage interval of fallow on soil water storage. N.D. Farm Res. 35(4):8–11.

Bauer, A., and H. L. Kucera. 1978. Effect of tillage on some soil physicochemical properties and on annually cropped spring wheat yields. North Dakota Agric. Exp. Bull. no. 506.

Bauer, A., E. B. Norum, J. C. Zubriski, and R. A. Young. 1966. Fertilizer for small grain production on summer fallow in North Dakota. North Dakota Agric. Exp. Stn. Bull. no. 461.

Black, A. L. 1968. Nitrogen and phosphorus fertilization for production of crested wheatgrass and native grass in northeastern Montana. Agron. J. 60:213–216.

Black, A. L. 1973. Crop residue, soil water, and soil fertility related to spring wheat production and quality after fallow. Soil Sci. Soc. Am. Proc. 37:754–758.

Black, A. L., P. L. Brown, A. D. Halvorson, and F. H. Siddoway. 1981. Dryland cropping strategies for efficient water-use to control saline seeps in the Northern Great Plains, U.S.A. Agric. Water Manage. 4:295–311.

Black, A. L., and L. L. Reitz. 1969. Row spacing and fertilization influences on forage and seed yields of intermediate wheatgrass, Russian wildrye, and green needlegrass on dryland. Agron. J. 61:801–805.

Black, A. L., and F. H. Siddoway. 1976. Dryland cropping sequence within a tall wheatgrass barrier system. J. Soil Water Conserv. 31:101–105.

Black, A. L., and F. H. Siddoway. 1971. Winter wheat recropping on dryland as affected by stubble height and nitrogen fertilization. Soil Sci. Soc. Am. Proc. 41:1186–1190.

Black, A. L., F. H. Siddoway, and P. L. Brown. 1974. Summer fallow in the Northern Great Plains (winter wheat). p. 36–50. *In* Summer fallow in the western United States. USDA-ARS Conserv. Res. Report no. 17.

Bond, J. J., J. F. Power, and W. O. Willis. 1971. Tillage and crop residue management during seedbed preparation for continuous spring wheat. Agron. J. 63:789–793.

Bond, J. J., and W. O. Willis. 1970. Soil water evaporation: First stage drying as influenced by surface residue and evaporation potential. Soil Sci. Soc. Am. Proc. 34:924–928.

Bond, J. J., and W. O. Willis. 1971. Soil water evaporation: Long term drying as influenced by surface residue and evaporation potential. Soil Sci. Soc. Am. Proc. 35:984–987.

Branson, F. A. 1956. Range forage production changes on a water-spreader in southeastern Montana. J. Range Manage. 9:187–191.

Brown, P. L. 1970. Dryland cereal production in Montana. p. 262–264. *In* W. C. Burrows et al. (ed.) Int. Conference on Mechanized Dryland Farming Proc. 11–15 Aug. 1969. Deere and Co., Moline, Ill.

Doering, E. J., and F. M. Sandoval. 1976a. Hydrology of saline seeps in the Northern Great Plains. Trans. ASAE 19:856–861, 865.

Doering, E. J., and F. M. Sandoval. 1976b. Saline seep development on upland sites in the Northern Great Plains. USDA Report no. ARS-NC-32.

Dunn, B. H., and H. M. Olson. 1978. Cow-calf beef production with irrigated forages. N.D. Farm Res. 36(1):13–21.

Emory, S. M. 1894. Organization—announcements. Montana Agric. Exp. Stn. Bull. no. 1.

Frank, A. B., and E. J. George. 1975. Windbreaks for snow management in North Dakota. p. 144–154. *In* Snow management on the Great Plains. Great Plains Agric. Coun. Pub. no. 73, Univ. of Nebraska, Lincoln.

Frank, A. B., D. C. Harris, and W. O. Willis. 1976. Influence of windbreaks on crop performance and snow management in North Dakota. p. 41–48. *In* R. Tinus (ed.) Shelterbelts on the Great Plains. Great Plains Agric. Coun. Pub. no. 78, Univ. of Nebraska, Lincoln.

Greb, B. W., and A. L. Black. 1971. Vegetative barriers and artificial fences for managing snow in the Central and Northern Plains. p. 96–111. *In* A. A. Hagen (ed.) Proc. Snow and Ice in Relation to Wildlife and Recreation Symp., Ames, Iowa. 11–12 February. Univ. of Iowa, Ames.

Haas, H. J., and W. O. Willis. 1962. Moisture storage and use by dryland spring wheat cropping systems. Soil Sci. Soc. Am. Proc. 26:506–509.

Haas, H. J., and W. O. Willis. 1968. Conservation bench terraces in North Dakota. Trans. ASAE 11:396–398, 402.

Haas, H. J., W. O. Willis, and J. J. Bond. 1974. Summer fallow in Northern Great Plains (spring wheat). p. 12–35. *In* Summer fallow in the western United States. USDA-ARS Conserv. Res. Report no. 17.

Halvorson, A. D., and A. L. Black. 1974. Saline-seep development in dryland soils of northeastern Montana. J. Soil Water Conserv. 29:77–81.

Halvorson, A. D., A. L. Black, F. Sobolik, and N. Riveland. 1976. Proper management—key to successful winter wheat cropping in Northern Great Plains. N.D. Farm Res. 33(4): 3–9.

Houston, W. R. 1960. Effects of waterspreading on range vegetation in eastern Montana. J. Range Manage. 13:289–293.

Howard, J. K. 1959. Montana high, wide and handsome. Yale Univ. Press, New Haven, Conn.

Lodge, R. W. 1970. Complementary grazing systems for the northern Great Plains. J. Range Manage. 23:268–271.

Lorenz, R. J., and G. A. Rogler. 1967. Grazing and fertilization affect root development of range grasses. J. Range Manage. 20:129–132.

Mathews, O. R., and T. J. Army. 1960. Moisture storage on fallowed wheatland on the Great Plains. Soil Sci. Soc. Am. Proc. 24:414–418.

Neff, E. L., and J. R. Wight. 1977. Overwinter soil water recharge and herbage production as influenced by contour furrowing on eastern Montana rangelands. J. Range Manage. 30: 193–195.

Nichols, J. T. 1969. Range improvement practices on deteriorated dense clay wheatgrass range in western South Dakota. South Dakota Agric. Exp. Stn. Bull. no. 552.

Rauzi, F. 1968. Pitting and interseeding native shortgrass rangelands. Wyoming Agric. Exp. Stn. Res. J. no. 17.

Rauzi, F. 1974. Mechanical and chemical range renovation in southwestern Wyoming. J. Range Manage. 27:48–52.

Rauzi, F. 1975. Snow management for water conservation in Wyoming. p. 180–186. *In* Snow management on the Great Plains. Great Plains Agric. Coun. Pub. no. 73, Univ. of Nebraska, Lincoln.

Rauzi, F., and C. L. Fly. 1968. Water intake on midcontinental rangelands as influenced by soil and plant cover. USDA Tech. Bull. no. 1390.

Ryerson, D. E., J. E. Taylor, L. O. Baker, H. H. Houlton, and D. W. Stroud. 1970. Clubmoss on Montana rangelands. Montana Agric. Exp. Stn. Bull. no. 645.

Schneider, R. P., F. Sobolik, and N. Riveland. 1978. No-till: Promise and problems. N.D. Farm Res. 35(4):12–14.

Smika, D. E., H. J. Haas, and J. F. Power. 1965. Effects of moisture nad nitrogen fertilizer on growth and water use by native grass. Agron. J. 57:483–486.

Thysell, J. C. 1938. Conservation and use of soil moisture at Mandan, N. Dak. USDA Tech. Bull. no. 617.

U.S. Department of Agriculture. 1970. Soil taxonomy of the national cooperative soil survey. USDA, U.S. Government Printing Office, Washington, D.C.

U.S. Department of Agriculture. 1974. Summer fallow in the western United States. USDA-ARS Conserv. Res. Report no. 17.

Van Haveren, B. P., and W. D. Striffler. 1976. Snowmelt recharge on a shortgrass prairie site. West. Snow Conf. Proc. 44:56–62.

Wendt, C. W., T. C. Olson, H. J. Haas, and W. O. Willis. 1970. Soil water evaporation. p. 207–227. *In* Evapotranspiration in the Great Plains. Great Plains Agric. Coun. Pub. no. 50, Univ. of Nebraska, Lincoln.

Wicks, G. A., and D. E. Smika. 1973. Chemical fallow in a winter wheat–fallow rotation. Weed Sci. 21:97–102.

Wight, J. R. 1973. Range improvement practices: Mechanical and fertilization. *In* The range beef cow. Symp. on Production III, Rapid City, S.D. 17–19 Dec.

Wight, J. R., and A. L. Black. 1978. Soil water use and recharge in a fertilized mixed prairie plant community. J. Range Manage. 31:280–282.

Wight, J. R., E. L. Neff, and F. H. Siddoway. 1975. Snow management on eastern Montana rangelands. p. 138–143. *In* Snow management on the Great Plains. Great Plains Agric. Coun. Pub. no. 73, Univ. of Nebraska, Lincoln.

Wight, J. R., and L. M. White. 1974. Interseeding and pitting on a sandy range site in eastern Montana. J. Range Manage. 27:206–210.

Willis, W. O. 1976. Dryland agriculture. p. 78–80. *In* R. E. Moreng (ed.) Western Resources Conference on Food Production and Resources of the West, Fort Collins, Colo. 15–16 July. Colorado State Univ., Fort Collins.

Willis, W. O. 1978. Snow on the Great Plains. p. 56–62. *In* S. C. Colbeck and M. Ray (ed.) Modeling of snow cover runoff. U.S. Army Corp. of Engineers, Cold Regions Res. and Engineering Lab., Hanover, N.H.

Willis, W. O., and J. J. Bond. 1971. Soil water evaporation: Reduction by simulated tillage. Soil Sci. Soc. Am. Proc. 35:526–529.

Willis, W. O., C. W. Carlson, J. Alessi, and H. J. Haas. 1961. Depth of freezing and spring run-off as related to fall soil-moisture level. Can. J. Soil Sci. 41:115–123.

Willis, W. O., and H. J. Haas. 1965. Dryland soil moisture losses in the Northern Great Plains. p. 49–53. *In* Proc. Great Plains Agric. Coun., Laramie, Wyo. 28–29 July. Great Plains Agric. Coun. Pub. no. 73, Univ. of Nebraska, Lincoln.

Willis, W. O., and H. J. Haas. 1969. Water conservation over winter in the Northern Great Plains. J. Soil Water Conserv. 24:184–186.

Willis, W. O., H. J. Haas, and C. W. Carlson. 1969. Snowpack runoff as affected by stubble height. Soil Sci. 107:256–259.

Willis, W. O., and J. F. Power. 1975. Soil temperature and plant growth in the Northern Great Plains. p. 209–219. *In* M. Wali (ed.) Prairie: A multiple view. Univ. of North Dakota Press, Grand Forks.

Young, R. A. 1962. Nitrogen fertilizer reduces the need for summerfallow. Fert. Solutions 6(4):12–15.

7

E. de Jong
University of Saskatchewan
Saskatoon, Saskatchewan
Canada

H. Steppuhn
Northern Illinois University
Dekalb, Illinois

Water Conservation: Canadian Prairies

7-1 SOILS, CLIMATE, AND MAJOR CROPS

Four major soil zones are recognized on the Canadian Prairies (Fig. 7-1): Brown (Aridic Borolls[1]), Dark Brown (Typic Borolls[1]), Black (Udic Borolls[1]), and Dark Gray (Boralfic Cryoborolls[1]). The Brown soil zone is the most arid, and the natural vegetation is typically a short grass mixed prairie. The Dark Brown soil zone is less arid, and the vegetation is of the mid-grass section of the mixed prairie. Black soils are typical of the fescue prairie-aspen grove (parkland) and true prairie grasslands, and the Dark Gray soils are found in transitional areas of grassland and forest. North of the Dark Gray soils, Gray soils (Alfisols[1]) have developed under forest.

Mean monthly values for precipitation, temperature $[(\overline{T}_{max} + \overline{T}_{min})/2]$, and estimated potential evapotranspiration for sites representative of the various soil zones are shown in Fig. 7-2. Precipitation increases from about 300 to 350 mm in the Brown soil zone up to 500 mm in parts of the Black and Gray soil zones; more than half of the precipitation falls from May to October. Potential evapotranspiration decreases from the Brown to the Gray soil zone. From mid-October to April, precipitation generally exceeds potential evapotranspiration; however, most of the precipitation comes as snow and is subject to considerable redistribution (Table 7-1). Assuming that 100 mm of precipitation has been stored in the soil after the growing

Contribution from the Department of Soil Science and the Division of Hydrology, University of Saskatchewan, Saskatoon, Saskatchewan, Canada.

[1] Equivalent in the U.S. Taxonomy.

Dryland Agriculture—Agronomy Monograph no. 23.

Table 7–1. Relative distribution of snow cover water equivalent in the Creighton Watershed, Bad Lake Basin (Gray et al., 1979).

Land type and use	Percent
Level plains	
Fallow	55
Stubble	71
Pasture	59
Hill tops	
Fallow	20
Stubble	48
Pasture	30
Gradual slopes	
Fallow	68
Stubble	69
Pasture and brush	79
Small draws	
Fallow	132
Stubble, pasture, and brush	138
Steep slopes	
Pasture and brush	253
Farm yards	150
Accumulated total snowfall (Nipher shielded gauge)	100
Average for entire watershed	77

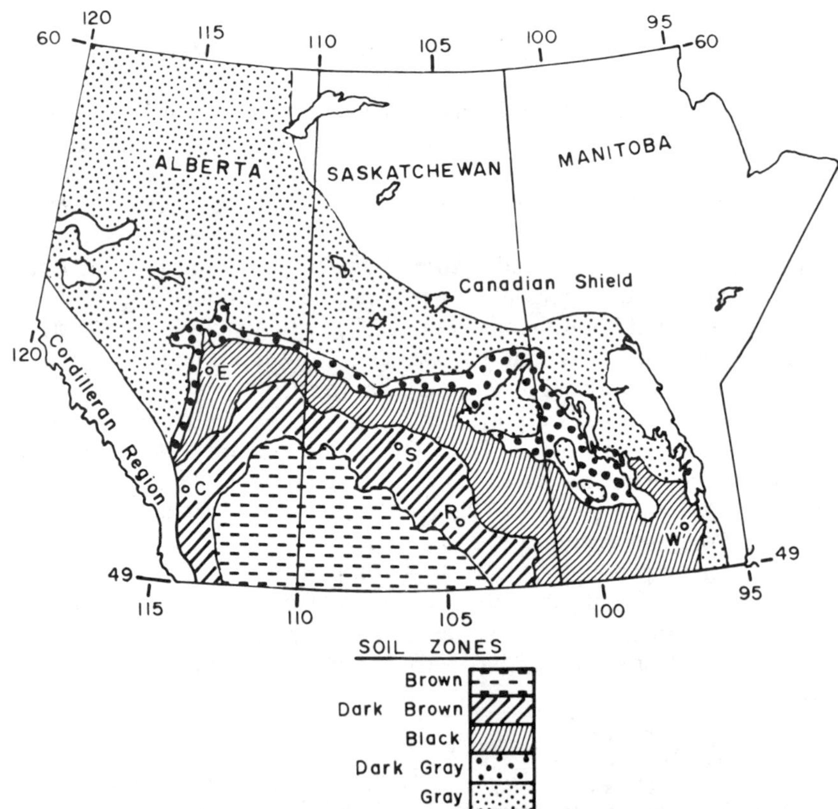

Fig. 7–1. Generalized soils map of the Canadian Prairie provinces.

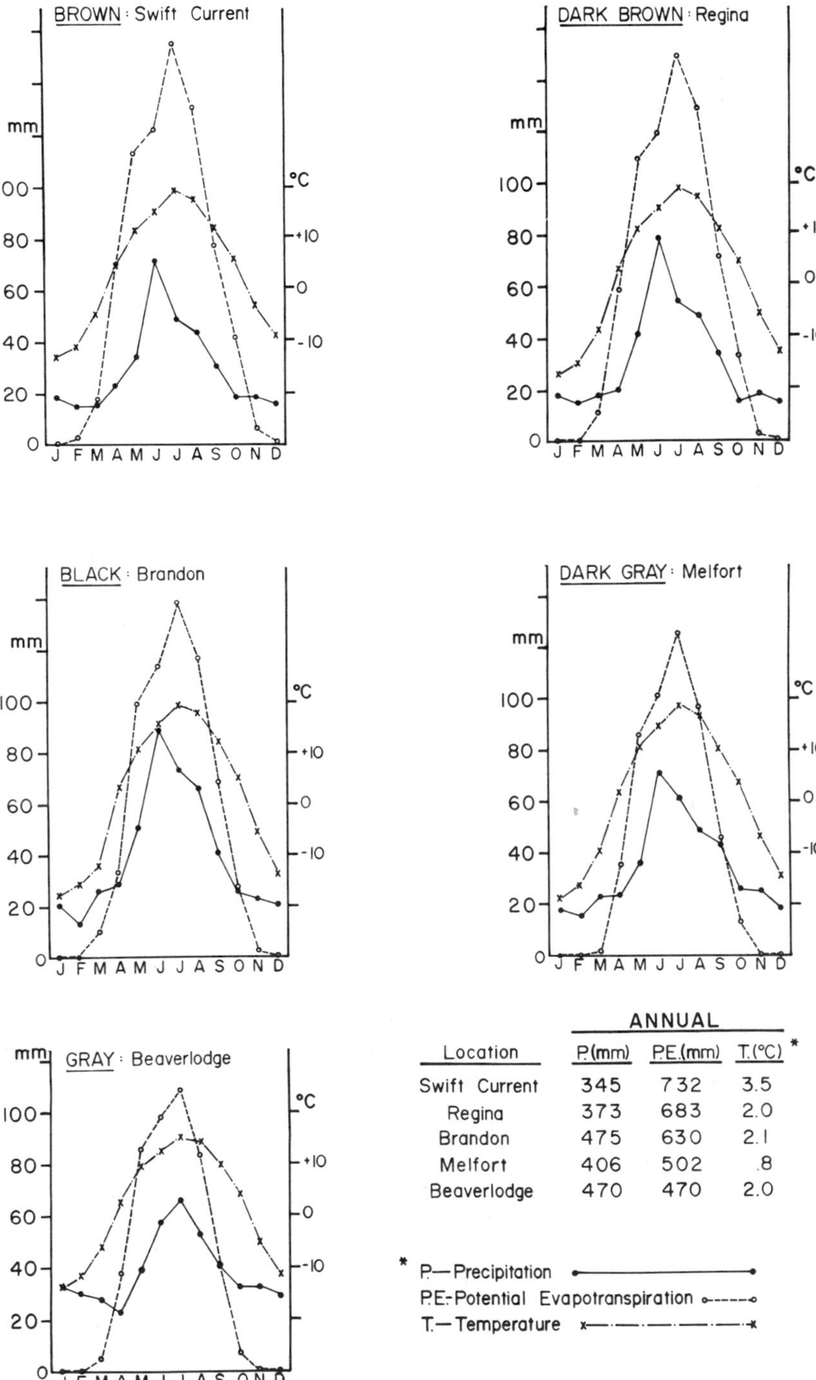

Location	ANNUAL P.(mm)	P.E.(mm)	T.(°C)*
Swift Current	345	732	3.5
Regina	373	683	2.0
Brandon	475	630	2.1
Melfort	406	502	.8
Beaverlodge	470	470	2.0

Fig. 7-2. Mean monthly climatic data for stations representative of the major soil zones in the Canadian Prairie provinces (Coligado et al., 1968).

Table 7–2. Use of improved land on the Canadian Prairies (mean and standard deviation for 1967 to 1976, Daviault, 1977).

Crop	Manitoba	Saskatchewan	Alberta	Prairie provinces
		10^6 ha		
Wheat	1.2 ± 0.3	6.2 ± 1.4	1.9 ± 0.5	9.3 ± 2.1
Oats	0.5 ± 0.1	0.8 ± 0.1	0.8 ± 0.1	2.0 ± 0.1
Barley	0.6 ± 0.2	1.5 ± 0.4	2.0 ± 0.2	4.1 ± 0.7
Rapeseed	0.2 ± 0.1	0.6 ± 0.3	0.5 ± 0.2	1.2 ± 0.5
Tame hay	0.5 ± 0.1	0.7 ± 0.2	1.3 ± 0.1	2.4 ± 0.3
Summer fallow	1.1 ± 0.2	7.3 ± 0.9	2.9 ± 0.3	11.4 ± 1.3

season and through the winter as available water, we can expect that the average water deficits for the next growing season will be from 400 mm down to 100 mm from the Brown to the Gray soil zones, respectively (Baier, 1976).

Spring wheat is the major crop on the Canadian Prairies, followed by barley, oat, rapeseed, and tame hay (Table 7–2). Summer fallow averages nearly 40% of the total cultivated acreage and increases from the Black and Gray soil zones to the Dark Brown and Brown soil zones. In the Gray soil zone, legumes and forages are often incorporated in the crop rotations, and barley is the most important grain crop (Hoyt et al., 1978).

7–2 WATER CONSERVATION PRACTICES

7–2.1 Water Conservation by Summer-fallowing

7–2.1.1 General Discussion

Early methods of summer fallowing involved repeated tillage, including plowing, to eliminate weeds and to minimize losses of soil water by evaporation (dust mulching). Dust mulching was discredited in part because wind erosion was a serious problem on the resulting "black" fallow. Studies on fallowing practices started in the 1920's (Anderson, 1967) and showed that plowed and plowless fallow produced equal yields if weeds were adequately controlled.

Efficiency of water storage during the summer fallow period was studied in detail in the Brown soil zone. The precipitation and conservation averages in Table 7–3 give no indication of the large annual variability. For example, in the field experiments, conservation during the first fall period ranged from a gain of 4.6 cm to a loss of 0.8 cm for an average of 2.1 cm. Occasional soil water losses occurred during the summer and second winter but never during the first winter.

Precipitation during the fall and winter following harvest is of major importance for recharging the depleted soil water reserves. Fall rain and winter snow are equally effective (Staple et al., 1960):

$$C_s = 0.035P + 0.20 \text{ or} \quad [1]$$

Table 7-3. Water conservation during the summer fallow period for spring wheat in the Brown soil zone in southern Saskatchewan.

Period (use)	Field data†			Lysimeters‡		
	Precipitation	Conservation		Precipitation	Conservation	
	mm	cm	%	mm	cm	%
Harvest-October (stubble)	55	2.1	36	174	5.6	32
November-April (stubble)	110	3.5	33			
May-October (fallow)	200	2.6	13	234	6.8	29
November-April (fallow)	111	1.9	16	97	1.1	12
Total	475	10.1	21	505	13.5	27

† Field data for 1939 to 1950 based on at least six fields/y (Staple and Lehane, 1952).
‡ Lysimeters for 1923 to 1947 (Soil Res. Lab., 1949).

$$C_s = 0.041S + 0.035R - 0.24F + 0.51 \qquad [2]$$

where C_s = conservation in stubble fields during the first fall and winter, cm
$P = R + S$
R = fall rain, mm
S = snow (water equivalent), mm
F = fall soil water (available water in 0- to 122-cm depth), cm.

About one third of the fall and winter precipitation is stored on stubble fields, but on fallow fields only about one tenth is stored. In over half of the years that data were obtained with the lysimeters (Table 7-3), there was no increase in soil water over the second fall and winter. The main factors determining the recharge efficiency of the snow are the amount of snow retained on the area (cf., Table 7-1) and the runoff losses of meltwater.

Runoff from snowmelt depends on the rate of thawing of the snowpack and the infiltration rate of the soil. Water content and temperature of the soil largely determine the infiltration rate (Fig. 7-3). The effect of soil water content on infiltration is illustrated by Eq. [2]; similar results were reported by Hobbs and Krogman (1971) in Alberta and by Willis et al. (1961) in North Dakota.

Water conservation is low during the summer mainly because of water loss by evapotranspiration. The importance of weed control is evident from Table 7-4 and is discussed later. Evaporation depends on amount of rain received and water content of the topsoil (Staple, 1964). In southern Saskatchewan, summer showers of less than 10 mm evaporate within 10 days unless they are associated with more rains, and in an average year fewer than 10 showers contribute to soil water storage in fallow fields between May and October.

Few data are available on soil water losses by deep drainage. Staple and Lehane (1952) claim that such losses were negligible for the field data shown in Table 7-3; however, drainage losses occurred on both fallow and stubble fields during the wet years from 1951 to 1956 (Staple et al., 1960). Van Schaik et al. (1976) calculated that between 1968 and 1972, fallow fields in southern Alberta lost 7.5 and 3.7 cm of water from the root zone in 1969

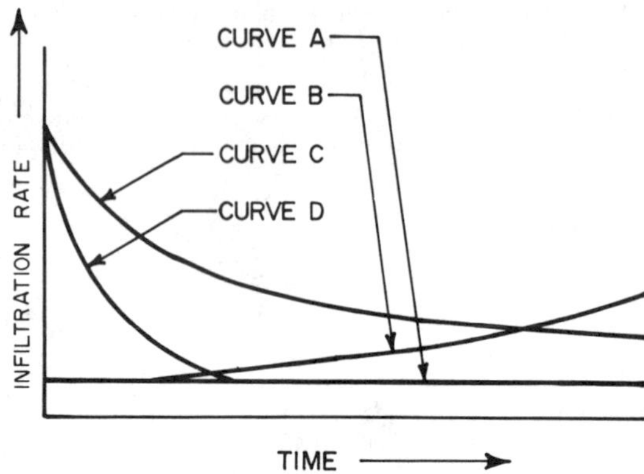

Fig. 7–3. Conceptual diagram of time variation infiltration rate of frozen soils (Gray et al., 1970). Curve *A* symbolizes the infiltration characteristics that prevail when the soil is frozen when saturated or when an impervious ice layer develops on the surface during the melting period. Curve *B* signifies the conditions that may exist when a soil is frozen at a high water content. Some of the meltwater is able to penetrate the soil and thus transfer heat to melt the ice. As the soil warms and the ice melts, the infiltration capacity increases. Curve *C* is the infiltration rate when the soil is frozen at a low water content and the soil temperature is near or above freezing. The ice thaws rapidly with the downward movement of water and infiltration proceeds as under unfrozen conditions. Curve *D* represents the condition in which the soil is frozen at a low water content but the soil temperature at the time of snowmelt is below freezing. Water entering the soil is frozen, and movement is inhibited.

and 1971 by deep percolation, respectively. These losses are similar to those calculated by Doering and Sandoval (1976) in North Dakota. No data are available for deep drainage losses in the Black and Gray soil zones, but they are likely larger than those in the Brown and Dark Brown soil zones.

On the Canadian Prairies most runoff is generated by the spring melt of the snowpack (Fig. 7–4). For two gauged watersheds in the Brown soil zone, annual runoff varied from about 5 to 90 mm (Nicholaichuk, 1967), whereas values from 0 to 50 mm were reported for a small watershed in the Dark Brown soil zone (Hall and Langham, 1970). On rolling topography (H. Steppuhn, D. E. L. Erickson, and W. Lin. 1978. Personal communication. Division of Hydrology, Univ. of Saskatchewan) 60% or more of the snow cover on fallow fields flows overland as runoff compared with 25% or less on stubble fields. The relatively greater runoff on fallowed fields is partly caused by the higher fall soil water contents than on stubble,

Table 7–4. Effect of weed growth on water conservation in summer fallow and on the yield of the following crop (3-year average) (Soil Res. Lab., 1949).

	Water stored in soil	Comparative yield of grain
	cm	%
Weed growth prevented	13.0	100
First cultivation 15 May	11.4	88
First cultivation 15 June	9.1	78
First cultivation 15 July	4.8	47

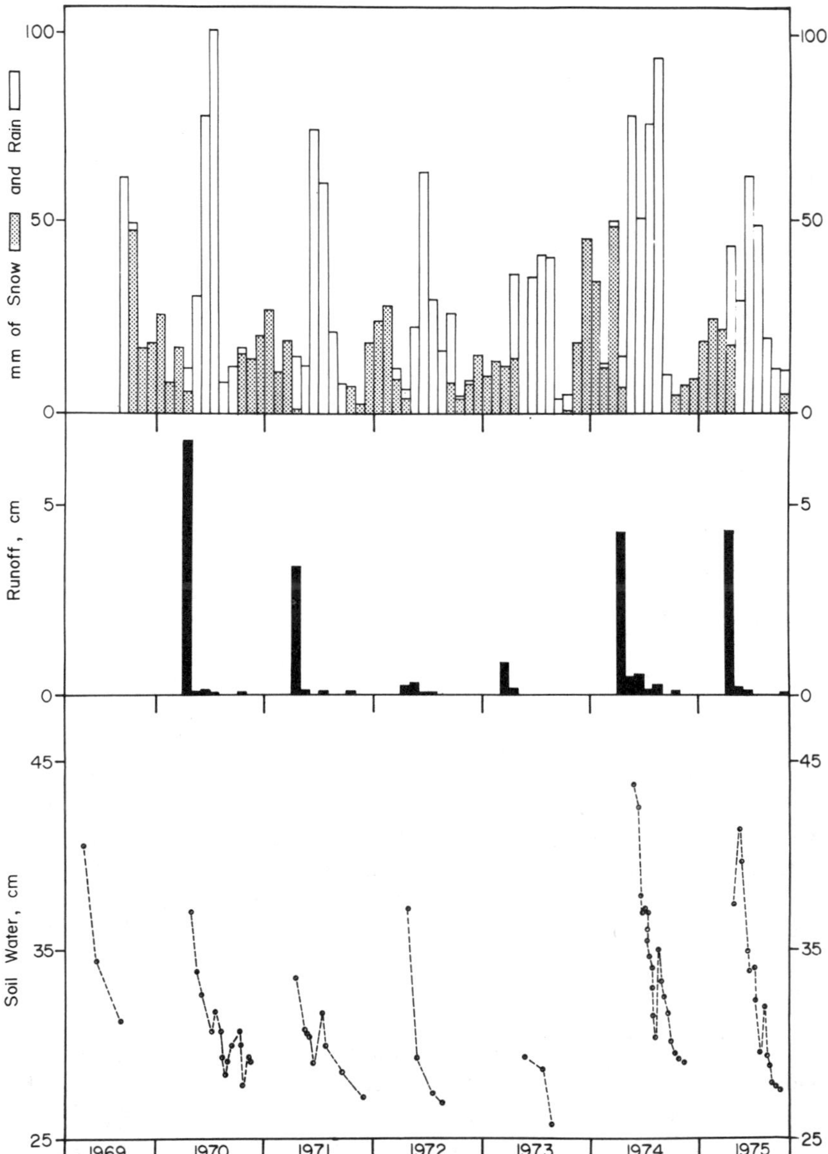

Fig. 7–4. Monthly precipitation and runoff, and chânges in soil water (in 0- to 120-cm depth) under grass in the Creighton Watershed, Bad Lake Basin (E. de Jong. 1978. Unpublished data. Division of Hydrology, Univ. of Saskatchewan).

but in addition the rate of snowmelt is faster on fallow fields than that on stubble or pasture (Erickson et al., 1978). During the growing season, runoff is small. On small plots (40 m^2) in the Black soil zone near Edmonton, runoff occurred on only 5 days during a typical summer and amounted to 1.4, 0.3, and 0.2 mm for fallow, cereal, and grass plots, respectively (Toogood, 1963).

Table 7-5. Changes in soil water during the fallow period in different soil zones.†

Zone	First winter‡	Fallow	
		Summer and fall	Winter
	cm		
Brown	5.7 ± 5.5§	1.5 ± 4.2§	2.5 ± 4.7§
Dark Brown	8.2 ± 3.5	1.1 ± 3.7	2.2 ± 3.7
Black	5.7 ± 4.5	1.3 ± 4.5	−0.7 ± 3.6
Dark Gray and Gray	2.0 ± 4.7	0.5 ± 3.5	−0.7 ± 2.5

† Calculated from data for 1971 to 1977 summarized by the Agrometeorology Research and Service Section, Agriculture Canada, Ottawa.
‡ The "winter" period includes some "fall" period as used in Table 7-3.
§ Water in the 0- to 122-cm depth (mean ± standard deviation).

Lysimeters present an ideal environment for soil water storage with weed growth prevented, no drainage losses, and the soil kept in a condition to minimize runoff. Under these conditions (Table 7-3) conservation efficiency during the fallow summer is twice as high as in the field, but conservation is lower the second winter for the reasons discussed above. The field data in Table 7-3 were obtained on nearly level fields. On rolling topography, soil water recharge over the winter period varies widely with slope position due to redistribution of snow or snowmelt water (de Jong and Cameron, 1980).

7-2.1.2 Water Conservation Efficiency by Soil Zone

Since 1970, soil water measurements have been taken in late fall and at seeding time at several sites in western Canada; because no data were gathered at harvest time, the results (Table 7-5) are not fully comparable with those in Table 7-3. In the Dark Gray and Gray soil zones, conservation during the first winter period is low because considerable soil water recharge usually had taken place between harvest and the late fall sampling. In these soil zones, precipitation exceeds potential evapotranspiration after mid-September (Fig. 7-2), and the measurements were usually taken later. Soil water storage varied widely from year to year, and the gain during the last 12 months of the summer fallow period is sometimes very small. Over the 7-year period, the differences in spring soil water content between fields that had been fallowed and those that had been cropped the previous year were (mean and standard deviation) 4.5 ± 4.5, 3.5 ± 3.3, 1.8 ± 4.5 and 0.3 ± 2.0 cm water in the 0- to 122-cm depth for fields in the Brown, Dark Brown, Black, and Dark Gray and Gray soil zones, respectively. Thus, even in the Brown and Dark Brown soil zones, stubble fields sometimes contain as much soil water as fallowed fields at seeding time. Summer fallowing is obviously of little benefit for soil water conservation in the Black and Gray soil zones. The differences between soil zones are attributable to differenes in climate (Fig. 7-2). The conservation efficiency over the 21-month fallow period drops from over 20% in the Brown and Dark Brown soil zones (Staple and Lehane, 1952; Lindwall and Anderson, 1981) to around 10% in the Black soil zones (Siemens, 1977; Bowren, 1977; Ferguson, 1963) to less than 10% in the Dark Gray and Gray soil zones (Bowren, 1977; Carder and Hennig, 1966).

7-2.1.3 Effect of Tillage on Soil Water Storage

One value of tillage is weed control (Table 7-4). On the average four timely tillage operations are sufficient to ensure adequate weed control during the summer fallow year, with three operations being sufficient in the Brown soil zone (Molberg et al., 1967). Additional tillage does not increase stored soil water or yields. Staple (1964) concluded that different tillage machines are equally effective in conserving soil water as long as weed growth is prevented. Because the soil tends to dry out to the depth of cultivation, especially when inverted, deep tillage or the use of disk or plow-type implements is not recommended. With chemicals that enable full weed control, chemical fallow stores as much or more water as mechanical fallow or combinations of both (Anderson, 1971; Bowren, 1977; Lindwall and Anderson, 1981). The use of herbicides has the advantage of leaving more trash on the surface and thus providing better erosion control.

Anderson's (1971) and Lindwall and Anderson's (1981) studies in southern Saskatchewan and Alberta showed no significant differences in overwinter soil water conservation between plots that had or had not received treatment for weed control after harvest. This agrees with an earlier study from which Staple et al. (1960) concluded that it was doubtful that different fall tillage methods had any effect on soil water conservation. Tests in the Black soil zone produced similar results (Bowren and Dryden, 1971). Apparently any gain in soil water from weed control was nullified by reduced snow accumulation, e.g., fall tillage of chemically fallowed fields with a blade tended to decrease the amount of soil water stored over the winter (Lindwall and Anderson, 1981). Fall tillage of fallowed fields with damming listers to hold snow and trap spring runoff resulted in no significant gains in soil water (Staple et al., 1960).

7-2.1.4 Advantages and Disadvantages of Summer Fallow

Data in Table 7-2 indicate the extent of summer fallowing on the Canadian Prairies. If summer fallowing is practiced primarily for water conservation, a positive correlation should exist between the area of stubble-seeded crops and precipitation during the previous fall and winter (cf., Eq. [2]). This is not the case, and in fact, summer fallowing has increased steadily over the 1930 to 1970 period (de Jong and Cameron, 1980; Michalyna and Hedlin, 1961). Other benefits of summer fallowing such as accumulation of NO_3-N, weed and pest control, spread of power and labor, economic factors, and a steady level of production are apparently at least as important to farmers as the need for water conservation.

Summer fallowing increases erosion hazards, leads to accelerated losses of organic carbon and soil structure, and possibly accelerates dryland salinity problems. The decrease in organic C content of surface soils, after initial cultivation of native lands, depends on the crop rotation practiced (Hill, 1954; Ridley and Hedlin, 1968). The greatest losses occur under the shortest rotation. A decrease in aggregate size occurs with the loss in organic C

(Emmond, 1971; Soil Res. Lab., 1949). Forages improve the aggregate size distribution (Soil Res. Lab., 1949), but in drier regions cereal yields are decreased following alfalfa or forages (Austenson et al., 1970), presumably because of greater soil water extraction by forages. In the wetter areas, especially in the Gray soil zone, the inclusion of forages and legumes in rotations is advantageous (Bowren, 1974; Bentley et al., 1971).

Recent reports of increased saline seeps lay the blame in part on increased leaching under summer fallow (Vander Pluym, 1978) compared with the original native grassland. No actual measurements of deep drainage losses under cultivated and noncultivated fields have been made in central and western Canada. Ferguson (1963) calculated that water losses by drainage, runoff, and evaporation outside the crop portion of the rotations increased markedly with the length of time that no crop grew, e.g., average annual water loss was 7 cm under continuous wheat and 21 cm under a wheat-fallow rotation.

7–2.2 Effects of Seeding Rates and Fertilization

Seeding rates for wheat range from 50 to 120 kg/ha, with the lower rates recommended for drier areas. Even lower seeding rates may be advantageous in the Brown soil zone when good agronomic practices are used, especially in dry years and on stubble (Pelton, 1969). Total water use was similar for seeding rates from 20 to 100 kg/ha, but rate of use was slower at the lower seeding rates. In the Black soil zone, yields of wheat on fallow and stubble increased as seeding rates increased from 67 to 168 kg/ha (Agric. Canada, 1949). Barley yields in the Dark Gray soil zone increased as the seeding rate increased from 54 to 134 kg/ha, yields in the Dark Brown soil zone were unchanged (Austenson, 1978). Similarly, doubling the row spacing reduced yields of wheat on stubble and fallow in the Black and Dark Brown soil zones (Rennie et al., 1966; Rennie, 1967).

Proper fertilization increases cereal yields on both fallow and stubble fields. In the Brown, Dark Brown, and Black soil zones, fertilizers did not affect total crop water use (de Jong and Rennie, 1969; Warder et al., 1963) and increased water use efficiency. In the Gray soil zone, fertilization increased the use of soil water and generally doubled the water use efficiency (Hoyt and Rice, 1977). Although soil water at harvest was lower under the fertilized treatments, these differences had largely disappeared by the next spring.

7–2.3 Other Water Conservation Practices

Terracing to conserve runoff water is not practiced in western Canada, although it increased fallow yields from 1410 to 1770 kg/ha and stubble yields from 470 to 700 kg/ha in an experiment in the Brown soil zone (Barnes, 1938). The yield increases can probably be attributed largely to better utilization of snowmelt water.

On fallow soil in shallow tanks, stubble mulches increased water conservation in an average summer by about 0.5 cm in the Brown soil zone

(Staple, 1964). Under more favorable conditions, increases of over 1 cm of stored soil water were obtained, but such gains are difficult to measure under field conditions. In southern Alberta zero-tillage increased soil water storage by nearly 3 cm over mechanical cultivation for weed control (Lindwall and Anderson, 1981); however, the highest yield was obtained on plots that were chemically fallowed and fall-tilled with a blade.

7-3 IMPROVING WATER CONSERVATION BY SNOW MANAGEMENT

7-3.1 General Discussion

Manipulation of the snow cover offers the greatest potential for increasing available water for dryland crops on the Canadian Prairies. The characteristics of snow that make it suitable for management are (i) its availability for manipulation while stored above ground, (ii) its susceptibility to redistribution by wind, and (iii) its compatibility with existing cropping practices. Approximately one third of the precipitation on the Canadian Prairies falls as snow, and a large portion of this snow usually persists through the winter. Snowfall varies widely from year to year, and the nature of the snow cover varies between regions that correspond roughly to the soil zones (McKay, 1964):

1. Brown soil zone: A snow cover from December through March, often discontinuous as a result of mid-winter melts.
2. Dark Brown soil zone: A generally continuous snow cover from December through March, subject to occasional mid-winter thaws.
3. Black and Gray soil zones: A continuous snow cover from November through mid-April, rarely subject to winter thaws.

The melting of the snow cover represents the single most important source of soil water recharge (cf., Fig. 7-4).

Snow covers at the time of maximum accumulation are rarely uniform (Table 7-1; Steppuhn and Dyck, 1974). The snow removed by wind from cultivated fields and pastures typically is deposited in ditches, fence rows, or adjacent to other physical structures. Snow harvesting seeks to prevent this transfer and to reduce evaporation from the snow as it is windborne. Data in Table 7-6 give some indication of the possibilities for harvesting snow in western Canada. The estimates of potential snow transport assumed an unlimited supply of snow and a threshold wind velocity of 7 m/s (at 10-m height) to initiate snowdrifting. The calculations of potential evaporation were based on the actual snow cover and show that during this particular winter evaporation caused a major decrease in the snow retained for soil water recharge. De Jong and Cameron (1980) estimated that on the average snow management might make an extra 3 cm of water available to the next crop.

The benefits from snow management depend on the specific cropping practices. Snow management is usually practiced on cropped fields scheduled to be reseeded the next spring. On such fields soil water recharge from snowmelt is more efficient than on fallowed fields.

Table 7-6. Snowfall, number of storms with wind speeds above the threshold for snowdrifting, and estimated snow transport and evaporation at six climatological stations for the 1974–1975 winter (Steppuhn, 1981).

Station	Snowfall water equivalent	Number of storms	Potential snow transported†	Potential snow evaporated‡
	mm		t/m	mm
Winnipeg, Manitoba	129	41	9.2	46
Regina, Saskatchewan	82	48	13.0	80
Saskatoon, Saskatchewan	78	40	5.9	29
Bad Lake, Saskatchewan	101	33	9.0	58
Edmonton, Alberta	108	27	2.6	27
Lethbridge, Alberta	137	39	21.7	103

† Based on an unlimited supply of snow; 1 t of snow deposited on an area of 10 m^2 is equal to 10 cm of water.

‡ Based on actual snow cover.

7-3.2 Snow Harvesting Practices

7-3.2.1 Stubble Management

The merits of tall stubble for trapping snow are well recognized (Janzen et al., 1960; Willis et al., 1969), and fall tillage operations should leave the stubble as undisturbed as possible. Sunflower or corn is sometimes seeded in rows 3 to 4 m apart on summer fallow fields to increase snow trapping and to reduce the wind erosion hazard. In reviewing studies on stubble management for snow trapping, Steppuhn (1981) concluded that tall stubble or stubble from crops such as sunflowers and corn is more efficient than short stubble.

Most cereals on the Canadian Prairies are cut and windrowed before combining. Cutting swaths at alternate heights increases the snow cover about 20 to 30% (Nicholaichuk and Norum, 1975) and available soil water by 1.3 cm over a 7-year period (Nicholaichuk, 1980).

7-3.2.2 Snowplowing

On the Canadian Prairies, the plowing of snow into ridges 2 to 3 m apart to trap windblown snow periodically receives attention. Matthews (1940) indicated that snowplowing increased the natural accumulation of snow by 100% in the ridges and 30% between them. Snow ridging did not affect yields of fallow-seeded cereals to any great extent, but grain yields on stubble were definitely improved (C. D. Matthews. 1940, 1941. Reports from Dominion Experimental Farm, Scott, Saskatchewan). Even better responses were obtained from hay and pasture fields, which also showed the highest overwinter gain in the soil water content.

Keys (Steppuhn, 1981) conducted snowplowing tests from 1963 to 1972 in the Brown and Dark Brown soil zones. In some years, ridges could not be established because of lack of snow. Ridging was thought to increase wheat yields by 10% at the site in the Brown soil zone and by 2% at the site in the Dark Brown soil zone. The results of Dyck et al. (1979) in the Brown soil zone over the 1972 to 1977 period also varied widely from year to year. In 3 of 5 years, snow ridging had no beneficial effect as the ridges were either blown away during blizzards, melted in mid-winter, or could not be formed because of insufficient snow cover. In the remaining years two ridging operations were completed each year and increased stored snow and soil water by at most 5 cm on the ridged fields.

7-3.2.3 Use of Vegetative Barriers

Various forms of vegetative barriers increase snow retention. For example, increased yields next to shelterbelts, originally planted to control wind erosion, can largely be attributed to increased soil water from trapped snow (Staple and Lehane, 1955), although reduced evaporative demand could also be a factor (Pelton, 1967).

Permanent grass barriers seeded in rows 15 m apart to trap snow are being tested on the Canadian Prairies (Nicholaichuk, 1980) and have been shown to be beneficial elsewhere (Black and Siddoway, 1976). However, vegetative barriers need not be permanent to be effective. Steppuhn and Erickson (Steppuhn, 1981) left 30-cm wide strips of unharvested wheat in cropped fields. At the time of maximum snow accumulation, snow volumes were increased by 88, 82, and 70% for strip spacings of 5.5, 11, and 16.5 m, respectively, compared with a stubble field with no strips. Special swather attachments that leave a strip of tall stubble but harvest the heads of the grain are now being tested with encouraging results (Nicholaichuk, 1980). Such temporary barriers allow easy adjustment from year to year to accommodate changes in machinery size and cropping practices.

7-3.2.4 Perspective

The methods of snow harvesting described above could perhaps increase overwinter soil water recharge by 3 cm or more. Although such gains are small, they are of the same order of magnitude as the gains in soil water during the summer fallow summer and winter (Table 7-5) and could reduce, or completely eliminate, the need for summer fallowing to conserve water. The practices are probably most effective on drier soils, i.e., on stubble fields in the Brown and Dark Brown soil zone, where the need for extra water conservation is the greatest. Techniques that involve simple stubble management procedures probably have the greatest potential (cf., Kirkland and Keys, 1981). Such procedures are cheap and interfere little with the farmers cropping system, hence they are usually quite cost effective (Nicholaichuk, 1980).

7-4 PROMISING RESEARCH

With the exception of small areas for winter rye and winter wheat, all cereals in western Canada are spring seeded. Winter wheat is grown successfully in southern Alberta and in some areas in southwestern Saskatchewan, but in other areas the risks of winter killing are too great. Winter wheat usually outyields spring wheat and probably uses water more efficiently because of differences in growth habits. The pattern of water use by winter wheat could reduce water losses by drainage and evaporation. Winter wheat also provides some erosion protection in fall and spring. The development of hardier varieties and new cultural practices, especially snow management, could increase the area in which winter wheat can be grown.

Snow management and tillage practices are being actively studied. Improved utilization of snow would reduce problems such as spring flooding and gully erosion. More research is needed on snowfall probabilities, their relationship to rain in the previous fall, and the redistribution of snow. Infiltration characteristics of frozen soils need further study.

To evaluate the salinity hazard associated with summer fallowing, studies must be conducted on the soil water regime of whole fields and the effects on the groundwater system of the area. Attention should be given to differences between slope positions within fields, and crops other than spring wheat should be included in these studies.

7-5 LITERATURE CITED

Agriculture Canada. 1949. Progress report 1936–1946. Melfort Experimental Farm, Agric. Canada, Ottawa.

Anderson, C. H. 1971. Comparison of tillage and chemical summerfallow in a semiarid region. Can. J. Soil Sci. 51:397–403.

Anderson, D. T. 1967. The cultivation of wheat. p. 338–355. *In* K. F. Nielsen (ed.) Canadian Centennial Wheat Symposium. Modern Press, Saskatoon, Saskatchewan, Canada.

Austenson, H. M. 1978. Principles of agronomy with particular reference to Saskatchewan conditions. Univ. of Saskatchewan, Saskatoon.

Austenson, H. M., A. Wenhardt, and W. J. White. 1970. Effect of summerfallowing and rotation on yield of wheat, barley and flax. Can. J. Plant Sci. 50:659–666.

Baier, W. 1976. Agroclimatic atlas Canada. Agrometeorology Res. and Service Section, Agric. Canada, Ottawa.

Barnes, S. 1938. Soil moisture and crop production under dryland conditions in western Canada. Agric. Canada Pub. no. 595, Ottawa.

Bentley, C. F., T. W. Peters, A. M. F. Hennig, and D. R. Walker (ed.) 1971. Gray wooded soils and their management. Dep. of Extension Bull. no. B71-1, Univ. of Alberta, Edmonton.

Black, A. L., and F. H. Siddoway. 1976. Dryland cropping sequences within a tall wheat grass barrier system. J. Soil Water Conserv. 31:101–105.

Bowren, K. E. 1974. Crop management on the Parkland soils of Saskatchewan. p. 6–14. *In* Proc. 1974 Soil Fertility Workshop, Saskatoon, Saskatchewan. February. Extension Division Pub. no. 244, Univ. of Saskatchewan, Saskatoon.

Bowren, K. E. 1977. The effect of reducing tillage on summerfallow when weeds are controlled in the Black soils zone in Saskatchewan. p. 5–11. *In* Proc. 1977 Soil Fertility and Crops Workshop, Saskatoon, Saskatchewan. February. Extension Division Pub. no. 328, Univ. of Saskatchewan, Saskatoon.

Bowren, K. E., and R. D. Dryden. 1971. Effect of fall and spring treatment of stubble land on yield of wheat in the Black soil region of Manitoba and Saskatchewan. Can. Agric. Eng. 13:32–35.

Carder, A. C., and A. F. M. Hennig. 1966. Soil moisture regimes under summerfallow, wheat and red fescue in the upper Peace River region. Agric. Meteorol. 3:311–331.

Coligado, M. C., W. Baier, and W. K. Sly. 1968. Risk analysis of weekly climatic data for agricultural and irrigation planning. Agrometeorology Section, Agric. Canada, Ottawa. Tech. Bull. no. 34 (Brandon), 38 (Melfort), 40 (Regina), 43 (Swift Current), 44 (Beaverlodge).

Daviault, R. 1977. Selected agricultural statistics for Canada. Agric. Canada Pub. no. 77/10, Ottawa.

de Jong, E., and D. R. Cameron. 1980. Efficiency of water use by agriculture for dryland crop production. Prairie Production Symp., Soils and Land Resources, Advisory Comm. to the Canadian Wheat Board, Saskatoon.

de Jong, E., and D. A. Rennie. 1969. Effect of soil profile type and fertilizer on moisture use by wheat grown on fallow or stubble land. Can. J. Soil Sci. 49:189–197.

Doering, E. J., and F. M. Sandoval. 1976. Hydrology of saline seeps in the Northern Great Plains. Trans. ASAE 19:856–861, 865.

Dyck, G. E., D. E. L. Erickson, and H. Steppuhn. 1979. Snow riding to increase all water. p. 1–12. *In* Proc. 1979. Soils and Crops Workshop, Saskatoon, Saskatchewan. February. Extension Division Pub. no. 403, Univ. of Saskatchewan, Saskatoon.

Emmond, C. S. 1971. Effect of rotations, tillage treatments and fertilizers on the aggregation of a clay soil. Can. J. Soil Sci. 51:235–241.

Erickson, D. E. L., W. Lin, and H. Steppuhn. 1978. Indices for prairie runoff from snowmelt. *In* Proc. Seventh Symp. of the Water Studies Inst. on Applied Prairie Hydrology. Water Studies Inst., Saskatoon.

Ferguson, W. S. 1963. Effect of intensity of cropping on the efficiency of water use. Can. J. Soil Sci. 43:156–165.

Gray, D. M., D. I. Norum, and J. M. Wigham. 1970. Infiltration and the physics of flow through porous media. p. 5.1–5.58. *In* D. M. Gray (ed.) Handbook on the principles of hydrology. Water Information Centre, Inc., Port Washington, N.Y.

Gray, D. M., H. W. Steppuhn, and F. L. Abbey. 1979. Estimating the areal snow water equivalent in the Prairie environment. p. 302–332. *In* Cold climate hydrology. Proc. Canadian Hydrology Symp.: 79, Vancouver. 10–11 May. Natl. Res. Coun. of Canada, Ottawa.

Hall, P. L., and E. J. Langham. 1970. Drainage basin study: Progress report no. 7 and program review. Pub. no. E70-4, Saskatchewan Res. Coun., Saskatoon.

Hill, K. W. 1954. Wheat yields and soil fertility on the Canadian Prairies after a half century of farming. Soil Sci. Soc. Am. Proc. 18:182–184.

Hobbs, E. H., and K. K. Krogman. 1971. Overwinter precipitation storage in irrigated and non-irrigated Chin loam soil. Can. J. Soil Sci. 51:13–18.

Hoyt, P. B., and W. A. Rice. 1977. Effects of high rates of chemical fertilizer and barnyard manure on yield and moisture use of six successive barley crops grown on three Gray Luvisolic soils. Can. J. Soil Sci. 57:425–435.

Hoyt, P. B., W. A. Rice, and A. M. F. Hennig. 1978. Utilization of Northern Canadian soils for agriculture. p. 232–247. *In* Symposia papers. Proc. 11th Int. Congr. Soil Sci., Vol. 3, Edmonton, Canada. 19–27 June. Int. Soc. of Soil Sci., Alberta Agric., Edmonton.

Janzen, P. J., H. A. Korven, G. K. Harris, and J. J. Lehane. 1960. Influence of depth of moist soil at seeding time and seasonal rainfall on wheat yields in southwestern Saskatchewan. Agric. Canada Pub. no. 1090, Ottawa.

Kirkland, K. J., and C. H. Keys. 1981. The effect of snowtrapping and cropping sequence on moisture conservation and utilization in west-central Saskatchewan. Can. J. Plant Sci. 61: 241–246.

Lindwall, C. W., and D. T. Anderson. 1981. Agronomic evaluation of minimum tillage systems for summerfallow in southern Alberta. Can. J. Plant Sci. 61:247–253.

Matthews, G. D. 1940. Snow utilization in prairie agriculture. Canada Dep. of Agric. Pub. no. 696, Ottawa.

McKay, G. A. 1964. Relationships between snow survey and climatological measurements for the Canadian Great Plains. p. 9–18. *In* Proc. 32nd Annual Meeting of the Western Snow Conference, Nelson, British Columbia. 21–23 April.

Michalyna, W., and R. A. Hedlin. 1961. A study of moisture storage and nitrate accumulation in soils as related to wheat yields on four cropping sequences. Can. J. Soil Sci. 41:5–15.

Molberg, E. S., E. V. McCurdy, A. Wenhardt, D. A. Drew, and R. D. Dryden. 1967. Minimum tillage requirements for summerfallow in western Canada. Can. J. Soil Sci. 47:211–216.

Nicholaichuk, W. 1967. Comparative watershed studies in southern Saskatchewan. Trans. ASAE 10:502–504.

Nicholaichuk, W. 1980. Snow management to provide additional water for agriculture. Prairie Production Symp., Soils and Land Resources, Advisory Comm. to the Canadian Wheat Board, Saskatoon.

Nicholaichuk, W., and D. I. Norum. 1975. Snow management on the Canadian Prairies. p. 118–127. *In* Snow management on the Great Plains, Bismarck, N.D. 29 July. Great Plains Agric. Coun. Pub. no. 73, Univ. of Nebraska, Lincoln.

Pelton, W. L. 1967. The effect of a windbreak on wind travel, evaporation and wheat yield. Can. J. Plant Sci. 47:209–214.

Pelton, W. L. 1969. Influence of low seeding rates on wheat yield in southwestern Saskatchewan. Can. J. Plant Sci. 49:607–614.

Ridley, A. D., and R. A. Hedlin. 1968. Soil organic matter and crop yields as influenced by frequency of summerfallowing. Can. J. Soil Sci. 48:315–322.

Siemens, L. B. (Chairman) 1977. Soils. p. 17–60. *In* Principles and practices of commercial farming. Univ. of Manitoba, Winnipeg.

Soil Research Laboratory. 1949. Soil moisture, wind erosion and fertility of some Canadian Prairie soils. Agric. Canada Pub. no. 819, Ottawa.

Staple, W. J. 1964. Dryland agriculture and water conservation. p. 15–30. *In* T. J. Army et al. (ed.) Research on water. ASA Spec. Pub. no. 4. Am. Soc. of Agronomy, Madison, Wis.

Staple, W. J., and J. J. Lehane. 1952. The conservation of soil moisture in southern Saskatchewan. Sci. Agric. (Ottawa) 32:36–47.

Staple, W. J., and J. J. Lehane. 1955. The influence of field shelterbelts on wind velocity, evaporation, soil moisture and crop yield. Can. J. Agric. Sci. 35:440–453.

Staple, W. J., J. J. Lehane, and A. Wenhardt. 1960. Conservation of soil moisture from fall and winter precipitation. Can. J. Soil Sci. 40:80–88.

Steppuhn, H. 1981. Snow and agriculture. p. 60–125. *In* D. M. Gray and D. H. Male (ed.) Handbook of snow: Principles, processes, management and use. Pergamon Press, Inc., Elmsford, N.Y.

Steppuhn, H., and G. E. Dyck. 1974. Estimating true basin snow cover. p. 314–324. *In* H. S. Santeford and J. L. Smith (compilers) Proc. of Symp. on Advanced Concepts and Techniques in the Study of Snow and Ice Resources, Monterey, Calif. 2–6 Dec. 1973. Natl. Res. Coun.-Natl. Academy of Sci., Washington, D.C.

Toogood, J. A. 1963. Water erosion in Alberta. J. Soil Water Conserv. 18:238–240.

Vander Pluym, H. S. A. 1978. Extent, causes and control of dryland saline seepage in the Northern Great Plains Regions of North America. p. 1.48–1.58. *In* Dryland-saline-seep control. Proc. Meeting of Subcommission on Salt-affected Soils, 11th Int. Congr. Soil Science, Alberta Agric., Edmonton.

van Schaik, J. C., D. S. Chanasyk, and E. H. Hobbs. 1976. An assessment of the capability of a computer program to éstimate fall soil moisture and deep percolation. Can. J. Soil Sci. 56:357–362.

Warder, F. G., J. J. Lehane, W. C. Hinman, and W. J. Staple. 1963. The effect of fertilizer on growth, nutrient uptake and moisture use of wheat on two soils in southwestern Saskatchewan. Can. J. Soil Sci. 43:107–116.

Willis, W. O., C. W. Carlson, J. Alessi, and H. J. Haas. 1961. Depth of freezing and spring run-off as related to fall soil moisture level. Can. J. Soil Sci. 41:115–123.

Willis, W. O., H. J. Haas, and C. W. Carlson. 1969. Snowpack run off as affected by stubble height. Soil Sci. 107:256–259.

8

R. E. Ramig and R. R. Allmaras
Agricultural Research Service
U.S. Department of Agriculture
Pendleton, Oregon

R. I. Papendick
Agricultural Research Service
U.S. Department of Agriculture
Pullman, Washington

Water Conservation: Pacific Northwest

The Pacific Northwest states of Idaho, Oregon, and Washington contain all or portions of the land resource regions (Austin, 1965) and dry-farmed cropland areas (Thomas and Grano, 1971) listed in Table 8-1. Because the Northwestern Forest, Forage, and Specialty Crop Region receives adequate precipitation, special water conservation practices are usually unnecessary. The Western Range and Irrigated Region and the Rocky Mountain Range and Forest Region both have little dry-farmed cropland. This chapter discusses water conservation practices used in the Northwestern Wheat and Range Region, which contains 78% of the dry-farmed cropland in the Pacific Northwest.

Three major land resource areas (Columbia Basin, Columbia Plateau, and the Palouse and Nez and Perce Prairies) within the Northwestern

Table 8-1. Land resource regions and dry-farmed cropland areas of Idaho, Oregon, and Washington.

Land resource region	Dry-farmed cropland, million hectares
Northwestern Forest, Forage, and Specialty Crop	0.58
Northwestern Wheat and Range	4.04
Western Range and Irrigated	0.14
Rocky Mountain Range and Forest	0.40
Total dry-farmed cropland	5.16

Contribution from the U.S. Department of Agriculture, Agricultural Research Service.

Dryland Agriculture—Agronomy Monograph no. 23.

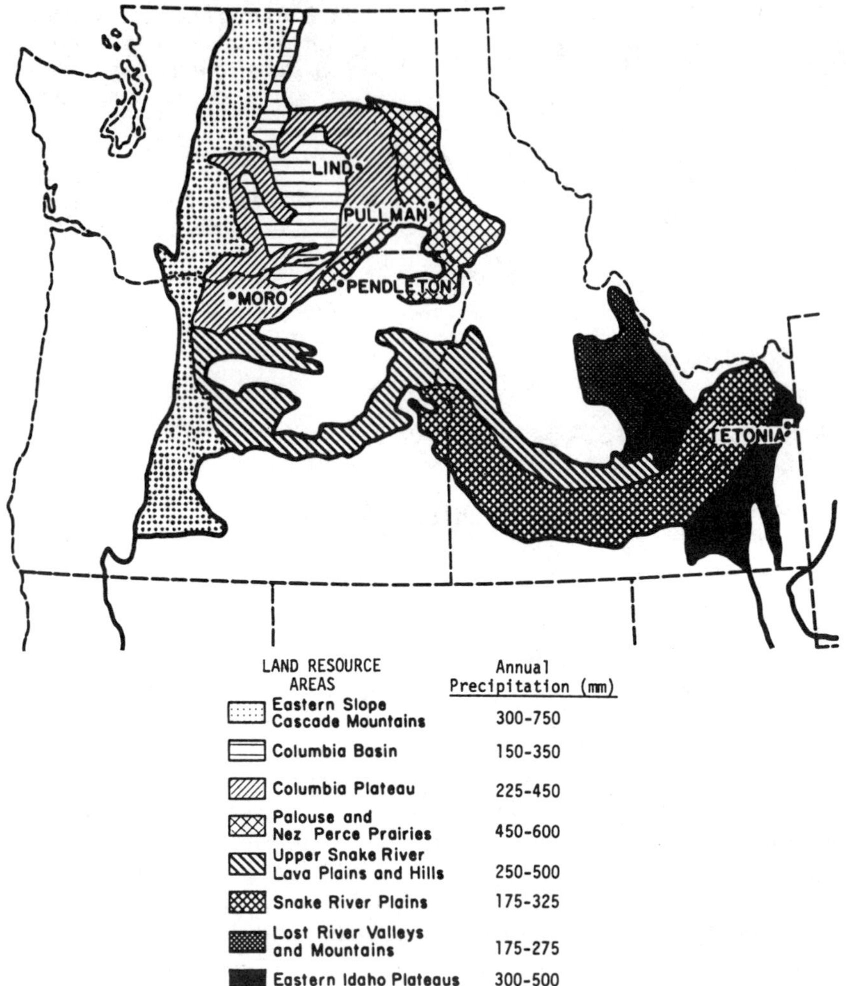

Fig. 8-1. Major land resource areas in the Northwestern Wheat and Range Region (Austin, 1965).

Wheat and Range Region form a contiguous body of land in northern Idaho, eastern Washington, and north central Oregon (Fig. 8-1) that totals 3.3 million ha of dry-farmed cropland (Thomas and Grano, 1971). A fourth major land resource area is the Eastern Idaho Plateaus in southeastern Idaho with 0.45 million ha of dry-farmed cropland (Fig. 8-1). Water conservation is paramount for economic production in croplands of these four major land resource areas. The remaining four major land resource areas in the Northwestern Wheat and Range Region are predominantly forest, range, or irrigated land. They contain only small areas of dry-farmed cropland (0.29 million ha), which are managed similarly to those in adjacent dry-farmed areas.

8-1 CLIMATE, TOPOGRAPHY, SOILS, AND CROPS

8-1.1 Climate

The Columbia Basin, Columbia Plateau, and Palouse and Nez and Perce Prairies major land resource areas (Fig. 8-1) occur at an elevation of 350 to 1400 m; summers are dry and winters are cool and wet. Climate is determined primarily by prevailing westerly winds from cyclonic storms that sweep in from the Pacific Ocean. Annual precipitation ranges from 150 to 600 mm, with 60 to 70% occurring during the November through April period. Rainfall from individual storms usually ranges from a few to 50 mm, but intensities are low, usually less than 4 mm/h (Horner et al., 1957). Six-hour rainfall amounts of 18 to 32 mm are expected to occur at least once in 2 years (Hershfield, 1961). The mild winter temperatures may freeze soils to depths ranging from a few to 40 cm. Soil freezing is more prevalent and deeper in the northern than southern parts of the region. Soils may freeze and snow may accumulate to a depth of 30 cm or more, especially at the higher elevations in the northern part of the region. Snowmelt or rain on frozen or saturated soil often produces runoff and severe soil erosion, especially in areas with higher precipitation and steeper topography.

In the Eastern Idaho Plateaus (1500 to 2000 m above sea level), the distribution of annual precipitation at Tetonia, Idaho, is transitional between the maritime climate of the Pacific Coast, represented by Pendleton, Oregon, and the continental climate of the Great Plains, represented by Akron, Colorado (Fig. 8-2). Annual precipitation ranges from 300 to 500 mm, with monthly distribution of about 25 mm except when summer rains increase the precipitation in May and June to 40 mm/month. The soil is frozen and snow covered during most of the winter, after which considerable water can be lost by runoff from rapid spring snowmelt on frozen soil.

8-1.2 Topography and Soils

Topography ranges from nearly level valleys to moderately and steeply sloping uplands. Much of the steeper land is found in the higher precipitation zones of eastern Washington and northern Idaho. Because the slopes are not only steep but also highly variable areally, soils of different land capability class (Klingebiel and Montgomery, 1973) are not managed separately within a field. Approximately four fifths of the cropland in this dry-farmed area has slopes of 8 to 30%, and slopes of some fields exceed 45%. Most of the soils are in the "erosion" land capability class because of the special practices needed to control soil erosion.

Soils are derived mainly from loess mixed with some volcanic ash. They belong to the order Mollisols, suborder Xerolls, and great groups Argixerolls and Haploxerolls (U.S. Geological Survey, 1970).

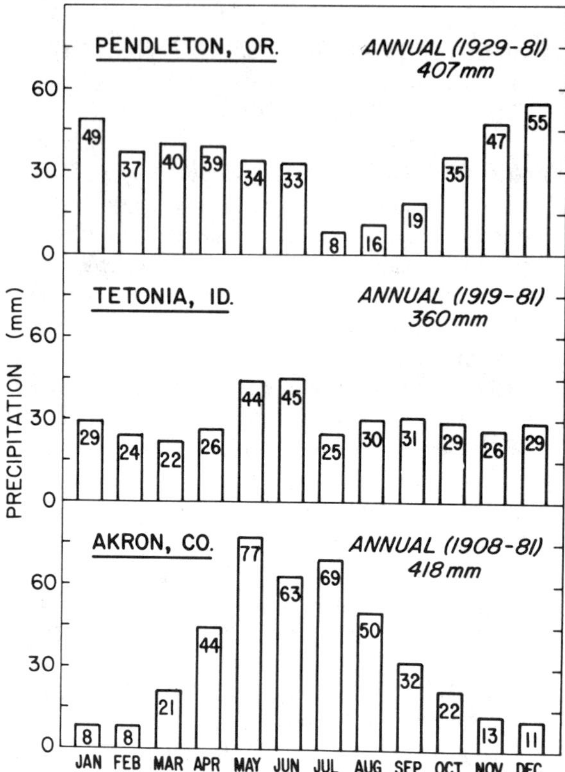

Fig. 8-2. Precipitation by months at Pendleton, Oregon, Tetonia, Idaho, and Akron, Colorado, to contrast the winter precipitation pattern at Pendleton and the uniform precipitation pattern at Tetonia (both in the Pacific Northwest) with the summer precipitation pattern at Akron in the Great Plains.

Haploxerolls are found in the Columbia Basin, Columbia Plateau, and Eastern Idaho Plateaus where summer fallow is practiced. Argixerolls are found in the Palouse and Nez and Perce Prairies and are often annually cropped to wheat, pea, lentil, alfalfa, and grass.

8-1.3 Cropping Practices

Cropping systems in the Pacific Northwest depend primarily on the amount of annual precipitation. Of the 3.3 million ha in the dry-farmed crop area in Oregon, Washington, and northern Idaho, approximately 1.7 million ha is in an alternate winter wheat-fallow zone, where annual precipitation ranges from 200 to 330 mm. Another 0.9 million ha is in an intermediate zone, where annual precipitation ranges from 330 to 380 mm. The usual cropping system in this intermediate zone is two or three crops between fallow periods (e.g., rotations of fallow-winter wheat-spring barley-spring barley or fallow-winter wheat-spring wheat). Cropping intensity depends largely on yearly precipitation but may also depend on economic factors and government farm policy. The remaining 0.7 million ha makes up the annual cropping zone, where annual precipitation exceeds 380 mm and where winter wheat is grown in rotation with pea, lentil, barley, or alfalfa.

The 0.45 million ha in the Eastern Idaho Plateaus, where annual precipitation ranges from 300 to 500 mm, largely is cropped in a rotation of winter wheat and fallow.

8-2 WATER CONSERVATION PRACTICES

Precipitation amount and distribution in the Pacific Northwest is especially favorable for soil water recharge. Daily potential evaporation during the recharge period is low enough that even small amounts of precipitation can percolate beyond the surface soil layers. Surface detention is a mechanism for delaying runoff. It is controlled by soil surface roughness, either random or oriented, and is an important factor for water conservation because much of the cultivated cropland is moderately to steeply sloping. In northeastern Oregon, over 90% of the cultivated cropland in the Columbia Plateau and Palouse and Nez and Perce Prairies major land resource areas has slopes ≥ 5% (Allmaras et al., 1979).

An unfavorable factor in some areas for soil water recharge is shallow soil depth. Frequency of this problem is reflected in the estimated average soil loss tolerance of 7.9 and 6.4 t ha^{-1} y^{-1} for cultivated land in the Columbia Plateau and Palouse and Nez and Perce Prairies major land resource areas, respectively (Allmaras et al., 1979). A soil loss tolerance less than 11.2 t ha^{-1} y^{-1} is assigned to a soil when the root zone is less than 150 cm deep. Soil water storage in a Ritzville silt loam soil (coarse-silty, mixed, mesic Calciorthidic Haploxerolls) from cereal harvest in August until March of the crop year 2 years later (19 months) was reduced as much as 3.8 cm, or 25%, when soil depth was less than 150 cm (Alban and Cushman, 1962).
period of greatest soil water recharge. During the November through March period, from 60 to 120 diurnal freeze-thaw cycles may be expected (Hershfield, 1974). Theoretical studies by Harlan (1973), Fuchs et al. (1978), and Cary et al. (1979) show that a freeze-thaw cycle redistributes water toward the surface. The severity of a diurnal freeze-thaw cycle, and thus displacement of water toward the surface, depends on soil water content, temperature, plant cover affecting the nocturnal radiation balance, and internal soil drainage as related to subsoil compaction. The magnitude of these effects is not well understood, but freeze-thaw cycles may decrease net water recharge by increasing evaporation and runoff losses.

Methods for reducing soil water losses are based primarily on suppressing unnecessary transpiration and evaporation. Weeds must be controlled rigorously during the March through August fallow period. During summer fallow, a tillage mulch is produced by a rod weeder (Papendick et al., 1973; Pikul et al., 1979) to conserve stored water and hold adequate amounts in the seed zone for wheat germination in the fall before the winter rains commence. There is usually a sharp interface between the soil mulch and underlying packed soil as produced by the rod weeder. Papendick et al. (1973) showed that water vapor and liquid transfers are restricted by the soil mulch, that liquid water moves upward to the packed soil–soil mulch interface, and that vapor flux then moves downward from the hot, dry soil mulch layer. These combined heat and water flows provide the overall water conservation and maintenance of moist soil at the anticipated seeding depth.

8–2.1 Summer Fallow

Summer fallow is the most widely and commonly used method of soil management to conserve water in the Pacific Northwest states.

There are 1.5 to 1.7 million ha (estimates by Mathews, 1951, and Leggett et al., 1974, respectively) of summer fallow in the Pacific Northwest states. Fallow appears essential for satisfactory long-term crop production where annual precipitation is less than 330 mm. In Oregon and Washington, where annual precipitation is 330 to 380 mm, fallowing may be needed depending on variation of annual precipitation. In the dry-farmed section of the Eastern Idaho Plateaus, summer fallow is usually necessary for stable crop production (Leggett et al., 1974). Factors that affect the extent of summer fallow in the Pacific Northwest are (i) the contribution to production stability, (ii) crop response on fallow, (iii) governmental programs controlling and limiting seeded "acreage," (iv) relations between production costs and commodity prices, and (v) soil erosion hazard.

During the first winter after wheat harvest, water storage efficiencies range from 50 to 75% of the precipitation (Leggett et al., 1974). Little, if any, precipitation is stored during March through October, and there is some net loss of soil water by evaporation. During the winter of the crop year, water storage varies from 10 to 50% of the precipitation. Thus, water storage is about 50% for the 17-month water storage period from September after harvest to March of the crop year. This storage efficiency contrasts with efficiencies of 15 to 33% reported by Evans and Lemon (1957) and Leggett et al. (1974) for the Great Plains, where 60 to 70% of the precipitation occurs from April through August (Fig. 8–2). Later research in the Central and Northern Great Plains (Smika and Whitfield, 1966; Smika and Wicks, 1968) have shown fallow water storage efficiencies from 25 to 45% with improved weed control and straw mulch management.

8–2.1.1 Fall Tillage

More than one system and time of initial (or primary) tillage is used for summer fallow in the Pacific Northwest because of geographic differences in amount, monthly distribution, and form (rain or snow) of winter precipitation. Latitude and elevation also affect the system by influencing duration of saturated or frozen soil surfaces.

Where annual precipitation is less than 280 mm, summer annual weeds such as Russian thistle and prostrate knotweed often infest wheat fields. The wide row spacing (30 to 40 cm) of the deep furrow drills required for mulch clearance and successful stand establishment keep wheat plants from competing sufficiently for light to prevent the late emerging weed growth. Herbicide sprays to control these weeds are only partially effective, because these weed seedlings are usually drought stressed from competing with vigorous wheat plants. If there are light postharvest rains, these weeds, especially the Russian thistle, may grow sufficiently large to interfere with tillage machine operation. Postharvest subsurface tillage in August with large sweeps (150 to 180 cm wide) at a depth of 10 to 12 cm undercuts the stubble and weeds, prevents formation of viable weed seeds, and leaves 80 to 90% of the residues erect to trap winter snows. The shattered soil surface improves water infiltration especially into frozen soils. Approximately 10% of the fallowed land in Oregon is fall-tilled in this manner to control weeds, improve infiltration, increase water conservation, and reduce wind and water erosion.

The low precipitation zone in Washington covers a greater range of latitude and elevation than the similar low precipitation zone in Oregon. Harvest is later and soils freeze more frequently than in Oregon. Nearly all of the stubble land in the northern areas of the wheat region is fall-chiseled with machines that have a chisel spacing of 30 cm working at a depth of 20 to 25 cm (Lindstrom, 1974). This tillage controls most of the summer weeds, leaves the soil surface in a rough condition to detain snowmelt or rain in depressions, and increases infiltration in the bottom of the chisel channels that frequently are not frozen. Lindstrom et al. (1974) reported that fall chiseling, fall disking, and no-tillage stored 87, 60, and 50%, respectively, of the September to May precipitation during cold winters (Fig. 8–3). During mild winters, 62% of the September to May precipitation was stored regardless of tillage (Fig. 8–4).

In the Eastern Idaho Plateaus, soils freeze every winter, and most of the winter precipitation occurs as snow. Standing stubble traps the snow, and in one comparison (Massee and McKay, 1979), standing stubble increased water storage by as much as 90 mm and grain yield by as much as 335 kg/ha compared with burned stubble. The deep snowpack in the standing stubble absorbs the snowmelt and forms a slush at the snow-soil interface. The temperature at this interface remains near 0°C for days, whereas the water slowly infiltrates into the soil. Soil temperature, water content, structure, and slope determine the extent of runoff; however, significant runoff is expected in 6 to 10 years (Massee and Siddoway, 1969). Fall chiseling 30 cm deep (below the frost zone), with shanks spaced 90 cm apart and with operations on the contour, reduced runoff and resulted in storage of 87% of the September through April precipitation (Massee and Siddoway, 1969); standing stubble alone resulted in storage of only 60% of the precipitation. Chiseling to a depth less than the frost zone was ineffective. Snowmelt ac-

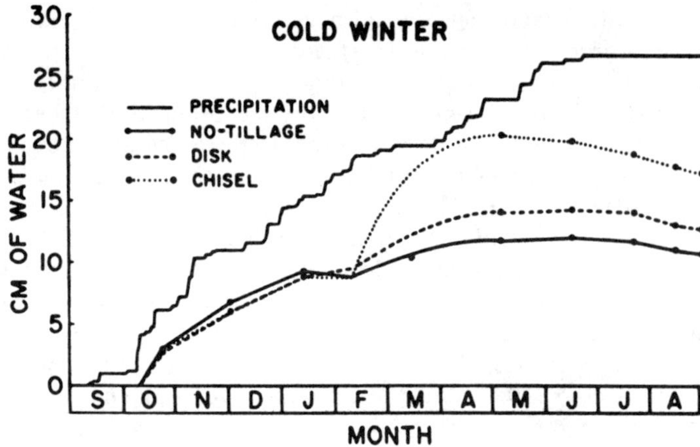

Fig. 8-3. Cumulative precipitation and soil water storage for three tillages during a year with a cold winter, frozen soil, and severe runoff at Lind, Washington (from Lindstrom et al., 1974).

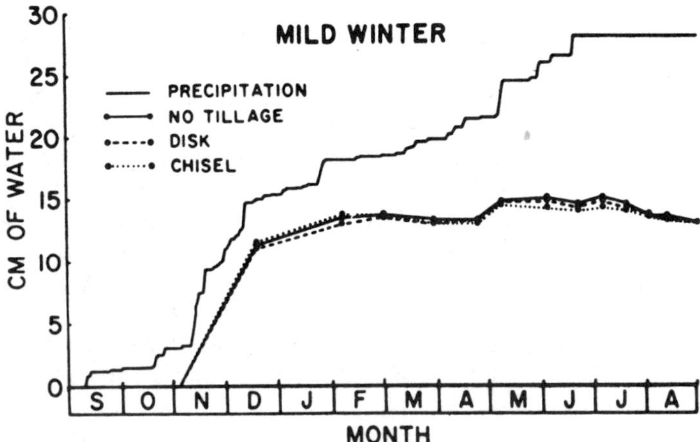

Fig. 8-4. Cumulative precipitation and soil water storage for three tillages during a year with a mild winter, nonfrozen soil, and no runoff at Lind, Washington (from Lindstrom et al., 1974).

cumulation in chisel furrows that were not on the contour accelerated erosion. Wheat stubble is left over the winter on about 50% of the land to be fallowed; 35% is fall-chiseled, and 15% is fall-disked.

Where annual precipitation ranges from 330 to 380 mm in Oregon, 90% of the harvested wheat stubble is left standing the first winter; some is grazed by cattle. Soil is frozen only intermittently and then to only shallow (less than 10 cm) depths. Fall tillage of weedfree stubble land after harvest has not proved beneficial. October through March water storage during the winters of measurement at each of two locations was 72% for standing stubble, 68% for flailed stubble, and 51% for burned or plowed stubble (R. E. Ramig. 1980. Unpublished data).

In the similar (330 to 380 mm) precipitation zone in Washington, from 50 to 95% of the stubble land is fall-tilled after harvest to (i) improve water detention and infiltration, (ii) control downy brome, and (iii) balance the seasonal workload of farm operations. The stubble is fall-chiseled where water detention and infiltration is the main objective. Stubble is fall-disked 2 to 10 cm deep with heavy offset disks when the objective is to kill germinated downy brome and volunteer wheat or to begin stubble decomposition by partial incorporation. The latter goal is desirable if there is more than 4500 kg/ha of straw present at harvest.

8-2.1.2 Spring and Summer Tillage

Spring tillage of summer fallow depends on the amount and type of fall tillage. When wheat stubble is fall-tilled, initial spring tillage is accomplished with sweeps, chisels, or disks (Lindstrom, 1974). Spring tillage should leave enough straw on the soil surface to protect the soil from wind erosion, develop a loosened soil mulch layer, and stop growth of weeds and volunteer grain. Papendick et al. (1973) observed no benefit and even detrimental effects from tillage deeper than 12 to 16 cm. In low precipitation zones where the amount of stubble is usually less than 3000 kg/ha, subsurface sweeps or field cultivators are used because they usually do not bury more than 20% of the residue. If the amount of residue exceeds 4500 kg/ha, the stubble is usually first disked, then chiseled, followed by skew treading. Massee et al. (1966), Massee and Siddoway (1970), Oveson and Appleby (1971), and Lindstrom (1974) all emphasize that the first tillage in the spring should be as early as possible after the soil surface has dried and before weeds and volunteer wheat have attained much growth. Subsequent tillage is usually one to five rod weedings during summer to control weeds and maintain a loose soil mulch about 8 to 12 cm deep. This layer provides a significant resistance to liquid and vapor water flow, thermal insulation, and a firm, moist seed zone in contact with the moist, undisturbed soil (Papendick et al., 1973).

In the Eastern Idaho Plateaus, spring tillage on standing stubble is initiated using a field cultivator or a disk (tandem, offset, or one-way) in middle to late April (Massee and McKay, 1979). Tillage depth is 8 to 10 cm deep. Three to four rod-weeding operations are performed during summer to control weed growth and to produce a firm seedbed.

Where growers choose not to till the stubble in the autumn in Oregon and Washington, summer fallow fields are either moldboard-plowed, subsurface-tilled, or chiseled in the spring as soon as the soil surface has dried and weeds begin to grow. The tillage date varies from early March to early April depending on elevation and latitude. After moldboard plowing, the land is immediately field-cultivated or spring-tooth-harrowed to close voids, bring some straw to the surface, and kill any remaining weeds. The subsurface tilled or chiseled land is field-cultivated. During summer the land is rod-weeded or field-cultivated, usually three to five times, as needed to control weed growth. Rod weeders are used if the soil is loose and friable enough so that the rod weeder can penetrate the soil. Field cultivators are used when the soil is too firm for a rod weeder.

8-2.1.3 Weed Control

Timely control of weeds, such as Russian thistle or prostrate knotweed, by subsurface sweep tillage in low precipitation zones (280 mm), immediately after wheat harvest reduces transpiration losses and prevents plants from producing viable seed. Vegetative growth, even without production of viable seed, must be suppressed to avoid production of a canopy that may not allow complete and thorough herbicide spray coverage or may catch and drag in stubble mulch tillage machines.

Herbicides offer the most feasible method of weed control during the winter and early spring when the soil is almost continuously wet and tillage is not possible or practical. Treatment usually consists of soil residual herbicides applied in November and December or contact herbicides applied in early spring. Herbicides control weeds during the wet period and eliminate excessive tillage that accelerates soil drying. Contact herbicides also are recommended when autumn rains germinate a flush of weeds before seeding and the soil remains too wet for cultivation control of the weed seedlings.

When viable seed development by annual broad-leaved weed species is suppressed for several years, herbicide treatment may no longer be necessary every year to control broad-leaved weeds in cereal crops. The situation with annual grassy weed species is similar but more difficult to achieve, especially with downy brome. When downy brome stands of 108 to 160 plants/m^2 were eliminated from wheat stands before March, Rydrych (1974) noted only a 6% reduction in wheat yield at Pendleton, Oregon. Compared with the control, when the competition was permitted until harvest, yield reduction was 40%.

8-2.1.4 Time of Planting

Time of seeding small grains in the fall is controlled by adequate soil water for germination and early growth. Because of dry summers, early (before 15 September) seeding of winter wheat is feasible only after fallow where water has been stored for adequate germination. Three problems limit success with early seedings: inadequate water for germination and emergence, increased susceptibility to disease, and late spring frost damage to early headed wheat. Summer fallow practices, which include rod weeding, usually retain water at the 8- to 15-cm depth in late August or early September in Oregon and Washington (Papendick et al., 1973). Early September seeding into this moist layer in the low precipitation zones usually provides good seedling establishment. Lindstrom (1974) showed that early establishment from these September seedings could increase yields as much as 650 kg/ha compared with October seedings, which usually require fall rains for germination. Early and uniform stands also provide better protection against soil erosion during winter (Cochran et al., 1970).

Rains (usually more than 5 mm) after deep seeding in moist soil with deep furrow drills may cause soil crusting and reduce emergence. Replanting may be necessary and is often done immediately after the rain. Hot, dry weather after shallow seeding in the moist soil layer will dry the soil and germinated seed, causing death of the plants.

Early planted wheat is susceptible to dryland foot rot in the low (less than 300 mm annual) precipitation zones. The disease is favored and most damaging when cultural treatments favor plant water stress (Papendick and Cook, 1974). Such cultural factors are high plant density and high N application, especially if continued long enough to build up disease innoculum in the soil. Wheat cultivars may escape damage if they have growth habits to avoid water stress. In intermediate rainfall zones, early seeding can favor the development of another foot rot, called straw breaker or eyespot, which may reduce yields by 50% or more. Bruehl and Cunfer (1972) found that a late fall or early spring application of the systemic fungicide benomyl [methyl 1-(butylcarbamoyl)-2-benzimidazolecarbamate; trade name Benlate, DuPont, Wilmington, Delaware] controls the disease. Earlier seeding can therefore be used if management safeguards are used to reduce damage from these two diseases.

In eastern Idaho, the time of seeding winter wheat on summer fallow depends on the soil water supply in the surface 8 cm and on the elevation, which determines length of growing season. Wheat is seeded with deep furrow drills to place seed in moist soil. Early seeding in August predisposes the plants to root, crown, and foot rots with attendant reductions in yield, just as in Oregon and Washington. Research has shown that optimum seeding date is early September using a 45 kg/ha seeding rate.

8-2.2 Special Conservation Tillage Practices in Fallow

Several new practices are being integrated into conservation tillage systems. Conservation tillage is defined (SSSA, 1979) as "any tillage sequence which reduces loss of soil or water relative to conventional tillage." In the dryland of the Pacific Northwest, stubble mulching and chemical weed control are both especially desired because their maintenance of residues on the surface is not only advantageous to soil erosion control but also water conservation.

8-2.2.1 Stubble Mulching

Stubble mulch tillage is defined (SSSA, 1979) as "tillage of the soil in such a way that plant residues or other materials are left to cover the surface." It is sometimes called mulch farming, trash farming, or plowless farming. Stubble mulch tillage is used extensively in the wheat-fallow areas for wind and water erosion control, but retention of residues on the surface also conserves water. Stubble mulching as a system of tillage is especially useful in the Pacific Northwest because it systematically encourages plant residue cover during the first winter after harvest, the fallow season, and even after winter wheat is seeded.

Surface residues are well known for their suppression of evaporation during the first stage of evaporation immediately after a precipitation event. Bond and Willis (1969) showed that this first-stage evaporation is extended by surface residues; more time is allowed for drainage deeper into the soil (Fig. 8-5). Idso et al. (1974) showed that first-stage evaporation, in which the evaporation rate depends on atmospheric potential, not soil conditions, predominated relatively more in an Arizona winter than in summer. One would expect even more predominance of first-stage evaporation in the Pacific Northwest winters, which explains why stubble mulching can be so successful for water conservation in the Pacific Northwest. The value of surface residues for water conservation is less during summer, when there is little precipitation. Compared with bare soil, however, surface residues cause slightly lower soil temperature during the day because of reflection of some solar radiation and provision of insulation on the soil surface.

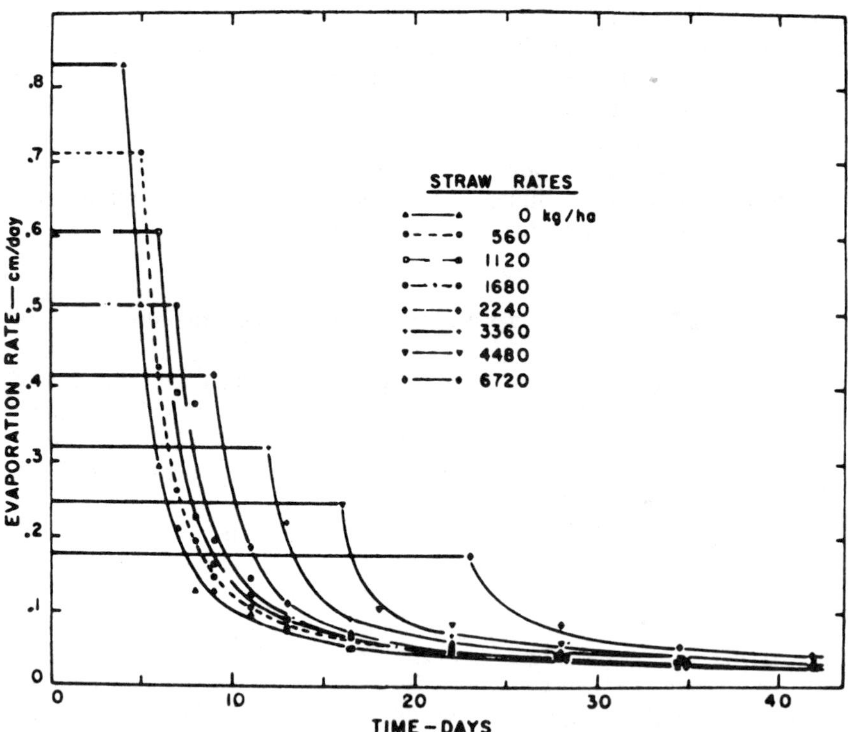

Fig. 8-5. Soil water evaporation rate with time as affected by rates of surface residue (from Bond and Willis, 1969).

Papendick and Miller (1977) reported that 0, 5000, and 11 000 kg/ha of residues on the soil surface from September through March, when evaporative demand was low at Lind, Washington, resulted in water storage of 8, 11, and 13 cm, respectively. These storages were 48, 66, and 81%, respectively, of the precipitation. At Tetonia, Idaho, standing stubble over the first winter after harvest stored 25 cm of water compared with 13 cm where the stubble was burned (Massee et al., 1966). One year later in the spring of the crop year, the respective storage amounts were 16 and 14 cm. The value of stubble mulch for increasing water conservation and storage during the cool, wet winters of the Pacific Northwest is apparent, but some of the additional stored water is lost during the dry summers. Oveson and Appleby (1971) found an optimum tillage system (at Pendleton, Oregon) to be control of weeds during the spring with nonpersistent herbicides and use of stubble mulch tillage before 1 June, followed with two or three rod weeder cultivations as needed during the summer. Seeding is then done in September.

In Idaho, Oregon, and Washington, approximately 45, 18, and 22%, respectively, of the cropland was in minimum tillage in 1978 to 1979 (Lessiter, 1979). Lessiter defined minimum tillage as "limited tillage, but where the total field surface is still worked by tillage equipment." About 83% of the minimum tilled land was stubble-mulched summer fallow. Cropland that was conventionally tilled in the same three states was 55, 80, and 78% of the total, respectively. Lessiter defined conventional tillage as "100% of the topsoil is mixed or inverted by plowing, power tilling, or multiple diskings."

8-2.2.2 Chemical Fallow

Exclusive use of herbicides to control weeds without any tillage during the summer fallow season (chemical fallow) has not been successful at continued field scale. Chemical fallow at Pendleton, Oregon, stored less water and accumulated fewer nitrates than tilled plots (Oveson and Appleby, 1971; Swan et al., 1974). In some years, the nontilled chemical fallow yielded significantly less grain than tilled plots, because the nontilled fallow had dried to such a depth that wheat could not be seeded into moist soil at seeding time. Emergence was then delayed until after fall rains. Late seeded wheat does not yield as well as timely sown wheat (Thill et al., 1978).

Rydrych and Muzik (1968) state that when herbicides are used to control weeds early in the spring (when the soils are often wet and tillage is not feasible), from three to four tillage operations can be eliminated with concurrent savings in energy. However, Oveson and Appleby (1971) found that fallow should be tilled at least once before June for greatest water conservation and successful seedbed preparation in the fall.

Herbicides can substitute for early tillage but sometimes fail to give adequate control of all weeds during drought periods. Some of these chemicals have extended residual activity that may damage crops seeded into coarse-textured soils. Recommended rates vary, depending on precipitation and soil type. Nonresidual contact herbicides can be used in the spring after all weeds have germinated and emerged, but they are expensive and sometimes applications must be repeated, limiting their feasibility. Local county extension agents and product labels should be consulted for necessary precautions and required waiting intervals before these chemicals are used.

8-2.3 Annual Cropping

Summer fallow is not necessary for crop production in zones where annual precipitation is greater than 400 mm. However, water conservation practices are needed in these areas to maximize crop yields. In the annual cropping systems of Oregon and Washington, the combination of water extraction by the crop and insufficient precipitation until late autumn requires either middle to late October seeding of winter cereals or seeding into dry soils. In eastern Idaho, the usual practice on annual cropping is to seed spring cereal varieties because the growing season in the autumn is too short and the winter too cold to permit establishing satisfactory stands of fall sown cereals, e.g., winter wheat.

8-2.3.1 Moldboard Plowing

Rotations used in the annual cropping systems of Oregon and Washington are winter wheat-pea, winter wheat-spring wheat, or winter wheat-spring wheat (or spring barley)-pea.

Tillage of water conservation in these rotations includes moldboard plowing the wheat stubble soon after harvest to cover about 90% of the stubble. Stubble protruding between the furrow slices traps drifting snow, and the random roughness of the plowed surface detains rain and snowmelt, increasing infiltration during the winter. The plowed surface warms and dries in the spring, permitting early field cultivation, fertilizer application, and seeding of the spring cereal crop or peas. After harvest, the spring cereal or pea stubble land is disked about 10 cm deep with a heavy offset disk to kill weeds and prepare a seedbed for the injection of fertilizers. The land is rod-weeded and seeded to winter wheat in October. If autumn rains have not occurred by 15 October in Washington or by 1 November in Oregon, the winter wheat is seeded in the dry seedbed with a double-disk opener drill to a depth of no more than 3 cm. When rains occur, the wheat germinates and emerges as the soil warms. Clods and residues on the dry surface enhance infiltration of winter precipitation and subsequent water storage; however, soil erosion control is only fair after a pea crop because surface plant residues are sparse and seedling growth small.

Best management practices adopted over the past 2 years help control soil erosion and clean up surface waters. Initial tillage of winter wheat stubble after harvest has changed from moldboard plowing to field chiseling or cultivating. The latter practices leave much of the crop residue on the soil surface to increase infiltration and decrease runoff and soil erosion. Similar changes are taking place after pea harvest.

8–2.3.2 Fall Chiseling

In the Eastern Idaho Plateaus, most of the overwinter precipitation occurs as snow on frozen soil, a combination that produces serious spring runoff in 6 of 10 years. Massee and Siddoway (1969) reported that fall-chiseling the land after wheat harvest to a depth of 25 to 30 cm with a chisel spacing of 90 cm increased the infiltration and essentially stopped runoff. With adequate fertilizer (22 kg of N/ha), annually cropped spring wheat yields were 820 and 1150 kg/ha, respectively, for no fall tillage and fall chiseling. Yield from fall-chiseled and fallowed plots was 1290 kg/ha. The value of fall chiseling for water conservation and annual grain production is readily apparent. Under these conditions, summer fallow has no advantage for spring wheat production.

A different situation exists in northeastern Oregon, where the winter climate is comparatively mild and soil freezes infrequently during winter. Allmaras et al. (1977) showed a significant reduction in the water content of the surface 40-cm layer of a fall-chiseled treatment compared with a non-chiseled treatment after only 2 days of internal drainage. This increased internal drainage should increase soil water storage accumulation during each of the many intermittent low-intensity rainstorms. Improved annual crop production is expected but has not been documented. Deep tillage of layered soils with dense subsoils in eastern Washington decreased the bulk density of the subsoil, increased root proliferation, and increased annual crop yields (Mech et al., 1967). The infiltration rate of the soil surface was decreased, but under the low-intensity natural precipitation of the area, water storage in the soil was increased. The high energy requirement of the deep tillage has prevented its adoption and use.

8–2.3.3 Cloddy Seedbeds and Standing Stubble

In the Palouse and Nez Perce Prairies Major Land Resource Area of northeastern Oregon and southeastern Washington, where annual precipitation exceeds 450 mm, winter wheat and green pea are alternately cropped. After harvest of green pea for processing, in June, the land is tilled before winter wheat is seeded in October. Horning and Oveson (1962) demonstrated the value of a cloddy seedbed. A single pass with either a moldboard plow or sweep through the dry ground after pea harvest produced large hard clods. Wheat was seeded in October either directly into these cloddy seedbeds or after two to four additional tillage operations were made to produce a conventional finely worked seedbed. The cloddy seedbeds reduced runoff and soil erosion. Unfortunately, water storage was not

Fig. 8-6. Seeding small grain into standing wheat stubble with an experimental no-till drill near Enterprise, Oregon (*above*), and view of seeded area (*below*).

measured. Stands of wheat plants in the cloddy seedbeds were irregular and relatively thin, yet produced as much grain as the uniform stands in the conventional seedbeds in 4 of 6 years. Cloddy seedbeds for wheat after pea are used by some farmers in northeast Oregon to reduce runoff and increase water conservation. They prefer to use chiseling rather than moldboard plowing or sweep tillage as their single tillage.

Water conservation benefits of standing stubble are illustrated by measurements made by Ramig and Ekin (1976). Leaving wheat stubble stand during the winter in a wheat-pea cropping sequence stored 252 of the 290 mm of precipitation that occurred from wheat harvest in July until 1

April. Only 185 mm of water was stored where the stubble was plowed after wheat harvest. The additional 67 mm of stored water where the stubble remained standing during the winter increased the yield of green peas 670 kg/ha. In those years when the shallow-rooted peas did not use all of the stored water, the following heat crop utilized the additional water to produce an extra 135 kg/ha of grain. The combined practices of (i) allowing wheat stubble to stand over the winter to store water for pea production and (ii) single-pass plowing to produce a cloddy seedbed for wheat after a pea crop not only increased water conservation and crop yields but also reduced runoff, soil erosion, and energy use for tillage. These two practices are gaining acceptance due to escalating energy costs and the Clean Water Act.

8-2.3.4 No-Tillage

No-tillage is a practice whereby a crop is planted directly in soil without any previous seedbed preparation since the harvest of the last crop (Fig. 8-6). Only the immediate seed zone is disturbed as the seed is planted. Some no-till seeding has been done in Washington since 1975. Lessiter (1979) reported less than 1% no-tillage planting in Idaho, Oregon, and Washington in 1979. First year research results at Pullman, Washington, indicated that the no-tillage system may enhance soil water storage. Seasonal runoff from winter wheat after summer fallow, recropped wheat after conventional tillage, and no-tillage recropped wheat was 140, 109, and 144 mm, respectively (K. E. Saxton. 1981. Personal communication. Pullman, Wash.). Many problems are associated with the system, e.g., the prevalence of cold and wet soils in the spring of the year, improper placement of fertilizers, high rodent populations in the residues, toxicity of the residues to germinating seedlings, and poor control of weeds, diseases, and possibly insects (Engle and Wetter, 1976). When spring crops are alternated with fall-seeded crops, acceptable weed control and crop yield can be obtained with expert use of herbicides.

8-2.3.5 Slot Mulching

Slot mulching is a subsoiling operation in which a mulching material, usually crop residues, is placed in the vertical slot in the soil immediately behind a soil-opening tool. This practice was first tried in the Pacific Northwest in 1959 and 1960 at Pendleton, Oregon, but failed to improve water conservation or reduce runoff and soil erosion. Subsequent tillage for summer fallow pulled some of the residue from the slot and then covered the slot and residue with loose soil that in effect prevented the residue from conducting water into the soil. Saxton et al. (1981) used mulched trenches on the contour, referred to as slot mulches, to reduce runoff and improve water conservation in the annual cropped areas. Straw was compacted into contour trenches 5 to 7.5 cm wide and 20 to 25 cm deep, spaced 3 to 6 m or more apart. Care was taken to avoid covering the trenches with loose soil.

Runoff from frozen soil flows into the straw-filled slots and infiltrates into unfrozen soil at the bottom of the slot. Slots must be deeper than the frost in the soil. This practice could be a valuable support practice to no-tillage. In early spring of 1979 when soil near Pullman, Washington, had deep frost, no-tillage plots planted to winter wheat had 114 mm of runoff and 1.12 t/ha of soil erosion. Adjacent no-tillage plots treated with slot mulch and planted to winter wheat had 10 mm runoff and 0.04 t/ha of soil erosion. The slot mulch increased infiltration about 100 mm, which may increase wheat yields 1300 to 2000 kg/ha.

8-3 RESEARCH DEVELOPMENT AND NEEDS

After 30 years of N fertilization, the surface soils (to depths as great as 40 cm) are becoming acid (Allmaras et al., 1979). Silica in the surface horizons is dissolved and leached to lower horizons where an "incipient pan" is developing. The influence of traffic is not distinguishable from the silica migration problem. Soil texture may determine the contribution of traffic on pan development. In time, water flow and root development may be restricted, thus reducing water infiltration and storage. Deep chiseling gave only temporary improvement of hydraulic properties in one soil (Allmaras et al., 1977). Additional research is needed on this complex problem to ascertain its prevalence and importance. Soil management guidelines are needed for controlling or reducing impacts of the problem in water conservation.

No-tillage planting reduced soil erosion by water in annual cropping (Saxton et al., 1981), but much water is still lost by runoff. Contour slot mulching controlled this runoff and essentially stopped all soil erosion. Should further research substantiate the value of no-tillage planting in contour slot mulched land for controlling runoff and erosion, it would be a major breakthrough for conservation farming in eastern Washington, north central Oregon, and northern Idaho. Additional research would be needed to design and develop machinery to use the slot-mulching practice on a field scale. Fertilizer requirements and application methods, plant phytotoxicity and disease control, rodent control, and proper seeding procedures would be needed to make the practice operational.

The amount of annually cropped dryland could be increased at the expense of summer fallow if gains in water conservation were as small as a 25% improvement over present practices. These gains in water conservation will require field experimentation coordinated with more use of process-oriented models of net soil water storage. Tillage effects on soil properties that control infiltration and evaporation must be represented along with weather input in these process-oriented models. Surface residue influences must also be included. Some progress on these process models is evident (Linden, 1979) as related to tillage-controlled soil properties. A much better representation of diurnal freeze-thaw cycles and associated internal drainage or water redistribution must be incorporated. Climatic variables must also be considered to project average gains and frequency of failures to store enough water for annual crop production. Thus risk can be assessed in adopting a new practice that is highly dependent on climate (Haan, 1979).

8-4 LITERATURE CITED

Alban, L. A., and H. E. Cushman. 1962. Fertilizer recommendations for wheat in the Columbia Basin of Oregon. Oregon State Univ. Extension Bull. no. 806.

Allmaras, R. R., S. C. Gupta. J. L. Pikul, Jr., and C. E. Johnson. 1979. Soil erosion by water as related to management of tillage and surface residues, terracing, and contouring in eastern Oregon. USDA-SEA Report no. ARR-W-10.

Allmaras, R. R., R. W. Rickman, L. G. Ekin, and B. A. Kimball. 1977. Chiseling influences on soil hydraulic properties. Soil Sci. Soc. Am. Proc. 41:796–803.

Austin, M. E. 1965. Land resource regions and major land resource areas of the United States. USDA Agric. Handbook no. 296.

Bond, J. J., and W. O. Willis. 1969. Soil water evaporation: Surface residue rate and placement effects. Soil Sci. Soc. Am. Proc. 33:445–448.

Bruehl, G. N., and B. Cunfer. 1972. Control of Cercosporella foot rot of wheat by benomyl. Plant Dis. Rep. 56:20–23.

Cary, J. W., R. I. Papendick, G. S. Campbell. 1979. Water and salt movement in unsaturated frozen soil: Principles and field observation. Soil Sci. Soc. Am. J. 43:3–8.

Cochran, V. L., R. I. Papendick, and C. D. Fanning. 1970. Early fall crop establishment to reduce winter runoff and erosion on Palouse slopes. J. Soil Water Conserv. 25:231–234.

Engle, C., and F. Wetter. 1976. No-till crop production—dryland eastern Washington. Washington Agric. Exp. Stn. Extension Bull. no. 674.

Evans, C. E., and E. R. Lemon. 1957. Conserving soil moisture in soil. p. 340–359. *In* Yearbook of agriculture. U.S. Dep. of Agric., U.S. Government Printing Office, Washington, D.C.

Fuchs, M., G. S. Campbell, and R. I. Papendick. 1978. An analysis of sensible and latent heat flow in a partially frozen unsaturated soil. Soil Sci. Soc. Am. J. 42:379–385.

Haan, C. T. 1979. Risk analysis in environmental modification. p. 30–51. *In* B. J. Barfield and J. F. Gerner (ed.) Modification of the aerial environment of plants. Am. Soc. of Agric. Engineers, St. Joseph, Mich.

Harlan, R. L. 1973. Analysis of coupled heat-fluid transport in partially frozen soil. Water Resour. Res. 9:1314–1323.

Hershfield, D. M. 1961. Rainfall frequency atlas of the United States for durations from 30 minutes to 24 hours and return periods from 1 to 100 years. U.S. Dep. of Commerce Tech. Paper no. 40.

Hershfield, D. M. 1974. The frequency of freeze-thaw cycles. J. Appl. Meteorol. 13:348–354.

Horner, G. M., W. A. Starr, and J. K. Patterson. 1957. The Pacific Northwest wheat region. p. 475–481. *In* Yearbook of agriculture. U.S. Dep. of Agric., U.S. Government Printing Office, Washington, D.C.

Horning, T. R., and M. M. Oveson. 1962. Cloddy and conventional seedbeds for wheat production in the northwest. Agron. J. 54:229–232.

Idso, S. B., R. J. Reginato, R. D. Jackson, B. A. Kimball, and F. S. Nakayama. 1974. The three stages of drying of a field soil. Soil Sci. Soc. Am. Proc. 38:831–837.

Klingebiel, A. A., and P. H. Montgomery. 1973. Land capability classification. USDA Agric. Handbook no. 210.

Leggett, G. E., R. E. Ramig, L. C. Johnson, and T. W. Massee. 1974. Summer fallow in the Northwest. p. 110–135. *In* Summer fallow in the western United States. USDA-ARS Conserv. Res. Report no. 17.

Lessiter, F. 1979. 1978-1979 No-till farmer acreage survey. No-Till Farmer, March issue, p. 5.

Linden, D. R. 1979. A model to predict soil water storage as affected by tillage practices. Ph.D. Dissertation. Univ. of Minnesota, St. Paul (Libr. Congr. Card no. Mic. NVC 80-331580). Univ. Microfilms, Ann Arbor, Mich. (Diss. Abstr. 40(11):5098b).

Lindstrom, M. J. 1974. Wheat-fallow management practices in the low rainfall areas of the United States Pacific Northwest. *In* Tillage and cultural practices for wheat under low rainfall conditions. Proc. Second Regional Wheat Workshop, Ankara, Turkey. 6–11 May. Rockefeller Foundation, New York.

Lindstrom, M. J., F. E. Koehler, and R. I. Papendick. 1974. Tillage effects on fallow water storage in the eastern Washington dryland region. Agron. J. 66:312–316.

Massee, T., and H. McKay. 1979. Improving dryland wheat production in eastern Idaho with tillage and cropping methods. Idaho Agric. Exp. Stn. Bull. no. 581.

Massee, T. W., and F. H. Siddoway. 1969. Fall chiseling for annual cropping of spring wheat in the Intermountain dryland region. Agron. J. 61:177–182.

Massee, T. W., and F. H. Siddoway. 1970. Summer fallow soil water losses on intermountain dryland and its effect on cropping winter wheat. Agron. J. 62:722–725.

Massee, T. W., F. H. Siddoway, H. C. McKay, and J. J. Mech. 1966. Moisture conservation for wheat production in the upper Snake River dryfarming area. USDA-ARS Conserv. Res. Report no. 10.

Mathews, O. R. 1951. Place of summer fallow in the agriculture of the western states. USDA Circ. no. 886.

Mech, S. J., G. M. Horner, L. M. Cox, and E. E. Cary. 1967. Soil profile modification by backhoe mixing and deep plowing. Trans. ASAE 10:775–779.

Oveson, M. M., and A. P. Appleby. 1971. Influence of tillage management in a stubble mulch fallow–winter wheat rotation with herbicide weed control. Agron. J. 63:19–20.

Papendick, R. I., and R. J. Cook. 1974. Plant water stress and development of *Fusarium* root rot in wheat subjected to different cultural practices. Phytopathology 64:358–363.

Papendick, R. I., M. J. Lindstrom, and W. L. Cochran. 1973. Soil mulch effects on seedbed temperature and water during fallow in eastern Washington. Soil Sci. Soc. Am. Proc. 37: 307–314.

Papendick, R. I., and D. E. Miller. 1977. Conservation tillage in the Pacific Northwest. J. Soil Water Conserv. 32:49–56.

Pikul, J. L., Jr., R. R. Allmaras, and G. E. Fischbacher. 1979. Incremental soil sampler for use in summer-fallowed soils. Soil Sci. Soc. Am. J. 43:425–427.

Ramig, R. E., and L. G. Ekin. 1976. Conservation tillage effects on water storage and crop yield in Walla Walla and Ritzville soils. Oregon Agric. Exp. Stn. Spec. Report no. 459.

Rydrych, D. J. 1974. Competition between winter wheat and downy brome. Weed Sci. 22: 211–214.

Rydrych, D. J., and T. J. Muzik. 1968. Downy brome competition and control in dryland wheat. Agron. J. 60:279–280.

Saxton, K. E., D. K. McCool, and R. I. Papendick. 1981. Slot mulch for runoff and erosion control. J. Soil Water Conserv. 36:44–47.

Smika, D. E., and C. J. Whitfield. 1966. Effect of standing wheat stubble on storage of winter precipitation. J. Soil Water Conserv. 21:138–141.

Smika, D. E., and G. A. Wicks. 1968. Soil water storage during fallow in the Central Great Plains as influenced by tillage and herbicide treatments. Soil Sci. Soc. Am. Proc. 32:591–595.

Soil Science Society of America. 1979. Glossary of soil science terms. Rev. ed. SSSA, Madison, Wis.

Swan, D. G., M. M. Oveson, and A. P. Appleby. 1974. Chemical and cultural methods for downy brome control and yield of winter wheat. Agron. J. 66:793–795.

Thill, D. C., R. E. Witters, and R. I. Papendick. 1978. Interactions of early- and late-planted winter wheat with their environment. Agron. J. 70:1041–1047.

Thomas, H., and A. M. Grano. 1971. Agricultural land and water resource base and productivity, Pacific Northwest. U.S. Dep. of Agric., Economic Res. Service, Natural Resource Economics Division, Corvallis, Ore.

U.S. Geological Survey. 1970. Distribution of principal kinds of soils: Orders, sub-orders, and great groups. Map no. 86. *In* National atlas of the United States of America. U.S. Dep. of Interior, U.S. Geological Survey, U.S. Government Printing Office, Washington, D.C.

9

R. E. Luebs
Consulting Agronomist
Lakewood, Colorado

Water Conservation: Pacific Southwest

Agricultural drylands of the Pacific Southwest are generally described as arid or semiarid. For discussion here, the Pacific Southwest includes the states of Arizona, California, Nevada, and that part of Utah south of an imaginary line extending from the southern Wyoming boundary to Nevada. Rainfall is below the threshold needed for dryland crop production over most of the four-state area, and evaporation is high. Approximately 0.6 million ha in California and approximately 20 000 ha in the relevant area of Utah are used for dryland agriculture in the Pacific Southwest (Luebs, 1974).

Small grain cereals predominate as dryland crops in the Pacific Southwest as they do in other areas of the western USA, although the particular cereals and varieties as well as growing season are different in the different regions.

9-1 SOILS

The majority of dry-farmed soils in California are classified at the order level as Alfisols and a much lesser number as Entisols and Mollisols. At the great group level, Haploxeralfs, Palexeralfs, Durixeralfs, Xerorthents, and Argixerolls occur in dryland areas. Location of occurrence is somewhat related to the geographic area of the state. In Utah most dry-farmed soils belong to the order Mollisols and the great group Argixerolls.

Soils affect water conservation for dryland agriculture in several ways: (i) surface structure effect on infiltration of water from rainfall; (ii) texture effect on plant-available water-holding capacity, evaporative soil water loss, and unsaturated drainage; (iii) depth effect on volume of soil for water-holding capacity; and (iv) soil slope effect on surface runoff. Poor

Dryland Agriculture—Agronomy Monograph no. 23.

infiltration characteristics of the surface are related to low organic matter content and excessive tillage.

Organic matter content and the distribution of soil separates are among the soil properties affecting soil structure. Soils of the Pacific Southwest are inherently low in organic matter, and many California soils have a tendency to form thick surface crusts on drying. Long-term cropping has markedly reduced organic matter content of these soils.

The effect of soil texture on available water-holding capacity was illustrated by unpublished results of R. E. Luebs, M. Brown, and A. E. Laag in Riverside County, California. Four soils of different textures were wetted to beyond the 120-cm depth and covered 5 days with polyethylene film before determining water content. At that time, measured plant-available soil water (tension less than 1.5 MPa) to the 120-cm depth was Hanford loamy sand, 8.9 cm; Hanford sandy loam, 14.5 cm; Hanford very fine sandy loam, 18.3 cm; and Porterville clay, 15.0 cm. The very fine sandy loam held twice as much plant-available water as the loamy sand, and the sandy loam held an intermediate amount. The clay soil, however, did not hold significantly more available water than the sandy loam. After 12 weeks of hot, rainless summer and no vegetative growth, the 120-cm depth had retained 2.5, 5, and 10 cm of available water for the loamy sand, sandy loam, and very fine sandy loam or clay, respectively.

About 50% of California dryland soils have a moderately fine texture, and the remainder are medium to moderately coarse, with about 24% being in the moderately coarse category. Dryland soils in Utah have a wide range of textures, but the medium-textured soils predominate in the central area and coarser-textured soils in the southeast.

Slightly less than one half of the California dryland soils are deeper than 90 cm. The large area of relatively shallow soils appears to be a considerable limitation in water conservation. Depth of Utah soils is probably greater on the average than California soils.

Soil slopes in California dryland areas are greater than 5% on nearly 75% of the acreage, whereas nearly the converse is true for Utah. Water conservation may be greater in Utah because of less runoff.

9-2 PRECIPITATION CHARACTERISTICS

The characteristics of precipitation are major factors in water conservation. Average annual rainfall amount and variability during the rainfall season and amounts of precipitation per storm all affect the water conservation process.

Dryland cropping areas of California receive an average of 200 to 500 mm of precipitation annually. Annual distribution of precipitation is distinctly seasonal in most of the state, with 70% of the average annual rainfall occurring during December through March. Precipitation in November and April may be significant for nonirrigated vegetation some years. Negligible amounts of precipitation are received during the remaining months.

Evaporative water losses from the soil and evapotranspiration are at a seasonal low during the winter rainfall period, allowing fairly efficient storage for the growing crop. However, if the land is not cropped during the rainfall season, high evaporative losses will occur during the hot, rainless summer before the next dryland crop can be grown.

Although rainmaking companies have been seeding clouds in areas of the western USA, little or no documentation of their effect on rainfall variability and amounts have been available. Atmospheric research on the principles of rainmaking is continuing.

Annual variability in precipitation increases as the total annual amount decreases for a land area or a period of time. At Riverside, California, where the 80-year average rainfall (1880 to 1960) was 281 mm, annual variability is high. For the 15-year period (1945 to 1960) when the mean was 231 mm, annual amounts range from 110 to 381 mm, or a decrease of 52% and an increase of 39% from the mean, respectively. The coefficient of variation for annual rainfall in the dry-farmed area of southern California is higher than for other semiarid areas in this country.

Seasonal variability in precipitation is also a major factor in dryland grain production. As with annual rainfall, seasonal rainfall variability in California is high. During the 1945–1960 period at Riverside, rainfall during one of the major rainfall months (December through March) was less than 20 mm for 11 of 15 years. Rainfall for each of 2 months was less than 20 mm one third of the years.

Seasonal distribution of rainfall is critical for dryland small grains, because as the crop progresses sequentially through its growth stages, a fraction of the final yield is being determined. Germination and emergence after a certain date result in a reduced yield increment. In California, the date of the beginning of the rainfall season after a rainless summer is important, because the seedbed will be dry and recent rain is necessary for germination. During approximately one half of the years the seedbed is dry on 1 December, which is considered about 1 week later than the optimum planting date.

Rainfall intensity and size of storm greatly affect water conservation. Generally, the winter rainfall storms in California are received at a low rate of intensity, which is ideal for infiltration into the soil. Runoff from the land is more a result of the length of the storm than the rate or intensity.

9-3 CONSERVATION OF WATER FROM PRECIPITATION

Conservation of water in the soil from precipitation on the land requires (i) movement of water into the soil and (ii) control of soil water losses. If water does not enter the soil at the rate of precipitation, it will run off or pond on the land surface. Runoff is the dominant pathway for water loss from the majority of land surfaces. Any runoff from a unit area of land receiving precipitation decreases water stored in the soil under dryland conditions and decreases precipitation use efficiency. Precipitation use efficiency is defined as the ratio of economic crop produced to the water used over a specified time. Water harvesting is a technique for concentrating normal or induced runoff to increase precipitation use efficiency.

Conservation of water from rainfall involves conserving water from a highly variable and unpredictable source. This variability is even more pronounced in low rainfall, dryland agriculture areas. As a consequence, a particular water conservation practice may have a relatively low probability of success in any one growing season. Where snowfall is a significant portion of precipitation in dryland farming areas, and snow accumulates on the ground surface during the winter as in Utah, the variability of precipitation has less effect on the results of conservation practices.

Water conservation for dryland agriculture must consider yearly rainfall distribution. California and Utah differ markedly in rainfall distribution, the former having a definite distribution of winter rainfall and the latter an ill-defined seasonal distribution, usually with more precipitation in the spring.

9-3.1 Soil Water Losses

Conservation of water in soil for dryland agricultural crops requires minimizing (i) water use by weeds, (ii) evaporative losses from the soil, and (iii) unsaturated drainage losses from the crop root zone. Eradication of weeds from the land during fallow periods is an obvious and easily accomplished practice. However, eradication of weeds from dryland crops may also be very important and is usually more difficult to accomplish because of plant population density. Aircraft spraying with herbicides can be effective. I am not aware of quantitative information on the effect of weed control on water conservation in the Pacific Southwest. Weed control by fallow tillage is a widespread practice, and significant unsaturated drainage from the root zone is limited to high rainfall seasons and coarser-textured soils. Although evaporative losses are the major pathway of soil water loss, they may be lower in dryland areas of California than in Utah, because nearly all rainfall occurs during the low temperature, low solar radiation period of the year in California.

Soil water losses from the 0- to 120-cm depth increment of several different-textured soils in the field are illustrated in Fig. 9–1. The magnitude of loss and the change in rate of loss over a 90-day period in the summer, with no rain in southern California, are shown in this figure. These bare soils were wetted to below the 120-cm depth and covered with polyethylene film after the last wetting. Five days after the soils were covered, water content measurements were made by the neutron method and plots were uncovered. Soil water contents at time zero, therefore, represented "field capacity." Total water losses for the 90-day period were similar for the loamy sand, sandy loams, and the very fine sandy loam, ranging from 6 to 7 cm. However, the clay had lost less than 4 cm of water during that period. At least three general conclusions can be made from these data. First, major water losses occur soon after wetting. In fact, 30 to 40% of the total loss over the

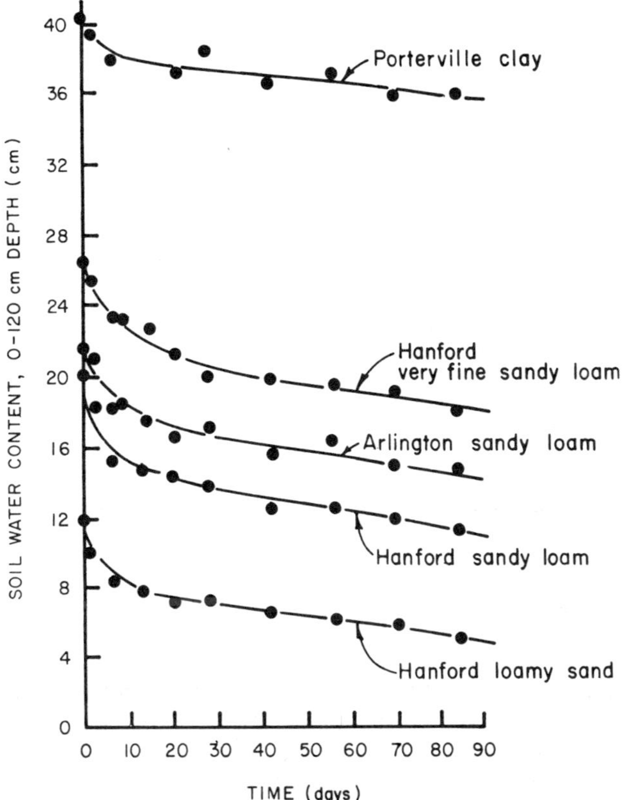

Fig. 9-1. Decrease in water content of wetted field soils following 5 days with no evaporation.

90 days was lost in the first 6 days of drying. Second, the rate of water loss reached relatively constant values in from 10 to 20 days. Third, losses were continuing at a measurable rate 90 days after measurements began. This continuing, constant rate of loss was similar for the coarse- and medium-textured soil and somewhat less for the clay.

9-3.2 Evaporation

Except for soil water losses by evapotranspiration from dense weed populations, evaporation from the soil is a major pathway for water loss. The relatively hot and rainless summer in southern California provided a good field situation for measuring evaporative losses from different-textured soils over a long period of time. Evaporative losses from a bare Hanford sandy loam totaled over 6.2 cm for an 85-day period (Fig. 9-2). A water loss of this general magnitude can be expected from sandy loam soil during the summer period in dryland areas having a winter rainfall distribution, assuming that the soils contain water to the field capacity level for a depth of 90 cm at the beginning of the summer.

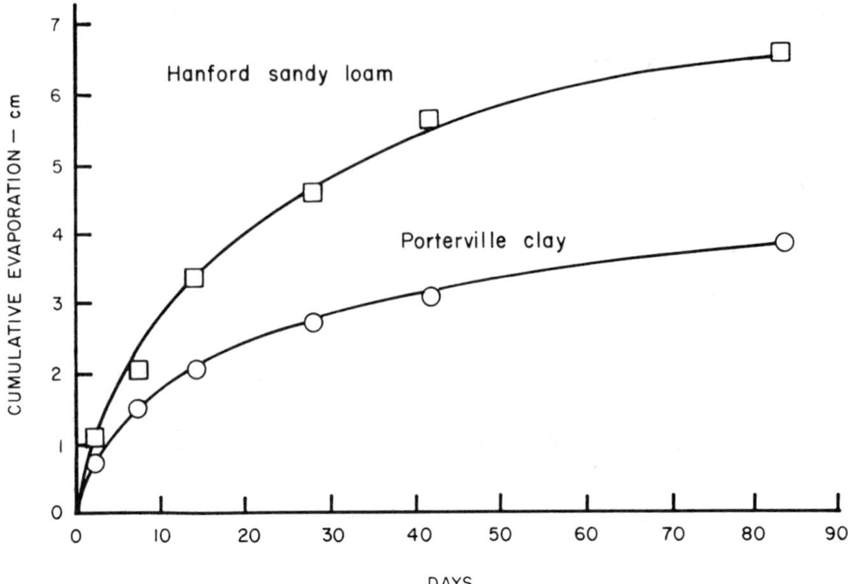

Fig. 9–2. Evaporative water losses from field soils after wetting.

A Porterville clay loam (fine, montmorillonitic, thermic Typic Chromoxererts) located several kilometers from the sandy loam site lost 35% less water to evaporation than did the sandy loam over the 85-day period. The difference in water loss between the two soils was much less early in the evaporative period. The longer period, however, is indicative of losses to be expected under summer fallow in the winter rainfall area of the Pacific Southwest. Data presented in Fig. 9–1 support the conclusion that soils high in clay content lose less water by evaporation.

The movement of water in soil by which losses occur is illustrated by data in Fig. 9–3. Tensiometer readings were taken to the 90-cm depth in a Hanford sandy loam (coarse-loamy, mixed, nonacid, thermic Typic Xerorthents) for 75 days after 20 mm of water were added in the field. Before this application, the soil had been wetted to beyond the 90-cm depth. The applied water decreased water tension in the soil at the 30-cm depth where previous drying had been occurring, but it had no discernible effect on the tension at the 60- and 90-cm depths, which were gradually increasing in tension from unsaturated drainage. One day after the application of water, the tension at the 30-cm depth no longer decreased, and after an additional day, it rose steeply under evaporative loss and continued to do so for the remaining 70 days of the test. Soil water tensions at the 60-cm depth increased slightly faster than at the 90-cm depth because of a slightly greater gradient for upward movement of water. About 55 days after the initiation of tensiometer readings, soil water tensions between the 60- and 90-cm depths were the same, indicating that the gradient for downward movement of water as unsaturated drainage was equal to the gradient for upward movement of water.

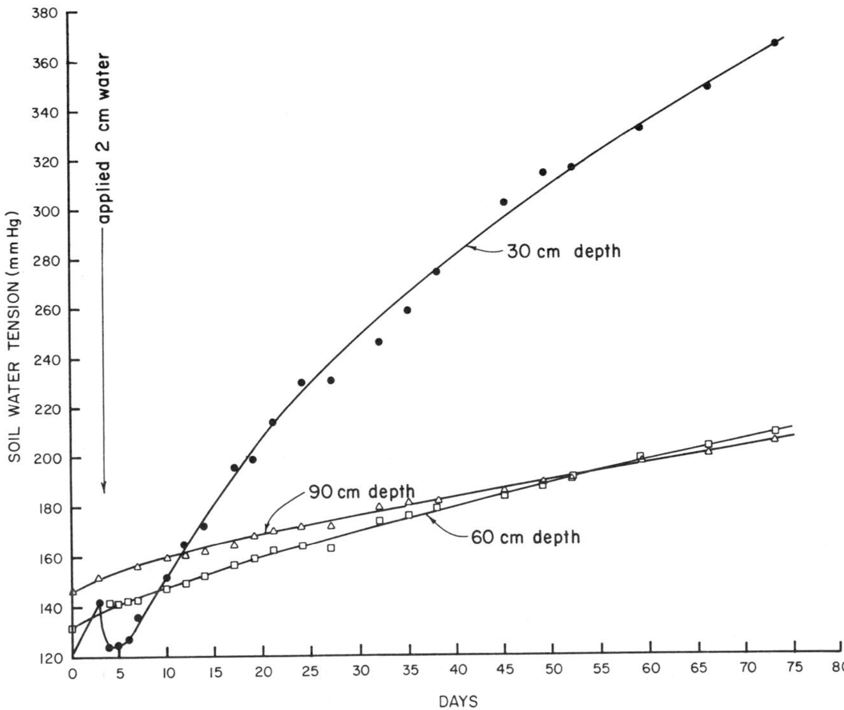

Fig. 9-3. Changing water tensions at different depths of a field soil during a drying period.

Under the high evaporative conditions existing in the Pacific Southwest, soil water losses by evaporation are likely to exceed those by unsaturated drainage in medium-textured soils. Soil water measurements by the neutron method for a period after a wetted Hanford very fine sandy loam was uncovered show that unsaturated drainage in 3 weeks had dropped to a rate too low to measure (Fig. 9-4), whereas evaporation was continuing at a relatively high rate of 0.8 mm/day. These data indicate the relative magnitude of water loss from these two pathways for the 120-cm depth of a very fine sandy loam during a rainless summer fallow period. However, unsaturated drainage can be considered a loss to the subsequent dryland crop only if the roots do not reach the depth to which the water is moving. Except for coarser-textured soils, it is unlikely that significant amounts of water are lost from the crop root zone by unsaturated drainage in the Pacific Southwest. This conclusion is based on many observations that the root zone is seldom filled to its water-holding capacity.

9-4 WATER CONSERVATION PRACTICES FOR DRYLAND AGRICULTURE IN PACIFIC SOUTHWEST

Water conservation practices for dryland crop production have generally not been extensively investigated in the Pacific Southwest. Most of the California research was done at or near Riverside, California, between 1955

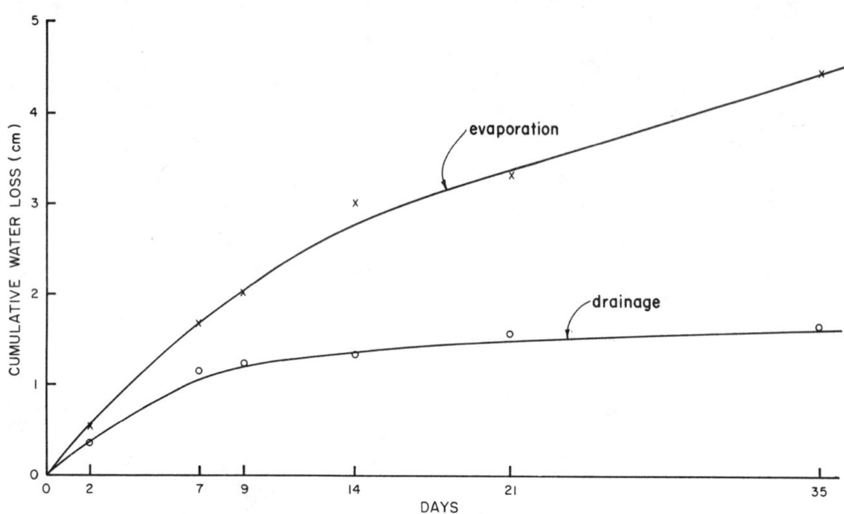

Fig. 9–4. Evaporative and unsaturated drainage losses from a wetted Hanford very fine sandy loam in the field following 5 days with no evaporation.

and 1970. Certain practices for water conservation, however, have been used over many years by farmers for dryland cropping.

9–4.1 Summer Fallow

Alternating fallow with cropping in a 2-year cycle in California has been termed "summer fallow," as in other areas of the West. However, the water storage period in California is during the winter. The following dry summer season before planting the crop is actually a soil water loss period.

The practice of summer fallow in southern California gained wide acceptance over 65 years ago and remains in use for nearly all dryland grain cropping in the state. Most years fallow increases the available soil water supply and assures economic grain production. On the other hand, because summer fallow also increases the soil N available for the crop, its value for water conservation has undoubtedly been overstated.

Shaw (1911) recognized summer fallow as a water conservation practice in California in 1911. Madsen (1916) showed that dryland crop yields were markedly increased by biennial cropping, which involved the summer fallow practice in alternate years. More recently, Luebs and Laag (1971) obtained data on water conservation with summer fallow for a Hanford sandy loam soil in southern California. Twenty-centimeter dikes at 7.5-m intervals on a 2 to 3% slope retained most of the runoff on annual cropped and summer fallow plots in their field studies. The fallow period in their studies averaged 16.5 months, from June until December of the next year. Over four fallow periods (1965 to 1969), an average of 5.7 cm of water was stored in the 0- to 180-cm depth increment of soil (Table 9–1). Essentially no water

Table 9-1. Soil water storage with summer fallow, Hanford sandy loam, Riverside County, California.†

Fallow period	Rainfall	Water stored	Efficiency of storage
	mm	cm	%
1965-1966	322	7.6	24
1966-1967	310	4.1	13
1967-1968	254	4.3	17
1968-1969	523‡	6.9	13
Average	352	5.7	17

† 180-cm depth.
‡ Approximately 150 mm lost as runoff.

was stored in the soil with continuous annual cropping, because available soil water was depleted at harvest, and little or no significant precipitation occurred between harvest and planting the next crop. Between 5 and 6 cm of stored soil water in the crop root zone is a highly effective amount for dryland grains in areas receiving an average of 350 mm rainfall. Luebs and Laag (1971) found that average barley grain yield was increased by 29% (from 1.58 to 2.04 t/ha) by this stored water. Incidentally, yield increases of this magnitude under the runoff control system used do not economically justify use of the summer fallow practice. Data also suggested that when runoff is not held on land with 2 to 3% slope, the yield advantage of summer fallow may be greater.

The average annual fraction of rainfall stored in soil in the studies of Luebs and Laag (1971) was 17%. This value, commonly termed fallow storage efficiency, is relatively low compared with values for the Great Plains.

Summer fallow as a water conservation practice is more than twice as effective in Utah as in California. The data of Bracken and Cardon (1935) for an 8-year period (1909 to 1917) showed the fallow storage efficiency to be 37.2%. During a later 7-year period (1925 to 1932), they were able to show an efficiency of 33.3%. Bennett et al. (1954) in Utah found that the average wheat yield increase from practicing summer fallow was 134%. Presumably, the major factor in this increase was the increased available soil water stored by using summer fallow.

9-4.2 Stubble Mulch

The stubble mulch system of tillage was investigated for dryland cropping in southern California over the 10-year period 1946 to 1956 (Luebs and Laag, 1964). Subsurface tillage representing the stubble mulch system was compared with moldboard plowing and disking as a primary fallow tillage method for water stored to a depth of 90 cm at the end of the summer fallow season in an alternate barley-fallow cropping system. In 1 of the 4 years that soil water measurements were made, more water was stored with the stubble mulch system. The increase in water stored was statistically significant and amounted to 4 cm more storage than with disking and 2 cm more than with plowing. During 3 of the 4 years there were no differences in water stored among the tillage systems. The average annual precipitation in the area during the period of these field studies was approximately 23 cm. The effect of the stubble mulch to conserve water in low rainfall areas may be less because of less straw production. The average annual straw production from dryland barley cropping over a 6-year period in the southern California studies was 760 kg/ha, a relatively light mulch.

9-4.3 Water Harvesting

Concentrating runoff water from areas adjacent to cropped areas by impounding on the surface is termed water harvesting. Water harvesting was studied for its possible water conservation benefits in dryland cropping near Riverside, California, from 1965 to 1969 (Luebs and Laag, 1971). It was compared with annual cropping and cropping after 1 year of summer fallow. Runoff was concentrated on leveled areas from adjacent sloping areas covered with polyethylene film. Runoff areas were the same size as cropped areas. As a result, the same total land area was involved for water harvesting as for summer fallow and twice the area as for annual cropping. Water harvesting plots were cropped annually. Plots for all cropping systems in this study area were diked to retain runoff water on the plot. As a consequence, concentration of water from natural runoff occurred within fallow and cropped plots that were not part of the water harvesting system. The system consistently conserved more water than annual cropping and summer fallow in plot areas where runoff occurred with the latter two systems. An average of 3 to 5 cm more available water was stored to the 135-cm soil depth by water harvesting at the barley culm elongation stage and 6 to 7 cm more water in the 180-cm depth at barley heading. Differences were consistent but less when the comparison was made with the annual and summer fallow plot areas in which concentration of natural runoff occurred. The amount of water conserved at heading was more than doubled by water harvesting even with runoff control in the other systems.

Soil water data obtained in these studies suggested that water harvesting resulted in unsaturated drainage from the root zone some years. Excessively high amounts of rainfall over short periods during one season also severely reduced the percentage of total seasonal runoff that could be impounded with the water harvesting system. The water harvesting principle appears to hold much promise for increased water conservation in dryland areas of the Pacific Southwest.

9-4.4 Terracing

Studies by Luebs and Laag (1971) at Riverside, California, with dryland barley cropping suggested that low ridges or terraces at short intervals on the contour for slightly sloping lands would be very effective for conserving water for dryland crops. They found that holding runoff water on 2 to 3% slopes of a Hanford sandy loam at intervals of approximately 8 m down slope was a very effective water conservation practice under continuous annual dryland cropping with annual precipitation of from 250 to 350 mm. They used earthen dikes about 20 cm high. As a result, an average of 3 cm of water was conserved in the soil to a depth of 135 cm fairly early in the barley growing season, and this water conservation was reflected in the 50% increase in barley yield over a 4-year period.

9-4.5 Snowmelt Control

Snow contributes significantly to precipitation in dryland areas. Large runoff losses frequently occur when the snow melts too rapidly in the spring for infiltration of water into the soil. A technique using fly ash for controlling melting of the snow in the early spring on dryland when the soil is not frozen is believed to result in more gradual snowmelt and, consequently, more water infiltration into the soil for storage in the root zone (R. Nielson. Personal communication. 1980. Utah State Univ., Logan). Fly ash is a by-product of coal-fired generating plants. Because it is dark, ash absorbs the sun's radiation and melts the snow when melting would not otherwise occur. The fly ash is spread on the snowpack covering dryland fields with soft-tracked, light ground-pressure vehicles. The magnitude of water conservation for dryland cropping with this technique is not yet fully known and probably deserves additional study.

9-5 FUTURE RESEARCH

Future research that merits consideration for increased water conservation under dryland farming depends on the particular area of the Pacific Southwest. In California and possibly in Utah, more field research on holding runoff on the land within relatively short distances, even on gentle slopes, and cropping annually appears to offer much promise. Small amounts of water conserved in this way are likely to have a marked positive effect on dryland grain yields. Judicious application of N fertilizer for annual cropping would no doubt be beneficial with additional available water.

In California, where evaporative losses from fallow are high during the hot, rainless summer period, a selection of economic crops and management practices that efficiently and economically use rainfall during the season of occurrence merits investigation. An analysis of optimum runoff areas and crops that could most efficiently use increased water over short periods during the growing season appears to be needed. In that area where rainfall is low, the rainfall season is short, and crop evapotranspiration is low, some portion of the land area may be more valuable for the water that falls on it than the dryland crop it could produce.

In Utah, where water from snowfall is a significant factor in dryland crop production, research to develop methods for greater storage of water in the soil may be needed. Documentation of the water conservation benefits of controlling snowmelt would be of considerable interest.

9-6 SUMMARY

General precipitation characteristics affecting water conservation for dryland farming in the Pacific Southwest are (i) lower annual precipitation with higher variability than in other dryland farming areas of the country, (ii) a distinctly winter rainfall area (California), and (iii) a year-round precipitation area (Utah).

The factors affecting water storage in the soil—runoff, soil texture, and soil depth—are believed to be well understood for the Pacific Southwest, as are those affecting the soil water losses—evaporation, unsaturated drainage, and evapotranspiration by weeds. Comprehensive field research on water conservation for dryland cropping, however, has been limited in the Pacific Southwest.

The practice of summer fallow is widely used throughout the dryland areas of the Pacific Southwest. In terms of the efficiency of water storage in the soil, summer fallow has been about twice as effective in Utah as it has been in California. As in other dryland areas of the country, a portion of the crop yield increases from summer fallow is attributed to increased available soil N.

Stubble mulch as a fallow tillage system conserves water for dryland crops in California occasionally, probably in those years with a particular type of rainfall distribution. The amount of crop residues available for mulch, however, is relatively low in California.

Data on water harvesting and runoff control research in California suggest that the general principles of water harvesting and runoff control with annual cropping should be emphasized in water conservation for dryland farming.

9-7 LITERATURE CITED

Bennett, W. H., D. W. Pittman, D. C. Tingey, D. R. McAllister, H. B. Peterson, and I. G. Sampson. 1954. Fifty years of dryland research at the Nephi Field Station, Utah Agric. Exp. Stn. Bull. no. 371.

Bracken, A. F., and P. V. Cardon. 1935. Relation of precipitation to moisture storage and crop yield. Agron. J. 27:8–20.

Luebs, R. E. 1974. Summer fallow in the Southwest. p. 136–148. *In* Summer fallow in the western United States. USDA-ARS Conserv. Report no. 17.

Luebs, R. E., and A. E. Laag. 1964. Tillage and nitrogen for dryland grain in a winter rainfall climate. California Agric. Exp. Stn. Bull. no. 804, Berkeley.

Luebs, R. E., and A. E. Laag. 1971. Rainfall use efficiency for dryland barley with three crop and water management systems. Soil Sci. Soc. Am. Proc. 35:336–340.

Madsen, B. A. 1916. A comparison of annual cropping, biennial cropping and green manures on the yield of wheat. California Agric. Exp. Stn. Bull. no. 270, Berkeley.

Shaw, G. W. 1911. How to increase the yield of wheat in California. California Agric. Exp. Stn. Bull. no. 211, Berkeley.

SECTION C

SOIL CONSERVATION

10

R. R. Allmaras
Agricultural Research Service
U.S. Department of Agriculture
Pendleton, Oregon

Soil Conservation: Using Climate, Soils, Topography, and Adapted Crops Information to Select Conserving Practices

Soil conservation in dryland agriculture consists mainly of protecting soil from erosion by wind and water. Salinization and saline seeps are sometimes hazardous to soil conservation in dryland agriculture and often require specialized practices to reduce them. Because these saline hazards are appropriately treated in each of the later chapters, they will not be extensively considered here.

Depending on a multitude of factors, including climate or weather, soil properties, topography, and cropping practices, each tract of land in the dryland of North America usually requires special management to protect against wind and water erosion. One soil erosion hazard may predominate for part of the year and the other during another part of the year. Hazard from either wind or water can also be significant during several seasons of the year. In some areas of the dryland, only one erosion hazard may be significant. Both forms of soil erosion may be controlled by the same soil and crop management practices but not always.

An overview of soil conservation in dryland agriculture is difficult without systematic use of resource information. This chapter illustrates the combined use of land resource information (including mapped soil information) and soil erosion equations to project soil erosion hazard and associated types of control practices in specific locales. The details are treated by geographic classification (land resource region) in subsequent chapters.

Soil conservation cannot always be practiced without considering other objectives in dryland agriculture. Water conservation is another important

Dryland Agriculture—Agronomy Monograph no. 23.

objective. In dryland agriculture there is nearly always a water deficit during the growing season for the major crop, and management practices must enhance soil water storage during the period when precipitation exceeds evapotranspiration (compare columns 6 and 8 in Table 10–1). Thus practices must satisfy both soil and water conservation during the nongrowing season. A common decision is whether a cereal grain straw should be incorporated or allowed to stand without disturbance after harvest. During the growing season, production- and water conservation–related practices may both compete with soil conservation practices. Some of these competing factors affect choice of planting time, surface residue treatment, weed and other pest control, and plant spacing to manage water stress.

Soil conservation is also significantly affected by economic, political or governmental, and social or cultural factors (Beatty, 1982). For instance, increased summer fallow and associated soil erosion is encouraged by governmental crop regulations that vary from year to year or are uncertain (Carlson et al., 1981). Land tenure relations also determine acceptance of erosion control practices (Dillman and Carlson, 1981).

10–1 LAND RESOURCE INFORMATION

The dryland agricultural area of the USA is contained mainly in 7 of the 20 land resource regions described by Austin (1981). These areas are identified in Fig. 10–1; the letter identification and corresponding name are given in Table 10–1. The dryland agriculture portion of the Canadian Provinces is shown in Fig. 18–1. Not all major land resource areas were included in land resource regions C (California Subtropical Fruit, Truck, and

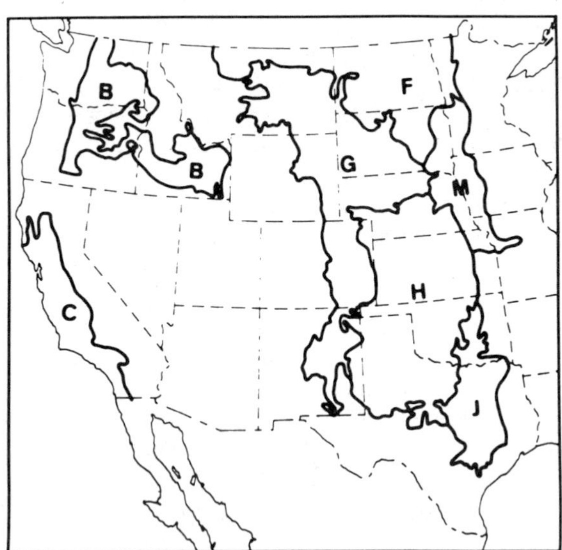

Fig. 10–1. Land resource regions in the U.S. portion of the North American dryland agriculture. (Letter designations and names are shown in Table 10–1.)

Table 10–1. Climate, topography, and major cultivated crop in the land resource regions of the USA and in Canadian Provinces within the dryland agricultural area of North America.

Letter identification in Fig. 10–1†	Land resource region†	Topography	Range of average annual temperature	Frost-free season	Average precipitation range	Major precipitation season	Range of annual pan evaporation‡	Erodible wind index§	Major cultivated dryland crops
			°C	Days	mm		mm		
--	Canadian Provinces	Smooth through knoll and depression to moderately rolling	2–5	90–120	250–420	Summer	1000–1400	--	Spring wheat and barley
B	Northwestern Wheat and Range	Smooth to deeply dissected plains and plateaus	7–10	120–200	250–600	Winter	900–1500	15	Winter wheat
C	California Subtropical Fruit, Truck, and Specialty Crop¶	Low mountains and broad valleys	13–19	200–365	300–1020	Winter	1100–1800	10	Winter and spring cereal grains
F	Northern Great Plains Spring Wheat	Smooth to rolling	4–7	100–145	250–600	Summer	700–1300	70	Spring wheat and barley
G	Western Great Plains Range and Irrigated	Moderately dissected Smooth to steep slopes	4–15#	100–200#	250–600††	Summer	1400–2800	50	Spring and winter wheat#
H	Central Grain Plains Winter Wheat and Range	Smooth to rolling	10–18	150–240#	500–750	Summer	1600–2800	50	Winter wheat and grain sorghum
J	Southwestern Prairies Cotton and Forage	Dissected and rolling Plains and Plateaus	16–21	200–280	600–1100	Spring, fall	1900–2700	20	Grain sorghum, cotton, and winter cereal grains
M	Central Feed Grains and Livestock‡‡	Smooth to steep Rolling	4–13	140–190#	500–900#	Spring, summer, fall	1000–1800	30	Corn and small grains

† Except for Canadian Provinces, the letter and name designations are as in Austin (1965).
‡ Source: Kohler et al. (1959).
§ Erodible wind index is the average monthly accumulation of wind force proportional to the cube of the wind velocity in meters per second (Skidmore and Woodruff, 1968).
¶ All of land resource region except for the Sierra Nevada Foothills Major Land Resource Area.
# North to south.
†† Wide fluctuation from year to year.
‡‡ Western part of land resource region including Loess, Till and Sandy Prairies, Nebraska and Kansas Loess-Drift Hills, and Iowa and Missouri Deep Loess Hills major land resource areas.

Specialty Crop Region; abbreviated as California Subtropical Fruit Region) and M (Central Feed Grains and Livestock Region; abbreviated as Central Feed Grains Region). No detailed discussion on soil conservation in the California Subtropical Fruit Region will appear in the monograph. However, similarity of natural features and crops suggests that information developed for the Northwestern Wheat and Range Region (Northwestern Wheat Region; region B) can be used to make inferences about soil conservation in that specific region.

The land resource region system of classification was designed to organize information about land as a resource for agriculture (and other commercial or recreational uses); the associated map is an updating and revision of an earlier 1950 map entitled "Problem Areas in Soil Conservation." The regions noted in Fig. 10–1 thus have a historical relation to soil conservation. The regions of Fig. 10–1 are made up of major land resource areas, which are in turn made up of land resource units. Land resource units are "geographic areas of land, usually several thousand acres (0.4 ha) in extent, that are characterized by particular patterns of soil (including slope and erosion), climate, water resources, land use, and type of farming" (Austin, 1965). At this level of characterization there is close similarity to general soil maps. A soil mapping unit is "a portion of the landscape that has similar characteristics and qualities whose limits are fixed by precise definitions" (Klingebiel and Montgomery, 1973). This definition indicates that the level and detail of description are similar for the soil mapping unit and the land resource unit. Later, as the water or wind erosion control planning guides are discussed (Chapt. 11 and 12), we will see that the land resource unit scale provides detailed supplemental information about what practices may be necessary and successful in soil erosion control.

Major land resource areas are made up of geographically contiguous and similar land resource units; land resource regions are made of geographically contiguous and similar major land resource areas. This hierarchal grouping undoubtedly averages over detailed information at the land resource unit level but preserves grouping of climatic, topographic, and major crop interactions for areas as large as the U.S. portion of the North American dryland.

Significant differences in soil erosion can be anticipated on the basis of climatic, topographic, and major cultivated dryland crop information listed in Table 10–1. For instance, water erosion is relatively more likely than wind erosion in the Northwestern Wheat Region; conversely, wind erosion is expected to dominate in the Western Great Plains Range and Irrigated Region (Western Great Plains Range Region; region G). Even though the ranges of average annual precipitation are the same in both regions, other characteristics (e.g., the summer vs. winter occurrence, the differing year-to-year variability, and precipitation intensity) may control the form of soil erosion. Examination of soils information for land resource units within each region would add significantly to the assessment regarding wind- or water-caused soil erosion. Information at this level is likely to locate lands in which both wind and water erosion are problems. Northern regions, where spring grains are grown, present much different opportunities and requirements for soil erosion control during the winter (or off growing season) periods than southern regions, where winter grains are grown. Similarities and differences in management based on information in Table 10–1 will be amplified in the subsequent chapters on soil conservation.

Soil conservation evaluation and planning for water and wind erosion control is facilitated by using, respectively, the universal soil loss equation (Wischmeier and Smith, 1978) and wind erosion equation (Chepil and Woodruff, 1963; Woodruff and Siddoway, 1965; Skidmore and Woodruff, 1968; Fisher and Skidmore, 1970). Not only do these equations guide the evaluation, but their use is also supported by resource data available from the Soil Conservation Service (SCS) and other sources. These data are available for areas as small as the land resource unit and soil mapping unit.

10–2 WATER EROSION

The universal soil loss equation (USLE), related to control of soil losses caused by sheet and rill erosion, contains a factor related to climate (*R,* rainfall erosion index), two factors related to a soil mapping unit (*K,* soil erodibility; and *S,* slope steepness), and three factors related to management (*P*, practice factor; *C,* cover and management factor; and *L,* uninterrupted length of slope). The detailed theory and background for the USLE is discussed in Chapter 11. Figure 10–2, taken from Wischmeier and Smith (1978) originally and modified by Foster et al. (1981), shows distribution of the *R* factor over the U.S. portion of the North American dryland. The modification by Foster et al. (1981) was to improve information in the western USA and to accommodate a change in units. The average annual rainfall erosion index ranges from about 200 to 600 in the Northwestern Wheat Region, whereas it may reach 5000 to 6000 MJ mm(ha h y)$^{-1}$ in the Southwestern Prairies Cotton and Forage Region (Southwestern Prairies; region J). This average annual rainfall erosion index is proportional to soil loss from a land in a bare, fallowed condition when all other factors (e.g., slope steepness, length of slope, practice, and soil erodibility) are each fixed at, for example, their mean value throughout the dryland area. Experience indicates that differences of slope steepness (a mapping unit characteristic) and permissible length of slope both must compensate for this difference in climatic factor. Of the six factors in the USLE, the *R* and the combined *SL* factors have ranges far greater than the others. In fact, the dryland agriculture west of the Rocky Mountains is characterized by steeper slopes and longer uninterrupted field lengths than east of the Rocky Mountains; note the correspondingly lower average annual rainfall erosion index west of the Rocky Mountains in Fig. 10–2. This comparison illustrates that a fixed value of the product of *R* and the *SL* factors can be used as a guide to predict where steep slopes and long uninterrupted fields can or cannot be farmed in the dryland.

Year-to-year variation (maximum divided by minimum value) in the annual rainfall erosion index is greater in an arid or semiarid than in a humid climate (Wischmeier and Smith, 1978); variation of *EI* (energy-in-

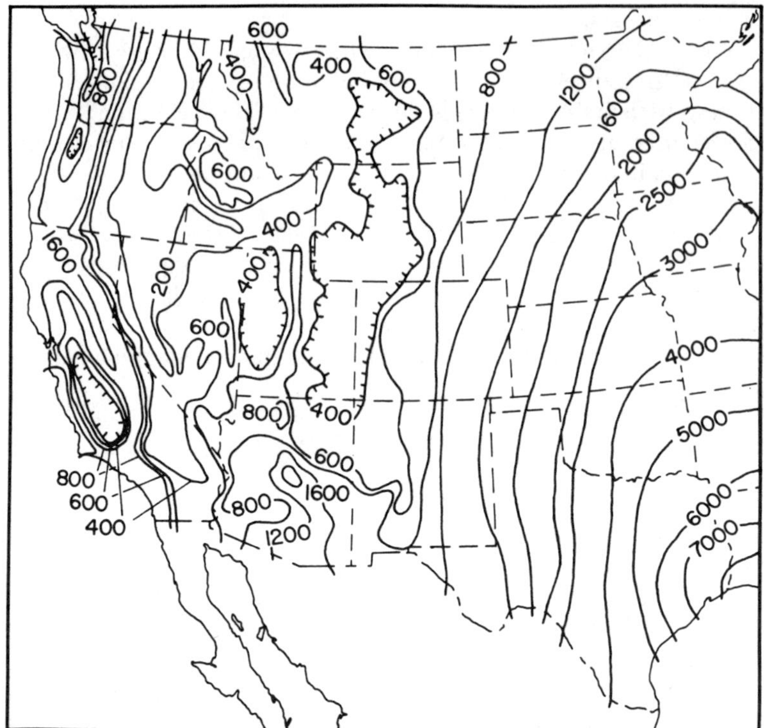

Fig. 10–2. Average annual rainfall erosion index for the U.S. portion of the North American dryland. Units are MJ mm(ha h y)$^{-1}$.

tensity product) among individual storms is also greater in an arid or semiarid climate. Wischmeier (1974) found that the *R* factor in the north central and western states of the USA could be approximated as a function of the amount of rainfall in a 6-h storm expected at least once in 2 years (Fig. 10–3). In the dryland area, rainfall amounts in a 10-year, 6-h storm are approximately double those in the 2-year, 6-h storm shown in Fig. 10–3 (Hershfield, 1961). Regardless of measures taken using the annual rainfall erosion index to plan the control of soil erosion by water, there is always a chance of severe storms during an average land tenure in dryland, because rainfall amounts in a 25-year, 6-h storm range from 2.5 to 5 times that in a 2-year, 6-h storm. According to an estimation procedure by Cooley (1980), the *R* value in a 25-year, 6-h storm can range from 340 in the Columbia Basin and portions of California, to 4600 in eastern Oklahoma and Texas. These values are approximately the same as the annual *R* (Fig. 10–2) for the same regions.

An improved assessment of soil erosion by water, beyond that afforded by the *R* value in Fig. 10–2, is achieved by considering the soil erodibility (*K*), slope steepness (*S*), and uninterrupted length of slope (*L*) factors. This information varies locally (among land resource units within land resource regions), is available from soil mapping units, and is elaborated on in Chapter 11.

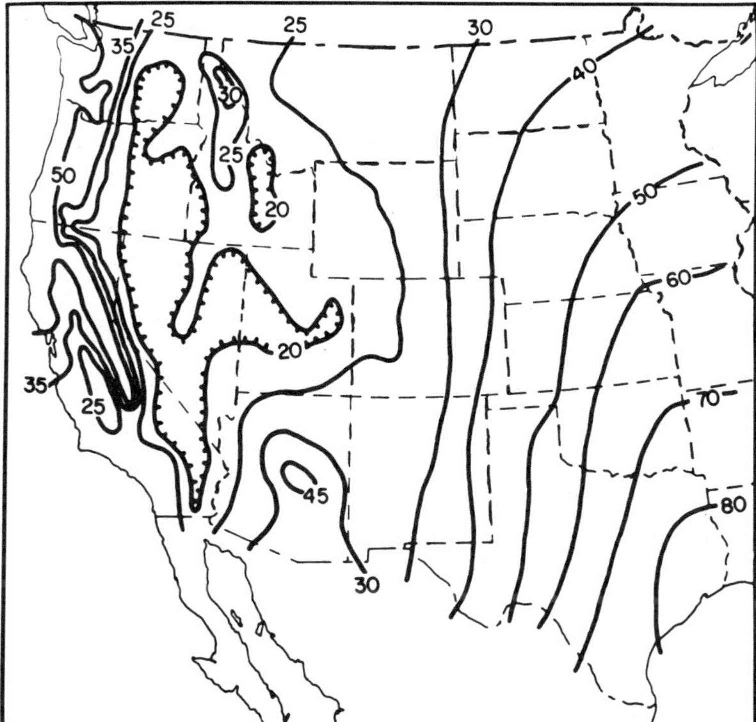

Fig. 10–3. Rainfall (in millimeters) expected for a 6-h period at least once in 2 years in the U.S. portion of the North American dryland.

An understanding of rill and sheet erosion from an upland site in a field is further improved by considering the factors under management control by the grower or farmer; these factors are the cover and management (*C*) and practice (*P*) factors. The *C* factor is influenced by the management of tillage and residue placement and by cover afforded by the crop itself. Cropping systems have a strong influence on available and useful options for managing the *C* factor; thus, the land resource region description and associated information (Table 10–1) is a logical grouping of managements used for discussing the *C* factor. Discussions in each of the subsequent chapters on soil conservation provide details within the various land resource regions. Most conservationists consider the cover and management factor as the most useful (now and potentially) means for control of soil erosion. Values of *C* may nominally range from 0.5 to 0.1 within any cropping system depending on use of excessive or reduced tillage methods. By definition, *C* is 1.0 for soil maintained in continuous bare fallow.

Any comparison of a field operation in an up-and-down direction as opposed to a tangent direction on the slope is included in the *P*, or practice, factor. Field operations in the direction of the contour, and alternate strips of different crops (or fallow vs. crop) on the contour are the main practice factors. Again, the cropping practice and soil mapping unit jointly provide basic information about expected soil erosion control from the practice factor.

Terraces are used for soil erosion control mostly in the Northwestern Wheat Region, Central Great Plains Winter Wheat and Range Region (Central Winter Wheat Region; region H), and Central Feed Grains Region (region M). Terraces are generally a combined channel and earthen embankment constructed across the slope; their purpose is to (i) reduce soil erosion, (ii) increase soil storage of water, (iii) convey water off of a slope, (iv) reduce sediment in runoff water, and (v) facilitate contour operation. Terraces may reduce sheet and rill erosion by reducing the length of slope (L) factor, but they also reduce gully erosion by reducing flow velocity and thus the ability of the runoff to transport sediment. The distinction between a rill and a gully is one of depth because both are water courses (Glossary Committee, 1970); a rill is generally intermittent and no deeper than several centimeters, whereas a gully is so deep that it cannot be obliterated by normal tillage operations. Others (Foster and Meyer, 1975) discriminate a gully and a rill based on concentrated flow; removal of one rill from the system would have no significant effect on flow, but removal of a gully would significantly affect flow over the slope. To use terraces as a viable erosion control practice depends on their compatibility with topography, normal mechanized farming, and a system for soil erosion and water management.

Terrace design and layout is a complex engineering procedure (SCS, 1977). However, the SCS curve number technique (SCS, 1972) may be used along with rainfall characteristics or storm-return frequency (Fig. 10–3) to project direct runoff and water volumes to be carried off of the slope. Such an estimation was done by Stewart et al. (1975, 1976) to project water quality responses to applied chemicals. Curve number projections are determined not only by the precipitation but also by internal soil drainage class (described in the soil mapping unit), cover afforded by the growing crop, contour or noncontour direction of field operations, and antecedant precipitation. Spacing of terraces is determined somewhat by the expected upland soil erosion (SCS, 1977).

10–3 WIND EROSION

Soil erosion by wind is affected by as many as 11 primary variables but is described (Woodruff and Siddoway, 1965) by several factors that combine variables. (Note: The older literature on wind erosion uses a prime sign, whereas the newer literature has the same unprimed letters. The prime notation is retained here to avoid confusion of wind and water erosion symbols.) They are: C', a local wind erosion climatic factor, I_s', soil and knoll erodibility, and three management related factors (K', soil ridge roughness as produced by cultivation; L', field length; and V', equivalent quantity of vegetative cover). The soil and topography factor, I_s', is related to the soil mapping unit but includes a management factor because the percentage of dry soil fractions greater than 0.84 mm is determined by field traffic as well as natural soil properties. Details regarding the wind erosion equation are contained in Chapter 12.

Several maps of the monthly wind erosion climatic factor are shown in Fig. 10-4 for the western USA. Some users may consider the lines on these maps as isoerodents. Maps for each month were developed by Skidmore and Woodruff (1968); those in Fig. 10-4 were chosen to illustrate the climate factor during critical postplanting periods for spring (April) and winter wheats (October) as examples. February and March could also be critical for winter wheat in both the Western Great Plains Range Region and the Central Winter Wheat Region. The climate factor, C', is derived from wind velocity and soil moisture deficit terms such that C' increases as the expected soil loss increases. The erodible wind index values in Table 10-1 confirm roughly the wind erosion hazard indicated by the C' in Fig. 10-4.

As with an earlier analysis of water erosion, the C' values in Fig. 10-4 may be considered proportional to soil loss on a bare, smooth soil surface that is unsheltered upwind (located at Garden City, Kansas) and with a similar proportion of dry soil fractions greater than 0.84 mm. Again, as with the R factor for water erosion, the C' factor varies widely within the dryland area. During February, March, and April wind erosion can be potentially severe in the North American dryland. It is especially serious in the Northern Great Plains Spring Wheat Region (Northern Spring Wheat Region; region F) and the Canadian Provinces, where aggregates are reduced to erodible size by precipitation and freezing over the winter, and where soils have been recently tilled in preparation for seeding spring wheat and are not yet covered by vegetation. The most serious period in the Central Winter Wheat Region occurs in February and March because winter wheat cover has not yet protected the soil weathered during winter. The C' value for October is generally lower than that for April (Fig. 10-4), but a wind erosion hazard is commonly experienced during this period in regions where winter wheat seedbeds have recently been prepared, e.g., the Northwestern Wheat Region, the Western Great Plains Range Region, and the Central Winter Wheat Region. Localized areas within these regions and even within the major land resource areas of these regions will show considerable variation in wind erosion based on I_S', a factor about which the soil mapping unit will provide much information. These generalizations are discussed in detail in subsequent chapters on soil conservation.

The climatic potential (C') for wind erosion shown by month in Skidmore and Woodruff (1968) only roughly depicts extremes of climates affecting potential for wind erosion. Recurrence frequency of wind velocities are not generally available to the same degree as recurrence frequency for rainfall storms; moreover, the climatic factor (C') is a joint function of wind velocity and soil moisture deficit. Zingg (1950) characterized the intensity frequency of various wind velocities in the wind erosion prone portion of the Central Winter Wheat Region and the Western Great Plains Range Region. Chepil et al. (1962) developed an intensity-frequency analysis of the annual climate factor for Garden City, Kansas; they showed that the severe

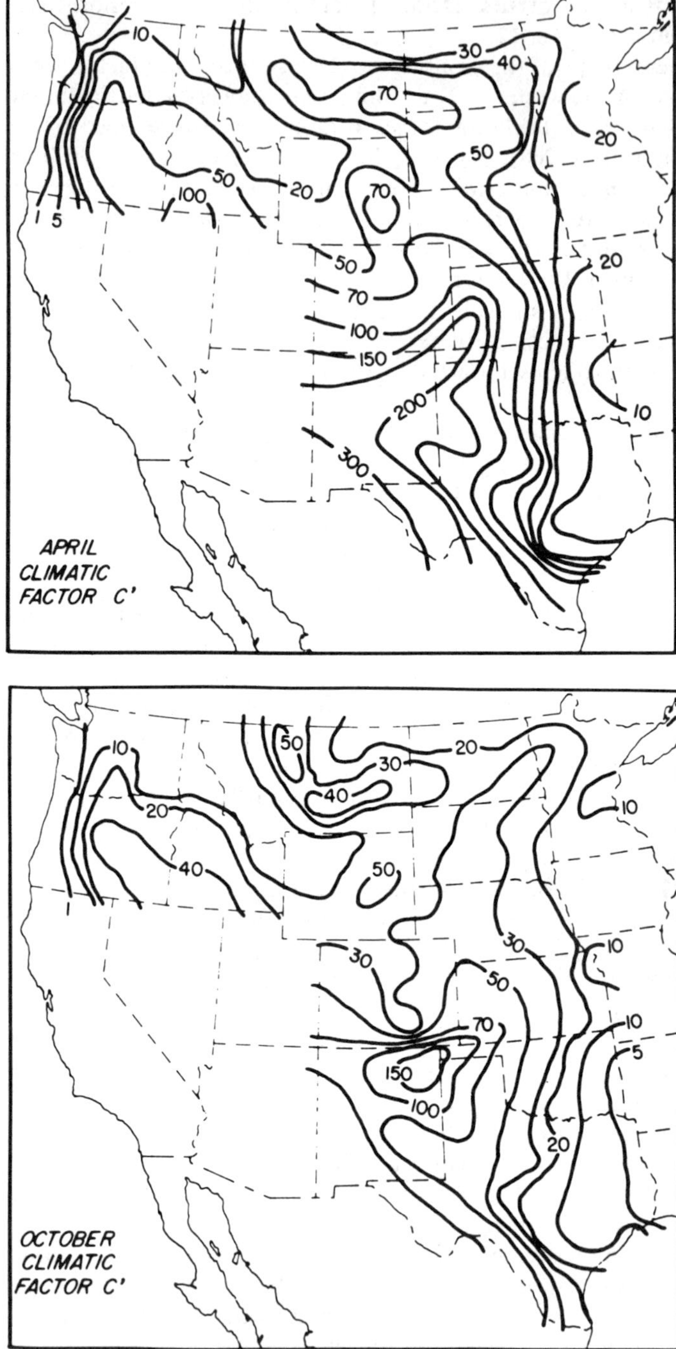

Fig. 10–4. Climate factor (isoerodents) for wind erosion for the months of April (*above*) and October (*below*) in the U.S. portion of the North American dryland.

wind erosion of the 1930's was, in fact, caused by unusually large magnitudes of the climate factor that occurred then. Woodruff and Siddoway (1965) state that the normal distribution of atmospheric wind velocities portends a greater probability of high winds as the average velocity increases.

Erosive winds are those that exceed about 5.3 m/s at the 2-m height (Skidmore, 1965). These winds have a velocity component, a prevailing direction, and a dominance in the prevailing direction, all three of which were computed for various locations in the USA (Skidmore and Woodruff, 1968). These computations confirm that the most erosive winds occur in spring and fall for most locations; however, high velocity winds also occur during winter in the Northern Spring Wheat Region and northern part of the Western Great Plains Range Region. During periods of high wind velocity, the prevailing direction changes little, and the most erosive winds occur from the prevailing direction. When most erosive winds come from the prevailing direction, field strips of alternate crops (or alternate crop and fallow) can be used to reduce the length of unprotected field. These alternate strips are ideally placed at right angles to the direction of the prevailing erosive winds.

The soil erodibility factor (I') decreases as the percentage of dry soil fraction greater than 0.84 mm increases. This factor can be estimated indirectly from information in the soil mapping unit; I' increases as the sand mechanical fraction increases, as organic matter decreases, when clay soils are exposed to freezing and thawing, and when fine-textured soils contain more than 3% $CaCO_3$. This factor has a numerical range based on soil texture information in soil mapping units within a major land resource area as great as the range of C' values in Fig. 10–4 over regions in the dryland.

Wind erosion hazard and associated management for control are elaborated in Chapters 14, 16, and 17. Specific measures available to the farmer or grower to reduce wind erosion in these regions include control of soil ridge roughness (K'), length of field (L') exposed to erosive wind, and vegetative cover (V'). Again, as with water erosion, these control measures depend on the cropping practice of the region.

10–4 VEGETATIVE COVER

Earlier discussions in this chapter identified vegetative cover as the most effective management practice for protection from wind and water erosion. In dryland agriculture, it is often difficult to maintain vegetative cover because drought may seriously reduce plant growth and ensuing crop residue. The hazard is accentuated also by the use of summer fallow. Systematic management of crop growth and residue is all the more critical under these circumstances.

Summer fallow is defined as "a farming practice wherein no crop is grown and all plant growth is controlled by cultivation or chemicals during a season when a crop might normally be grown. Thus, production for one season is forfeited in anticipation that there will be at least partial compensation by increased crop production the next season'' (Haas et al., 1974a). Without judicious management to maintain soil cloddiness and crop residues on the surface, the soil condition during this fallow period is a smooth and bare surface. This is the surface considered earlier in the discussion of climatic forces for water and wind erosion, respectively, in Fig. 10–2 and 10–4. Successful control of soil erosion thus requires ample crop residue at harvest and infrequent tillage to maintain crop residues on the surface during the summer fallow season. This potentially erosive condition extends until the newly seeded crop is large enough to again protect the soil surface from wind and water erosive forces. In a typical winter wheat–fallow sequence, green vegetative wheat covers the land for only 3 of the 24 months in a typical spring wheat–fallow sequence.

Several of the following chapters discuss estimation of surface crop residue in a wheat–fallow system based on type of tillage implement and number of times it will be used in a tillage management scheme. This estimation provides a basis to predict the number and kind of operations needed to maintain soil erosion (wind and water) below some specified level, given the average local climate. This concept includes evaluation of cover needed at various critical periods during a typical fallow period and the subsequent period for wheat germination and early growth. First consideration in fall-seeded wheat must be given to the critical period after germination; for spring-seeded wheats, first consideration must be given to the second winter period of the fallow year. Earlier critical periods occur during the first winter and spring following wheat harvest. Once the crop residue is buried, it is not practical to move it back to the surface. Critical cover may be estimated using similar concepts for a nonfallow cropping sequence.

Probabilities of achieving straw yields below a specified maximum value are shown in Table 10–2. Straw production in a wheat-fallow cropping system was compared with that in a continuous cropping of the same wheat type. Interpretation of the information is illustrated for continuous spring wheat in the Northern Spring Wheat Region. In 8 out of 10 years straw yields are expected to be less than 4350 kg/ha; in 2 out of 10 years they are expected to be less than 930 kg/ha; in half the years they are expected to be less than 2320 kg/ha. Climatic variability affected straw production least in the Northwestern Wheat Region and most in the Central Winter Wheat Region; other regions experienced variability of straw production between these two extremes.

Straw production in the wheat–fallow system was always greater than in the continuous wheat system (Table 10–2). Even after an adverse crop production year (depicted by maximum production for a probability = 0.2), straw production in the wheat fallow system was about 1000 kg/ha. With 825 kg/ha of straw after wheat seeding, soil loss is expected to be 34% of that from a fallowed bare soil surface exposed to the same rainfall and runoff energy (Wischmeier and Smith, 1978); the same amount of straw after wheat seeding would reduce wind erosion to about half that from a bare and smooth soil surface (Woodruff and Siddoway, 1965). Even in a

Table 10-2. Probabilities for production of wheat straw in selected land resource regions where summer fallow is practiced in the U.S. portion of the North American dryland.

Land resource region and portion of region	Wheat type	Probability of producing less than indicated amount of straw†					
		Wheat-fallow			Continuous wheat		
		0.2	0.5	0.8	0.2	0.5	0.8
		kg/ha					
Northwestern Wheat and Range	Winter	6710	7580	8750	2100	2530	3090
Western Great Plains Range and Irrigated							
Western	Winter	1180	4130	8760	150	1520	4060
Eastern	Winter	4750	7300	10320	450	2850	5300
Northern Great Plains Spring Wheat	Spring	1830	4290	6890	930	2320	4350
Central Great Plains Winter Wheat							
North central	Winter	2300	5280	9580	670	2790	5610
Southwestern	Winter	1210	2980	5020	410	1650	3390

† Estimates based on original grain production probabilities (Haas et al., 1974b; Greb et al., 1974; Johnson et al., 1974; Leggett et al., 1974) adjusted using residue to grain weight ratios of 1.7 for winter wheat and 1.3 for spring wheat. A ratio of 1.6 was used for winter wheat in the Northwestern Wheat and Range Region. Yields adjusted to present production levels based on grain yield trends shown by Bond and Umberger (1979).

summer fallow management system conservative with tillage, at least half of the crop residue at wheat harvest is either buried or decomposed (Greb et al., 1970). Thus, at least 1650 kg/ha of straw at harvest would be required to be certain of retaining at least 825 kg/ha on the surface after wheat seeding. Because in at least 3 in 10 years harvest residues in continuous wheat (Table 10-2) can be expected to be less than 1650 kg/ha, there is a serious risk of soil erosion with continuous wheat in dryland areas; the risk is much reduced in wheat-fallow.

The information in Table 10-2 was developed from long-term experiments completed before many of the modern methods of conservation (or reduced) tillage were developed. As new conservation tillage systems are developed along with improved varieties (better than those used in Table 10-2), more continuous small grain will be planted. Later chapters on soil conservation emphasize this change. Examples of improved crop residue production and maintenance on the surface using herbicides along with current tillage and planting technology are given by Black and Siddoway (1977), Soil Conservation Society of America (1977), Fenster et al. (1977), and von Hulst (1971).

Probabilistic crop residue productions in Table 10-2 show that tillage and residue management requires machinery and timing flexibility. For wheat-fallow in the eastern part of the Western Great Plains Range Region, as an example, residue productions greater than 10 320 kg/ha are expected for 2 of 10 years, between 7300 and 10 320 kg/ha for 3 of 10 years, between 4750 and 7300 kg/ha for 3 of 10 years, and less than 4750 kg/ha for 2 of 10 years. If approximately 825 kg/ha must be on the surface after wheat is seeded, then there must be considerable variation in the amount of crop residue buried. Such an objective requires a systematic flexibility in the type of tillage implement and number of field operations. In years when residue production is less than 4750 kg/ha, herbicides would provide a means to reduce tillage and burial of crop residues. These innovations in tillage, seeding, and residue management are sensitive to natural features and cropping systems of adaptable crops. They are discussed in subsequent chapters on soil conservation.

10–5 LITERATURE CITED

Austin, M. E. 1981. Land resource regions and major land resource areas of the United States. Rev. ed. USDA Agric. Handbook no. 296.

Beatty, M. T. (task force chairman) 1982. Soil erosion: Its agricultural, environmental, and socioeconomic implications. Coun. for Agric. Science and Technology Report no. 92, Ames, Iowa.

Black, A. L., and F. H. Siddoway. 1977. Winter wheat recropping on dryland as affected by stubble height and nitrogen fertilization. Soil Sci. Soc. Am. J. 41:1186–1190.

Bond, J. J., and D. E. Umberger. 1979. Technical and economical causes of productivity changes in U.S. wheat production, 1949–76. USDA Tech. Bull. no. 1598.

Carlson, J. E., D. A. Dillman, and W. R. Lassey. 1981. The farmer and erosion: Factors influencing the use of control practices. Idaho Agric. Exp. Stn. Bull. no. 601.

Chepil, W. S., F. H. Siddoway, and D. V. Armbrust. 1962. Climatic factor for estimating wind erodibility of farm fields. J. Soil Water Conserv. 17:162–165.

Chepil, W. S., and N. P. Woodruff. 1963. The physics of wind erosion and its control. Adv. Agron. 15:211–302.

Cooley, K. R. 1980. Erosivity values for individual design storms. J. Irrig. Drain. Div. Am. Soc. Civ. Eng. 106:135–145.

Dillman, D. A., and J. E. Carlson. 1981. Influence of absentee landlords on soil erosion control practices. J. Soil Water Conserv. 37:37–41.

Fenster, C. R., H. I. Owens, and R. H. Follett. 1977. Conservation tillage for wheat in the Great Plains. USDA Cooperative Extension Service Bull. no. PA-1190.

Fisher, P. S., and E. L. Skidmore. 1970. WEROS: A Fortran IV program to solve the wind-erosion equation. USDA Report no. ARS-41-174.

Foster, G. R., D. K. McCool, K. G. Renard, and W. C. Moldenhauer. 1981. Conversion of the universal soil loss equation to SI metric units. J. Soil Water Conserv. 36:355–359.

Foster, G. R., and L. D. Meyer. 1975. Mathematical simulation of upland erosion by fundamental erosion mechanics. p. 190–207. *In* Present and prospective technology for predicting sediment yields and sources. USDA Report no. ARS-S-40.

Glossary Committee. 1970. Resource conservation glossary. Soil Conserv. Soc. of Am., Ankeny, Iowa.

Greb, W. B., D. E. Smika, and A. L. Black. 1970. Water conservation with stubble mulch fallow. J. Soil Water conserv. 25:58–62.

Greb, W. B., D. E. Smika, N. P. Woodruff, and C. J. Whitfield. 1974. Summer fallow in the Central Great Plains. p. 51–85. *In* Summer fallow in the western United States. USDA-ARS Conserv. Res. Report no. 17.

Haas, H. J., W. O. Willis, and J. J. Bond. 1974a. Introduction. p. 1–11. *In* Summer fallow in the western United States. USDA-ARS Conserv. Res. Report no. 17.

Haas, H. J., W. O. Willis, and J. J. Bond. 1974b. Summer fallow in the Northern Great Plains (spring wheat). p. 12–35. *In* Summer fallow in the western United States. USDA-ARS Conserv. Res. Report no. 17.

Hershfield, D. M. 1961. Rainfall frequency atlas of the United States for durations from 30 minutes to 24 hours and return periods from 1 to 100 years. U.S. Dep. of Commerce Tech. Paper no. 40.

Johnson, W. C., C. E. Van Doren, and E. Burnett. 1974. Summer fallow in the Southern Great Plains. p. 86–109. *In* Summer fallow in the western United States. USDA-ARS Conserv. Res. Report no. 17.

Klingebiel, A. A., and P. H. Montgomery. 1973. Land-capability classification. USDA Agric. Handbook no. 210.

Kohler, M. A., T. J. Nordenson, and D. R. Baker. 1959. Evaporation maps for the United States. U.S. Dep. of Commerce Tech. Paper no. 37.

Leggett, G. E., R. E. Ramig, L. C. Johnson, and T. W. Massee. 1974. Summer fallow in the Northwest. p. 110–135. *In* Summer fallow in the western United States. USDA-ARS Conserv. Res. Report no. 17.

Skidmore, E. L. 1965. Assessing wind erosion forces: Directions and relative magnitudes. Soil Sci. Soc. Am. Proc. 29:587–590.

Skidmore, E. L., and N. P. Woodruff. 1968. Wind erosion forces in the United States and their use in predicting soil loss. USDA Agric. Handbook no. 346.

Soil Conservation Service. 1972. Hydrology. Natl. Engineering Handbook, Section 4. USDA-SCS, U.S. Government Printing Office, Washington, D.C.

Soil Conservation Service. 1977. S.C.S. engineering practice standards for terraces. Natl. Engineering Handbook, Section 2. USDA-SCS, U.S. Government Printing Office, Washington, D.C.

Soil Conservation Society of America. 1977. Conservation tillage problems and potentials. J. Soil Conserv. Soc. Am. 32:9–65.

Stewart, B. A., D. A. Woolhiser, W. H. Wischmeier, J. H. Caro, and M. H. Frere. 1975. Control of water pollution from cropland. Vol. I: A manual for guideline development. USDA Report no. ARS-H-5-1.

Stewart, B. A., D. A. Woolhiser, W. H. Wischmeier, J. H. Caro, and M. H. Frere. 1976. Control of water pollution from cropland. Vol. II: An overview. USDA Report no. ARS-H-5-2.

von Hulst, H. 1971. Tillage and seeding practices and machines for crop production in semiarid areas. Agric. Development Paper no. 92. Food and Agric. Organization of the United Nations, Rome.

Wischmeier, W. H. 1974. New developments in estimating water erosion. p. 179–186. *In* Land use: Persuasion or regulation? Proc. 29th Annual Meeting of the Soil Conserv. Soc. of Am., Syracuse, N.Y. 11–14 Aug. Soil Conservation Soc. of Am., Ankeny, Iowa.

Wischmeier, W. H., and D. D. Smith. 1978. Predicting rainfall erosion losses—a guide to conservation planning. USDA Agric. Handbook no. 537.

Woodruff, N. P., and F. H. Siddoway. 1965. A wind erosion equation. Soil Sci. Soc. Am. Proc. 29:602–608.

Zingg, A. W. 1950. The intensity-frequency of Kansas winds. USDA Soil Conserv. Service Tech. Pub. no. 88.

11

Kenneth G. Renard
Agricultural Research Service
U.S. Department of Agriculture
Tucson, Arizona

George R. Foster
Agricultural Research Service
U.S. Department of Agriculture
West Lafayette, Indiana

Soil Conservation: Principles of Erosion by Water

The study of soil erosion by water is dynamic, with many advances developing since the advent of digital computers. Although scientists and engineers have been investigating the factors affecting erosion for many decades, most of the early work was limited to storms or, even more commonly, seasonal phenomena. Efforts to incorporate known physical laws in such early work led to the solution of problems. Current computer capability allows solutions of known physical relationships that facilitate description of spatial and temporal variability.

The hydraulic principles involved in soil erosion by water were applied primarily to larger river systems where the flow depths are large relative to the roughness elements. Thus, the transfer of such technology to upland soil erosion problems must be made with complete understanding of the assumptions involved. Students of soil erosion problems, and practitioners, need to be aware that many technological improvements being made will require continual upgrading of the subject matter. The material presented subsequently is intended to provide an introduction to soil erosion theory and some of the strategies for the control of erosion.

11-1 WATER EROSION

Erosion and sedimentation by water involve processes of detachment, transport, and deposition of soil particles. The major forces are from raindrop impact and water flowing over the land surface. Erosion may be un-

Dryland Agriculture—Agronomy Monograph no. 23.

noticed on a smooth soil surface where raindrops are eroding large quantities of sediment, whereas it can be dramatic where concentrated flow creates an extensive rill and gully system. Factors affecting erosion can be expressed in an equation of the form

$$E_r = f(C_l, S_p, T_o, SS, M) \qquad [1]$$

where

E_r = erosion,
f = function of ( ),
C_l = climate,
S_p = soil properties,
T_o = topography,
SS = soil surface conditions, and
M = Human activities.

Sediment yield should not be confused with erosion. Sediment yield is the amount of eroded sediment that is delivered to a point in the watershed remote from the origin of the detached particles. In a watershed, sediment yield includes the erosion from slopes and channels, mass wasting, and deposition of eroded sediment on both slopes and in channels.

The universal soil loss equation (USLE) is essentially an expression of the functional relationship shown in Eq. [1] (Wischmeier and Smith, 1965, 1978). It was developed from extensive field experimentation and has been widely accepted by conservation planners. The USLE, a product of six terms representing the major factors in Eq. [1], gives an estimate of the average annual erosion expected from a field-sized area. The USLE is

$$A = RKLSCP \qquad [2]$$

where

A = the computed soil loss per unit area expressed in the units selected for K and for the time period selected for R. In practice, these are usually so selected that they compute A in metric tons per hectare per year, but other units can be selected.
R = the rainfall and runoff factor, which is the number of rainfall erosion index units for a given time period, usually a year, plus a factor for runoff from snowmelt or applied water where such runoff is significant.
K = the soil erodibility factor, which is the soil loss rate per erosion index unit for a specified soil as measured on a unit plot, defined as a 22.1-m length of uniform 9% slope continuously in clean-tilled fallow.
L = the slope-length factor, which is the ratio of soil loss from the field slope length to that from a 22.1-m length under identical conditions.
S = the slope-steepness factor, which is the ratio of soil loss from the field slope gradient to that from a 9% slope under otherwise identical conditions.

C = the cover and management factor, which is the ratio of soil loss from an area with specified cover and management to that from an identical area in tilled continuous fallow.

P = the support practice factor, which is the ratio of soil loss with a support practice such as contouring, stripcropping, or terracing to that with straight-row farming up and down the slope.

The material presented below is not intended to replace the detailed material in Agriculture Handbook no. 537 (Wischmeier and Smith, 1978). The details contained in the handbook must be familiar to the user of the USLE. The material we presented here is a general overview of the water erosion problem.

11-2 CLIMATE

Climate directly but obscurely influences erosion. For example, the semiarid western USA, with generally low rainfall, has high sediment yield per unit area (Langbein and Schumm, 1958), because the low rainfall generally supports a poor vegetative cover to protect the soil from raindrops and because erosion from channels is generally large. Similarly, land slopes are often greater in the mountainous areas of the West. Thus, climate interacts with other factors such as topography and soil cover. Climate is often described by annual rainfall in an area, although the seasonal distribution of the precipitation, the intensity of the precipitation, the form of precipitation, and temperature at critical periods are also important climatic indicators. Erosive rains are not uniformly distributed over a typical year except in some areas in the Southeast. In western Nebraska, 24% of the annual rainfall erosivity potential normally occurs in the 1-month period from 15 May to 15 June. Because this period coincides with seedbed preparation for crops such as corn, when there is little ground cover to protect against erosion, the hazard of erosion is usually severe. In contrast, a negligible amount of average annual rainfall erosivity occurs during the same period in the Palouse area of the Northwest.

The erosivity of precipitation coming as snow is less than that for an equivalent rainfall because there is no detachment by raindrops. However, warm rain on snow can produce large rates of runoff that may be quite erosive when coupled with a highly erodible soil produced by frequent freeze-thaw cycles. Rain on snow with frozen ground, although it produces large amounts of runoff, will not produce erosion.

The "concentration of energy" in rainstorms is also important. Large amounts of rainfall at low intensities (low energy concentration) may cause much less total erosion than small amounts of rainfall at high intensities (high energy concentration). Such high-energy rainfall is characteristic of Arizona, where a single 2-h thunderstorm commonly produces over 50% of the annual R value (Renard et al., 1974).

The rainfall erosivity term R in the USLE expresses the important characteristics of rainfall influencing erosion. Analysis of experimental data showed that erosion is highly correlated with rainfall energy, rainfall amount, and rainfall intensity. These factors may be combined as the product (*EI*) of a storm's total kinetic energy and its maximum 30-min intensity. Since erosion is linearly related to *EI,* values for the individual storms are summed for an annual erosivity *R* value. Average annual values for *R* range from a low in the West of 200 to almost 10 000 MJ mm(ha h y)$^{-1}$ (megajoule-millimeters per hectare hour per year) in the Southeast. If one assumes that all factors except *R* of the USLE are the same, erosion in the Southeast will be over 25 times that in the West. However, for undisturbed conditions, the high rainfall will produce very good ground cover, and thus little erosion, whereas ground cover will be poor in the low rainfall area, allowing very high erosion rates. Human activities that disturb the ground cover can greatly change this natural situation.

Two types of information are needed to represent the seasonal pattern of rainfall erosivity: (i) the average annual *R* value and (ii) its distribution throughout the year. This information is available in Agriculture Handbook no. 537 (Wischmeier and Smith, 1978), and a map of the annual *R* factor expressed by using the International System of Units (SI) was presented by Foster et al. (1981b).

The USLE estimates expected erosion over a long period. It must be modified to estimate erosion for individual storms. The modifications usually involve changing the *R* term to reflect the greater variability of erosion on a storm-by-storm basis. Foster et al. (1977) replaced *R* with a term defined as

$$E = (0.5\, R_{st} + 3.5\, V_u\, \sigma^{1/3}) \qquad [3]$$

where

E = storm erosivity from rainfall and runoff MJ mm(ha h)$^{-1}$
R_{st} = single storm *EI* erosivity value MJ mm(ha h)$^{-1}$
V_u = storm runoff (mm)
σ = storm peak runoff rate mm h^{-1}

In effect, Eq. [3] reflects the erosion contribution by both raindrop impact and runoff. In watersheds where significant deposition occurs, sediment yield is almost entirely related to flow characteristics, and the coefficient of R_{st} in Eq. [3] can be set to zero. Williams (1975) used this approach with the equation

$$Y = 90.5\, (Q\, q_p)^{0.56}\, K\, L\, S\, C\, P \qquad [4]$$

where

Y = computed soil loss (tonnes) for the runoff event,
Q = storm runoff (m^3), and
q_p = peak runoff rate for the storm (m^3 s^{-1}).

The variables Q and q_p are estimated using commonly available techniques for estimating runoff (SCS, 1972). Equation [4] has been used and widely tested for estimating sediment yield.

The USLE and a delivery ratio (Roehl, 1962) to account for deposition are often used to estimate sediment yield. Better ways of estimating sediment yield using a runoff-based equation such as Williams' modification of the USLE (Eq. [4]) or recent fundamental models that route sediment through the watershed system are being developed (Knisel et al., 1980). Such models are derived from more basic erosion relationships that consider continuity of mass and the processes of detachment, transport, and deposition by raindrop impact and the flow of water over the land surface and in the channels (see Appendix). The models also calculate erosion rates for individual storms. They can consider the spatial distribution of soils, the land use, and the topography in the field or watershed.

11-3 SOIL PROPERTIES

Some soils are much more susceptible to erosion than others. In fact, during the same rainfall, a highly erodible soil can erode 10 times faster than a nonerodible soil. Factors affecting a soil's resistance to erosion include its organic matter content, primary particle-size distribution, structure, iron and aluminum oxide content, electrochemical bonds, moisture content, and aggregation (Wischmeier et al., 1971; Grissinger, 1966; Partheniades, 1972). The K term in the USLE quantifies a soil's erodibility. The U.S. Department of Agriculture Soil Conservation Service and Agricultural Research Service have tabulated the values of K for most major soils found in the continental USA and have presented a nomograph for estimating a value where they are unavailable (Fig. 11-1).

Good management practices (e.g., improving organic matter content, soil tilth, and soil structure) may decrease erodibility. Continuous tillage destroys soil structure and decreases organic matter content, thereby increasing erodibility. These management influences are described in other terms of the USLE. In concept, the K factor is a fixed soil property that remains constant over a long period. In practice, some soils are more susceptible to rill erosion than other soils. Since rill erosion is a function of runoff volume and rate (see Appendix), erosion control on these soils requires maintenance of high infiltration rates and prevention of large accumulation of runoff by, e.g., installing terraces and diversions.

11-4 TOPOGRAPHY

Topography affects erosion on two scales—macro- and micro-. Slope steepness, length, and shape express the macroscale. Steepness and row side slopes and roughness configurations left by tillage implements are expressions of microtopography.

11-4.1 Macrotopography

Topography, as expressed by slope steepness, length, and shape, influences erosion. The energy of the eroding agents increases with slope steepness. Additional runoff accumulates with longer slope lengths, which in-

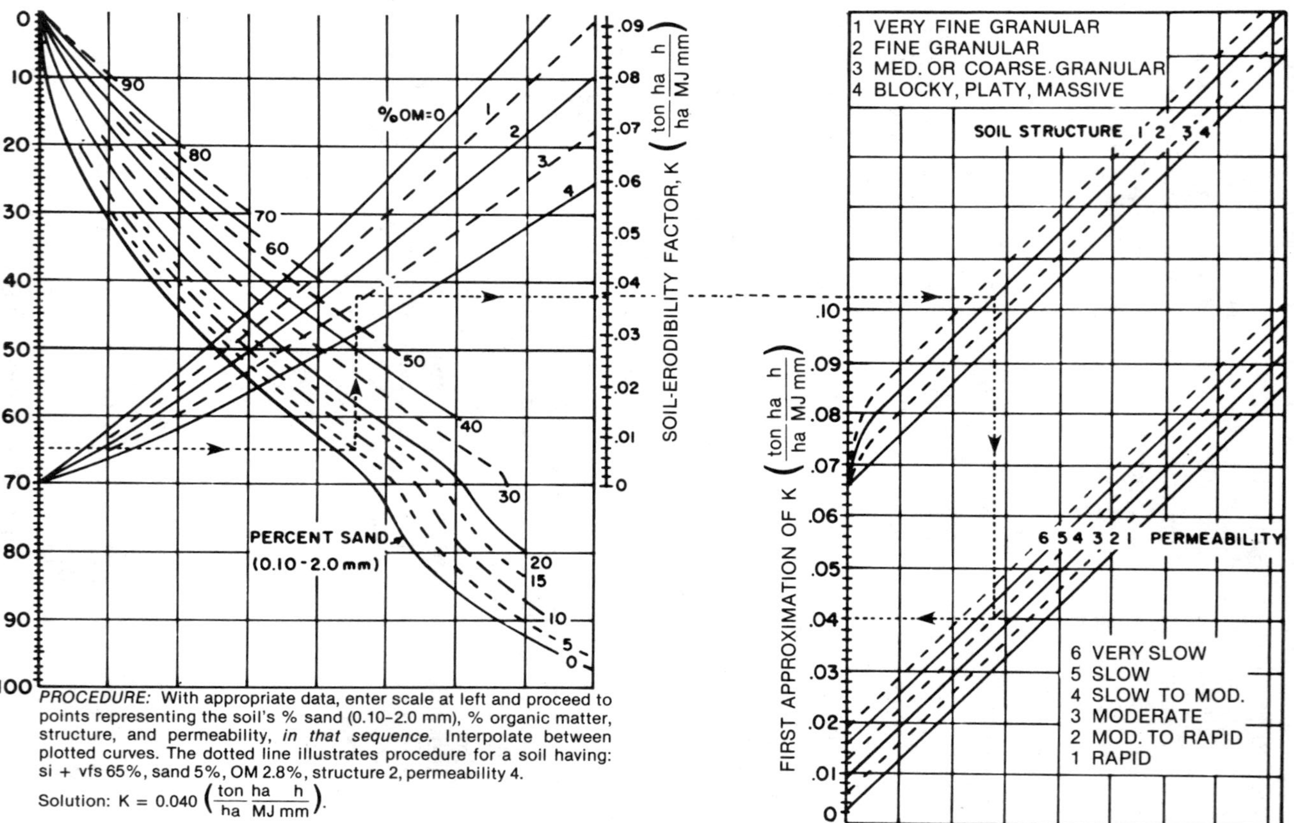

Fig. 11–1. The soil erodibility nomograph in SI units (Wischmeier and Smith, 1978).

creases the capacity of flow to detach and transport soil particles. The effect of slope shape on erosion depends on the locations of steepest portions along the slope. Erosion is greatest when steep slopes occur at the end of the slope where maximum runoff has accumulated. The options available for modifying the macrotopography of a field or watershed are limited except for construction, mine reclamation, and similar sites where extensive land reshaping is possible.

In the USLE, L and S are dimensionless factors accounting for the effects of slope length and steepness. Their product is the topographic factor that is the expected ratio of soil losses from a given length and steepness of slope to corresponding losses from the standardized plot dimensions used in other terms of the USLE (22.1-m long plot with 9% gradient). Nomographs that assume uniform slopes are available for obtaining the LS term (e.g., Wischmeier, 1977; Wischmeier and Smith, 1978). The formula for computing the combined slope length-steepness factor is

$$LS = (\lambda/22.1)^m (65.41 \sin^2 \theta + 4.56 \sin \theta + 0.065) \quad [5]$$

where λ = the slope length in meters; m = 0.5 if the slope is steeper than 5%, 0.4 on slopes of 3.5 to 4.5%, 0.3 for slopes of 1 to 3%, and 0.2 on uniform gradients of less than 1%; and θ = the angle of slope. The LS value can be modified to account for the effects of slope irregularities using the procedure detailed by Foster and Wischmeier (1974). An example of one such computation is given in Agriculture Handbook no. 537 (Wischmeier and Smith, 1978). Equation [5] has been modified somewhat for use in the Pacific Northwest (Chapt. 19).

Watershed topography and geomorphologic characteristics may greatly influence runoff and, hence, erosion. Watersheds with dense channel networks (i.e., a high proportion of channels per unit watershed area), steep channels, basin shapes, and geology that increase runoff volumes and rates all tend to increase erosion. Analytical and experimental results have shown the effect of slope shape on erosion (Fig. 11–2).

Slopes of the sediment load curves in the upper part of the figure indicate the rate of erosion as a function of position on the slope. Erosion losses from the convex slope increase very rapidly near the bottom of the slope, whereas the losses from the concave slopes are greater near the top, indicating deposition may occur at the bottom. The magnitude of the erosion and the location where deposition begins are functions of several factors, including the susceptibility of the soil to detachment by flow relative to detachment by raindrop impact, slope curvature, and runoff rate.

11–4.2 Microtopography

Microtopography scale refers to the roughness and surface configurations that remain after tillage. The microtopographic influence is described by the cover management and support practice factor of the USLE. The

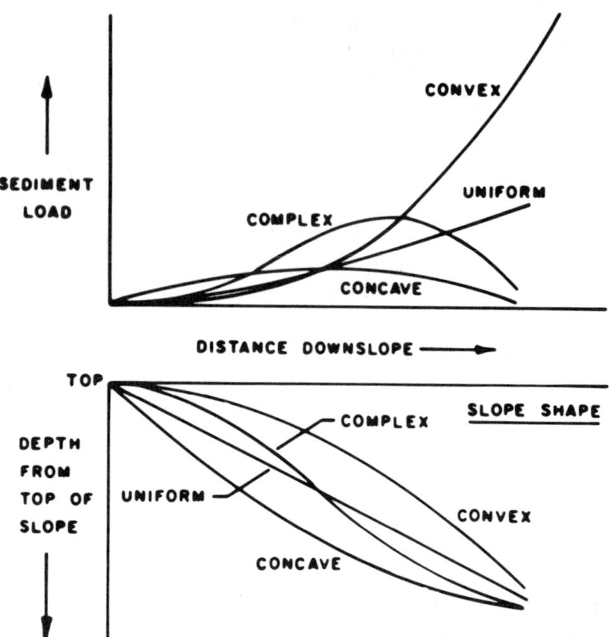

Fig. 11-2. Influence of slope shape on sediment load and rate of erosion. Steepness of sediment load curves indicates rate of erosion, and negative slope indicates deposition (from Meyer et al., 1975).

steep side slope of a row (microtopography) may produce large quantities of detached soil particles, but if the main slope gradient (macrotopography) is flat, little sediment may leave the field. On the other hand, if the main slope gradient is steep and rows run uphill and downhill, the steep side slopes of rows will increase erosion.

Depression roughness can store large quantities of detached sediment, greatly reducing the amount of sediment leaving a field. However, as the depressions fill, sediment delivery increases to a rate comparable to that from a relatively smooth surface.

Roughness can often eliminate or greatly reduce runoff from small to moderate rainstorms. Preventing runoff from a storm completely eliminates erosion for that storm. Large storms with high runoff water can cut rills through the depression storage areas, reducing or even eliminating such storage capacity. Thus, roughness must be reestablished periodically to maintain its effectiveness.

11-5 SURFACE COVER

Surface cover includes vegetative canopies and material in contact with the ground surface. A vegetative canopy intercepts raindrops above the soil surface and ground surface covers (mulches, plant residue, and erosion pavement), and tends to dissipate some of the raindrop energy and reduce runoff velocity.

The number of raindrops intercepted by a canopy depends on its density. Some of the intercepted water reforms drops, which then fall to the soil surface or ground cover. The energy of the reformed drops is usually much less than that of the nonintercepted drops because their height of fall from the canopy is usually too low for them to regain terminal velocity. Some of the intercepted water becomes stemflow. Consequently, a canopy does not reduce runoff erosivity so much as it reduces rainfall erosivity, especially during larger storms. Material in contact with the soil surface is more effective than a canopy in reducing erosion. The energy of raindrops striking this cover is completely dissipated, whereas drops falling from a canopy regain some energy. Furthermore, material in contact with the soil slows the runoff, which increases the flow depth, providing a greater buffer for reducing the hydrodynamic impact forces of the raindrop (Mutchler and Young, 1975). Also, slowing the flow increases infiltration and reduces runoff erosivity.

Wischmeier and Smith (1978) used a cover management term C in the USLE to account for the effects of surface cover. Values of C presented in Agriculture Handbook no. 537 (Wischmeier and Smith, 1978) consider different canopy heights, densities, and different ground surface cover rates. Relationships are also available for various rangeland and woodland conditions (Wischmeier, 1975; Dissmeyer and Foster, 1981). Because cover varies greatly throughout the year, C values that reflect the crop growth stage must be used. At planting time, the seedbed for a conventionally planted corn is three times as susceptible to erosion as it is just before harvest with a full canopy. The distribution of cover interacts with the distribution of rainfall erosivity over the year. Erosion may be reduced by planting crops or managing land use so that soil-disturbing activities are performed during periods of less erosive rains.

Agriculture Handbook no. 537 (Wischmeier and Smith, 1978) gives the procedure for deriving local C factors in the USLE. A weighted C value is developed that reflects the distribution of the crop stages, the crop rotation, and the distribution of the average annual EI at the specific location where erosion is being estimated. Such a computation is illustrated in Table 8 of the handbook.

Erosion pavements reduce erosion under desertlike conditions. Where plant cover is limited, gravel-covered surfaces are very effective in absorbing the impact energy of raindrops. Renard et al. (1974) suggested that the percentage of the soil surface covered by gravel or larger material should be included with the plant cover in determining a C value for use in the USLE.

11-6 HUMAN ACTIVITIES

Human activity affects erosion by use of the land and installation of supporting erosion control practices. Adequate erosion control on land that is especially susceptible to erosion may restrict land use to forestry or grassland. Land having fewer restrictions may be cropped, but may require special care, including the use of crop rotations. In many modern agricultural systems, continuous corn and soybean have replaced rotations of meadow, small grain, and corn. Adequate erosion control for continuous row cropping often requires use of conservation tillage systems. Most conversation tillage systems control erosion by leaving all or a part of the previous year's crop residue on the surface. Management practices that incorporate crop residues to improve soil tilth and structure are effective in increasing infiltration and therefore in decreasing runoff and erosion. Removing too much of the crop residue for energy production, animal feed, or building materials can greatly reduce the effectiveness of these systems.

These types of management effects are described by the USLE *C* factor. Supporting practices such as contouring, stripcropping, terraces, diversions, and detention ponds are accounted for in the *P* factor of the USLE.

11-7 SUPPORT PRACTICES

Most support practices control erosion by reducing the erosivity and transport capacity of runoff. Often the gradient and flow velocity along the runoff path are reduced enough that erosion by flow in rills and waterways is eliminated. The reduction in runoff erosivity is often great enough that deposition occurs, which reduces sediment yield from the field.

The sedimentation processes usually involved are detachment from areas where runoff erosivity has not been reduced and deposition in the area where it has been reduced. The sediment load entering this area must exceed the transport capacity of the flow for deposition to occur. The ratio of sediment leaving the deposition area to that entering depends on the relationship between the amount of sediment entering the area and the transport capacity of flow in the area (Neibling and Foster, 1977). This ratio is the *P* factor of the USLE.

11-7.1 Contouring, Graded Furrows, and Contour Ridges

Flow may travel along the contour before breaking over the row ridges, depending on the ridge height and the runoff characteristics. Flow along the contour is on a gradient considerably less than the direct downslope gradient. If the gradient along the row middles—the low area between higher plant row ridges—is steep, rill erosion may occur. If the gradient along the row is flat, deposition will occur in row middles. Where the flow breaks over the row ridges at low points, considerable rill erosion may occur. If the rows run vertically on the slope, transport capacity of flow in the middles increases rapidly as slope increases. On steep slopes where no deposition occurs, erosion from the ridges could be much greater than if the field were not ridged (Mutchler and Murphree, 1980).

11-7.2 Grass Strips

Detachment occurs on tilled areas upslope from the strips of close-growing vegetation, so that the sediment load at the upper edge of the vegetative strip is relatively high. The upper edge of the strip slows the runoff and reduces its transport capacity. If the transport capacity is reduced to less than the incoming sediment load, deposition occurs, which reduces the sediment load. Dense vegetation also eliminates most erosion within the strip.

11-7.3 Terraces

Terraces reduce rill erosion on overland flow areas by reducing slope length, but the total reduction in erosion is not great because interrill erosion per unit area is unaffected. For example, reducing slope length to one fourth of its original value cuts soil loss only in half according to the USLE. The greatest benefit of properly designed terraces is safe conveyance of concentrated surface runoff from the field without channel erosion.

If the sediment load reaching the terrace channel from the interterrace area is greater than the transport capacity of flow in the terrace channel, deposition occurs in the channel at a rate dependent on grade and other factors (Foster and Ferreira, 1981). If terrace channels are steep, they may erode with no deposition.

11-7.4 Waterways

Many farm fields contain natural or constructed waterways that convey runoff from the field. Tillage can significantly increase the erodibility of the soil in the waterway (Foster et al., 1980a). If an intense storm with large runoff rates occurs soon after tillage, erosion may be severe. These tilled waterways may erode downward to a relatively nonerodible layer at the depth of secondary or primary tillage; after reaching the nonerodible layer, they will widen. While they widen, erosion rate decreases. Erosion may be low during a storm if a previous, more intense storm has formed a wide channel.

Eroding waterways may be protected by reshaping to a broader, shallower channel and establishing grass. A good grass stand can almost completely eliminate erosion, and depending on the amount of incoming sediment and flow hydraulics, can cause deposition in the waterway. This deposition may reduce sediment yield at the edge of the field, but deposition in waterways frequently causes maintenance problems. Much of the deposition will be along the edge of the waterway, hampering the entrance of runoff into the waterway. The flow will move along the edges of a poorly maintained waterway, causing serious erosion problems.

11-7.5 Impoundments

Natural impoundments such as backwater at field outlets and constructed impoundments such as tile outlet terraces can trap large amounts of sediment by deposition and significantly reduce sediment yield at the edge of the field. Impoundment deposition depends on the opportunity for the sediment to settle to the bottom before runoff carries it away. Therefore, fewer fine particles than coarse particles are trapped. Likewise, impoundments that drain rapidly are less efficient sediment traps. Simple models, based on the settling velocity of particles, are available for estimating deposition and amount of sediment leaving impoundments (Laflen et al., 1978; Ward et al., 1977).

11-8 SEDIMENT CHARACTERISTICS

Sediment yield refers to delivery of sediment to a point in the watershed remote from the origin of the detached particles. The quantity of sediment yield and the sediment's size characteristics are important in analyses of nonpoint source pollution where the effectiveness of best management practices (those practices that minimize off-site pollution) is being evaluated. Sediment delivery is a transport process. When sediment available from the detachment processes exceeds the flow's transport capacity, deposition results. The USLE and a delivery ratio to account for deposition are often used to estimate sediment yield.

Fine soil particles cause the greatest water quality problems, yet these are the most difficult particles to remove from the flow. Practices that control sediment yield by inducing deposition remove the coarse particles first, leaving the fine ones. Therefore, the most effective sediment control for water quality maintenance is prevention of most erosion in the first place, thus eliminating the need for removal of sediment from the flow by deposition. Furthermore, control of sediment yield by control of erosion helps to maintain soil productivity.

Rates and amounts of deposition depend on the diameter and the density of the sediment particles. Sediment is typically a mixture of primary particles and aggregates. Diameters range from that for primary clay (0.002 mm) to that for coarse sand and aggregates (1.0 mm). Specific gravity ranges from 2.65 for primary particles to 1.6 for aggregates (Foster, 1982).

The distribution of sizes and densities of sediment particles depends on soil properties, previous management, tillage, and upslope deposition. On poorly aggregated soils the size distribution of the sediment at the point of detachment may be almost that of the primary particles of the soil mass. On other soils, most of the sediment may be eroded as aggregates. For example, a soil of 20% clay may yield sediment that is only 5% primary clay (Foster, 1982).

11-9 EROSION CONTROL STRATEGIES

Erosion is controlled in one or both of two ways: (i) reduce the erosive agents' energy (i.e., raindrop impact and runoff) or (ii) increase the soil's resistance to detachment. The second way is long term and highly desirable, but its impact is usually less than that of the first. Maintenance of cover in contact with soil is the most effective single factor for reducing erosion; a 90% mulch cover will cut erosion by 93% (Wischmeier and Smith, 1978). This type of soil cover reduces the energy of both raindrop impact and runoff. Often, it increases infiltration and reduces evaporation as well. Canopy, especially that near the soil surface, can reduce erosion caused by raindrop impact if the cover is dense. Minimizing the degree of soil disturbances leaves more crop residue on the soil surface and produces few soil particles that are easily detached and transported. Maintenance of soil roughness and contour tillage store detached soil particles and reduce runoff for small to moderate rainstorms. Shortening of slope length with terraces and diversions reduces rill erosion but not erosion by raindrop impact. Terraces and detention structures control sediment yield more effectively than they control erosion.

Good management practices such as crop rotations and incorporation of organic matter from crop residue and manure improve soil structure and soil tilth. These practices increase the soil's resistance to detachment, increase infiltration, and reduce runoff. Over the long term, the net result is a significant reduction in erosion. Meadow rotations are effective in the short run. Obviously, erosion is greatly reduced during the rotation period with the meadow. Moreover, a residual effect from the period with the meadow lasts for about 2 years for subsequent crops, such as corn.

11-10 SUMMARY

Soil erosion by water occurs when raindrops impact on bare soil or when water moves over the soil surface creating a shearing force. Erosion exceeding the natural regeneration rate may reduce the capacity of a soil to produce crops. Excessive erosion by surface runoff causes gully erosion that inconveniences farming operations and reduces land value. Climate, soil, topography, cover, and human activities on the land all affect erosion. Empirical equations based on field experiments are used to plan conservation practices to adequately control erosion.

Runoff transports the eroded sediment from the fields. If the transport capacity of the runoff is less than the sediment detached upslope, deposition that reduces the amount of sediment leaving the field occurs. Large and dense particles are the first to deposit during a storm event, leaving an enriched amount of fine sediment in the flow. These fine sediments are the carriers of many agricultural chemicals. Sediment may cause considerable off-site damage by filling water conveyance and storage structures, being a pollutant itself, and being a carrier of pollutants such as some fertilizer and pesticide compounds. Sediment yield estimates are required in water quality analysis to assess the importance of on-farm erosion to downstream degradation of water quality.

To date, most erosion and sediment yield prediction techniques have been empirical, but fundamentally based equations and models are being developed that will significantly improve prediction accuracy and will apply to a broad range of conditions.

11–11 APPENDIX

11–11.1 Basic Mechanics of Erosion by Water

Concepts of basic erosion mechanics were understood by the 1930's (Cook, 1936). Although equations describing these concepts had been formulated by the 1940's (Ellison, 1947), empiricism rather than theory yielded the universal soil loss equation (USLE) (Wischmeier and Smith, 1978). More recently, applied erosion models based on fundamental concepts and theory have been developed (Foster, 1982, Foster et al., 1980a). An understanding of basic erosion mechanics is useful for application of fundamental models as well as the USLE and extension of the USLE to new situations where data are limited.

11–11.2 Basic Principles

Erosion of soil by water is a process of detachment of particles (sediment) from the soil mass by the erosive agents rainfall and runoff. Detachment, the removal of sediment from the soil mass at a point, adds particles to the sediment load transported downslope by runoff. Deposition, the reverse of detachment, is the removal of soil particles from the sediment load in the runoff and their addition to the soil mass at a location remote from where the sediment originated. Deposition occurs where the transport capacity of the runoff is less than the sediment load in the runoff (Foster and Meyer, 1977).

For illustration, consider a concave slope. On the steep upper end of the slope, transport capacity often exceeds the sediment load (Fig. 11–3). Downslope, sediment load has increased due to detachment. Transport capacity tends to increase as runoff accumulates along the slope and tends to decrease with the steady decrease in steepness along the slope. The net effect is that transport capacity increases to a point part way down the slope and then decreases. If the slope becomes flat enough, transport capacity will become less than the sediment load, and deposition will begin.

Runoff as overland flow on many soils tends to concentrate in many small channels. Erosion occurring in these channels is defined as rill erosion (Fig. 11–4). Erosion occurring on areas between the rills is defined as interrill erosion (Foster and Meyer, 1977). Flow in rills transports the sediment downslope, and flow on interrill areas moves sediment laterally toward the rills. Interrill flow is generally along the microslopes, whereas flow in rills is along the macroslope.

As previously mentioned, erosion is dynamic, it varies with time within a storm and varies from storm to storm. Since erosion dynamics are complex, we have assumed a quasi-steady state to simplify this analysis. The main governing equation is continuity expressed by (Foster and Meyer, 1975):

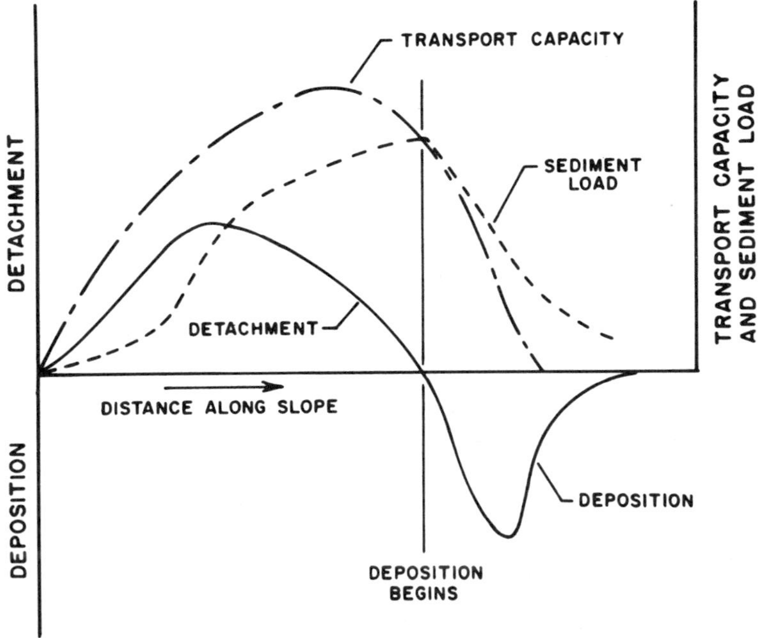

Fig. 11-3. Detachment, transport, and deposition processes along a concave slope that flattens enough to cause deposition.

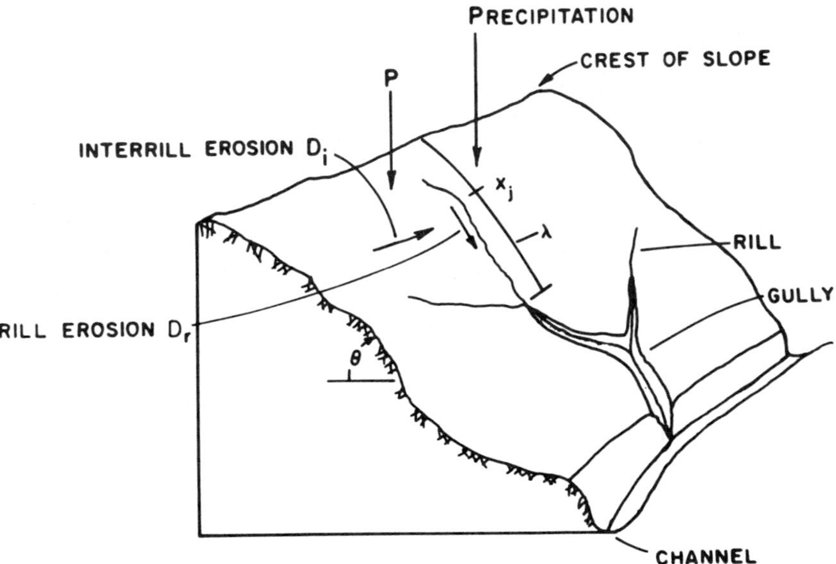

Fig. 11-4. Erosion on interrill areas, rills, gullies, and channels.

$$dg/dx = D_f + D_i \qquad [6]$$

where g = sediment load (mass per unit width per unit time), D_f = rate of detachment or deposition by flow in rills (mass per unit area per unit time), D_i = rate of delivery of sediment to rills by interrill erosion (mass per unit area per unit time), and x = distance downslope. Equation [6] is integrated to give sediment load at the location λ as

$$g = \int_0^{\lambda} (D_f + D_i)dx. \qquad [7]$$

The next step is to write relationships for D_f and D_i.

11–11.3 Relationships for Detachment and Deposition

Sediment delivery after detachment from interrill areas is closely related to raindrop impact and may be written as (Foster et al., 1977)

$$D_i = K_i\, i^2\, (as + b)\phi_i \qquad [8]$$

where D_i = rate of delivery of sediment from interrill areas, K_i = soil erodibility factor for interrill erosion, i = rainfall intensity, s = sine of slope angle θ, a and b = coefficients, and ϕ_i = soil loss ratio for interrill erosion. Soil loss ratio expresses the effect of such factors as vegetative canopy, surface cover in direct contact with the soil, tillage, residue incorporation, roughness, and previous land use. Soil loss ratio is the ratio of soil loss with the given condition to that of a base or unit condition where the soil loss ratio is defined as 1.0. Clean tilled, continuous fallow soil is the unit condition used here and in the USLE (Wischmeier and Smith, 1978).

Detachment by rill erosion, which is closely related to flow hydraulics, may be written as (Foster et al., 1977)

$$D_r = 2K_r\, \sigma\, x\, (cs^2)\phi_r \qquad [9]$$

where D_r = rate of detachment by rill erosion, K_r = soil erodibility factor for rill erosion, σ = rate of runoff per unit area, c = a coefficient, and ϕ_r = a soil loss ratio for rill erosion. Equation [9], based on the concept that detachment by rill erosion is a function of shear stress of the runoff as it flows over the soil surface, was derived using hydraulic relationships for shear stress as a function of flow depth and slope steepness (Foster et al., 1977).

Rate of deposition is a function of the flow hydraulics, the diameter and density of sediment particles, and the sediment load. It is expressed by (Foster and Meyer, 1975) as

$$D_d = \alpha\, (T_c - g) \qquad [10]$$

where D_d = rate of deposition and T_c = transport capacity of the flow. The deposition coefficient α for overland flow is given by (Foster et al., 1980b) as

$$\alpha = 0.5\, V_f/\sigma x \qquad [11]$$

where V_f = fall velocity of sediment particles. The coefficient α is large for large and dense particles, expressing the rapid deposition of these particles when transport capacity becomes less than sediment load. The coefficient α is small for small and light particles expressing their slow rate of deposition and distant travel before deposition once transport capacity becomes less than sediment load.

11-11.4 Derivation of Erosion Equations

Erosion equations can be derived for uniform slopes where the independent variables can be assumed to be constant along the slope.

11-11.4.1 Detachment Along Entire Slope

The equation for this case is obtained by substitution of Eq. [8] and [9] into Eq. [7] to give

$$g = i^2 K_i(as + b)\phi_i \lambda + \sigma K_r(cs^2)\phi_r \lambda^2. \qquad [12]$$

The average soil loss A for the slope length is

$$A = g/\lambda \qquad [13]$$

or

$$A = i^2 K_i (as + b)\phi_i + \sigma K_r (cs^2)\phi_r\lambda. \qquad [14]$$

Erosion equations are usually written for soil loss A for the storm. Equation [14] is intergrated with respect to time to obtain (Foster et al., 1977)

$$A = E_i K_i(as + b)\phi_i + E_r K_r(cs^2)\phi_r\lambda \qquad [15]$$

where E_i and E_r = interrill and rill erosivity factors, respectively.

11-11.4.2 Deposition Along Entire Slope

One example of deposition along the length of a uniform slope is when interrill erosion on row side slopes exceeds the transport capacity of runoff in the row middles. Another example is deposition of sediment eroded from soil projecting above the flow in depressions of a rough surface. Deposition rate along the slope is given by (Foster et al., 1980b) as

$$D = [\psi/(1 + \psi)](dT_c/dx - D_i) \qquad [16]$$

where dT_c/dx = the change of transport capacity with respect to distance and may be taken as a constant for uniform slopes and

$$\psi = 0.5 V_f/\sigma. \qquad [17]$$

The sediment load at λ is

$$g = (\psi dT_c/dx - D_i)[\lambda/(1 + \psi)], \qquad [18]$$

and soil loss is

$$A = (\psi dT_c/dx + D_i)[\lambda/(1 + \psi)]. \qquad [19]$$

For a storm, T_c, D_i, and ψ are averages for the storm. The interpretation of Eq. [19] is as follows. If ψ is large, as for particles that are readily deposited, soil loss is approximately equal to dT_c/dx, the change in transport capacity with distance, and sediment load is approximately equal to transport capacity at the end of the slope. If ψ is small, as for particles that are not readily deposited, soil loss is approximately equal to the interrill erosion rate D_i, with sediment load approximately equal to $D_i\lambda$.

Since the equations describing deposition are nonlinear, a typical diameter and density for the sediment cannot be assumed. The sediment distribution is broken into classes, and the governing equations are applied to each class separately. However, interaction between the classes is required in the computation of transport capacity (Foster, 1982).

11-11.5 Universal Soil Loss Equation

The USLE is given by Foster and Wischmeier (1974) as

$$A_{st} = R_{st} K (\lambda/\lambda_u)^m [65.41 \sin^2 \theta + 4.56 \sin \theta + 0.0654]\phi \qquad [20]$$

where A_{st} = soil loss for a storm, R_{st} = storm erosivity factor, K = soil erodibility factor, m = slope length exponent, λ_u = length of unit plot (22.1 m), θ = angle of the slope, and ϕ = soil loss ratio. The USLE in Eq. [20] is on a storm basis, which is its most basic form. In practice, the USLE is intended to estimate average soil loss for many storm occurrences over an extended time, e.g., a crop stage or a year. The average value for erosivity R for the period, the average sum of R_{st}, is used in such calculations. Soil loss ratio is replaced by a cover management factor C, which is soil loss ratio weighted according to the interaction of nonuniform distributions of R_{st} and ϕ.

11-11.5.1 Theoretical Interpretation of USLE Factors

11-11.5.1.1 Equation Form. The USLE, like most empirical erosion equations, is of the form

$$A \propto \lambda^m. \qquad [21]$$

Equations [20] and [21], as approximations to Eq. [14], lump rill and interrill erosion together. Equations [20] and [21] estimate no soil loss at a zero slope length, whereas Eq. [14] gives some soil loss, which can be quite large. There is a limit to the shortness of a slope length to which the USLE applies. The USLE may not apply to slopes shorter than 5 m (Foster et al., 1981b) and certainly does not apply to slopes as short as 1 m, such as row side slopes. Soil loss at a zero slope length is from interrill erosion, which is assumed to be constant over a uniform slope.

If Eq. [20] and [21] are fitted to Eq. [14], the exponent m varies for different conditions. The exponent equals (Foster et al., 1977)

$$m = 1/(1 + \beta) \qquad [22]$$

where

$$\beta = [E_i K_i (as + b)\phi_i]/[E_r K_r (cs^2)\phi_r \lambda]. \qquad [23]$$

In this equation, the numerator is soil loss caused by interrill erosion, and the denominator is soil loss caused by rill erosion. Equations [22] and [23] show that the slope length exponent increases toward a maximum of 1 as rill erosion increases. This qualitatively agrees with Wischmeier and Smith's (1978) recommendation for an increase in m for slopes up to 5% such as was used in Eq. [5]. Other factors affecting the slope length exponent are the rate and amount of runoff relative to rainfall, interrill erodibility relative to rill erodibility, interrill soil loss ratio relative to that for rill erosion, and slope length (Meyer et al., 1975). Experimental data have indicated variability in m but have not been adequate to define the relationship (Wischmeier et al., 1958).

11-11.5.1.2 Irregular Slopes. Theory permits application of the USLE to irregular slopes where deposition does not occur and runoff rate per unit area is relatively uniform along the slope (Foster and Wischmeier, 1974). Detachment at a point, as described by the USLE, is

$$D = (1 + m)\, R\, K\, (x/\lambda_u)^m\, S\, C\, P \quad [24]$$

where S = the USLE slope factor given by

$$S = (65.4s^2 + 4.56s + 0.0654). \quad [25]$$

The factors K, S, C, and P are based on conditions at x. Note that erosion rate at λ is $(1 + m)$ times the soil loss for the slope length λ.

Irregular slopes may be analyzed by dividing the slope into segments so that the uniformity of K, S, C, and P is maintained with a segment but variability from segment to segment can be asumed. By Eq. [7], the sediment ΔG_j produced on a segment is

$$\Delta G_j = \int_{x_{j-1}}^{x_j} D\, dx \quad [26]$$

or

$$\Delta G_j = (R/\lambda_u{}^m) K_j\, S_j\, C_j\, P_j (x_j^{m+1} - x_{j-1}^{m+1}) \quad [27]$$

where Δx_j = the segment length.

The total sediment produced on the slope composed of k segments is

$$G = \sum_{j=1}^{k} \Delta G_j. \quad [28]$$

The average soil loss for the entire slope is given by

$$A = G/\lambda. \quad [29]$$

If the slope segments are of equal length,

$$x_j = j\lambda/k \quad [30]$$

average soil loss for the slope is

$$A = R \sum_{j=1}^{k} (\lambda/\lambda_u)^m\, S_j\, K_j\, C_j\, P_j\, \omega_j \quad [31]$$

where

$$\omega_j = (j/k)^{m+1} - [(j - 1)/k]^{m+1}. \quad [32]$$

Equations [31] and [32] are the algorithm given by Wischmeier and Smith (1978) for analyzing irregular slopes. Note that $(\lambda/\lambda_u)^m\, S_j$ is the LS factor for the entire slope length using the slope of the jth segment. Average soil loss on the jth segment is given by

$$A_j = R\,K_j\,C_j\,P_j\,(\lambda/\lambda_u)^m\,S_j\,\omega_j\,k. \qquad [33]$$

Average soil loss A_j on a particular segment may be several times the average for the entire slope.

11–11.5.1.3 Erosivity. Since the USLE lumps interrill and rill erosion together, the USLE factor for rainfall erosivity, R_{st}, represents rainfall erosivity for both interrill and rill areas.

Erosivity R_{st} is

$$R_{st} = EI \qquad [34]$$

where E = storm energy and I = maximum 30-min intensity. Total storm energy is obtained by integrating energy concentration e over the total rainfall duration as

$$E = \int_o^T e\,i\,dt \qquad [35]$$

where T = duration of rainfall, i = rainfall rate, and dt = differential of time. In application, Eq. [35] is usually written in discrete form as

$$E = \sum_{k=1}^{n} e_k\,\Delta V \qquad [36]$$

where ΔV = an incremental depth of rainfall, and e_k = rainfall energy per unit of rainfall for the kth increment of rain and is given by

$$e_k = 0.119 + 0.0873\,\log_{10} i_k \qquad [37]$$

where e = unit rainfall energy in megajoules per hectare per millimeter [MJ(ha mm)$^{-1}$] of rain, and i_k = intensity in millimeters per hour for the kth increment. Since e varies little with i, especially above 100 mm/h,

$$E = a_e\,V\,i_{ch} \qquad [38]$$

where a_e = a coefficient, V = depth of rainfall, and i_{ch} = a characteristic rainfall rate. Assuming that maximum 30-min intensity I is a good measure of i_{ch} gives

$$EI = a_e\,T\,I^2 \qquad [39]$$

where T is a rainfall duration defined by $T = V/I$. Note that I^2 is similar to i^2 in Eq. [8] for interrill erosion. The i_{ch} term gives a rate of interrill erosion per unit of rainfall, which, multiplied by volume of rainfall, gives total interrill erosion for the storm.

For runoff, characteristic runoff rate (σ_{ch}), is related to I by $\sigma_{ch} = I - f_{ch}$ where f_{ch} is a characteristic infiltration rate. Likewise, depth of runoff is related to depth of rainfall. The I term can be viewed as giving a rate of rill erosion per unit of runoff for a characteristic runoff rate, which, multiplied by the depth of rainfall corresponding to the depth of runoff, gives total detachment for rill erosion for the storm. The factor EI, although originally derived empirically (Wischmeier and Smith, 1958), is a simple variable for erosivity that contains the important variables that should be in an erosivity factor. These variables are the intensity and volume of rainfall and the rate and volume of runoff.

11-11.6 Fundamental Models

Detailed models based on fundamental concepts of erosion mechanics and sediment transport are being developed that give the user power beyond the USLE (Foster, 1982). However, they typically require a computer and may require considerable effort to assemble input data. Also, some require measured data to calibrate model parameters. These models calculate erosion and sediment yield on a storm-by-storm basis, which is necessary where both detachment and deposition processes are important. Analysis of complex, spatially varied areas requires detailed models of this type. Also, these models more accurately consider hydrologic processes than simple relationships like the USLE. They will be used more and more in the future.

11-12 LITERATURE CITED

Cook, H. L. 1936. The nature and controlling variables of the water erosion process. Soil Sci. Soc. Am. Proc. 1:487–494.

Dissmeyer, G. E., and G. R. Foster. 1981. Estimating the cover-management factor (C) in the universal soil loss equation for forest conditions. J. Soil Water Conserv. 36:235–240.

Ellison, W. D. 1947. Soil erosion studies. Agric. Eng. 28:145–146, 197–201, 245–248, 297–300, 349–351, 402–405, 442–444.

Foster, G. R. 1982. Modeling the erosion process. p. 297–380. *In* C. T. Haan (ed.) Hydrologic modeling of small watersheds. Am. Soc. of Agric. Engineers, St. Joseph, Mich.

Foster, G. R., and V. A. Ferreira. 1981. Deposition in uniform grade terrace channels. p. 185–197. *In* Crop production with conservation in the 80's. Am. Soc. of Agric. Engineers, St. Joseph, Mich.

Foster, G. R., L. J. Lane, and J. D. Nowlin. 1980a. A model to estimate sediment yield from field sized areas: Selection of parameter values. p. 193–281. *In* CREAMS—a field scale model for chemicals, runoff, and erosion from agricultural management systems, Vol. II: User manual. USDA-SEA Conserv. Res. Report no. 26.

Foster, G. R., L. J. Lane, J. D. Nowlin, J. M. Laflen, and R. A. Young. 1980b. A model to estimate sediment yield from field sized areas: Development of model. p. 36–64. *In* CREAMS—a field scale model for chemicals, runoff, and erosion from agricultural management systems, Vol. I: Model documentation. USDA-SEA Conserv. Res. Report no. 26.

Foster, G. R., D. K. McCool, K. G. Renard, and W. C. Moldenhauer. 1981a. Conversion of the universal soil loss equation to SI metric units. J. Soil Water Conserv. 36:355–359.

Foster, G. R., and L. D. Meyer. 1975. Mathematical simulation of upland erosion by fundamental erosion mechanics. p. 190–207. *In* Present and prospective technology for predicting sediment yield and sources. USDA Report no. ARS-S-40.

Foster, G. R., and L. D. Meyer. 1977. Soil erosion and sedimentation by water—an overview. p. 1–13. *In* Proc. Natl. Symp. on Soil Erosion and Sedimentation by Water, Chicago 12–13 Dec. Am. Soc. of Agric. Engineers. Pub. no. 4-77, St. Joseph, Mich.

Foster, G. R., L. D. Meyer, and C. A. Onstad. 1977. A runoff erosivity factor and variable slope length exponents for soil loss estimates. Trans. ASAE 20:683–687.

Foster, G. R., J. R. Simanton, K. G. Renard, L. J. Lane, and H. B. Osborn. 1981b. Discussion of "Application of the universal soil loss equation to rangelands on a per-storm basis," by Trieste and Gifford in Journal of Range Management 33:66–70, 1980. J. Range Manage. 34:161–164.

Foster, G. R., and W. H. Wischmeier. 1974. Evaluating irregular slopes for soil loss prediction. Trans. ASAE 17:305–309.

Grissinger, E. H. 1966. Resistance of selected clay systems to erosion by water. Water Resour. Res. 2:131–138.

Knisel, W. G. (ed.) 1980. CREAMS—a field scale model for chemicals, runoff, and erosion from agricultural management systems. USDA-SEA Conserv. Res. Report no. 26.

Laflen, J. M., H. P. Johnson, and R. O. Hartwig. 1978. Sedimentation modeling of impoundment terraces. Trans. ASAE 21:1131–1135.

Langbein, W. B., and S. A. Schumm. 1958. Yield of sediment in relation to mean annual precipitation. Trans. Am. Geophys. Union 39:1076–1084.

Meyer, L. C., G. R. Foster, and M. J. M. Romkens. 1975. Source of soil eroded by water from upland slopes. p. 177–189. *In* Present and prospective technology for predicting sediment yield and sources. USDA Report no. ARS-S-40.

Mutchler, C. K., and C. E. Murphree. 1980. Modeling erosion and sediment yield from flatland watersheds. CREAMS—a field scale model for chemicals, runoff, and erosion from agricultural management systems, Vol. III. Supporting documentation. USDA-SEA Conserv. Res. Report no. 26.

Mutchler, C. K., and R. A. Young. 1975. Soil detachment by raindrops. p. 113–117. *In* Present and prospective technology for predicting sediment yields and sources. USDA Report no. ARS-S-40.

Neibling, W. H., and G. R. Foster. 1977. Estimating deposition and sediment yield from overland flow processes. p. 75–86. *In* D. T. Kao (ed.) Symposium on urban hydrology and sediment control. Proc. Int. Symp. on Urban Hydrology, Hydraulics, and Sediment Control, 18–21 July. Lexington, Ky. Univ. of Kentucky, Lexington.

Partheniades, E. 1972. Results of recent investigations on erosion and deposition of cohesive sediments. p. 1–39 (Chapt. 20). *In* H. W. Shen (ed.) Sedimentation (Einstein). Published by editor, Fort Collins, Colo.

Renard, K. G., J. R. Simanton, and H. B. Osborn. 1974. Applicability of the universal soil loss equation to semiarid rangeland conditions. p. 18–32. *In* Hydrology and water resources in Arizona and the Southwest, Vol. 4. Univ. of Arizona, Tucson.

Roehl, J. W. 1962. Sediment source areas, delivery ratios, and influencing morphological factors. Int. Assoc. of Scientific Hydrology Pub. no. 59, p. 202–213.

Soil Conservation Service. 1972. Hydrology. Natl. Engineering Handbook, Section 4, USDA-SCS, U.S. Government Printing Office, Washington, D.C.

Ward, A. D., C. T. Haan, and B. J. Barfield. 1977. Simulation of the sedimentology of sediment basins. Tech. Report no. 103, Univ. of Kentucky, Water Resources Inst., Lexington.

Williams, J. R. 1975. Sediment-yield prediction with universal equation using runoff energy factor. p. 244–252. *In* Present and prospective technology for predicting sediment yields and sources. USDA Report no. ARS-S-40.

Wischmeier, W. H. 1975. Estimating the soil loss equation's cover and management factor for undisturbed areas. p. 118–124. *In* Present and prospective technology for predicting sediment yields and sources. USDA Report no. ARS-S-40.

Wischmeier, W. H. 1977. Soil erodibility by rainfall and runoff. p. 45–56. *In* T. J. Toy (ed.) Erosion: Research techniques, erodibility and sediment delivery. GEO Abstracts, Ltd., Norwich, England.

Wischmeier, W. H., C. B. Johnson, and B. V. Cross. 1971. A soil erodibility monograph for farmland construction sites. J. Soil Water Conserv. 26:189–193.

Wischmeier, W. H., and D. D. Smith. 1958. Rainfall energy and its relationship to soil loss. Trans. Am. Geophys. Union 39:285–291.

Wischmeier, W. H., and D. D. Smith. 1965. Predicting rainfall erosion losses from cropland east of the Rocky Mountains. USDA Agric. Handbook no. 282.

Wischmeier, W. H., and D. D. Smith. 1978. Predicting erosion losses: A guide to conservation planning. USDA Agric. Handbook no. 537.

Wischmeier, W. H., D. D. Smith, and R. E. Uhland. 1958. Evaluation of factors in the soil loss equation. Agric. Eng. 39:458–462, 474.

12

Leon Lyles, L. J. Hagen, and E. L. Skidmore
Agricultural Research Service
U.S. Department of Agriculture
Manhattan, Kansas

Soil Conservation: Principles of Erosion by Wind

Interest in the fluid transport of solids—in this special case, wind erosion—continues not only because erosion affects soil productivity and damages plants but also because wind erosion is an important contributor of atmospheric aerosols.

The most comprehensive summaries on the movement of surface material by wind action have been done by Bagnold (1941) for desert sands and Chepil and Woodruff (1963) for agricultural lands. Control strategies have been discussed by Woodruff et al. (1977).

12-1 MEAN VELOCITY PROFILES

The fluid flows of interest to the agricultural community—those sufficient to cause wind erosion—are classified as turbulent boundary layer flows over relatively rough surfaces. For such flows over stable (noneroding) surfaces in the "constant stress layer" (the lower 10 to 20% of the boundary layer depth), the following form of the logarithmic law is often used to describe the mean velocity profile:

$$\bar{u}_z = \frac{u_*}{k}\left[\ln\left(\frac{Z - D}{Z_0}\right) + \phi_{(z)}\right] \qquad [1]$$

where $\bar{u}_z$ is the mean velocity at height z from some reference plane; u_* is the friction velocity defined as $(\tau_0/\varrho)^{1/2}$, where τ_0 is the shear stress at the surface and ϱ is air density; k is von Karman's constant (0.4); D is a zero

Contribution from Agricultural Research Service, U.S. Department of Agriculture, in cooperation with the Kansas Agricultural Experiment Station, Department of Agronomy Contribution 79-230-A.

Dryland Agriculture—Agronomy Monograph no. 23.

plane displacement; Z_0 is a roughness parameter; and $\phi_{(z)}$ is the integral diabatic influence function (Stearns, 1970). For most wind tunnel flows and winds of high velocity and turbulence, we can safely assume that near the surface $\phi_{(z)} = 0$ (the function $\phi_{(z)}$ is zero for adiabatic conditions).

For flows over relatively smooth—yet aerodynamically rough—stable surfaces, D is omitted because it becomes very small relative to Z and $Z_0 = 1/30\, d_p$ (Nikuradse, 1950), where d_p is particle diameter, so that Eq. [1] becomes

$$\frac{\bar{u}_z}{u_*} = \frac{1}{k} \ln\left(\frac{Z}{d_p}\right) + 8.5, \qquad [2]$$

which is also found in the literature.

Particles in transport near the surface alter the mean velocity profile. Based on work by Bagnold (1941) and Andres (1970), White and Schulz (1977) gave the following form of the profile equation for particle-laden flows:

$$\frac{\bar{u}_z}{u_*} = \frac{1}{k} \ln\left(\frac{Z}{d_p}\right) - 2.29 + 10.79\frac{u_{*t}}{u_*} \qquad [3]$$

where u_{*t} is the threshold friction velocity. If u_* approaches u_{*t} (the stable surface case), Eq. [3] becomes Eq. [2].

12-2 PARTICLE DYNAMICS

Most published information concerning the movement of particles by wind distinguishes between three principal types of particle movement: suspension, saltation, and surface creep. However, the way the first particles are moved has received less attention. Before 1962, most writers were satisfied by Bagnold's (1941) description of particles rolling along the surface by direct wind pressure for about 30 cm before starting to bounce off the ground. Bisal and Nielsen (1962) concluded, after observing particles in a shallow pan mounted on the viewing stage of a binocular microscope, that most erodible particles vibrated (oscillated) with increasing intensity as wind speed increased and then left the surface instantly (as if ejected). Lyles and Krauss (1971), from wind tunnel observations, reported that as mean wind speed approached the threshold value, some particles (0.59 to 0.84 mm in diameter) began to rock back and forth (oscillate) and hypothesized that the oscillation frequency was related to the frequency band containing the maximum energy of the turbulent motion. Although average peak frequency of the longitudinal energy spectra and particle oscillation frequency were of same order of magnitude, more comprehensive research on particle oscillation is needed. Interestingly, Azizov (1977), who was studying soil moisture effects on wind erosion, noted a critical wind speed at which individual particles began oscillatory forward motions to a distance less than or equal to their diameters.

12-2.1 Suspension

Suspension of particulates by wind erosion often causes loss of productivity on the eroding field (Lyles, 1975), transport causes visibility and health hazards, and deposition causes chemical and sediment pollution. The suspended particles can range in size from about 2 to 100 μm, with mass median diameter of about 50 μm in an eroding field (Chepil, 1957a; Gillette and Walker, 1977). However, in long-distance transport, particles less than 20 μm in diameter predominate, because the larger particles have significant sedimentation velocities (Gillette, 1977). Some suspension-size particles are present in the soil, but many are created by abrasive breakdown during erosion. Some of the smallest are carried in crevices on the larger particles and are shed on impact (Rosinski et al., 1976).

Vertical flux (F_v) of suspension-size particles is often described by

$$F_v = F_0(u_*/u_{*0})^P \qquad [4]$$

where u_* is friction velocity, F_0 is a reference flux at a reference friction velocity u_{*0}, and P is a power that can range from about 2 to 6 (Englemann and Sehmel, 1976). Some evidence shows that P is about 3 on sandy soils (i.e., vertical flux increases at the same rate as horizontal), but P may be larger on finer-textured soils (Gillette, 1977). The value of F_0 is also uncertain. Chepil (1945) reported that 3 to 38% of the eroding soil could be carried in suspension, depending on soil texture. Generally, the vertical flux is less than 10% of the horizontal (Gillette, 1977, 1978).

Concentration profiles of suspended particulates can be easily predicted only for equilibrium conditions. When net vertical flux is zero,

$$F_v = -V_s\chi(Z) - K(Z)\,\partial\chi/\partial Z = 0 \qquad [5]$$

where V_s is settling velocity and χ is concentration. If particulate eddy diffusivity (K) is equal to that for momentum, $K = u_*kZ$, where $k \cong 0.4$ and Z is height above the surface, the solution of Eq. [5] is

$$\chi = \chi_1\,(Z/Z_1)^{-V_s/u_*k}. \qquad [6]$$

Measurements during soil erosion confirm that these concentration profiles frequently occur (Chepil and Woodruff, 1957).

Widespread wind erosion causes dust storms, whose climatology in the contiguous USA has been studied by Orgill and Sehmel (1976). Highest dust frequency (with visibility less than 11 km) occurs in the Southern Great Plains, where most of the area is affected by dust 1% of the time. A small part of the area is affected by dust 3% of the time. Secondary dust frequency maxima occur in the western states, Northern Great Plains, Southern Coastal Pacific and inland valleys, and the Southeast. Highest frequencies occur in the afternoon between 1200 and 2000 local standard time. In the Great Plains, Hagen and Woodruff (1973) found the average dust storm to last 6.6 h and estimated the median dust concentration to be 4.83 mg/m^3 at observer height. Dusty hours vary widely from year to year, but the 2-year mean precipitation was correlated (r^2 = 0.76) with the annual hours of dust in the Great Plains. Pollard (1977), who investigated the meteorology of Southern Great Plains dust storms, found that wind speed and direction were correlated with their occurrence but that antecedent moisture was not.

From visibility and wind speed data, approximate calculations showed that in the Great Plains during the 1950's, 221 million t of dust was suspended annually and in the 1960's, 70 million t (Hagen and Woodruff, 1975). These calculations correlated well with dust depositions measured by Smith et al. (1970). These results suggest that particulate suspension from wind erosion exceeds that from all other sources (both natural and artificial) —the latter source accounts for about 31.9 million t annually in the USA (Walther, 1972).

The primary ambient air quality standard for total suspended particulates is 75 μg/m^2 (annual geometric mean) and 260 μg/m^3 (maximum 24-hour average) (*Federal Register,* 1971). Because of the high concentration of coarse particles in dust storms, rural areas often fail to meet this standard (USEPA, 1976), and several investigators (e.g., Corn, 1971; Husar, 1976) have suggested that they be changed to reflect the health hazards posed by the size and chemical composition of the particulates.

Suspended particulates are best reduced by good wind erosion control practices. Tall vegetation can sometimes be used to trap particulates already suspended (Hagen and Skidmore, 1977; Smith, 1977).

12-2.2 Saltation

The characteristics of saltation (jumping) particles in wind have been described by several research workers (Free, 1911; Bagnold, 1941; Chepil, 1945; Zingg, 1953). However, theoretical understanding of the saltation process is at present incomplete (White and Schulz, 1977). In saltation, individual particles lift off the surface (eject) and follow distinctive trajectories under the influence of air resistance and gravity. Such particles (100 to 500 μm) rise at fairly steep angles but are too large to be suspended by the flow; they return to the surface either to rebound or to embed themselves and initiate movement of other particles. The bulk of total transport, roughly 50 to 80%, is by saltation. Saltating particles rise less than 120 cm; most rise less than 30 cm.

Chepil (1945) reported lift-off angles of 75 to 90° and impact angles of 6 to 12° from the horizontal. White and Schulz (1977), using microbeads to study particle trajectories in a wind tunnel, reported an average lift-off angle of 50 ± 20° for 57 filmed observations and average impact angle of 14 ± 3° for 43 observations.

Many investigations have shown the importance of lifting forces in saltation (White and Schulz, 1977). Because of the large differences between densities of air and sand (soil) particles, Owen (1964) suggested that the lift force due to interaction between particle motion and environmental vorticity does not play a significant part in determining the motion of a particle (see Tennekes and Lumley, 1972, for the role of vorticity in turbulence dynamics). He did note that a lift force could result from the spin of a particle. White and Schulz (1977) have shown that considering the lift associated with particle rotation—called the Magnus effect—greatly improves the agreement between theoretical trajectories calculated from the equations of particle motion and those filmed in a wind tunnel. From photographs of saltation flow, Chepil (1954) found appreciable particle rotation of 200 to 1000 revolutions/s. From changes in light reflection of microbeads on film, White and Schulz (1977) estimated spinning rates of 115 to 500 revolutions/s.

Chepil (1945) reported that ratios of height of rise (h) to length of path (L) for saltating soil particles were about 1:10 for all the agricultural soils studied. From solutions of the equations of particle motion, White and Schulz (1977) indicated that h/L is a function of particle diameter (at constant u_*), with larger ratios associated with the smaller diameters in the saltation size range and decreasing to approach those of Chepil at 500 μm.

Finally, Owen (1964) stated that Bagnold deduced that lift-off velocities ($\bar{u}_L$) of saltating particles were comparable with u_*. White and Schulz (1977) reported $\bar{u}_L = 69.3$ cm/s when $u_* = 39.6$ cm/s, although about 18% of the particles examined from filmed trajectories had velocities between 20 and 40 cm/s. Thus, Bagnold's deduction is not unrealistic based on limited data.

12–2.3 Surface Creep

Mineral soil (sand) particles 500 to 1000 μm in diameter, too large to leave the surface in ordinary erosive winds, are set in motion by the impacts of saltating particles. In high winds, the whole sand surface appears to be creeping slowly forward at speeds $\ll 2.5$ cm/s—pushed and rolled (driven) by the saltation flow. The rippling of windblown sand has been attributed to unevenness in surface creep flow (Bagnold, 1941). Reportedly, surface creep constitutes 7 to 25% of total transport (Bagnold, 1941; Chepil, 1945; Horikawa and Shen, 1960). Factors affecting the proportion of total transport in surface creep seem to be wind speed, particle size distribution, and surface geometry (roughness).

12–3 THRESHOLD CONDITIONS

Bagnold (1941) used an experimental coefficient, A, to describe the threshold friction velocity, u_{*t} (the minimum at which the flow has sufficient energy to initiate particle movement). The equation is

$$u_{*t} = A(\alpha g\, d_p)^{1/2} \qquad [7]$$

where α is immersed density ratio, $(\varrho_p - \varrho)/\varrho$, g is the gravitational acceleration, d_p is particle diameter, ϱ_p is particle density, and ϱ is air density. In air, A has a value of 0.08 to 0.12, perhaps as great as 0.2 without saltation flow. Iversen et al. (1973) and Wood et al. (1974) noted that A is a function of the particle friction Reynolds number $R_f = u_* d_p/\nu$, where ν is kinematic viscosity of air. In Wood's summary of previous research, A ranged from 0.08 to 0.22 for R_f greater than 0.7; however, most values were between 0.08 and 0.12. Iversen et al. (1976) extended Bagnold's equation to include interparticle forces (due to moisture, electrostatic effects, and other forces of cohesion) for estimating threshold values for small particles and in low-density atmospheres (extraterrestrial). Their estimates resulted in somewhat lower minimum values for u_{*t} in the fine particle range than those given by Eq. [7]. Those estimates and recent low air-density experiments (unpublished) also showed that the coefficient A is not a unique function of R_f for small particles, but because of interparticle forces, it is smaller for larger values of α. Some question remains concerning what value to use for d_p in materials of wide particle size range. Also, no standard method of determining u_{*t} has been used or specified. Theoretically, $u_{*t} \equiv u_{*\max}$ for $q \equiv 0$ (where q is particle flux), but it is difficult to measure experimentally (Lyles, 1977).

We suggest that threshold consideration be limited to flow and particle properties and exclude boundary geometry. There is a growing tendency to extend the threshold concept to include the presence of nonerodible roughness (elements)—thus making it dependent on many physical dimensions and geometric patterns of those elements. A clearer approach would be to partition the shear stress between erodible and nonerodible material and thus avoid confounding the threshold concept.

12-4 PARTICLE FLUX

When considering particle flux (flow rates), we must distinguish between cases in which all the erodible particles are less than 1000 μm and cases in which there are mixtures of erodible particles and nonerodible elements, e.g., aggregates or clods greater than 1000 μm and vegetative materials. Before we proceed, however, a brief review of soil avalanching (Chepil, 1957b) is appropriate.

On an unprotected eroding field, the particle flux is zero on the windward edge and increases with distance downwind until, if the field is large enough, the flux reaches the maximum amount that a given wind can sustain. This increase of particle flux with distance downwind has been called "soil avalanching." The more erodible the surface, the shorter the distance in which maximum flux is reached.

12-4.1 All Erodible Particles

The simplest case involves all erodible particles where several equations have been developed to predict soil (sand) flux from an area under specific soil and wind conditions. Most equations, empirically developed, relate mass of soil moved to surface-shear stress or friction velocity of the wind and erodibility characteristics of the soil.

The functional form of those equations is

$$q = f[(\text{soil properties})^a, (\text{flow properties})^b] \qquad [8]$$

where q is particle flux (mass per unit width per unit time) and a and b are constants. The soil properties term may include size, density, and shape; the flow properties term may include mean wind speed at some reference height ($\bar{u}_z$) or more commonly the friction velocity (u_*) for the flow in question, air density, and, rarely, a turbulence parameter. Specific flux equations for all erodible particles are contained in reports by O'Brien and Rindlaub (1936), Bagnold (1941); Kawamura (1951), Zingg (1953), Owen (1964), Kadib (1965), Makaveev (1967), and Iversen et al. (1976). The flux varies considerably among the equations because of different values found for the constants and for coefficients introduced in explicit equations. Because it contains a turbulence term, the Makaveev's (1967) equation is the only one given here:

$$q = C \varrho_p d_p \frac{\sigma_u}{\bar{u}_t}\left(\frac{\bar{u}_z}{\bar{u}_t}\right)^2 (\bar{u}_z - \bar{u}_t) \qquad [9]$$

where $\bar{u}_t$ is threshold mean wind speed at reference height z, σ_u is the root mean square (RMS) of the longitudinal velocity fluctuations (turbulence), and other terms are as previously defined. Substituting Eq. [1] into Eq. [9] and reducing give

$$q = C \varrho_p d_p \frac{\sigma_u}{u_{*t}}\left(\frac{u_*}{u_{*t}}\right)^2 (u_* - u_{*t}), \qquad [10]$$

which shows the well-known cubic relationship between friction velocity and particle flux and further supports why u_* is so often used to indicate the wind's capability to erode soil particles. Substituting the relationship $\sigma_u = C_1 u_*$ (Lumley and Panofsky, 1964) into Eq. [10] gives

$$q = C_2 \varrho_p d_p \left(\frac{u_*}{u_{*t}}\right)^3 (u_* - u_{*t}), \qquad [11]$$

which indicates particle flux is proportional to u_*^4 and illustrates the ability of u_* to characterize some turbulent properties of flow. All the previous particle flux equations assume steady or stationary *mean* flow. Fan and Disrud (1977) suggested that mean wind velocities are not generally stationary and that steady-flow flux equations may underpredict actual erosion.

12-4.2 Mixtures of Erodible Particles and Nonerodible Elements

Field soils seldom contain only erodible-size particles; mixtures of erodible and nonerodible elements are the more common and more complex case. An equation for transport of field soils is complicated by factors other than erodible particle size gradation, e.g., proportion and size of nonerodible fractions, field roughness, vegetation, and soil moisture content. No reliable equations have been developed to express field soil flux for short time periods or for single erosive windstorms.

The protective role of nonerodible elements in the erosion process has been characterized by Lyles et al. (1974) in wind tunnel studies. For example, a soil initially with buried nonerodible elements like clods is eroded by a wind of characteristic friction velocity, u_*. As the erodible material is removed, the roughness is increased (because of exposure of portions of the nonerodible material), which increases the total friction velocity. The friction velocity (or drag) may be thought of as being divided between the nonerodible elements and the erodible soil (intervening surface). Thus, as more erodible material is removed from the initially smooth, erodible surface, more drag is absorbed by nonerodible elements and less is absorbed by the erodible soil. After sufficient time, enough soil is eroded so that the intervening surface drag decreases to the threshold where erosion ceases and the soil is stabilized. Stabilizing agricultural fields by nonerodible elements is complicated by variation in speed, direction, and duration of winds plus possible generation of erodible-size particles from larger aggregates by abrasion. However, the role of nonerodible elements is clearly to absorb part of the total wind drag—reducing the drag on erodible particles.

From wind tunnel studies, Lyles and Allison (1976) published a regression equation that predicts the degree of protection provided by standing crop residues and nonerodible soil aggregates:

$$\left(\frac{u_*}{u_{*t}}\right)_s = 1.638 + 17.044\frac{NA_s}{A_t} - 0.117\frac{L_y}{L_x} + [(1.0236)^c - 1] \qquad [12]$$

where $(u_*/u_{*t})_s$ is called the critical friction velocity ratio, because erosion begins when this value is exceeded; the larger the ratio, the greater the wind erosion protection. The term u_* is the total friction velocity for a stable surface at a given free stream velocity, and u_{*t} is the threshold friction velocity for the erodible particles in question. The other parameters are N/A_t, number of stalks in area A_t; A_s, average silhouette area (projected area facing flow) of a single stalk; L_y, distance (center to center) between stalks normal to wind direction; L_x, the corresponding distance in the wind direction; and c, percentage of dry soil aggregates greater than 1.0 mm in diameter (limited to $\leq$ 50%). Research is needed to test Eq. [12] under field conditions.

12-5 PRINCIPLES AND GENERAL STRATEGIES OF CONTROL

Two general principles of wind erosion control are obvious: (i) reduce wind forces on erodible particles or (ii) create particles resistant to wind forces. From knowledge of erosion processes and mechanics, four specific principles of wind erosion control have been identified (Woodruff et al., 1977):

1. Establish and maintain vegetation or vegetative residues.
2. Produce or bring to the soil surface nonerodible aggregates or clods.
3. Reduce field width along prevailing wind erosion direction.
4. Roughen the land surface.

These principles are unchanging, but the control *practices* that grow out of them (which are discussed in subsequent sections) vary in space (from place to place) and may change over time along with cropping and management systems. Any present or proposed (future) control strategy should be evaluated on how it influences the four principles. Those four principles plus a factor for climate were used by Woodruff and Siddoway (1965) to develop the now well-known wind erosion equation that predicts potential annual erosion rates:

$$E = f(I, K, C, L, V) \qquad [13]$$

where E is the potential annual soil loss rate, I is the soil erodibility, K is the soil ridge roughness factor, C is the climatic factor, L is the unsheltered median travel distance of the wind across a field, and V is the equivalent vegetative cover. Briefly, the soil erodibility index I is the potential soil loss in $t(\text{ha y})^{-1}$ from a wide unsheltered, isolated field with a bare, smooth, non-crusted surface based on climatic conditions near Garden City, Kansas. It is related to surface soil cloddiness greater than 0.84 mm and may be determined by dry sieving and use of a table given by Woodruff and Siddoway (1965). Values range from 0 to 695 t/ha as surface aggregates greater than 0.84 mm decrease to 1%. For convenience, the U.S. Soil Conservation Service uses wind erodibility groups based primarily on soil texture as a guide for selecting I factors.

Soil ridge roughness factor K is a measure of the effects of ridges on erosion amounts relative to a smooth surface. It is determined from the height-spacing measurement of ridges caused by tillage implements and then consulting a chart given by Woodruff and Siddoway (1965). It varies from 0.5 for ridged surfaces to 1.0 for smooth surfaces.

The local wind erosion climatic factor C characterizes the erosive potential of climate (wind speed, precipitation, and air temperature) at a particular location relative to Garden City, Kansas, which has an annual value of 100% based on long-term data. Equations for computing the factor and a map giving general ranges of annual values for the western half of the USA are given by Chepil et al. (1962). Monthly climatic factors for most of the USA are presented by Skidmore and Woodruff (1968).

The equivalent field width L in the equation recognizes that the rate of soil flow increases with distance downwind across an eroding field until reaching the transport capacity of a given wind and that winds have a prevailing direction and a preponderance in the prevailing direction. It may be determined from field size and orientation and data on prevailing wind direction and preponderance in the prevailing wind erosion direction given by Skidmore and Woodruff (1968).

The vegetative cover factor V considers the quantity, kind, and orientation of vegetation or vegetative residues on erosion amounts. All vegetative materials must be converted to an equivalent small-grain standard before use in the equation. The standard is absolutely flat, small-grain stubble with straw aligned parallel with wind direction on smooth ground and in rows 25 cm apart at right angles to wind direction. Data for converting various crops to their small-grain equivalent are given by Craig and Turelle (1964), Woodruff and Siddoway (1965), and Lyles and Allison (1980, 1981).

Relations among variables in the equation are complex, and a single equation that expresses E as a function of the dependent variables has not been devised. The equation was initially solved in a stepwise procedure involving tables and graphs (Woodruff and Siddoway, 1965; Skidmore and Woodruff, 1968), but its solution has been computerized (Skidmore et al., 1970).

The equation was developed to estimate the potential erosion from a particular field or to estimate the field conditions necessary to reduce potential erosion to tolerable amounts and has been used extensively for design and evaluation of control systems (practices).

12-6 LITERATURE CITED

Andres, R. M. 1970. The mechanics of dust lifting with particular emphasis on the planet Mars. Ph.D. Dissertation, Library, Saint Louis Univ., St. Louis, Mo.

Azizov, M. T. 1977. Influence of soil moisture on the resistance of soil to wind erosion. (In Russian.) Pochvovedenie 1:102–105.

Bagnold, R. A. 1941. The physics of blown sand and desert dunes. Methuen, London.

Bisal, F., and K. F. Nielsen. 1962. Movement of soil particles in saltation. Can. J. Soil Sci. 42: 81–86.

Chepil, W. S. 1945. Dynamics of wind erosion: I. Nature of movement of soil by wind. Soil Sci. 60:305–320.

Chepil, W. S. 1957a. Sedimentary characteristics of duststorms: III. Composition of suspended dust. Am. J. Sci. 255:206–213.

Chepil, W. S. 1957b. Width of field strips to control wind erosion. Kansas Agric. Exp. Stn. Tech. Bull. no. 92.

Chepil, W. S., F. H. Siddoway, and D. V. Armbrust. 1962. Climatic factor for estimating wind erodibility of farm fields. J. Soil Water Conserv. 17:162–165.

Chepil, W. S., and N. P. Woodruff. 1957. Sedimentary characteristics of duststorms: II. Visibility and dust concentration. Am. J. Sci. 225:104–114.

Chepil, W. S., and N. P. Woodruff. 1963. The physics of wind erosion and its control. Adv. Agron. 15:211–302.

Corn, M. 1971. Particle size: Relationship to collector performance, emission standards, and ambient air quality. p. 965–969. *In* H. M. England and W. T. Beery (ed.) Proc. Second Int. Clean Air Congr., Washington, D.C. Academic Press, Inc., New York.

Craig, D. G., and J. W. Turelle. 1964. Guide for wind erosion control on cropland in the Great Plains States. U.S. Dep. of Agric., Soil Conserv. Service, U.S. Government Printing Office, Washington, D.C.

Englemann, R. J., and G. A. Sehmel (ed.) 1976. Proceedings of the conference on atmospheric-surface exchange of particulates and gaseous pollutants. Richland, Wash. 4–6 Sept. 1974. Energy Res. Development Administration Symp. Series 38, CONF-740921, Dep. of Commerce, Natl. Tech. Information Service, Springfield, Va.

Fan, L. T., and L. A. Disrud. 1977. Transient wind erosion: A study of the nonstationary effect on rate of wind erosion. Soil Sci. 124:61–65.

Federal Register. 1971. National primary and secondary ambient air quality standards. Fed. Regist. 36:8186–8201.

Free, E. E. 1911. The movement of soil material by the wind. USDA Bureau of Soils Bull. no. 68.

Gillette, D. A. 1977. Fine-particle emissions due to wind erosion. Trans. ASAE 20:890–897.

Gillette, D. A. 1978. A wind tunnel simulation of the erosion soil: Effect of soil texture, sandblasting, windspeed, and soil consolidation on dust production. Atmos. Environ. 12: 1735–1743.

Gillette, D. A., and T. R. Walker. 1977. Characteristics of airborne particles produced by wind erosion of sandy soil, high plains of West Texas. Soil Sci. 123:97–110.

Hagen, L. J., and E. L. Skidmore. 1977. Wind erosion and visibility problems. Trans. ASAE 20:898–903.

Hagen, L. J., and N. P. Woodruff. 1973. Air pollution from duststorms in the Great Plains. Atmos. Environ. 7:323–332.

Hagen, L. J., and N. P. Woodruff. 1975. Particulate loads caused by wind erosion in the Great Plains. J. Air Pollut. Control Assoc. 25:860–861.

Horikawa, K., and H. W. Shen. 1960. Sand movement by wind action. Beach Erosion Board, Corps of Engineers, Tech. Memo. no. 119.

Husar, R. B. 1976. Fine particle standard: An assessment of the need and an attempt of its formulation. p. 1–32. *In* ASME Air Pollution Control Div., Fourth Natl. Symp., St. Louis, Mo.

Iversen, J. D., R. Greeley, J. B. Pollack, and B. R. White. 1973. Simulation of Martian eolian phenomena in the atmospheric wind tunnel. p. 191–213. *In* Seventh Conf. on Space Simulation, NASA, SP-336.

Iversen, J. D., R. Greeley, B. R. White, and J. B. Pollack. 1976. The effect of vertical distortion in the modeling of sedimentation phenomena: Martian crater wake streaks. J. Geophys. Res. 81:4846–4856.

Iversen, J. D., J. B. Pollack, R. Greeley, and B. R. White. 1976. Saltation threshold on Mars: The effect of interparticle force, surface roughness, and low atmospheric density. Icarus 29:381–393.

Kadib, A. A. 1965. A function for sand movement by wind. Univ. of California Hydraulic Engineering Laboratory Tech. Report no. HEL-212.

Kawamura, R. 1951. Study on soil movement by wind. Inst. of Sci. and Tech. Report no. 5, Tokyo, p. 95–112.

Lumley, J. L., and H. A. Panofsky. 1964. The structure of atmospheric turbulence. Interscience, New York.

Lyles, L. 1975. Possible effects of wind erosion on soil productivity. J. Soil Water Conserv. 30:279–283.

Lyles, L. 1977. Wind erosion: Processes and effect on soil productivity. Trans. ASAE 20: 880–884.

Lyles, L., and B. E. Allison. 1976. Wind erosion: The protective role of simulated standing stubble. Trans. ASAE 19:61–64.

Lyles, L., and B. E. Allison. 1980. Range grasses and their small grain equivalents for wind erosion control. J. Range Manage. 33:143–146.

Lyles, L., and B. E. Allison. 1981. Equivalent wind-erosion protection from selected crop residues. Trans. ASAE 24:405–408.

Lyles, L., and R. K. Krauss. 1971. Threshold velocities and initial particle motion as influenced by air turbulence. Trans. ASAE 14:563–566.

Lyles, L., R. L. Schrandt, and N. F. Schmeidler. 1974. How aerodynamic roughness elements control sand movement. Trans. ASAE 17:134–139.

Makaveev, V. M. 1967. The movement of solid particles in turbulent stream flow. (In Russian.) Tr. Koord. Soveshch. Gidrotekh. 36:11–18.

Nikuradse, J. 1950. Law of flow in rough pipe. Natl. Advisory Comm. for Aeronautics Tech. Memo. no. 1292. (Transl. of 1933 paper of Nikuradse.)

O'Brien, M. P., and B. P. Rindlaub. 1936. The transport of sand by wind. Civ. Eng. (N.Y.) 6:325.

Orgill, M. M., and G. A. Sehmel. 1976. Frequency and diurnal variation of duststorms in the contiguous USA. Atmos. Environ. 10:813–825.

Owen, P. R. 1964. Saltation of uniform grains in air. J. Fluid Mech. 20:225–242.

Pollard, M. C. 1977. An investigation of duststorm generation in the southern Great Plains. Air Force Inst. of Technology, Wright-Patterson Air Force Base, Ohio, Report no. AFIT-CI-78-46.

Rosinski, J., G. Langer, C. T. Nagamoto, and J. S. Bogard. 1976. Generation of secondary particles from single large soil particles upon impact. p. 638–647. *In* R. J. Engleman and G. A. Sehmel (ed.) Proc. Conf. on Atmospheric-Surface Exchange of Particulate and Gaseous Pollutants, Richland, Wash. 4–6 Sept. 1974. Energy Res. Development Administration Symp. Series 38, CONF-740921, Dep. of Commerce, Natl. Tech. Information Service, Springfield, Va.

Skidmore, E. L., P. S. Fisher, and N. P. Woodruff. 1970. Wind erosion equation: Computer solution and application. Soil Sci. Soc. Am. Proc. 34:931–935.

Skidmore, E. L., and N. P. Woodruff. 1968. Wind erosion forces in the United States and their use in predicting soil loss. USDA-ARS Agric. Handbook no. 346.

Smith, R. M., P. C. Twiss, R. K. Krauss, and M. J. Brown. 1970. Dust deposition in relation to site, season, and climatic variables. Soil Sci. Soc. Am. Proc. 34:112–117.

Smith, W. H. 1977. Removal of atmospheric particulates by urban vegetation: Implications for human and vegetative health. Yale J. Biol. Med. 50:185–197.

Stearns, C. R. 1970. Determining surface roughness and displacement height. Boundary-Layer Meteorol. 1:102–111.

Tennekes, H., and J. L. Lumley. 1972. A first course in turbulence. The M.I.T. Press, Cambridge, Mass.

U.S. Environmental Protection Agency. 1976. Monitoring and air quality trends report, 1974. USEPA Report no. EPA-450/1-76-001, Research Triangle Park, N.C.

Walther, E. G. 1972. A rating of the major air pollutants and their sources by effect. J. Air Pollut. Control Assoc. 22:352–355.

White, B. R., and J. C. Schulz. 1977. Magnus effect in saltation. J. Fluid Mech. 81:497–512.

Wood, G. P., W. R. Weaver, and R. M. Henry. 1974. The minimum free-stream windspeed for initiating motion of surface material on Mars. NASA Report no. TMX-71959.

Woodruff, N. P., L. Lyles, F. H. Siddoway, and D. W. Fryrear. 1977. How to control wind erosion. Rev. ed. USDA Agric. Information Bull. no. 354.

Woodruff, N. P., and F. H. Siddoway. 1965. A wind erosion equation. Soil Sci. Soc. Am. Proc. 29:602–608.

Zingg, A. W. 1953. Wind-tunnel studies of the movement of sedimentary material. p. 111–135. *In* Proc. Fifth Hydraulogy Conference Bull. no. 34, State Univ. of Iowa Studies in Engineering.

13

C. W. Richardson and Earl Burnett
Agricultural Research Service
U.S. Department of Agriculture
Temple, Texas

Soil Conservation: Southern Great Plains (Southeastern)

The southeastern part of the Southern Great Plains includes the Southwestern Prairies Cotton and Forage Land Resource Region (Southwestern Prairies Region), which lies in Oklahoma and Texas (Fig. 13-1). The Southwest Prairies Region includes the Texas Blackland Prairies, Grand Prairie, Cross Timbers, and Texas Claypan major land resource areas (Austin, 1965). The climate is characterized by high summer temperatures and relatively mild winters. The mean annual frost-free period ranges from about 200 days in the northern part to about 280 days in the southern part. The mean annual precipitation ranges from about 620 mm along the western edge to about 1100 mm near the eastern side. The distribution by months at most locations shows a peak in April or May, followed by a dry period in July and August and a secondary peak in September and October. Thus, even though the annual precipitation at most locations within the region is usually adequate for good crop production, the distribution within the year causes severe moisture stress during most growing seasons.

Water erosion is a major concern during the spring and fall. The average annual rainfall erosion index in this region varies from 250 to 350 (Wischmeier and Smith, 1978). Much of the erosive rains occur at a time when soil surface conditions are highly susceptible to erosion losses. About 35% of the erosive rains occur within 2 months after the average planting dates of spring-planted crops when the plant canopy is small and the soil is loose (Stewart et al., 1975). An additional 20% of the erosive rains occurs in the fall after the plant canopy has been removed. Many of the soils are inherently highly erodible. Major erosion producing rainfall events are characterized by high intensities. Good soil conservation practices are essential to prevent serious water erosion losses.

Wind erosion is insignificant in most of the area due to a high clay content of the soils. However, a severe wind erosion hazard does exist in some

Dryland Agriculture—Agronomy Monograph no. 23.

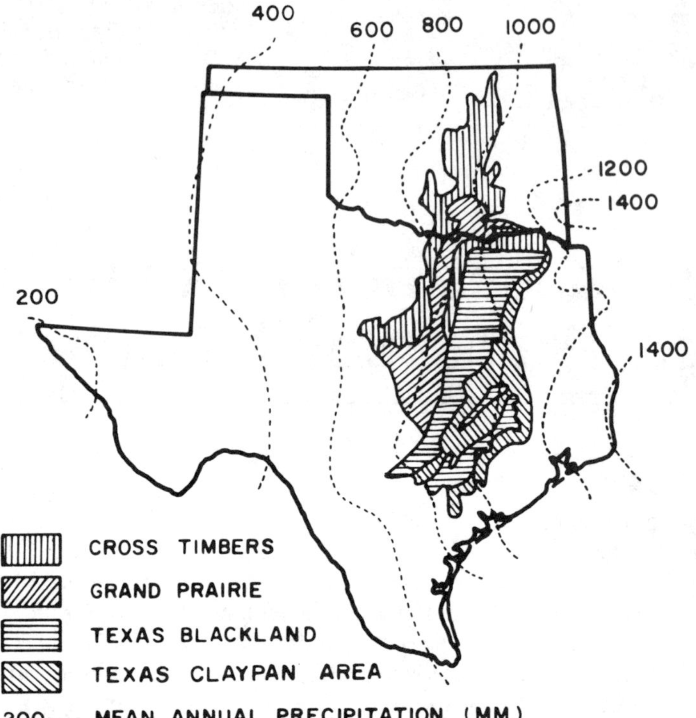

Fig. 13-1. Southern Prairies Cotton and Forage Region, including major land resource areas no. 84-87.

areas along the western edge of the region that have sandy soils and produce crops that provide little soil protection during the spring when wind velocities are highest.

13-1 SOILS AND CROPS

The soils and crops of the Southwestern Prairies Region are highly variable among the four major land resource areas within the region. The Texas Blackland Prairie occurs in the central part of the region and includes about 5.1 million ha extending from northeast Texas to the south central part of the state. About one half of this land resource area is cropland, with grain sorghum, cotton, cereal grains, and forages as the major crops. Pastures and meadows occupy about one fourth of the land area. The general surface relief is undulating to gently rolling (Godfrey et al., 1967). The soils are dark, highly calcareous, montmorillonitic clays with a high water-holding capacity and a high shrink-swell potential. The soil depths vary from moderately deep to shallow. In some places along the western part of the Blackland Prairie, spots occur where the soil has been washed away and the underlying chalk is exposed. The major soil series include Houston Black (fine, montmorillonitic, thermic Udic Pellusterts), Austin (fine-silty, carbonatic, thermic Entic Haplustolls), and Heiden (fine, montmorillonitic, thermic Udic Chromusterts). The eastern one third of the area, commonly called the graylands, has soils that are neutral to slightly acid clays to sandy loams. The major soil series in this part of the area are Burleson (fine, montmorillonitic, thermic Udic Pellusterts), Wilson (fine, montmorillonitic, thermic Vertic Ochraqualfs), and Crockett (fine, montmorillonitic, thermic Udertic Paleustalfs) (Carter, 1931; Godfrey, 1964; SCS, 1978).

The Grand Prairie occupies 3.3 million ha in Texas and Oklahoma. The soils are black to brown calcareous clays. Much of the surface is rolling, and the soils are shallow with limited water-holding capacity. In much of the area, water erosion has cut deeply. Large areas of stony soils occur, and in places bare rock is exposed to the surface. The major cash crops are grain sorghum, forage sorghum, and cereal grain with small amounts of cotton. Much of the cereal grains are grown for winter livestock grazing, and forage sorghum is grown for summer grazing to supplement range or permanent pasture grazing.

The Cross Timbers area occurs in Texas and Oklahoma and includes 4.6 million ha. The soils are highly variable. The upland areas have light, slightly acid, loamy sands and sandy loams. The bottomlands have dark, neutral to calcareous, clayey and loamy soils along minor streams. The surface ranges from gently rolling to very rolling, with smooth undulating broad divides, and considerable areas of hilly and rough stony lands in some sections. Large portions of the area has been subject to severe erosions. On the steeper slopes much of the soil has been removed, leaving stony and shallow soils. On the gentle slopes that have been cultivated the soil has been thinned by erosion with an accompanying formation of gullies (Carter, 1931). Most of the land is used for range or pasture. About one fifth of the land is used for crops. The major crops grown include peanuts, grain sorghum, forage sorghum, corn, cotton, and cereal grains. The cereal grains are grown primarily for cattle grazing but some are harvested for grain.

The Texas Claypan area contains approximately 3.4 million ha of nearly level to gently rolling, moderately dissected, postoak savannah. Much of the land was once cultivated, but severe erosion has caused some cultivated land to be returned to grass or trees. The major crops are cereal grains for livestock grazing with lesser amounts of grain sorghum, forage sorghum, and cotton. Corn and vegetables are locally important. The soils are characterized by a relatively thin loamy to sandy loam gray surface over an indurated claypan that restricts water movement and root penetration. There are some deep sandy soils with less clayey subsoils. The soils are generally acid. The major soil series are Lufkin (fine, montmorillonitic, thermic Vertic Albaqualfs), Tabor (fine, montmorillonitic, thermic Udertic Paleustalfs), Freestone (fine-loamy, siliceous, thermic Glossaquic Paleudalfs), and Axtell (fine, montmorillonitic, thermic Udertic Paleustalfs) (SCS, 1978).

13-2 EROSION PROBLEMS AND CONTROL

Water erosion is a major problem throughout much of the Southwestern Prairies Region due to high rainfall intensities, erodible soils, rolling topography, and clean tillage of crops. Erosion control research has been conducted in the Blackland Prairie for more than 50 years. Hill et al. (1944) summarized some of the early research dealing with terrace design and spacing, gully surveys, and soil movement lines. Smith et al. (1954) described the effects of climate, cropping systems, and slope length and steepness on erosion losses. The effectiveness of graded terraces for controlling erosion was described by Smith and Henderson (1961), and the application of parallel graded terraces were described by Smith et al. (1959). The effect of surface mulches on erosion control was reported by Adams (1966).

Soil conservation practices that include terraces, contour cultivation, and grassed waterways significantly reduced sediment that moved into streams (Baird, 1964). Using 1960 price relationships Magee (1962) found that these conservation practices are profitable in the Blackland Prairie but that about 12 years are required for increased earnings to pay all costs of establishing the program.

Graded terraces combined with contour farming, grassed waterways, and crop rotations have been proved in practice as well as research to be excellent means of controlling water erosion in this region. Most of the cropland is terraced or has been terraced at one time. Terrace and waterway maintenance are required to ensure adequate erosion protection.

Difficulties in using large farm equipment on the irregular contours of this region has caused many farmers to abandon contour farming on terraced land. Point rows, which were merely inconvenient 10 or 15 years ago, have become operationally impossible with large modern equipment. As a result, some farmers, particularly in the Blackland Prairie, have begun to till across terraces to more efficiently utilize wide equipment. This practice began about 1970 and has since become widespread. Tilling across terraces accelerates the erosion of soil into terrace channels and necessitates extensive annual terrace maintenance. The installation of parallel terraces in recent years seems to have reduced the use of this practice, but in some instances farmers are also farming across parallel terraces (Swanson et al., 1974. Terraces and farming operations. Unpublished report).

Wind erosion is a relatively minor problem in most of this region. Significant wind erosion is confined to areas in the western Cross Timbers where peanuts are grown on sandy soils. In these areas strip cropping peanuts with sorghums and cover crops of cereal grains are used to help control wind erosion. These practices are excellent means of controlling wind erosion but are not used extensively.

13-3 CROPPING AND TILLAGE PRACTICES

Cropping and tillage practices used in the Southwestern Prairies Region have a major effect on erosion problems and erosion control practices. Most of the crops grown produce small amounts of residue. Conventional weed control practices that rely on clean tillage and maximum surface disturbance are in direct conflict with good erosion control practices. The conventional cultural practices will be described to illustrate this problem.

Cropping systems used for grain sorghum and cotton are similar. Seedbed preparation usually begins with the forming of lister beds in the fall after the erosive rains. The beds are cultivated once or twice during the winter for weed control. Fertilizer is usually surface-applied in late winter. Many farmers apply a preemergence herbicide. Grain sorghum is normally planted about mid-March, and cotton is planted in April; however, these dates vary with latitude within the region. Most farmers use conventional bedder planters, but use of planters with double-disk openers on flat seedbeds are becoming more popular. Row spacings of about 100 cm are common. The timing of planting and the row widths are such that the crop canopy provides minimal soil protection during the erosive rains of April and May. Both grain sorghum and cotton are cultivated one to three times during the growing season with a conventional sweep cultivator. Johnsongrass is the most severe weed problem and frequently limits crop yields. Most of the tillage operations are intended to control johnsongrass. Johnsongrass control, particularly in grain sorghum, is a major impediment to the use of reduced tillage and maintenance of surface residue.

Grain sorghum is normally harvested in midsummer with a conventional combine. Following harvest, stalks are chopped off near the soil surface, and the plant roots are plowed out with sweeps or disks to prevent regrowth and further moisture use. This practice leaves the soil loose and essentially bare for the erosive fall rains.

Cotton is harvested in the fall. The plant is usually sprayed with a desiccant to remove the leaves before harvest. The lint and boll are then stripped from the stalk with a cotton stripper. The stalks are chopped and plowed out similar to the practices with grain sorghum. Very little plant residue remains on the soil after cotton harvest.

Both winter wheat and oats are grown throughout much of the region. Seedbed preparation usually begins in August or September before the fall erosive rains. Planting is done in the fall using a conventional grain drill, with drill rows spaced about 15 to 20 cm apart. Fertilizer may be incorporated during seedbed preparation or applied to the surface after plant emergence. Some of the cereal grains, particularly oats, are grown solely to provide winter livestock grazing. Then livestock are allowed to feed on the grain until no plant material remains for grazing. More often, however, the cereal grains are grown both to provide winter livestock grazing and to produce grain for harvest. When the grain is to be harvested, the livestock will be removed about 1 March so that the plant can produce seed. The grain will normally be harvested in June using a conventional combine.

Peanut production is basically confined to the western Cross Timbers. Seedbed preparation for peanuts, if the land is to be deep plowed, will normally begin in late February or early March. The deep plowing is done with a moldboard about 20 cm deep and buries most of the soil-protecting vegetation. The purpose of the deep plowing is to combat southern blight, a serious disease of peanuts. Southern blight control by deep plowing directly conflicts with residue management for soil conservation.

Not all of the land to be planted to peanuts is deep plowed. When the land is not to be deep plowed, the seedbed preparation typically begins in May and is done with chisels, sweeps, and disks. This system has the advantage of maintaining protective cover for a longer period.

Fertilizers and preplant herbicides are applied on the surface and incorporated with a disk or rolling cultivator. The peanuts are usually planted in June on 100-cm rows. They may be planted flat or by breaking open beds. Either a conventional bedder planter or a double-disk opener planter may be used. The double-disk opener planter requires less power and can plant successfully into plant residue. However, the conventional bedder planter permits a wider range of planting depths and allows the seed to be placed in moist soil.

The peanuts are usually cultivated from one to three times using sweep or rolling cultivators. Nitrogen is often added to the growing crop by top-dressing or side-dressing.

Harvesting is usually in October and includes digging, windrowing, turning as needed for drying, and threshing with a combine. Following harvest, most farmers bale the peanut vines for hay. These practices result in very little plant residue on the surface. Some farmers rotate peanuts with high residue-producing crops to reduce the erosion potential.

13-4 NEW DEVELOPMENTS AND RESEARCH NEEDS

Research conducted in this region at Temple, Texas, has shown an alternative to the problem of using wide equipment on terraced fields. A system of graded furrows were evaluated for controlling runoff and erosion (Richardson et al., 1969). With the system, crops were grown on ridges, and the furrows between the crop rows were used to convey runoff to a grassed waterway. The grades within the furrows were carefully controlled to permit surface drainage at nonerosive velocities. The system was shown to control erosion equal to a good terrace system while permitting 20% faster tillage (Richardson, 1973). The graded furrow system has not been accepted by farmers because of land smoothing required to install the system and the change in management required to maintain the ridges during the crop growing season. Difficulties in harvesting cotton grown on the ridges has also contributed to the lack of acceptance.

The use of closer spaced rows for grain sorghum and cotton has also shown promise as a method for both increasing crop yields and reducing water erosion. Grain sorghum yields were 20% greater (Adams et al., 1976) and soil loss during the growing season was 30% less (Adams et al., 1978) with 50-cm row spacing than with the conventional 100-cm spacing. The yield increase with closer row spacing was attributed to more efficient interception of incoming radiant energy by the plant canopy (Arkin et al., 1978). The reduction in soil loss with closer spacing was due to greater protection of the soil provided by the plant canopy during periods of vegetative growth. Some farmers in the Blackland Prairie are beginning to use closer row spacing. The major problems that have limited the use of closer row spacing seem to be difficulty of cotton harvest with available equipment and weed control. The most practical row spacing that permits interrow cultivation and retains the yield and erosion advantages is about 60 cm. Grain sorghum in closely spaced rows present no special harvest problems. Some farmers are using closely spaced rows for grain sorghum and conventionally spaced rows for cotton. Broadcast cotton strippers that are now available may cause the use of closer row spacing for cotton to increase.

Opportunities exist in the Southwestern Prairies Region for reducing erosion through improved cultural practices. The management of crop residues from cereal grains and grain sorghum offers excellent opportunities for erosion control. However, research needs to be done before residue management can be used extensively to control erosion without detrimental effects on crop production. The primary technical barriers to good residue management systems are weed control problems (primarily johnsongrass) and the lack of adequate mechanical equipment for residue management on heavy clay soils. Good residue management in this region requires a change in tillage and planting equipment and new technology in the use of herbicides for weed control. Research is underway to develop tillage and planting equipment designed for residue management on the major soils of the region (Morrison et al., 1980). New techniques are also being developed for incorporating herbicides into the soil with minimal disturbance of the residue on the surface. Developments such as these should enable more effective residue management for erosion control in the future.

13–5 LITERATURE CITED

Adams, J. E. 1966. Influence of mulches on runoff, erosion, and soil moisture depletion. Soil Sci. Soc. Am. Proc. 30:110–114.

Adams, J. E., G. F. Arkin, and E. Burnett. 1976. Narrow rows increase dryland grain sorghum yields. Texas Agric. Exp. Stn. Miscellaneous Pub. no. 1248.

Adams, J. E., C. W. Richardson, and E. Burnett. 1978. Influence of row spacing of grain sorghum on ground cover, runoff, and erosion. Soil Sci. Soc. Am. Proc. 42:959–962.

Arkin, G. F., J. T. Ritchie, and S. J. Maas. 1978. A model for calculating light interception by a grain sorghum canopy. Trans. ASAE 21:304–308.

Austin, M. E. 1965. Land resource regions and major land resource areas of the United States. USDA Agric. Handbook no. 296.

Baird, R. W. 1964. Sediment yields from blackland watersheds. Trans. ASAE 7:454–456.

Carter, W. T. 1931. The soils of Texas. Texas Agric. Exp. Stn. Bull. no. 431.

Godfrey, C. L. 1964. A summary of the soils of the Blackland Prairies of Texas. Texas Agric. Exp. Stn. Miscellaneous Pub. no. 698.

Godfrey, C. L., C. R. Carter, and G. S. McKee. 1967. Land resource areas of Texas. Texas Agric. Exp. Stn. Bull. no. 1070.

Hill, H. O., W. J. Peevy, A. S. McCall, and F. G. Bell. 1944. Investigation of erosion control and reclamation of eroded land at the Blackland Conservation Experiment Station, Temple, Texas, 1931–41. USDA Tech. Bull. no. 859.

Magee, A. C., R. W. Baird, and J. B. Pope. 1962. Conservation pays in the Blacklands. Texas Agric. Exp. Stn. Miscellaneous Pub. no. 602.

Morrison, J. E., L. H. Wilkes, and E. B. Hudspeth. 1980. Equipment for conservation tillage systems. *In* B. L. Harris and A. E. Colburn (ed.) Conservation tillage in Texas. Texas Agric. Extension Service Bull. no. 1290.

Richardson, C. W. 1973. Runoff, erosion, and tillage efficiency on graded furrows and terraced watersheds. J. Soil Water Conserv. 28:162–164.

Richardson, C. W., R. W. Baird, and D. W. Fryrear. 1969. Graded furrows for water erosion control. J. Soil Water Conserv. 24:60–63.

Smith, R. M., A. C. Bennett, and R. C. Henderson. 1959. Researchers report on parallel-type terraces. Soil Water 8(5):7, 17.

Smith, R. M., and R. C. Henderson. 1961. Some interpretations of runoff and erosion from terraces on blackland soil. USDA Report no. ARS-41-42.

Smith, R. M., R. C. Henderson, and O. J. Tippit. 1954. Summary of soil and water conservation research from the Blackland Experiment Station, Temple, Texas, 1942–53. Texas Agric. Exp. Stn. Bull. no. 781.

Soil Conservation Service. 1978. Classification of soil series in the United States. USDA-SCS, U.S. Government Printing Office, Washington, D.C.

Stewart, B. A., D. A. Woolhizer, W. H. Wischmeier, J. H. Caro, and M. H. Frere. 1975. Control of water pollution from cropland, Vol. 1. A manual for guideline development. USDA Report no. ARS-H-5-1.

Wischmeier, W. H., and D. D. Smith. 1978. Predicting Rainfall Erosion Losses—a guide to conservation planning. USDA Agric. Handbook no. 537.

14

Wendell C. Johnson
Agricultural Research Service
U.S. Department of Agriculture
Bushland, Texas

E. L. Skidmore
Agricultural Research Service
U.S. Department of Agriculture
Manhattan, Kansas

B. B. Tucker
Cooperative Extension Service
Oklahoma State University
Stillwater, Oklahoma

Paul W. Unger
Agricultural Research Service
U.S. Department of Agriculture
Bushland, Texas

Soil Conservation: Central Great Plains Winter Wheat and Range Region

The Central Great Plains Winter Wheat and Range Region (Central Winter Wheat Region) of the U.S. Great Plains includes major land resource areas no. 71 through 80 (Austin, 1965) and lies within the states of Nebraska, Kansas, Colorado, New Mexico, Oklahoma, and Texas. This region, which covers parts of the Central and Southern Great Plains, is semiarid (Thornthwaite, 1931) and is characterized by limited and erratic precipitation, high summer temperatures, high average wind speeds, and high evaporation.

Contribution from the U.S. Department of Agriculture, Agricultural Research Service, Bushland, Texas, and Manhattan, Kansas, and Oklahoma State University, Stillwater, Oklahoma.

Dryland Agriculture—Agronomy Monograph no. 23.

The above conditions result in generally low dryland crop yields, especially in the western and southern portions of the region. They also result in potentially high wind erosion. The wind erosion potential increases during prolonged dry periods when residue production by crops is low or when weather, tillage, and other management practices interact or act singly to decrease or destroy surface residues or soil roughness. Wind erosion usually is a problem only in localized areas in the eastern portion of the Central and Southern Great Plains. Water erosion predominates in the eastern portion of the Great Plains where precipitation is more frequent and intense rainstorms often occur. In this portion of the Great Plains, rain may fall on wet soil that has little capacity of additional storage. Also, some soils are on relatively steep slopes and have low infiltration rates, which often leads to runoff and potentially high water erosion. Because of intense storms, runoff and erosion are also problems in the western and southern portions of these Great Plains, even though the soils have less slope than in the eastern portion.

Through the years, many practices that have an impact on soil erosion have been developed. One practice that has had a major impact on reducing erosion has been the development of irrigation, which has its major impact on controlling wind erosion by providing adequate water for more reliable and greater crop production. With irrigation, annual cropping is possible, and crop residue levels are usually adequate to control erosion if the residues are properly managed. Most water for irrigation in the region is obtained from the Ogallala Aquifer. Unfortunately, the aquifer is limited and being depleted, especially in the southern portion. Therefore, strategies for erosion control must allow for the effect that the disappearance of this water source will have on the region.

As an aid in the discussion of the soil erosion problems in the vast Central and Southern Great Plains region, this region has been divided into Northern and Southern subregions. The major land resource areas included in the two subregions are indicated by number and name in Table 14–1. Also indicated in the table are the major soils, along with their classification, occurring in each major land resource area.

14–1 NORTHERN SUBREGION

The Northern Subregion includes most of Kansas, the southern half of Nebraska, and a small area in northeastern Colorado (Fig. 14–1*A*).

14–1.1 Soils

Because soils are classified according to natural relationships, their classification gives clues about their behavior, need for conservation practices, and how they will respond under alternative management systems.

Table 14-1. Major land resource areas (numbered) and major soils of the Central and Southern Great Plains.

Number	Name	Series	Classification
		Northern Subregion	
71	Central Nebraska Loess Hills	Colby	Fine-silty, mixed (calcareous), mesic Ustic Torriorthents
72	Central High Tableland	Keith	Fine-silty, mixed, mesic Aridic Argiustolls
		Richfield	Fine, montmorillonitic, mesic Aridic Argiustolls
		Ulysses	Fine-silty, mixed, mesic Aridic Haplustolls
73	Rolling Plains and Breaks	Harney	Fine, montmorillonitic, mesic Typic Argiustolls
74	Central Kansas Sandstone Hills	Lancaster	Fine-loamy, mixed, mesic Udic Argiustolls
75	Central Loess Plains	Hastings	Fine, montmorillonitic, mesic Udic Argiustolls
76	Bluestem Hills	Irwin	Fine, mixed, mesic Pachic Argiustolls
79	Great Bend Sand Plains	Naron	Fine-loamy, mixed, thermic Udic Argiustolls
		Southern Subregion	
77	Southern High Plains	Acuff	Fine-loamy, mixed, thermic Aridic Paleustolls
		Amarillo	Fine-loamy, mixed, thermic Aridic Paleustalfs
		Brownfield	Loamy, mixed, thermic Arenic Aridic Paleustalfs
		Dalhart	Fine-loamy, mixed, mesic Aridic Haplustalfs
		Olton	Fine, mixed, thermic Aridic Paleustolls
		Patricia	Fine-loamy, mixed, thermic Aridic Paleustalfs
		Pullman	Fine, mixed, thermic Torrertic Paleustolls
		Ulysses	Fine-silty, mixed, mesic Aridic Haplustolls
78	Central Rolling Red Plains	Abilene	Fine, mixed, thermic Pachic Argiustolls
		Altus	Fine-loamy, mixed, thermic Pachic Argiustolls
		Brownfield	Loamy, mixed, thermic Arenic Aridic Paleustalfs
		Enterprise	Coarse-silty, mixed, thermic Typic Ustochrepts
		Foard	Fine, montmorillonitic, thermic Typic Natrustolls
		Hollister	Fine, mixed, thermic Pachic Paleustolls
		Miles	Fine-loamy, mixed, thermic Udic Paleustalfs
		Pratt	Sandy, mixed, thermic Psammentic Haplustalfs
		Rowena	Fine, mixed, thermic Vertic Calciustolls
		Sagerton	Fine, mixed, thermic Typic Paleustolls
		Tillman	Fine, mixed, thermic Typic Paleustolls
		Tivoli	Mixed, thermic Typic Ustipsamments
80	Central Rolling Red Prairies	Bethany	Fine, mixed, thermic Pachic Paleustolls
		Bonti	Fine, mixed, thermic Ultic Paleustalfs
		Chigley	Fine, mixed, thermic Udic Paleustalfs
		Grant	Fine-silty, mixed, thermic Udic Argiustolls
		Kirkland	Fine, mixed, thermic Udertic Paleustolls
		Lindy	Fine, mixed, thermic Udic Haplustalfs
		Norge	Fine-silty, mixed, thermic Udic Paleustolls
		Pond Creek	Fine-silty, mixed, thermic Pachic Argiustolls

(continued on next page)

Table 14–1. Continued.

Number	Name	Series	Classification
		Port	Fine-silty, mixed, thermic Cumulic Haplustolls
		Reinach	Coarse-silty, mixed, thermic Pachic Haplustolls
		Yahola	Coarse-loamy, mixed (calcareous), thermic Typic Ustifluvents
		Zaneis	Fine-loamy, mixed, thermic Udic Argiustolls

The soils of this subregion are generally medium-textured Ustolls, with Udic, Typic, and Aridic subgroups in the eastern, central, and western portions, respectively (Table 14–1; Soil Survey Staff, 1975).

The Ustolls are in a subhumid, semiarid climate; drought is frequent and may be severe. During a drought, soil erosion by wind becomes a problem. Wind erosion is the dominant conservation problem on the Aridic Ustolls. Water erosion is more frequent on the Udic Ustolls. The major soils are cohesive, medium to fine textured, and of montmorillonitic or mixed mineralogy. They will form clods and likely respond to emergency tillage for wind erosion control. If the Entisol (Colby silt loam) were a Psamment instead of an Orthent, one would expect severe wind erosion problems and sand dunes, especially with the hot, dry climate of the region.

Soils of the Northern Subregion developed under climax vegetation of tall and mid-height grasses, except for a small area of short grasses in eastern Colorado and southwestern Kansas (Shantz, 1923).

14–1.2 Climate

The climate becomes drier from east to west (average annual precipitation 900 to 400 mm) and winds become stronger (Fig. 14–1*C*). The wind erosion climatic factors are 73 and 20 for major land resource areas no. 72 and 76, respectively, whereas the rainfall erosion index increases from about 1200 in the northwest to about 4000 in the southeast portion of this subregion (Fig. 14–1*D*). The concentration of high values of the wind erosion climatic factor as centered in the Oklahoma panhandle is striking. This area was termed the windiest inland area in the USA (Flora, 1948) and was the center of the "dust bowl" in the 1930's. The 1950's drought was also severe in this area.

14–1.3 Conservation Practices

14–1.3.1 Summer Fallow

Variation in climate is reflected in the type of adapted farming operation. The high climatic factor values in Fig. 14–1*C* indicate a high wind erosion hazard in major land resource area no. 72. The *R* value, averaging

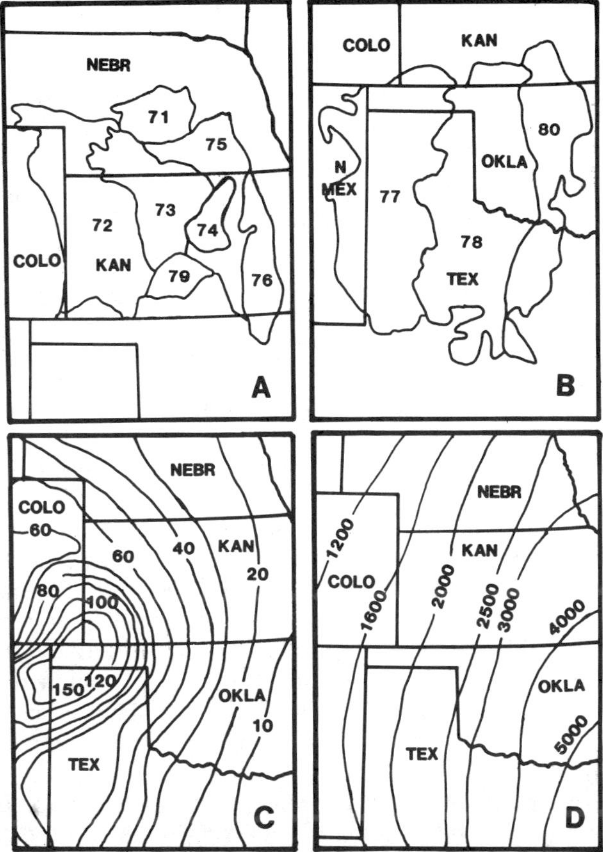

Fig. 14-1. Major land resource areas of Central and Southern Great Plains: *A,* Northern Subregion; *B,* Southern Subregion; *C,* Wind erosion climatic factor, *C* (in percent relative to that at Garden City, Kansas) (Woodruff and Siddoway, 1965); and *D,* Isoerodent (water erosion) factor, *R* [in MJ mm (ha h y)$^{-1}$] (Foster et al., 1981).

about 1600 for water erosion, suggests a low to moderate water erosion hazard. The low annual precipitation and high evaporative potential mandates fallow-wheat; between 75 and 85% of the dryland wheat occurs on fallow (Greb et al., 1974). Both early and late summer fallowing are practiced in the western Kansas part of major land resource area no. 72. The first tillage operation during fallow usually is done with a chisel, sweep, or disk implement, with the type used depending on the amount of residue produced by the preceding crop. The first tillage usually is performed immediately after harvest. In delayed fallowing, the first fallow tillage is performed about 1 May when weed growth begins to markedly reduce stored soil water. Delayed fallowing and the associated reduced tillage has an economic advantage; in addition, the reduced rate of soil organic matter decomposition could raise the ultimate equilibrium level of soil humus (Hobbs and Thompson, 1971).

Chemical weed control during the summer fallow period has the advantage of controlling weeds with fewer tillage operations, thereby preserving the residues for wind erosion protection, slowing down the loss of soil humus, and conserving water ordinarily lost in cultivation (Black and Power, 1965; Fenster et al., 1965; Smika and Wicks, 1968). More research is needed, however, before chemical fallow can be recommended for general use in western Kansas (Banbury et al., 1977).

Moving eastward from major land resource area no. 72 to areas no. 71, 73, and 79, one enters a more humid region that Greb et al. (1974) called a "fallow-transition" area where fallow is often used but is not always needed. In the Rolling Plains and Breaks area (no. 73), the major cropping system is wheat-sorghum-fallow. In this much-tested system, two crops are produced in 3 years rather than one crop in 2 years as in the wheat-fallow system. The wheat-sorghum-fallow system offers a farmer greater flexibility in responding to extremely erratic precipitation that may be below or above average. If, in the fall of the year, lack of soil water makes it unwise to seed wheat, the farmer can forego wheat seeding in favor of raising sorghum the following summer. Getting the land back into wheat again, however, involves an 11-month fallow period because the sorghum is normally harvested too late to allow seeding wheat at an optimum time even if soil water conditions were adequate. The situation is different from that in Montana or North Dakota where, if water is lacking for fall seeding of winter wheat, the opportunity still exists to seed spring small grain (Haas and Boatwright, 1960). The wheat-sorghum-fallow system has a high wind erosion protection potential because relatively large amounts of residue are produced and the soil usually is rougher and cloddier than after a wheat crop (Ramig and Smika, 1964).

Still farther east, with a further increase in rainfall in major land resource areas no. 75 and 76, summer fallowing is not needed, and corn, sorghum, wheat, and soybeans are raised by continuous cropping.

14-1.3.2 Stubble Mulch Tillage

Seedbed preparation for winter wheat, before the dust bowl years, was done with the moldboard or one-way disk plow, which inverted the soil and left very little stubble on the soil surface. During the 1930's, the urgency of the wind erosion problem and recognition that crop residues left on the soil surface were a major deterrent to wind erosion gave rise to stubble mulch tillage (Duley, 1959; Zingg and Whitfield, 1957). Experiments with stubble mulch tillage began as early as 1937 in the Great Plains. Stubble mulch tillage implements perform subsurface tillage, mainly with sweeps, blades, or rods that are drawn beneath the surface without inverting the soil or covering plant residues. Improvised implements were used at first, but presently a number of excellent commercial sweep and blade implements are on the market whose size has kept pace with the increasing size of farm tractors. In the 1940's, a typical sweep plow had five 0.75-m sweeps, was pulled by a 37 kW tractor, and had a cutting width of 3 m. Today, a typical sweep plow will have five or six sweeps covering a lateral width of 6 to 9 m and is pulled by a 75 kW to 110 kW tractor. Still larger plows of 9- to 18-m width are used with the larger four-wheel-drive tractors in the 150 kW to 370 kW range. These plows fold hydraulically in three to seven sections for easy transporting. The greater flexibility also allows them to do a better job on uneven ground. Stubble mulch tillage has become recognized as the key to controlling wind erosion in wheat fields of western Kansas (Norwood and Herron, 1977). In addition to protecting against wind erosion, the surface residues increase water infiltration and reduce evaporation, thereby increasing crop production. Even stubble mulch tillage, however, reduces the amount of surface residues by about 20% with each operation (Anderson, 1961; Fenster, 1960; Woodruff and Chepil, 1958; Woodruff et al., 1965), and some cropland may be difficult to protect adequately against wind erosion, even though noninverting tillage is used (Skidmore et al., 1979; Skidmore and Siddoway, 1978).

14-2 SOUTHERN SUBREGION

The Southern Subregion is comprised of major land resource areas no. 77, 78, and 80, which lie mainly in the Texas Panhandle and western Oklahoma but also extend into eastern New Mexico, southern Kansas, and extreme southeastern Colorado (Fig. 14-1*B*).

14-2.1 Southern High Plains

The Southern High Plains is on a smooth, gently sloping, outwash plateau comprised of calcareous materials eroded from the Rocky Mountains. These materials were reworked by wind at a time when the prevailing wind direction was southerly. As a result of this movement by wind, surface soil texture grades from sand at Big Spring to clay loam at Amarillo (Lotspeich and Coover, 1962; Fig. 14-2*A*). Due to variation in soil and climate, the south to north sequence of adapted land use is rangeland; cot-

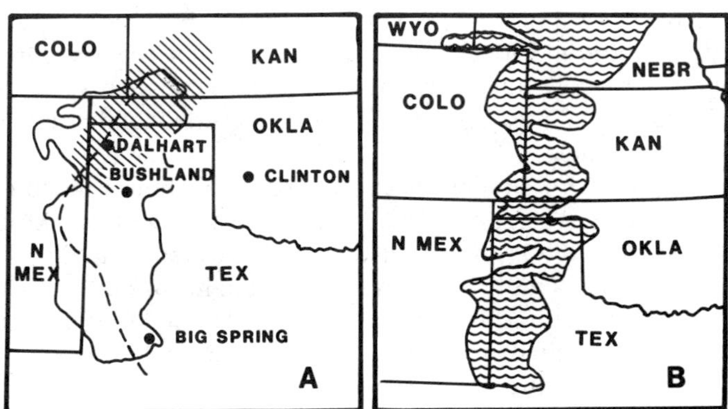

Fig. 14-2. *A,* Southern Subregion showing location of Dalhart sandy soil problem area (*shaded*); effective rainfall line, western limit of cultivation of fine-textured soil (Dale, 1941); and places mentioned in the text. *B,* Location of Ogallala aquifer (Frye, 1970).

ton and sorghum; cotton, sorghum, and wheat; and wheat and sorghum (Bonnen, 1960; Rule, 1939). Wind is the primary erosion hazard with C values ranging from 65 to 150 (Fig. 14–1*C*).

14–2.1.1 Soils

Soils of the Southern High Plains are mainly loamy Ustolls with Aridic and Torrertic subgroups in the northern and eastern sections; Calcic subgroups of Ustolls and Aridic subgroups of Ustalfs dominate the southern and western areas. Aridic subgroups of Ustochrepts are found on slopes along draws with Fluventic subgroups of Ustolls on flood plains. The Ustolls developed under short grasses and the Ustalfs under mid-height grasses.

14–2.1.2 Problem Areas

The soil erosion problem will be discussed for three distinct areas in major land resource area no. 77 (Fig. 14–2*A*). These are (i) the sandy row cropland at Big Spring, Texas; (ii) the fine-textured wheat and sorghum land at Bushland, Texas; and (iii) the Dalhart sandy soil problem area. Finally, the anticipated effect of depletion of the extensive Ogallala Aquifer on wind erosion will be considered under section 14–2.1.2.4.

14–2.1.2.1 Wind Erosion Control on Sandy Row Cropland. The type location for the sandy row cropland is at the Big Spring Field Station, Big Spring, Texas (Fig. 14–2*A*).

The southern part of the Southern High Plains has 1.5 million ha of cotton, 100 000 ha of sorghum, and 70 000 ha of wheat with interspersed areas of rangeland. In the sandy area, tillage is used to control wind erosion because cropland is bare during the winter and spring months when the *C′* factor ranges from about 30 to 70 (Fig. 14–1*C*; Skidmore and Woodruff, 1968). The growing season is long, and the climate is characterized by high-intensity storms—a single one of which may produce 10 or even 20% of the annual precipitation. These large storms are often separated by several weeks without rain. Droughts extending over two consecutive years tend to reduce the cultivated area because the severely eroded land is either planted to grass or abandoned.

Cotton production under the existing soil and climatic conditions presents a real challenge for wind and water erosion control. After cotton harvest, less than 1000 kg/ha of surface residue usually remains. This rules out stubble mulch tillage and leaves the soil roughening effect of tillage as the only practical method for erosion control. Highly erodible sandy soils with a clay layer within 45 to 60 cm of the surface can be deep plowed, and the clods brought to the surface by this tillage will resist breakdown for several years. If there is sufficient soil water after harvest in November or December, the soils may be moldboard or lister plowed (Fryrear, 1969).

With some soil series, moldboard plowing will increase surface aggregation. However, the lister is the primary wind erosion control implement. Listing is effective if it is done before the soil has started to erode and if there is adequate soil water to result in clod formation.

Ideally, a high-residue crop should be grown in rotation with cotton to provide some surface residues and to increase soil organic matter. However, herbicide carryover can be hazardous to cotton, and care must be taken if cotton is rotated with grain sorghum or wheat. The area is too dry for soybeans, but guar holds promise as a crop to be grown in rotation with cotton.

Terraces are used in the southern extremity of the Southern High Plains for water conservation. Nearly all new terrace systems are of the level, parallel type to accommodate large, mounted, row crop equipment. With a properly designed and maintained terrace system, all rain can be held on the field and stored in the soil. However, the traditional practice of changing row direction each year to avoid hummocking and blowouts cannot be used, and special care must be taken to keep the field smooth enough for good implement performance.

To hold evaporation to a minimum, fine sandy loam soils, which dry very quickly, should be cultivated as soon as possible after a rain. Also, raindrop impact separates the coarse and fine soil fractions and leaves a thin film of sand grains on the soil, which, if not covered, may be eroded by wind within a few hours after a rain. If the crop is planted and rain occurs before emergence, the soil may crust and prevent emergence. In this case, a rotary hoe or sand fighter will aid emergence by breaking the crust and will also reduce wind erosion.

14–2.1.2.2 Wind Erosion Control on Fine-textured Wheat and Sorghum Land. The type location for fine-textured wheat and sorghum land is at the USDA Conservation and Production Research Laboratory, Bushland, Texas (Fig. 14–2*A*).

Dryland agriculture in the northern part of the Southern High Plains is practiced on smooth uplands characterized by the Pullman clay loam (0 to 3% slope). Although 80% of this soil occurs on slopes less than 1% and water erosion is not a major problem, 5 to 8 cm of surface soil was removed by wind erosion during the 1930's. Thirty-five years of research with conservation tillage in this area conclusively shows that stubble mulch tillage effectively controls wind erosion. Also, compared with clean tillage with the one-way disk plow, stubble mulch tillage increased wheat yields by 17%, slowed the decline of soil humus by 11%, did not increase weed and insect problems, and increased water storage by summer fallowing by 20%. With either tillage method, however, summer fallowing to store water was not very efficient. An average of only about 15% of the fallow season precipitation was stored as soil water (Johnson and Davis, 1972).

Stubble mulch tillage can be practiced with comparatively simple equipment in this area. At the Conservation and Production Research Laboratory, dryland wheat and sorghum were produced annually for many years using the stubble mulch plow with 75-cm wide sweeps as the only tillage tool. Only once in 35 years was it necessary to shred wheat straw preparatory to sweep tillage. A rod weeder did not perform well in the Pullman clay loam soil because of too much compaction. Also, there is not a great need for this implement because it usually is not necessary to control winter annual grasses. Wheat is usually seeded with a high-clearance, deep-furrow drill in 35-cm spaced rows.

Less land is summer-fallowed in this area than in major land resource area no. 72, because wheat yields are not increased as much by fallowing. As a result, somewhat less than 25% of the wheat is produced on summer-fallowed land. The small yield increase is attributable to soil and climatic factors that lower the water storage efficiency and to higher native soil fertility, due in part to a shorter time in cultivation, which makes increased fertility from summer fallowing a lesser factor in yield. The soils typically have a high water-holding capacity and low permeability. The growing season is long, the evaporation rate is high, and the amount of snow received is small compared with that in major land resource area no. 72.

Delayed summer fallowing is not used by farmers in the Texas Panhandle, probably because of the large amount of weed seed produced during the long weed-growing period from wheat harvest to first frost. This interval, for example, is about 6 to 7 weeks longer in the panhandle of Texas than in the panhandle of Nebraska. However, delayed stubble mulch fallowing on small experimental plots at Bushland over a period of years produced about the same average wheat yield as early fallowing with a 30% saving in number of cultivations. Delayed fallowing affords excellent erosion protection, conserves soil humus, and could be used for soil conservation if a way were found to reduce the amount of weed seed produced. Excellent protection against erosion is also provided when residues are maintained on the surface in no-tillage cropping systems (Unger et al., 1971; Unger and Wiese, 1979). Although delayed-fallowed land is usually thoroughly dried out by summer and fall weed growth and develops large shrinkage cracks, it usually is in a very receptive state for storing water from spring rains. If a concentration of good rains occurs in May or June before much cultivation has been performed, delayed fallow land rapidly stores as much water as conventionally summer-fallowed land.

14–2.1.2.3 Wind Erosion in Dalhart Sandy Soil Problem Area. An area in which soil conservation has especially benefited from the development of irrigation is the 0.8 million ha Dalhart sandy soil area in the northwest corner of major land resource area no. 77 (SCS, 1963; Fig. 14–2*A*). This was a critical wind erosion problem area in the 1930's (Bennett, 1939; Eby and Whitfield, 1940). Currently in this area, 0.2 million ha of land that is highly susceptible to wind erosion is producing sufficient residues for erosion control by means of sprinkler irrigation. On a smaller area, excellent erosion control is being obtained from corn production, or corn and small grain production, using sprinkler-irrigation systems combined with minimum tillage methods (Unger et al., 1977). In Dallam and Hartley counties (Texas) taken together, in the heart of the sandy land area, there are more than 2 ha of irrigated land/ha of land in dryland crops. However, the aquifer is not being replenished, and wind erosion will return in its former severity when irrigation is no longer feasible unless suitable control measures are used.

14–2.1.2.4 Depletion of Ogallala Aquifer as a Factor in Future Wind Erosion Control in Southern Great Plains. An important threat to wind erosion control in the Central and Southern Great Plains is the prospect of depletion of the nonrenewing Ogallala Aquifer whose areal extent is shown in Fig. 14–2*B*. Interdispersal of irrigated and dry-farmed land helps protect the dry-farmed land from erosion; when the irrigated area ultimately decreases, the risk of wind erosion will increase. Currently, in the Texas High Plains as a whole, there is a hectare of irrigated land for each hectare of cultivated dryland.

Irrigation from the Ogallala Aquifer increased rapidly during the 1950's largely because of the drought of 1952 to 1956. In the Texas High Plains, the irrigated area increased fivefold from 1958 to 1976, or from 0.46 to 2.51 million ha (New, 1976). In western Kansas, the irrigated area increased from about 0.68 million ha in 1950 to 1.1 million ha in 1975 (Kansas Water Resources Board, 1967; Hay, 1976). The present land area irrigated from the Ogallala formation in Texas, Kansas, and the Oklahoma panhandle is over 3.6 million ha or about the same as the irrigated area in California (*Irrigation Journal,* 1976). The land irrigated from the Ogallala Aquifer has the potential of becoming dryland again in the next 50 years.

Estimating the rate and pattern of depletion of the Ogallala Aquifer is very difficult because unpredictable factors, e.g., the cost of power and the variability of the aquifer itself, are involved. One prediction is that the recoverable volume of water in storage under the Texas High Plains as a whole in the year 2020 will be about 35% of the 1970 volume (Walker and Taylor, 1975). An earlier study (Hughes and Harman, 1969) indicated an even more rapid rate of depletion. Gutentag (1978) estimated that water in the groundwater reservoir in western Kansas, not all of which is recoverable, was about 50 times the annual volume being pumped.

14–2.2 Central Rolling Red Plains and Central Rolling Red Prairies

Along its eastern edge, the Southern High Plains ends in a cap rock escarpment beyond which the land surface drops abruptly through an elevation of several hundred meters to an erosional plain, the Central Rolling Red Plains (major land resource area no. 78). This area, as well as the Central Rolling Red Prairies (major land resource area no. 80) with which it merges to the east, derives its name from Permian redbed materials that have been exposed by severe geologic erosion. In some places, high, caliche-topped buttes indicate the original land surface. Bennett (1939) stated that the eastern half of major land resource areas no. 78 and 80, which were then called the "Red Plains," had received damage from erosion that was exceeded in few other areas in the USA. He also noted that even before settlement the area was classified as an eroded plains.

14-2.2.1 Climate-Soil Interactions

The Central Rolling Red Prairies contain some of the most productive wheatland soils in the Southern Plains. The Central Rolling Red Plains are also quite productive but receive less rainfall, are more rolling, contain more rangeland, and have a more diversified agriculture. In the northern portion of the Central Rolling Red Plains, wheat and forages are the predominant crops. In the southern portion, such row crops as cotton, peanuts, and grain sorghums are also widely grown.

Rainfall in the area varies from 500 to 800 mm annually, with wide fluctuations from year to year. Long periods of adequate rainfall are frequently followed by periods of abnormally low rainfall. Drought is common, and even when the area as a whole is in a wet cycle, drought occurs in localized areas. In the center of the area, at Clinton, Oklahoma, the average annual rainfall is 690 mm. The 80% quantile is 870 mm and the 20% quantile is 530 mm, meaning that less than 870 mm is expected 80% of the time and that less than 530 mm is expected 20% of the time (Curry, 1972).

In the Central Rolling Red Plains, wind as well as water erosion has been severe and has affected the productivity of the coarse-textured soils. High winds can occur any time of the year in connection with thunderstorms but are most common in early spring months when weather fronts are most frequent. Winter precipitation is usually low. Therefore, soils are vulnerable to wind erosion in the early spring when wind velocities are high and the soil water content is low. Crop damage from wind erosion can also be severe in early summer whenever high winds follow high-intensity rainfall, which disperses clean-tilled soils. This occurs on both coarse- and fine-textured soils because the soil peds are not resistant to breakdown by the impact of raindrops. High-intensity storms often occur when soils are water saturated, resulting in considerable runoff with accompanying water erosion of unprotected soils.

14-2.2.2 Soils

Soils in the Central Rolling Red Prairie and Plains (Gray and Roozitalab, 1976; Austin, 1965; Table 14-1) have mixed mineralogy, thermic temperature, and Ustic moisture regimes, and they vary in texture from clay loams to loamy sands. With the exception of stream terraces, most upland soils in the Central Rolling Red Prairies are silt loams and clay loams. They tend to be rather deep Ustolls, e.g., the Pond Creek, Grant, Bethany, and Kirkland series. However, on the more sloping areas, only a few centimeters of topsoil overlay sandstones or shales. Alluvial soils vary from sands to clay loams, the most productive of which are the sandy and silt loams, with the Norge, Reinach, Port, and Yahola series predominating. The alluvial soils and terraces are very important agriculturally to the area.

Soils of the Central Rolling Red Plains are quite diverse, ranging from sands to clays. Fortunately, the sandy loams and loamy sands are frequently underlain by fine-textured clays and clay loams, making them rather productive considering the climate. The clayey layers occur at depths from only a few to 150 cm or more from the surface. Sandy soils underlain by this finer-textured material at 45 to 90 cm seem to be the most productive. These are characterized by the Miles, Altus, Brownfield, and Pratt series. Although wind erosion is more severe on these soils than water erosion, they are extremely erosive by high-intensity rains when the soil profile is supersaturated with water.

Some of the alluvial-aeolian sandy soils in the Central Rolling Red Plains, e.g., the Tivoli and Enterprise series, do not contain the fine-textured layers at helpful depths and are, therefore, very droughty.

Medium- to fine-textured soils occur in large areas of the Central Rolling Red Plains on both smooth upland flats and sloping areas. In general, the silt loams and clay loams are found on the smooth flats and are moderately resistant to extreme water erosion. They are represented by the Foard-Tillman-Hollister Association. The most sloping areas are occupied by soils that are rather thin or have weakly expressed profiles in deep sandy materials, unconsolidated shales, or caliche. Alluvial soils in the area are less extensive than in the Central Rolling Red Prairie, although they are locally quite important. Narrow strips of very productive soils occur along terraces of each of the major streams traversing the area. Bottomland is less extensive than terrace land, and on the bottomland, soils are not well developed because of the high sediment load of the streams and frequent flooding.

14-2.2.3 Practices for Wind and Water Erosion Control

Settlers first used this area for grazing livestock. Before that, it supported populations of bison and other wildlife. Evidence points to widespread overgrazing during drought periods, even before settlement. Much geologic erosion, especially by water, is apparent and continues today. Climate and quite erosive soil materials teamed up and still plague the area. Whereas crop production in Kansas and Texas commenced with widespread plowing of the prairie in the 1870's, settlement and farming date only from the 1890's in Oklahoma. The far western counties of Oklahoma were not settled until after the turn of the century. In Texas, the land was managed primarily in rather large units for cattle ranching. Plowing and crop production were slow to develop and occurred first on the best soils. In Oklahoma, land runs and homesteading restricted each farmer to 65 ha. To obtain enough cash for survival, many sold one half of their land, resulting in many 32.5-ha homesteads. In efforts to survive, farmers plowed most of the land on a farm and attempted crop production regardless of slope or soil depth. Without modern conservation methods, erosion took its toll, soils became less productive, and the plight of the farmers continually increased, culminating in the severe drought and depression of the 1930's (McPheters, 1938).

14-2.2.3.1 Revegetation of Eroded Land. The first problem facing the newly developed Soil Erosion Service in 1933 was permanent revegetation of land unsuitable for crop production (Harper, 1951). In those early days, revegetation was a mammoth job. Little or no research information was available. After experimentation with various species, including introduced legumes, it was learned that native grass mixtures were best for revegetation (Daniel et al., 1942) but that the process was slow and expensive. The first major obstacle was the development of a grass seed industry to obtain needed quantities of seed. Lack of harvesting and processing methods as well as seeding methods were major obstacles. Eventually, specialized equipment was developed to handle the problems. Presently, over 2 million ha have been revegetated to native grasses in Oklahoma, with another 0.8 million ha planted to permanent improved species. The improved species are primarily bermudagrass and weeping lovegrass. Improved varieties of native grasses have also been released, and these add to forage production. Revegetation of lands unsuited for cultivation has added substantially to the livestock economy of the region. Many farmers and ranchers now plant combinations of native bluestem, bermudagrass, and weeping lovegrass to extend the grazing season and to obtain more flexibility in grazing systems. However, each time more grain or cotton production is needed, more marginal land is plowed and returned to crop production. This occurred during World Wars I and II and again in the early 1970's when crop prices soared. Each plow-out period was followed by a need for revegetation. Better varieties, improved equipment, and availability of fertilizers are now making revegetation more successful and profitable in a shorter time period.

14-2.2.3.2 Terraces for Erosion Control. The major obstacle in managing terraced fields in the Central Rolling Red Plains and Prairies is terrace maintenance. Modern farming equipment is quite large and not suited to small fields. It is, therefore, no longer practical to farm terrace intervals as separate fields if terraces are closely spaced. Farming over terraces necessitates reshaping the terraces periodically. Also, loam and sandy loam soils have a tendency to move downhill when saturated with water, resulting in extremely difficult terrace maintenance problems. Terraces, however, are vitally needed on these soils.

Modern terraces in the Central Rolling Red Plains and Central Rolling Red Prairies are broad enough to be crossable with large farming equipment. They are constructed with soil material from the back side of the terrace with only a shallow, saucer-shaped channel. Many nearly level terraces that successfully conserve water and reduce erosion have been constructed. Most level terraces are constructed with open ends that allow runoff. On permeable soils, the ends are closed to impound runoff and allow more time for water infiltration.

A relatively new idea in water management and erosion control is construction of level benches where one third of the area is leveled to receive runoff water from the other two thirds (Hauser and Cox, 1962; Zingg and Hauser, 1959). In this plan, the more drought-resistant crops are grown on the upper two thirds, and those requiring more soil water are planted on the level, water-receiving, lower one third. Even though the advantages of this system are obvious, certain disadvantages have resulted in limited acceptance. Construction costs are high, and the necessary severe land scraping reduces soil productivity. Several years of rather intense management along with soil amendments are often required to restore productivity. Increased yields on the lower, water-accepting area occur only in years with substantial runoff. Runoff in the region averages only 3 to 10% of the average annual rainfall. In drought years, no runoff may occur.

Other innovations in terrace construction in the region include (i) increasing terrace intervals by rigidly practicing supplementary conservation measures, e.g., stubble mulching and conservation tillage, and (ii) constructing parallel terraces that have rectangular intervals of approximate equal size and that can easily be farmed as fields, thus allowing for easier planning of rotations and cropping systems. Parallel terraces are especially useful in row cropping because they avoid the need for point rows and uneven lands. Parallel terraces are not generally successful on very erosive soils and require that the slopes of the field be gentle and rather uniform across the field.

In the Central Reddish Prairie, solonetz spots (high sodium soils) occur in some associations. Soils of these "slick spots" are disperse, impermeable, and extremely erodible (Plice and Emerson, 1947). Terraces commonly fail where they traverse such areas, and special care must be taken to avoid them.

14-2.2.3.3 Waterways and Gully Shaping. Gully shaping, leveling (or both) is often necessary on fields before constructing terraces. The success or failure of terraces depends on the waterway system. A properly designed waterway conducts water from the terraces off the field and into a satisfactory stream without creating a gully. Waterways must be adequately protected with vegetation, which usually must be established before terrace construction.

Waterways take land out of crop production and are hard to maintain if siltation from the cultivated field occurs. In some cases, insects are harbored in waterway vegetation and spread to the crop in the field.

14-2.2.3.4 Contouring. Nearly all contour farming in the Central Rolling Red Plains and Prairies is done along with terracing. In row crop areas, contouring is a common practice. There is a growing trend toward tilling the soil without respect to the contour because of the use of larger equipment, but rows frequently run parallel to the terraces when crops are planted. Contouring helps terrace systems perform better and is an effective water conservation and erosion control practice. The use of large equipment is the biggest deterrent to contouring. Point rows resulting from irregular terrace intervals are difficult to plant, cultivate, and harvest with large equipment. Contouring is not widely practiced in wheat production.

14-2.2.3.5 Tillage. Many changes in tillage methods and frequency of tillage have occurred since cultivation of crops began. Stubble mulching has received considerable attention in the region and is practiced to some degree on a relatively large area, especially in the Rolling Red Plains. The choice of tools depends on the amount of residue produced, the soil, and soil water conditions. Leaving sufficient residue on the surface almost always involves some use of sweeps or large blades, but rod weeders and mulch treaders are also used. However, many wheat producers also use chisels and, if the stubble is heavy, disk plows. The residue should remain anchored to the soil for proper drill operation. Hoe-type drills perform better than disk opener drills in stubble-mulched soil. Rolling coulters are not widely used ahead of the openers primarily because of the extra expense involved.

The practice of stubble mulching has declined during the past decade because of failure to control annual winter grasses, e.g., downy brome, Japanese brome, cheat, and hairy chess. Wheat that is planted early for increased grazing yields is particularly vulnerable to winter annual grass infestation. In addition, these winter grasses are more tolerant to soil acidity than wheat and are especially aggressive competitors with wheat where surface soils are acid. These grasses can be partially controlled by moldboard plowing at about 4-year intervals and are effectively controlled by crop rotations. Certain herbicides have been labeled for controlling the *Bromus* sp. in wheat, but they are rather restrictive as to soils and varieties (Runyan et al., 1982).

Stubble mulching is effective for reducing both wind and water erosion and for improving soil structure (Tanchandrphongs and Davidson, 1970). However, in the short run, unlike in major land resource area no. 72 where soil fertility is higher, yields often are not increased by stubble mulch tillage (Knisel et al., 1961).

Conventional clean tillage operations for wheat production in the Central Rolling Red Prairie generally include the use of (i) a chisel, (ii) a moldboard plow, (iii) a disk or field cultivator, and (iv) a spring-tooth harrow (two times). The moldboard plow is not used for clean tillage in the Central Rolling Red Plains, and the second tillage is often done with a disk plow or field cultivator. Moldboard plowing is too slow and expensive for the very extensive farming regions where farms are large.

In addition to stubble mulch tillage, chisels and field cultivators, which leave the soil surface rough and cloddy, are used for erosion control. Disk tillage tends to leave the soil surface too smooth and without adequate-sized clods for controlling erosion (Fenster et al., 1965).

Large areas in the Central Rolling Red Plains are sandy on the surface but contain clayey substrata (Gray and Roozitalab, 1976). These soils are extremely erodible by wind and are difficult to manage. Deep plowing to bring sufficient clay to the surface to increase the cloddiness helps immensely in wind erosion control. An area of about 1 million ha in the Rolling Red Plains has been deep plowed with large (90-cm) disk plows. The operation is expensive but results in a permanent improvement. Experiments have shown rather dramatic yield increases after deep plowing these soils (Harper and Bensing, 1950).

14–2.2.3.6 Stripcropping. Stripcropping is practiced to some extent in the region on the most wind-erodible soils. In general, the strips are too wide to adequately control wind erosion. The greater efficiencies resulting from farming larger blocks causes a compromise that results in the strips being too wide. Strips of small grains, cotton, sorghum, and alfalfa are grown in nearly every combination. The rows run east and west, which is at right angles to the prevailing winds. On terraced soils, the rows cross the terraces because wind erosion is a greater threat than water erosion.

14–2.2.3.7 Windbreaks. Trees for windbreaks were extensively planted for providing protection against wind erosion in the late 1930's. In fact, the first shelterbelt planted in the USA was in Greer County, Oklahoma. Shelterbelts provide protection on wind erosive soils (Stoeckeler and Williams, 1949). Many quarter section fields were protected by two or three shelterbelts. However, the shelterbelts took considerable land out of crop production because the trees occupied space and because the tree species that were used competed with crops for a considerable distance from the tree row. For this reason, many shelterbelts were removed in the 1960's when land costs soared and production was needed on as much land as possible. Trees in some older plantings also died during the drought of the early 1950's.

Recent plantings of trees for shelterbelts have been greatly improved. Fewer tree rows are used, thus taking less land out of production, and taprooted trees, which extract water and nutrients for only a short distance, are used. This objective was achieved by compromising with aerodynamics of the shelterbelt because of tree shape but without apparent loss of effectiveness. In spite of the improvements, no widespread trend toward increased plantings has developed.

14–2.2.3.8 Cover Crops. Cover crops are frequently planted on wind-erosive soils in the region. Rye and other small grains are often planted in cotton either before or immediately after harvest. The cover crop can be grazed in the spring while it is protecting the soil. A disadvantage of this system is the extraction of soil water by the cover crop. This water must be replenished or the subsequent summer crop will suffer (Murphy, 1956). During extremely dry spring months, soil preparation for the summer crop can be hampered. To overcome the disadvantage of excessive water extraction by the cover crop, there is a trend toward killing the crop with contact herbicides early enough to give an opportunity for restoring soil water while the residue still protects the soil.

Summer cover crops are also used on some very sandy soils to protect against wind erosion during the period in which a fall-sown crop is being established. Sudan grasses and other forage sorghums can be seeded in late summer and, if soil water is adequate, sufficient growth obtained to protect the soil during fall and winter. The growth of sudangrass can be stopped either by a contact herbicide or by frost. Alfalfa is traditionally fall-seeded on very sandy soils by using this approach. This is an improvement over using small grain as a nurse crop for alfalfa because it eliminates the competition between small grain and the alfalfa during early spring growth.

14-2.2.3.9 Catch Crops. Catch crops are important in the region, especially in the Rolling Red Plains. Flexibility is a vital part of crop production in the region. Because of both wind and water erosion hazards, soils cannot be left bare for extended periods. When crop failures occur, it is important to change plans and seed a crop whenever water conditions are favorable. Any adapted crop can be used as a catch crop, but those most commonly used in the region include sorghum, sudan, and oat. Often plans are changed because of the weather to switch the cash crops to fit the growing conditions.

14-3 TECHNOLOGICAL NEEDS

Technological and management practices that allow the production of large amounts of food and fiber under dryland conditions in the Central and Southern Great Plains have been developed. Although many practices aid in controlling wind and water erosion, erosion remains a major problem in this highly important crop production region. Therefore, an urgent need to develop improved practices or to hasten application of effective erosion control measures to practical crop production systems remains.

Much research has been conducted in relation to the factors that influence wind and water erosion. As a result of this research, generalized equations for wind and water erosion have been developed (Woodruff and Siddoway, 1965; Wischmeier and Smith, 1978). These equations are applicable to many situations but need improvements, especially where conservation tillage is practiced.

Conservation tillage, a rather recent development, includes those farming practices that retain plant residues from previous crops on the soil surface. Stubble mulch tillage, which is widely used in the Great Plains, is a type of conservation tillage. Other types, including no-tillage, have been developed, and some are applicable to the Great Plains. Conservation tillage has the potential to control erosion on many soils, because plant residues shield the surface from high winds and raindrops and prevent the loosening and carrying away of soil particles (Cruse and Erbach, 1981).

Although conservation tillage controls erosion in many situations, it has not been adapted to all conditions and crops because the technological and management practices developed for clean-tilled soils are not always applicable to conservation-tilled soils. Therefore, research involving conservation tillage is needed to develop practices that will maintain or improve crop yields and control erosion and yet be practical economically.

Finally, a strong extension program to educate and encourage producers to use effective conservation practices in their crop production systems is needed. Many presently available practices aid in controlling erosion, and these should be used until improved practices are available. Only through the use of effective conservation of our soil resource can we hope to indefinitely maintain crop productivity in the highly important Central and Southern Great Plains.

14-4 LITERATURE CITED

Anderson, D. T. 1961. Surface trash conservation with tillage machines. Can. J. Soil Sci. 41: 99-114.

Austin, M. E. 1965. Land resource regions and major land resource areas of the United States. USDA Agric. Handbook no. 296.

Banbury, E. E., L. R. Reinhordt, and W. L. Barker. 1977. Reducing soil erosion with residue management. Kansas Agric. Extension Service Circ. no. 403.

Bennett, H. H. 1939. Soil conservation. McGraw-Hill Publications, New York.

Black, A. L., and J. F. Power. 1965. Effect of chemical and mechanical methods on moisture storage, wheat yields, and soil erodibility. Soil Sci. Soc. Am. Proc. 29:465-468.

Bonnen, C. A. 1960. Types of farming in Texas. Texas Agric. Exp. Stn. Bull. no. B-964.

Cruse, R. M., and D. C. Erbach. 1981. Conservation tillage. p. 65-68. *In* Yearbook of Agriculture. U.S. Dep. of Agric., U.S. Government Printing Office, Washington, D.C.

Curry, B. R. 1972. Monthly precipitation quantiles in Oklahoma. Oklahoma Agric. Exp. Stn. Miscellaneous Pub. no. MP-88.

Dale, T. 1941. Conservation farming for the hardlands of the Southern Great Plains. U.S. Dep. of Agric., Soil Conserv. Service, U.S. Government Printing Office, Washington, D.C.

Daniel, H. A., H. M. Elwell, and H. F. Murphy. 1942. Conservation and better land use for Oklahoma. Oklahoma Agric. Exp. Stn. Bull. B-257.

Duley, F. L. 1959. Progress of research on stubble mulching in the Great Plains. J. Soil Water Conserv. 14:7-11.

Eby, J. K., and C. J. Whitfield. 1940. Soil and erosion changes on the Dalhart sand dune area. Agron. J. 32:290-296.

Fenster, C. R. 1960. Stubble mulching with various types of machinery. Soil Sci. Soc. Am. Proc. 24:518-523.

Fenster, C. R., O. C. Burnside, and G. A. Wicks. 1965. Chemical fallow studies in winter wheat fallow rotations in western Nebraska. Agron. J. 57:469-470.

Fenster, C. R., N. P. Woodruff, W. S. Chepil, and F. H. Siddoway. 1965. Performance of tillage implements in a stubble mulch system: III. Effects of tillage sequences on residues, soil cloddiness, weed control, and wheat yield. Agron. J. 57:52-55.

Flora, S. D. 1948. Climate of Kansas. Kansas State Board of Agric. Report no. 285.

Foster, G. R., D. K. McCool, K. G. Renard, and W. C. Moldenhauer. 1981. Conversion of the universal soil loss equation to SI metric units. J. Soil Water Conserv. 36:355-359.

Frye, J. C. 1970. The Ogallala formation—a review. Int. Center for Arid and Semi-Arid Land Studies Spec. Report no. 39, Texas Technological Univ., Lubbock.

Fryrear, D. W. 1969. Reducing wind erosion in the Southern Great Plains. Texas Agric. Exp. Stn. Miscellaneous Pub. no. MP-929.

Gray, F., and M. Roozitalab. 1976. Benchmark and key soils of Oklahoma. Oklahoma State Univ. Miscellaneous Pub. no. MP-97.

Greb, B. W., D. E. Smika, N. P. Woodruff, and C. J. Whitfield. 1974. Summer fallow in the Central Great Plains. p. 51-85. *In* Summer fallow in the western United States. USDA-ARS Conserv. Res. Report no. 17.

Gutentag, E. D. 1978. Relationship between ground water and energy in the Ogallala aquifer in western Kansas—a preliminary evaluation. Mid-Continent Res. and Development Coun., Fort Collins, Colo.

Haas, H. J., and G. O. Boatwright. 1960. Let's take another look at summer fallow in the Northern Plains. J. Soil Water Conserv. 15:176-179.

Harper, H. J. 1951. Problems and progress of soil conservation. Adv. Agron. 3:265-322.

Harper, H. J., and O. H. Brensing. 1950. Deep plowing to improve sandy land. Oklahoma Agric. Exp. Stn. Bull. no. B-362.

Hauser, V. L., and M. B. Cox. 1962. Evaluation of Zingg conservation bench terrace. Agric. Eng. 43:462–464, 467.

Hay, D. R. 1976. 1975 Kansas irrigation survey engineering newsletter. USDA Cooperative Extension Service, Kansas State Univ., Manhattan.

Hobbs, J. A., and C. A. Thompson. 1971. Effect of cultivation on the nitrogen and organic carbon contents of a Kansas Argiustoll (Chernozem). Agron. J. 63:66–68.

Hughes, W. F., and W. L. Harman. 1969. Projected economic life of water resources, subdivision no. 1. High Plains underground water district. Texas Agric. Exp. Stn. Tech. Monogr. no. 6.

Irrigation Journal. 1976. 1976 Irrigation survey, survey issue. Irrig. J. 26(6):23–30.

Johnson, W. C., and R. G. Davis. 1972. Research on stubble-mulch farming of winter wheat. USDA Conserv. Res. Report no. 16.

Kansas Water Resources Board. 1967. Irrigation in Kansas. Kansas Water Resource Board Pub., Planning for Development, 701 Project no. Kansas P-43, Report no. 16(e), Part I, Topeka.

Knisel, W. G., Jr., M. B. Cox, and B. B. Tucker. 1961. Effect of land treatment on runoff at Cherokee, Oklahoma. Oklahoma Agric. Exp. Stn. Bull. no. B-583.

Lotspeich, F. B., and J. R. Coover. 1962. Soil forming factors on the Llano Estacado. Tex. J. Sci. 14:7–17.

McPheters, W. H. 1938. Soil erosion control in Oklahoma. Oklahoma Extension Circ. no. 218, Series 56.

Murphy, H. F. 1956. Green manure and cotton in a double-cropping system on a fine-textured soil. Oklahoma Agric. Exp. Stn. Bull. no. B-472.

New, L. 1976. High Plains irrigation survey. Texas Agric. Extension Service, College Station.

Norwood, C. A., and G. M. Herron. 1977. Use stubble mulch to control erosion in western Kansas. Kansas Agric. Exp. Stn. Circ. no. 411.

Plice, M. J., and R. L. Emerson. 1947. Some bronomic and other characteristics of slick-spot soils in Oklahoma. Proc. Okla. Acad. Sci. 27:97–101.

Ramig, R. E., and D. E. Smika. 1964. Fallow-wheat-sorghum: An excellent rotation for dryland in central Nebraska. Nebraska Agric. Exp. Stn. Pub. no. SB-483.

Rule, G. K. 1939. Land facts on the Southern Plains. USDA-SCS Miscellaneous Pub. no. 334.

Runyan, T. J., W. K. McNeal, and T. F. Peeper. 1982. Differential tolerance of wheat cultivars to metribuzin. Weed Sci. 30:94–98.

Shantz, H. L. 1923. The natural vegetation of the Great Plains region. Ann. Assoc. Am. Geogr. 13:81–107.

Skidmore, E. L., M. Kumar, and W. E. Larson. 1979. Crop residue management for wind erosion control in the Great Plains. J. Soil Water Conserv. 34:90–98.

Skidmore, E. L., and F. H. Siddoway. 1978. Crop residue requirements to control wind erosion. p. 17–33. *In* W. R. Oschwald (ed.) Crop residue management systems. ASA Spec. Pub. no. 31, Am. Soc. of Agron., Madison, Wis.

Skidmore, E. L., and N. P. Woodruff. 1968. Wind erosion forces in the United States and their use in predicting soil loss. USDA Agric. Handbook no. 346.

Smika, D. E., and G. O. Wicks. 1968. Soil water storage during fallow in the Central Great Plains as influenced by tillage and herbicide treatments. Soil Sci. Soc. Am. Proc. 32:591–595.

Soil Conservation Service. 1963. The Dalhart soils: A compilation of benchmark data pertaining to the series. USDA-SCS mimeo, U.S. Government Printing Office, Washington, D.C.

Soil Survey Staff. 1975. Soil taxonomy—a basic system of soil classification for making and interpreting soil surveys. USDA-SCS Agric. Handbook no. 436.

Stoeckeler, H., and R. A. Williams. 1949. Use of shelter belts for controlling wind erosion. p. 191–199. *In* Yearbook of agriculture. U.S. Dep. of Agric., U.S. Government Printing Office, Washington, D.C.

Tanchandrphongs, S., and J. M. Davidson. 1970. Bulk density, aggregate stability, and organic matter content as influenced by two wheatland soil management practices. Soil Sci. Soc. Am. Proc. 34:302–305.

Thornthwaite, C. W. 1931. The climates of North America according to a new classification. Geogr. Rev. 21:633–655.

Unger, P. W., R. R. Allen, and A. F. Wiese. 1971. Tillage and herbicides for surface residue maintenance, weed control, and water conservation. J. Soil Water Conserv. 26:147–150.

Unger, P. W., and A. F. Wiese. 1979. Managing irrigated winter wheat residues for water storage and subsequent dryland grain sorghum production. Soil Sci. Soc. Am. J. 43:582–588.

Unger, P. W., A. F. Wiese, and R. R. Allen. 1977. Conservation tillage in the Southern Plains. J. Soil Water Conserv. 32:43–48.

Walker, L., and H. Taylor. 1975. Texas Water Development Board High Plains study shows 340 million acre-feet of water in 45-county area. p. 20–22. *In* Water for Texas, no. 1 and 2. Texas Water Development Board, Austin.
conservation planning. USDA-SEA Agric. Handbook no. 537.

Woodruff, N. P., and W. S. Chepil. 1958. Influence of oneway disk and subsurface-sweep tillage factors affecting wind erosion. Trans. ASAE 1:81–85.

Woodruff, N. P., C. R. Fenster, W. S. Chepil, and F. H. Siddoway. 1965. Performance of tillage implements in a stubble mulch system: I. Residue conservation. Agron. J. 57:45–49.

Woodruff, N. P., and F. H. Siddoway. 1965. A wind erosion equation. Soil Sci. Soc. Am. Proc. 29:602–608.

Zingg, A. W., and V. L. Hauser. 1959. Terrace benching to save potential runoff for semiarid land. Agron. J. 51:289–292.

Zingg, A. W., and C. J. Whitfield. 1957. Stubble-mulch farming in the western states. USDA Tech. Bull. no. 1166.

15

R. A. Young, M. J. Lindstrom, and R. F. Holt
Agricultural Research Service
U.S. Department of Agriculture
Morris, Minnesota

Soil Conservation: Central and Northern Great Plains (Eastern)

The eastern Central and Northern Great Plains include the western part of the Central Feed Grains and Livestock Land Resource Region (Central Feed Grains Region) (Fig. 1-1). Major land resource areas no. 102, 106, and 107 (Fig. 15-1) are located within this resource region (Austin, 1965). These resource areas extend from west central Minnesota through eastern South Dakota, western Iowa, eastern Nebraska, northeastern Kansas, and into northwestern Missouri and encompass about 16.6 million ha. The average annual precipitation ranges from about 510 mm in the northwestern portions to 910 mm in the southeast extremities. The rainfall factor (R) in the universal soil loss equation for water erosion varies from about 100 in the northern portion to about 250 in the southern (Wischmeier and Smith, 1978); the climatic factor (C') in the wind erosion equation decreases from 20 in the western portion to about 10 in the eastern portion of the area shown in Fig. 15-1 (Woodruff and Siddoway, 1965). Major problems associated with agriculture are limited soil moisture for crop growth during periods when evapotranspiration exceeds precipitation and intense summer storms that severely erode unprotected soils.

The loess mantle soils, covering a large portion of the area, generally have a high water-holding capacity and are highly erosive. However, the combination of high intensity, summer rainfall with the subsequent runoff, and low noncropping season precipitation contributes to shortages in stored soil water for crop growth. Conservation practices that control erosion while maintaining high levels of crop production can help solve some of these agricultural problems.

Contribution from the North Central Soil Conservation Research Laboratory, Agricultural Research Service, USDA, Morris, Minn., in cooperation with the Minnesota Agricultural Experiment Station, Science Journal Series 10,902.

Dryland Agriculture—Agronomy Monograph no. 23.

Fig. 15-1. Dryland portion of the Central Feed Grains and Livestock Region.

Stewart et al. (1975) have estimated that the relative potential contribution of cropland to watershed sediment yield ranges from high to very high and that on over 40% of the major land resource areas erosion is the dominant limitation to agricultural use. The amounts of cultivated land in major land resource areas no. 102, 106, and 107 are 68, 55, and 65%, respectively. Figure 15-2 shows the amount of cultivated land used for the major crops in these areas.

15-1 PRACTICES

By themselves, individual conservation practices are seldom sufficient to reduce soil losses to tolerance levels. A combination of practices is usually required, e.g., terracing and conservation tillage. Figure 15-3 summarizes the amount of cultivated area protected by terraces, diversions, field windbreaks, strip cropping, and contouring in the approximately 9 million ha of the dryland portion of the Central Feed Grains Region.

15-1.1 Terracing

Terracing, a method of controlling erosion by constructing broad channels across the slope of rolling land, is used where other cropping or management practices are inadequate to provide the desired amount of pro-

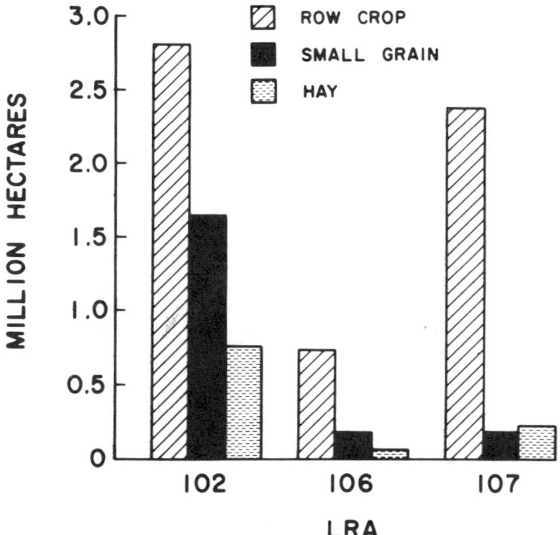

Fig. 15-2. Amounts of cultivated land used to grow major crops in the various major land resource areas (Soil Conservation Service. 1979. Unpublished data).

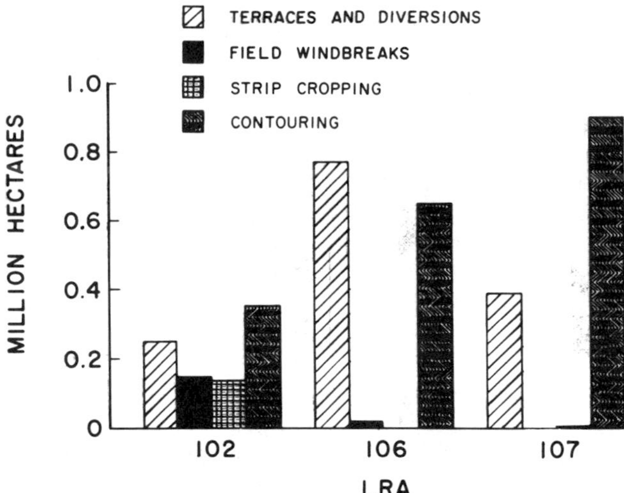

Fig. 15-3. Area protected by various conservation practices in the major land resource areas (Soil Conservation Service. 1979. Unpublished data).

tection. Terraces can stabilize cropland on extremely steep slopes; however, they are expensive to construct, require a fair amount of technical expertise, and are not always compatible with modern farming methods. Their construction usually requires the use of specialized equipment, e.g., bulldozers, graders, and other heavy earth movers, although if conditions are suitable, some small terraces may be constructed on gently sloping land with a moldboard plow or disk. Once terraces are installed, they require more maintenance than most other conservation methods. Proper maintenance, which

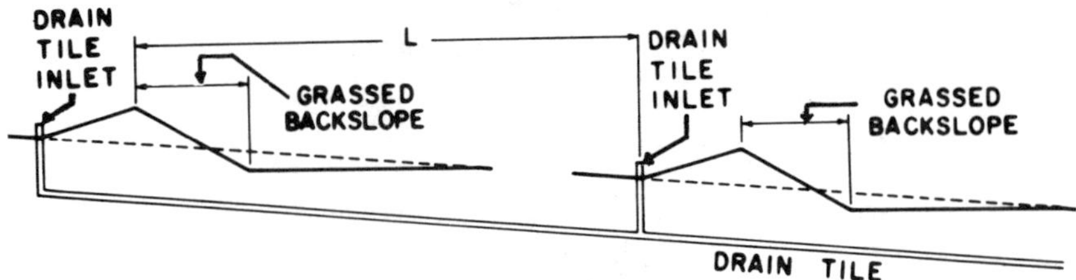

Fig. 15-4. Steep seeded back slope terrace.

usually can be accomplished with standard farm equipment, is very important, especially for the first 1 or 2 years after construction.

The use of terraces is limited in the northern part of the region and increases further south (Fig. 15-3). Major land resource area no. 102, which includes the northern portion of the region, is dominated by rolling hills with highly irregular topography that is not conducive to terrace construction. Further south, slopes tend to be longer and more uniform, making their use more practical.

Several different types of terraces, as described by Laflen and Moldenhauer (1971), are used in the area. All have the same primary purpose—reducing erosion by shortening the slope length to control erosive concentrated flow. In the 1950's and 1960's, most terraces constructed in this region were broad-based, level terraces on the deep loess soils and broad-based, gradient terraces on the more shallow soils. In more recent years, farmers have straightened the row pattern to accommodate the use of multirow machinery. As a result, the main purpose of terraces, reducing slope length, has been negated in some instances. In some extreme cases, the trend toward larger tractors and other farm equipment has caused some terrace systems to be abandoned because they restrict their operation. In the 1960's and 1970's, steep, seeded, back slope terraces became the most common type of terrace built. Most of these were built from the lower side, which tended to reduce the slope between terraces. Also, they were typically drained by an underground tile system that did not need to follow the contour (Fig. 15-4). They can be parallel and nearly straight, thus eliminating point rows and irregular areas, making them much more amenable to modern farming methods. Research has shown that these terraces reduce soil losses from the areas above them by about 95% (Laflen et al., 1972).

15-1.2 Diversions

Diversions are usually used to divert all runoff away from a site and to channel it to a suitable outlet. Like terraces, they usually require heavy equipment for their construction. They are not used for erosion control to any great extent in the region, because less than 0.2% of the total cultivated area is protected by diversions.

15–1.3 Stripcropping

Strip cropping is a relatively limited conservation practice in the dryland portion of the Central Feed Grains Region. Only 1.5% of the cultivated area is protected by this method. This practice is usually used in combination with other conservation practices, e.g., contouring or crop rotations. If the strips are formed on the contour, they protect slopes from water erosion by limiting their length and reducing runoff. However, if they are oriented normally to the direction of the prevailing winds, they protect cropland against wind erosion. They help to conserve soil moisture, especially when formed on the contour, by increasing snow catch and decreasing runoff from snowmelt and rainfall. However, the narrow strips are not always compatible with modern farming methods that use large multirow equipment. Also, the greater number of exposed crop borders between strips can present a risk of increased insect damage (Schwab et al., 1966).

15–1.4 Contouring

Contouring, the practice of carrying out all tillage practices as closely as possible following the natural contour of the land, tends to form small ridges and furrows that decrease the velocity of surface runoff and effectively decreases erosion. However, if no other conservation practice (e.g., terracing or strip cropping) is used, these contour ridges can potentially overtop or break and release relatively large volumes of water suddenly from the row, possibly increasing erosion (Schwab et al., 1966). On steeper slopes, these breaks can cause gulleying.

Contouring requires no special equipment other than hand levels or simple levels unless it is done with terraces; then more accurate leveling systems are required.

Almost 20% of the cultivated land in this region is contoured. Most of this land is in the southern portion of the region, where 65 and 32% of the cultivated land in major land resource areas no. 106 and 107 are contoured, respectively. Further north, major land resource area no. 102 has only about 7% of the cultivated area contoured, again, in agreement with observations regarding terracing, as a result of the highly irregular topography with generally short and variable slopes not conducive to this practice.

15–1.5 Windbreaks

Field windbreaks, or shelterbelts, are any barriers formed to protect land from wind, but they usually consist of shrubs or trees planted to protect fields and conserve moisture. They reduce wind velocities in their vicinity, thus decreasing evaporation and reducing severe burning of crops, increasing yields (Woodruff et al., 1963). They also provide farm woodlots and wildlife habitat.

Only about 3% of the cultivated acres in the region is protected by field windbreaks. Most of this area is in the northern part of the region where strong, erosive winds are more common. Further south, field windbreaks are used less as a conservation practice.

Research has shown that vegetative barriers obstructing the path of wind reduce the wind speed and momentum of air flow and thus afford wind erosion protection (Woodruff et al., 1972). Properly oriented barriers will reduce wind erosion forces by more than 50% from the barrier leeward to 20 times its height (Skidmore and Hagen, 1977). Tall-growing crops, e.g., rows of corn planted at regular intervals in fields of low-growing crops, also dissipate wind energy and increase crop yields more effectively than do more dense wind barriers. Studies by Radke (1976), Radke and Burrows (1970), Radke and Hagstrom (1973, 1974, 1976), and Brown and Rosenberg (1972) have shown that crops sheltered by annual wind barriers grow taller, produce more dry matter, and have larger leaf area indices and higher yields. Yields of sheltered soybean have been increased as much as 25% (Radke and Hagstrom, 1973).

15-2 MINIMUM TILLAGE AND CONSERVATION CROPPING SYSTEMS

Figure 15-5 shows data on the hectares of cultivated land, cultivated land adequately treated, land with conservation cropping systems, and land under minimum tillage. The working definition of conservation cropping

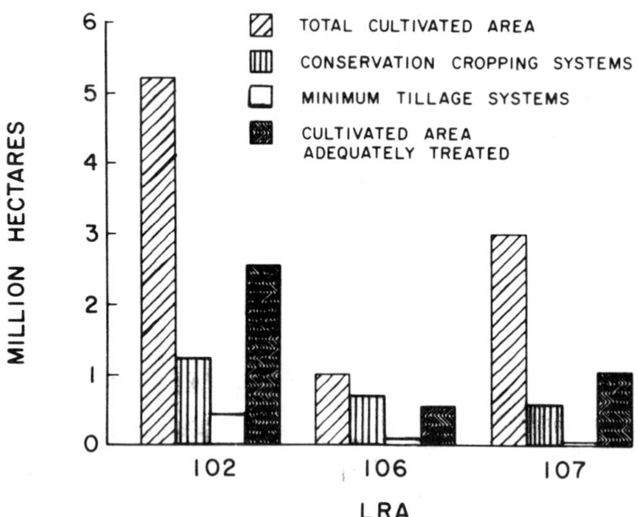

Fig. 15-5. Total cultivated area and area protected by conservation systems for the major land resource areas (Soil Conservation Service. 1979. Unpublished data).

systems varies throughout the USA, but, in general, it includes those management systems that are designed to reduce soil loss more than that reduced by conventional management systems. They may include residue management systems, reduced tillage systems, crop rotations with a sod crop, and contour or stripcropping. The reported amount of land under minimum tillage was 6% of the cultivated area for the three major land resource areas. The large area reportedly under conservation systems indicates that farmers are becoming more aware of water runoff and soil erosion problems. However, a comparison of adequately treated cultivated land with the total cultivated land shows that conservation efforts must continue.

According to Amemiya (1977), of the many conservation tillage systimes used in the western Corn Belt, disking and chisel plowing are the most common. Till and no-till planting are being tried throughout the area; however, at present, these systems are used on only a small percentage of the total cultivated area.

Amemiya (1977) presented yields of corn and soybean from many tillage trials conducted in the western Corn Belt. In general, yields from conservation tillage systems were comparable with conventional clean-tilled systems. Griffith et al. (1977), however, reported yield reductions with conservation and no-till systems in the eastern Corn Belt on soils with poor drainage. Other possible problems (Amemiya, 1977) with conservation tillage systems are cooler soil temperatures in early spring (especially on heavy-textured, imperfectly drained soils), weed control (in the northern part of the Corn Belt), insect and disease problems, poor seed placement, and lower nutrient availability. Control of erosion by conservation tillage practices has been well documented, but many factors relating to crop production require additional research.

15-3 DEVELOPMENTS IN TECHNOLOGY

The critical period for soil erosion is between primary tillage, either fall or spring, to the time when the crop canopy is established. During this period, the soil is bare from conventional tillage and susceptible to raindrop impact that dissipates most of its contained energy at the soil surface, detaching soil particles, breaking down soil aggregates, reducing surface roughness, and causing surface sealing. The total effect is reduced infiltration and increased potential for runoff and soil erosion.

Crop residue management systems that leave residue on the soil surface are a promising method for reducing runoff and erosion. Surface coverage by residue dissipates raindrop energy, which increases infiltration and thus reduces the amount of potential runoff. To show the effectiveness of residue cover in reducing soil loss, Wischmeier (1973) used a mulch factor (ratio of soil loss with various percentages of mulch cover to loss with no mulch). Gupta et al. (1979) and Lindstrom et al. (1979) demonstrated, by applying the universal soil loss equation (Wischmeier and Smith, 1978), that tillage systems that leave residue on the soil surface can substantially reduce soil loss. For example, Lindstrom et al. (1979) calculated for major land resource areas no. 102, 106, and 107, respectively, that soil losses would be 14, 28, and 45 t ha^{-1} y^{-1} for conventional (fall plow, spring disk) systems when all residue is removed and 7, 11, and 19 t ha^{-1} y^{-1} for conservation (fall chisel, spring disk, or sweep) systems when 3.9 t/ha of residue is left on the soil after harvest.

Residue remaining on the soil surface will increase the amount of soil moisture available for crop growth both by increasing infiltration and by reducing evaporation of stored water (Onstad and Otterby, 1979). Miller and Shrader (1976) developed a yield response curve to soil moisture and evaluated by computerized modeling techniques the potential effects of conservation tillage systems on corn yields in western Iowa. This study showed that when spring soil moisture levels were normal or below normal, conservation tillage systems effectively increased yield estimates over conventional tillage systems.

Agricultural pollutants are the main source of nonpoint water pollution, with runoff from cropland contributing about 50% of the total sediment entering the nation's waterways (Wadleigh, 1968). To select best management practices for controlling erosion, maintaining soil productivity, and reducing sediment yield to prevent excess degradation of water quality, we must estimate nutrient and pesticide losses from farm fields, in addition to erosion and sediment yield, since the farm field is the basic management unit. Since it is impossible to monitor the runoff from all farm fields, soil loss or sediment-yield-prediction models provide the only means of obtaining these estimates. Through the use of established soil loss tolerances for given areas, models enable planners to choose the best management practices that will keep soil losses within allowable limits.

Today, the universal soil loss equation is probably the most commonly used soil loss prediction model (Wischmeier and Smith, 1960). However, new, more fundamental models based on the physical principles of the erosion process have been developed (Frere, 1978; Knisel, 1978; Knisel, 1980; Foster et al., 1977). These models simulate basic hydrologic, erosion, sedimentation, and chemical transport processes from field-size areas and relate them to land use and management. Such models are probably the single most significant development in erosion research today. By using allowable soil loss values and desired end products as inputs, they enable the planner to select the best management practice to obtain the desired goals. Because models are only as accurate as their weakest component, these new generation models can be used to determine the particular aspects of the erosion process that require more information and, thus, show where further research is needed.

15-4 TECHNOLOGY CONCERNS AND NEEDS

In various regions, research is needed to develop cropping management systems that will maximize water storage throughout the year. Several studies have investigated soil water recharge during the nongrowing season in the northern Corn Belt and Great Plains. At Swift Center, Saskatchewan,

Staple et al. (1960) found that 37% of precipitation stored over the winter was conserved in stubble fields compared with 9% in fallow. Willis et al. (1961) reported that snow-covered soils gained soil water over the winter, whereas bare soils (fallow) lost water at Mandan, North Dakota. Timmons and Holt (1968) demonstrated that recharge during the fall (crop maturity to freeze-up) in western Minnesota and eastern South Dakota was most efficient. However, all of these investigators have shown an inverse relationship between soil water content at freeze-up and soil water increase from freeze-up to thaw. In general, therefore, we can assume that some management practices are considerably more effective than others in increasing soil water storage during the season when the soil is frozen. A real challenge exists to develop systems that will maximize soil water storage during the noncropping season to assure adequate moisture for the following crop.

Even though the problems associated with soil water storage during a 5-month frozen soil period are not as critical in the southern parts of the region as in the northern parts, adequate soil water is still required to assure good crop yields. Only about 25% of the annual precipitation for the three major land resource areas discussed occurs during the noncropping season (i.e., 1 October through 31 March), potentially limiting soil water storage. The relationship between stored soil water and crop yields has been demonstrated by several investigators (Denmead and Shaw, 1962; Dale, 1964; Holt et al., 1964). Soil conservation measures are effective in preventing surface runoff as well as erosion and therefore afford an added incentive for soil conservation.

The fact that a soybean crop tends to increase erodibility has been documented as far back as the early 1940's (Browning et al., 1942), but researchers have been unable to explain the reason for it. Since 1960, soybean hectarage has increased sharply, about 200% in the region. Most of this increase was in the southwestern portion in major land resource area no. 106, where now more than seven times as many hectares are planted to soybeans as in 1960 (USDA, 1978, 1960). With this large increase in soybean hectarages, the potential for soil erosion is also greatly increased, and additional research is needed on the problems resulting.

More research is also needed on the effect of various amounts of residue on erosion from various slopes, the effects of soil properties on erosion under similar conditions, the effects of tillage on wet soils, and the effects of tillage and ridge planting of row crops. Also, in general, research is needed on ways to encourage users to adopt proved soil and water conservation cropping systems and practices.

15-5 LITERATURE CITED

Amemiya, M. 1977. Conservation tillage in the western Corn Belt. J. Soil Water Conserv. 32: 29-36.

Austin, M. E. 1965. Land resource regions and major land resource areas of the United States. USDA Agric. Handbook no. 296.

Brown, K. W., and N. J. Rosenberg. 1972. Shelter-effects on micro-climate, growth and water use by irrigated sugar beets in the Great Plains. Agric. Meteorol. 9:241-263.

Browning, G. M., M. B. Russell, and J. R. Johnston. 1942. The relation of cultural treatment of corn and soybeans to moisture conditions and soil structure. Soil Sci. Soc. Am. Proc. 7: 108–113.

Dale, R. F. 1964. Changes in moisture stress days since 1933. p. 23–43. *In* Weather and our food supply. Center for Agric. and Economic Development Report no. 20, Iowa State Univ., Ames.

Denmead, O. T., and R. H. Shaw. 1962. Availability of soil water to plants as affected by soil moisture content and meteorological conditions. Agron. J. 54:385–390.

Foster, G. R., L. D. Meyer, and C. A. Onstad. 1977. An erosion equation derived from basic erosion principles. Trans. ASAE 20:678–682.

Frere, M. H. 1978. Models for predicting water pollution from agricultural watersheds. p. 501–509. *In* G. C. Vansteenkiste (ed.) Modeling, identification and control in environmental system. Elsevier North-Holland, Inc., New York.

Griffith, D. R., J. V. Mannering, and W. C. Moldenhauer. 1977. Conservation tillage in the eastern Corn Belt. J. Soil Water Conserv. 32:20–28.

Gupta, S. C., C. A. Onstad, and W. E. Larson. 1979. Predicting the effects of tillage and crop residue management on soil erosion. J. Soil Water Conserv. 34:77–79.

Holt, R. F., D. R. Timmons, W. B. Voorhees, and C. A. Van Doren. 1964. Importance of stored soil moisture to the growth of corn in the dry to moist subhumid climatic zone. Agron. J. 56:82–85.

Knisel, W. G., Jr. 1978. A system of models for evaluating nonpoint source pollution—an overview. Int. Inst. for Applied Systems Analysis, CP-78-11, A-2361, Laxenburg, Austria.

Knisel, W. G. (ed.) 1980. CREAMS: A field scale model for chemicals, runoff, and erosion from agricultural management systems. USDA Conserv. Report no. 26.

Laflen, J. M., H. P. Johnson, and R. C. Reeve. 1972. Soil loss from tile-outlet terraces. J. Soil Water Conserv. 27:74–77.

Laflen, J. M., and W. C. Moldenhauer. 1971. Soil conservation on agricultural land. J. Soil Water Conserv. 26:225–229.

Lindstrom, M. J., S. C. Gupta, C. A. Onstad, W. E. Larson, and R. F. Holt. 1979. Tillage and crop residue effects on soil erosion in the Corn Belt. J. Soil Water Conserv. 34:80–82.

Miller, E. L., and W. D. Shrader. 1976. Moisture conservation potential with conservation tillage treatments in the thick loess area of western Iowa. Agron. J. 68:374–378.

Onstad, C. A., and M. A. Otterby. 1979. Crop residue effects on runoff. J. Soil Water Conserv. 34:94–96.

Radke, J. K. 1976. The use of annual wind barriers for protecting row crops. p. 79–86. *In* R. Tinus (ed.) Shelterbelts on the Great Plains. Great Plains Agric. Coun. Pub. no. 78, Univ. of Nebraska, Lincoln.

Radke, J. K., and W. C. Burrows. 1970. Soybean plant response to temporary field windbreaks. Agron. J. 62:424–429.

Radke, J. K., and R. T. Hagstrom. 1973. Plant-water measurements on soybeans sheltered by temporary corn windbreaks. Crop Sci. 13:543–548.

Radke, J. K., and R. T. Hagstrom. 1974. Wind turbulence in a soybean field sheltered by four types of wind barriers. Agron. J. 66:273–278.

Radke, J. K., and R. T. Hagstrom. 1976. Strip intercropping for wind protection. p. 201–222. *In* R. I. Papendick. Multiple cropping. ASA Spec. Pub. no. 27, Am. Soc. of Agron., Crop Sci. Soc. of Am., and Soil Sci. Soc. of Am., Madison, Wis.

Schwab, G. O., R. K. Frevert, T. W. Edminster, and K. K. Barnes. 1966. Soil and water conservation engineering. ed. 2. J. Wiley & Sons, Inc., New York.

Skidmore, E. L., and L. J. Hagen. 1977. Reducing wind erosion with barriers. Trans. ASAE 20:911–915.

Staple, W. J., J. J. Lehane, and A. Wenhardt. 1960. Conservation of soil moisture from fall and winter precipitation. Can. J. Soil Sci. 40:80–88.

Stewart, B. A., D. A. Woolhiser, and W. H. Wischmeier, J. H. Caro, and M. H. Frere. 1975. Control of water pollution from cropland, Vol. I. USDA-ARS Report no. ARS-H-5-1.

Timmons, D. R., and R. F. Holt. 1968. Soil water recharge during the dormant period in the northwest Corn Belt. J. Soil Water Conserv. 23:177–180.

U.S. Department of Agriculture. 1960. Agricultural statistics. USDA, U.S. Government Printing Office, Washington, D.C.

U.S. Department of Agriculture. 1978. Agricultural statistics. USDA, U.S. Government Printing Office, Washington, D.C.

Wadleigh, C. H. 1968. Wastes in relation to agriculture and forestry. USDA Miscellaneous Pub. no. 1065.

Willis, W. D., C. W. Carlson, J. Alessi, and H. J. Haas. 1961. Depth of freezing and spring runoff as related to fall soil-moisture level. Can. J. Soil Sci. 41:115–123.

Wischmeier, W. H. 1973. Conservation tillage to control water erosion. p. 133–141. *In* Conservation tillage. Proc. Natl. Conference, Des Moines, Iowa. 28–30 March. Soil Conserv. Soc. of Am., Ankeny, Iowa.

Wischmeier, W. H., and D. D. Smith. 1960. A universal soil-loss estimating equation to guide conservation farm planning. Int. Congr. Soil Sci. Trans. Seventh (Madison, Wis.) 1:418–425.

Wischmeier, W. H., and D. D. Smith. 1978. Predicting rainfall-erosion-losses—a guide to conservation planning. USDA Agric. Handbook no. 537.

Woodruff, N. P., D. W. Fryrear, and L. Lyles. 1963. Reducing wind velocity with field shelterbelts. Kansas Agric. Exp. Stn. Tech. Bull. no. 131.

Woodruff, N. P., L. Lyles, E. H. Siddoway, and D. W. Fryrear. 1972. How to control wind erosion. USDA Agric. Information Bull. no. 354.

Woodruff, N. P., and F. H. Siddoway. 1965. A wind erosion equation. Soil Sci. Soc. Am. Proc. 29:602–608.

16

F. H. Siddoway
Agricultural Research Service
U.S. Department of Agriculture
Sidney, Montana

C. R. Fenster
University of Nebraska
Scottsbluff, Nebraska

Soil Conservation: Western Great Plains

The Western Great Plains Range and Irrigated Land Resource Region (Western Great Plains Range Region) extends across the western portion of the Great Plains dry farming area (Austin, 1965) between 32° N Lat in southern New Mexico to 48.5° N Lat in north central Montana, a distance of about 1600 km (Fig. 1-1). From north to south, the region covers significant portions of Montana, Wyoming, Nebraska, South Dakota, Colorado, and New Mexico. In an area covering such extensive length and breadth, one may find many soil and climatic diversities. In spite of these diversities one finds similar dryland farming and the associated crop production and erosion hazards. Thus, unfavorable soils, strong slopes, or low moisture supplies make dry farming very uncertain (Austin, 1965). Table 16-1 lists three of the major land resource areas within the region from north to south to illustrate climatic and physiographic variations. The region is subdivided into 13 major land resource areas, which are defined as geographically associated units characterized by particular patterns of soil (including slope and erosion), climate, water resources, land use, and type of farming.

Current land use reflects the general adaptive extent of dry farming. Largely through trial and error, those lands that could sustain profitable dryland productivity over time were selected. Marginal lands were brought

Contribution from the Agricultural Research Service, U.S. Department of Agriculture, in cooperation with the Montana Agricultural Experiment Station, Sidney, Montana (Journal Series no. 1028), and the University of Nebraska, Agricultural Experiment Station, Scottsbluff, Nebraska.

Dryland Agriculture—Agronomy Monograph no. 23.

Table 16–1. Pertinent physical characteristics of selected major land resource areas in the Western Great Plains and Irrigated Region (Austin, 1965).

Major land resource area and number	Land use	Climate	Dryland topography	Predominant dryland soils
Northern Rolling Plains (no. 58), 172 000 km^2	Rangeland 80%; dryland 4–5%; principal crops: winter wheat, spring wheat, and barley	Avg annual temperature 4.4–7.2°C; avg annual precipitation 300–410 mm, highest in May and June; snowfall 900 mm; potential evaporation 700–1000 mm; growing season 120–140 days.	Elevation 915–1830 m; gently sloping	Mollisols Aridisols
Central High Plains (no. 67), 85 500 km^2	Rangeland 75%; dryland 25%; principal crops: winter wheat and sorghum	Avg annual temperature 7.2–10°C; avg annual precipitation 330–430 mm, highest in April, May, and June; snowfall 610–910 mm; potential evaporation 1100–1300 mm; growing season 120–160 days.	Elevation 915–1830 m; rolling to undulating plains	Aridisols Mollisols
Pecos-Canadian Plains and Valleys (no. 70), 74 900 km^2	Rangeland 75%; dryland 2%; Principal crops: small grains, sorghum, and beans	Avg annual temperature 10–15.6°C; avg annual precipitation 300–410 cm, highest in July; snowfall 150–310 mm; potential evaporation 1500–1800 mm; growing season 160–200 days.	Elevation 1220–1830 m; gentle to rolling slopes	Mollisols

under cultivation when an above-average precipitation series of years coincided with high prices (e.g., 1940's) but abandoned or seeded back to grass when these favorable conditions ended (Chepil, 1957a; Zingg, 1954). Where water resources could be developed, choice lands with relatively deep and level soils were irrigated. Land too steep, rocky, shallow, coarse, or otherwise unsuitable for cultivation remained in range. Dry farming developed on land that could not easily be brought under irrigation but that was otherwise amenable to cultivation and crop production. However, with new technology, considerable land is being brought under irrigation. Except for the relatively small major land resource areas no. 59, 63, and 64, dryland farming comprises less than 20% of the land. Major land resource areas no. 62 and 65 have insignificant areas of dryland farming.

Annual precipitation generally decreases from east to west, ranging from 560 mm in the eastern part of the region in Nebraska and South Dakota to 300 mm annually in the western part. Most of the area receives between 300 and 400 mm/y. The highest proportion of annual precipitation falls during the growing season, and almost without exception, June is the peak month. Months on either side of June gradually taper off to the relatively dry winter period. Snowfall averages about 90 mm of water over most of the region, with lesser amounts in the southern part. Average annual temperatures of 4 to 7°C in the north increase to 10 to 16°C in the south with respective frost-free periods of 120 to 180 days. Winds are relatively strong (Fig. 10–5) compared with areas outside the Great Plains, especially during the late winter and early spring months (Baldwin, 1968; Skidmore and Woodruff, 1968).

Wind speed averages about 4.9 m s^{-1} y^{-1} at a 9.1-m height over most of the region but is as high as 6.3 m s^{-1} y^{-1} in the southeastern corner of Colorado (Chepil et al., 1962). Annual lake evaporation varies almost directly with latitude, averaging from about 900 mm/y in Montana to 1800 mm/y in New Mexico.

Annual wind erosion from cropland on class IIIe land in the region's states averages about 12.3 t/ha, and water erosion averages 4.3 t/ha (SCS. 1978. National erosion inventory estimates. Unpublished). The total exceeds the annual erosion tolerance of 11.2 t/ha by 5.4 t/ha. Wind erosion is most serious during the dry late winter and early spring months, whereas water erosion generally corresponds to the higher rainfall spring and summer months. In the northern part of the region, runoff from melt of the winter snow pack in February and March causes moderate erosion (Haas et al., 1974b).

16–1 WIND EROSION HAZARD

In terms of climate, wind erosion varies with wind speed and surface soil moisture content. When the wind reaches a velocity of about 5.4 m/s at a height of 30 cm above the ground surface, loose, dry, and unconsolidated soil starts to erode (Skidmore, 1965). Above this threshold level, the capacity of the wind to cause movement of soil is proportional to the wind speed cubed times the duration of wind (Bagnold, 1943; Chepil, 1945; Zingg, 1953). For example, a 10.7 m/s wind will erode soil at a rate eight times higher than a 5.4 m/s wind. Precipitation quantity and frequency affects the amount of time that the soil surface is wet, and radiation, temperature, humidity, and wind affect the rate of drying. A moist soil surface will not erode but becomes increasingly erodible as it dries.

The climatic factor

$$C = 34.5\ V^3/(\text{P-E})^2$$

has been formulated to indicate the relative mean rate of wind erosion that would occur as a function of wind velocity, V, and Thornthwaite's precipitation-evaporation index (or P-E index), as a percentage of the reference location, Garden City, Kansas (Chepil et al., 1962). Practically all of the Western Great Plains Range Region falls within a climatically highly hazardous annual wind erosion classification—C values between 72 and 150. The C values have also been calculated on a monthly basis, C', (Woodruff and Armbrust, 1968) and are shown by months for the USA by Skidmore and Woodruff (1968). The climatic hazard is generally highest during late winter and early spring, a period of high winds and low precipitation. To further compound the problem during this time, vegetative cover is at the yearly minimum, and the soil is more granular and thus more erodible because clods have been weathered by cycles of freezing, thawing, wetting, and drying.

In southeastern Colorado and New Mexico, C' is as high as 200 for March, April, and May (Skidmore and Woodruff, 1968). In general, the climatic hazards of wind erosion decrease from west to east across the region because this is the direction of increasing precipitation in the Great Plains, which in turn decreases the magnitude of the climatic factor, C. Northern areas are more protected during the winter months by snow cover, although drifting snow and soil erosion can occur simultaneously, as evidenced by "dirty" drifts.

States in the region have 1.13 million ha of wheat cropland that lose in excess of 22 t ha^{-1} y^{-1} from wind erosion, 990 000 ha that lose between 11 and 22 t ha^{-1} y^{-1}, and 1.9 million ha that lose between 4.5 and 11 t ha^{-1} y^{-1}. About 11 t ha^{-1} y^{-1} is generally considered the maximum quantity of soil that can be removed and still maintain the productive capacity of most soils (Chepil et al., 1962). This quantity, termed the T value (tolerable soil loss), is associated with no visible soil movement and no injury to plants. Approximately 2 million ha in this region exceed 11 t ha^{-1} y^{-1}. Losses from other cropland in the region occur on fewer hectares but are nevertheless of serious proportion (USDA, 1980).

In a procedure evaluating the effect of wind erosion on soil productivity, Lyles (1975) cited studies showing that wind erosion removes the finer soil fractions, leaving behind the coarser sands and gravel, and other studies document losses in productivity. Using estimated rates of wind erosion that would occur with representative amounts of residue cover, he estimated that annual wheat yield reductions would range between 0.7 and 140 kg/ha depending on soil type and climatic hazard. This loss in productivity can be partly compensated for by supplying extra fertilizer and improving management technology. However, where excessive soil loss has reduced the water-holding or infiltration capacities that affect the quantity of water stored in the root zone, the productive potential of the soil is irreversible under dryland culture.

16–2 WATER EROSION HAZARD

The water erosion hazard (annual erosivity) from rainfall, which is a function of raindrop impact and rate of runoff likely to be associated with the rain, has been quantified in terms of rainfall erosion index (R) values for the USA in the universal soil loss equation (Wischmeier and Smith, 1978). The R values in this region are relatively low, ranging from 20 to 50 in the two northern states to as high as 100 in the extreme eastern portion of the region (Fig. 10–2). The water erosion hazard increases rapidly east of the Western Great Plains Range Region because of the higher rainfall storm energy and intensity. Erosion from snowmelt is common in Montana, North Dakota, and Wyoming, where the snowpack gradually builds up from December to late February or early March and the melt rate often exceeds the infiltration rate of still frozen soils (Haas et al., 1974a). South of these states erosion from snowmelt is less serious, because snow melts intermittently over the entire winter and infiltrates relatively dry soils (Greb, 1979). For equal quantities of water, erosion is less serious from snowmelt than from rainfall because rate of melt is usually lower than rainfall intensity, and the impact force of rainfall that causes particle detachment is absent with snowmelt. Although, on the average, the water erosion hazard is relatively low in the region, it is nevertheless an insidious phenomenon, and erosion from intensive spring and summer storms is sometimes devastating in localized areas. Average annual potential runoff in the region is less than 2.5 cm, excluding major land resource area no. 63, where it is 2.5 to 3.2 cm (Stewart et al., 1975). Most of the potential runoff occurs during the growing season, and cropland is generally a low contributor to watershed sediment yields.

The states in the region have only 150 000 of wheat cropland that lose in excess of 22 t ha^{-1} y^{-1} from water erosion, 0.31 million ha that lose between 11 and 22 t ha^{-1} y^{-1}, and 1.68 million ha that lose 4.5 to 11 t ha^{-1} y^{-1} (USDA, 1980).

16–3 SOILS AND LAND FORMS

Texture is the dominant soil characteristic relative to inherent erodibility. A soil surface of stable dry aggregates of about 70% clods greater than 0.84 mm in diameter resists erosion from all but the highest winds (Chepil, 1958). Clods of this size are not easily moved by wind, and they protect smaller clods and particles in their lee. In general, susceptibility to wind erosion increases as texture becomes coarser. Based on a relative soil erodibility index, very fine, fine, and medium sands have an index value of 310; loamy sands, 134; sandy loams, 86; clays and silty clays subject to granulation, 86; loams and sandy clays, 56; silt loams and clay loams, 48; and silty clay loams, 38 (Craig and Turrelle, 1964). When the clay content is greater than about 20% of the sand-silt-clay soil composition, soil clods under the influence of wetting and drying and especially freezing and thawing tend to disintegrate into small granules that are as erodible as fine grains of sand (Chepil, 1955). Sandy or coarse-textured soil classes lack sufficient amounts of silt and clay to bind sand particles together and form clods resistant to wind erosion. Consequently, such soils form a "single grain" structure or weakly cemented clods, conditions that are quite susceptible to erosion by wind (Chepil, 1953). Extensive areas of sandy soils occur in the region in the states of Nebraska, Colorado, and Montana and to a lesser extent in Wyoming, New Mexico, and South Dakota (Chepil and Burnett, 1963). In contrast, sandy soils are less subject to water erosion than finer-textured soils because of their greater water infiltration capacity. A soil is generally more susceptible to water erosion as the silt fraction increases (Wischmeier and Mannering, 1969). Most of the dryland soils in the region fall within the medium texture classification. With good cultural practices, a fairly stable clod structure that is resistant to wind erosion can be maintained.

Although most of the dry cultivated land is on gentle to moderate slopes where water erosion can usually be controlled, three characteristics of dry farming are especially typical of this region and contribute to erosion: (i) large fields, (ii) large powerful tractors with multiple hitching of tillage equipment and drills, and (iii) summer fallowing to conserve water for the next crop. Wind erosion increases from zero at the windward edge of a field and increases with distance downwind until a maximum rate of soil removal is reached that depends on soil texture and wind speed. Sands reach a maximum rate of flow at a distance of 30 m; loamy sands, 50 m; sandy loams, 280 m; silt loams, 810 m; and silty clay loams, 1290 m (Chepil, 1957b). Water erosion is likewise related to field size and increases with steepness and length of slope (Wischmeier and Smith, 1978). For example, the potential soil loss on a unit hectare basis for a 2% 300-m slope would be twice as high as that for a 30-m slope. The corresponding comparison for a 5% slope would be over three times as great. Large tractors with multiple hitch systems are more conveniently accommodated by large fields, and more powerful tractors allow faster operating speeds, which tend to pulverize the soil and bury protective residue, especially during the first two tillage operations (Anderson, 1967). However, higher speeds with subsurface tillage implements result in more soil agitation and better weed control, especially under residues. Summer fallow, a necessity with present technology in most of this region, is the overriding cause of both wind and water erosion. Soil losses are directly related to the amount of vegetative cover that can be maintained on the soil surface during the period of fallow up to the time that the seeded crop provides the cover needed to control erosion. The region has about 1.75 million ha of summer fallow divided among the states in descending order as follows: Colorado, 0.9 million ha; Montana,

0.55 million ha; Nebraska, 0.27 million ha; South Dakota, 0.255 million ha; Wyoming, 0.122 million ha; North Dakota, 53 000 ha; and New Mexico, 12 000 ha (Haas et al., 1974a).

16–4 EROSION CONTROL PRACTICES

Practices that are both effective and adaptable for controlling wind erosion are generally also useful in controlling water erosion. Erosion is caused by a force acting on the soil to detach and transport particles from one point to another. Principles of control involve preventing the force (moving air or water) from coming in contact with the soil or by moderating the impact of the force. Various management practices have been devised for control, and the extent of application has been largely determined by costs and relative convenience of their use.

16–4.1 Conservation Tillage

Stubble mulch farming uses tillage and planting operations that leave the stubble or crop residues essentially in place on the land as a surface cover before and during the preparation of a seedbed and at least partly during the growing of a succeeding crop. The extended drought and wind erosion that occurred during the 1930's "dust bowl" seriously damaged an estimated 43% of a 6.5 million ha area (Joel, 1937). At the end of this era, which has been classed the most serious man-made environmental problem the USA has ever seen (Lockeretz, 1978), it became evident that the methods of tilling dryland would have to be changed. The moldboard plow was replaced by the disk (a surface mixing implement), which in turn was replaced by subsurface tillers. The latter two implements, depending on the amount of original residue and speed and number of operations, leave a proportion of postharvest residue on the surface during fallow and crop establishment to protect against both wind and water erosion. The most effective practical method of preventing wind erosion (and water erosion) is an adequate cover of either growing plants or residues from the previous crop (Woodruff et al., 1977). Although other conservation practices are employed on a limited basis on the region's drylands, stubble mulch farming is the most universal.

There is no fixed system of providing a mulched soil surface condition during the period of summer fallow. Implements used in the first tillage operation range from 5-cm chisels up to 2-m V-blade sweeps. The most commonly used tool that has evolved after 30 or more years of experience has been the 46- to 61-cm sweep. After the first tillage, from three to six secondary operations are done with sweep plows, field cultivators, rod weeders with semipoint chisels or shovel attachments, or plain rod weeders. When residues are especially heavy (more than 4500 kg/ha), a disk type implement may be used for one or more tillage operations to reduce stubble

Table 16–2. Kilograms of residue required per hectare to hold wind erosion to a tolerable level of 11 kg ha^{-1} y^{-1} (Woodruff et al., 1977).

	Wheat residue		Sorghum residue		Growing wheat	
Soil texture†	Standing	Flattened	Standing	Flattened	In furrow	On smooth ground
	kg/ha					
Silts	500	1040	2020	2910	560	480
Clay and silty clay	900	1790	3700	5320	1090	920
Loamy fine sands	1180	2380	4700	6940	1340	1120

† Silts with 50% nonerodible fractions greater than 0.84 mm in diameter; clay and silty clay with 25% nonerodible fractions; loamy fine sand with 10% nonerodible fractions.

to a more manageable quantity (Fenster et al., 1977). When the normal tillage period terminates during the first part of September, either winter wheat is seeded or the field is left fallow through the winter until a spring cereal (wheat, barley, or oat), sorghum, or other spring annual is seeded. The critical period for potential wind erosion occurs after the end of tillage until the growing crop provides adequate cover. For winter wheat, this period usually extends from September to April in the northern part, from September to March in the central part, and from September to February in the southern part. For spring-seeded crops, the period extends from September to May.

The goal of stubble mulching for wind erosion control is to maintain an adequate protective cover and a cloddy-rough soil surface at all times, even after the crop is planted. Different crops vary in the effectiveness of their residues to provide sufficient cover, as illustrated in Table 16–2 (Woodruff et al., 1977).

Table 16–3 shows that certain implements also bury various amounts of residue during each tillage operation (Woodruff et al., 1977). By know-

Table 16–3. Typical residue reduction by various tillage implements (Woodruff et al., 1977).

Tillage machine	Percent of residue reduction by each tillage operation
Stirring or mixing machines	
One-way disk (61- to 66-cm disks)	50
One-way disk (46- to 56-cm disks)	40
Tandem or offset disks	50
Power disk	60
Field cultivator (41- to 46-cm sweeps)	20
Chisel plow (5-cm chisels 30 cm apart)	25
Mulch treader (spade-tooth)	25
Mulch treader (spike-tooth)	30
Sidewinder rotary tiller (30 cm tilled on 102-cm center)	30
Subsurface machines	
Blades (91 cm or wider)	10
Sweeps (30- to 41-cm)	15
Rod weeders with plain rod	10
Rod weeders with semichisels or shovels	15

ing the amount of residue before tillage and estimating the number of tillage operations normally required during the fallow season with various combinations of machines, one can design a tillage system to preserve the quantity of residue needed to control erosion. Regardless of type, each operation buries some of the remaining residue; therefore, it is important to keep the number of tillage operations to the absolute minimum needed for weed control to preserve as much residue as possible going into the critical wind erosion period. By proper selection of combinations of implements, the quantity of residue needed for erosion control can be achieved. The data in Table 16–3, along with knowledge about the type and quantity of residue after harvest and the quantity of residue needed for wind erosion control, can be used to select implement combinations and operations to preserve needed residue cover. Among dryland crops, only winter wheat occasionally produces so much residue that tillage and drilling operations are hindered. With this exception, the objective should be to preserve as much residue as possible. If the residue is not needed for erosion control, water conservation is increased by maintaining more residue on the surface (see Chapters 2 through 9). Generally the greatest problem a farmer encounters is preserving adequate residue to control erosion while still controlling weeds. Such winter annual weeds as downy brome and volunteer winter wheat are difficult to kill with sweep implements if the spring season is late or wet (Fenster et al., 1977). Their early spring growth forms a sod that sweep tillage will not adequately break up to achieve a weed kill, and timely follow-up tillage may be required.

Research on water erosion control in this region has been very meager in terms of direct measurements of runoff and soil loss, but Table 16–4 illustrates the range of losses that can be expected from stubble mulch fallow land compared with bare fallow (Fenster et al., 1977).

Table 16–5 lists the *C* factor from the universal soil loss equation for various quantities of mulch in eastern Colorado (SCS, 1977). This value is a fraction of the annual soil loss from water erosion that would occur relative to bare areas. It illustrates the effectiveness of mulch in controlling soil losses from fields.

During years when rainfall and consequent crop production are slightly below to above average, crop residues can be maintained on the surface in sufficient amounts to provide satisfactory erosion control. After a dry year, and especially after a series of dry years, serious wind erosion can occur because clod structure deteriorates and the quantity of residues are inadequate for control (Chepil et al., 1963).

Table 16–4. Runoff and soil loss produced in wheat-fallow rotation at Alliance, Nebraska, from 1958 to 1965 (Fenster et al., 1977).

	During fallow period		During growing period	
	Stubble mulch	Bare	Stubble mulch	Bare
Runoff, cm of water	4.8	11.9	11.4	19.0
Soil loss, t/ha	3.8	24.5	10.0	39.1

Table 16–5. Factor *C* for various quantities of mulch (SCS, 1977).

Straw mulch†	*C* factor
Bare area, t/ha	1.00
0.6	0.52
1.1	0.35
1.7	0.24
2.2	0.18
3.4	0.10
4.5	0.06
6.7	0.03
9.0	0.02

† Adequately anchored in soil.

16–4.2 Field Stripcropping

Stripcropping, the arrangement of alternate narrow strips of crop and fallow to reduce the avalanching effect of wind erosion, is used quite extensively in the northern part of the region. Strips are usually oriented north-south or east-west, whichever direction is most perpendicular to prevailing erosive winds. Very narrow strips (6 to 30 m) are required for control on sandy soils, whereas widths can be as much as 130 m wide on a silty clay loam. For medium-textured soils, widths are intermediate between these two extremes (Chepil, 1957b).

Control is improved by combining mulching with stripcropping, but both methods lose effectiveness when drought limits the quantity of residue. Objections to strips are associated with difficulties in cultivating up to the crop-fallow border and killing weeds. With narrow strips, accommodating the various widths of tillage, planting, and harvest machinery occasionally can be a problem for most efficient operations. Over time on extremely erodible sandy soils, the trapping of soil in standing residue can eventually create an undulating or wavelike topography. In spite of these problems, stripcropping is an effective erosion control method and is used extensively in the northern part of the region where spring-planted crops predominate. Field strips provide some water erosion control when the strips fall on the contour, but this is merely coincidental, because wind control strips should ideally be oriented perpendicular to prevailing erosive winds.

16–4.3 Contour Stripcropping

Contour stripcropping is designed mainly for water erosion control, and the objective is to shorten the length of erosive slopes. Strips are more effective when cultivated strips are alternated with strips of perennial grasses or close-growing crops that filter out sediment and disburse concentrations of channeled runoff. Because they break up large fields, contour strips afford some wind erosion protection, but perpendicular orientation of strips to prevailing winds (a desirable feature for wind erosion control) is coincidental. Contour strips have the additional advantages of forcing contour tillage and seeding, which on their own merits reduce water erosion by about 50% on slopes of 3 to 8% over a slope length of 60 to 910 m (Wischmeier and Smith, 1978). Contour stripcropping has been applied on only a limited basis in the region. The principal disadvantage is the inconvenience of operating large machinery in narrow fields. Irregular and rolling topography is incompatible with equal-width strips and odd-sized areas seeded to grass occupy land that could otherwise be used for crop production.

16–4.4 Shelterbelts and Wind Barriers

Trees, shrubs, perennial grasses, and annual crops—when planted in single or multiple rows at appropriately spaced intervals across a field—provide multiple benefits, one of which is the protection against wind erosion through the mechanism of reducing wind speed. A typical one-row tree shelterbelt effectively reduces wind speed a distance of about 10 to 15 times the height of the trees. Within this distance range, wind speed approaches 80% of open field wind speed and then gradually increases to nearly 100% at a distance of about 40 times the height of the trees (Stoeckeler, 1962; Woodruff et al., 1963). Although shelterbelts were probably the first widely studied and promoted method of solving wind erosion problems in the Great Plains (Davis, 1976), their application in the Western Great Plains Range Region has been quite limited. Introduced species of trees and shrubs are often not adapted to the semiarid climate of the region. Shelterbelts also compete with an adjacent crop for moisture and nutrients and thereby reduce production (Greb and Black, 1961). One advantage of windbreaks over most other types of wind erosion control is their relative permanency. During a series of drought years, shelterbelts may be the only effective and persistent control measure. Some concern has been voiced in recent years about the removal of windbreaks. Reasons often cited are that they occupy too much land, sap moisture and nutrients from adjacent cropland, obstruct irrigation systems, are damaged by drift of broadleaf herbicides, and harbor disease organisms and insects (Goldsmith, 1976).

A system of perennial tall wheatgrass barriers (Aase et al., 1976; Black and Siddoway, 1976; Siddoway, 1970) increases water conservation by trapping enough snow to make annual cropping feasible in the Northern Plains. By holding snow on the land, soil water storage is increased during the overwinter period after harvest to the extent that summer fallow may not be needed between cropping periods. A single- or double-row planting of tall wheatgrass forms a barrier about 1.2 m tall, and when the barriers are spaced about 15 m apart, they reduce potential wind erosion by about 90% (F. H. Siddoway et al. 1977. Unpublished information). In turn, annual cropping provides almost continuous vegetative cover, with the net effect of eliminating wind erosion except during periods of consecutive annual droughts. Application of this barrier system is progressing slowly because of its relative newness. Two farmer objections to application are the division of large fields into narrow (approximately 15-m) land parcels and the inconvenience of tilling near the barriers with large-scale equipment. As tillage implements increase in width, it becomes more difficult for the tractor operator to precisely till next to a field border or, in this case, a grass barrier. If the land is tilled away from the barrier, weeds may grow in the space and produce seed. Nevertheless, this practice, where adaptable, provides effective wind erosion control and water conservation.

16–4.5 Emergency Tillage

Emergency tillage is a last resort wind erosion control operation that can provide a degree of relief when because of climate (drought and high winds) or mismanagement the soil is blowing or conditions threaten imminent erosion. Implements such as listers, chisels, shovels, and "sand fighters" traverse the fields at a right angle to erosive winds to roughen the soil surface and create clods (Woodruff et al., 1957, 1977). The hectarage treated varies with the severity of erosion. For example, in 1977 (the fourth worst erosion year in the Great Plains since 1936), over 1.4 million ha was emergency tilled (SCS, 1977).

16–4.6 Terracing

Level bench terraces have been researched in the region with the dual objective of conserving water and controlling erosion (Black, 1968; Michelson, 1968). The benches effectively increased soil water storage and, by reducing slope length and quantity of runoff, they also decreased water erosion potential. However, application of this terrace system has been very limited. The high cost of construction and difficulty of performing tillage, planting, and harvest operations on small fields with large equipment have precluded their routine use.

16–5 TECHNOLOGICAL NEEDS

Because erosion is aggravated by large field size and because climate is marginal for successful crop production, erosion control options are somewhat limited. Many of the measures that control erosion such as terraces, closely spaced shelterbelts, and narrow strips cannot easily be applied with the machinery that is needed to farm large economic units. Machinery manufacturers have not standardized on uniform width of units for multiple hitching. Nevertheless, stubble mulch tillage is one erosion control practice that is almost universally applied to dry-farmed cropland. The combined technology that has resulted in large productivity increases during the last 20 years has, in turn, enhanced erosion control with mulch tillage because the residue remaining for protection is proportional to the yield of the harvested crop.

In spite of technological advances, the degree of erosion control that can be achieved depends on the vagaries of climate. One of the most promising and relatively recent developments for dryland erosion control, the no- and minimum-till concept, is sometimes termed "ecofallow," or chemical fallow. Ecofallow is a system of controlling weeds and conserving soil moisture in a crop rotation with a minimum disturbance of crop residue and soil (Wicks, 1976). It involves the use of herbicides to eliminate most of the tillage operations. Tillage operations may be needed to prepare the seedbed or to manage residues for proper planting. Researchers at North Platte, Nebraska, used a wheat-fallow rotation during a 6-year period (1963 to 1969) to make a comparison between no-tillage with herbicides and conventional tillage (Wicks and Smika, 1973). The plots receiving no-tillage (weeds controlled by herbicides) had the least weed growth, the most soil water stored, and the highest amount of surface residues retained. No-till also had the highest grain yields of all fallow treatments. The bare fallow system yielded 2690 kg/ha; stubble mulch–fallow, 2880 kg/ha; and no-till fallow 3160 kg/ha. This concept preserves more protective residue cover for longer periods by maintaining the stubble aftermath in a vertical position.

Good and Smika (1978) stated that control of both wind and water erosion is almost complete in the central Great Plains with chemical fallow. Calculated potential soil loss from a large block of land with stubble mulch tillage was 22 t/ha and with chemical fallow only a trace. Research and limited application of the practice has shown the merits of the concept (Anderson, 1976; Black and Siddoway, 1977; Phillips et al., 1976). In the Northern Plains, where snow management can increase water conservation so that crops can be grown annually, the land can be protected by vegetative cover except for the short period of crop establishment (Black and Siddoway, 1976). Any research advances that increase water conservation, water use efficiency, or crop production in general also benefit erosion control. For example, through control of the runoff, more water can be stored in the soil; water availability increases crop production, and the larger crop, in turn, provides more residue for protective vegetative cover.

No-till, a relatively new practice for drylands, can go beyond the proved erosion control benefits of stubble mulch tillage by increasing the quantity and extending the duration of residue cover. Although progress has been made, many associated problems remain to be solved, including weeds, plant diseases, and insect pests, development of large-scale seeding equipment, nutrient cycling, phytotoxic effects, and the breeding of varieties that are compatible with the soil and air environment created by higher quantities of cereal residues on the soil surface. In the absence of tillage, herbicides are the foundation of good weed control, and several wide-spectrum contact and residual types compatible with particular crops and environments are needed. Precision equipment and techniques for herbicide application need to be developed.

Successfully developing this erosion control practice until its application is routine will require a multidisciplinary approach. The erosion control principles are sound and compatible with large-scale dry farming, and the potential benefits may be the long-range salvation of dry farming in this region.

Water is the most limiting factor in crop production, and wind and water erosion are sensitive to water supply. Any advances in water conservation and water use efficiency can help in minimizing erosion losses by increasing dry matter production and decreasing the need for summer fallow. Fallow is a very inefficient method of storing water, and any improvements in water storage over the first winter period after harvest could decrease the need for fallow. Snow management, improved infiltration, suppression of evaporation, and breeding of drought tolerant crops are pertinent areas of research.

We also need to know the effects of erosion and cropping on the productive capacity of soil. These influences have been masked by the application of improved technology. Historically, this land has been cropped for only a relatively short period. Is the land deteriorating, improving, or maintaining the status quo? If this information were known, we could recommend proper action.

16-6 LITERATURE CITED

Aase, J. K., F. H. Siddoway, and A. L. Black. 1976. Perennial grass barriers for wind erosion control, snow management, and crop production. p. 69–78. *In* R. Tinus (ed.) Shelterbelts on the Great Plains. Great Plains Agric. Coun. Pub. no. 78, Univ. of Nebraska, Lincoln.

Anderson, C. H. 1976. Eco-fallow in the Northern Great Plains area of Canada and the United States. p. 33–45. *In* Conservation tillage. Great Plains Agric. Coun. Pub. no. 77, Univ. of Nebraska, Lincoln.

Anderson, D. T. 1967. Factors affecting trash conservation with the wide blade cultivator. Can. Agric. Eng. 9:98–99.

Austin, M. E. 1965. Land resource regions and major land resource areas of the United States. USDA Agric. Handbook no. 296.

Bagnold, R. A. 1943. The physics of blown sand and desert dunes. Methuen and Co. Ltd., London.

Baldwin, J. L. 1968. Climatic atlas of the United States. U.S. Dep. of Commerce, Environmental Sci. Services Admin., Environmental Data Service.

Black, A. L. 1968. Conservation bench terraces in Montana. Trans. ASAE 11:393–395.

Black, A. L., and F. H. Siddoway. 1976. Dryland cropping sequences within a tall wheatgrass barrier system. J. Soil Water Conserv. 31:101–105.

Black, A. L., and F. H. Siddoway. 1977. Winter wheat recropping on dryland as affected by stubble height and nitrogen fertilization. Soil Sci. Soc. Am. J. 41:1186–1190.

Chepil, W. S. 1945. Dynamics of wind erosion: III. The transport capacity of the wind. Soil Sci. 60:475–480.

Chepil, W. S. 1953. Factors that influence clod structure and erodibility of soil by wind: I. Soil texture. Soil Sci. 75:473–483.

Chepil, W. S. 1955. Factors that influence clod structure and erodibility of soil by wind: IV. Sand, silt, and clay. Soil Sci. 80:155–162.

Chepil, W. S. 1957a. Dust bowl: Cause and effects. J. Soil Water Conserv. 12:108–111.

Chepil, W. S. 1957b. Width of field strips to control wind erosion. Kansas Agric. Exp. Stn. Tech. Bull. no. 92.

Chepil, W. S. 1958. Soil conditions that influence wind erosion. USDA Tech. Bull. no. 1185.

Chepil, W. S., and E. Burnett. 1963. Management of sandy lands in the central United States. USDA Farmers Bull. no. 2195.

Chepil, W. S., F. H. Siddoway, and D. V. Armbrust. 1962. Climatic factor for estimating wind erodibility of farm fields. J. Soil Water Conserv. 17:162–165.

Chepil, W. S., F. H. Siddoway, and D. V. Armbrust. 1963. Climatic index of wind erosion conditions in the Great Plains. Soil Sci. Soc. Am. Proc. 20:449–452.

Craig, D. G., and J. W. Turrelle. 1964. Guide for wind erosion control in the Great Plains. U.S. Dep. of Agric., Soil Conserv. Service, U.S. Government Printing Office, Washington, D.C.

Davis, R. M. 1976. Great Plains windbreak history and overview. p. 8–11. *In* R. Tinus (ed.) Shelterbelts on the Great Plains. Great Plains Agric. Coun. Pub. no. 78, Univ. of Nebraska, Lincoln.

Fenster, C. R., H. I. Owens, and R. H. Follett. 1977. Conservation tillage for wheat in the Great Plains. USDA Extension Service Pub. no. PA-1190.

Goldsmith, L. 1976. Action needed to discourage removal of trees that shelter cropland in the Great Plains. p. 12–18. *In* R. Tinus (ed.) Shelterbelts on the Great Plains. Great Plains Agric. Coun. Pub. no. 78, Univ. of Nebraska, Lincoln.

Good, L. G., and D. E. Smika. 1978. Chemical fallow for soil and water conservation in the Great Plains. J. Soil Water Conserv. 33:89–90.

Greb, B. W. 1979. Reducing drouth effects on cropland in the west-central Great Plains. USDA Information Bull. no. 420.

Greb, B. W., and A. L. Black. 1961. Effects of windbreak plantings on adjacent crops. J. Soil Water Conserv. 16:223–227.

Haas, H. J., W. O. Willis, and J. J. Bond. 1974a. Introduction. p. 1–11. *In* Summer fallow in the western United States. USDA-ARS Conserv. Res. Report no. 17.

Haas, H. J., W. O. Willis, and J. J. Bond. 1974b. Summer fallow in the northern Great Plains (sring wheat). p. 12–35. *In* Summer fallow in the western United States. USDA-ARS Conserv. Res. Report no. 17.

Joel, A. H. 1937. Soil conservation reconnaissance survey of the southern Great Plains wind erosion area. USDA Tech. Bull. no. 556.

Lockeretz, W. 1978. The lessons of the dust bowl. Sci. Am. 66:560–569.

Lyles, L. 1975. Possible effects of wind erosion on soil productivity. J. Soil Water Conserv. 30:279–283.

Michelson, R. H. 1968. Conservation bench terraces in eastern Colorado. Trans. ASAE 11: 389–395.

Phillips, W. M., P. W. Unger, and H. A. Greer. 1976. Eco-fallow in the southern Great Plains. p. 56–63. *In* Conservation tillage. Great Plains Agric. Coun. Pub. no. 77, Univ. of Nebraska, Lincoln.

Siddoway, F. H. 1970. Barriers for wind erosion control and water conservation. J. Soil Water Conserv. 25:180–184.

Skidmore, E. L. 1965. Assessing wind erosion forces: Direction and relative magnitudes. Soil Sci. Soc. Am. Proc. 29:587–590.

Skidmore, E. L., and N. P. Woodruff. 1968. Wind erosion forces in the United States and their use in predicting soil loss. USDA-ARS and Kansas Agric. Exp. Stn., Agric. Handbook no. 346.

Soil Conservation Service. 1977. Severe erosion hits plains. Soil Conserv. 42(12):22.

Soil Conservation Service. 1977. Universal soil loss equation. USDA-SCS Colorado Tech. note no. 50.

Stewart, B. A., D. A. Woolhiser, W. H. Wischmeier, J. H. Caro, and M. H. Frere. 1975. Control of water pollution from cropland, Vol. 1. USDA-ARS and ORD-EPA Report no. ARS-H-5-1, U.S. Government Printing Office, Washington, D.C.

Stoeckeler, J. H. 1962. Shelterbelt influence on Great Plains field environment and crops. USDA Production Res. Report no. 62.

U.S. Department of Agriculture. 1980. Major uses of nonfederal land. p. 103–188. *In* 1980 Appraisal: Part I. Soil, water, and related resources in the United States: Status, condition, and trends. Soil and Water Resources Conservation Act. USDA-SCS, Washington, D.C.

Wicks, G. 1976. Eco-fallow: A reduced tillage system for the Great Plains. Weeds Today 7(2):20–23.

Wicks, G., and D. E. Smika. 1973. Chemical fallow in a winter wheat fallow rotation. Weed Sci. 21:97–102.

Wischmeier, W. H., and J. V. Mannering. 1969. Relation of soil properties to its erodibility. Soil Sci. Soc. Am. Proc. 33:131–137.

Wischmeier, W. H., and D. D. Smith. 1978. Predicting rainfall erosion losses—a guide to conservation farming. USDA Agric. Handbook no. 537.

Woodruff, N. P., W. S. Chepil, and R. D. Lynch. 1957. Emergency chiseling to control wind erosion. Kansas Agric. Exp. Stn. Tech. Bull. no. 90.

Woodruff, N. P., D. W. Fryrear, and Leon Lyles. 1963. Reducing wind velocity with field shelterbelts. Kansas Agric. Exp. Stn. Tech. Bull. no. 131.

Woodruff, N. P., and D. V. Armbrust. 1968. A monthly climatic factor for the wind erosion equation. J. Soil Water Conserv. 23:103–104.

Woodruff, N. P., L. Lyles, F. H. Siddoway, and D. W. Fryrear. 1977. How to control wind erosion. USDA Agric. Information Bull. no. 354.

Zingg, A. W. 1953. Wind-tunnel studies of the movement of sedimentary material. p. 111–135. *In* Proc. Fifth Hydraulogy Conference Iowa Inst. of Hydraulogy Bull. no. 34.

Zingg, A. W. 1954. The wind erosion problem in the Great Plains. Trans. AGU. 35:252–258.

17

A. L. Black and A. Bauer
Agricultural Research Service
U.S. Department of Agriculture
Mandan, North Dakota

Soil Conservation: Northern Great Plains

In the Northern Great Plains, the practices employed to conserve water, as well as most other cropping practices, were developed to be compatible with and to complement soil resource conservation practices. This chapter discusses major soil conservation practices that are being used, those being tested, and new soil conservation techniques for the Northern Great Plain Spring Wheat Land Resource Region (Northern Spring Wheat Region) (Fig. 1-1), which includes major land resource areas no. 52 to 57.

17-1 CLIMATE

Climate of the Northern Great Plains is distinctly continental, ranging from semiarid to subhumid. The most common agent of erosion in this region is wind, although water erosion can, at times, be a problem. Wind erosion is a dry weather phenomenon; thus, in a sense it is a water problem. Potential for wind erosion is highest in late fall and early spring, because these seasons have the greatest incidence of potentially erosive winds, cultivated soils usually have the least vegetative cover, and water content of the soil usually is low. Presence or absence of snow cover is the principal factor affecting surface soil water content in the spring. Potential for water erosion is highest during the high-intensity summer rainfall periods, but water erosion also occurs during the spring thaw period because soil often is frozen or has reduced permeability because of low temperatures when snowmelt occurs.

Although progress has been made in adapting soil conservation practices to reduce erosion potential, soil losses to wind and water erosion continue to occur because of extremes in weather factors, e.g., prolonged drought, windstorms of high velocity, runoff from high intensity summer

Dryland Agriculture—Agronomy Monograph no. 23.

storms, and runoff from rapid snowmelt. Severe drought was the fundamental cause of the great dust clouds (originating in the Northern Great Plains) that reached the Atlantic coast in May 1934 (Kellogg, 1935). The drought of the 1930 era, later to be referred to as the "dirty Thirties," was the result of successive years of extreme rainfall deficiencies coinciding with extremes in summer and winter temperatures. Soils became increasingly drier until root zone water reserves were exhausted. Because of the prolonged drought, little or no vegetation was produced, and thus soils were essentially devoid of any protective vegetative cover. In the 50 years since the 1930's, successive years of extreme drought have not occurred in the Northern Great Plains even though some 1-year periods of drought have been recorded. The dirty Thirties were followed by the "fabulous Forties," occasioned by a combination of timely average to above-average precipitation, high grain yields, and high prices for farm products.

17-2 SOIL EROSION

Two basic methods used to prevent or control wind erosion are discussed in detail in Chapter 12. The basic principles are to:

1. Reduce the wind velocity near the soil surface below that required to initiate soil movement. Roughening the soil surface and providing obstructions, e.g., vegetative covers, are common approaches used.
2. Increase the size of the soil aggregates or increase the proportion of nonerodible soil aggregates at the soil surface so as to increase the minimum threshold velocity to initiate soil detachment. This can be achieved with various cropping and tillage practices.

To prevent or control water erosion requires that detachment of soil particles be eliminated. A major energy source for detachment is raindrop impact. Maintaining residues on the soil surface is a way of absorbing the direct impact. Intercepting the direct impact of raindrops also prevents disintegration of soil aggregates, and thereby the soil maintains a higher infiltration rate.

17-2.1 Wind Erosion Control Practices

Summer fallow is identified as a major source of sediment from wind and water erosion (Young, 1958; Haas and Boatwright, 1960). Soil loss and sediment could be reduced 60% by eliminating summer fallow from the rotation (Lewis and Clark, 1978). However, summer fallow is considered to add a degree of stability to crop production in the drier regions of the Northern Spring Wheat Region. With careful management of summer fallow, and particularly seed zone soil water supply, the probability of crop establishment is improved, thereby ensuring the continuation of sufficient crop residue to protect the land (Haas et al., 1974).

17-2.1.1 Wind Stripcropping

Wind stripcropping, the first wind erosion control method adopted in the Northern Great Plains (Howard, 1959), minimizes the avalanche phenomenon, the tendency for erosion rate to increase downwind across a field, by breaking large fields into narrow alternate strips of crop and summer fallow. This is the field length, which is the L' factor in the wind erosion equation (Woodruff and Siddoway, 1965). Individual strip widths vary from as narrow as 15 m for coarse-textured to over 90 m for least erosive medium-textured soils. Wind erosion forces are high in the Northern Great Plains from October through April (Skidmore, 1965). During the winter months snow cover may offer intermittent protection on residue-covered fields, but a bare surface soil often loses snow cover and then becomes especially susceptible to wind erosion because of aggregate breakdown caused by thawing, freezing, and drying (Chepil and Woodruff, 1963; Bisal and Nielsen, 1964). Also, stripcropping does not control wind erosion when erosive winds (4.5 m/s) occur parallel to the direction of the strips (Skidmore, 1965).

17-2.1.2 Residue Management

Stubble mulch tillage has augmented stripcropping in the last 20 years for controlling wind erosion. The level of protection afforded by stubble mulching is proportional to kind, quantity, and orientation of crop residue maintained on the soil surface (Woodruff and Siddoway, 1965). Straw production by spring wheat varies with grain yield (Bauer and Zubriski, 1978). Assuming a straw/grain ratio of 1.45 for the Northern Great Plains (USDA, 1977), straw production averages 2400 kg/ha. Weight loss of untilled spring wheat stubble after the first overwinter period is about 33% (Black, 1973a). Typically, therefore, 1600 kg/ha of surface residue are present at the beginning of the summer fallow period. Subsurface tillage operations during fallow reduce surface residue quantities about 50% after four operations (Anderson, 1961). Since the summer fallow period for spring wheat extends through the second winter, an estimated 800 kg/ha of crop residue remains on the soil surface at the beginning of that period. Therefore, the potential wind erosion hazard on summer fallow increases progressively for the second winter and through spring seedbed preparation and crop establishment. In addition, each tillage operation increases the proportion of erodible soil aggregates (Black and Power, 1965; Siddoway, 1963). Mean weight-diameter of aggregates is also decreased during summer fallow independently of tillage frequency (Bauer and Conlon, 1974).

Combinations of chemical and mechanical fallow methods conserve more residue and more nonerodible soil clods than conventional tillage (Black and Power, 1965). Under drought conditions during summer fallow, however, adverse residual chemical effects on subsequent spring wheat yields were recorded with the herbicides then available. Interest and opportunities for developing minimum- or no-till crop–summer fallow systems have been renewed. The recent advent of several new herbicides having less residual activity is at least partially responsible for the interest.

17–2.1.3 Barriers

Most of the field windbreak plantings, about 0.5 million ha, are in the more humid portions of the Northern Great Plains (major land resource areas no. 55, 56, and 57) where the relative need for wind erosion control and water conservation is less. Windbreaks are not used extensively in the semiarid region (major land resource areas no. 52, 53, and 54) because of few adaptable tree and shrub species, the difficulties of establishment, and the susceptibility to damage by diseases, insects, and herbicides.

Historically, the principal field barrier has been a multirow planting of trees and shrubs. The spacing between these belts usually is too wide in itself to be a complete erosion control system for a field (George et al., 1963). In recent years, multirow belts have given way to single-row belts. The design of the single-row belts has, with time, evolved from a close, within-row spacing of trees to more permeable, wider spacings (Frank and George, 1975). Experience and research have shown that barriers effectively reduce the wind speed blowing perpendicular to the barrier for a leeward distance of 8 to 12 times the height of the barrier (Chepil and Woodruff, 1963).

The direction of erosive winds is not always from a right angle to the barrier, and any deviation from perpendicularity leaves a portion of the land unprotected. This fact becomes significant in the choice and design of barriers. When wind direction is not perpendicular to the barrier, the proportion of protected vs. unprotected land does not differ among barrier heights, but the distance paralleling the wind in the unprotected portions (between barriers spaced to provide protection at 10 times their height) is larger for tall barriers than for short barriers (Siddoway, 1970).

Vegetative barriers of various annuals planted in strips provide wind erosion protection to summer fallow only when their growth is adequate. All annuals have shortcomings that limit their adoption, however. Crops such as barley or oat prostrate early in the winter (Wilkinson and Schroer, 1966), reducing the effectiveness of protection. Flax will remain more erect than small grains, but stand establishment is more difficult, hence less reliable, because of the required shallower depth for flax seedings. Strips of double rows of unharvested corn or corn planted on summer fallow at various width intervals, although effective in wind erosion control (Geiszler, 1961), can contribute to water erosion problems if runoff from snowmelt occurs when row orientation is parallel to slope direction. Also, if corn matures, volunteer corn in succeeding cereal crops presents a problem of control. Strips of double rows of sunflower spaced as much as 21 m apart provided wind erosion protection without the adversities of a volunteer crop (Hoag and Geiszler, 1971). In addition to their individual shortcomings in erosion control, annual vegetative barriers can lower the yield potential of the following crop by reducing the available N, the soil water supplies, or both (Wilkinson and Schroer, 1966; A. Bauer. 1981. Unpublished data). Water storage through snow catchment can offset the soil water removed by the barrier, whereas snowmelt water can transport nitrates downward within the root zone.

Perennial vegetative barriers of single- or double-row tall wheatgrass barriers spaced at 10- to 15-m intervals across a field have the potential for controlling wind erosion (Black and Siddoway, 1971, 1976). With an average annual erosive wind frequency as occurs over a 20-year period at Great Falls, Montana, the grass barriers would reduce the duration of exposure of lands and lower the rate of erosion with a net affect of reducing erosion potential about 94% (F. H. Siddoway. 1981. Personal communication). Additional advantages of these types of barriers are ease of establishment, low cost, and lack of potential injury from herbicides used for broadleaf weed control.

17–2.2 Water Erosion Control Practices

Water erosion is a problem on the gently to moderately rolling cultivated soils of the Northern Spring Wheat Region. Summer fallow land is susceptible to water erosion from both snowmelt runoff and intense summer rains, but soil erosion losses are generally small from land with a vegetative cover. Erosion from snowmelt runoff occurs whenever the rate of snowmelt exceeds the infiltration capacity of frozen soils (Willis et al., 1961). Water erosion may not be severe in any one year, and what erosion occurs is usually limited to natural, unprotected field drainageways. However, intense summer rainstorms cause erosion damage, not limited to drainage ways, even though they occur rather infrequently at a given site. Rainfall amounts are less than 1.3 cm/day in 60 to 90% of the days with rain (Ramirez, 1974), but recorded rainfall intensities in the Northern Great Plains have been as high as 7 to 10 cm/h (Linsley et al., 1949).

17–2.2.1 Land Forming

Land forming techniques have been developed to control water erosion on favorable sites of uniform slopes and deep soils in dryland areas of the Southern and Central Great Plains. Terracing is the primary erosion control technique employed. This involves graded terraces for gradual export of excess water and flat channel terraces or level conservation bench terraces to collect, store, and use excess water (Zingg and Hauser, 1959). Although level bench terracing affords nearly complete protection against water erosion, the practice has limited acceptance in the Northern Great Plains. Land slope is the primary limiting factor because this determines the bench width. Resulting bench widths become too narrow for modern farming equipment, and the construction costs too high in the rolling Northern Great Plains (Black, 1968). However, level bench terracing does control water erosion from snowmelt and increases soil water supplies nearly twofold for subsequent crop production (Black, 1968; Haas and Willis, 1968).

Grassed waterways are natural drainageways usually reshaped and seeded to adaptable grasses that serve to nonerosively transport concentrations of runoff water from fields. Although grassed waterways are used extensively, they do nothing to control sheet and rill erosion on a field basis even though they do prevent deep gullying of natural drainageways. Unless associated with a complete erosion control system for a field, grassed waterways are of limited benefit.

17-2.2.2 Conservation Cropping

Contour stripcropping is usually designed with equal-width strips of crop and summer fallow with variable-width buffer areas of grass in between to compensate for topography irregularities. Contour strips shorten slope length, which reduces sheet and rill erosion, reduces total runoff, and may filter sediment in upright stubble. Although contour farming is effective when used in combination with stubble mulching, terraces, and water diversion, the land area devoted to the practice is relatively small.

Stubble mulching is by far the principal practice relied on to reduce or prevent water erosion. Mulches intercept the direct impact of raindrops and prevent dispersion of soil aggregates (sealing) and, thereby, maintain a relatively higher infiltration rate (Barnes and Bohmont, 1958). As wheat straw levels increase in a stubble mulch system, the percentage of dry soil aggregates greater than 0.84 mm increases at the soil surface, and the bulk density of the soil decreases (Black, 1973b). Seemingly, larger pores develop, and these larger pores, associated with increasing wheat-straw mulch levels, favor greater infiltration and less water erosion.

17-3 TILLAGE AND EQUIPMENT

The basic tillage implements used are the chisel plow with individual 30- to 45-cm sweeps, disk (tandem or one-way), rod weeder, and moldboard plow. Tillage is not traditionally performed in the fall on stubble land to be summer-fallowed the next year unless precipitation after the harvest causes excessive weed growth. A typical sequence of spring tillage operations for summer fallow in the semiarid region (major land resource areas no. 52, 53, and 54) include (i) disking when surface residue levels exceed about 3000 kg/ha and weed growth is moderate to heavy or (ii) tilling with a sweep cultivator when residue levels are less than 3000 kg/ha. Subsequent fallow operations, usually three to four, are performed with sweep cultivators with or without rod weeder, or harrow, attachment. Even though this type of tillage sequence is commonly referred to as conventional stubble mulch tillage, relatively small amounts (about 800 kg/ha) of surface residue will be present at the beginning of the second winter of the summer fallow period. In the subhumid region (major land resource areas no. 55, 56, and 57), the moldboard plow is typically used to initiate summer fallow, with subsequent tillage with a field cultivator. Summer-fallowing coarse- and moderately coarse-textured soils is not common in the subhumid region.

After the second winter of the summer fallow period (spring wheat planting time), surface soil conditions are extremely variable, ranging from wet to dry, powdery to crusted, and cloddy to smooth. Therefore, seedbed preparation may proceed with different tillage implements, or combinations of implements, as dictated by weather and soil conditions. The objective is to prepare a seedbed without pulverizing the soil to the extent of creating an erosion hazard and to maintain surface soil water content at a level necessary to obtain an adequate stand.

In a crop-crop sequence, the moldboard plow is the most frequently used implement for preparing a seedbed on land cropped the previous year. Fall moldboard plowing is a common practice on medium-, moderately fine-, and fine-textured soils in the subhumid region (major land resource areas no. 55, 56, and 57). Percentage of nonerodible-sized aggregates in these soils usually is high enough to prevent wind erosion unless they are subjected to wetting and freeze-drying (Hinman and Bisal, 1968; Bisal and Nielsen, 1964) or moistened by snowmelt and followed by exposure to widely fluctuating air temperatures (Anderson and Bisal, 1969). The moldboard plow, because it inverts the soil especially when used in simultaneous combination with a packer-pony press drill, is a reliable method of ensuring placement of seed into moist soil, thus enhancing relatively rapid crop establishment. This is a till-plant system. Surface soil cloddiness may deteriorate rapidly after moldboard plowing of moderately coarse-textured soils to increase potential wind erodibility compared with tillage with implements that leave residue on the surface (Power et al., 1958).

Minimum- and no-till methods are in the developmental stage for the Northern Spring Wheat Region. Research has shown that potential wind erodibility of soils in a spring wheat–summer fallow system could be reduced 10-fold with a no-till summer fallow system compared with conventional sweep cultivation (Black and Power, 1965). Limited tillage and no-till cropping systems that conserve crop residue on the soil surface increase dry aggregate soil structure, reduce soil erosion, and increase soil water storage. The combined influence of these factors is manifested in increased crop yields, greater crop residue production, and extension of periods of protective cover. Because of the increase in crop residue production and all of the subsequent beneficial effects it has on soil properties when returned to the soil (Black, 1973a, b), conservation tillage, or no-till cropping systems, intuitively would become progressively more beneficial with time, thereby improving the chemical and physical properties of soils.

17-4 CROP ROTATIONS

The principal cropping system in the region is spring wheat–summer fallow. (The preceding discussion of wind and water erosion control practices centered around this sequence.) However, other crops are grown either on summer fallow or following another crop. These include winter wheat, spring-sown barley and oats, oilseed crops (safflower, sunflower, mustard, and flax), and alfalfa.

The spring wheat–winter wheat–summer fallow cropping system (Black and Siddoway, 1976, 1977) affords superior soil protection against wind and water erosion compared with conventional system of spring wheat–summer fallow. Winter wheat seeded into upright spring wheat (or spring barley) stubble provides both plant and soil protection from wind. Upright stubble provides the added bonus of conserving additional water through snow trapping. Winter wheat produces more crop residue per unit of grain yield than spring wheat and therefore potentially provides better soil protection during the subsequent summer fallow period than spring grain residues.

In a crop-crop sequence, e.g., spring barley or oats after wheat, the greatest soil erosion hazards associated with the second crop occur during seedbed preparation and seedling establishment because the soil surface is devoid of residues. Summer fallowing after spring barley or oats presents the same problems as discussed for spring wheat in a 21-month summer fallow sequence, namely, the inadequacy of sufficient crop residue production to protect the soil for such a long period of time.

Oilseed crops grown in cropping systems, usually after wheat without summer fallow, increase the wind erosion hazard for several reasons. Preemergence herbicides must be incorporated properly with tillage operations before seeding. These operations reduce soil aggregate size, thereby increasing potential soil erodibility. Oilseed crops produce relatively less crop residue than wheat, and the residues decompose faster. On the same soil, dry aggregate size decreased more on summer fallow when the crop previous to the summer fallow was an oilseed crop than when the previous crop was spring wheat (A. L. Black. 1964. Unpublished data). Therefore, summer fallowing after oilseed crops is not recommended unless stored soil water the following spring is less than about 8 cm. If the area is summer-fallowed following an oilseed crop, the use of a contact broadleaf herbicide should be used as a substitute for one or more tillage operations to conserve as much crop residue as possible for soil protection.

Alfalfa is an important cultivated crop for dryland and shows a noteworthy increase in hectarage in recent years. Alfalfa provides year-round soil protection and a rather dependable supply of hay for livestock. When planted on recharge areas, alfalfa will provide hydrologic control of saline seep areas because its root system extracts water to depths of 6 m (Krall and Brown, 1976). Alfalfa supplies additional N to the soil for succeeding crops, especially if it is not harvested for forage.

17-5 PROMISING RESEARCH DEVELOPMENTS

Properly designed and placed barriers and the conversion of cultivated land producing annual crops to permanent, perennial vegetation are the only two measures that provide year-round protection against wind erosion. No-till planting of winter wheat into upright spring grain stubble in flexible cropping sequences also provides complete soil protection (Black and Siddoway, 1977). The use of tall wheatgrass barriers for wind erosion control has an added benefit of providing additional stored soil water through snow trapping to make intensive or annual cropping feasible (Black and Siddoway, 1976). Likewise, upright small grain stubble also controls wind erosion and traps snow to increase soil water supplies for a crop (e.g., winter wheat) seeded without tillage (Black and Siddoway, 1977). Adequate weed control and proper fertilization practices are prerequisites to the success of these types of erosion control practices.

17-6 RESEARCH NEEDS

Quantitative data of soil and water losses due to runoff from snowmelt and high-intensity rainstorms in the Northern Great Plains are limited to such a study on surface-mined sites, which included an undisturbed range site (Gilley et al., 1977). A data base is urgently needed to provide the means for making soil and water loss comparisons on cropland among existing and approved soil conservation practices. Specific interactions of topography, climate, soil type, cropping systems, and crop residue management as related to soil erosion by water need to be identified to enable more effective planning of appropriate soil conservation practices in this region.

Cropping systems involving no-till and minimum tillage concepts have been researched and adopted in many areas of the humid and subhumid regions of the USA (SCSA, 1973). However, in the semiarid regions of the Northern Great Plains, few have adopted such minimum- or no-till systems with the exception of no-till seeding of winter wheat into spring grain stubble in a 3-year cropping system (Black and Siddoway, 1977). With the emphasis on energy (fuel) conservation, interest has been renewed in minimum- or no-till summer fallow methods to partially or completely eliminate the necessity of tillage. Herbicides now available have improved the potential for developing a successful no-till summer fallow program. Also, proper fertilization and no-till planting techniques need to be developed to assure optimum yields and water use efficiency. Other problems, real and anticipated, involving crop adaptation, cultivars, weeds, insects, and diseases also need to be evaluated simultaneously. Tillage equipment that incorporates lesser quantities of residues during the summer fallow period than those now used is urgently needed. Better machinery for proper seed placement of fertilizer, herbicides, and crop seeds in one operation in no-till cropping systems for large-scale dryland operations in the Northern Great

Plains must be designed and developed. Year-round soil protection requires better utilization of crop residue and vegetative cover; therefore, concentrated interdisciplinary research is needed, combined with knowledge of the climatic restraints imposed in semiarid agriculture.

17-7 LITERATURE CITED

Anderson, C. H., and F. Bisal. 1969. Snow cover effect on the erodible soil fraction. Can. J. Soil Sci. 49:287–296.

Anderson, D. T. 1961. Surface trash conservation with tillage machines. Can. J. Soil Sci. 41: 91–114.

Barnes, O. K., and D. W. Bohmont. 1958. Effect of cropping practices on water intake rates in Northern Great Plains. Wyoming Agric. Exp. Stn. Bull. no. 358.

Bauer, A., and T. J. Conlon. 1974. Effect of tillage interval of fallow on available soil nitrogen, some soil physical properties, and wheat yields. North Dakota Agric. Exp. Stn. Res. Report no. 51.

Bauer, A., and J. C. Zubriski. 1978. Hard red spring wheat straw yields in relation to grain yields. Soil Sci. Soc. Am. J. 42:777–781.

Bisal, F., and K. F. Nielsen. 1964. Soil aggregates do not necessarily break down over-winter. Soil Sci. 98:345–346.

Black, A. L. 1968. Conservation bench terraces in Montana. Trans. ASAE 11:393–395.

Black, A. L. 1973a. Crop residue, soil water, and soil fertility related to spring wheat production and quality on fallow. Soil Sci. Soc. Am. Proc. 37:754–758.

Black, A. L. 1973b. Soil property changes associated with crop residue management in a wheat-fallow rotation. Soil Sci. Soc. Am. Proc. 37:943–946.

Black, A. L., and J. F. Power. 1965. Effect of chemical and mechanical fallow methods on moisture storage, wheat yields, and soil erodibility. Soil Sci. Soc. Am. Proc. 29:465–468.

Black, A. L., and F. H. Siddoway. 1971. Tall wheatgrass barriers for soil erosion control and water conservation. J. Soil Water Conserv. 26:107–111.

Black, A. L., and F. H. Siddoway. 1976. Dryland cropping sequences within a tall wheatgrass barrier system. J. Soil Water Conserv. 31:101–105.

Black, A. L., and F. H. Siddoway. 1977. Winter wheat recropping on dryland as affected by stubble height and nitrogen fertilization. Soil Sci. Soc. Am. Proc. 41:1186–1190.

Chepil, W. S., and N. P. Woodruff. 1963. The physics of wind erosion and its control. Adv. Agron. 15:211–302.

Frank, A. B., and E. J. George. 1975. Windbreaks for snow management in North Dakota. p. 144–154. *In* Snow management on the Great Plains. Great Plains Agric. Coun. Pub. no. 73, Univ. of Nebraska, Lincoln.

Geiszler, G. N. 1961. Corn rows control wind erosion. N.D. Farm Res. 21(11):13–14.

George, E. J., D. Broberg, and E. L. Worthington. 1963. Influence of various types of field windbreaks on reducing wind velocity and depositing snow. J. For. 61:345–349.

Gilley, J. E., G. W. Gee, A. Bauer, W. O. Willis, and R. A. Young. 1977. Runoff and erosion characteristics of surface-mined sites in western North Dakota. Trans. ASAE 20:697–704.

Haas, H. J., and G. O. Boatwright. 1960. Let's take another look at summer-fallow in the Northern Plains. J. Soil Water Conserv. 15:176–179.

Haas, H. J., and W. O. Willis. 1968. Conservation bench terraces in North Dakota. Trans. ASAE 11:396–398.

Haas, H. J., W. O. Willis, and J. J. Bond. 1974. Summer fallow in the Northern Great Plains (spring wheat). p. 12–35. *In* Summer fallow in western United States. USDA-ARS Conserv. Res. Report no. 17.

Hinman, W. C., and F. Bisal. 1968. Alterations of soil structure upon freezing and thawing and subsequent drying. Can. J. Soil Sci. 48:193–197.

Hoag, B. K., and G. N. Geiszler. 1971. Sunflower rows to protect fallow from wind erosion. N.D. Farm Res. 28(5):7–12.

Howard, J. K. 1959. Montana high, wide, and handsome. Yale Univ. Press, New Haven, Conn.

Kellogg, C. E. 1935. Soil blowing and dust storms. USDA Miscellaneous Pub. no. 221.

Krall, J. L., and P. L. Brown. 1976. Cultural practices for the control of saline seep in the Northern Plains. p. 64–78. *In* Conservation tillage. Great Plains Agric. Coun. Pub. no. 77, Univ. of Nebraska, Lincoln.

Lewis and Clark 1805 Regional Council for Development. 1978. Control of water pollution from cropland. Lewis and Clark 1805 Regional Coun. for Development Tech. Supplement no. 4, Mandan, N.D.

Linsley, R. K., M. A. Kohler, and J. L. H. Paulhus. 1949. Applied hydrology. McGraw-Hill Publications, New York.

Power, J. F., T. S. Aasheim, and G. P. Hartman. 1958. Spring wheat production and soil and water conservation as influenced by methods of summer fallowing on a chestnut soil in northeastern Montana. Soil Sci. Soc. Am. Proc. 22:460–463.

Ramirez, J. M. 1974. The effects of added rainfall during the growing season in North Dakota. North Dakota Agric. Exp. Stn. Res. Report no. 52.

Siddoway, F. H. 1963. Effects of cropping and tillage methods on dry aggregate soil structure. Soil Sci. Soc. Am. Proc. 27:452–454.

Siddoway, F. H. 1970. Barriers for wind erosion control and water conservation. J. Soil Water Conserv. 25:180–184.

Skidmore, E. L. 1965. Assessing wind erosion forces: Directions and relative magnitudes. Soil Sci. Soc. Am. Proc. 29:587–590.

Soil Conservation Society of America. 1973. Conservation tillage. Proc. Natl. Conserv. Tillage Conference, Des Moines, Iowa. 28–30 March 1973. SCSA, Ankeny, Iowa.

U.S. Department of Agriculture. 1977. Agricultural statistics. USDA, U.S. Government Printing Office, Washington, D.C.

Wilkinson, G. E., and F. W. Schroer. 1966. Water, nitrogen removal by cover crops on fallow. N.D. Farm Res. 24(6):4–7.

Willis, W. O., C. W. Carlson, J. Alessi, and H. J. Haas. 1961. Depth of freezing and spring run-off as related to fall soil-moisture level. Can. J. Soil Sci. 41:115–123.

Woodruff, N. P., and F. H. Siddoway. 1965. A wind erosion equation. Soil Sci. Soc. Am. Proc. 29:602–608.

Young, R. A. 1958. Can nitrogen fertilizer and modern weed control methods eliminate summerfallow? N.D. Agric. Exp. Stn. Bimon. Bull. 20(3):3–9.

Zingg, A. W., and V. L. Hauser. 1959. Terrace benching to save potential runoff for semiarid land. Agron. J. 51:289–292.

18

W. Earl Johnson
University of Saskatchewan
Saskatoon, Saskatchewan, Canada

Soil Conservation: Canadian Prairies

The extent of the dryland area in the prairies of Canada is indicated in Fig. 18–1. Table 18–1 explains areas labeled in Fig. 18–1. It includes the Aridic Borolls, Typic Borolls, and a portion of the Udic Borolls, corresponding approximately to the region with a soil water deficit of 120 mm or greater. This continental climate is characterized by extremes with mean January temperatures down to −9°C and July mean temperatures up to 18°C. Monthly wind speeds range up to an average of 21 km/h. These high monthly average wind speeds occur in April and May, which comprise a highly vulnerable period for wind erosion. Annual precipitation varies tremendously with Swift Current (Fig. 18–1) located in the Aridic Boroll zone, ranging from a low of 210 mm in 1937 to a high of 620 mm in 1891. A wide range in soil texture and topography is characterized by extensive hectarages of smooth level plains, a moderate hectarage of gently rolling topography with long smooth slopes, and a major hectarage of glaciated topography characterized by a high frequency of knolls and ridges in an immature drainage pattern. These features and others (to be discussed in this chapter) contribute to serious soil conservation hazards and have given rise to many practices peculiar to the Canadian dryland.

18–1 FOUR INTERRELATED SOIL CONSERVATION PROBLEMS

18–1.1 Wind Erosion

Wind erosion has been a recurring serious problem on cultivated land in the dryland area of western Canada since the early days of settlement. The first severe cycle occurred in the period 1917 to 1920 and prompted establishment of the first soil surveys and research programs seeking methods for erosion control.

Dryland Agriculture—Agronomy Monograph no. 23.

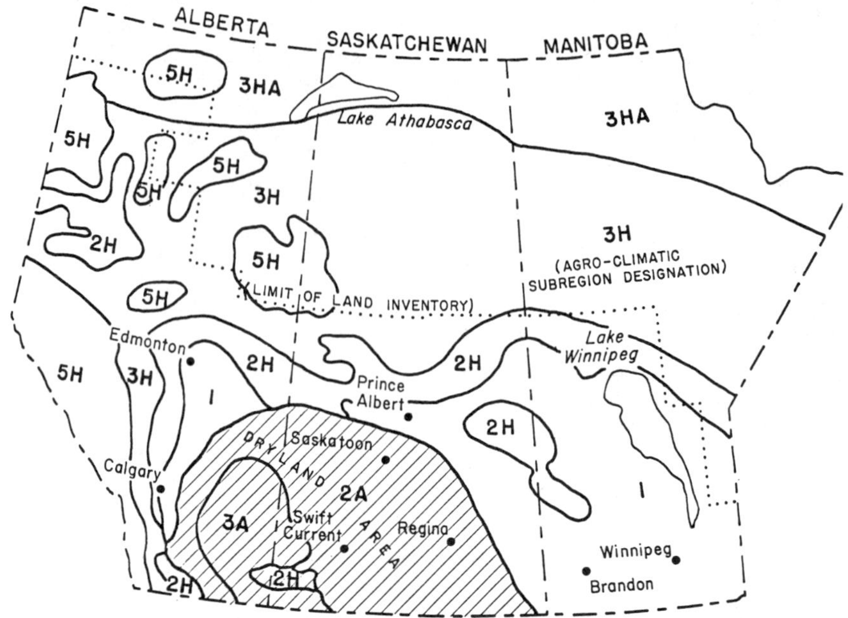

Fig. 18–1. Canadian dryland area. For a key to the subregions, see Table 18–1.

Intensive summer fallowing has been a major cause of extensive wind erosion. The problem was further aggravated by an erroneous adoption of the "dust mulch" theory to conserve moisture (Bracken, 1915; Baver, 1948). Moldboard plowing, combined with excessive harrowing to achieve the dust mulch, left bare soil with a large proportion of very fine aggregates. This highly erodible surface, combined with drought, and intensified by adverse economic conditions, led to the severe wind erosion of the 1930's. Programs, problems, and people have been well described by Gray (1967).

General adoption of combine harvesting and development of subsurface tillage during the 1930's and 1940's provided the two main requirements for wind erosion control. Crop residues were left in place on the land and maintained on the surface.

18–1.2 Water Erosion

Water erosion has been more sporadic and did not receive attention in the literature or extension publications until 1950. Snowmelt and local thunderstorms are the main causes of water erosion. The frequency of high-intensity precipitation is quite low. At Swift Current, over the period from 1922 to 1954, measurable precipitation occurred on 2301 days out of a total of 7062 days in the April–October period (Soil Res. Lab., 1956). Precipitation for a 24-h period exceeded 25 mm only 49 times, and precipitation of 20 to 25 mm occurred only 46 times in the 33 years. Water erosion generally produces small to moderate gully development, which has been acceler-

Table 18–1. Explanation of the subregions in Alberta, Saskatchewan, and Manitoba shown in Fig. 18–1.

Subregion designation	Precipi-tation	Frost-free period	Remarks
	mm	Days	
1	400–450	> 90	Both adequate for most dryland crops
2A	300–400	--	Frost-free period usually long enough to mature wheat; precipitation usually limiting 50% of the years
2H	400–450	75–90	Frost damage to wheat 30% of the years
3A	300	110–115	Precipitation usually severely limiting for most dryland crops
3H	--	--	Precipitation adequate; severe frosts render wheat production impractical; precipitation in subregion no. 3HA about 300 mm
5H	--	< 60	Precipitation usually adequate; frost-free period inadequate for cereals

ated significantly by disruption of natural drainage. Water erosion is much more severe in the more moist regions of the province where moisture deficits are less than 120 mm.

18–1.3 Deterioration of Soil Physical Condition

Another soil conservation problem is deterioration of soil tilth caused by long-term use of summer fallow. As noted in Chapter 26, yields of spring wheat per crop hectare (a yield basis that includes the hectarage of summer fallow used for the crop) have shown only a moderate increase in the past 50 years. Yields per seeded hectare have increased markedly, but the proportion of land in summer fallow each year has increased substantially in the last 35 years. Wheat yields for the 5-year periods 1924 to 1928 and 1974 to 1978 for Alberta and Saskatchewan are shown in Table 18–2. Increases per seeded hectare were 547 and 643 kg/ha, but increases per crop hectare, when the hectarage of summer fallow was taken into account, were only 206

Table 18–2. Wheat yields in Saskatchewan and Alberta for the two periods 1924 to 1938 and 1974 to 1978 and distribution of seeded hectarage on summer-fallowed land.

Province and period	Wheat yield (kg/ha) for indicated basis		Percent of wheat hectarage seeded on summer fallow
	Per seeded hectare	Per crop hectare†	
Alberta			
1924–1928	1353	862	59
1974–1978	1900	1088	74
Saskatchewan			
1924–1928	1183	831	42
1974–1978	1826	1008	81

† A crop hectare includes hectarage of the seeded wheat plus the hectarage of the summer fallow of the previous year seeded to wheat and not other crops.

and 177 kg/ha. Statistics for 1974 to 1978 show the actual hectarage seeded on summer fallow. This figure is not available for 1924 to 1928, and the total previous year's summer fallow hectarage has been allocated to wheat, which would overestimate hectarage slightly since some summer fallow hectarage would be used for other crops. Consequently, yield increases are more favorable than would be totally accurate. Hedlin and Rigaux (1976) reported that fertilizer use and varietal improvement should provide a yield increase of 350 kg/ha. Weed control, tillage equipment, insect control, and seeding equipment are other technological changes that would add further to yield potential.

It is reasonable to conclude that the natural productivity of prairie soils has declined and that yields per seeded hectare do not reflect the total potential increase from current production inputs. The increases per hectare over 50 years are very small. Over 50 years ago, Shutt (1925) showed that prairie soils had lost about 40% of the original N and organic matter. Considerably less than 50% of the N loss could be accounted for by crop removal. More recent studies have shown that many soils have lost 50% or more of the original N and organic matter (Rennie et al., 1976). Although the rate of loss has decreased, the decline is continuing. Erosion losses probably account for less than 20% of the N lost (Moss, 1935; Nicholaichuck and Read, 1973; Toogood, 1963). The remaining large disappearance of N has occurred through deep percolation or denitrification.

The intensive use of summer fallow is a major contributor to these latter losses. In cycles when surface moisture is adequate, organic N rapidly mineralizes. This N is then susceptible to leaching losses or denitrification if the moisture regime allows.

These documented losses reduce potential soil N supplies and severely deteriorate physical condition. This loss in physical condition decreases moisture efficiency, increases susceptibility to water erosion, and, of course, decreases productivity. Fully definitive studies are generally lacking, but analysis by Mack (1977) does indicate the highest relative yield reduction in years of low moisture stress. This would relate well to the larger effect of deterioration in soil physical condition under moist or saturated soil moisture levels.

18-1.4 Soil Salinization

The practice of summer fallow in a standard 2-year program produces saturated soil moisture conditions during good precipitation cycles. Surface runoff and water movement through the soil are both increased. A 23-year study at Swift Current showed an average 30% of the November–April precipitation stored in the soil after harvest (Soil Res. Lab., 1956). After the summer period of the summer fallow, only 9% of precipitation for the November–April period was stored. Consequently, over 300% more water goes to evaporation and runoff on summer fallow as on cropland during the fall-to-spring period.

It was generally considered that little moisture moved below the root zone in the drier regions on summer fallow. Deeper sampling to 240 and 300 cm has shown that moisture will move below the root zone about 3 years in 10 (Soil Res. Lab., 1956). Nitrate nitrogen (NO_3-N) levels have increased down to 240 cm or more. This data indicates the potential for movement of salts by seepage.

There has been a large increase in land moderately to seriously affected by salinity in the past 20 years. Intensive use of summer fallow has been an important causative factor. Salinity encroachment thus represents one of the main threats to productivity on a large hectarage.

Increases in area affected by soil salinity, deterioration in soil condition, or both without soil erosion are equally important as erosion control for long-range soil productivity. Soil conservation must have a broader scope then mere erosion control. Significant changes in cropping systems and production inputs are required to halt and, we hope, reverse the decline in basic soil productivity.

18-2 SOIL FACTORS AND WIND EROSION

Soil characteristics important in wind erosion were established largely by the Soil Research Laboratory (1949, 1956). The proportion of surface aggregates greater than 0.42 mm in diameter is a fairly accurate indication of resistance to soil erosion. In general, a high percentage of dry aggregates greater than 0.83 mm has been used as a satisfactory measure of resistance to wind erosion.

Soil texture is the most important soil characteristic related to wind erosion. Sandy soils, which do not form cohesive aggregates, are extremely susceptible. Fine-textured clays have temporarily resistant aggregates, but the aggregates break down rapidly during wetting and drying, freezing and thawing, and particularly on freeze-drying (Hinman and Bisal, 1968). Soils of medium texture have the most stable aggregates and are most resistant to wind erosion (Soil Res. Lab., 1949).

Calcium carbonate content of the soil has a pronounced effect on wind erosion (Soil Res. Lab., 1949). From 2 to 3% (wt/wt) $CaCO_3$ decreased aggregate stability, and the amount of soil eroded in wind tunnel studies was increased by 17 to 22 t of soil/ha over an 18-min period. The content of $CaCO_3$ in the Haploborolls and the Calciborolls in knoll positions is moderately high to high. The increase in wind velocity over knolls, combined with the high carbonate content, makes the knolls major focal points for initiation of wind erosion.

Decomposing organic matter is a most important agent in development of soil aggregates (Rennie et al., 1954). On the other hand, decomposed organic matter or humus may produce aggregates too small to resist wind erosion (Soil Res. Lab., 1949).

Soil water content significantly affects aggregation or pulverization by soil cultivation (Fig. 18-2; Soil Res. Lab., 1949). It is inadvisable to culti-

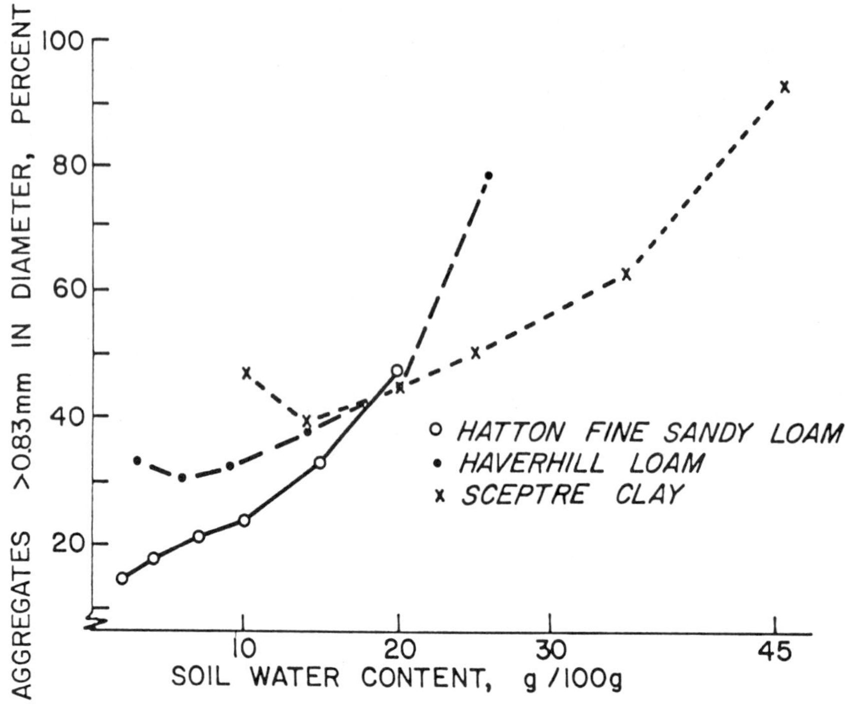

Fig. 18-2. Soil structure response to soil water content at tillage time (adapted from Soil Res. Lab., 1949).

vate sandy soils when they are dry. For loam and clay soils, tillage close to the sticky point produces the best effect on stable aggregates. For these soils, tillage at a slightly moist content is most damaging to aggregation, and tillage is better when dry than when slightly moist. Timely tillage is an important management factor in maintaining soil tilth.

18-3 CROP RESIDUE MANAGEMENT

Current assessment of soil water reserves and associated requirements of fertilizers and weed control provides a basis for extended cropping instead of continuing summer fallow in an uninterrupted rotation. In the severe moisture deficit regions and in drier cycles, 5 million ha or more will continue in summer fallow. Markets and cost-price economics will be influencing factors in summer fallow hectarage.

Erosion control during summer fallow is largely dependent on stubble mulch. Maintenance of adequate residues anchored in the soil depends first on production inputs to grow sufficient residue. Subsurface tillage, seeding machinery that will retain some surface residues, and maximum substitution of herbicides for tillage are the other requirements.

Table 18-3. Average reduction in wheat crop residue for each operation with various tillage machines (Anderson, 1967).

Machine	Residue reduction per operation	Expected range†
	%	
Wide-blade cultivator	10	−13 to 30
Rod weeder	10	−15 to 20
Heavy-duty cultivator with rod weeder	10	−12 to 30
Heavy-duty cultivator	20	−12 to 30
One-way disk	50	10 to 70
Tandem disk	50	10 to 75

† Negative figures reflect the machine's ability to return some residue to the surface.

Bisal and Ferguson (1970) found that 840 kg/ha of wheat residues protect medium-textured soils for wind speeds up to 1200 cm/s. For fine-textured soils, their studies indicated a need for amounts of residue somewhat greater than the 1400 kg/ha suggested by other workers. They concluded that 2000 kg/ha would protect fine-textured soil against all but the most extreme conditions. On coarse-textured soils, adequate residue protection was the most critical problem in the wheat cropping system.

For crop seeded on summer fallow recommended rates of fertilizer increase wheat straw yields an average of 400 kg/ha or more on Aridic Borolls at Swift Current. Data for wheat on land cropped the previous year indicate an average increase of 300 kg/ha or more using recommended fertilizer rates with satisfactory moisture rserves (F. G. Warder. 1966. Unpublished data. Research Station, Swift Current). With straw yields frequently at the 1000 kg/ha level or lower, these increases in straw yield are important in growing sufficient crop residue.

Johnson (1977) reviewed conservation tillage programs for the prairie provinces of Canada, including the subhumid as well as the dryland area. Anderson (1961, 1964, 1965, 1967) has given detailed study to the characteristics of tillage machines for saving surface residues. Table 18-3 shows the percentage reduction in residues for each operation of various machines (Anderson, 1967). One operation with either type of disk reduces an initial 1500 kg/ha straw yield below the 840 kg/ha needed on medium-textured soils. A wide-blade cultivator or the rod weeder would leave at least the required amount (850 kg/ha) after five operations. It is obvious that the use of disk implements must be restricted except on fields with very heavy residue. Detailed evaluations (Anderson, 1967) for the combined cultivator–rod weeder implement show that the rod weeder tends to return residues to the surface. An important requirement for the rod weeder is operation at a minimum depth of 5 to 6 cm to keep residues anchored. Operation of the rod weeder at a shallow depth pulverizes surface soil and leaves residues loose on the surface.

Anderson (1971) reported the percentage residue conserved with varying degrees of herbicide substitution for tillage (Table 18-4). Reduction in

Table 18–4. Surface retention of residue and wheat yields as related to summer fallow, chemical assist, or tillage-only summer fallow.

Fall treatment	Summer fallow program	Residue conserved	Wheat yield
		% of original	kg/ha
Nil	Herbicide	62	1542
Nil	Herbicide, tillage†	38	1494
Nil	Tillage, herbicide‡	43	1484
Nil	Tillage	35	1497
Herbicide	Tillage	37	1559
Tillage	Tillage	24	1510

† First spring treatment was herbicide.
‡ First spring treatment was tillage.

use of tillage had no significant adverse effect on yields in this study. Complete chemical summer fallow retained 62% of the original residue compared with 24% for the most intensive tillage program. Tillage during the first fall of the summer fallow period produced the largest reduction of surface residues. Herbicide use after harvest gave good control of winter annuals and preserved standing stubble to trap snow.

Most of the wheat crop requires two operations of swathing and combining for harvesting. Large combines in current use can pick up and thresh windrows collected from swaths of 9 m or greater width; the straw is not redistributed uniformly over the original swath especially if the swath is 5 m or more wide. Nonuniform redistribution of combined crop residues, therefore, complicates surface residue management. Straw choppers and modifications of baffles on the straw spreader will ultimately improve straw distribution and allow efficient operation of tillage machinery. Although the concentrated residue may serve as a barrier to prevent soil loss from the field, the intervening areas may have insufficient residue to prevent drifting. The result is frequently the development of drift soil ridges in the field.

Straw choppers must be used with discretion in the dryland area before the summer fallow year. Straw chopping is necessary for tillage where residues are over 3000 kg/ha, but chopped residues disappear more quickly.

Many soil-applied herbicides introduced in recent years require severe soil manipulation for adequate incorporation. Moreover, crop residues in excess of 25% ground cover interfere with efficacy of some herbicides. Herbicides that control weeds shortly after germination offer a real advantage to moisture efficiency. For good residue maintenance, however, new methods of injecting or placing soil-applied herbicides or use of surface-applied herbicides is necessary.

Anderson and Russel (1964) found that straw mulch rates up to 4500 kg/ha had little to no deleterious effect on spring or winter wheat yields in southern Alberta. There is no reason to retain less than maximum percentages of residues for erosion control. Satisfactory seeding methods and other production inputs should be used to give maximum potential residue production. Many types of drills are more suitable than the one-way disk seeder combination for seeding because they retain more residue for protection during the vulnerable period after seeding.

18-4 FOCAL AREAS, SHELTERBELTS, AND OTHER SUPPORT MEASURES

Calcareous knolls are an important focal point for wind erosion control. Erosion from these sites increases the $CaCO_3$ content of neighboring areas and enlarges the susceptible area. Knolls should be given special attention, including application of barnyard manure, waste straw, or other materials. Deep tillage should be avoided, because it serves only to mix the tilled layer with deeper horizons that contain a greater $CaCO_3$ content than the original surface soil.

Knoll positions are frequently low in both P and N. High rates of P fertilizer broadcast and incorporated will increase the residues needed to improve erosion control.

Stripcropping is a proved method for wind erosion control. If summer fallow becomes necessary on coarse-textured soils, only a combination of stubble mulch and rigorous stripcropping will control soil drifting. The width of strip required varies from 10 m on loamy sands to 100 m on clay loams. Stripcropping is less effective on fine-textured clays than on clay loam to sandy soils. On the fine-textured soils, however, crop production is generally large enough to give adequate residues for protection. For water erosion control, which occurs mainly on medium-textured soils, strips should be across the slope. When this direction is parallel to the wind, more attention to residue retention is needed because stripcropping gives best wind control when oriented perpendicular to the prevailing wind.

Wind speed and turbulence is affected by knolls, barriers, soil ridges, buildings, and road grades. These effects produce focal areas for drifting that need special attention.

Some areas are of variable texture; coarse-textured soil occurs as small local spots in medium- to fine-textured soil areas. These areas require special protection measures, including permanent grass to prevent spread of damage to neighboring land.

Field shelterbelts have proved effective for soil drifting control in areas with systematic planting. Shelterbelts reduce wind velocity by 20% at distances of 20 times the height of the shelterbelt to the leeward side. Studies have shown an overall wheat yield increase of 47 kg/ha in shelterbelt areas over that obtained in exposed areas (Soil Res. Lab., 1956). The area occupied by the trees was included in the calculation.

Systematic shelterbelts, at spacings of 95 m for coarse-textured soils and increasing to 200 m on medium-textured soils, are effective and practical on these soil types on smooth topography where salinity hazards are slight.

Much more attention must be given to the design of shelterbelts on steep or long slopes, on fine-textured soils, and in areas of moderate to high salinity hazard. Shelterbelts up and down slopes or planted across intermittent water channels may increase water erosion. The additional surface water from snow trapped by shelterbelts is a factor that aggravates salinity in salinity hazard areas (Sommerfelt, 1973). The caragana shelterbelts have relatively low porosity. Their low porosity often causes heavy snow drifts near the shelterbelt, drifts that seriously delay field operations in the spring. Weed control and seeding are also not timely compared with the more exposed area. Shelterbelts with slightly greater porosity or some change in planting density would improve the potential for extended use.

18-5 FACTORS INCREASING WATER EROSION HAZARD

Although wind erosion damage has occurred on a much larger area, water erosion has become a significant problem in the past 30 years. Water runoff not only causes soil damage but represents a loss of stored water, the major limiting factor in dryland crop production. Sediment from water erosion is the most important contributor to lower surface water quality.

As with wind erosion, most water erosion occurs on land that has been summer-fallowed. Hectarage in summer fallow increased rapidly after a wheat acreage reduction program in 1941 and is currently double the hectarage in summer fallow in 1940. Toogood (1963) reported soil losses from runoff plots in Alberta, in a region somewhat more humid than the true dryland area. Losses from a 2-year wheat-fallow rotation averaged almost 2400 kg ha^{-1} y^{-1}. On a 5-year rotation of continuous grain and hay, soil losses were reduced to 140 kg ha^{-1} y^{-1}.

The severe reduction in organic matter and the resulting deterioration in physical condition of the soil are important causes of water erosion. The Soil Research Laboratory (1949) showed that water stable aggregates greater than 0.83 mm ranged from 10.14% on an uncultivated virgin clay to 52.5% on a virgin loam soil. On cultivated soils the range was 1.1 to 31.9%. Generally the percentage of water stable aggregates greater than 0.83 mm in cultivated soils was less than one half of the percentage in prairie sod.

Soils of clay to silt loam texture on steep or long slopes are most susceptible to water erosion. Detachability and transportability of soil particles are much greater than on soils of loam texture on glacial till. Solonetzic soils (Argiborolls) with low organic matter content in surface horizons have low infiltration rates; their dense B horizons also produce a low internal permeability. These soils are, therefore, quite susceptible to water erosion.

Disruption of natural drainage patterns has been an important cause of gully erosion. Insufficient culvert capacity in road grade crossings and imbalance in capacity concentrates runoff. The need for care in designing shelterbelt patterns has been discussed above.

Stubble mulch and stripcropping are basic practices in water erosion control. The complexity of topography prevents large-scale application of contour stripcropping. It is feasible, however, to have field strips across the general slope of the land.

A major amount of water erosion is caused by runoff from snowmelt in the spring. Willis et al. (1961) report from small plot studies in North Dakota that as much as 90% of the snowmelt water is lost as surface runoff when plots were frozen at high moisture levels. Gray (1965) listed mean annual runoff coefficients for the dryland region largely in the range of 100 to 200 ha cm km^{-2} for the dryland area. Snowmelt therefore represents a major part of the average annual runoff. Standing stubble, grass barriers, shelterbelts, or cover crop on summer fallow all retain a blanket of snow that reduces frost penetration (Staple et al., 1960). Water erosion is thereby reduced because the soil thaws and restores permeability more quickly.

Coefficients for the universal soil loss equation (Wischmeier and Smith, 1978) have not been developed for the Canadian Prairies. Complexity of topography with frequent ridges and knolls is one problem. Soil series occur in complex patterns governed largely by slope position and aspect. Snowmelt runoff characteristics vary considerably with the intermittent occurrence of freezing cycles. The latter prevents satisfactory use of terraces and interferes with channel flow.

Accelerated runoff and saturated soil conditions during fallow periods both contribute significantly to the increase in salinity of poorly drained soils. Saline seep characteristics are treated in some detail in Chapter 7. Salinity encroachment and the increasing susceptibility to water erosion constitute proof that the cropping system must be modified to maintain and increase productivity.

18-6 CROPPING SYSTEMS

Although marketing and other economic constraints present problems, it is agronomically feasible to reduce the proportion of crop land in summer fallow each year from 40 to 50% down to 30 to 40% or less. These choices of cropping system have a significant influence on soil conservation (Toogood, 1963).

The highly erodible coarse-textured soils store little moisture and should not be summer-fallowed. Aridic and Typic Borolls coarser than very fine sandy loam and Udic Borolls coarser than fine sandy loam are best retired to long-term pasture or hay production. With good management, erosion is fully controlled. Just as important, these highly permeable soils contribute significantly to ground water movement and salinization of neighboring soils. Medium-textured soils also depend in large degree on growing season precipitation and should be cropped when spring reserves of available water are 8 cm or more (Experimental Farm, 1961). Fine-textured soils can store up to 20 cm of water in the rooting zone. Crops make more efficient use of soil water than precipitation on fine-textured soils (Soil Res. Lab., 1949, and 1956). Crop production is generally heavy enough to give adequate residue for erosion control on summer fallow.

It is now well established that much of the summer fallow contribution to grain yield is the N mineralized during the summer fallow year. Another factor, of course, is the positional availability of the mineralized N, because much of it is located below 15 cm, where it is available to the crop when the surface soil is dry.

The soil test for NO_3-N in the upper 60 cm of soil in late fall or early spring gives a reasonably accurate assessment of N requirements. Herbicides have reduced dependence on summer fallow for weed control. The de-

Table 18–5. Total crop residue and percentage of original residue remaining at the end of the summer fallow year.

Crop	Cultivar	Crop residue, kg/ha	Percent residue conserved
Oat	Ajax	1149	21.6
Wheat	Chinook	1068	24.2
Barley	Vantage	1033	20.9
Brown mustard	--	578	15.9
Flax	Norland	380	14.8
Peas	Chancellor	350	20.8
Polish rapeseed	--	251	6.4

velopment of soil testing, herbicides, and equipment to negotiate through surface residues can allow for a significant increase in extended cropping based on assessment of current moisture reserves.

Wheat currently occupies in excess of 75% of the crop hectarage over much of the dryland area. Diversification to reduce this percentage considerably would allow for use of a wider spectrum of herbicides, better labor distribution, and more opportunity for spring or fall operations. Sunflower, mustard, flax, lentil, *campestris* varieties of rapeseed, silage corn, and some increase in oat, and barley all could be used. The oilseeds and grain legumes will not provide sufficient residues to control erosion on summer fallow. Anderson (1968) has shown the poor residue retention for these crops (Table 18–5). They must be followed by cereal grains that will provide sufficient residue if summer fallow becomes necessary.

Winter wheat is being grown successfully in southern Alberta, and winter rye has been an important crop on coarse-textured soils. Winter cereals represent an important addition to erosion control and soil maintenance. The early spring growth is beneficial in saline hazard areas and gives soil erosion control in the vulnerable period of April and May. Winter cereals offer the best potential to follow low-residue annuals if moisture reserves are too limited for spring seeding. Fallow using minimum tillage but with good weed control until August and early September will allow some buildup of moisture reserves.

For adequate soil maintenance or improvement, forage legumes, alfalfa, yellow-fallowered alfalfa, and sweetclover, or in mixtures with grasses, are needed in the cropping systems. Intensive use is required on strongly saline and contiguous areas, steep slopes, and permanent use as grassed waterways.

The benefits of legume-grass mixtures or legumes to soil condition and productivity are indicated in Chapter 26. The main requirement for successful inclusion of legumes and grasses in the cropping program is flexibility in the cropping system to adjust for moisture cycles.

In all cropping systems, all production inputs should be maintained at the full economic optimum. Soil erosion control and soil productivity maintenance in the dryland area depends on a vigorous growing crop or the residues from that crop.

18-7 NEW DEVELOPMENTS AND RESEARCH NEEDS

Reduced tillage with a major substitution of herbicides for tillage is a well-established management to control erosion. Zero tillage, using herbicides only for weed control, is showing promise on both summer fallow and continuous cropping.

Residue management and seeding equipment for zero tillage requires further research. Research and development is required to develop fertilizer sources and application methods under zero tillage. Timing of operations is important, but development of soil moisture guidelines and simple test methods for moisture are needed for best seeding operations.

Basic soil studies to assess the effect of zero tillage on soil physical and chemical characteristics are essential. With extended cropping, the annual application of herbicides requires assessment of residues and soil effects.

Plant materials research on the winter cereals rye and wheat is of major importance. Improved winter hardiness, marketable winter wheat, and rye with improved nutritional and feed utilization characteristics would add a major dimension to production and soil conservation in western Canada.

18-8 LITERATURE CITED

Anderson, C. H. 1968. A comparison of annul crops for seed and residue in the semi-arid region of western Canada. Can. J. Plant Sci. 48:287-291.

Anderson, C. H. 1971. Comparison of tillage and chemical summer fallow in a semi-arid region. Can. J. Soil Sci. 51:397-404.

Anderson, D. T. 1961. Surface trash conservation with tillage machines. Can. J. Soil Sci. 41: 91-114.

Anderson, D. T. 1964. Some factors affecting trash conservation with disc-type tillage implements. Can. Agric. Eng. 6:11-13, 19.

Anderson, D. T. 1965. Some factors affecting trash conservation with the heavy duty cultivator. Can. Agric. Eng. 7:45-46, 69, 56.

Anderson, D. T. 1967. The cultivation of wheat. p. 338-355. *In* K. F. Nielsen (ed.) Canadian Centennial Wheat Symp., Saskatoon, Saskatchewan. 10-12 May. Western Co-operative Fertilizer Ltd., Calgary.

Anderson, D. T., and G. C. Russell. 1964. Effects of various quantities of straw mulch on the yield and growth of spring and winter wheat. Can. J. Soil Sci. 44:109-118.

Baver, L. D. 1948. Soil physics. ed. 2. John Wiley & Sons, Inc., New York.

Bisal, F., and W. S. Ferguson. 1970. Effect of non-erodible aggregates and wheat straw on initiation of soil drifting. Can. J. Soil Sci. 50:31-34.

Bracken, J. 1915. The tillage of stubble land. Field Husbandry Circ. no. 7, Dep. of Agric., Government of Saskatchewan.

Experimental Farm, Lethbridge, Alberta. 1961. Progress report 1953-1958. Information Division, Agric. Canada, Ottawa.

Gray, D. M. 1965. Climate and surface water. p. 46-66. *In* Water and Climate Symp., Report no. 2. Water Studies Inst., Univ. of Saskatchewan and Saskatchewan Res. Coun., Saskatoon.

Gray, J. H. 1967. Men against the desert. Modern Press, Saskatoon, Saskatchewan, Canada.

Hedlin, R. A., and L. R. Rigaux. 1976. Crop yield changes in the Prairie Provinces, 1958–1976. p. 73–86. *In* Technical and scientific papers. Manitoba Agronomists' Conference, Winnipeg. 14–15 December. Manitoba Dep. of Agric., Winnipeg.

Hinman, W. C., and F. Bisal. 1968. Alterations of soil structure upon freezing and thawing and subsequent drying. Can. J. Soil Sci. 48:193–198.

Johnson, W. E. 1977. Conservation tillage in western Canada. J. Soil Water Conserv. 32:61–65.

Mack, A. R. 1977. Long-term variations in moisture-stress index, Section 3.2. p. 1–8. *In* Climatic variability in relation to agricultural productivity and practices. Canada Comm. on Agrometeorology, Res. Branch, Agric. Canada, Ottawa.

Moss, H. C. 1935. Some field and laboratory studies of soil drifting in Saskatchewan. Sci. Agric. 15:665–679.

Nicholaichuk, W., and D. W. Read. 1973. Nutrient contribution to surface water from agricultural lands. CSAE Mimeo, Canada Soc. of Agric. Engineers Annual Meeting, Victoria, British Columbia. 20–22 August. CSAE, Victoria.

Rennie, D. A., E. Truog, and O. N. Allen. 1954. Soil aggregation as influenced by microbial gums, level of fertility and kind of crop. Soil Sci. Am. Proc. 18:399–403.

Rennie, D. A., G. J. Racz, and D. K. McBeath. 1976. Nitrogen losses. Section II: Nitrogen inputs to the plant system. p. 325–353. *In* Proc. Western Canada Nitrogen Symp., Calgary, Alberta. 20–21 January. Sponsored by Proc. Alberta Soil Science Workshop. Western Canada Soils Coordinating Comm., Edmonton.

Shutt, F. T. 1925. Influence of grain growing on the nitrogen and organic matter content of western prairie soils of Canada. Canada Dep. of Agric. Bull. no. 44. New Series.

Soil Research Laboratory. 1949. Soil moisture, wind erosion and fertility of some Canadian prairie soils. Agric. Canada Tech. Bull. no. 71, Pub. no. 819, Ottawa.

Soil Research Laboratory. 1956. Progress report 1948–54. Agric. Canada, Ottawa.

Sommerfeldt, T. G. 1973. Dryland salinity investigation at Nobleford. p. 47–49. *In* Proc. Alberta Dryland Salinity Workshop, Lethbridge, Alberta. 18–19 January. Alberta Dep. of Agric. Soils Branch of Plant Industry Division, Edmonton.

Staple, W. J., J. J. Lehane, and A. Wenhardt. 1960. Conservation of soil moisture from fall and winter precipitation. Can. J. Soil Sci. 40:80–88.

Toogood, J. A. 1963. Water erosion in Alberta. J. Soil Water Conserv. 18:238–240.

Willis, W. O., C. W. Carlson, J. Alessi, and H. J. Haas. 1961. Depth of freezing and spring runoff as related to fall soil-moisture level. Can. J. Soil Sci. 41:115–124.

Wischmeier, W. H., and D. D. Smith. 1978. Predicting rainfall erosion losses—a guide to conservation planning. USDA Agric. Handbook no. 537.

19

R. I. Papendick and D. K. McCool
Agricultural Research Service
U.S. Department of Agriculture
Pullman, Washington

H. A. Krauss
Soil Conservation Service
U.S. Department of Agriculture
Spokane, Washington

Soil Conservation: Pacific Northwest

Dryfarmed cropland in the Pacific Northwest covers about 4 million ha in the Northwestern Wheat and Range Land Resource Region (Northwestern Wheat Region) (Thomas and Grano, 1971). About 3.3 million ha is in a contiguous area in eastern Washington, northern Idaho, and north central Oregon. Most of the hectarage is in the Columbia Basin, Columbia Plateau, and the Palouse and Nez Perce Prairies major land resource areas (Austin, 1965). The remaining 0.7 million ha is largely in southern and southeastern Idaho with some in eastern Oregon in the Snake River Plains, Upper Snake River Lava Plains and Hills, and Eastern Idaho Plateaus major land resource areas (Austin, 1965).

Water and wind erosion are the dominant soil conservation problems. The type and amount of erosion are influenced by climate, topography, soils, and cropping systems. These factors vary widely from place to place and affect the type of practices used for erosion control. This chapter describes how these factors relate to the erosion hazard and practices for control.

19-1 CLIMATE, TOPOGRAPHY, SOILS, AND CROPS

19-1.1 Climate

The dryland of Oregon, Washington, and northern Idaho is characterized by humid winters and dry summers. Climate ranges from semiarid in the west to subhumid in the east near the mountainous areas. Precipitation

Dryland Agriculture—Agronomy Monograph no. 23.

ranges from 200 to 600 mm annually, with slightly over 60% occurring during November through March. During most winters, freezing weather and snow cover are interrupted several times by rainy weather and complete thawing. Rainfall from individual storms ranges from a few to 50 mm, but intensities are low—usually less than 4 mm/h. Between rainy periods, soils may freeze to 40 cm deep, and snow may accumulate to depths of 30 cm or more, especially at the higher elevations. Soil freezing and its effects on erosion is a problem in those higher elevation portions of the wheat region that have insufficient snow cover for insulation. This includes much of the Columbia Plateau and the Palouse and Nez Perce Prairies. Elevation plays an unusually important role in climate of the dryland; agriculture may be found at elevations ranging from 100 to 1000 m.

In southern and southeastern Idaho, the precipitation is more uniformly distributed throughout the year; it is slightly higher in May and June than in other months. Precipitation ranges from 250 to 500 mm/y but is highly variable over relatively short distances from elevation influences. Soil is snow covered and frozen most of the winter (Leggett et al., 1974), although the extent of this varies with elevation.

High winds can occur in any season of the year. However, winds of soil-eroding velocity are most frequent and prolonged in the fall and spring (Skidmore and Woodruff, 1968), the time when crop plants are small and most susceptible to wind damage. Winds are predominantly westerly (Skidmore and Woodruff, 1968).

19–1.2 Topography and Soils

Topography ranges from nearly level valleys to moderately and steeply sloping uplands. Most of the steeper land is in the Palouse and Nez Perce Prairies located in eastern Washington, eastern Oregon, and northern Idaho (Naffziger and Horner, 1958). Because the slopes in the Palouse are not only steeper but also highly irregular within small blocks of land, managing fields according to capability class is extremely difficult. Approximately 71% of the cropland in dry-farmed areas has slopes of 8 to 30% (preliminary data from the 1977 National Resource Inventories; J. J. Goebel. 1980. Personal communication. Statistical Laboratory and Dep. of Statistics, Iowa State Univ., Ames). Slopes on some fields exceed 50%.

Soils in the dry-farmed areas are derived mainly from loess. Some in north central Washington and northern Idaho are derived from glacial drift. Nearly all upland soils contain varying amounts of volcanic ash in their profiles. Most soils are permeable, well drained, and deep enough to store precipitation, but some (e.g., in parts of north central Oregon) are shallow and have limited water storage capacity. Soils in the low and intermediate precipitation zones typically are fine sandy loams or silt loams. They grade from silt loams to silty clay loams in the higher precipitation zone.

Estimated soil loss tolerance to water erosion ranges from 2.2 to 11.2 t/ha (Allmaras et al., 1979; SCS, 1975, 1976a, b). The higher tolerances are assigned to deep, renewable soils, whereas lower tolerances are assigned to shallower soils or those with nonrenewable subsoil material.

19–1.3 Cropping Practices

The cropping systems depend primarily on the amount of precipitation. Dry-farmed areas are divided into summer fallow zones, the intermediate zones, and annual crop zones. In the 3.3 million ha contiguous area in Oregon, Washington, and northern Idaho, about 1.7 million ha is in alternate winter wheat–fallow (200 to 350 mm of annual precipitation). Another 0.9 million ha is in the intermediate zone (350 to 450 mm of annual precipitation), and annual cropping is practical when precipitation is above normal. The usual system is to produce two or three crops between fallow periods (e.g., a fallow–winter wheat–spring barley rotation). The intensity of cropping varies with fluctuations in yearly precipitation, conservation considerations, economic factors, and government policy. The remaining 0.7 million ha makes up the annual crop zone, where average annual precipitation exceeds 450 mm. Winter wheat is grown in rotation with pea, lentil, and spring cereals such as wheat and barley. Winter wheat–fallow is practiced in most of the intermountain area because moisture is inadequate for annual cropping or precipitation fluctuates widely from year to year.

19–2 EROSION PROBLEMS

Wind, water, and tillage erosion (individually or in combination) are major problems on nearly all the dry-farmed croplands. Water erosion is most prevalent and causes the greatest soil loss; however, wind erosion and "tillage erosion" can also be serious over widespread areas.

19–2.1 Wind Erosion

Wind erosion is most serious on coarse-textured soils and where annual precipitation is less than 300 mm (Leggett et al., 1974). Most affected are the Columbia Basin and Columbia Plateau. The erosion hazard is greatest in early spring and fall when the soil surface is dry and unprotected and winds are highest (Skidmore and Woodruff, 1968). Summer-fallowed fields are especially vulnerable to blowing where surface residues are inadequate or winter wheat makes little fall growth. Small, highly pulverized areas of a field often contribute huge volumes of dust during a storm; these areas include field corners, roads, and boundaries where extra tillage or traffic has occurred (Leggett et al., 1974).

Procedures for selecting erosion control practices for wind erosion are based on the wind erosion equation (Chepil and Woodruff, 1963; Woodruff and Siddoway, 1965). The equation is written as

$$E = 2.242 f(I', K', C', L', V)$$

where

E = the potential average annual soil loss in metric tons per hectare per year,
I' = a soil erodibility index,
K' = a soil ridge roughness factor,
C' = a climatic factor,
L' = field length along the prevailing wind erosion direction,
V = equivalent quantity of vegetative cover (Chepil and Woodruff, 1963; Woodruff and Siddoway, 1965).

Factor values for the equation and soil loss tolerance can be obtained from local wind erosion control guides developed by the U.S. Soil Conservation Service.

19–2.2 Water Erosion

Water erosion is a major conservation problem on about 72% of the dry-farmed croplands in Oregon, Washington, and northern Idaho (CAST, 1975). It is also serious on fallow land in the intermountain area. The most severe erosion occurs in the steeply sloping areas of the Palouse and Nez Perce Prairies in eastern Washington, eastern Oregon, and northern Idaho. Soil losses from croplands in eastern Washington may average 10 to 50 t/ha annually and in some years may exceed 450 t/ha in certain fields (ESCS-FS-SCS, 1978). The high soil loss is a result of a combination of winter precipitation, steep slopes, and management practices that leave the soil pulverized and unprotected during the rainy season. Erosion is most severe from rain and snowmelt on thawing soil. Virtually every major runoff event or flood is the result of runoff on frozen soil and a single storm event can account for 75% of the annual erosion. Although erosion occurs with rain and snowmelt, runoff is the chief eroding agent. This dominant influence of runoff contrasts with dominance of rainfall energy in areas east of the Rocky Mountains (McCool et al., 1976).

The universal soil loss equation (USLE) (Wischmeier and Smith, 1978) is being used to formulate conservation practices for erosion control (Brooks et al., 1974). The relationships for the R_T, LS, and C factors have been adapted to conditions of the Pacific Northwest (McCool et al., 1976). Tentative values for the rainfall and runoff (R_T) factor have been determined for the Oregon-Washington-Idaho croplands (McCool et al., 1976). These values are adjusted to account for frozen soil and snowmelt runoff effects on soil loss; they range from 700 [MJ mm $(\text{ha h y})^{-1}$] in the high precipitation zone in northern Idaho to 200 in the fallow areas (Fig. 19–1).

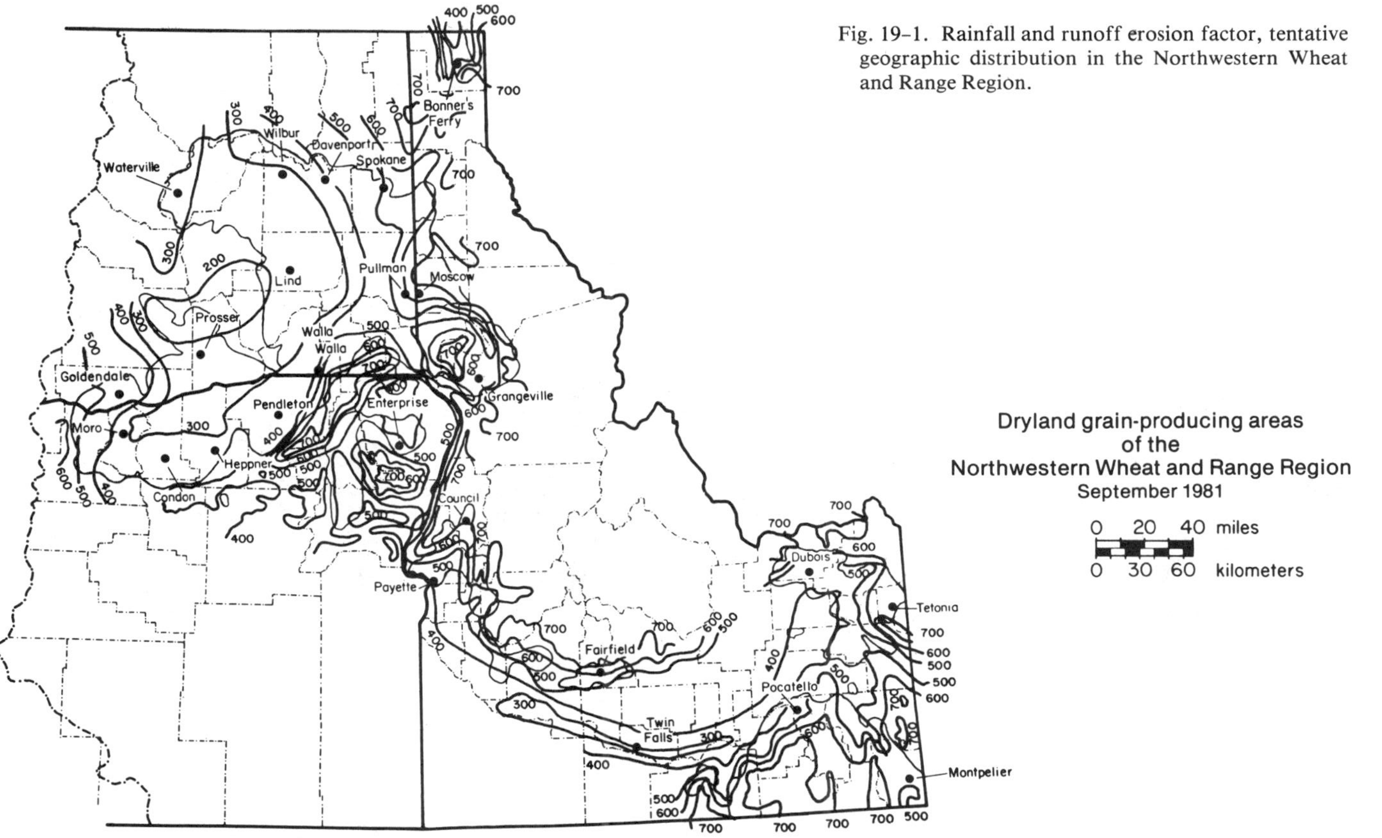

Fig. 19–1. Rainfall and runoff erosion factor, tentative geographic distribution in the Northwestern Wheat and Range Region.

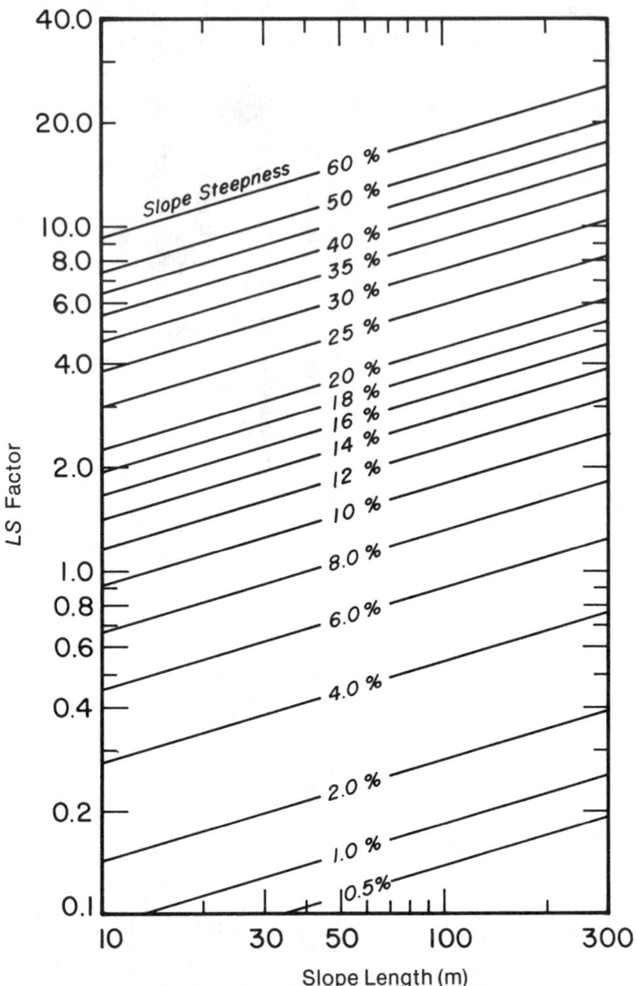

Fig. 19–2. Tentative *LS* factors for the Northwestern Wheat and Range Region.

These values are relatively low compared with those of other regions in the USA where rainfall intensity is much higher (Foster et al., 1981; Wischmeier and Smith, 1978). In the contiguous Oregon, Washington, and northern Idaho area, over 80% of the R_T factor value occurs in November through March. Seasonal distribution in southern and southeastern Idaho will vary somewhat because of the higher May and June precipitation.

The slope length and steepness factor relationships shown in Fig. 19–2 were developed specifically from Pacific Northwest data (McCool et al., 1976) and are different from Midwest-derived relationships. For slope characteristics given by Kaiser et al. (1951), the *LS* values range from 2 to 8 for much of the land in the higher precipitation zones and from 0.9 to 6 in the fallow areas. On some areas with long, steep slopes, *LS* values may exceed 12.

Table 19-1. USLE crop management factors for erosion control with different cropping systems.

Cropping zone	Crop sequence	C value†
Wheat-fallow (200 to 350 mm)	Winter wheat/fallow	0.08–0.30
	Winter wheat/spring grain/fallow	0.08–0.28
Intermediate (350 to 450 mm)	Winter wheat/spring grain/fallow	0.08–0.35
	Winter wheat/spring grain/spring grain/fallow	0.08–0.29
	Winter wheat/spring grain	0.08–0.17
	Winter wheat/spring grain/alfalfa + grass/alfalfa + grass	0.06–0.11
	Spring grain	0.04–0.11
Annual crop (greater than 450 mm)	Winter wheat/pea or lentil	0.20–0.30
	Winter wheat/spring grain	0.08–0.17
	Winter wheat/spring grain/pea or lentil	0.15–0.24
	Winter wheat/spring grain/alfalfa + grass/alfalfa + grass	0.06–0.11
	Spring grain	0.04–0.11
	Winter wheat/winter wheat	0.12–0.24

† Range of C values for a variety of conservation tillage practices; values may double for very intensive tillage. These C values when used in the USLE predict an average soil loss in metric tons per hectare per year.

Tentative USLE C values for different conservation cropping systems applicable to the region range from 0.04 to 0.35 (Table 19–1). The higher values are generally for more intense tillage practices; the lower values apply where reduced tillage practices are used. For very intensive tillage practices, the values may be double those shown. The C value is lowest with alfalfa or grass in the rotation or with continuous spring grain cropping. With winter wheat, the C value increases because there is less soil protection at the time when the erosion hazard is the greatest.

Estimates of USLE P values have been made by the U.S. Soil Conservation Service in the state of Washington for a number of practices in current use (Table 19–2). These values were obtained either directly from Wischmeier and Smith (1978) or were interpolated from their data. The P factor values increase as slope steepness increases, i.e., practices are less effective as slope steepness increases. Contour stripcropping, contour strips of winter wheat alternating with rough-tilled stubble, and field strips of winter wheat alternating with rough-tilled stubble are among the most effective practices (Table 19–2). Cross-slope farming provides the least protection against erosion, and the effectiveness of the divided slope practice is intermediate.

19–2.3 Tillage Erosion

Tillage erosion is the result of movement of soil downslope by mechanical implements. It is most severe on steep slopes, particularly where the moldboard plow is used (Mech and Free, 1942), and it is recognized as a hazard because much of the cropland has steep slopes. Farmers find it impractical to turn the furrow slice uphill on slopes steeper than about 25%. Repeated downhill plowing can move massive amounts of soil downslope

Table 19–2. USLE practice (*P*) factors for water erosion control.

Percent slope	Up and down hill†	Cross-slope farming‡		Contour farming¶		*P*			Contour stripcropping¶¶		
	P	*P*	Maximum slope length limits§	*P*	Maximum slope length limits#	Divided slopes††	Field strips W–F‡‡	Contour strips W–F§§	*P*	Strip width limit##	Maximum slope length
			m		m					m	m
1–2	1.0	0.80	60	0.60	120	0.52	0.45	0.38	0.30	40	245
2–5	1.0	0.75	45	0.50	90	0.44	0.38	0.32	0.25	30	180
6–8	1.0	0.75	30	0.50	60	0.44	0.38	0.32	0.25	30	120
9–12	1.0	0.80	20	0.60	35	0.52	0.45	0.38	0.30	25	75
13–16	1.0	0.85	12	0.70	25	0.61	0.52	0.44	0.35	25	50
17–20	1.0	0.90	10	0.80	20	0.70	0.60	0.50	0.40	20	35
21–25	1.0	0.95	8	0.90	15	0.79	0.68	0.56	0.45	15	30

† Farming around the field without regard to slope during fall seeding and fall tillage.
‡ Tentative interpolation between contour farming and uphill and downhill farming when SCS cross-slope specifications are followed.
§ Tentative limits. May be increased 50% if SCS cross-slope specifications are followed.
¶ Following the contour on all major slopes with fall seeding and fall tillage. Usually not feasible without permanent guideline such as level terrace. Not to exceed deviations in SCS contour farming specification.
# Limit may be increased by 25% if small grain residue cover after fall seeding will regularly exceed 1.1 t/ha (50% cover).
†† Tentative interpolation between wheat/rough-rilled stubble field strips and contour farming. Not to exceed 180-m slope length unless there is a grass strip.
‡‡ Tentative interpolation between contour farming and contour stripcropping; W–F = wheat–fallow.
§§ Tentative interpolation between contour stripcropping and wheat/rough-tilled stubble field strips; W–F = wheat–fallow.
¶¶ Annual cropping as per Wischmeier and Smith (1978, p. 36, Table 14, column A).
Adjust strip-width limit, generally downward, to accommodate widths of farm equipment.

and accounts for the major loss of soil from hilltops in the steeply sloping areas. Evidence of tillage movement are soil banks 3 to 4 m high formed along field borders where the upslope operator turns the furrow toward the border and the downslope operator turns the furrow away from the border (Papendick and Miller, 1977). In the Palouse of eastern Washington, the original topsoil has long been removed from most hilltops, primarily by tillage erosion. Today's powerful tractors accelerate this movement, especially with the moldboard plow, but they also move other types of implements faster than before, magnifying downslope movement with almost all tillage implements.

19-3 PRACTICES FOR EROSION CONTROL

Practices for erosion control are most conveniently discussed according to the cropping system zones. The erosion control practices (described in the sections on cultural and supporting practices in each cropping zone; sections 19-3.1 through 19-3.3) reflect recommended practices based on the best technology currently available. These do not always coincide with practices or systems in widespread use on the land.

19-3.1 Winter Wheat-Fallow Zone

The standard cropping system is alternate winter wheat-fallow, although two crops may be grown in sequence when above-normal precipitation allows. In most areas the soil is deep enough to store two seasons' precipitation. The topography in most areas is only moderately sloping; nevertheless, water erosion is still a hazard. Wind erosion is an ever-present problem, especially on coarse-textured soils. Based on average wheat yields, residue production ranges from 2.3 to 7.5 t/ha.

19-3.1.1 Cultural Practices

Practices that control water erosion also generally control wind erosion. The most effective control measures are stubble-mulching of fallow, maintaining cloddy surfaces, and early fall planting of winter wheat. Because of irregular topography, it is difficult to practice contour chiseling or seeding without permanent guidelines, e.g. level terraces or contour strips. Where runoff from frozen soil is a problem, fall chiseling of stubble across the slope is recommended for water erosion control (Lindstrom et al., 1974). Fall tillage is also used to control early weed growth (Horning, 1962).

Fall chiseling of wheat stubble is normally done to a depth of 20 to 30 cm, with narrow point shanks spaced 30 to 60 cm apart. Chiseling produces large clods and helps water penetrate through surface-exposed fractures as well as through the tool channel when soils are frozen (Lindstrom et al.,

1974). Sweep cultivation or disking may be used to control downy brome and Russian thistle. Chiseling of sandy soils when very dry is not always recommended, because the weakly aggregated dry soil may flow back and fill the chisel track. The operation is more effective for reducing runoff and increasing moisture intake of frozen soils if done after some fall rains occur.

Stubble mulching is accomplished most satisfactorily by using the high-clearance chisel or subsurface sweep for the initial spring tillage (Horning, 1962). The one-way disk is not recommended unless initial residue amounts exceed 8 t/ha (Horning, 1962), which is an upper limit in most of the wheat-fallow areas. After fall seeding, surface residue for wind or water erosion control should range from 1.1 to 2.2 t/ha or should be not less than 30% of the initial residue.

Tillage normally commences in the spring after weeds begin to grow and the soil surface has dried. Secondary tillage with a rod weeder or spring-tooth cultivator is performed two to four times during the summer to control weeds and to break capillary continuity to the surface and insulate the seedzone layer (Papendick et al., 1973). Early fall wheat stands are seeded between mid-August and early September, early enough so that the plants have three to seven tillers by 1 December. Planting is done across the slope with deep furrow drills capable of penetrating the dry tillage layer and passing through the surface residue mulch. Earliest seeding is generally done with 40 to 46 cm–spaced openers; opener spacing narrows to about 30 cm with later planting.

19–3.1.2 Supporting Practices

Stripcropping at right angles to the direction of erosive winds reduces wind erosion. Stripcropping is also recommended for water erosion control, particularly on shallow soils with low soil loss tolerances. Construction of level, broad-based terraces is proving to be effective and practical where needed and where topography permits. Terraces have proved to be more practical than stripcropping, where crop residues are utilized in a winter grazing program. Field stripcropping, cross-slope planting, early fall planting, and terracing combined with stubble mulching and fall chiseling provide adequate control systems under most conditions where an erosion hazard exists (Allmaras et al., 1979).

19–3.2 Intermediate Zone

In this zone, water erosion is the major soil conservation problem. Some of the steepest topography in the region occurs in this zone. Three- or 4-year rotations that include a year of fallow are more feasible economically and more effective for erosion control than alternate winter wheat–fallow that is sometimes practiced. Typical rotations include 2 or 3 years of grain and 1 year of fallow. Annual cropping is possible, but with current practices, some years it is not economically feasible. It is more feasible to annually crop on moderately deep to shallow soils, because they lack the water storage capacity to store more than one season's precipitation.

19-3.2.1 Cultural Practices

Stubble mulching and early fall planting usually provide adequate erosion control on winter wheat after fallow for land capability classes and subclasses I, IIe, and IIIe (Klingebiel and Montgomery, 1961) agricultural land. On steeper land subclasses IVe and VIe, soil losses with these practices may exceed the soil loss tolerance. In this case, alfalfa, grass, or both after 2 or 3 years of grain are necessary in the crop rotation for erosion control. In many areas, from 15 to 25% of the landscape is subclass IVe and VIe land, but the areas are often isolated in the field, making special treatment difficult. To treat the small, steep, erosive areas, additional portions of the field must be sown to grass to prevent undercutting the toe of the grassed area by tillage and buildup of overburden at the top of the grassed area. If this happens, a soil slip may occur, and large sections of the hillside may avalanche.

During cropping, retention of 25 to 33% of the residues on the surface and maintenance of cloddy seedbeds are effective for erosion control most years. Weed control and seedbed preparation are usually accomplished with a chisel or disk, followed by tillage with a rod weeder or spring-tooth cultivator. It is recommended that date of planting on fallow be early enough (before 15 September) for plants to reach the five-tiller stage by 1 December (CAST, 1975). This stage provides good winter hardiness and erosion control.

19-3.2.2 Supporting Practices

Fall seeding is done across the slope. Stripcropping, gradient terraces, or both are effective for support of stubble mulch fallow, crop residue use, and reduced tillage methods. Unfortunately, much of the topography is highly irregular and does not lend itself to some of these supporting practices. Level terraces are being constructed for use on long slopes in the intermediate zone in Oregon. Grassed waterways are also used as a supporting practice to control erosion from concentrated flows and to filter sediments from runoff as long as adjacent sheet and rill erosion is not too heavy.

Divided slope farming for erosion control has become popular with some operators who farm the short, steep slopes in eastern Washington. However, there is little or no research information on the effectiveness of divided slope farming on soil erosion in the area. With this practice, the slopes are divided horizontally into different cropping sequences (e.g., winter wheat vs. chiseled stubble) to reduce or intercept runoff and eroded soil from upper areas. The divided areas do not necessarily follow the contour. The advantage of divided slope to farmers is that point corners, which sometimes occur with true contour strips, are avoided.

19–3.3 High Precipitation Zone

Water erosion is almost the exclusive erosion hazard. Considerable areas consist of steep topography with short, highly irregular slopes. Longer slopes are common on foothills near mountainous areas. The steepest areas in the high precipitation zone, just as in the intermediate zone, contain numerous isolated pieces of subclass IVe and VIe land in which special erosion control measures are needed or in which erosion would be above tolerance limits if cropped under conventional management.

19–3.3.1 Cultural Practices

Annual cropping is essential for adequate erosion control. For maximum erosion control, the moldboard plow is not recommended except possibly on level lowland areas with heavy residue cover. Crops that produce low amounts of residues, e.g., pea and lentil, are not recommended more than 1 out of 3 years unless supporting practices are used (CAST, 1975). On land to be seeded to winter wheat after pea or lentil, the seedbed is prepared with a chisel or disk, left rough (not more than one harrow operation), and planted.

With spring crops, either the land is rough-tilled in the fall or the stubble of the previous crop is allowed to stand over the winter. Disking and spring-tooth cultivation are used for seedbed preparation.

For winter wheat after small grain, the stubble is chiseled or disked followed by secondary tillage that leaves the surface cloddy and covered with residue. If chisels are used in heavy stubble, they should be equipped with curved shanks, because these pass through and bury more residue than straight-point tools. Heavy surface residues interfere with planting operations, because the double-disk drill in common use is not designed to operate in such seedbeds.

On subclass IVe land, alfalfa, grass, or both should be included in the rotation after 3 years (grain-grain-pea or grain-grain-lentil) unless erosion can be controlled by the above practices. Subclass VIe land should be sown to permanent grass.

19–3.3.2 Supporting Practices

Planting is done across the slope. Divided slope farming is used on slopes no longer than about 180 m unless the additional distance is in an intervening grass strip, a hilltop seeding of permanent vegetation, or a third strip (i.e., strips not to exceed approximately 90-m width). Terracing and stripcropping are often of limited use, because they are impractical on the steep, irregular topography. However, they have potential on long, regular slopes. Because runoff occurs nearly every year, grassed waterways are essential for controlling erosion from concentrated flow along drainageways.

19-4 NEW TECHNOLOGICAL DEVELOPMENTS

Major emphasis on erosion research is being given to conservation tillage and residue management systems, more intensive annual cropping, and improved methods for prediction of soil erosion. Often application of the research is limited to specific cropping zones because of the great diversity in climate, soils, and topography.

19-4.1 Conservation Tillage and Residue Management

New approaches being researched include no-till fallow and delayed spring tillage (both of which require use of herbicides) as alternatives to intensive tillage fallow, no-till with annual cropping, and the slot mulch. The latter is a new approach for residue management to reduce runoff and erosion on frozen soil and steep hillsides.

19-4.1.1 No-till Fallow

With no-till fallow, weeds must be controlled throughout the season with herbicides, without tillage. Planting is accomplished with equipment capable of placing the seed and fertilizer in the previously undisturbed soil and stubble. Currently, two major problems restrict the use of no-till fallow. First, we do not have a satisfactory herbicide program for cereal cropping systems that reliably and economically controls weeds without some tillage. Primary limitations are (i) lack of chemicals that have an adequate residual effect but that do not carry over and affect the subsequent crop and (ii) high cost of short-residual chemicals that prohibits their use. Second, research shows that with no-till fallow, the surface layers of some soils dry deeper during the summer than with tilled fallow, which may make it difficult or impossible to establish early fall stands (Lindstrom et al., 1974). The first limitation may be surmountable; the second may not be overcome.

Early fall stands are essential for maximum yields from the available water supply. Fine-textured soils (silt loam or finer) apparently dry deeper than coarse-textured soils (fine sandy loam or coarser), which do not form shrinkage cracks and tend to "self-mulch" without tillage (Papendick and Campbell, 1974). Thus, no-till fallow may be feasible for some of the coarse-textured soils in the Columbia Basin but may not be feasible for the silt loams in the Columbia Plateau. The soil mulch produced by summer fallow tillage reduces seed zone drying during the summer by restricting liquid water flow to the surface and thermally insulating the seed zone. The net effect is a reduction of vapor loss (Papendick et al., 1973). A detailed mathematical model has been developed of soil water flow in the shallow layers; the model accounts for thermally induced vapor flow (Hammel, 1979) and will be used to assess the feasibility of no-till fallow for different parts of the wheat-fallow areas.

Because of higher precipitation, no-till fallow may be feasible in certain areas of the intermediate zone where it may be desirable to fallow if weeds are adequately controlled and no-till planting equipment can be developed for seeding on steep slopes. Weed pressure, particularly from grass species, is considerably greater in this zone than in the low precipitation zone.

19–4.1.2 Delayed Spring Tillage

A possible alternative to no-till fallow for silt loam soils is delayed spring tillage. Normally, spring tillage in the wheat-fallow areas begins in March and is followed by several rod weedings to control weeds and to establish a soil mulch. With delayed tillage, the primary spring tillage is not performed until late May or early June before the onset of hot weather. Subsequent tillage is limited to one or two rod weeding operations. Weeds are controlled in early spring with short- or no-residual chemicals. This practice especially protects the soil from wind during the critical months of March and April. Furthermore, fewer tillage operations keep more residues and soil clods on the soil surface at planting time, a practice that aids erosion control the following winter. Preliminary field research showed that with good weed control, water storage and water content in the seed zone at planting time were not different from conventional fallow (R. I. Papendick and J. E. Hammel. 1978. Unpublished research).

19–4.1.3 No-till in Annual Cropping Systems

Under investigation is direct drilling of crops into stubble of cereals, pea, lentil, and killed grass sod. Also under development are reduced tillage drills capable of placing seed and fertilizer in stubble or killed sod or in a narrow, minitilled zone. Growers are expressing considerable interest in these practices, and a few are beginning to use both commercial and improvised planting equipment. On the steeper areas, erosion occurs with almost any type of cultivation. Even if water erosion is controlled, tillage erosion removes the soil from the slopes (Mech and Free, 1942; Papendick and Miller, 1977). With no-till or minimum tillage planting into adequate residues, erosion is prevented or is very low in all precipitation areas (ESCS-FS-SCS, 1978). Thus, if technical problems can be solved, crop production appears feasible on even the steepest slopes.

With direct drilling into the heavy cereal stubble typical of the high precipitation zone, crop yields are often 10 to 30% lower than yields obtained with conventional tillage planting (Papendick and Miller, 1977). This low yield so far has been a major restraint to no-till planting of cereals. Causes of reduced yields are high crown set, phytotoxicity to seedlings from surface residues, fertility imbalance, increased weed infestation, increased disease, and, in some cases, lowered spring soil temperature (Papendick and Miller, 1977). Yields of winter wheat with no-till planting following pea or lentil crops, which produce low amounts of residues, are usually equal to or higher than yields with conventional planting; however, erosion control may be no better than with conventional seeding if residue levels are too low. Limited research data from small plots indicate runoff under no-till seeding in the annual cropping zone may be as high or higher than under conventional seeding (D. K. McCool. 1982. Unpublished data).

Experiments have been underway for 4 years in eastern Washington investigating the feasibility of direct drilling cereals and pea into killed sod of established bluegrass grown for seed production. The results appear promising with a spring crop (pea, spring barley, or spring wheat) following sod kill with 0.5 kg/ha of glyphosate [N-(phosphonomethyl)glycine] applied in mid- to late April. Subsequently, a winter wheat–spring crop rotation is followed with glyphosate applied only before the spring crop (alternate years) to kill volunteer winter wheat and regrowth of the bluegrass. Yields have been equal to or superior to those from tilled seedbeds. Broadleaf weeds have not been a problem, and erosion has been completely controlled (L. F. Elliott and R. I. Papendick. 1979. Unpublished data). Direct sod seeding greatly extends the erosion control benefits of a sod crop in the crop rotation. These researchers are now considering controlling erosion on steeper slopes by direct sod seeding of small grain into killed field grasses.

A major concern by many people to the no-till approach is the potential for increased use of pesticides. They feel that any increase in pesticide use could lead to greater pollution of water by these chemicals; increased resistance of weeds, insects, and pathogens to pesticide applications; and increased destruction of wildlife as well as greater danger to human and animal life from pesticide contamination (USDA, 1980). Thus, future research should take into account the potential hazards from increased use of pesticides as well as the benefits of no-till methods for erosion control.

19–4.1.4 Slot Mulch

A new concept of tillage and residue management termed "slot mulch" is under investigation for control of runoff and erosion during major runoff events. Conceptually, the approach is to compact residues of a grain crop into a narrow continuous slot approximately 5 to 10 cm wide by 20 to 25 cm deep, cross slope, with the soil ridged on the downslope side. The most effective spacing between slots has not been determined; however, preliminary results suggest that a field combine swath width may suffice.

Unlike the older vertical mulch concept where the mulched trench is sealed by soil, the straw in the slot mulch is left well exposed at the surface. The mulch maintains good macroporosity for water infiltration and insulates the slot so that the bottom portion remains unfrozen during prolonged cold weather. Field studies have confirmed that the slot mulch is highly effective in increasing water infiltration and reducing erosion where surface sealing and frozen soils otherwise restrict water intake (Saxton et al., 1981). The mulch concept needs to be further developed and incorporated into conservation tillage methods where heavy runoff is a potential problem. It has great potential as part of a no-till conservation system.

19–4.2 Annual Cropping in Drier Areas

Annual cropping is recognized as an effective erosion control practice because more residues can be retained on the surface, the soil is normally tilled less between crops, and the soil has greater water storage capacity during the winter compared with cropping after fallow. The main emphasis of research is to extend annual cropping into drier areas of the intermediate cropping zone and, if possible, into the wheat-fallow areas in years of higher than normal precipitation.

Experience indicates that annual cropping generally is economically feasible 5 to 7 out of 10 years with conventional practices over much of the intermediate cropping zone. Complete elimination of conventional winter wheat–fallow would reduce soil erosion to 30 to 50% of current levels (Table 19–1). Success of annual cropping depends on improved water conservation and development of drought-resistant crop varieties. The potential for improved water conservation lies in surface residue management and no-till planting of spring crops. To maximize water retention, all the crop residues must be retained on the surface over the winter, preferably as upright stubble, and weeds must be controlled in the fall and after spring warm-up. Spring crops (spring wheat or barley) are planted no-till while the soil is still moist but able to support equipment. A preplant application of 0.4 kg/ha of glyphosate and a late spring application of 0.4 kg/ha of bromoxynil (3,5-dibromo-4-hydroxybenzonitrile) have shown good potential for weed control. (Mention of a pesticide does not constitute a recommendation for use by USDA or imply registration.) Two years of research indicate that crop yields for no-till planting into residue levels typical of the intermediate precipitation zone can exceed conventional planting yields by 20% or more (R. D. Schirman. 1978. Unpublished results). The yield increase is attributed to (i) additional moisture available to the crop because with no-till the soil is not disturbed and is less exposed to drying in the spring and (ii) quicker emergence because planting is not delayed by the usual seedbed preparation. Crops with improved drought tolerance are needed to reduce yield losses during high-moisture stress periods.

19–4.3 Erosion Prediction

Research is being conducted to adapt the USLE to Pacific Northwest conditions for use as a tool in erosion control planning. Improvements are especially needed in the R_T and C factors for different parts of the region (McCool et al., 1977). Research on the components of the C factor such as clod size, residue decomposition, antecedent moisture, and seeding date cover relationships is underway. Limited research is also underway to improve USLE *P* values for practices such as slot mulching, deep furrow seeding, cross-slope farming, contour farming, divided slopes, and wheat-fallow, wheat-pea, and winter wheat-spring grain strips.

19-5 LITERATURE CITED

Allmaras, R. R., S. C. Gupta, S. L. Pikul, Jr., and C. E. Johnson. 1979. Tillage and plant residue management for water erosion control on agricultural land in eastern Oregon. J. Soil Water Conserv. 34:85–90.

Austin, M. E. 1965. Land resource regions and major land resource areas of the United States. USDA Agric. Handbook no. 296.

Brooks, F. L., J. W. Turelle, W. H. Wischmeier, and D. K. McCool. 1974. Universal soil loss equation. USDA-SCS Tech. Notes Conserv. Agron. no. 32, Portland, Ore.

Chepil, W. S., and N. P. Woodruff. 1963. The physics of wind erosion and its control. Adv. Agron. 15:211–302.

Council for Agricultural Science and Technology. 1975. Erosion and sedimentation in the loessial region of Washington, Idaho, and Oregon. CAST Report no. 42, Iowa State Univ., Ames.

Economics, Statistics, and Cooperative Service, Forest Service, and Soil Conservation Service. 1978. Palouse cooperative river basin study. USDA-ESCS-FS-SCS, U.S. Government Printing Office, Washington, D.C.

Foster, G. R., D. K. McCool, K. G. Renard, and W. C. Moldenhauer. 1981. Conversion of the universal soil loss equation to SI metric units. J. Soil Water Conserv. 36:355–359.

Hammel, J. E. 1979. Modeling tillage effects on evaporation and seed zone water content during fallow in eastern Washington. Ph.D. Dissertation. Washington State Univ., Pullman. Univ. Microfilms, Ann Arbor, Mich. (Diss. Abstr. 40(1):19b).

Horning, T. R. 1962. Stubble mulching in the Northwest. USDA Information Bull. no. 253.

Kaiser, V. G., W. A. Starr, and S. B. Johnson. 1951. Types of topography as related to land use in Whitman County, Washington. Northwest Sci. 25:69–75.

Klingebiel, A. A., and P. H. Montgomery. 1961. Land-capability classification. USDA Agric. Handbook no. 210.

Leggett, G. E., R. E. Ramig, L. C. Johnson, and T. W. Massee. 1974. Summer fallow in the Northwest. p. 110–135. *In* Summer fallow in the western United States. USDA-ARS Conserv. Res. Report no. 17.

Lindstrom, M. J., F. E. Koehler, and R. I. Papendick. 1974. Tillage effects on fallow water storage in the eastern Washington dryland region. Agron. J. 66:312–316.

McCool, D. K., M. Molnau, R. I. Papendick, and F. L. Brooks. 1977. Erosion research in the dryland grain region of the Pacific Northwest: Recent developments and needs. p. 50–59. *In* Soil erosion: Prediction and control. Soil Conserv. Soc. of Am. Spec. Pub. no. 21, Ankeny, Iowa.

McCool, D. K., R. I. Papendick, and F. L. Brooks. 1976. The universal soil loss equation as adapted to the Pacific Northwest. p. 135–147. *In* Proc. Third Fed. Interagency Sediment Conference, Symp. 2, Denver, Colo. 22–25 March. Water Resour. Coun., Washington, D.C.

Mech, S. J., and G. R. Free. 1942. Movement of soil during tillage operations. Agric. Eng. 23:379–382.

Naffziger, L. M., and G. M. Horner. 1958. Effect of cropping and tillage practices on runoff and erosion in the Palouse area of Washington and Idaho. Trans. ASAE 1:34–35.

Papendick, R. I., and G. S. Campbell. 1974. Wheat-fallow agriculture: Why, how, when? *In* Tillage and cultural practices for wheat under low rainfall conditions. Proc. Second Regional Wheat Workshop, Ankara, Turkey. 6–11 May. Rockefeller Foundation, New York.

Papendick, R. I., M. J. Lindstrom, and V. L. Cochran. 1973. Soil mulch effects on seedbed temperature and water during fallow in eastern Washington. Soil Sci. Soc. Am. Proc. 37: 307–314.

Papendick, R. I., and D. E. Miller. 1977. Conservation tillage in the Pacific Northwest. J. soil Water Conserv. 32:49–56.

Saxton, K. E., D. K. McCool, and R. I. Papendick. 1981. Slot mulch for runoff and erosion control. J. Soil Water Conserv. 36:44–47.

Skidmore, E. L., and N. P. Woodruff. 1968. Wind erosion forces in the United States and their use in predicting soil loss. USDA Agric. Handbook no. 346.

Soil Conservation Service. 1975. The universal soil loss equation for Idaho. USDA-SCS Tech. Notes Conserv. Agron. no. 19, Boise, Idaho.

Soil Conservation Service. 1976a. Sheet and rill erosion control guide, state of Washington. USDA-SCS, Spokane, Wash.

Soil Conservation Service. 1976b. Universal soil loss equation. USDA-SCS Tech. Notes Conserv. Agron. no. 29, Portland, Ore.

Thomas, H., and A. M. Grano. 1971. Agricultural land and water resource base and productivity, Pacific Northwest. U.S. Dep. of Agric., Economic Res. Service, Natural Resource Economics Division, Corvallis, Ore.

U.S. Department of Agriculture. 1980. Report and recommendations on organic farming. A special report prepared for the Secretary of Agriculture. USDA, U.S. Government Printing Office, Washington, D.C.

Wischmeier, W. H., and D. D. Smith. 1978. Predicting rainfall-erosion losses from cropland east of the Rocky Mountains—a guide to conservation planning. USDA Agric. Handbook no. 537.

Woodruff, N. P., and F. H. Siddoway. 1965. A wind erosion equation. Soil Sci. Soc. Am. Proc. 29:602–208.

SECTION D

CROPPING PRACTICES

20

D. E. Smika
Agricultural Research Service
U.S. Department of Agriculture
Akron, Colorado

Cropping Practices: Introduction

Cropping practices for dryland agriculture have been developed on the basis that water is generally the factor limiting production. Successful cropping practices make the most efficient use of water and provide maximum conservation of soil and water resources. This concept has not always been practiced; farmers who settled in semiarid areas of the western part of the USA starting in the late 1840's brought with them familiar cropping practices from the humid east. Many of these practices were actually a carry-over from the native country of immigrant farmers—also humid areas. When precipitation was better than normal in dryland areas, these practices of the early settlers produced satisfactory results. In normal and below-normal precipitation years, production failed. When crops failed in successive years, the settlers generally moved elsewhere. By the late 1800's, there was positive evidence that the semiarid areas of the USA were going to be permanently settled. Stable, economic crop production from these areas was urgently needed but was not actually achieved until the 1940's.

In 1862, the Morrill Bill, passed in the U.S. Congress, established agricultural colleges and experiment stations throughout the USA. Research from these institutions helped the early dryland farmer, but much was yet to be learned. As the drier areas became more intensively cropped and experiment station results became available, a need arose to exchange ideas among those directly involved in dryland cropping. To meet this need, the first Dryland Farming Congress was held in 1907. The organization that followed was important in promoting early dry farming practices. It often awarded prizes for outstanding production to stimulate interest. From research and promotional schemes, cropping practices evolved that provided the necessary economic stability from dryland production in the semiarid areas.

Contribution from Soil, Water, and Air Sciences, U.S. Department of Agriculture, Agricultural Research Service, Akron, Colorado.

Dryland Agriculture—Agronomy Monograph no. 23.

Table 20–1. Hectares of fallow and major crops produced on dryland in each geographic area, 1977 statistics.

Geographical area	Total fallow	Crop: Wheat	Corn	Rye	Barley	Oat	Sorghum	Cotton	Special crops†	Oilseed‡
	hectares × 1000									
Southern Great Plains	1645.6	5862.0	--	--	--	--	2241.2	2922.4	--	0.2
Central Great Plains	5080.0	7961.2	119.6	75.6	216.8	190.0	2581.2	--	64.8	--
Northern Great Plains	5682.4	7606.0	1484.0	--	2008.0	2072.8	136.8	--	--	506.0
Canadian Prairie Provinces	7077.6	10 111.0	730.0	242.0	4646.0	2132.0	--	--	--	1421.0
Pacific Northwest	1730.4	2391.6	--	--	685.6	121.6	--	--	70.6	--
Pacific Southwest	280.4	436.8	--	--	494.4	165.2	--	--	--	--
Total for all areas	21 496.4	34 368.6	2333.6	317.6	8050.8	4681.6	4959.2	2922.4	135.4	1927.2

† The special crop in the Central Great Plains is millet, in the Pacific Northwest pea.

‡ Oilseed crops in the Southern Great Plains are sunflower and guar, in the Northern Great Plains safflower and flax, and in the Canadian Provinces rapeseed and mustard.

One of the practices that evolved for dryland crop production was the use of fallow. This practice is common to all semiarid dryland cropping areas. Because of the different roles that fallow plays in various areas and its potential interaction with characteristics that are unique to individual areas within the USA and Canada, each of the different geographic areas will be discussed individually in the following chapters: Southern Great Plains, Central Great Plains, Northern Great Plains, Canadian Prairie Provinces, Pacific Northwest, and Pacific Southwest.

Wheat is the major crop throughout all dryland cropping areas. Winter wheat predominates in all areas except the Northern Great Plains and Canadian Prairie Provinces, where spring wheat and durum are most common. Most of the wheat in all areas is produced on medium- to fine-textured soils, with only limited plantings on the sandy soils. Planted hectarages of crop-fallow for each geographic area are presented in Table 20–1.

Since most wheat is produced on soils capable of storing considerable amounts of water, fallow plays an integral role in the overall wheat production system in all areas. The value of fallow for this purpose ranges from minimal in some areas to absolutely necessary in others (Greb et al., 1974). Where fallow is needed, its major attribute is the stabilization of economic production. This stabilization occurs in the entire Central Great Plains area (Smika, 1970) and in other areas where annual precipitation ranges from about 250 to 400 mm in the Northern Great Plains, Canadian Prairie Provinces, and the Pacific Northwest to about 400 to 450 mm in the Pacific Southwest and the Southern Great Plains.

The concept and importance of fallow must be thoroughly understood and appreciated before further study of dryland cropping systems can be made. This information was presented in detail earlier in this book. For the purposes of this chapter, however, the necessity of fallow for wheat production can be defined as the capacity of fallowed soil to supply water to the crop during prolonged periods without rainfall. With winter wheat in the Great Plains, this period occurs in the fall after emergence and after spring growth begins—typically 6 to 8 weeks before spring rains are expected. In the Pacific Northwest, fallow is most important for the fall stand establishment of winter wheat. Other benefits of fallow include control of problem weeds, disease, and insects; storage of 18 months of nitrate mineralization; and production of enough crop residue to prevent wind erosion.

20–1 LITERATURE CITED

Greb, B. W., D. E. Smika, N. P. Woodruff, and C. J. Whitfield. 1974. Summer fallow in the central Great Plains. p. 51–85. *In* Summer fallow in the western United States. USDA-ARS Conserv. Res. Report no. 17.

Smika, D. E. 1970. Summer fallow for dryland winter wheat in the semiarid Great Plains. Agron. J. 62:15–17.

21

T. L. Jackson
Oregon State University
Corvallis, Oregon

A. D. Halvorson
Agricultural Research Service
U.S. Department of Agriculture
Sidney, Montana

B. B. Tucker
Oklahoma State University
Stillwater, Oklahoma

Soil Fertility in Dryland Agriculture

Essentially all of the dryland soils have supported the same cultivated crops, i.e., wheat, sorghum, and oil crops, since they were plowed out of prairie grasses. The ecology of the grass prairie provided a reserve of readily decomposable soil organic matter that supplied N for crops for a number of years. However, production of nonlegume crops combined with summer fallow (where the crop is grown once in two seasons) for maximum release of soil N and storage of soil water reduced this reserve supply of soil N. Present yields of most nonlegume crops grown every year depend on use of commercial fertilizers.

Sievers and Holtz (1922, 1926) identified the importance of an incubation period to release available N from soil organic matter during the summer fallow season. They estimated that in 39 years of cropping, mostly in a wheat–summer fallow system, silt loam soils of eastern Washington had lost 22% of the N and 34% of the humus.

Soil N and C losses have been large with cropping systems studied in Oklahoma. Soil organic matter losses on a Pratt fine sandy loam (Psammentic Haplustalf) at Woodward, Oklahoma, averaged 60% by 1957 (Haas and Evans, 1957). Organic matter losses were significantly faster in the Southern Great Plains than in northern regions because of higher temperatures. This loss in soil organic matter has resulted in increased dependence

Dryland Agriculture—Agronomy Monograph no. 23.

on fertilizer to supply the crop N requirement. The deep prairie soil (Kirkland silt loam Udertic Paleustoll) at Stillwater, Oklahoma, supplied enough native soil N to produce about 1344 kg of wheat/ha, or approximately 45 kg of N ha^{-1} y^{-1} for more than 86 years. Fallow and row crop farming accelerated the loss of soil N and C when compared with continuous cropping (Haas and Evans, 1957).

Pittman (1977) found that six of seven nonfertilized dryland crop rotations in southern Alberta, including combinations of fallow, wheat, alfalfa, and grass had depleted several major plant nutrients in the soil within 20 years. Average contents of organic matter, total N, and exchangeable K were decreased by 14.5, 10.1, and 26.7%, respectively, in the 0- to 15-cm soil depth and by 24.1, 13.3, and 25.7%, respectively, in the 15- to 37-cm depth. Total P content of the soil changed little during 20 years of cropping.

In general, soil organic matter content is highest in the eastern part of the Northern Great Plains area and decreases from east to west. Soil test summaries show that these soils are inherently low in available soil P and high in available K (Bauer et al., 1966; Ward and Carson, 1975). Therefore, yield responses to P fertilization are common but not to K. Nitrogen fertilizer requirements vary with the amount of N mineralized from the time of crop removal until the time the next crop is seeded and with the yield potential of the seeded crop under the prevailing moisture conditions.

In the semiarid Northern Great Plains, water, available soil N, and P often limit plant growth. In addition, soil temperatures are frequently below optimum during the early growth stages, and ambient air temperatures are above optimum during heading and ripening stages.

A summary of long-term cropping studies in the Pacific Northwest shows (Horner et al., 1960) that soil organic matter ranged from 3 to 4% in the Pullman, Washington area (about 500 mm of rainfall), 2.0% at Pendleton, Oregon (about 350 mm of rainfall), and 1.3% at Lind, Washington (about 200 mm of rainfall) when experiments were started in the 1920's. Marked decreases have been observed in all areas since that time. Two major factors were responsible for the rapid decrease in soil organic matter: (i) limited return of crop residues with little or no addition of N to the fallow system and (ii) loss of topsoil by soil erosion from both wind and water during the summer fallow period.

Most of the soils in Washington, Oregon, and adjacent dryland areas of Idaho had adequate levels of available soil P to supply crop needs when cropping started. The prairie wheat-producing areas of Canada and associated soils along the Canadian-U.S. border that are largely glacial till soils and inherently low in available P were some of the first soils to show response to applied P. Erosion of hilltops has resulted in both P and K deficiencies.

Soils of the Southern Great Plains vary markedly in their capacity to supply crops with adequate P; some soils have adequate supplies of P, whereas others have extremely low supplies. There seems to be little relationship between the type of cropping system and P availability. However, soil erosion has played an important role in reduced P availability.

This chapter discusses soil associations specifically related to nutrient availability and fertilizer response. The major soil associations throughout the dryland area and the relationships between chemical and physical characteristics, as well as environmental and management practices, are outlined in Chapter 1.

21-1 NUTRIENT DEFICIENCIES IN CALCAREOUS SOILS

Calcareous soils generally have apatite compounds, and dicalcium phosphate ($CaHPO_4$) and tricalcium phosphate [$Ca_3(PO_4)_2$] are the major inorganic P components (Olsen, 1953; Seats and Stanberry, 1963). As levels of organic P have diminished with cropping, P deficiency has become more widespread. The alkaline pH of calcareous soils favors the dominance of 2:1 type clay minerals (Grim, 1953). Low leaching losses of K have contributed to high soil K levels and to K fixation when hydrous mica minerals are present.

The carbonate chemistry of soil systems and soil pH influence the availability of cations as well as P. The influence of soil pH, CO_2, carbonates, and P compounds on solubility of $H_2PO_4^-$ is discussed by Lindsay and Moreno (1960). The solubility of $H_2PO_4^-$ or HPO_4^{2-} in soils is a complex system. Calcium carbonate, CO_2, and soil pH are the critical factors controlling P in soil solution in neutral to alkaline soils. The effect of calcareous soil systems on plant availability of soil P has been reviewed (Grunes, 1959; Seats and Stanberry, 1963; Miller, 1974).

An understanding of the reactions taking place when water-soluble P fertilizers are added to these soils is important. The $H_2PO_4^-$ ion reacts with Ca^{2+} to form dicalcium phosphate ($CaHPO_4$), which has a solubility of 0.2 g/L. As these ions react in a calcareous soil, the contact with particles of lime ($CaCO_3$) result in formation of tricalcium phosphate [$Ca_3(PO_4)_2$], which can become a surface coating on limestone granules. An equilibrium exists between the limited amounts of $H_2PO_4^-$ and HPO_4^{2-} ions in the soil solution and these less soluble P compounds. Thus, in fertilizer management, practices that either minimize intimate contact between P fertilizers added and limestone particles present in these soils or that place P fertilizer in intimate contact with acid-forming materials, e.g., ammonium sulfate [$(NH_4)_2SO_4$], increase the solubilization of compounds formed.

Potassium deficiencies are being identified on eroded hilltops in the Pacific Northwest and on sandy soils in the Northern and Southern Great Plains. The widespread occurrence of hydrous mica minerals (Grim, 1953) throughout these areas has ensured a relatively high reserve of slowly available soil K.

The solubility of heavy metal micronutrients generally decreases as soil pH increases, and higher pH vlaues are more prevalent where erosion or land leveling has exposed calcareous subsoils. Zinc, Mn, and Fe deficiencies can be expected on crops that are susceptible to deficiencies of these nutrients. Fortunately, wheat and barley, two of the most widespread crops produced under dryland conditions, are either efficient at taking up these nutrients from calcareous soils or have low requirements. Deficiencies are seldom found in these crops.

Deficiencies of Zn and Fe are more frequently encountered on sorghum, corn and flax crops in the Great Plains. Practices followed to identify and correct these problems will be discussed in later sections of this chapter.

Sulfur deficiency is widespread throughout the Pacific Northwest, isolated areas of the Great Plains, and western prairie provinces of Canada, principally on noncalcareous soils where gypsum ($CaSO_4 \cdot 2H_2O$) is not present in subsoils. This deficiency was seldom a problem in earlier production of cultivated crops in these areas because breakdown of soil organic matter during a summer fallow season supplied enough S for crop requirements. All of these practices increase the S requirement of the cropping system. The widespread use of commercial N with increased yields and increased incorporation of crop residue to help control erosion has started to increase equilibrium soil organic matter levels (Horner et al., 1960) and has thus increased total S needs. In the Pacific Northwest, these effects have been combined with the production of legumes, which have higher S requirements, and a return to annual cropping or cropping 2 out of 3 years to wheat and barley in many areas. Thus, where $(NH_4)_2SO_4$ or other S-carrying fertilizers have not been used, S deficiency has become a more common occurrence (Leggett et al., 1959; Ramig et al., 1975).

21-2 NUTRIENT DEFICIENCIES IN ACID SOILS

Some dryland soils in the western states are moderately acid, and some have acidic A horizons with alkaline B and C horizons. Acid soils are found around the foothills of the mountains in the Pacific Northwest, some Rocky Mountain areas, Turtle Mountains in North Dakota, and higher rainfall areas in Oklahoma, Texas, and Kansas. The rainfall is high enough, especially when combined with lower evapotranspiration rates at higher elevations, to support a stand of timber. Increased use of ammonium fertilizers in the last 30 years has also contributed to soil acidity in these areas.

Similar areas can be found in Alberta where environmental conditions resulted in the development of moderately acid soils. However, a review of literature has indicated that soil acidity has probably not been severe enough to result in Al toxicities in these areas. Application of lime on soil with pH values below 5.0 has resulted in increased wheat yields in Oklahoma and the Pacific Northwest states.

Sulfur deficiency is more widespread on these moderately acid soils, probably because of greater leaching potential. Molybdenum deficiencies have been identified on legumes (Reisenauer, 1956) on moderately acid soils in the Pacific Northwest and may be more widespread on land cleared of trees.

21-3 IDENTIFYING NUTRIENT DEFICIENCIES

Researchers, extension workers, and farmers generally rely on three methods to identify nutrient deficiencies and to predict the need for fertilizers. The two most widespread are (i) visual observation of response to fertilizer applications and (ii) the identification of specific visual deficiency symptoms. The search for more dependable methods of recommending specific rates of fertilizer has led to the development of a third method—chemical analyses of both soil and plant samples to identify nutrient deficiencies and predict plant response to fertilizers quantitatively. An extensive review of assumptions underlying use of soil and plant analyses and methods employed throughout the USA has been edited by Walsh and Beaton (1973).

21-3.1 Use of Soil Analyses to Predict Fertilizer Response

The use of soil analyses to predict plant response to fertilizer is based on a correlation between some fraction of a nutrient extracted from a representative soil sample and a measured response from fertilizer application. Soil analyses-fertilizer response relationships, except for NO_3-N, are generally based on an analysis of "plow layer" soil samples. This procedure has generally given a reliable basis for predicting response where adequate and up-to-date field calibration data are available. Levels of available P, K, and micronutrients are generally lower in subsoil horizons than in the "plow layer"; therefore, chemical analyses of surface soil samples provide a measure of these nutrients that can be related to crop response. A notable exception to this rule has been in the wheat-growing areas of Utah (Neilson and Lamborn, 1970; Lamborn, 1970) where buried profiles sometimes contain more available P than surface soils.

21-3.1.1 Nitrogen Soil Analysis

Dahnke and Vasey (1973) have reviewed some of the important problems associated with use of different N soil test procedures. Recent studies have emphasized that NO_3-N in the surface 1.2 m of soil is more important than the NO_3-N in the soil below a depth of 1.5 m. Soil testing laboratories in the Great Plains are using NO_3-N soil analysis to measure carryover or residual and mineralized N. These laboratories recommend soil sampling to a minimum of 60 cm, with samples to 1.2 m preferred where plants have unrestricted rooting to this depth.

Leggett (1959) emphasized that the N required for anticipated optimum yield was supplied by (i) NO_3-N present in the root zone at planting, (ii) N that should be released from soil organic matter during the spring of the year, and (iii) commercial N to be added. A reasonably accurate measure of the NO_3-N in the root zone can be made at planting. The amount of N released from soil organic matter during the spring of the year for a specific soil and environmental condition is reasonably constant. Major errors in these predictions are associated with the marked fluctuation in yield potential caused by variation in environmental conditions between years. Summer fallowing usually ensures that the soil water profile is near field capacity to a depth of 120 cm by the start of the growing season, but rainfall during the 3 or 4 months preceding crop maturity and temperature, especially 4 weeks before crop maturity, can combine to cause wide fluctuations in yields from year to year. Nitrogen removed by livestock grazing small grain crops in the Southern Great Plains can have a marked effect on crop nutrient requirements. This practice will be discussed in latter sections of this chapter.

Yield increases from fertilizer applications are not always assured. In dryland farming regions in the Great Plains and west, available soil water varies and often limits yields. Therefore, yields fluctuate widely. Growers are faced with the problem of deciding on realistic yield goals for specific soil conditions that are most profitable over a period of years.

21-3.1.2 Phosphorus Soil Analysis

Soil P solubility on neutral and alkaline pH soils is controlled by the formation of $CaHPO_4$ and $Ca_3(PO_4)_2$ as previously discussed. The sodium bicarbonate procedure to extract P from calcareous soils, developed by Olsen et al. (1954), is in widespread use (Thomas and Peaslee, 1973).

The sodium bicarbonate extract has been used to predict P response on moderately acid soils in western Oregon (Jackson et al., 1964), indicating a wide range of adaptation for this extracting procedure if adequate field calibration data are available.

The Bray-1 weak acid extract is used (Halvorson, 1970) to measure soil P in western Oregon and in acid soils of northern Idaho. Oklahoma and Kansas have used the Bray-1 weak acid with soil solution ratio (up to 1 to 100) to extract P on acid and calcareous soils. This increased amount of extracting solution ensures that a moderate amount of carbonate in a soil system will not neutralize the acid-extracting solution. The range of extracting solutions used to measure some fraction of "plant available P" emphasizes that different extraction procedures can be used on many soils *if good field calibration data are available* and if different critical levels are established for different environmental conditions.

The question of P levels in subsoils should be established for different soil formations since it is generally assumed that subsoil levels of plant available P are low where surface soils are deficient. Lamborn (1970) found this was a factor affecting calibration of P soil tests in Utah.

21-3.1.3 Potassium Soil Analysis

Plant response to application of K has been limited in dryland agricultural production in the western U.S. Removal of K by dryland crops has been relatively low. However, responses to K fertilization have been measured under some conditions. Differences in clay content and types of minerals present in soils can affect established critical K soil test values because these soil properties affect the reserve K supply.

Ammonium acetate has been used extensively to extract cations and is a suitable solution for use in analytical flame instruments, because the NH_3 and acetate volatilize with heat; thus, salt problems on burners are minimized. The NH_4 ion, however, tends to "trap" K ions in nonexpanding hydrous mica minerals and does not provide a good measure of slowly available or "reserve" K.

Good characterization of "plant-available" and "reserve" soil K plus a quantitative identification of the hydrous mica component of clay minerals will be an important part of future soil chemistry studies where K response becomes more widespread.

Doll and Lucas (1973) have pointed out that good correlations have been obtained between some measure of exchangeable K, crop response, and K removed by crops where adequate field calibration data have been available.

21-3.1.4 Sulfur Soil Analysis

Response from S fertilization has generally been limited to legumes or annual cropping in higher rainfall areas. Use of soil analyses to predict S response presents problems comparable with the use of NO_3-N soil analyses (Reisenauer et al., 1973). Sulfur is released from soil organic matter during incubation periods that favor microbiological activity, with oxidation producing SO_4-S that is subject to leaching from most soils. Quantitative chemical problems associated with measuring low SO_4-S levels has limited use of S soil tests to predict S response.

Plant analyses for S offers one technique for evaluating S fertilizer programs (Pumphrey and Moore, 1965b); however, critical S levels vary with crop maturity so that sampling procedures must be specifically defined. Nitrogen/sulfur ratios offer a possible solution to this problem (Pumphrey and Moore, 1965a) because both N and S decrease with crop maturity. An N/S ratio wider than about 16:1 in forage and grain crops has been suggested as an indication of S deficiency.

21-3.1.5 Micronutrient Soil Analysis

Heavy metal micronutrient deficiencies occur most often on irrigated soils where leveling for surface irrigation has exposed calcareous subsoils and where corn, potato, bean, and other crops are produced that are more susceptible to Zn, Mn, and Fe deficiency.

Chelate-extracting solutions developed by Lindsay and Norvell (1978) have more widespread use than other extractants for heavy metal micronutrients (Viets and Lindsay, 1973).

Boron deficiency can be a problem in dryland legume production but is seldom observed for wheat, barley, or grass crops. Boron deficiency is often associated with a deficiency of growing season rainfall because low soil water limits B availability (Tisdale and Nelson, 1975); thus, B deficiency on alfalfa varies considerably from year to year. The hot water extraction method of Berger and Truog (1939) is often used to assess soil B. However, because a large number of factors (soil pH, soil water, soil organic matter, clay minerals, and soil texture) affect B availability, relationships between B soil tests and response to B application are difficult to establish.

21–3.2 Use of Plant Analyses

Chemical analyses of plant samples provide a basis for quantitative evaluation of deficiencies, toxicities, imbalances, and other crop production problems. Plant part sampled (i.e., leaf vs. stem), stage of maturity, proper sample handling, possible sources of contamination, and other factors can affect the level of nutrients found in the plant; failure to consider these factors has sometimes resulted in misuse of this tool to diagnose soil fertility–plant nutrition problems.

The use of plant analyses as a basis for predicting fertilizer needs in dryland agriculture has limitations. Many times fertilizers should be applied at or before planting if special application methods, like banding, are required for optimum fertilizer use efficiency. Since a specific growth stage must be achieved before a proper plant sample can be taken, it is often too late to get full benefit of fertilizer applied at this late growth stage. Phosphorus and cations that are not easily moved with rainfall into the root feeding zone should almost always be applied at or before planting for efficient uptake. Thus, for these nutrients, plant analyses for dryland agricultural production becomes more of a postmortem and a method to assess current year's fertilizer program before the next crop is planted.

Limited progress has been made in establishing critical nutrient levels for small grains because of problems in using plant analyses information for adjusting current fertilizer needs. Brown and Jones (1977) reported that plant samples must be taken at very early stages of growth (probably the 3 to 4 leaf stage) to obtain a reliable estimate of N needed for winter wheat in irrigated areas of southern Idaho. Unpublished data by the senior author suggests that leaf samples should be taken during tillering to evaluate the P status of winter wheat under western Oregon growing conditions. Other researchers have emphasized the need for plant sampling during tillering in assessing the P nutrition of winter wheat in Oklahoma (Baker and Tucker, 1973). Leaf samples taken at later growth stages have limited diagnostic value.

21–4 NUTRIENT REQUIREMENTS BY DIFFERENT CROPS AND CROPPING SYSTEMS

In general, as the level of plant available soil water increases, the yield response to added fertilizers increases. Thus, the yield potential from available water, stored in the soil or expected as rainfall, and the amount of plant available N required to achieve that yield become the two most critical factors in balancing fertilizer requirements.

The influence of stored soil water at seeding and growing season precipitation on the response of spring wheat to N and P fertilization is shown in Tables 21–1 and 21–2 (Snider et al., 1968). Because yield responses to fertilization are governed by available water, most of the states in the Great Plains and Canadian provinces consider the amount of available water

Table 21-1. Effect of amount of stored soil water at seeding and growing season rainfall on spring wheat yields and yield responses to N fertilizer on nonfallowed soils, North Dakota (Snider et al., 1968).

Stored water†	Rainfall‡	Number of trials	Yield of check	Yield response to added N (kg of N/ha)§ 22	45	67
cm	cm		kg/ha			
Less than 5	Less than 15	5	640¶	150	130	190
	15 to 20	11	1110	170	120	100
	More than 20	10	1360	320	370	310
		26	1040	220	220	200
5 to 10	Less than 15	9	1270	190	270	250
	15 to 20	5	1450	380	540	620
	More than 20	6	1870	300	540	420
		20	1490	290	420	390
More than 10	Less than 15	2	1610	350	390	230
	15 to 20	5	1590	450	790	1040
	More than 20	13	1700	560	870	940
		20	1660	520	800	890

† To 120-cm depth or dry zone above 120 cm.
‡ From seeding to 5 days before harvest. Less than 15 cm is below average rainfall during the spring wheat growing season in North Dakota; 15 to 20 cm from slightly below to average; more than 20 cm from average to above average.
§ A constant rate of P (0–46–0) was applied by drill attachment.
¶ Figures rounded from original publication.

Table 21-2. Effect of amount of stored soil water and growing season rainfall on spring wheat yields and yield responses to P fertilizer in fallowed soils of "very low" P test level, North Dakota (Snider et al., 1968).

Stored water†	Rainfall‡	Number of trials	Yield of check	Yield response to added P (kg of P/ha)§ 7	22	17	22
cm	cm		kg/ha				
Less than 5	Less than 15	1	280	50	70	110	140
	15 to 20	2	1140	280	280	290	300
	More than 20	2	1130	350	470	340	380
		5	960	260	330	270	300
5 to 10	Less than 15	5	1270	120	190	210	210
	15 to 20	3	1940	390	400	420	400
	More than 20	3	1860	540	790	840	840
		11	1610	310	410	440	440
More than 10	Less than 15	5	1610	340	430	500	550
	15 to 20	8	2220	440	550	620	660
	More than 20	4	2110	560	790	830	810
		17	1940	440	570	630	670

† To 120-cm depth or dry zone above 120 cm.
‡ From seeding to 5 days before harvest. Less than 15 cm is below average rainfall during the spring wheat growing season; 15 to 20 cm from slightly below to average; more than 20 cm from average to above average.
§ Phosphorus (0–46–0) was applied by drill with the seed.

when making fertilizer recommendations. In North Dakota, fertilizer recommendations are based not only on soil test data obtained in the laboratory but also on stored-soil water at seeding time, crop to be grown, and indirectly, on probable growing season rainfall and yield potential (Wagner and Vasey, 1970). In Saskatchewan, generalized N and P fertilizer recommendations are made for each soil zone (e.g., Brown, Dark Brown Soils), soil texture, and crop species based on long-term average yield data (Saskatchewan Advisory Council on Soils, 1977).

21–4.1 Wheat and Other Small Grains

Wheat is grown on more acres than any other crop in the dryland areas of the western USA and Canada. Approximately 16 million ha are devoted to wheat production in the dryland area of the USA. In the prairie provinces of Canada over 9 million ha of wheat are grown. Barley is an important crop in the western states with more than 0.5 million ha seeded each year. Oat and rye are grown for both grain and forage, but hectarages have declined in recent years.

Wheat yields have increased substantially during the last 20 years, and it is no coincidence that yields have increased simultaneously with increased fertilization of that crop.

Most of the early fertilizer experiments on wheat in the Great Plains showed that P was the most limiting nutrient. Yield increases from P fertilizers were reported from several locations (Bauer et al., 1966; Eck et al., 1957; Harvey et al., 1961; Power et al., 1961; Russel et al., 1958). With continued cropping, N soon became limiting, and N deficiency symptoms were widespread. Now, N fertilizer is applied to a majority of the dryland wheat fields.

Wheat requires approximately 30 kg of available soil N for each 1000 kg of grain produced up to a yield of about 2200 kg/ha, or to the point where the yield curve starts to level off. Above that, slightly more N is required to produce a unit of grain because of decreased N efficiency. Balancing N requirements with available water has been discussed in section 21–3.1.1.

In the Southern Great Plains, much of the winter wheat is grazed by cattle in the fall and winter. Grazing removes N, and additional N must be supplied if wheat is harvested for both forage and grain. Each 1800 kg of dry weight of small grain forage produced requires about 36 kg of N (Tucker, 1977). The amount of wheat forage potential depends on many factors, including soil water, temperatures, and seeding dates. Grazing yields of 800 to 1600 kg of forage/ha are common for the fall and winter, and grazing can continue until the jointing stage of growth without seriously affecting grain yields (Staten and Heller, 1949).

The N requirements of other small grains (barley, oat, and rye) are nearly identical to wheat on a kilogram of grain basis (Johnson and Tucker, 1979; Table 21–3). Growers should recognize that barley and oat are much

Table 21-3. Nitrogen requirements for small grain and oilseed crops being used in North and South Dakota to make N fertilizer recommendations (Wagner and Dahnke, 1977; Gelderman et al., 1977).

Crop	North Dakota	South Dakota
	g of N/kg of grain yield	
Winter wheat	--	40
Spring wheat	33-47	40
Feed barley	33-48	40
Malting barley	27-35	33
Oats	38-47	38
Rye	36-50	39
Flax	54	54
Safflower	40-53	50
Sunflower	50	50
Mustard	60	60
Rape	60	60

more susceptible to lodging with excess N than the commonly grown wheat cultivars. In addition, N rates on malting barley are usually about 30% lower to control protein levels and preserve malting quality.

Wheat is sensitive to low soil P levels. Most experiments show P fertilizer requirements for broadcast applications to be 40 to 60 kg of P/ha on extremely deficient soils, with lower rates or a "starter" application on soils testing moderately high.

Wheat and other small grain crops do not remove large amounts of soil K, and a large majority of soils used for wheat production in the West contain adequate exchangeable K. Responses of wheat to K fertilization can be substantial when soil P levels are low, but yield increases are generally lower than for forage and row crops. As with P, K fertilizer needs for small grains are determined by soil tests. Even with soil tests as low as 75 kg of exchangeable K/ha, 55 to 60 kg of K fertilizer will give maximum yields (Tucker, 1977).

21-4.2 Oil Crops

Oilseed crops occupy about 1.5 million ha of dryland soils in the Northern Great Plains and larger areas in the dryland areas of Canada, but only small hectarages in other dryland areas of western USA. The N requirements for flax, safflower, sunflower, mustard, and rape are reported in Table 21-3 (Bergman et al., 1975; Cobia and Zimmerman, 1975; Gelderman et al., 1977; Wagner and Dahnke, 1977). Mustard and rape generally require slightly more N than sunflower, safflower, or flax per kilogram of grain produced. Adams et al. (1978) reported that flax required 7 g of P and 40 g of K/kg of grain produced. In the Northern Great Plains, minor element deficiencies are not common in these crops. However, Moraghan (1978) reported that "chlorotic dieback" of flax results from a Zn deficiency. Addition of Zn corrected the nutritional problem. Although B deficiencies are rare, sunflower has a relatively high requirement; where B responses have been observed, 1 to 2 kg of B/ha has corrected the deficiency. Recommended rates of N, P, and K fertilizers vary with soil test levels and yield goals. Recommended rates of P and K for oilseed crops are similar to those recommended for small grain crops for similar soil test levels.

21–4.3 Grain Sorghum and Corn

Grain sorghum is grown extensively in the dryland areas of the Central and Southern Great Plains. Corn grown on dryland in the Northern Great Plains is used primarily for forage.

Nutrient requirements for grain sorghum are slightly less than for corn because of its more extensive fibrous root system and the lower grain-forage ratio for grain sorghum. The fertilizer needed for both crops depends on the soil's supply of each nutrient. The potential yield depends on the supply of N. Estimated N requirements range from 17.9 to 26.8 g of N/kg of grain or 33 g of N/kg of forage (Wagner and Dahnke, 1977). After a yield goal has been established, the preplant soil NO_3-N in the 60 cm soil profile can be subtracted from the total N requirement to determine fertilizer application.

Average yields of dryland sorghum and corn range from 3000 to 4000 kg/ha on most soils. Yield levels of 2000 to 3000 kg/ha are common in the drier regions where average annual precipitation is less than 500 mm. Nitrogen fertilizer rates commonly used vary from 18 to 60 kg/ha.

Rates of P on dryland grain sorghum and corn vary from 20 to 30 kg/ha on soils tested low in available P. Sorghum and corn respond to starter band applications of P especially if planted in cool soils (less than 10°C).

Sorghum and corn respond to K fertilizer on soils low in that nutrient. Lodging near maturity is often the first symptom of K deficiency on sorghum. The lodging may be a direct result of secondary disease resulting from low K levels. Whitney and Murphy (1974) showed the relationship between K rates, yield, and lodging on a K-deficient soil in Kansas. Practical K fertilizer levels for grain sorghum range from 20 to 70 kg K/ha on soils testing very low to moderately high K levels.

Most of the sorghum produced in the Southern Great Plains are grown on calcareous soils with resultant high pH values. Therefore, the availability of certain micronutrients may be restricted. Iron deficiency is common, with Zn deficiencies being reported (Thompson, 1969).

Soil-applied chelated Fe is effective in correcting Fe deficiencies, but rates required are sufficiently high to render them uneconomical. Sorghum seed coated with chelated Fe has been effective in reducing Fe deficiency, and foliar spraying various Fe compounds has also been effective. It is important to apply foliar sprays before the deficiency becomes too severe.

Zinc deficiencies in grain sorghum and corn have been observed in Kansas (Thompson, 1969) and Texas (Fenn, 1969) on high pH soils simultaneously having high soil P levels. Soil tests for Zn can be used to predict the need for Zn application. Zinc deficiencies can be corrected with the application of 3 to 8 kg of Zn/ha as $ZnSO_4$. Chelated Zn is available and lower rates can be used, but low rates of Zn need to be banded close to the seed to be effective (Bauer, 1968).

Some nutrient responses to S have been noted on grain sorghum in Texas, Nebraska, and Kansas (Tucker and Bennett, 1968). These responses have occurred on low organic matter coarse-textured soils. Sulfur requirements can be met by applying sulfate sources such as $CaSO_4$, K_2SO_4, and $(NH_4)_2SO_4$. Elemental S sources are effective if applied well ahead of planting. Where S is deficient, applying 30 to 40 kg of S/ha is generally sufficient.

21-4.4 Legumes

The principal legumes grown in the dryland areas are alfalfa and sweetclover. Some summer legumes are grown in the southern Plains, including mung bean, cowpea, guar, and Austrian winter pea.

Alfalfa production in the Southern and Central Great Plains has been variable throughout this century; plantings increase during periods of favorable rainfall and generally decrease with adverse climate and pest outbreaks. Also, the high quality of alfalfa hay has resulted in increased demands from the large cattle feedlots that have been established in the region. These increased demands for hay have also increased seed requirements and thus production of seed for new plantings.

Alfalfa generally removes more Ca, P, and K from the soil than most other field crops. Each metric ton of alfalfa hay will remove about 20 to 30 kg of N, 5 kg of P, 20 kg of K, 10 kg of Ca, 5 kg of Mg, and 5 kg of S.

Alfalfa yields are reduced from soil acidity; thus lime must be applied for optimum production to some acid soils along the eastern portions of the Southern Great Plains and few areas in eastern Washington and northern Idaho. Most states recommend application of lime when the soil pH is less than 6.2 to obtain good production and to ensure stand longevity.

Legume crops like alfalfa and sweetclover should obtain their N needs through rhizobium fixation of atmospheric N. Therefore, the major limiting nutrients for legume production are P, K, and S; Mo deficiency occurs on a small number of moderately acid soils. Small amounts of N are often applied at or just prior to seeding to help establish seedlings before the rhizobia actively supply N to the plant. Soils low in available P frequently require 75 to 100 kg of P/ha to maximize alfalfa yields (Tucker, 1977). Corresponding P needs for alfalfa production in the Northern Great Plains region would be about one-half those reported for the Southern Great Plains. Phosphorus fertilization is just as important in increasing alfalfa seed yields as hay yields on P-deficient soils.

Potassium removal by alfalfa cut for hay is large. Therefore, many fields with a history of high alfalfa yields are becoming deficient in K. Potassium fertilizer recommendations range from 200 to 250 kg of K/ha in the Southern Great Plains and 60 to 90 kg of K/ha in the Northern Great Plains on soils testing very low in K.

Both Zn and Fe deficiency symptoms have been identified on alfalfa in the Southern Great Plains, and even though the symptoms have been corrected by Zn and Fe applications, data are not available showing yield increases from these micronutrients.

The summer legumes, cowpea, mungbean, and guar, all have similar soil nutrient requirements, with P being the most limiting nutrient. Much smaller quantities of P are necessary to reach optimum yields than for alfalfa. Maximum yields of these crops have been obtained with the application of 40 to 80 kg of P/ha. Most Southern Great Plains soils contain adequate K for maximum yields unless these crops are harvested as hay over a period of time.

Legumes contribute to the N economy of the soil, leaving some residual N for subsequent crops. However, total N_2 fixation while the crop is growing can be higher than N left for subsequent crops. Some comparisons of annual N fixation and residual, or carryover, N show 220 and 90 kg N/ha for alfalfa, 110 and 45 kg of N/ha for cowpea, 90 and 25 kg of N/ha for mungbean, and 135 and 45 kg of N/ha for guar, respectively. Tucker et al. (1971) concluded that over 90 kg of N/ha was available to a wheat crop the first year after alfalfa, and 50 to 70 kg of N/ha was available for the second wheat crop. If wheat is planted after alfalfa, stubble mulching the alfalfa sod will reduce the N released the first year. The practice tends to distribute the N released over 2 years after the alfalfa crop, thus helping to avoid excessive vegetative growth the first year.

The economic value of using legumes to supply the N needs of crops in the Northern Great Plains is questionable with the current costs of commercial N fertilizers. Brown (1964) concluded that legumes contributed 34 to 45 kg of N/ha per year to following crops; however, residual N from the legume is not enough to meet N requirements of grain crops in a 3- and 4-year rotation. Therefore, supplemental N is needed for optimum crop production, especially when forage is removed as hay (Haas et al., 1976; Jensen and Weiser, 1971; Bauer and Zimmerman, 1975). The N contribution varies with the legume growing conditions. Reduced subsoil moisture rather than inadequate N supply is a serious problem that accompanies the growing of deep-rooted legumes in drier areas.

21-4.5 Grasses

There are approximately 72 million ha of rangeland in the Northern Great Plains. Nitrogen deficiency has been identified as an important limiting factor to herbage production on these rangelands (Lorenz and Rogler, 1973; Power, 1972; Rogler and Lorenz, 1969; Wight and Black, 1979). Power (1972) concluded that adding rates up to 540 kg of N/ha to grasslands saturates the capacity of the soil-plant system to immobilize N. The system can then be maintained in a N-saturated condition if

$$\text{Annual fertilizer additions} + \text{Mineralization}$$

$$= \text{Immobilization} + \text{Irreversible losses.}$$

Thus, N can be eliminated as a growth-limiting factor to provide grass production from the available water supply. Responses to P fertilization are obtained after N requirements have been satisfied (Lorenz and Rogler, 1973; Wight and Black, 1979). Rogler and Lorenz (1969) report a 72% increase in animal carrying capacity on fertilized pasture compared with an unfertilized pasture. For optimum production, 35 to 56 kg of N/ha is required annually as fertilizer to satisfy the needs of rangelands.

Grasses in the Great Plains area consist of both warm and cool season species. Native grasses account for the largest hectarage, but several thousand hectares of improved species and strains have been planted. Improved pastures respond more to fertilizer applications of N and P in many situations (Houston, 1957; Williamson et al., 1977; Ahring et al., 1973; Moser and Anderson, 1965). Native ranges and meadows receive very little fertilizer, although some N and P responses have been reported (Harper, 1957; Grable et al., 1965; Elwell and Daniel, 1953). Native ranges have low total production, and competition from less desirable species often limits the economic returns from fertilization. Inefficient utilization of forage by grazing animals also limits returns on many ranches. Palatability of grasses is frequently improved by N application.

Cool season grasses respond to P application more than summer-growing species. The amount of P needed to reach maximum yields, however, appears to be about the same for both cool and warm season species (Elder and Tucker, 1968).

One kilogram of N fertilizer per hectare will increase dry matter forage production 12 to 16 kg/ha on cool season grasses and 20 to 25 kg/ha on warm season forages. In both cases, with grazing steers, this quantity of forage can add 1 kg to a beef animal with average management. Therefore, 1 kg of added N is expected to produce 1 kg of beef within the responding range of the yield curve.

21–4.6 Cotton

Cotton is grown on 1 million ha of dryland in the Southern Great Plains. Soils on which cotton is grown in the Southern Great Plains are more often deficient in N than other plant nutrients (Tucker and Tucker, 1968). Because of the rather warm climate, soil organic matter content has been severely depleted through oxidation during years of cultivation.

An understanding of the effects of N on growth and fruiting behavior of cotton is essential for N fertilizer management. Cotton fruit is borne on a fruiting branch that originates at the leaf axil on the main stem or on a vegetative branch (Tharp, 1960). Because N does promote plant growth, it increases the number of fruiting sites and yield potential (Tucker and Tucker, 1968). However, the cotton plant remains in the vegetative stage whenever insect infestations or climatic conditions cause fruit abortion and whenever excessive N rates in the above situation create too large a stalk for proper insect and disease control. Excessive available soil N too late in the season often reduces the plant's ability to mature and causes indeterminate growth. Therefore, enough soil N must be present early in the season to produce sufficient fruiting sites for adequate yield, but excessive N must be avoided in case a period of fruit abortion occurs. Rather high rates of available soil N are needed during rapid fruiting, but best yields and grades of cotton are produced if soil N is depleted during the maturing period. This program of N management requires special skill and at best is difficult to accomplish.

Several schemes to accomplish the above have been devised, the most successful being soil NO_3-N analyses at planting coupled with monitoring plant NO_3-N in petioles during the growing season (Tucker, 1965). This plan is being employed in some irrigated areas and in higher rainfall belts. It is not extensively used for dryland cotton production in the Southern Great Plains, where the common practice is to apply all the N prior to planting. Effects of N on plant type and fruiting are usually favorable if N rates are not excessive.

Two bales (450 kg) of cotton/ha is a good yield for dryland conditions in the Southern Great Plains. A bale of cotton contains approximately 18 kg of N in the seed and burs (Tucker and Tucker, 1968) and represents about 60% of the N contained in the plant. The soil N required for a two-bale yield is about 60 kg of N/ha.

Phosphorus will increase yields on deficient soils and frequently increases earliness and maturity. Early fruit set and maturation reduces vegetative production. One bale of cotton removes about 5.2 kg of P (Jones and Bardsley, 1965). In many experiments, 85 to 100% of the maximum yield increases are obtained with 20 kg of P/ha on soils with relatively low P levels. Soils of the Southern Great Plains generally have adequate K for cotton production, and limited K fertilizer is recommended. However, some of the sandy upland soils in the 600 to 800 mm rainfall area have responded to K whenever soil test K drops below 250 kg/ha. With a moderately high soil test (200 to 250 kg of K/ha) only 15 to 30 kg K/ha is suggested. Higher amounts are suggested at lower soil test levels.

Zinc has been deficient on a few cotton fields in the southern Great Plains, and B deficiencies have been suspected on a limited number of soils. Researchers in the Texas Blacklands (McLean and Niles, 1968) demonstrated that excess P applications reduced Zn availability. Where Zn is deficient, broadcast applications of 6 to 8 kg of Zn/ha as $ZnSO_4$, or band applications of 2 to 4 kg of zinc chelate is recommended.

21–4.7 Summer Fallow Cropping Systems

The term "summer fallow" has traditionally described a system in which a crop is grown once in two seasons. Fallow has been widely practiced in the dryland wheat-growing regions of the western USA (USDA-ARS, 1974). The popularity of summer fallow in dry regions is primarily attributed to water conservation leading to increased stability of production. Other benefits include better weed control and increased mineralization of N with 60 to 100 kg of N/ha accumulation from microbial decomposition during the summer fallow period. Summer fallow hectarage has fluctuated widely during the past few decades because of government agricultural programs but in general has declined rapidly in recent years.

The great advantage for summer fallowing has been ascribed to increased N mineralization (Haas et al., 1976). However, increased N fertilizer use has diminished this advantage, and many soils do not supply adequate mineralizable N with summer fallowing to meet the crop needs. Nitrogen fertilization is based on established yield potential and levels of soil NO_3-N present to achieve projected yield. Need for other nutrients does not change materially because of fallowing, and nutrents are applied commensurate with soil test levels.

In general, P is the major limiting nutrient under summer fallow conditions in the Northern Great Plains (Bauer et al., 1966; Black, 1969). Increases in grain yield from N and P fertilization are generally greater for early and medium seeding dates than for late-seeded spring wheat. Therefore, N and P fertilization of early-seeded spring wheat increases grain yields and economic returns more than the same N and P fertilizer levels for late seeding (Black and Siddoway, 1977b; Saskatchewan Advisory Council on Soils, 1977).

Most crops in the Central and Southern Great Plains are grown in a continuous (one crop each year) cropping system, and the nutrient demands have generally been established under this system. Where crops are grown every year on a field, the length of the fallow period (or period between harvest and planting) and soil water determine quantity of N mineralized. For small grains, the fallow period occurs during the summer, which is warm and is the period of greatest rainfall, with optimum conditions for mineralization of nutrients. Indigenous soil N varies considerably in the Southern Great Plains, but on large hectarages only 20 to 40 kg of available N/ha is mineralized.

In the Pacific Northwest, the amount of additional N needed to reach optimum yields on fallowed soils varies with location and precipitation. Leggett et al. (1959) found that in areas with less than 250 mm of growing season rainfall, 20 to 40 kg of N/ha was needed; in areas of 250 to 380 mm of growing season rainfall, N needs were 30 to 60 kg/ha; and in areas of more than 380 mm of growing season rainfall, 60 to 80 kg of N/ha was needed. They concluded that S fertilization often resulted in vegetative responses, usually without increased grain yield, in wheat grown on fallowed land.

21-4.8 Flexible Cropping Systems

Under flexible cropping systems in the Northern Great Plains, the farmer assesses the available soil water and expected growing season precipitation at seeding time before deciding to either crop or summer fallow (Halvorson and Kresge, 1982). If the farmer decides to crop, a crop is selected that should yield well on the available water supply. Good weed control and crop rotation are essential under this system. Nitrogen is gener-

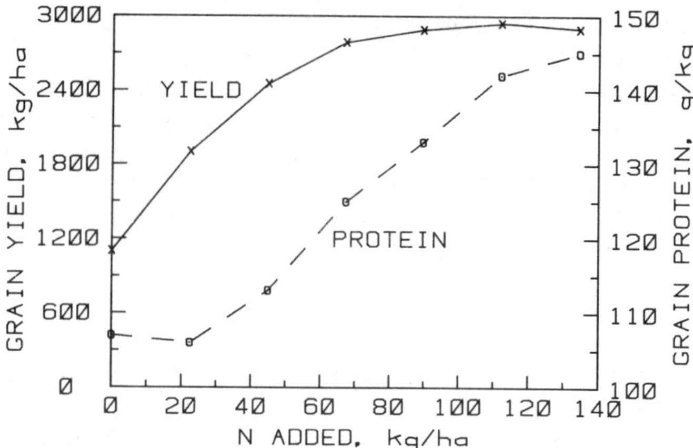

Fig. 21-1. Average grain yield and protein content of three winter wheat varieties (Winalta, Froid, and Centurk) (Halvorson et al., 1976).

ally the major limiting nutrient under a recrop situation. In the Northern Great Plains, approximately 34 kg of N/ha remains in 120 cm of the soil profile after a small grain crop (Black and Ford, 1976; Halvorson et al., 1976). Generally, N fertilizer needs to be added to attain maximum yields. A typical response of winter wheat to added N under recrop conditions is shown in Fig. 21-1 (Halvorson et al., 1976). Black and Siddoway (1976) found 34 kg/ha of additional fertilizer N generally provides adequate N for recrop conditions and near maximum yield. Crop rotation and chemical weed control are essential for maximizing returns on fertilizer dollars invested.

The P fertilizer requirements for crops grown on recrop land are essentially the same as for fallow conditions; an adequate soil P supply is necessary. Adequate P levels can be established either by broadcasting a relatively high rate of P to supply adequate P for several crops or by supplying each crop with the amount of P needed at seeding time. Recommended broadcast rates of P are approximately 1.5 to 2 times higher than rates recommended for banding with the seed.

Most N recommendations made for recropping in the Northern Great Plains are based on soil water available to the crop to be grown. Greater responses to fertilization are obtained as water supplies increase (Bauer et al., 1967; Russell et al., 1958). Larger supplies of soil water can be provided for crop growth by using snow management practices, such as tall wheatgrass barriers or standing stubble to trap winter snow (Black and Siddoway, 1976).

Because of the variable climate, especially rainfall, in the Central and Southern Great Plains, the most economical cropping systems over a period of time seem to be those that take advantage of soil water conditions at a given planting date (Harvey et al., 1961). These flexible cropping systems have been labeled as "opportunity rotations" (Luebs, 1962). Alfalfa is the most important legume used in cropping systems in the Southern Great Plains. Small grains do not yield well after alfalfa, but substantial yield increases have been reported for sorghum and cotton after alfalfa (Murphy and Tucker, 1969). Sweetclover, although productive in some years, is poorly adapted for use in short rotations. Cropping sequence studies in the Southern Great Plains have shown that all crops do well after cotton, provided soil water was replenished. Generally crops do poorly after grain sorghum.

Fertilizer requirements for crops in various cropping systems in the Southern Great Plains are computed on the same basis as in continuous monoculture with adjustments in the N fertilizer application being made for: (i) changes in yield goals, (ii) amounts and kinds of plant residues at seeding, and (iii) types of tillage during the fallow period.

In addition to N, S is often limiting during annual cropping in the Pacific Northwest. Application of S fertilizers will generally increase wheat yields under annual cropping provided that high levels of N fertilizer have been applied and no S has been applied in recent years.

Legumes in the crop sequence supply some N to subsequent crops and thereby reduce N fertilizer needs.

21-5 FERTILIZER AND WATER USE EFFICIENCY

The classic example of fertilizer increasing the efficiency with which water is used by plants is illustrated by response to N fertilizer. Assuming other nutrients are not limiting, we find that N can frequently double or triple the yield when adequate water, stored soil water, and growing season rainfall are available to grow a crop. Under most conditions, and unless affected by disease, insects, and other adversities, a crop will remove most of the available water from the rooting zone during a summer's growing season. Water in the surface 30 cm not used by the crop will probably be lost by evaporation during the hot, dry August weather in the Southern Great Plains and areas with extended summer drought.

Use of the summer fallow system of farming to store subsoil water, reduce weed problems, and increase the supply of available soil N in areas of Montana and the Dakotas has contributed to the development of saline seeps (Halvorson and Black, 1974; Doering and Sandoval, 1976a, b) on rolling topography as excess soil water has moved laterally to hillside seep areas. Systems are being developed to establish annual cropping that will use more of the potentially available water. Fertilizer and herbicides will be essential in replacing the traditional summer fallow system of farming.

21-6 FERTILIZER APPLICATION—SOURCE AND METHOD

Urea, $(NH_4)_2SO_4$, NH_4NO_3, anhydrous ammonia, solution 32, ammonium phosphates, and monocalcium phosphate (concentrated superphosphate) are the major sources of N, P, and N-P fertilizers used in this area of the USA. Low costs have increased the amount of urea and anhydrous ammonia applied in recent years. Anhydrous ammonia is used extensively on large farms in many wheat-producing areas and accounts for over 80% of the N used in the Southern Great Plains.

Ammonium sources of N have some special advantages and disadvantages for this area. The NH_4 ion is positively charged and is not subject to leaching until after biological oxidation to NO_3-N. Ammonium nitrogen can be applied as soil temperatures approach 10°C when biological activity is reduced; under low temperature (less than 10°C) conditions, these materials will stay in the NH_4 form until the next spring. Nitrification inhibitors can be applied with NH_4 sources of N to provide additional delays to the nitrification process. Ammonia volatilization from N fertilizers has been studied extensively in laboratory experiments, but difficulties in measuring NH_3 volatilization has limited the experiments conducted under field conditions. Fenn and Kissel (1973) devised a procedure for measuring NH_3 volatilization directly; using this technique, they found large losses of N from $(NH_4)_2SO_4$ when it was applied to the surface of a calcareous Houston Black Clay (Udic Pellustert). Losses from NH_4NO_3 were much smaller. In these studies, high N rates (140 and 280 kg of N/ha) were applied in single applicataions. Laboratory studies (Fenn and Kessel, 1975) showed losses from surface-applied $(NH_4)_2SO_4$-N to be substantially lower at low application rates. Surface applications of urea were inferior to other N sources in Oklahoma whenever single applications exceeded 90 kg of N/ha (Bohl, L. G. 1976. The effect of nitrogen sources, rates, and application methods on bermudagrass production. M.S. Thesis. Oklahoma State Univ., Stillwater). Losses ranged from 19 to 50% with rates of 550 kg of NH_4-N/ha. Nitrogen losses increased as soil temperatures increased.

The application of urea to perennial grass crops is of concern to agronomists. Field experiments in the Southern and Central Great Plains indicate volatilization losses of N from surface applications of urea to grass crops. Significant losses occur whenever: (i) the urea was applied to wet sods followed by no rain and rapid drying conditions (generally high temperatures, low humidity, and windy), (ii) application rates exceeded 90 kg of N/ha, and (iii) the sod was dense enough to intercept applied urea. Grazing or haying the forage prior to application helped reduce potential NH_3 losses. Also, burning destroys the urease enzyme and reduces these losses.

Urea and anhydrous ammonia are becoming more prominent N sources in the Northern Great Plains. Beaton (1978) forecast a significant reduction in production and consumption of dry NH_4NO_3 with considerable expansion in manufacture and sale of solid urea. Power (1974) points out that at high N rates, urea is generally 5 to 40% less effective than NH_4NO_3. He suggests that efficiency may not be as important as the fact that maximum yields obtained with urea may be less than with NH_4NO_3. Black and Siddoway (1977a) also found that urea was less effective than NH_4NO_3 when applied at rates above 45 kg of N/ha or applied in the late spring. In general, rates of N banded with the seed should not exceed 11 kg of N/ha in most Northern Great Plains areas (Black et al., 1980). The rate of N that results in salt or NH_3 toxicity varies with the source of N, soil pH, exchange capacity of the soil (clay and humus), soil water, and crop.

Chemical reactions take place as urea reacts with water in the soil. The initial hydrolyses forms ammonium carbonate or bicarbonate, which raises the pH in the area around a fertilizer granule. Thus, NH_3 can be released to the atmosphere; if urea is banded with the seed, the NH_3 released can reduce the stand of germinating seedlings (Black et al., 1980).

Problems resulting from banding NH_3-producing fertilizers are generally associated with placing the fertilizer near, or in contact with, the seed. Poor seedling emergence and subsequent root damage may result when fertilizers with high salt index are placed too close to the seed row under low soil water conditions. Seedling damage is accentuated when the soil has less than optimum water because salt concentrations increase with decreasing soil water. High rates or band applications of urea, $NH_4H_2PO_4$, and urea ammonium phosphate near the seed row may cause seedling damage from free NH_3, especially in dominantly neutral and calcareous soils of the Northern Great Plains. Pairintra (1973) found that damage to spring wheat seedlings from applications of various $(NH_4)_2PO_4$ fertilizers increased in the order ammonium polyphosphate < monoammonium phosphate < diammonium phosphate < urea ammonium phosphate. For each fertilizer, the damage was more severe with higher amounts of $CaCO_3$ in the soil.

As long as reasonable precautions are followed with special attention to local recommendations for time, source, and method of application, growers should expect efficient results from commercial N fertilizers.

Water-soluble P sources (monocalcium phosphate [$CaH_4(PO_4)_2$] or ammonium phosphates) have advantages over $CaHPO_4$ or $Ca_3(PO_4)_2$, especially where the soil pH is neutral or above. The increased rate at which water-soluble P materials supply HPO_4 or H_2PO_4 to the soil solution has prompted their wide use in these areas. Phosphorus is relatively immobile in the soil. Therefore, all P materials should be applied at or before planting; they should either be banded with or near the seed, mixed with the seedbed, or plowed down.

Increased efficiency in use of P and most micronutrient fertilizers, when banded rather than mixed or broadcast, has been observed under a wide range of conditions (Ellington, 1978; Mortvedt, 1976; Richards, 1977), particularly in the Northern Great Plains where spring soil temperatures are often 5 to 15°C. As soil temperatures increase, the advantages from banding P decrease; in areas with higher soil temperatures, banding decreases P response (Baker et al., 1970). Also, as the P rate needed to alleviate deficiencies increases, differences in effectiveness between broadcast and band applications decreases. At very high rates (100 kg of P/ha), broadcast applications of P are just as effective as band placement (Ellington, 1978; A. L. Black, 1975. Unpublished data. USDA-ARS, Sidney, Mont.). Broadcast application requires less labor and time.

When using broadcast applications, Bauer and Kucera (1969) found that moldboard plowing incorporated P fertilizer to a deeper depth than a double disk or Noble blade. As minimum tillage and flexible systems become more prominent in the Northern Great Plains, placement for effective P fertilizer use becomes more important.

Mixing an acid-forming source of N, e.g., $(NH_4)_2SO_4$, with P fertilizers when banding with or near the seed on calcareous soils will increase the uptake of P and micronutrients (Miller, 1974; Jackson and Carter, 1976).

Sulfur is generally supplied as $(NH_4)_2SO_4$, 16–20–0 ammonium phosphate-sulfate, single superphosphate (10% S), K_2SO_4, $CaSO_4 \cdot 2H_2O$, and elemental S. All SO_4 sources are readily available for plant uptake and the SO_4 anion moves downward in the soil; thus, method of application is not critical. Elemental S must be finely ground so that oxidation to the plant available SO_4 ion will proceed at an adequate rate.

Micronutrients (Zn, Mn, and Cu) are generally applied as sulfates, with Fe applied as a chelate in most applications. Banding these nutrients with an acid-forming N fertilizer (Viets et al., 1957) can increase micronutrient uptake and is a practical way of getting soil applied Mn into a plant. Foliar application is an efficient method of applying most of these nutrients and is a practical way of correcting Fe deficiencies without lowering the pH of the soil system. Sorghum seed coated with chelated Fe, followed by foliar sprays whenever chlorosis appears, has been used to correct Fe deficiencies in the Southern Great Plains. High seeding rates have limited the use of Fe seed treatment with small grains.

Foliar application of chelated micronutrients reduces the risk of foliar burn, and uptake per unit of nutrient applied can be more efficient than with sulfate sources. These advantages help justify the higher cost per kilogram of nutrient applied for chelated materials. Generally, chelates of Zn, Mn, and Cu have limited advantage for soil application.

Application of B for dryland agriculture is generally limited to alfalfa and other legumes with high B requirements. Boron is almost always broadcast with some other material or applied as a foliar spray. Limited materials are on the market, thus water soluble B is generally used. Boron applications should not exceed 1.5 kg/ha annually in nonirrigated fields because the margin is small between deficiency from one crop and toxicity for a sensitive crop that might follow.

21-7 RESIDUAL CARRYOVER OF FERTILIZERS

Residual carryover of N, P, and S are the most important to evaluate because they have the largest impact on cost and plant growth. The carryover can be estimated by soil testing.

21-7.1 Measuring Residual or Naturally Occurring Levels of Nutrients

Chemical analysis of soil samples offers the best quantitative method of measuring residual carryover from fertilizer application and the naturally occurring levels of available nutrients. The measurement is an important aspect of managing different cropping systems. It becomes especially critical after drought periods, as in 1977, and periods of higher than normal rainfall, when N can be leached out of the root zone.

Carryover of excessive N from fertilization is probably minimal in dryland areas of the northern Great Plains because of the relatively low rates of N applied (less than 40 kg/ha) on intensively cropped lands and the even lower rates on summer fallowed land. Black and Halvorson (1976. Unpublished data. USDA-ARS, Sidney, Mont.) found that with the application of 34 and 67 kg of N/ha annually, residual N slightly accumulated in the 120- to 180-cm soil depth. The accumulation was greater for the 67 kg of N/ha treatment and under summer fallow system. Where safflower (a deep-rooted crop) was grown, the accumulation of residual N was negligible and in most cases no greater than in the area not receiving N fertilizer for at least 8 years. Under recrop conditions in the Northern Great Plains, 34 kg of N/ha is a common level of residual NO_3-N remaining to a depth of 120 cm after a small grain crop (Brown, 1964; Halvorson et al., 1976).

Residual P carry-over from P fertilization in the Northern Great Plains is detectable with the sodium bicarbonate soil test. Phosphorus carry-over work by Read et al. (1977) and Bailey et al. (1977) showed that a single large application of 100 kg of P/ha was effective over an 8-year period in increasing wheat and flax yields. Higher applications of 200 and 400 kg of P/ha were expected to be effective for several more years because of the higher level of extractable P remaining in the soil from these treatments. Even in the eighth year after application, yield increases were 11 to 60% greater than the check.

Tucker (1977) has observed marked carryover or buildup of both N and P in the Southern Great Plains. Use of P fertilizers has resulted in a marked reduction of P responses being observed in the field.

In a recent series of wheat experiments, Gardner et al. (1975) found significant accumulations of NO_3-N on wheat-producing farms in the Columbia Basin counties of Oregon.

21-7.2 Losses of Nutrients by Leaching

Leaching of nutrients is restricted to NO_3-N and SO_4-S, and these losses are probably restricted to summer fallow areas during periods of above normal rainfall.

Over 70% of the area in the Northern Great Plains is in grass. Under present management, Power (1970) concluded that the possibility of NO_3-N leaching through soils in grass is remote. In soils cropped annually to small grains, movement of NO_3-N has been generally restricted to the upper 2 m of soil. He concludes that NO_3-N could possibly move below the root zone in spring grains in a fallow rotation during the 21 continuous months the soil supports no growing crop.

Sulfur will be subject to leaching or accumulation in the subsoil whenever $(NH_4)_2SO_4$ is used as an N source, and an optimum N application results in applying 10 times or more S than the crop needs.

21-7.3 Adverse Effects From Excess Nutrients

Adverse effects from excess nutrients are generally limited to (i) naturally occurring concentrations, (ii) excess application of fertilizer, and (iii) accumulations of by-product contamination from industry.

Naturally occurring concentrations of B have been found in some areas of California and Oregon. Selenium has accumulated on shale soils in

Wyoming and the Dakotas. Excess Se is not a problem limiting plant growth, but can be a problem for animal health. Whereas Se can be toxic to animals in some areas of the Northern Great Plains, Se deficiency for livestock can be a problem in parts of the Pacific Northwest.

Some soils have naturally occurring accumulations of Mo and a soil pH >6.5 results in maximum availability of Mo. Molybdenum toxicity does not limit plant growth, but excess concentrations of Mo in animal diets can interfere with Cu utilization, especially in ruminants (Allaway, 1971, 1975).

Few instances of fertilizer applications have been high enough in dryland farming to result in toxic or adverse accumulations of nutrients.

21-8 EFFECTS OF EROSION ON SOIL FERTILITY AND NUTRIENT REQUIREMENTS

Soil erosion has caused a depletion of soil organic matter and loss of organic N and P in the surface 0 to 15 cm of soil. The "topsoil" also has higher concentrations of plant-available K under many prairie soils. Thus, reduced production on eroded hilltops can be caused from reduced availability of N, P, and K as much as from reduced storage of soil moisture. Reductions in soil fertility, as measured by losses in total N and organic C, have been greater under fallow than continuous cropping in the Northern Great Plains (USDA-ARS, 1974).

Eck et al. (1965) reported that many researchers have found that deficiencies of N, P, and K were primarily responsible for reduced yields where soils have eroded to the subsoils. On a Pullman silty clay loam (Torrertic Paleustoll) at Bushland, Texas, Eck et al. (1957) found that only N was required for maximum yield on the undisturbed soil, but both N and P were required when 10.16 cm or more of topsoil was removed.

The loss of A horizon and exposure of more calcareous horizons in subsoils presents another problem. The solubility, and thus plant availability, of Zn, Mn, and Fe is largely controlled by soil pH (Lindsay, 1972). As layers of $CaCO_3$, which accumulate under limited rainfall, and grassland prairies are exposed, the soil pH approaches 8.0 to 8.3. This high pH results in a marked reduction in availability of heavy metal micronutrients.

21-9 ENVIRONMENTAL IMPACT OF FERTILIZER

Nutrients from commercial fertilizer find their way into streams and groundwater as an environmental contaminant. Viets (1971a, b) has provided an excellent review of the different viewpoints surrounding this problem.

Most of the N and essentially all of the P, the two nutrients of greatest concern in water eutrophication, moves from cultivated farmland into streams by soil erosion. The use of commercial fertilizer has been an important factor in reducing soil erosion. Fertilizer has increased yields of both grain and crop residue. Proper management of crop residues during tillage is an important factor reducing soil erosion. Soil erosion is much more severe with summer fallowing or alternate year cropping than with annual cropping if residues are not left on the surface. Most erosion occurs during the winter after summer fallow, when crop residues have decomposed, and during early season growth, before crops provide ground cover. Commercial fertilizers, especially N, have been essential in the adoption of annual cropping in lower rainfall areas, thereby further reducing erosion.

Our present levels of crop production would require much larger hectarages if commercial fertilizer were not used. Farmers generally retire the steep lands, which are more costly to till and more subject to erosion than level land, as cultivated hectarages have been reduced. Expansion of cultivated hectarage to meet present levels of crop production would thus require cultivation of many hectares subject to more serious erosion than much of our present farmland.

Viets (1971a, b) has concluded that the overall impact of using commercial fertilizer has been to reduce environmental contamination. The impact is greater in dryland areas where yields are lower than with irrigation, and fertilizer is used in smaller quantities.

21-10 EFFECTS OF FERTILIZERS ON CROP QUALITY

The application of fertilizers can have important effects on quality of the crop produced. Quality is important for both human and animal nutrition.

One of the most obvious effects of fertilizers on crop quality is that of N fertilization on protein content of grain (Olson and Koehler, 1968). McGuire et al. (1974) found that fertilization not only increased yield and protein content of wheat, but also resulted in improved baking qualities, namely larger loaf volume, higher grain and texture scores, and greater dough stability. The first increments of added N generally give the largest yield increase with limited protein increase per unit of N. As the yield increase per unit of applied N decreases, the increase in protein per unit of applied N increases. These trends will vary somewhat with different wheat cultivars.

Nitrogen fertilization can be a critical factor in maintaining protein content of hard red winter wheats above 13%, the point where protein premiums frequently start. Millers can blend lower protein bread wheats with higher protein bread wheats to maintain the minimum desirable gluten strength for bread flour (Finney and Yamazaki, 1967).

The desirability of high protein for bread wheat is in contrast with medium to low protein content of soft white wheats from the Pacific Northwest. This wheat is used for pastry flour (7 to 8% protein) and exported to Asia for noodle flour (9 to 10% protein). Rainfall and temperature can have a marked effect on wheat quality in the Pacific Northwest. A soft white club wheat used for pastry flour fertilized for a yield potential of 4 t/ha in the Palouse area can have this yield potential decreased 20 to 30% in dry years with about the same amount of protein produced per hectare. This decrease can eliminate this class of wheat from the pastry trade but makes the wheat a desirable product for livestock feed.

Protein content is most frequently the factor limiting the quality of livestock feed. The addition of N fertilizers to grain crops for livestock not only increases both the grain yield and protein content, but generally increases the nutritional value of the grain. Similar effects are noted when forage grasses are fertilized with N for livestock pasture, silage, and hay. Literature showing possible effects of P fertilization on increasing the feed value of hay for livestock has been reviewed by Reid and Jung (1974). Legume hay, especially alfalfa, generally has a wide Ca/P ratio (often 10:1) while dairy cows require a Ca/P ratio of about 4:1 for optimum milk production. This ratio can be balanced by feeding grain with higher P contents and by adding inorganic P supplements to the diet. Thus, a frequently heard statement is that P fertilizer should be applied only to increase the yield of forage crops and not to influence nutritional value. However, recent work in Australia (Ozanne and Howes, 1970, 1971; Ozanne et al., 1976) has suggested that the P concentrations of subterranean clover (*Trifolium subterraneum* L.) required for maximum digestibility, especially as clovers start to mature, is higher than P required for optimum yield. Reduced digestibility was not corrected by supplementing the diet with inorganic P supplements in these studies. This question needs further investigation.

Bauer and Vasey (1964) found that K fertilization improved the quality of malting barley by increasing the percent of plump kernels, decreasing grain protein, and increasing yield. Jackson et al. (1964) showed that both P and K fertilization increased yield, test weight, and kernel size of spring-planted barley in western Oregon; these increases were more pronounced in years when yellow dwarf virus was present.

All of the previously cited effects might be considered direct effects because either N or P was being added as fertilizer and the N or P content of the crop was being increased. The major problem comes in evaluating and identifying a wide range of indirect effects, such as: (i) possible reductions in Se from S fertilization; (ii) increases in protein content of legumes from S, Mo, or P fertilization; and (iii) possible adverse effects of N and S fertilizers on micronutrient concentrations.

Probably the easiest beneficial effect to document is the increase in N content, and thus, indirectly, protein, of legumes when nutrients are added that increase N fixation by *Rhizobium.* Sulfur, P, Mo, and Ca are probably needed more luxuriantly by *Rhizobium* than by many of the host legume plants. These indirect beneficial effects are widely recognized. However, possible adverse effects are often more difficult to explain.

Possible toxicities and deficiencies of Se in the western states presents a complex problem. Selenium is not required for plants but is required for animals. Present experiments with plants have not established beneficial effects. However, in forage plants the range in Se concentrations that will cause toxicity or deficiency in ruminants is relatively small. Levels below 0.10 mg/kg of Se result in white muscle disease and frequent death of calves and lambs, and levels above 1.0 mg/kg start to cause Se toxicity (Allaway, 1971). Excess Se is accumulated in some plants growing on shale soils in parts of the Dakotas and Wyoming.

Sulfur fertilization can increase the yield of many legumes 5 to 10 times in many areas of the Pacific Northwest where Se levels are low (Westermann and Robbins, 1974). Thus an alfalfa hay yield of 1 t/ha with 0.10 mg/kg of Se could be increased to 6 t/ha and have the Se concentration diluted below 0.03 mg/kg without actually decreasing the total Se removed by the crop. A level of 0.03 mg/kg of Se and lower usually produces widespread white muscle disease.

Also, leaching a soil with sulfate salts will hasten the leaching losses of Se (Brown and Carter, 1969). With these effects, S fertilization, an essential practice for producing alfalfa hay and other legumes in the northwest states, frequently results in biologically significant reduced concentrations of Se in the forage produced.

In contrast to S fertilization, P fertilization frequently results in increased levels of Se in the forage (Carter et al., 1972). Rock phosphate frequently contains small amounts of Se, and P fertilization may increase Se uptake (Robbins and Carter, 1970).

21-11 RESEARCH DEVELOPMENT AND NEEDS

Today, farmers are making expert use of knowledge that took years to learn through research and development. Research effort in soil fertility must continue to provide answers to present and future crop production problems. The identification of specific nutrient deficiencies and establishing a basis for predicting response from fertilizer must remain in high priority for soil fertility-plant nutrition related research. At present, soil and plant analyses are the main tools for identifying deficiencies and predicting fertilizer responses, but refinements in those techniques are needed. Some fundamental studies on ion activities and concentrations are likely to improve soil test procedures. Minimum threshold levels of ion concentrations in soil solution need to be established for each crop throughout the growing season, and the nutrient release rates for soils need to be determined. Calibration of soil tests is a continuing process for assuring efficiency of the test employed as soils change under the impact of varied cropping and tillage practices, increased yield potential, and prior fertilizer treatments.

Soil fertility investigations in the future must be conducted by utilizing all crop cultural practices that can maximize crop yields. Many changing cultural practices are rendering some of today's knowledge on fertilizer use obsolete. Examples of needed research are given below.

21-11.1 Use of Fertilizer in Conservation Tillage

Conservation tillage is a management alternative currently gaining in popularity and slowly being adopted to reduce soil loss and conserve fuel on many agricultural soils in dryland farming regions throughout the West.

The reduced tillage and protective soil cover, however, are known to affect other soil processes, e.g., soil temperature, water infiltration rates, and the distribution of decayed plant material and applied fertilizer in the surface soil layer. Crop residues incorporated in the surface soil can immobilize N, can reduce nitrification rates, and might be phytotoxic to seedlings. Aeration is thought to change with a shift toward less oxidation and less mineralization of N with conservation tillage systems. Rooting density may also be altered, resulting in a soil plant system greatly altered from conventional tillage. The amount, placement, and timing of fertilizer application with these new systems must be determined.

21–11.2 Nitrogen Transformation and Movement in Soils and Improvement in Nitrogen Fertilizer Efficiency

Nitrogen fertilization is necessary for continued optimum crop production. Even though N is one of the most efficiently used plant nutrients, management can be vastly improved. Under fertilization results in low yields and restricts production. Overfertilization increases expenses without increasing returns, reduces yields, and poses a potential pollution hazard.

Most farmers in the semiarid region use less than optimum rates of N because of the uncertainties associated with rainfall and other factors. Field research must provide the information needed to predict optimum fertilization rates. Associated laboratory studies are also needed to better understand transformations and movements of N. These changes can cause unproductive losses of fertilizer N from soils. Research is a prerequisite to finding practices to minimize nutrient losses. In addition, research data concerning the use, transformation, and movement of N in soils provides assurance that only justified and rational measures will be instituted to protect the environment from excessive nitrates in water supplies.

21–11.3 Establishment of Secondary and Micronutrient Needs and Correction of Deficiencies

Previous research in soil fertility has emphasized N, P, and K. Less emphasis was placed on secondary nutrients and micronutrients. Yields of crops and forages have been increased by N, P, and K fertilization and superior cultivars. As a consequence a greater demand has been placed on soils to provide an adequate supply of other nutrients. Crop removal of available secondary nutrients and micronutrients, a decline in soil organic matter content, soil loss by erosion, and purer fertilizer materials have accentuated potential problems and increased the possibility that these elements may be limiting crop yields.

Intensified field research involving sources, rates, timing, and methods of application with different crops, management systems, and nutrients are vitally needed.

21-11.4 Effect of Fertilizer Use on Crop Diseases

Fertilizers can affect crop diseases. Even though the effect is generally beneficial, increased susceptibility of plants to diseases by fertilizer application has been reported. A luxuriant vegetative growth of wheat and other small grains resulting from fertilizer use can provide more favorable conditions for incubation of fungi and bacteria. However, optimum nutrition can also increase resistance of plants to diseases.

Huber et al. (1968) and Walker (1975) have shown that maintaining N in the ammonium form as long as possible will increase the percentage of NH_4-N taken up by a plant. Increased excretion of acidic ions accompanies uptake of N into the root rhizosphere; this increased uptake, plus increased acidification in the root rhizosphere, reduces the infection of take-all root rot. Younts and Musgrave (1958) showed that chloride can influence nutrient absorption and stalk rot in corn. Recent research by Powelson and Jackson (1978) has shown that the chloride ion can also influence the susceptibility of winter wheat to take-all root rot.

Keisling (1978) reported bacterial leaf spot in bermudagrass was reduced by increasing K rates. Potassium deficiency has also been reported to accentuate charcoal stalk rot in grain sorghum.

These examples indicate that the area of plant disease–plant nutrition relationships warrants expanded research.

21-11.5 Crop Quality

In a previous section, several effects of fertilizers on crop quality were discussed. The effects of increased N, P, and K seem to be well documented, but the effects of other essential nutrients on crop composition and nutritive value are not well understood. Measurement of quality factors are not well defined and require an integrated evaluation by scientists from a range of disciplines.

21-11.6 New Cultivars

As genetic yield capabilities are expanded, the continued need for fertilizer studies on new or improved cultivars is self-evident. Plant scientists are accumulating new information concerning the ability of genetic materials to better utilize specific nutrients. Iron-efficient lines of grain sorghum and soybean are now available. High protein hard red winter wheat lines are being utilized in developing improved cultivars. Aluminum-tolerant wheats, which endure very acid soils, have been identified. Possibilities for improved plant nutrition resulting from cooperative efforts between plant geneticists and plant nutritionists seem unlimited.

21-11.7 Energy Conservation

Fertilizer mining, processing, manufacturing, transportation, and application consumes considerable energy. As energy costs increase, fertilizer prices will probably increase. Some sources of fertilizers require less energy input than others. As new fertilizer materails are developed, they must be evaluated for efficiency and effectiveness both from a monetary and energy standpoint.

21-11.8 Impact of Fertilizer on Other Soil Processes

The long-term productivity of our soil resources depends on the maintenance of the soils in a desirable chemical, physical, and biological condition. Any deleterious effect can reduce crop production, whereas any improvement in soil processes can increase crop production and fertilizer efficiency.

Fertilizer salts, when used over a long period of time, can contribute to increased salinity. In addition, continued N use can result in potential soil acidity. Obviously, these factors need to be studied and predictive models developed.

The effects of continued, long-term fertilizer use on soil physical properties and biological characteristics of soils must be determined.

21-12 LITERATURE CITED

Adams, E. P., E. J. Williamson, P. Carson, and R. Gelderman. 1978. Fertilizing flax. USDA Cooperative Extension Service, USDA-FS 680, South Dakota State Univ., Brookings.

Ahring, R. M., C. M. Taliaferro, and L. G. Morrill. 1973. Effects of cultural and management practices on seed production of "Plains" bluestem. J. Range Manage. 26:143-146.

Allaway, W. H. 1971. Feed and food quality in relation to fertilizer use. p. 533-556. *In* R. A. Olson et al. (ed.) Fertilizer technology and use. Soil Sci. Soc. of Am., Madison, Wis.

Allaway, W. H. 1975. The effect of soils and fertilizers on human and animal nutrition. USDA-ARS-SCS and Cornell Univ. Agric. Exp. Stn. Information Bull. 378.

Bailey, L. D., E. D. Spratt, D. W. L. Read, F. G. Warder, and W. S. Ferguson. 1977. Residual effects of phosphorus fertilizer. II. For wheat and flax grown on Chernozemic soils in Manitoba. Can. J. Soil Sci. 57:263-270.

Baker, J. M., and B. B. Tucker. 1973. Critical N, P, and K levels in winter wheat. Commun. Soil Sci. Plant Anal. 4:347-358.

Baker, J. M., B. B. Tucker, and L. G. Morrill. 1970. Effects of sources of phosphorus under varying soil temperature and moisture regimes on the emergence of winter wheat. Soil Sci. Soc. Am. Proc. 34:694-697.

Bauer, A. 1968. Effects of zinc fertilizer on corn yield in southeastern N.D. N.D. Farm Res. 25(1):3-6.

Bauer, A., and H. L. Kucera. 1969. Influence of tillage method on incorporation of fertilizer phosphorus. N.D. Farm Res. 56(5):4-6.

Bauer, A., E. B. Norum, J. C. Zubriski, and R. A. Young. 1966. Fertilizer for small grain production on summer fallow in North Dakota. North Dakota Agric. Exp. Stn. Bull. no. 461. Rev. Jan. 1970. North Dakota State Univ., Fargo.

Bauer, A., and E. H. Vasey. 1964. Potassium fertilizer for barley. N.D. Farm Res. 23(4):19-22.

Bauer, A., E. H. Vasey, R. A. Young, and J. L. Oxbun. 1967. On non-fallowed soil—stored soil moisture best guide to nitrogen needed. N.D. Farm Res. 24(11):15-24.

Bauer, A., and L. Zimmerman. 1975. Alfalfa, harvested for hay, as a source of available nitrogen for wheat. North Dakota Agric. Exp. Stn. Res. Report no. 57.

Beaton, J. D. 1978. Urea: Its popularity grows as a dry source of nitrogen. Crops Soils 30(6): 11-14.

Berger, K. C., and E. Truog. 1939. Boron determination in soils and plants. Ind. Eng. Chem. Anal. Ed. 11:540-545.

Bergman, J. W., G. P. Hartman, and A. L. Black. 1975. Safflower production guidelines. Montana Agric. Exp. Stn., Capsule Information Series no. 8, Montana State Univ., Bozeman.

Black, A. L. 1969. Nitrogen topdressing on winter wheat? Check P needs first. Mont. Farmer-Stockman 56(14):12.

Black, A. L., and R. H. Ford. 1976. Available water and soil fertility relationships for annual cropping systems. p. 286-290. *In* A. Shaw et al. (ed.) Proc. Regional Saline-Seep Control Symp., Boseman, Mont. 9-11 Dec. 1975. Montana Cooperative Extension Service Bull. no. 1132. Montana State Univ., Bozeman.

Black, A. L., A. D. Halvorson, L. L. Reitz, and C. A. Reule. 1980. Spring wheat stand and yield losses from applying urea-N fertilizer with the seed. N.D. Farm Res. 37(3):8-12.

Black, A. L., and F. H. Siddoway. 1976. Dryland cropping sequences within a tall wheatgrass barrier system. J. Soil Water Conserv. 31:101-105.

Black, A. L., and F. H. Siddoway. 1977a. Winter wheat recropping on dryland as affected by stubble height and nitrogen fertilization. Soil Sci. Soc. Am. J. 41:1186-1190.

Black, A. L., and F. H. Siddoway. 1977b. Hard red and durum spring wheat responses to seeding date and NP-fertilization on fallow. Agron. J. 69:885-888.

Brown, B. D., and J. P. Jones. 1977. Nitrogen tissue test procedures for irrigated winter wheat. Agron. Abstr., p. 154.

Brown, M. J., and D. L. Carter. 1969. Leaching of added selinium from alkaline soils as influenced by sulfate. Soil Sci. Soc. Am. Proc. 33:563-565.

Brown, P. L. 1964. Legumes and grasses in dryland cropping systems in the northern and central Great Plains. USDA-ARS Miscellaneous Pub. no. 952.

Carter, D. L., C. W. Robbins, and M. J. Brown. 1972. Effect of phosphorus fertilization on the selenium concentration in alfalfa (*Medicago sativa*). Soil Sci. Soc. Am. Proc. 36: 624-627.

Cobia, D., and D. Zimmerman. 1975. Sunflowers, production, pests, and marketing. North Dakota State Univ. Extension Bull. no. 25.

Dahnke, W. C., and E. H. Vasey. 1973. Testing soils for nitrogen. p. 97-114. *In* L. M. Walsh and J. D. Beaton (ed.) Soil testing and plant analysis. Soil Sci. Soc. of Am., Madison, Wis.

Doering, E. J., and F. M. Sandoval. 1976a. Saline seep development on upland sites in the northern Great Plains. USDA Report no. ARS-NC-32.

Doering, E. J., and F. M. Sandoval. 1976b. Hydrology of saline seep development on upland sites in the northern Great Plains. Trans. ASAE 19:856-861, 865.

Doll, E. C., and R. E. Lucas. 1973. Testing soils for potassium, calcium and magnesium. p. 133-152. *In* L. M. Walsh and J. D. Beaton (ed.) Soil testing and plant analysis. Soil Sci. Soc. of Am., Madison, Wis.

Eck, H. V., V. L. Hauser, and R. H. Ford. 1965. Fertilizer needs for restoring productivity on Pullman silty clay loam after various degrees of soil removal. Soil Sci. Soc. Am. Proc. 29:209-213.

Eck, H. V., B. A. Stewart, and G. O. Boatwright. 1957. Wheat fertilization studies in western Oklahoma. Oklahoma Agric. Exp. Stn. Bull. no. B-493.

Elder, W. C., and B. B. Tucker. 1968. Pasture management: Report of progress on forage crop grazing studies and fertilizer tests. Oklahoma Agric. Exp. Stn. Processed Series no. P-583.

Ellington, C. P. 1978. Crop yield responses to phosphorus. p. 25–41. *In* Phosphorus for agriculture. Potash/Phosphate Inst., Atlanta.

Elwell, H. M., and H. A. Daniel. 1953. Native grass fertilization. Oklahoma Agric. Exp. Stn. Progress Report no. M-249.

Fenn, L. B. 1969. Residual effect of heavy phosphorus fertilization with and without zinc on grain sorghum yields. Texas Agric. Exp. Stn. Progress Report no. 2663.

Fenn, L. B., and D. E. Kissel. 1973. Ammonia volatilization from surface applications of ammonium compounds on calcareous soils: I. General theory. Soil Sci. Soc. Am. Proc. 37: 855–859.

Fenn, L. B., and D. E. Kissel. 1975. Ammonium volatilization from surface applications of ammonium compounds on calcareous soils: IV. Effect of calcium carbonate content. Soil Sci. Soc. Am. Proc. 39:631–633.

Finney, K. F., and W. T. Yamazaki. 1967. Quality of hard, soft, and durum wheats. *In* K. S. Quisenberry and L. P. Reitz (ed.) Wheat and wheat improvement. Agronomy 13:471–504. Am. Soc. of Agron., Madison, Wis.

Gardner, E. H., T. W. Thompson, H. Ken, J. Hesketh, F. Hagelstein, M. Zimmerman, and G. Cook. 1975. The nitrogen and sulfur fertilization of dryland wheat in Oregon's Columbia Plateau. p. 99–104. *In* Proc. 26th Annual Pacific Northwest Fertilizer Conference, Salt Lake City. 15–17 July. Pacific Northwest Plant Food Assoc., Portland, Ore.

Gelderman, R., E. Adams, E. Williamson, and P. Carson. 1977. Fertilizer guide for soil test—computerized recommendations, June, 1977. South Dakota Agric. Exp. Stn. Plant Sci. Pamphlet no. 34.

Grable, A. R., F. M. Willhite, and W. L. McCuistion. 1965. Hay production and nutrient uptake at high altitudes in Colorado with different grasses in conjunction with alsike clover and nitrogen fertilization. Agron. J. 57:543–547.

Grim, R. E. 1953. Clay mineralogy. McGraw-Hill Publications, New York.

Grunes, D. L. 1959. Effect of nitrogen on the availability of soil and fertilizer phosphorus to plants. Adv. Agron. 11:369–396.

Haas, H. J., and C. E. Evans. 1957. Nitrogen and carbon changes in Great Plains soils as influenced by cropping and soil treatments. USDA Tech. Bull. no. 1146.

Haas, H. J., J. F. Power, and G. A. Reichman. 1976. Effect of crops and fertilizer on soil nitrogen, carbon, and water content, and on succeeding wheat yields and quality. USDA Report no. ARS-NC-38.

Halvorson, A. D., and A. L. Black. 1974. Saline-seep development in dryland soils of northeastern Montana. J. Soil Water Conserv. 29:77–81.

Halvorson, A. D., A. L. Black, F. Sobolik, and N. Riveland. 1976. Proper management: Key to successful winter wheat recropping in northern Great Plains. N.D. Farm Res. 33(4):3–9.

Halvorson, A. D., and P. O. Kresge. 1982. FLEXCROP: A dryland cropping systems model. USDA-ARS Production Res. Report no. 180.

Halvorson, A. R. 1970. Current status of soil test standardization in the five northwest states. p. 135–139. *In* Proc. 21st Annual Pacific Northwest Fertilizer Conference, Salt Lake City. 14–16 July. Pacific Northwest Plant Food Assoc., Portland, Ore.

Harper, H. J. 1957. Effects of fertilization and climatic conditions on prairie hay. Oklahoma Agric. Exp. Stn. Bull. no. B-492.

Harvey, C., D. L. Jones, and C. E. Fisher. 1961. Dryland crop rotations on the southern high plains of Texas. Texas Agric. Exp. Stn. no. MP-544.

Horner, G. M., M. M. Oveson, G. O. Baker, and W. W. Pawson. 1960. Effect of cropping practices on yield, soil organic matter and erosion in the Pacific Northwest wheat region. USDA-ARS and Agric. Exp. Stn. of Idaho, Oregon, and Washington, Bull. no. 1.

Houston, W. R. 1957. Renovation and fertilization of crested wheatgrass stands in the northern Great Plains. J. Range Manage. 10:9–11.

Huber, D. M., C. G. Painter, H. C. McKay, and D. L. Peterson. 1968. Effect of nitrogen fertilization on take-all of winter wheat. Phytopathology 58:1470–1472.

Jackson, T. L., and G. E. Carter. 1976. Nutrient uptake by Russet Burbank potatoes as influenced by fertilization. Agron. J. 68:9–12.

Jackson, T. L., S. Vacharotayan, L. A. Alban, and R. Peterson. 1964. Phosphorus availability in reddish brown lateritic soils. II. Relationships between field, greenhouse, and soil analyses. Agron. J. 56:558–560.

Jensen, L. A., and V. L. Weiser. 1971. North Dakota crop rotations for profit. USDA Cooperative Extension Service Bull. no. 14, North Dakota State Univ., Fargo.

Johnson, G. V., and B. B. Tucker. 1979. Fertilizer recommendation guides. Oklahoma Agric. Extension Fact Sheet no. 2225.

Jones, U. S., and C. E. Bardsley. 1965. Phosphorus nutrition. p. 213–254. *In* F. C. Elliot et al. (ed.) Advances in production and utilization of quality cotton: Principles and practices. Iowa State Univ. Press, Ames.

Keisling, T. 1978. Potassium helps maintain stand, prevent winter kill, promote strong root growth in coastal bermudagrass. Better Crops Plant Foods 62(Summer):10–12.

Lamborn, R. E. 1970. Variability in subsoil phosphorus reserves in Utah soils. p. 140–145. *In* Proc. 21st Annual Pacific Northwest Fertilizer Conference, Salt Lake City. 14–16 July. Pacific Northwest Plant Food Assoc., Portland, Ore.

Leggett, G. E. 1959. Relationships between wheat yield, available moisture and available nitrogen in eastern Washington dry land areas. Washington Agric. Exp. Stn. Bull. no. 609.

Leggett, G. E., H. M. Reisenauer, and W. L. Nelson. 1959. Fertilization of dry land wheat in eastern Washington. Washington Agric. Exp. Stn. Bull. no. 602.

Lindsay, W. L. 1972. Inorganic phase equilibria of micronutrients in soils. p. 41–58. *In* J. J. Mortvedt et al. (ed.) Micronutrients in agriculture. Soil Sci. Soc. of Am., Madison, Wis.

Lindsay, W. L., and E. C. Moreno. 1960. Phosphate phase equilibria in soils. Soil Sci. Soc. Am. Proc. 24:177–182.

Lindsay, W. L., and W. A. Norvell. 1978. Development of a DTPA soil test for zinc, iron, manganese, and copper. Soil Sci. Soc. Am. J. 42:421–428.

Lorenz, R. J., and G. R. Rogler. 1973. Interaction of fertility level with harvest date and frequency on productiveness of mixed prairie. J. Range Manage. 26:50–54.

Luebs, R. E. 1962. Investigations of cropping systems, tillage methods, and cultural practices for dryland farming at the Fort Hays Branch Exp. Stn. Kansas Agric. Exp. Stn. Bull. no. 449.

McGuire, C. F., J. R. Sims, F. H. McNeal, and P. L. Brown. 1974. Fertilizing Montana wheats to improve grain yield and milling and baking quality. Montana Agric. Exp. Stn. Bull. no. 674. Montana State Univ., Bozeman.

McLean, G. W., and G. A. Niles. 1968. Synergistic effect of irrigation, fertilization, micronutrients, and starter fertilizer on yield and quality of cotton grown on Blackland prairie soil. Texas Agric. Exp. Stn. Res. Report—1968, PR-2571.

Miller, M. H. 1974. Effects of nitrogen on phosphorus absorption by plants. p. 643–668. *In* E. W. Carson (ed.) The plant root and its environment. Univ. Press of Virginia, Charlottesville.

Moraghan, J. T. 1978. Chlorotic dieback in flax. Agron. J. 70:501–505.

Mortvedt, J. J. 1976. Band fertilizer placement—how much and how close? Fert. Solutions 20(6):90–96.

Moser, L. S., and K. L. Anderson. 1965. Nitrogen and phosphorus fertilization of bluestem range. Trans. Kans. Acad. Sci. 67:613–616.

Murphy, H. F., and B. B. Tucker. 1969. Long-term fertilizer studies on Kirkland-Bethany silt loam. Oklahoma Agric. Exp. Stn. Bull. no. B-667.

Neilson, R. F., and R. E. Lamborn. 1970. A progress report on the phosphorus status of dryland soils in northern Utah. p. 146–149. *In* Proc. 21st Annual Pacific Northwest Fertilizer Conference, Salt Lake City. 14–16 July. Pacific Northwest Plant Food Assoc., Portland, Ore.

Olsen, S. R. 1953. Inorganic phosphorus in alkaline and calcareous soils. *In* W. H. Pierre and A. G. Norman (ed.) Soil and fertilizer phosphorus. Agronomy 4:153–188. Am. Soc. of Agron., Madison, Wis.

Olsen, S. R., C. V. Vole, F. S. Watanabe, and L. A. Dean. 1954. Estimation of available phosphorus in soils by extraction with sodium bicarbonate. USDA Circ. no. 939.

Olson, R. A., and F. E. Koehler. 1968. Fertilizer use on small grains. p. 253–272. *In* L. B. Nelson et al. (ed.) Changing patterns in fertilizer use. Soil Sci. Soc. of Am., Madison, Wis.

Ozanne, P. G., and K. M. W. Howes. 1970. The effects of grazing on the phosphorus requirement of an annual pasture. Aust. J. Agric. Res. 22:81–92.

Ozanne, P. G., and K. M. W. Howes. 1971. Preference of grazing sheep for pasture of high phosphate content. Aust. J. Agric. Res. 22:941–950.

Ozanne, P. G., D. B. Purser, K. M. W. Howes, and I. Southey. 1976. Influence of phosphorus content on feed intake and weight gain in sheep. Aust. J. Exp. Agric. Anim. Husb. 16:353–360.

Pairintra, C. 1973. Influence of NH_3 from ammonium phosphate fertilizers on germination, seedling growth, and small plant yield of wheat (*Triticum aestivum* L.). Ph.D. Dissertation. Montana State Univ., Bozeman. (Diss. Abstr. 73:27,490.)

Pittman, U. J. 1977. Crop yields and soil fertility as affected by dryland rotations in southern Alberta. Commun. Soil Sci. Plant Anal. 8:391–405.

Powelson, R. L., and T. L. Jackson. 1978. Suppression of take-all (*Gaeumanomyces graminis*) root rot of wheat with fall applied chloride fertilizers. *In* Proc. 29th Annual Pacific Northwest Fertilizer Conference, Beaverton, Ore. 11–13 July. Pacific Northwest Plant Food Assoc., Portland, Ore.

Power, J. F. 1970. Leaching of nitrate-nitrogen under dryland agriculture in the northern Great Plains. p. 111–112. *In* Relationship of agriculture to soil and water pollution. Cornell Univ. Press, Ithaca, N.Y.

Power, J. F. 1972. Fate of fertilizer nitrogen applied to a northern plains rangeland ecosystem. J. Range Manage. 25:367–371.

Power, J. F. 1974. Urea as a nitrogen fertilizer for Great Plains grasslands. J. Range Manage. 27:161–164.

Power, J. F., P. L. Brown, T. J. Army, and M. G. Klages. 1961. Phosphorus responses by dryland spring wheat as influenced by moisture supplies. Agron. J. 53:106–108.

Pumphrey, F. V., and D. P. Moore. 1965a. Sulfur and nitrogen content of alfalfa herbage during growth. Agron. J. 57:237–239.

Pumphrey, F. V., and D. P. Moore. 1965b. Diagnosing sulfur deficiency of alfalfa (*Medicago sativa* L.) from plant analysis. Agron. J. 57:364–366.

Ramig, R. E., P. E. Rasmussen, R. R. Allmaras, and C. M. Smith. 1975. Nitrogen-sulfur relations in soft white winter wheat. I. Yield response to fertilizer and residual sulfur. Agron. J. 67:219–224.

Read, D. W. L., E. D. Spratt, L. D. Bailey, and F. G. Warder. 1977. Residual effects of phosphorus fertilizer. I. For wheat grown on four Chernozemic soil types in Saskatchewan and Manitoba. Can. J. Soil Sci. 57:255–262.

Reid, R. L., and G. A. Jung. 1974. Effects of elements other than nitrogen on the nutritive value of forage. p. 395–435. *In* D. A. Mays (ed.) Forage fertilization. Am. Soc. of Agron., Madison, Wis.

Reisenauer, H. M. 1956. Molybdenum content of alfalfa in relation to deficiency symptoms and response to molybdenum fertilization. Soil Sci. 81:237–242.

Reisenauer, H. M., L. M. Walsh, and R. G. Hoeft. 1973. Testing soils for sulfur, boron, molybdenum, and chlorine. p. 173–200. *In* L. M. Walsh and J. D. Beaton (ed.) Soil testing and plant analysis. Soil Sci. Soc. of Am., Madison, Wis.

Richards, G. E. (ed.). 1977. Band application of phosphatic fertilizers. Agric. Products Dep., Olin Corp., Little Rock, Ark.

Robbins, C. W., and D. L. Carter. 1970. Selenium concentrations in phosphorus fertilizer materials and associated uptake by plants. Soil Sci. Soc. Am. Proc. 34:505–509.

Rogler, G. A., and R. J. Lorenz. 1969. Pasture productivity of crested wheatgrass as influenced by nitrogen fertilization and alfalfa. USDA-ARS and North Dakota Agric. Exp. Stn. Tech. Bull. no. 1402.

Russell, G. C., A. D. Smith, and U. J. Pittman. 1958. The effect of nitrogen and phosphorus fertilizers on the yield and protein content of spring wheat grown on stubble fields in southern Alberta. Can. J. Plant Sci. 38:139–144.

Saskatchewan Advisory Council on Soils. 1977. Saskatchewan fertilizer and cropping practices. Published by authority of Minister of Agric. Hon. Edgar Kaeding.

Seats, L. F., and C. O. Stanberry. 1963. Advances in phosphate fertilization. p. 155–188. *In* M. H. McVickar et al. (ed.) Fertilizer technology and usage. Soil Sci. Soc. of Am., Madison, Wis.

Sievers, F. J., and H. F. Holtz. 1922. The silt loam soils of eastern Washington. Bull. Wash. Agric. Exp. Stn. 116:1–62.

Sievers, F. J., and H. F. Holtz. 1926. The significance of nitrogen in soil organic matter relationships. Bull. Wash. Agric. Exp. Stn. 206:1–43.

Snider, A. E., A. Bauer, E. B. Norum. 1968. Growing season precipitation probabilities in North Dakota. USDA Cooperative Extension Bull. no. 4, North Dakota State Univ., Fargo, p. 1–26.

Staten, H. W., and V. G. Heller. 1949. Winter pasture for more feed and better feed at lower costs. Oklahoma Agric. Exp. Stn. Bull. no. 333.

Tharp, W. H. 1960. The cotton plant—how it grows. USDA Handbook no. 178.

Thomas, G. W., and D. E. Peaslee. 1973. Testing soils for phosphorus. p. 115–132. *In* L. M. Walsh and J. D. Beaton (ed.) Soil testing and plant analysis. Soil Sci. Soc. of Am., Madison, Wis.

Thompson, C. A. 1969. Grain sorghum fertilization research in Kansas fertilizer research progress report—1970. Mimeo Report, Dep. of Agron., Kansas State Univ., Manhattan, p. 5–26.

Tisdale, S. L., and W. L. Nelson. 1975. Soil fertility and fertilizers. ed. 3. MacMillan Inc., New York.

Tucker, B. B. (ed.). 1977. Oklahoma soil fertility handbook. Cooperative Extension Service, Oklahoma State Univ., and Oklahoma Plant Food Educational Soc., Inc., Stillwater.

Tucker, B. B., and W. F. Bennett. 1968. Fertilizer use on grain sorghum. p. 189–220. *In* L. B. Nelson et al. (ed.) Changing patterns in fertilizer use. Am. Soc. of Agron., Inc., Madison, Wis.

Tucker, B. B., M. B. Cox, and H. V. Eck. 1971. Effect of rotations, tillage methods, and N fertilization on winter wheat production. Agron. J. 63:699–702.

Tucker, T. C. 1965. The cotton petiole, guide to better fertilization. Plant Food Rev. 11:9–11.

Tucker, T. C., and B. B. Tucker. 1968. Nitrogen nutrition. p. 183–211. *In* Advances in production and utilization of quality cotton: Principles and practices. Iowa State Univ. Press, Ames.

U.S. Department of Agriculture, Agricultural Research Service. 1974. Summer fallow in the western United States. USDA-ARS Conserv. Res. Report no. 17.

Viets, F. G., Jr. 1971a. Fertilizer use in relation to surface and groundwater pollution. p. 517–532. *In* R. A. Olson et al. (ed.) Fertilizer technology and use. ed. 2. Soil Sci. Soc. of Am., Madison, Wis.

Viets, F. G., Jr. 1971b. Water quality in relation to farm use of fertilizer. BioScience 21:460–466.

Viets, F. G., Jr., L. C. Boawn, and C. L. Crawford. 1957. The effect of nitrogen and types of nitrogen carrier on plant uptake of indigenous and applied zinc. Soil Sci. Soc. Am. Proc. 21:197–201.

Viets, F. G., Jr., and W. L. Lindsay. 1973. Testing soils for zinc, copper, manganese and iron. p. 153–172. *In* L. M. Walsh and J. D. Beaton (ed.) Soil testing and plant analyses. Soil Sci. Soc. of Am., Madison, Wis.

Wagner, D. F., and W. C. Dahnke. 1977. NDSU soil testing programs. p. 54–81. *In* The 1978 crop production guide. North Dakota Agric. Assoc. and North Dakota State Univ., Fargo.

Wagner, D. F., and E. H. Vasey. 1970. Refining fertilizer recommendations by state areas, stored moisture conditions and rainfall probabilities. USDA Cooperative Extension Service, North Dakota State Univ., Fargo.

Walker, J. 1975. Take-all diseases of gramineae: A review of recent work. Rev. Plant Pathol. 54:113–144.

Walsh, L. M., and J. D. Beaton (ed.) 1973. Soil testing and plant analysis. Soil Sci. Soc. of Am., Madison, Wis.

Ward, R. C., and P. L. Carson. 1975. Soil fertility levels of South Dakota soils: A summary of soil tests. South Dakota Agric. Exp. Stn. Bull. no. B-624.

Westermann, D. T., and C. W. Robbins. 1974. Effect of SO_4-S fertilization on Se concentrations of alfalfa (*Medicago sativa* L.). Agron. J. 66:207–208.

Whitney, D., and L. F. Murphy. 1974. Potassium fertilization and lodging. Better Crops Plant Food 58(3):22–26.

Wight, J. R., and A. L. Black. 1979. Range fertilization: Plant response and water use. J. Range Manage. 32:345–349.

Williamson, E. J., P. Carson, R. Gelderman, J. Goos, and B. Nettleton. 1977. Fertilizing big bluestem grass. Eastern South Dakota Agric. Res. Progress Report 1978.

Younts, S. E., and R. B. Musgrave. 1958. Chemical composition, nutrient absorption, and stalk rot of incidence of corn as affected by chloride in potassium fertilizers. Agron. J. 50:426–429.

22

Jerry D. Eastin
University of Nebraska
Lincoln, Nebraska

T. E. Dickinson
DeKalb-Pfizer Genetics, Inc.
DeKalb, Illinois

D. R. Krieg
Texas Tech University
Lubbock, Texas

A. Bruce Maunder
DeKalb-Pfizer Genetics, Inc.
Lubbock, Texas

Crop Physiology in Dryland Agriculture

Dryland agriculturalists, both producers and researchers, naturally place heavy emphasis on water deficit problems. Production problems, however, arise from a number of additional environmental and climatic factors, such as transient temperatures, season length (dictated by heat units and available water), wind (advected heat and cold), altitude, latitude, and radiation. Essentially, these factors are either water and temperature dependent or can be analyzed in terms of their effects on crop production. For example, altitude and latitude both influence temperature and water availability. Winds are both created by temperature differentials and are responsible for large quantities of advected energy that may either favorably or adversely influence crop production. Rarely, if ever, can water effects on production be separated from temperature effects. Therefore, mainly water

Contribution of the Nebraska Agricultural Experiment Station. Published as Paper no. 7208, Journal Series, Nebraska Agricultural Experiment Station, Lincoln, Nebraska.

and temperature will be considered in this chapter on crop physiology and plant improvement in dryland agriculture. Photoperiod effects will rarely be discussed.

Achieving an understanding of how plants avoid or tolerate the environmental rigors of dryland agriculture is important. However, research information remains of limited value unless mechanisms of stress resistance can be incorporated into production systems through cultural/genetic means (the two are essentially inseparable). For this reason a genetic–plant breeding emphasis must be pursued along with a discussion of physiology.

Furthermore, acquiring a basic understanding of limiting physiological and developmental processes usually requires the researcher to manipulate and use divergent genetic materials expressing physiologically related characteristics. Closer interaction between physiological and genetic research should permit substantially faster gains in the rate of plant improvement. Such interfaces are poorly developed, probably because of our inability to state, in any given environment, the sequence of the most limiting physiological processes for grain yield. Such inability is not surprising in view of the many complex processes bearing on yield. The fact that one cannot usually make positive correlation between grain yield and so primary a process as photosynthesis—as evidenced by the flurry of photosynthetic research in the 1960's and 1970's—suggests that we should reconsider the kind of research to undertake.

22-1 YIELD AND WATER USE EFFICIENCY

Agronomists often express water use efficiency as the ratio of yield (Y) to evapotranspiration (ET) in field experiments. The WUE in greenhouse and growth chamber experiments may be grams of dry matter produced per liter of water (or other suitable dimensions). Short-time WUE evaluations on any plant can be expressed as net CO_2 uptake (mass) per unit mass of water transpired (Arnon, 1975; Rawson et al., 1978). Some authors talk of transpiration ratio, which is the inverse of WUE, and others mention transpiration/photosynthesis (Slatyer, 1964). Generally we use the definition

$$\mathrm{WUE} = \mathrm{Y}/\mathrm{ET}.$$

Obviously WUE may be increased by increasing Y, holding ET constant, reducing ET, and increasing ET at a slower rate than the increase in Y. Alternatively, WUE rises if Y can be held constant or reduced at a slower rate than ET is reduced. In crop plants WUE frequently is increased by increasing yield without increasing ET. The objective is to consider the influences of water and temperature on selected essential physiological and developmental processes that bear on (i) yields and (ii) evapotranspiration. The literature is voluminous, and because of space limitations, citations in this chapter will be selective.

22-2 RELEVANT DEVELOPMENT AND PHYSIOLOGICAL CONSIDERATIONS

We mentioned earlier our inability to determine the order of yield-limiting physiological processes in the production of cereal grain or grain legumes. Despite our general knowledge and recognition of the complexity of dry matter production, much research has been designed to relate photosynthesis or another primary physiological process to grain yields. Such research usually is not very informative (although photosynthesis, for example, may correlate well with forage or total dry matter production). A more realistic approach might be to (i) analyze a particular grain yield in terms of seed size and seed number, (ii) consider plant developmental processes that limit those components, and (iii) try to evaluate primary physiological processes during the limiting developmental sequences (Eastin et al., 1983). This approach accomplishes two things. First, it greatly shortens the time spans (critical periods) over which intensive physiological analyses need to be done. Second, it puts the thinking of the physiologist in common with that of the plant breeder who is used to analyzing yield limitations in terms of the seed size and seed number.

22-2.1 Components of Yield

Examples of water and temperature effects on yield components of crops suited to the Great Plains will be given for cereal grains (chiefly sorghum and wheat), soybean as a grain legume, and cotton where fiber rather than grain is the economic product. To develop examples requires determining the most stress-sensitive periods in a crop's life cycle. Salter and Goode (1967) review earlier references on sensitivities of crop growth stages.

Seed number per unit land area usually correlates more positively with grain yield than does seed size (weight), as for sorghum (Kambal and Webster, 1966; Stickler et al., 1961; Blum, 1967, 1970; Quinby, 1963; Doggett and Jowett, 1967; Beil and Atkins, 1967; Fischer and Wilson, 1975; Eastin and Sullivan, 1974; Ogunlela, 1979). Warrington et al. (1977) and Fischer (1980) refer to a portion of similar literature for wheat. Seed size appears to be a valuable yield stability component under marginal conditions. Plant development will be considered as it relates to seed number and size.

22-2.2 Crop Development

22-2.2.1 Wheat

Warrington et al. (1977) further defined sensitive periods in wheat by manipulating temperature during three growth stages (Table 22-1). Growth stage 1 (GS1) is the vegetative stage extending from planting to inflores-

Table 22-1. Sensitivity of wheat to temperature during three growth stages.

Growth stage	Number of grains/ear Day temperature (°C)†		
	15	20	25
GS1	57.8	59.8	56.8
GS2	70.5	59.8	19.5
GS3	57.2	59.8	48.2

† Night temperature is 5°C lower.

cence initiation. Growth stage 2 (GS2) is the inflorescence development period ending at bloom. Growth stage 3 (GS3) is the grain-filling period ending at physiological maturity. Day/night temperatures used were 25/20, 20/15, and 15/10°C. Plants grown at each temperature regime in GS1 were subsequently subdivided and grown at the same temperatures during GS2. The same system was used to test for temperature effects during GS3. Total grain number per main stem ear is given below for each growth stage held at 15/10, 20/15 and 25/20°C when the two other growth stages were held at 20/15°C. The 15/10, 20/15, and 25/20°C treatments are listed by day temperature only (i.e., 15, 20, and 25°C). Temperatures during GS1 had little influence on seed number. Temperature effects during GS2 were pronounced, with the 15/10°C regime giving the highest seed number produced. A 15% reduction occurred at 20/15°C and a 72% reduction at 25/20°C. The respective lengths of GS2 at the 15, 20, and 25°C day temperatures were 62, 37, and 29 days. Respective rates of main stem kernel production at 15, 20, and 25°C were 1.14, 1.62, and 0.67, indicating that production at 20/15°C was the most efficient. The seed number was higher at 15/10°C than at 20/15°C, but the 18% extra seeds required 40% more growing time to develop.

An excellent field study by Cackett and Wall (1971) illustrates similar crop response in natural field environments. They grew irrigated field wheat at similar latitudes but at altitudes of 438 m (mean) and 1505 m, which resulted in an average lower daily mean temperature of 3.5°C at the higher elevation over 2 years (19.7 vs. 16.2°C). GS2 covered 71 days, producing 15.5 fertile spikelets per plant at the high elevation compared with 50 days producing 11.8 spikelets at the low elevation. Thus, a 42% longer GS2 produced 31% more fertile spikelets associated with an 86% higher yield. Average grain fill lengths were 55 and 48 days for the low and high elevations. Respective average high and low elevation yields were 8020 and 4332 kg/ha of which 60% of the difference was attributed to grain number and 40% to grain weight.

Concerning water stress, Fischer (1973) noted about a 25% grain number reduction when xylem water tension reached about 1.5 MPa at 5 to 15 days before ear emergence. Wheat is highly sensitive to both temperature and water stress during spike development.

Wheat is well adapted to the Great Plains climate primarily because of its growth habit. The plant avoids the worst stress by maturing just as the hottest and driest period usually begins. Because sorghum, by contrast, is planted in spring and grows through the hot summer, its growth habit and resistance to suboptimal water and temperature must be greater. The temperature optima for sorghum is generally on the order of 5°C or more higher (Eastin, 1976) than for wheat (Warrington et al., 1977). While direct comparisons appear unavailable, their root systems probably differ in extension and water scavenging characteristics if for no other reason owing to different temperature optima.

22-2.2.2 Sorghum

The GS1, GS2, and GS3 growth stages reported by Eastin (1972a) will be used here. Terminology is the same as for wheat. Lee et al. (1974) reviewed literature on inflorescence development and characterized the chronology of events during GS2 because that period was obviously critical in a number of cereals (Salter and Goode, 1967; Langer, 1967). Also in 1970, field experiments began on three height line isogenes; the work related to both panicle development (seed number) and seed size (Eastin, 1972a). Three isogenic height hybrids (2 × 3-, 3 × 3- and 4 × 3-dwarf RS 626 hybrids) grown under irrigation yielded 17 940, 15 080, and 14 501 kg/ha total dry matter, respectively. Grain yields did not differ significantly, but the tall 2 × 3-dwarf hybrid had a 29.5% larger seed size than the others, indicating that seed number per unit land area was about 29.5% smaller. This difference was not expected because the hybrids seemed to possess the same genetic potential for seed size. The difference in seed number apparently resulted from the difference in total dry matter production (16 and 19% higher vegetative dry matter in the 2 × 3-dwarf over the 3 × 3 and 3 × 4) between panicle initiation and bloom. Presumably, competition between simultaneously expanding vegetative and floral parts for available assimilates resulted in a reduced seed number in the tall (2 × 3-dwarf) hybrid. Grain yield, however, was not reduced because the 2 × 3-dwarf was able to compensate by increasing seed size. Obviously seed size in sorghum is potentially larger than is normally realized.

The experiment on isogenic height hybrids underscores the reality of competition between simultaneously expanding floral and vegetative parts during GS2. This competition likely shifts more in favor of vegetative expansion (thereby reducing seed number potential) as the environment becomes suboptimal in the drier and hotter dryland. Fischer (1980) reviewed wheat literature in which the higher yielding Mexican wheats are reputed to yield more by virtue of higher seed number related to shorter plant stature (presumably lower vegetative production).

Castleberry (1973) conducted experiments bearing on the stage sensitivity of sorghum during GS2 and on the capacity of seed size compensation. Three genotypes were adjusted to stands of 0.25 million plants/ha; the stands were subsequently thinned to 0.167 million plants/ha at about weekly intervals beginning about 2 weeks after emergence. Total light interception was about the same for all stands for most of the season; therefore, thinning increased the amount of light available per plant. Plants were able to use the extra light energy to increase seed number per plant sufficiently to avoid grain yield losses from thinning up to about mid-GS2, which coincides with differentiation and expansion of pistil and stamen primordia. Thinning beyond that time reduced yield, because seed size did not fully compensate for reduced seed number per unit land area. From these data, the floret differentiation (FD) stage appears critical, and the plants seem readily able to use extra assimilates for setting a high seed number up to that time. Among genotypes, however, marked differences exist in capacities to partition assimilate to developing florets.

While the Castleberry (1973) experiments were being run, Eastin (1976) extended the work of Downes (1972) in evaluating temperature effects on grain sorghum. Downes tested temperate, intermediate, and cool tolerant sorghums beginning at panicle initiation (PI) with day/night combinations of 29/17, 29/22, 29/37, and 34/22°C. The first significant point is that all hybrid yields were reduced 25 to 36% when night temperatures were raised about 5°C above their optimum and at 10°C above optimum yields were reduced about 50%. The yield reductions were usually paralleled by reductions in seed number. This result implicates a deleterious effect of temperature during GS2 because floret abortion after bloom was not observed. The second point is that the night temperature optima of the temperate hybrid (29/22°C) was about the same as that of a cool tolerant hybrid (29/17°C) (Eastin, 1976, 1980).

These data point to a fairly high sensitivity in sorghum, as in wheat, to moderately suboptimal temperatures when seed is being set. The temperature optima for wheats appear to be at least 5°C lower than for sorghums as a group. It is clear that temperature optima within sorghums vary considerably, so selection for genotypes to fit divergent temperatures is feasible. Unfortunately, rapid selection techniques for temperature typing are not available.

Growth room work is valuable, but one should supplement it with field observations when possible. Ogunlela (1979), using special equipment designed to control night temperature in the field at fixed amounts, followed temperature effects during GS2 and briefly after bloom. His treatments of ambient +5°C were carried for 1-week successive intervals during GS2 plus the first week after bloom for 2 years. The RS 671 sorghum was the most sensitive near the differentiation of floret components (pistil and stamens) and their subsequent expansion. He showed a 28% yield reduction for a modest 5°C above ambient temperatures. In all treatments the yield reductions were closely paralleled by both seed number reductions and grams of grain weight per GS3 day per plant (production efficiency). Gipson et al. (1979) also reported reduced yield from elevated night temperatures.

Monitoring temperature effects on seed number is relatively easy compared with evaluating water stress effects on seed number. Whiteman and Wilson (1965) reported on drought effects on sorghum through delayed flowering. Hultquist (1973) evaluated drought effects on DeKalb C-42Y (considered drought resistant) and RS 626 (normal) sorghums in greenhouse pots. Both hybrids were sensitive to water stress but reacted differently to the severe stress imposed at peduncle elongation (about FD). When RS 626 was stressed severely enough to kill a portion of its expanding panicle (in the boot), the partial panicle was exerted on relief of the stress. Also nodal and basal tillers appeared. By contrast, when C-42Y was stressed severely enough to kill any portion of the panicle, the entire plant died. Photosynthetically fixed $^{14}CO_2$ transport in the two hybrids was different. The C-42Y hybrid, compared with RS 626, translocated a higher proportion of ^{14}C exported from the fed leaf to the developing panicle and a lower proportion to the lower nodes. These contrasting characteristics may at first appear incongruous in that C-42Y is considered more drought resistant in the portion of the Great Plains from south central Nebraska extending south through the Texas Panhandle. Climatic characteristics of the region must be considered.

Unpredictable, intermittant rainfall is normal. However, the level of stress imposed by Hultquist is seldom realized naturally in the Great Plains even though water stress may be common during the season. The ability of C-42Y to partition a greater portion of available assimilates to the developing panicle probably relates to its superior yielding ability. In the chronology of sorghum hybrid development in the United States, RS 610 (RS 626 is a later sister hybrid differing mostly in head smut resistance) was one of the first two or three hybrids produced widely, and C-42Y was developed several years later. The RS 626 type germplasm appears to have a growth habit more closely tuned to survival than to maximum production as evidenced by (i) ability to recover from stress by producing small amounts of seed from partially damaged panicles and (ii) translocating greater portions of available assimilate under stress to lower nodes and roots, such translocation permitting both basal and modal tillering once stress is removed. The survival habit of RS 626 may be more similar to that of sorghums in some of their native or wild habitats. Probably there are production areas where the growth habit of RS 626 may be the more desirable than C-42Y. Also, the type or level of drought or heat resistance of a genotype best suited for one area is not likely to be the same for all areas.

Lewis et al. (1974) used selective irrigation to demonstrate high sensitivity to water stress at the boot stage. The results agreed with Hultquist's (1973) results (see also Eastin, 1980; Krieg, 1983).

Bennett (1979) applied water stress treatments gradually to water stress sorghum using osmotica in nutrient solutions in two experiments. In this first experiment a 0.65-MPa stress for 9 days at panicle development reduced seed number 47%, grain yield 39%, and photosynthesis 27%. The 0.4-MPa stress had no significant yield effects. Bloom stress of 0.5 MPa had little effect, but 0.75 MPa reduced photosynthesis about 85% and yield 33% (28% seed number reduction). Water use efficiency of total dry matter production was reduced 10% in the 0.65-MPa treatment during panicle development. Reduction in WUE for the 0.75-MPA bloom treatment was 9%.

Their second experiment used a preconditioning treatment of 0.6 MPa beginning on 10-day-old plants followed by stress at panicle development

(0.65 MPa) and at full bloom (0.75 MPa). The 0.65-MPa panicle treatment reduced photosynthesis an average of 19.5% on nonprestressed plants and only 5% on prestressed plants. The average panicle yield and seed number reductions were 54 and 42%, respectively. A yield reduction trend (not significant) was noted at the 0.75 MPa bloom treatment, but photosynthesis was not reduced significantly. The absence of deleterious effects may be attributed to a decrease in osmotic potential as leaf water potential decreased, but turgor potential remained about the same. These grain yield–seed number reduction results are consistent with others reported previously.

The seed number component of yield has been emphasized with only incidental reference to seed size (weight). Seed weight may be relatively more important in semiarid than in subhumid environments. Once the environmental stress have been exerted on processes that fix seed number potential, the only way plants can compensate to a return to favorable water and temperature conditions is through an increase in seed size. Therefore, seed size potential is critical in semiarid regions.

Reference to sorghum and wheat suffice to illustrate seed size compensation in cereals. Eastin (1972b) reported a 29% greater seed size in a tall 2 × 2 dwarf sorghum hybrid than in its 3 × 3 and 3 × 4 dwarf counterparts. Tufail (1971) removed 23% of the seeds in sorghum to get a 24% seed size increase with no yield reduction. Removal of 51% of the seeds reduced yield because seed size increased only 38%. Fischer (1980) detailed seed number and seed size considerations in wheat.

22–2.2.3 Soybean

Grain legumes are not grown widely in the Great Plains. The soybean is no exception but will be discussed as representing legume limitations in general. Even under irrigation in low humidity areas west of the western Corn Belt (eastern Nebraska) the crop does not produce consistently as well as in the Corn Belt. Reasons for this behavior are not well understood.

Soybean growth and developmental descriptions differ considerably from those of cereal grains. Whigham (1980) reported three types of growth habit as determinate (vegetative development nearly complete before flowers appear), indeterminate (flowering begins before vegetative production ceases and continues to end of life cycle), and semideterminate (intermediate between the other types). The growth stage system of Fehr and Caviness (1977) appears adequately descriptive to help analyze environmentally influenced yield-related (dependent) events. They characterize development by subdividing events in the vegetative and reproductive stages. The vegetative period is defined by emergence (VE), cotyledon unfolding (VC), and subsequent node appearance (Vi. . .Vn). Noting node number is critical because it related to potential pod number (eventual seed number). Reproductive development is divided into eight stages from beginning bloom (R1) to full maturity (R8). Staging is accomplished by selecting 10 representative plants from various sites in the field and determining the number of plants at the most advanced stage. The appropriate stage is named when more than 50% of the plants are at that given stage. Some researchers report growth stage fractions. When vegetative and reproductive growth are proceeding simultaneously, both stages are detailed. Consequently, the system adapts well to both determinate and indeterminate growth habits.

The utility of the system lies in a positive relationship between node number and pod number and, further, pod number is associated with seed number and yield. Therefore, if the soybean crop were staged every few days during the entire life cycle, yield reductions could possibly be related to changes in developmental pace or peculiarities that influence pod number and other characteristics (seed size). The changes, in turn, can be related to the limiting environmental conditions prevalent at each stage. Knowing both the seasonal time of yield loss and the yield reduction (reduction of node number, pod number, seed number, and seed size) should be useful in subsequent efforts to screen for plants resistant to particular environmental stresses. The developmental times when node number, pod number, seed number, and seed size are most influenced are obviously different (Fehr and Caviness, 1977) for detailed staging.

For purposes of analyzing yield limitations, a staging system detailing nodal development similar to the one used for soybeans is generally useful for grain legumes and cotton because of the frequent correlation between node number and fruiting bodies (pods or bolls).

Several persons have investigated the influence of water stress during different stages of development. A single irrigation recharging the soil profile at full flower in central Nebraska (dry conditions) gave a yield nearly as high as did six subsequent irrigations (Howell, 1963). Doss et al. (1974) found the pod fill stage to be the most sensitive to water stress. Adequate watering during pod fill gave from 24 to 55% higher grain yields over natural rainfall in Alabama. A positive linear relationship was demonstrated between yield and amount of water applied soon after flowering began until maturity. Sionit and Kramer (1977) stressed two soybean cultivars in growth chamber pots (vermiculite-gravel mix) for 7 to 10 days before rewatering during flower induction, flowering, pod formation, and pod filling. Lowest seed yields occurred after pod formation and pod fill stresses. Pod formation stress reduced pod number an average of only 16%, but seed number was reduced 39% to give a 36% yield reduction. Pod filling stress reduced neither pod number nor seed number, but the yield reduction still averaged 36%. Ashley and Ethridge (1978) in Georgia compared unirrigated plants with plants irrigated at bloom, pod fill, and both. They found that irrigation at pod fill frequently produced yields equal to bloom or full irrigation treatments. J. A. Specht (1980. Personal communication) irrigated eight soybean genotypes with a factorial arrangement of no irrigation (0) or one irrigation (1) at flowering, podding, and pod filling. The treatment arrangements and mean seed yield in kilograms per hectare were:

Irrigation treatment							
000	001	010	011	100	101	110	111
Avg grain yield, kg/ha							
2570	3550	3140	3420	2660	3350	2900	3380
C	A	AB	A	C	A	BC	A

The multiple range test shows (*A* values) that any irrigation combination with irrigation at pod filling equaled the full irrigation treatment yield, and none exceeded the pod filling irrigation alone. Sensitivity during pod filling is well established. It should be investigated further if progress is to be made in better adapting soybeans to the Great Plains. The same is likely true for other potential legume crops. Pod filling occurs generally from mid-August or a little earlier through September. The period represents a portion (August) of the drier part of the year in the Great Plains. Teare et al. (1973) demonstrated that soybeans in Kansas began using water more heavily than sorghum about 15 August as the canopy was closed. A higher stomatal resistance in sorghum was apparently associated with sorghum's lower water use even though sorghum had twice the root weight per volume of soil as soybeans. One factor limiting the spread of soybeans in the Great Plains is inability of Corn Belt–bred soybeans to control water loss.

Ashley (1983) recently reviewed water relations in soybeans.

22–2.2.4 Cowpea

No generally adapted grain legume is being grown extensively in the drier parts of the Great Plains. Dry bean culture exists in low rainfall areas such as eastern Colorado and western Nebraska, but the hectarage is extremely small compared with that for major cereals like wheat and sorghum. Market limitations are partly responsible for holding hectarage down. Cowpea perhaps should be evaluated as an alternative because it is considered to be a drought-tolerant legume (Morse, 1920; Rachie and Roberts, 1974; Zablotowicz et al., 1981). Elston and Bunting (1980) point out that cowpeas are grown in some very dry climates in the Sahel of Africa and apparently avoid stress by maturing quickly. Indeterminate, restricted branching types can mature within 70 days from sowing when rainfall is about 250 mm with mean potential evapotranspiration of 210 mm. Soil storage of 100 mm can offset deleterious effects of the often erratic rainfall in such dry areas.

Summerfield et al. (1980) provided an excellent review of cowpea, including vegetative and reproductive details plus ecological adaptation. Those from the modern centers of production (India and Brazil) indicate that cowpea is more tolerant of "tropical hazards" than dry bean.

Growth habit characteristics of cowpea are not unlike those of some other grain legumes. Determinate, intermediate, and indeterminate types exist with crop durations from 60 to 140 days or more. Reported sole crop and mixed cropping yields range from 100 to 3800 kg/ha with U.S. yields ranging from 1000 to 2800 kg/ha (seldom irrigated). However, some reports give U.S. yields as lower and hectarage declining.

Cowpea has a strong tap root plus well-branched secondary and higher order roots that host populations of primary, secondary, and tertiary nodules capable of generally exceeding N_2 fixation capacity/ha of soybeans with proper *Rhizobium* inoculation.

Growth habits include erect, semierect, trailing prostrate, and climbing. Quite apart from degree of vertical or horizontal habit, the number of inflorescences (reproductive nodes) appears to be the most important grain yield attribute. The number of reproductive nodes relates to the number of mature fruits and the number of seeds per unit land area, both of which determine yield. Frequently there is a positive correlation between seed number and seed weight. The critical factor is assessing ecological adaptation for an area may be whether or not environmental conditions permit developing a sufficiently high reproductive node number consistent with the water and nutrient supply available to mature the crop. Such an assessment will need to be made if genotypes available within the side range of variability are secured for specific Great Plains environments.

Rachie and Roberts (1974) consider cowpea to be well adapted to hot, semiarid environments and to be "highly drought resistant" with good growth capabilities at 35°C and above (Rachie and Rawal, 1976). Whereas it appears likely that cowpea drought resistance may largely be a drought avoidance mechanism through early maturity, with 70 days from sowing (Elston and Bunting, 1980), the crop appears to maintain a higher leaf water potential under stress than soybean. However, day temperatures of 33°C can reduce yields by 50% compared with 27°C days (Summerfield et al., 1978). Also drought stress during the reproductive stage can reduce yield sharply (Hall et al., 1979). Wien et al. (1979) showed that N_2 fixation was severely reduced with water stress but that fixation levels exceeded control plant levels when stress was removed.

As with most crops, the ultimate influence on grain yield of stress-induced aberrations in individual physiological processes is usually not well understood. For cowpea, however, successful breeding for drought resistance likely will involve selection for earlier maturity (A. E. Hall and D. A. Garrity. 1980. Personal communication). Cowpea possesses some apparent good traits for adaptation to the Great Plains, but it also has definite weaknesses to temperature and water stress. Nonetheless, perhaps there is merit in exploratory trials for the Great Plains in view of the crop's diversity exposed by recent research, primarily done in the semiarid and subhumid tropics.

22-2.2.5 Cotton

New World cottons originated in Central America and the northern portion of South America, respectively, and thus thrive on warm to hot weather. Largely due to its indeterminate growth habit, cotton is widely adapted to a range of rainfall, soil, and culture that allow the plant "to wait until conditions are right" before it sets a fruit load. Cotton in the Great Plains is grown in the southern portion only and, therefore, usually is subjected to both heat and drought stress at some time in its life cycle.

For convenience, the life cycle will be divided into three phases. They are germination and seedling establishment, vegetative development, and reproductive growth.

Establishing a uniform stand of vigorous seedlings is one of the most critical steps in cotton production and thus subject to environmental control. Adequate water for imbibition can be obtained from soils with low water potentials if good seed–soil contact is attained (Wanjura and Buxton, 1972) because of the high concentration of colloidal material (~20% protein) in the cotton seed. The initial imbibition phase is largely insensitive to temperature within biological limits. Resumption of metabolic activity following hydration initiates a temperature-sensitive phase of the germination process. Physiological zero is assumed to be approximately 15°C. In addition, the embryo is especially sensitive to short periods of chilling (~5°C) during this initial phase of germination. Significant damage to the radicle can result when the imbibed seed is exposed at temperatures of 5 to 10°C for 12 to 24 hours (Christiansen, 1979).

Seed hydration is rather insensitive to soil water potential, whereas seedling emergence is highly sensitive to both the temperature and water status of the soil environment. Both hypocotyl and radicle elongation rates exhibit a temperature optimum near 35°C, with a minimum of 14°C and a maximum of 41.9°C (Wanjura et al., 1970). Under nonlimiting temperatures, hypocotyl elongation is more sensitive to soil water stress than is radicle growth.

Vegetative development and growth habit has been well described by Mauney (1979). The growth habit of cotton is characterized as indeterminate and dimorphic, meaning that the apical meristem remains vegetative throughout the life of the plant. Fruit are produced on axillary branches called sympodia. In agronomic types, the apical meristem produces leaves at a rate of one every 3 days until fruit production rate reaches a peak, at which time the apical meristem becomes dormant. The plastochron index (leaf appearance interval) is subject to slight modification by genetics and environment, especially temperature. The initiation rate is approximately twice as fast at 30°C compared with 18°C (Hesketh et al., 1972). The number of mean stem nodes is critical because this factor determines the potential number of fruiting sites, a number that is similarly important in grain legumes.

Leaf area development is a function of rate and duration of leaf initiation and expansion rate and final size of individual blades. Water stress affects the duration of the leaf initiation process and reduces plant height through effects on both total number of leaves produced (nodes) and internode length. In addition, water stress has a pronounced effect on blade size primarily through effects on cell division rather than cell expansion (Berlin et al., 1981). Diurnal expansion rates are similar in stressed and nonstressed leaves except for the very young, small leaves (Krieg, 1981). Total plant leaf area therefore is reduced by mild water stress, which reduces the production potential of the plant (Jordan, 1979).

Root development appears to be less sensitive to soil water stress than does leaf development possibly because of a change in photosynthate partitioning.

Photosynthetic rates are also rather tolerant of plant stress. A 0.4- to 0.5-MPa differential between a well-watered control and a plant experiencing soil water stress is required to cause a significant reduction in single leaf photosynthetic rates (Krieg and Sung, 1979). Ratios of gross to net photosynthetic rates are apparently unaffected by water stress and are only sensitive to temperature (Perry and Krieg, 1981). Changes in photosynthetic rate attributable to water stress or leaf age are largely caused by reduced RuBP carboxylase/oxygenase activity, which is directly related to soluble protein concentrations (Krieg, 1981).

Respiratory activity is a function of growth rate and maintenance requirements, which, in turn, are related to prevailing temperatures. When photosynthetic productivity is reduced by reductions in leaf area and photosynthetic rate, growth rates are reduced proportionately. Maintenance respiration is largely a function of temperature and is probably not affected until stress is extreme.

Floral bud initiation from axillary meristems signals the onset of reproductive growth, usually beginning at the fifth or sixth node above the cotyledonary nodes. As the developing fruit load becomes a strong sink, new leaf growth on the main axis essentially ceases, apparently because of lack of available assimilate to meet the demands of all competing sinks. The result in cessation of main axis leaf initiation and sympodial floral bud initiation. If the growth rate of the cotton plant is limited by water stress, fewer nodes are available for fruit branch development (Jordan, 1979). Leaf area and thus assimilation capacity are reduced by water stress; a similar response is noted in sympodial development, resulting in two or three floral buds per branch instead of five or six.

Normally, production of fruiting sites is not a major limitation in cotton production because usually less than 20% of the flowers and only 40 to 60% of the young fruit actually result in harvestable yield (Longenecker and Eric, 1968). Shedding of potential fruit (squares) and actual fruit (young bolls) is therefore a mechanism the plant utilizes to match sink activity with source activity (McMichael, 1979). Fruit shedding is apparently limited to bolls less than 15 days after anthesis.

The rate of dry matter accumulation by the fruit is subject to control by assimilate supply. Both current photo assimilate and existing reserves stored in leaf and stem tissue are used in boll filling. The boll period (time from anthesis to physiological maturity) may also be slightly increased under stress. The final result is a mature boll unless stress is so extreme as to essentially cause plant death. The vegetative to reproductive ratio remains fairly constant.

In overall management of cotton, the seedling establishment phase is considered one of the most critical developmental periods and is subject to both temperature and water stress. Although some compensation can take place within each plant, yields are often limited by plant population, especially irregular plant spacing. Once plant establishment has been achieved, no single growth stage appears to be any more sensitive than any other to environmental stresses. However, because early season growth rates determine potential fruiting sites and leaf area index, it is essential not to allow excessive stress during this developmental period. The total growing season length must be considered, however. If the season is short from cool, late season conditions, if drought can be expected in late summer, and if insect problems are known to exist at certain times, excessive vegetative growth is not desirable.

A very high percentage of total yield results from fruit set on the first fruiting position on each sympodial branch; therefore, it is critical to manage the water resources so that minimal stress exists during the early flowering period. Most fruit that attains an age of 10 to 15 days post-anthesis will probably mature. Yield will be largely controlled by plant population and number of mature fruit per plant.

22-2.2.6 Pearl Millet

Rachie and Majmudar (1980) recently completed a comprehensive review of *Pennisetum americanum,* known as *P. typoides* L. in much of the older literature. Pearl millet is more efficient in water utilization and has higher heat tolerance than corn or sorghum. It is common in the drier areas of India and the southern peripheries of the Sahara Desert, and it also is grown in some warm, subhumid areas. However, the drier adaptation characteristics and heat tolerance of pearl millet suggest that considering growing the crop in the drier Great Plains may be of value. Until now pearl millet has been used as a forage crop in the Southeast. Only in the past few years has there been an effort to consider grain production. A modest research effort at U.S. public institutions is in progress, mostly within USDA at Tifton, Georgia, and at Kansas State, Texas A&M, and Colorado State Universities, and the University of Nebraska.

Pennisetum probably originated in Africa (Staph, 1934), where a large range in diversity can be found. Werth (1943), however, suggests India may be the place of origin, with subsequent movement to Africa. In both regions the crop is often found in areas dry enough that sorghum production is marginal (Burton, 1951).

Maiti and Bisen (1978) characterized growth and development of pearl millet in terms of GS1, GS2, and GS3 as detailed earlier with wheat and sorghum. Further, they characterized development in detail from panicle initiation to physiological maturity on two genotypes. Although their description of millet development cannot be directly compared with that of sorghum by Lee et al. (1974) at Nebraska, some important differences seem to exist. Temperature differences were not specified in the two studies, but India was probably warmer, which would hasten developmental pace. Averaged over summer and winter seasons, the two Indian millets matured in 67 and 83 days, respectively, demonstrating large maturity differences. Sorghums frequently would be later maturing in the Great Plains, although early maturing genotypes are available. The second apparent species difference that might relate to yield capacity is the relative amount of time spent developing potential seed number during GS2. Our experience is that about one third of the sorghum life cycle is spent in each growth stage. By contrast the Indian data show that GS2 accounts for only about one fourth of the life cycle in pearl millet. Powers et al. (1980) provided exceptional pictures of panicle development stages but gave a less detailed account of the GS2 time course, which appeared to be about 25 days compared with 35 days for GS1. Reduced GS2 days usually means reduced seed number and reduced grain yield in sorghum (Eastin, 1976). Pearl millet reaction to reduced GS2 days appears not to be documented in the literature, but results of Fussell et al. (1980) suggest it is likely to be similar to that in both wheat and sorghum.

Fussel and Pearson (1978), Fussel and Dwarte (1980), and Witt (1981, 1979; Personal communication, 1981) demonstrated the black layer in pearl millet to be a good indicator of grain physiological maturity. Witt further tested sorghums and millet under a range of water levels at Garden City, Kansas, where the grain-filling period in millet was 15 to 20% shorter than in sorghums. This shorter period would normally appear to be a detriment except as production resources become progressively more limiting. Ferraris (1973) considered the broad adaptation of pearl millet. Cultivation in rainfall areas ranging from 125 to 800 mm was reported by Hill (1969), Aiyer (1947), and Jain and Mathur (1961). Krishnaswamy (1962) reported production for grain to occur in areas as low as 80 mm. The generally recognized superiority of pearl millet over corn and sorghum under dry, hot conditions in sandier soils relates heavily to its short duration of growth (Rachie, 1966). Long maturity types are also available. A wide range of maturities and rainfalls exist in possible adaptation areas of the Great Plains. Sufficient genetic variability for those traits exists and is being utilized in evaluating pearl millet potential in the Great Plains.

22–2.3 Physiological Processes

The literature is replete with references to environmental influences on essential physiological processes, especially photosynthesis. References will be made to selected reviews and papers. General plant reactions will be described, and subsequently consideration will be given to problems in physiological processes compared with problems in morphology and development.

22–2.3.1 Photosynthesis and Transpiration

Ludlow (1980) referred to several reviews on stomatal control of water loss and pointed to the pivotal role of stomata regarding assimilation and transpiration. Many, including Boyer (1976), have considered various resistances to the influx of CO_2 from air outside the leaf to carboxylation sites. Cuticular resistance is very large compared with all other resistances, and stomatal resistance is large compared with boundary layer and mesophyll resistances. Discussion of stomatal activity can be conveniently handled in terms of stomatal conductance (the inverse of resistance) because CO_2 and water vapor fluxes are proportional to conductance (Ludlow, 1980; Boyer, 1976; Turner et al., 1978). Reduction in stomatal conductance is probably the first and major water-induced resistance (Slatyer, 1973) to influence photosynthesis, whereas photosystem 2 appears to be affected also about the time when whole leaf photosynthesis is affected (Boyer, 1976). Photosystem 1 is affected adversely about the same time but to a lesser degree (Keck and Boyer, 1974).

A major characteristic of the Great Plains is its low humidity. Ludlow (1980) cited evidence showing that physiological changes make stomata more or less responsive to differences in water vapor pressure. Generally, stomates of leaves under water stress are less responsive than stomates of nonstressed leaves to changes in vapor pressure deficits, and the responsiveness declines further with increasing cycles of stress. He also cited a report on Negev Desert apricot, in which stomatal conductance declined faster (was more sensitive) to increasing leaf/air vapor pressure differences in a nonirrigated (stress) vs. irrigated situation. This increased sensitivity would seem to be a desirable trait where total water is limited and its distribution is erratic, especially if the crop has a low dehydration tolerance. Plant selection for both types of response appears possible. Less is known regarding ability to select for both types of response within a species.

Papers describing stomate adjustment under water stress by Turner and Jones (1980), Ludlow (1980), and Morgan (1980) cite many references implicating cellular osmotic adjustment as the main responsible mechanism. This adjustment does not fully account for the lowering of leaf water potential for stomate adjustment. The osmotic and turgor potential components of water potential are of chief interest. Osmotic potential lowering can result from concentration of osmotic solutes from cell shrinkage as water is lost or from active accumulation of solutes in the cell (Turner and Jones, 1980). The latter authors cite references on osmoregulation or turgor regulation as regulating cell osmotic potential by adding or removing solutes from solution until the osmotic potentials of the intracellular solution and solution bathing the cell are about equal. Plant reaction to osmoregulation or turgor regulation may be responses to either cell turgor or solute concentration. This variability has been reflected in the interchangeable use of the two terms by scientists. Turner and Jones (1980) suggest the use of the term osmotic adjustment to indicate net solute increases by means of active accumulation of solutes rather than passive solute concentration; and, they suggest the terms osmoregulation and turgor regulation to describe only osmotic potential responses to salinity in higher plants, lower plants, and microorganisms. Also turgor maintenance, which can be partial or full, results from osmotic adjustment responses to water deficits. They cited numerous references of water stress–induced osmotic adjustment of plants in leaves, hypocotyls, roots, and reproductive organs. *Triticum aestivum* L. did not maintain turgor pressure by osmotic adjustment from about 0 to 1.5 MPa, whereas *Triticum dicoccum* L. maintained full turgor over the same range by osmotic adjustment (Morgan, 1980). A sorghum experiment (Fereres et al., 1978) showed osmotic adjustment to maintain partial turgor pressure. Slow drying permitted much greater osmotic adjustment than rapid drying. Turner (1975) and Wenkert et al. (1978) showed osmotic adjustment (up to 0.6 MPa) in corn in response to diurnal fluctuations of leaf water status.

Ludlow (1980) considered the relative merits of stomatal adjustment as opposed to closure to depend on factors such as species dehydration tolerance, length, and severity of dry periods, the importance of production over survival, mode of reproduction (survival), and annual or perennial growth habit. Stomatal adjustment appears beneficial in dehydration-tolerant species (often perennials) where emphasis is on survival. Dehydration-tolerant plants often lack good dehydration avoidance and are slow growers. Therefore, Sullivan (1972) suggested crop plants intermediate in both dehydration avoidance and tolerance would be desirable. Stomatal adjustment in moderately dehydration tolerant crops (e.g., sorghum) in semiarid areas appears desirable. Higher dehydration tolerance of sorghum compared to corn (Ludlow, 1976; Turner, 1979) is probably associated with the greater stomatal adjustment (closure) in corn (Ludlow, 1980). Ludlow further considers reduced stomatal adjustment advantages for annual grain crops in arid areas to avoid water loss and undue stress at critical growth stages (inflorescence development). Stomatal closure not only reduces water loss and helps keep internal water status high but also shuts down CO_2 uptake. Shutdown reduces ability of the plant to adjust osmotically, reduces energy available for root exploration for water, and reduces energy available for cellular maintenance. In addition, stomatal closure raises plant temperature, thus requiring a higher heat tolerance for survival. Sullivan et al. (1977) demonstrated that corn and pearl millet, whose stomates close more readily than sorghum, are both more heat tolerant than sorghum. Whereas corn may be more heat tolerant than sorghum, however, tighter closing of stomates forces temperatures to rise above lethal temperatures more easily than in sorghum, which maintains at least slow transpirational cooling and survives longer under severe stress. Obviously the best combination of desirable plant characteristics is highly dependent on amount and frequency of rainfall and temperature. Judgment on the best selection criteria is difficult.

An example is the relative merits of sorghum over pearl millet in the drier parts of the Great Plains. Witt (1981) compared sorghum and millet in terms of the assimilation efficiency ratio (AER)—photosynthesis/relative transpiration. The AER for sorghum declined from 285 to 42 going from 18 to 42°C, whereas the AER for millet began at 210 and progressed to 270 over the same temperature range. Both had an AER of about 240 at 32°C, but above that temperature millet was clearly superior. Such response to temperature needs to be considered along with developmental habits of the two species when their potential in breeding programs for their use in moderate and very dry areas with differing temperatures is evaluated. Developmental properties influenced by osmotic adjustments will be considered briefly later.

22-2.3.2 Respiration

Hsiao's (1973) review reveals that dark respiration generally declines with moderate to severe stress. Brix (1962) and Shearman et al. (1972) noted that moderate stress increased respiration, and severe stress reduced it below controls. McCree and van Bavel (1977) and Wilson et al. (1980) suggest the growth component of respiration is not affected much by stress but growth rates are drastically reduced. Maintenance respiration, however, declines under severe stress (Wilson et al., 1980). Krieg (1981) suspects mild stress increases maintenance respiration, to be followed by a maintenance respiration reduction under prolonged severe stress (drastic substrate reduction), which seems logical in view of respiration trends noted by Shearman et al. (1972). With photosynthesis reduced and growth halted under severe stress, maintenance respiration consumes a greater proportion of the daily respiration (McCree and van Bavel, 1977).

22-2.3.3 Morphological Considerations

Hsiao (1973) and Hsiao and Acevedo (1975) presented a list of plant processes in order of generalized sensitivity to water stress. Cell growth, wall synthesis, nitrate reductase activity, and protein synthesis are far more sensitive to water stress than stomatal opening and CO_2 assimilation. Probably these sensitive processes (i.e., growth or direct components of growth and development) should command relatively more attention than they do as yield-limiting processes. Hsiao and Acevedo's (1975) definition of cell growth as "irreversible enlargement or expansion of cells" is excellent in terms of yield considerations. The definition, though seemingly simple, is broadly encompassing, because sustained cell enlargement requires "concomitant synthesis of membranes, organelles, proteins, and cell wall material and is almost always accompanied by differentiation at the subcellular level." Cellular synthetic activities culminate in expansive growth only when sufficient internal hydrostatic or turgor pressure is generated to physically expand the cell.

22-2.3.4 Cell Division and Expansion

Green (1968) showed that the slightest change in turgor pressure caused an immediate change in growth rate in *Nitella.* Green et al. (1971) further showed that cell extensibility and threshold turgor for growth can change with stress level in a manner to permit growth under stress. Such changes require osmotic adjustment as referred to in the physiological processes section. The review by Turner and Jones (1980) contains numerous references to osmotic adjustments maintaining turgor as water potential declines. Cases are also cited where stress generally reduces cell elasticity but also can increase cell elasticity.

Extensive reviews by Begg and Turner (1976) and Begg (1980) show that earlier literature reported cell division as being less sensitive to water deficits than cell expansion. However, later references suggest more or less equal sensitivities for cell division and cell enlargement in many species. The net stress effect can be greatly reduced leaf area in keeping with limited water in semiarid areas.

22-2.3.5 Differentiation

Hsiao and Acevedo (1975) intimated that cell enlargement could not be sustained without many concomitant synthetic activities, which are almost always accompanied by differentiation at the subcellular level. One would suspect that differentiation may be about as sensitive to stress as other growth processes. That differentiation associated with development of cereal inflorescences and grain legume seeds are indeed very sensitive to stress is common knowledge (Salter and Goode, 1967; Begg and Turner, 1976). As pointed out earlier, the seed number component of yield is especially sensitive to environmental stresses. Therefore, after a damaging stress during prebloom inflorescence, the seed size component of yield lends stability to yield (compensation for seed number loss) in sorghum genotypes that are the least influenced by adverse environments (G. Heinrichs. 1981. Unpublished data). Ogunlela (1979) raised night temperature 5°C above ambient in sorghum for a week beginning with the appearance of stamen and pistil primordia in the absence of water stress. The result was to reduce seed number per plant 30%, resulting in a 28% grain yield loss. This effect of the heat treatment was reflected during grain fill in that grain production efficiency (grams of grain per plant per day) was also reduced 30% in the absence of stress during grain filling.

In summary, thoughts regarding physiological, morphological, and developmental processes center on the extreme sensitivity of cell division, cell expansion, and differentiation. These growth-related processes are far more sensitive to water deficits than is CO_2 assimilation even though the former ultimately require the latter to persist. It would appear that a more complete understanding of environmental effects on cell division, cell expansion, and differentiation of floral components—and the response ranges available within a species for these traits—would be useful in improving the rate of crop improvement in semiarid areas. This is not to suggest diminishing research efforts in CO_2 assimilation, because it is the ultimate driving force, especially under stress conditions.

22-2.4 Water Consumption

Water requirement data in this section are selective rather than comprehensive and will generally, but not always, apply to the Great Plains.

22-2.4.1 Sorghum

Musick et al. (1963) applied several regimes of water and N levels at Garden City, Kansas. Evapotranspiration (ET) ranged from 400 to 635 mm, and WUE ranged from 112 to 164 kg ha^{-1} cm^{-1}. Daily ET ranged from 2.5 to 7.9 mm. Hanks et al. (1969) reported an ET of 0.48 cm/day at Akron, Colorado. Jones (1975) tested sorghum in the Texas High Plains under different tillage systems to get a mean WUE of 34.02 kg ha^{-1} cm^{-1} (range of 29.6 to 42.6). Ohmes and Penas (1977) reviewed several years of their own WUE work emphasizing limited irrigation. Water use efficiencies were on the order of 112 kg of grain ha^{-1} cm^{-1} for two irrigations and 73 kg ha^{-1} cm^{-1} for four irrigations. J. D. Eastin and M. D. Witt (1981. Unpublished data) also noted higher WUE for limited water levels compared with full irrigation in sorghum, whereas the opposite was true in corn. Bordorosky and Hoy (1975) noted water use efficiencies at Colby, Kansas, to be about 145 kg of grain ha^{-1} cm^{-1} and about 113 kg ha^{-1} cm^{-1} at 18 and 23 ha/cm of water used, respectively. Eck and Musick (1979) did not measure WUE on sorghum at Amarillo, Texas, but determined the early boot to heading period to be the most sensitive to field water stress.

A discussion of corn in the Great Plains is conspicuously absent because sorghum is generally felt to be a more promising feed grain. However, with the development of ecofallow technology to enhance precipitation capture and conservation, corn is perhaps more promising in certain areas. Castleberry and Lorette (1979) reviewed some of the promising activities in developing stress-tolerant corn. The lack of synchrony under stress between the tassel and silk is a major problem but can be selected through breeding.

22-2.4.2 Wheat

Many of the papers cited for sorghum also have WUE data for winter wheat. Examination illustrates the differences between C_3 and C_4 plants. Jones (1975), in Texas research on WUE, shows an average of 15.2 kg ha^{-1} cm^{-1} for wheat and 34.0 kg ha^{-1} cm^{-1} for sorghum. Thompson (1977) reported consumptive use of dryland wheat and sorghum at Hays, Kansas, from 1966 to 1976. Years grouped by average water use levels of about 42, 48, and 51 cm resulted in 44, 50, and 50 kg of grain ha^{-1} cm^{-1} from wheat. Years grouped by average water use levels of about 36 and 39 cm resulted in 75 and 81 kg of grain ha^{-1} cm^{-1} from sorghum. The WUE for preirrigated wheat (15 cm of preirrigated with 40 cm of total use) at Garden City, Kansas, was 74 compared with 521 kg of grain ha^{-1} cm^{-1} when irrigated three times with a total water use of 71 cm.

The most critical time to obtain maximum benefit from irrigation is near the heading stage, according to van Brackle (1980). Fischer's (1973) water stress treatments suggest a sensitive period may occur a little earlier than heading. His data indicate the deleterious effects are clearly on spike development (seed number). Kirkham and Kanemasu (1983) have a comprehensive review on water relations in wheat.

A comparison on WUE of C_3 and C_4 plants was mentioned earlier. Many references are available, some of which were mentioned above. Begg and Turner (1976) summarized the thoughts of many on the subject partly by quoting research from Akron, Colorado, of Shantz and Piemeisel (1927) as rearranged by Downes (1969). Water use efficiencies (grams of water per kilogram of dry matter) were 3.44 and 3.14 for C_4 species dicots and grasses, respectively, contrasted with 1.59 and 1.49 for C_3 species dicots and grasses, respectively. Such data are of particular interest in the Great Plains. Water use efficiencies in C_4 plants of twice those in C_3 plants are not unusual. More detailed discussions on WUE are given by Nobel (1980) and Ludlow (1976, 1980) for reference.

22-2.4.3 Soybeans

Data of Teare et al. (1973) are pertinent to consideration of soybeans in the Great Plains because a sorghum comparison was made. Total ET for soybeans and sorghum was about 32 and 38 cm, respectively. Higher water use by soybeans was related to a higher leaf area from mid-July on. Sorghum root weight was about double the soybean root weight, but soil water potential at 90 cm depth in mid-August reached 700 kPa under soybeans compared with 300 kPa under sorghum. Sorghum obviously uses soil water more effectively than does soybean, a factor reflected by its relative importance as a crop in the Great Plains. Water use efficiency of the sorghum was about three times as high as soybeans.

Garden City, Kansas, data from 1975 and 1976 (Ohmes and Penas, 1977) are more representative of performance in the drier Great Plains areas. Over a range of available water levels from 67 to 108 cm, yield levels ranged from 1500 to 2125 kg/ha with respective WUE's of 24 and 22 kg of grain kg^{-1} cm^{-1}. Ashley (1983) and Whigham (1980) provide reviews on soybean relating to factors affecting water use.

22-2.4.4 Cowpea

Water use data for cowpea in the Great Plains are not available. We mention cowpea because they are said to be "highly drought resistant" (Rachie and Roberts, 1974). Nonetheless, Cutler (1979) considers them to be very sensitive to mild midday water deficits in subhumid parts of Nigeria. Perhaps their drought resistance stems more from short growth duration, which escapes drought. Elston and Bunting (1980) report the crop to mature in 70 days before substantial stress results in areas where rainfall is less than 300 mm.

22-2.4.5 Cotton

Approximately 40% of the total U.S. cotton production occurs in the semiarid southwestern states where water use efficiency is a concern because of increasing costs of irrigation and lack of available water supplies. Daily water use is dependent on both the plant and the environment. Ritchie and others (Ritchie, 1971; Ritchie and Burnett, 1971; Ritchie and Henderson, 1972; Ritchie and Jordan, 1972) have described the influence of soil water availability and plant and atmospheric factors in determining the evapotranspiration rates of cotton. According to their reports, daily actual evapotranspiration (ET_a) is basically a function of the degree of ground cover and the leaf area index when soil water supplies are relatively high. When LAI exceeds 3.0 with a ground cover of approximately 75%, ET_a is equal to potential ET when soil water supply exceeds 25 to 30% of the "available water" holding capacity. Total seasonal water use is therefore a function of ET, length of growing season and the prevailing leaf area index. In the desert Southwest, e.g., in Arizona and California, seasonal water use will exceed 125 cm due to both excessive evaporative demand and long growing seasons (Grimes et al., 1969; Kittock, 1973). On the Texas High Plains, seasonal water use is normally 50 to 60 cm (Longenecker and Eric, 1968). The major difference between the two environments is the length of the growing season. Under the shorter growing season of the Texas High Plains, irrigation water management is critical in maximizing yield and water use efficiency. Optimum supplies of soil water are required in the early season for seedling establishment and maximum rates of leaf area development. Once fruiting begins, water supply must be managed to maximize fruit development while minimizing continued vegetative development (Longenecker and Erie, 1968). Because of the indeterminate growth habit of the cotton plant, soil water and N supplies determine the relative amounts of vegetative and reproductive growth under short growing season (Grimes et al., 1969; Longenecker and Eric, 1968).

Water use efficiency, defined as kilograms of lint produced per centimeter of water consumed, appears to be fairly constant in the range of 7 to 9. In the Arizona-California area, yields range from 800 to 1000 kg/ha with 125 cm of water, whereas in the Texas High Plains, yields of 400 to 500 kg/ha result from 50 to 60 cm of water. Jordan (1983) has recently reviewed the water relations and water use literature for cotton.

22-3 FUTURE RESEARCH

Projections of future research based on current research efforts that attempt to integrate physiology and breeding are tenuous at best. We do not know enough about the influence of water and temperature on both individual and integrated physiological and morphological events as they relate to yield. However, expanding our understanding should permit determining something of the limiting order of essential plant processes. Subsequently, genetic and cultural manipulation can be attempted to attenuate the effects of the limiting process. As one example of genetic manipulation, we will discuss the commericial sorghum breeding program at DeKalb AgResearch, Inc., where a modified reciprocal recurrent selection program is used to incorporate desirable plant reactions to stress into commercial hybrids (A. B. Maunder and T. E. Dickinson. 1980. Personal communication).

Charcoal rot can result in stalk lodging in the Great Plains when heat and drought stress occur simultaneously in the early stages of grain filling. To screen for resistance, plants are fully watered until well after panicle initiation and then left to suffer whatever heat and drought stress a dry field environment imposes. The best resistant genotypes are then incorporated into populations or pools of otherwise adapted germplasm.

DeKalb researchers test for high diffusive resistance under drought and associated heat tolerance necessary for survival when stomates partially close and evaporative cooling is diminished. Lines selected from the modified reciprocal recurrent selection program are crossed during the F_4 generation with tester parents known to have heritable drought characteristics. It is at this time that the lines are tested for heat tolerance along with known tolerant hybrids by using Sullivan's (1972) leaf disc test. The next year the heat test is rerun on lines and hybrids to determine heritability. During the same season, diffusive resistance readings are taken at middle or late GS2 under stress and irrigated conditions using a procedure similar to that of Kanemasu et al. (1969). Lines having desirable traits from the diffusive resistance and heat tests, as well as good yield data and field performance, are crossed onto a wider range of established lines to be evaluated as potential hybrids for commercial use. A limited number of the original lines with the best heat tolerance are recombined and recycled in the program.

Hybrid and advanced line tests for stress-induced dormancy and recovery during late season rains are conducted in west Texas, where the probability of stress is high. Notes are taken on color, stunting, and leaf rolling. Field tests are supplemented in greenhouse pots, where plants are water stressed during GS2 and then evaluated after stress is relieved for both 24-hour and long-term recovery.

Root function is checked using gypsum blocks in dryland sorghum tests to monitor water uptake to compare with top growth performance. This checking gives an idea of water uptake and use. Selected genotypes are also grown in 1.83-m long polyvinyl chloride pipes in a soil-sand-peat moss mixture that is washed free for direct root observation.

Selected lines from all tests enter the yield testing program. In the final analyses, wide range testing under dryland and irrigation provides the only sure way of evaluating cultivars for farmer use. Obviously the complexity of drought performance for a hybrid does not allow researchers to use a single criteria to evaluate for drought and heat tolerance. With the system previously described, a new dryland hybrid entering the commercial market will have been bred and selected for a number of traits. This testing ensures a farmer of a crop that is adapted for a particular area and farming practices and is able to yield respectably in a dry year.

Readers are referred to Hurd (1975) and Townley-Smith and Hurd (1979) for procedures currently being used in a wheat breeding program for stress to supplement thoughts presented on the above sorghum breeding program. We will now discuss plant research that may be of value in expanding genetic-cultural improvement of the type discussed in the preceding commercial program.

Development of breeder screening techniques should be enhanced through a better understanding of basic plant processes. Based on the relative sensitivities of essential plant processes, the first factor deserving attention is growth as characterized by Hsiao and Acevedo (1975). In other words, cell elongation, cell division, and differentiation should be evaluated more completely in terms of stress effects on each process and should include evaluation of genotype differences. Generating an adequate seed number component of yield, which normally correlates positively with yield, must largely be a function of those processes. Begg and Turner (1976) and Begg (1980) reviewed literature pertinent to the relative sensitivities of cell elongation and cell division. Possibly less attention has been given to differentiation at the cellular level than to cell elongation and division. The reasons why vegetative growth appears less influenced by stress than reproductive growth do not seem to be well researched and need attention.

Genotype differences in stress damage to reproductive growth in sorghum were demonstrated by Hultquist (1973). Differences in assimilate partitioning characteristics of stress and nonstress resistant sorghums have been shown (Hultquist, 1973; Rice, 1980), but mechanisms dictating the differences remain unknown. Further, it is not known whether assimilate supply was closely related to limitations regarding differentiation of floral components or if assimilate distribution patterns noted might have been a function of differentiation control by mechanisms less closely or only indirectly related to assimilate supply. The latter seems more likely.

Barlow et al. (1977) noted the extreme sensitivity of apical meristem growth to stress even though apices can survive severe stress more readily than many other tissues. Barlow et al. (1980) reported wheat apices surviving nicely at 6 MPa. Survival probably related to apex enclosure by leaves (partly cutinized) to inhibit water loss by osmotic adjustment partly from translocated carbohydrates. The question that looms much larger than survival is the high sensitivity of the apex to expansive growth as opposed to dry weight increases (Barlow et al., 1980) at the time of floret differentiation. Specifically, they found wheat apices stop growing between 6 and 8 days after stress is imposed even though dry weight increased for another 7 days. Growth response (emphasizing cell division) in actively dividing tissue may be a turgor-mediated response because some cells require that some minimal size be attained before cell division can occur (Doley and Leyton, 1968). Hence the impact of osmotic adjustment characteristics are again obvious. Similarly, because roots in cereal expand rapidly at the time floret components are being developed, balances in metabolism between roots and aerial portions need further evaluation. Rice's work (1979) seems pertinent since he demonstrated an average of 31% greater assimilate requirement in nutrient solutions for roots of a nonstress-tolerant over a stress-tolerant hybrid. Relative turgor may relate to the balance achieved by different genotypes between root exploration to scavenge for water and the simultaneous development of floret number, which correlates positively with grain yield. The importance of amplifying such knowledge is obvious from Ogunlela's sorghum data (1979), which show that simply elevating night temperature 5°C above ambient (no water stress) during floret differentiation resulted in about 28% reductions in both seed number and yield. As a result, metabolic efficiency during grain filling was reduced 28%.

The concerns expressed about growth and differentiation were largely in cereals. The same considerations are probably equally critical in grain legume crops but will not be discussed.

A matter of concern during grain filling is how to lengthen the grain fill period and prevent premature stress-induced senescence, which cuts off assimilate translocation for grain growth. Also, once stress during inflorescence development has reduced seed number, the ability of the plant to compensate by increasing seed size becomes critical (Eastin, 1980). Concerning the latter, Sanders (1955) and Paulson (1962) detailed sorghum seed development. Beginning with anthesis, free nuclear division occurs with endosperm cell wall formation beginning on day 3. By day 8, cell division essentially stopped (cell number determined) and starch formation commences. Panicle heat treatments between days 7 and 9 definitely limit seed size (Dickinson, 1976), possibly by limiting starch granule size or by rendering cell walls less expandable. Research is needed because of the unpredictable occurrence (time) and intermittent nature of stresses in the Great Plains. The same research need applies generally to other grain crops for the region.

The emphasis placed on growth and development is not meant to detract from needed research on photosynthesis, because all growth eventually depends on the conversion of light energy to chemical energy. However, numerous discussions on photosynthesis are available elsewhere. Finally, the requirements for plant survival are not necessarily the same as the requirements for optimizing grain production in adverse environments (Fischer and Turner, 1978; Turner, 1979). Hultquist's data (1973), discussed earlier to compare differences between stress-tolerant and nonstress-tolerant sorghums, seem to present a good example of survival mechanisms. Developing the plant characteristics that permit survival in a given stress environment required centuries. By contrast, understanding plant characteristics required for optimum grain production rather than survival in different environments has been pursued for only a few decades. Our modest level of understanding is not surprising. It seems obvious that pursuit of a clearer understanding will come partially, at least, from a better understanding of mechanisms controlling growth and differentiation.

Physiologists and breeders have created an awareness of critical dryland traits that the breeder can use as references when selecting from germplasm. The techniques for screening are far from perfect, but combined efforts of the breeder and physiologist can now partially characterize germplasm according to a few usable drought and temperature responses. The genetic complexity of each stress mechanism and the existence of several mechanisms operating simultaneously suggest that quantitative breeding approaches will be required to obtain favorable gene frequencies in potential cultivars or parental lines. Obviously, crops that can be hybridized will benefit more quickly from this physiologist–geneticist team approach. For more detail, Eslick and Hockett (1975) discuss some of the problems plant breeders face in obtaining needed gene frequencies for WUE. They use the example that from 1 to 10 hypothetical gene pairs could be involved, a range that helps to describe the complexity of the problems.

22–4 LITERATURE CITED

Aiyer, A. K. Y. N. 1947. Field crops of India with special reference to Mysore. Government Press, Bangalore, p. 84–90.

Arnon, I. 1975. Physiological principles of dryland crop production. p. 3–145. *In* U.S. Gupta (ed.) Physiological aspects of dryland farming. Oxford & IBH Publishing Co., New Dehli.

Ashley, D. A. 1983. Soybeans. p. 389–422. *In* I. D. Teare and M. M. Peet (ed.) Crop water relations. Interscience, New York.

Ashley, D. A., and W. J. Ethridge. 1978. Irrigation effects on vegetative and reproductive development of three soybean cultivars. Agron. J. 70:467–471.

Barlow, E. W. R., R. E. Munns, and C. J. Brady. 1980. Drought response of opical meristems. p. 191–206. *In* N. C. Turner and P. J. Kramer (ed.) Adaptation of plants to water and high temperature stress. John Wiley & Sons, Inc., New York.

Barlow, E. W. R., R. E. Munns, N. S. Scott, and A. H. Reisner. 1977. Water potential, growth, and polyribosome content of the stressed wheat apex. J. Exp. Bot. 28:909–916.

Begg, J. E. 1980. Morphological adaptations of leaves to water stress. p. 33–42. *In* N. C. Turner and P. J. Kramer (ed.) Adaptation of plants to water and high temperature stress. Interscience, New York.

Begg, J. E., and N. C. Turner. 1976. Crop water deficits. p. 161–217. *In* N. C. Brady (ed.) Advances in agronomy. Academic Press, Inc., New York.

Beil, G. M., and R. E. Atkins. 1967. Estimates of general and specific combining ability in F_1 hybrids for grain yield and its components in grain sorghum, *Sorghum vulgare*. Crop Sci. 7:225–228.

Bennett, J. M. 1979. Responses of grain sorghum [*Sorghum bicolor* (L.) Moench] to osmotic stresses imposed at various growth stages. Ph.D. Dissertation. Univ. of Nebraska, Lincoln.

Berlin, J. D., J. E. Quisenberry, B. L. McMichael, M. Bloodworth, and F. Bailey. 1981. Morphometric analyses of water stressed cotton leaves. p. 45. *In* J. M. Brown (ed.) Proc. Beltwide Cotton Production Res. Conference, New Orleans. 5–8 Jan. Natl. Cotton Coun., Memphis, Tenn.

Blum, A. 1967. Effect of soil fertility and plant competition on grain sorghum panicle morphology and panicle weight components. Agron. J. 59:400–403.

Blum, A. 1970. Heterosis in grain production by sorghum. Crop Sci. 10:28–31.

Bordorosky, P., and D. Hoy. 1975. Irrigating grain sorghum in North Kansas. Colby Branch Exp. Stn. Bull. no. 592, Kansas State Univ., Manhattan.

Boyer, J. S. 1976. Water deficits and photosynthesis. p. 154–190. *In* T. T. Kozlowski (ed.) Water deficits and plant growth, Vol. 4. Academic Press, Inc., New York.

Brix, H. 1962. The effect of water stress on the rates of photosynthesis and respiration in tomato plants and loblolly pine seedlings. Physiol. Plant 15:10–20.

Burton, G. W. 1951. The adaptability and breeding of suitable grasses for the southeastern states. Adv. Agron. 3:197–241.

Cackett, K. E., and P. C. Wall. 1971. The effect of altitude and season length on the growth and yield of wheat (*Triticum aestivum* L.) in Rhodesia. Rhod. J. Agric. Res. 9:107–120.

Castleberry, R. M. 1973. Effects of thinning at different growth stages on morphology and yield of grain sorghum [*Sorghum bicolor* (L.) Moench.]. Ph.D. Dissertation. Univ. of Nebraska, Lincoln (Diss. Abstr. 74-632).

Castleberry, R. M., and R. J. Lorette. 1979. Latente: A new type of drought tolerance? p. 45–56. *In* H. D. Loden and D. Wilkinson (ed.) Report of the 34th Annual Corn and Sorghum Res. Conference. American Seed Trade Assoc., Washington, D.C.

Christiansen, M. N. 1979. Influence of atmospheric parameters. p. 279–283. *In* J. M. Brown (ed.) Cotton physiology: A treatise. Section I. Flowering, fruiting, and cutout. Proc. Beltwide Cotton Production Res. Conference, Phoenix, Ariz. 7–11 Jan. Natl. Cotton Coun., Memphis, Tenn.

Cutler, J. M. 1979. Adaptation of cowpea to semi-arid environments in West Africa. Unpublished report to the Grain Legume Improvement Programme, Int. Inst. of Tropical Agric., Ibadan, Nigeria.

Dickinson, T. E. 1976. Caryopsis development and the effect of induced high temperatures in [*Sorghum bicolor* (L.) Moench]. M.S. Thesis. Univ. of Nebraska, Lincoln.

Doggett, H., and D. Jowett. 1967. Yield increase from sorghum hybrids. Nature (London) 216:798–799.

Doley, D., and L. Leyton. 1968. Effects of growth regulating substances and water potential on the development of xylem in *Fraxinus.* New Phytol. 67:579–584.

Doss, B. D., R. W. Pearson, and H. T. Rogers. 1974. Effect of soil water stress at various growth stages on soybean yield. Agron. J. 66:297–299.

Downes, R. W. 1969. Differences in transpiration rates between tropical and temperate grasses under controlled conditions. Planta 88:261–273.

Downes, R. W. 1972. Effect of temperature on the phenology and grain yield of *Sorghum bicolor.* Aust. J. Agric. Res. 23:585–594.

Eastin, J. D. 1972a. Efficiency of grain dry matter accumulation in grain sorghum. p. 7–17. *In* J. I. Sutherland and R. J. Falasca (ed.) American Seed Trade Assoc., Washington, D.C.

Eastin, J. D. 1972b. Photosynthesis and translocation in relation to plant development. p. 214–216. *In* N. G. R. Rao and L. R. House (ed.) Sorghum in seventies. Oxford & IBH Publishing Co., New Delhi.

Eastin, J. D. 1976. Temperature influence on sorghum yield. p. 19–23. *In* H. P. Loden and D. Wilkinson (ed.) Proc. of the 31st Annual Corn and Sorghum Res. Conference. American Seed Trade Assoc., Washington, D.C.

Eastin, J. D. 1980. Sorghum development and yield. *In* S. Yoshida (ed.) Symp. on Potential Productivity of Field Crops Under Different Environments, Los Baños, Laguna, Philippines. 22–26 Sept. Int. Rice Res. Inst., Los Baños, Laguna, Philippines. (In press.)

Eastin, J. D., R. M. Castleberry, T. J. Gerik, J. H. Hultquiest, V. Mahalakshmi, V. B. Ogunlela, and J. R. Rice. 1983. Physiological aspects of high temperature and water stress. p. 91–112. *In* C. D. Raper, Jr., and P. J. Kramer (ed.) Crop reactions to water and temperature stresses in humid, temperate climates. Westview Press, Boulder, Colo.

Eastin, J. D., and C. Y. Sullivan. 1974. Yield consideration in selected cereals. p. 871–877. *In* R. L. Bieleski et al. (ed.) Mechanisms of regulation of plant growth. Bull. no. 12, Royal Soc. of New Zealand, Wellington.

Eck, H. V., and J. T. Musick. 1979. Plant water stress effects on irrigated grain sorghum. I. Effects on yield. Crop Sci. 19:589–592.

Elston, J., and A. H. Bunting. 1980. Water relations of legume crops. p. 37–42. *In* R. J. Summerfield and A. H. Bunting (ed.) Advances in legume science. Her Majesty's Stationary Office, London.

Eslick, R. F., and E. A. Hockett. 1975. Genetic engineering on a key to water-use efficiency. p. 13–23. *In* J. F. Stone (ed.) Plant modification for more efficient water use. Elsevier North-Holland, Inc., New York.

Fehr, W. R., and C. E. Caviness. 1977. Stages of soybean development. USDA Cooperative Extension Service and Agric. and Home Economics Exp. Stn. Spec. Report no. 80, Iowa State Univ., Ames.

Fereres, E., E. Acevedo, D. W. Henderson, and T. C. Hsiao. 1978. Seasonal changes in water potential and turgor maintenance in sorghum and maize under water stress. Physiol. Plant 44:261–267.

Ferraris, R. 1973. Pearl millet (*Pennisetum typhoides*). Commonwealth Agric. Bureau Rev. Series 1/97376, London.

Fischer, K. S., and G. L. Wilson. 1975. Studies of grain production in [*Sorghum bicolor* (L.) Moench]. V. Effect of planting density on growth and yield. Aust. J. Agric. Res. 26:31–41.

Fischer, R. A. 1980. Wheat. *In* S. Yoshida (ed.) Symp. on Potential Productivity of Field Crops Under Different Environments, Los Baños, Laguna, Philippines. 22–26 Sept. Int. Rice Res. Inst., Los Baños, Laguna, Philippines. (In press.)

Fischer, R. H. 1973. The effect of water stress at various stages of development on yield in-processes in wheat. p. 233–241. *In* R. O. Slatyer (ed.) Plant response to climatic factors. United Nations Educational, Scientific, and Cultural Organization, Paris.

Fischer, R. A., and N. C. Turner. 1978. Plant productivity in the arid and semiarid zones. Ann. Rev. Plant Physiol. 29:277–317.

Fussell, L. K., and D. M. Dwarte. 1980. Structural changes of the grain associated with black region formation in *Pennisetum americanum.* J. Exp. Bot. 31:645–654.

Fussell, L. K., and C. J. Pearson. 1978. Course of grain development and its relationship to black region appearance in *Pennisetum americanum.* Field Crops Res. 1:21–31.

Fussell, L. K., C. J. Pearson, and M. J. T. Norman. 1980. Effect of temperature during various growth stages on grain development and yield of *Pennisetum americanum.* J. Exp. Bot. 31:621–633.

Gipson, J. R., J. W. Johnson, and D. T. Rosenow. 1979. Effect of night temperature on phenotypic periods and yield components of grain sorghum. Agron. Abstr., p. 88.

Green, P. B. 1968. Growth physics in *Nitella*: A method for continuous in vivo analyses in extensibility based on a micromanometer technique for turgor pressure. Plant Physiol. 43: 1169–1184.

Green, P. B., R. O. Erickson, and J. Bugg. 1971. Metabolic and physical control of cell elongation rate. Plant Physiol. 47:423–430.

Grimes, D. W., W. L. Dickens, and W. D. Anderson. 1969. Functions for cotton (*Gossypium hirsutum* L.) production from irrigation and nitrogen fertilization variables: I. Yield and evapotranspiration. Agron. J. 61:769–773.

Hanks, R. J., H. R. Gardner, and R. L. Florian. 1969. Plant growth–evapotranspiration relations for several crops in the Central Great Plains. Agron. J. 61:30–34.

Hesketh, J. D., D. W. Baker, and W. G. Duncan. 1972. Simulation of growth and yield in cotton: II. Environmental control of morphogenesis. Crop Sci. 12:436–439.

Hill, G. D. 1969. Performance of forage sorghum hybrids and Katherine Pearl Millet at Bubia. Papua New Guinea Agric. J. 21:1–6.

Howell, R. W. 1963. Physiology of the soybean. p. 107. *In* A. G. Norman (ed.) The soybean. Academic Press, Inc., New York.

Hsiao, T. C. 1973. Plant responses to water stress. Annu. Rev. Plant Physiol. 24:519–570.

Hsiao, T. C., and E. Acevedo. 1975. Plant responses to water deficits, water-use efficiency and drought resistance. p. 59–84. *In* J. F. Stone (ed.) Plant modification for more efficient water use. Elsevier North-Holland, Inc., New York.

Hultquist, J. H. 1973. Physiologic and morphologic investigation ofsorghum [*Sorghum bicolor* (L.) Moench]: I. Vascularization. II. Response to internal drought stress. Ph.D. Dissertation. Univ. of Nebraska, Lincoln (Diss. Abstr. 73-25451).

Hurd, E. A. 1975. Phenotype and drought tolerance in wheat. p. 39–56. *In* J. F. Stone (ed.) Plant modification for more efficient water use. Elsevier North-Holland, Inc., New York.

Jain, S. V., and C. M. Mathur. 1961. Efficiency of different nitrogenous fertilizers for bajra production in desert soils of Rajasthan. Indian J. Agron. 5:151–156.

Jones, O. R. 1975. Yields and water-use efficiencies of dryland winter wheat and grain sorghum production systems in the Southern High Plains. Soil Sci. Soc. Am. Proc. 39:98–103.

Jordan, W. R. 1979. Influence of edaphic parameters on flowering, fruiting, and cutout. A role of plant water deficit. p. 297–301. *In* J. M. Brown (ed.) Cotton physiology: A treatise. Section I. Flowering, fruiting, and cutout. Proc. Beltwide Cotton Production Res. Conference, Phoenix, Ariz. 7–11 Jan. Natl. Cotton Coun., Memphis, Tenn.

Jordan, W. R. 1983. Cotton. p. 213–254. *In* I. D. Teare and M. M. Peet (ed.) Crop water relations. Interscience, New York.

Kambal, A. E., and O. J. Webster. 1966. Manifestations of hybrid vigor in grain sorghum and the relations among the components of yield, weight per bushel, and height. Crop Sci. 6: 513–515.

Kanemasu, E. T., G. W. Thurtell, and C. B. Tanner. 1969. Design, calibration and field use of a stomatal diffusion porometer. Plant Physiol. 44:881–885.

Keck, R. W., and J. S. Boyer. 1974. Chloroplast response to low leaf water potentials: III. Differing inhibition of electron transport and photophosphorylation. Plant Physiol. 53: 474–479.

Kirkham, M. B., and E. T. Kanemasu. 1983. Water relations of wheat. p. 481–520. *In* I. D. Teare and M. M. Peet (ed.) Crop water relations. Interscience, New York.

Kittock, D. L. 1973. Pima cotton lint yield as influenced by irrigation schedule, cultivar and altitude. Agron. J. 65:498–501.

Krieg, D. R. 1981. Leaf development and function as related to water stress. p. 41–42. *In* J. M. Brown (ed.) Proc. Beltwide Cotton Production Res. Conference, New Orleans. 5–8 Jan. Natl. Cotton Coun., Memphis, Tenn.

Krieg, D. R. 1983. Water relations of sorghum. p. 351–388. *In* I. D. Teare and M. M. Peet (ed.) Crop water relations. Interscience, New York.

Krieg, D. R., and F. J. M. Sung. 1979. Source-sink relations of cotton as affected by water stress during boll development. p. 302–305. *In* J. M. Brown (ed.) Cotton physiology: A treatise. Section I. Flowering, fruiting, and cutout. Proc. Beltwide Cotton Production Res. Conference, New Orleans. 5–8 Jan. Natl. Cotton Coun., Memphis, Tenn.

Krishnaswamy, N. 1962. Bajra: *Pennisetum typhoides* S & H. Indian Coun. of Agric. Res. Cereal Crop Series no. 11, New Delhi.

Langer, R. H. M. 1967. Physiological approaches to yield determination in wheat and barley. Field Crop Abstr. 20:101.

Lee, K.-W., R. C. Lommasson, and J. D. Eastin. 1974. Developmental studies on the panicle initiation in sorghum. Crop Sci. 14:80–84.

Lewis, R. B., E. A. Hiler, and W. R. Jordan. 1974. Susceptibility of grain sorghum to water deficit at three growth stages. Agron. J. 66:589–591.

Longenecker, D. E., and L. J. Eric. 1968. Irrigation water management. p. 321–346. *In* F. D. Elliot et al. (ed.) Advances in production and utilization of quality cotton: Principles and practices. Iowa State Univ. Press, Ames.

Ludlow, M. M. 1976. Ecophysiology of C_4 grasses. p. 364–380. *In* O. L. Lange et al. (ed.) Water and plant life. Springer-Verlag New York, Inc., New York.

Ludlow, M. M. 1980. Adaptive significance of stomatal responses to water stress. p. 123–138. *In* N. C. Turner and P. J. Kramer (ed.) Adaptation of plants to water and high temperature stress. John Wiley & Sons, Inc., New York.

Mahalakshmi, V. 1978. Temperature influence on respiration and starch synthetase in sorghum [*Sorghum bicolor* (L.) Moench]. Ph.D. Dissertation. Univ. of Nebraska, Lincoln (Diss. Abstr. 79-10507).

Maiti, R. K., and S. Bisen. 1978. Studies on growth and development of panicles and grain in two contrasting genotypes of pearl millet. p. 115–125. *In* C. P. Malik et al. (ed.) Physiology of sexual reproduction in flowering plants. Kalyani Publishers, New Delhi.

Mauney, J. R. 1979. Production of fruiting points. p. 256–261. *In* J. M. Brown (ed.) Cotton physiology: A treatise. Section I. Flowering, fruiting, and cutout. Proc. Beltwide Cotton Production Res. Conference, Phoenix, Ariz. 7–11 Jan. Natl. Cotton Coun., Memphis, Tenn.

McCree, K. J., and C. H. M. van Bavel. 1977. Respiration and crop production: A case study with two crops under water stress. p. 199–216. *In* J. J. Landsberg and C. V. Cutting (ed.) Environmental effects on crop physiology. Academic Press, Inc., New York.

McMichael, B. L. 1979. The influence of plant water stress on flowering and fruiting in cotton. p. 301–302. *In* J. M. Brown (ed.) Cotton physiology: A treatise. Section I. Flowering, fruiting, and cutout. Proc. Beltwide Cotton Production Res. Conference, Phoenix, Ariz. 7–11 Jan. Natl. Cotton Coun., Memphis, Tenn.

Morgan, J. M. 1980. Differences in adaptation to water stress within crop species. p. 369–382. *In* N. C. Turner and P. J. Kramer (ed.) Adaptation of plants to water and high temperature stress. John Wiley & Sons, Inc., New York.

Morse, W. J. 1920. Cowpeas: Culture and varieties. USDA Farmers Bull. no. 1148.

Musick, J. T., D. W. Grimes, and G. M. Herron. 1963. Irrigation water management and nitrogen fertilization of grain sorghums. Agron. J. 55:295–298.

Nobel, P. S. 1980. Leaf anatomy and water use efficiency. p. 43–55. *In* N. C. Turner and P. J. Kramer (ed.) Adaptation of plants to water and high temperature stress. John Wiley & Sons, Inc., New York.

Ogunlela, V. B. 1979. Physiological and agronomic responses of grain sorghum [*Sorghum bicolor* (L.) Moench] hybrid to elevated night temperatures. Ph.D. Dissertation. Univ. of Nebraska, Lincoln (Diss. Abstr. 80-10871).

Ohmes, F., and P. Penas. 1977. Consumptive water-use determinations for crops at Garden City. p. 31–62. *In* Proc. First Annual Workshop on Water-use Res., Manhattan, Kan. (Mimeo.) Sponsored by Kansas State Univ., Manhattan.

Paulson, I. W. 1962. Embryology and seedling development to floral transition of *Sorghum vulgare* Pers. Ph.D. Dissertation. Iowa State Univ., Ames.

Perry, S. W., and D. R. Kreig. 1981. Gross:net photosynthetic ratios of cotton as affected by environment and genotype. p. 51–52. *In* J. M. Brown (ed.) Proc. Beltwide Cotton Production Res. Conference, New Orleans. 5–8 Jan. Natl. Cotton Coun., Memphis, Tenn.

Powers, D., E. T. Kanemasu, P. Singh, and G. Kreitner. 1980. Floral development of pearl millet [*Pennisetum americanum* (L.) K. Schum]. Field Crops Res. 3:245–265.

Quinby, J. R. 1963. Manifestation of hybrid vigor in sorghum. Crop Sci. 3:288–291.

Rachie, K. O. 1966. Aridity and salinity; a survey of soils and land use. p. 42–127. *In* H. Boyko (ed.) Salinity and aridity: New approaches to old problems. Junk Publ., The Hague.

Rachie, K. O., and J. V. Majmudar. 1980. Pearl millet. Pennsylvania State Univ. Press, University Park.

Rachie, K. O., and K. M. Rawal. 1976. Integrated approaches to improving cowpea, *Vigna unguiculata* L. Walp. Int. Inst. of Tropical Agric. Tech. Bull. no. 5, Ibadan, Nigeria.

Rachie, K. O., and L. M. Roberts. 1974. Grain legumes of the lowland tropics. *In* N. C. Brady (ed.) Advances in Agronomy 26:1–132. Academic Press, Inc., New York.

Rawson, H. M., N. C. Turner, and J. E. Begg. 1978. Agronomic and physiological responses of soybean and sorghum crops to water deficits. IV. Photosynthesis, transpiration and water-use efficiency of leaves. Aust. J. Plant Physiol. 5:179–194.

Rice, J. R. 1980. Physiological investigations of grain sorghum [*Sorghum bicolor* (L.) Moench] subjected to water stress conditions. Ph.D. Dissertation. Univ. of Nebraska, Lincoln. Diss. Abstr. 40(1–2):527B. Univ. Microfilm no. 7918697, Ann Arbor, Mich.

Ritchie, J. T. 1971. Dryland evaporative flux in a subhumid climate: I. Icrometeorological influences. Agron. J. 63:51–55.

Ritchie, J. T., and E. Burnett. 1971. Dryland evaporative flux in a subhumid climate: II. Plant influences. Agron. J. 63:56–62.

Ritchie, J. T., and R. C. Henderson. 1972. Dryland evaporative flux in a subhumid climate: III. Soil water influence. Agron. J. 64:168–173.

Ritchie, J. T., and W. R. Jordan. 1972. Dryland evaporative flux in a subhumid climate: IV. Relation to plant water status. Agron. J. 64:173–176.

Salter, P. J., and J. E. Goode. 1967. Crop responses to water at different growth stages. East Malling, Maidstone, Kent, Commonwealth Bureau of Horticulture and Plantation Crops (Resources Res. Series 2).

Sanders, E. H. 1955. Developmental morphology of the kernel in grain sorghum. Cereal Chem. 32:12–25.

Shantz, H. L., and L. N. Piemeisel. 1927. The water requirement of plants at Akron, Colo. J. Agric. Res. (Washington, D.C.) 34:1093–1190.

Shearman, L. L., J. D. Eastin, C. Y. Sullivan, and E. J. Kinbacher. 1972. Carbon dioxide exchange in water stressed sorghum. Crop Sci. 12:406–409.

Sionit, N., and P. J. Kramer. 1977. Effect of water stress during different stages of growth of soybean. Agron. J. 69:274–276.

Slatyer, R. O. 1964. Efficiency of water utilization by arid zone vegetation. Ann. Arid Zone 3:1–12.

Slatyer, R. O. 1973. Effects of short periods of water stress on leaf photosynthesis. p. 211–276. *In* R. O. Slatyer (ed.) Plant response to climatic factors. Proc. Uppsala Symp., 1970. United Nations Educational, Scientific, and Cultural Organization, Paris.

Staph, O. 1934. *In* D. P. Thisleton (ed.) Flora of tropical Africa. Dyer Gramineae 9:954–1090.

Stickler, F. C., A. W. Pauli, H. H. Laude, H. D. Wilkins, and J. L. Mings. 1961. Row width and plant population studies with grain sorghum at Manhattan, Kansas. Crop Sci. 4:297–300.

Sullivan, C. Y. 1972. Mechanisms of ehat and drought resistance in grain sorghum and methods of measurement. p. 247–264. *In* N. G. P. Rao and L. R. House (ed.) Sorghum in seventies. Oxford & IBH Publishing Co., New Dehli.

Sullivan, C. Y., N. V. Norcio, and J. D. Eastin. 1977. Plant responses to high temperatures. p. 301–318. *In* A. Muhammed et al. (ed.) Genetic diversity in plants. Plenum Publishing Corp., New York.

Summerfield, R. J., F. R. Minchin, E. Roberts, and P. Hadley. 1980. Cowpea. *In* S. Yoshida (ed.) Symp. on Potential Productivity of Field Crops Under Different Environments, Los Baños, Laguna, Philippines. 22–26 Sept. Int. Rice Res. Inst., Los Baños, Laguna, Philippines. (In press.)

Summerfield, R. J., F. R. Minchin, K. A. Stewart, and B. J. Ndunguru. 1978. Growth, reproductive development and yield of effectively nodulated cowpea plants in contrasting aerial environments. Ann. Appl. Biol. 90:277–291.

Teare, I. D., E. T. Kanemasu, W. L. Powers, and H. S. Jacobs. 1973. Water-use efficiency and its relation to crop canopy area, stomatal regulation and root distribution. Agron. J. 65:207–210.

Thompson, C. 1977. Consumptive water-use determinations for sorghum and wheat under dryland conditions at Hays. p. 1963–1967. *In* Consumptive water-use by crops in Kansas. Proc. First Annual Workshop on Water Use Res., Manhattan, Kan. Kansas State Univ., Manhattan.

Townley-Smith, T. F., and E. A. Hurd. 1979. Testing and selecting for drought resistance in wheat. p. 447–464. *In* H. Mussell and R. C. Staples (ed.) Stress physiology in crop plants. John Wiley & Sons, Inc., New York.

Tufail, M. 1971. Studies on components of yield in relation to plant structure in [*Sorghum bicolor* (L.) Moench]. M.S. Thesis. Univ. of Nebraska, Lincoln.

Turner, N. C. 1975. Current comparisons of stomatal behaviour, water status, and evaporation of maize in soil at high or low water potential. Plant Physiol. 55:932–936.

Turner, N. C. 1979. Drought resistance and adaptation to water deficits in crop plants. p. 343–372. *In* H. Mussell and R. C. Staples (ed.) Stress physiology in crop plants. John Wiley & Sons, Inc., New York.

Turner, N. C., J. E. Begg, H. M. Rawson, S. D. English, and A. B. Hearn. 1978. Agronomic and physiological responses of soybean and sorghum crops to water deficits: III. Components of water leaf potential, leaf conductance, $^{14}CO_2$ photosynthesis and adaptation to water deficits. Aust. J. Plant Physiol. 5:169–177.

Turner, N. C., and M. M. Jones. 1980. Turgor maintenance by osmotic adjustment: A review and evaluation. p. 87–104. *In* N. C. Turner and P. W. Kramer (ed.) Adaptation of plants to water and high temperature stress. John Wiley & Sons, Inc., New York.

Wanjura, D. F., and D. R. Buxton. 1972. Water uptake and radicle emergence as affected by soil moisture and temperature. Agron. J. 64:427–431.

Wanjura, D. F., D. R. Buxton, and H. N. Stapleton. 1970. A temperature model for predicting initial cotton emergence. Agron. J. 62:741–743.

Warrington, I. J., R. L. Dunstone, and L. M. Green. 1977. Temperature effects at three development stages on the yield of the wheat ear. Aust. J. Agric. Res. 28:11–27.

Wenkert, W., E. R. Lemon, and T. R. Sinclair. 1978. Water content–potential relationship in soya bean: Changes in component poentials for mature and immature leaves under field conditions. Ann. Bot. (London) 42:295–307.

Werth, E. 1943. Geography and history of millets. Angew. Bot. 19:41.

Whigham, D. K. 1980. Soybean. *In* S. Yoshida (ed.) Symp. on Potential Productivity of Field Crops Under Different Environments, Los Baños, Philippines. Int. Rice Res. Inst., Los Baños, Philippines. (In press.)

Whiteman, P. C., and G. L. Wilson. 1965. Effects of water stress on the reproductive development of *Sorghum vulgare* Pers. Univ. Queensl. Pap. 14:233–239.

Wien, H. C., E. J. Littleton, and A. Ayanaba. 1979. Drought stress of cowpea and soybean under tropical conditions. p. 294–301. *In* H. Mussell and R. Staples (ed.) Stress physiology in crop plants. John Wiley & Sons, Inc., New York.

Wilson, J. R., M. M. Ludlow, M. J. Fisher, and E.-D. Schulze. 1980. Adaptation to water stress of the leaf water relations of four tropical forage species. Aust. J. Plant Physiol. 7: 207–220.

Wilson, J. R., C. H. M. van Bavel, and K. J. McCree. 1980. Carbon balance of water-deficient grain sorghum plants. Crop Sci. 20:153–158.

Witt, M. D. 1981. Physiological studies of pearl millet and grain sorghum subjected to high temperature and moisture stress. Ph.D. Dissertation. Univ. of Nebraska, Lincoln (Diss. Abstr. 18-18188).

Zablotowicz, R. M., D. D. Focht, and G. H. Cannell. 1981. Nodulation and N fixation of field-grown California cowpeas as influenced by well irrigated and droughted conditions. Agron. J. 73:9–12.

23

Ordie R. Jones and Wendell C. Johnson
Agricultural Research Service
U.S. Department of Agriculture
Bushland, Texas

Cropping Practices: Southern Great Plains

Research on dryland crops and cropping sequences has been conducted at various locations throughout the Southern Great Plains. Results have shown that only a limited number of crops can be produced profitably on drylands in the area (Daniel and Finnell, 1939; Harvey et al., 1961; Keating and Mathews, 1957; Mathews and Barnes, 1940). The major adapted crops are winter wheat, sorghum, and upland cotton. Other crops grown are forages of various types, sunflower, and guar. Cropping practices and systems employed to produce these crops are discussed below. Discussion is limited to the Southern High Plains, Rolling Red Plains, and Reddish Prairie land resource areas of the Southern Great Plains.

23-1 CROPS

23-1.1 Winter Wheat

23-1.1.1 General Cropping Practices

The distribution and area occupied by winter wheat in 1980 is shown in Fig. 23-1. Most wheat in the Rolling Red Plains and Reddish Prairie is cropped annually. In the Southern High Plains, and particularly in the Oklahoma panhandle, from 40 to 50% of wheat is grown in an alternate crop and fallow sequence commonly referred to as "summer fallowing."

Compared with continuous cropping, summer fallowing reduces total grain production in all areas of the Southern Great Plains. However, in the

Contribution from Agricultural Research Service, U.S. Department of Agriculture, in cooperation with The Texas Agricultural Experiment Station, Texas A&M University, College Station, TX 77843.

Dryland Agriculture—Agronomy Monograph no. 23.

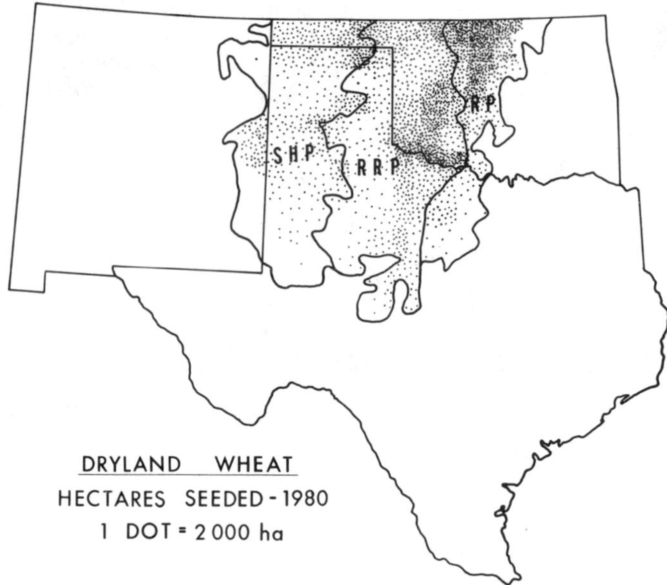

Fig. 23-1. Distribution of dryland wheat in Southern High Plains (SHP), Rolling Red Plains (RRP), and Reddish Prairie (RP) majorland resource areas of the Southern Great Plains. (Drawn from data supplied by Crop and Livestock Reporting Services of Texas, Oklahoma, and New Mexico.)

drier western area of the Southern Great Plains, fallowing greatly decreases the probability for crop failure. At Bushland, Texas, with average annual precipitation of 47 cm, average wheat yields (1942 to 1970) were 650 kg/ha for continuous wheat and 1000 kg/ha for wheat grown on summer fallow; thus, fallowing increased yield per harvested hectare by 54%. The probability of producing a 1000 kg/ha wheat crop was 0.50 with summer fallow and only 0.25 for continuous cropping (Johnson et al., 1974). On a Pratt fine sandy loam (sandy, mixed Thermic Psammentic Haplustalfs) at Woodward, Oklahoma, with average rainfall of 60.5 cm/y, summer fallowing increased yield only 28% (1160 kg/ha to 1490 kg/ha) in comparison with continuous cropping (Locke and Mathews, 1953). Thus, on the Rolling Red Plains and Reddish Prairie, summer fallowing is not recommended, because yields are not increased sufficiently by fallowing and the soil erosion hazard may be increased during the 15-month fallow period, even when stubble mulch tillage is used (Johnson et al., 1974).

Winter wheat is produced on most soil types in the Southern Great Plains. The largest wheat areas are on fine-textured soils such as loams and clay loams. With larger water-holding capacities, these soils produce greater yields than sandier soils during favorable years. However, during dry years, there are fewer crop failures on coarser soils, and yields are better than on fine-textured soils (Mathews and Brown, 1938). In marginal rainfall areas, as in the Southern High Plains resource area, coarse-textured soils such as sands and sandy loams are not adapted to wheat production because they provide insufficient soil water storage to sustain plants during winter and early spring when precipitation is low and plants are subject to damage from wind erosion.

Yields of continuous wheat increase from west to east across the Southern Great Plains in response to increasing precipitation. Yields average 650 kg/ha at Bushland, Texas (Johnson et al., 1974) and Goodwell, Oklahoma (H. E. Reeves. 1978. Personal communication), increase to 1140 kg/ha at Woodward, Oklahoma (Locke and Mathews, 1953), and to 1750 kg/ha in Garfield County in north central Oklahoma. In the Reddish Prairie of Oklahoma, sometimes called the "breadbox" of Oklahoma, 80 to 90% of all cultivated land is continuously cropped to wheat, with average county yields ranging from 1340 to 2020 kg/ha (D. M. Fain. 1978. Personal communication). In the drier areas (Southern High Plains Major Land Resource Area), commercial fertilizers are not normally required because water is the factor that limits yield, whereas in the Rolling Red Plains and Reddish Prairie, both N and P are necessary because nutrient supply rather than water often limits yield.

Continuous annual cropping of land to one crop is not a generally recommended practice. Problems may develop with weeds, diseases, insects, and soil fertility (Arnon, 1972). Continuous cropping of winter wheat is not objectionable, except possibly with respect to weed control. At Stillwater, Oklahoma, continuous wheat has been grown on the same plot of land since 1893. Organic matter and N in the topsoil have gradually decreased, but the average yield has not declined appreciably even where no fertilizer has been applied. Unfertilized wheat has averaged 850 kg/ha, whereas manured treatments have averaged 1380 kg/ha. A similar yield increase in response to the addition of commercial fertilizer indicates that soil fertility rather than water is the primary factor limiting wheat production in the Reddish Prairie (Harper, 1959; Webb et al., 1980).

Current fertilizer recommendations for the Reddish Prairie of Oklahoma are based on a minimum requirement of 112 kg/ha of N, 32 kg/ha of P, and 230 kg/ha of K. Normally only N and P are added because sufficient K is usually present in the soil. Since most wheat grown in Oklahoma is grazed, additional fertilizer beyond the requirement for grain production may be applied.

23-1.1.2 Seeding Date

The best time to seed winter wheat on the Southern Great Plains varies between years. On the Rolling Red and Southern High Plains, highest grain yields can be expected when wheat is seeded from about 20 September to 10 October (Porter et al., 1952). Highest grain yields is the Reddish Prairie result from seeding in mid- to late October.

Winter wheat, however, can be seeded from late August until mid-November. Major factors determining seeding date are soil water supply, diseases, and whether or not the wheat will be grazed. Early seeding of wheat (20 August to 10 September) has the advantages of providing forage for grazing and reducing the wind erosion hazard, but it has several disadvantages. Early seeding of wheat, followed by warm weather, may subject plants to infections by root rot, particularly in southern parts of the Rolling Red Plains and the Reddish Prairie. Early seeding encourages heavy growth of plants, which can suffer winter injury unless grazed. Heavy growth also depletes soil water reserves, making the plants totally dependent on precipitation to sustain spring growth. Late seeding (late October to mid-November) may result in poorly tillered, shallow-rooted plants that are subject to winter injury. Winter wheat seedlings in the three- to four-leaf stage survive cold temperatures better than more developed plants or those in the one- to two-leaf stage (Schlehuber and Tucker, 1967).

23-1.1.3 Seedbed Preparation

Stubble mulch tillage is used by many farm operators to maintain crop residues on the surface for soil protection against erosion and for increased water storage. The initial tillage after harvest should be the deepest, with subsequent operations at shallower depths. Chisel plows or stubble mulch sweep plows with large V-shaped blades are used in initial operations at depths of about 15 cm, with subsequent operations at 5- to 10-cm depths. Tillage should be used only as often as required to control weeds. Shallow tillage with V-shaped blades or rod weeders should be performed immediately before seeding to eliminate weeds and compact the seedbed to aid in seedling establishment.

In the Rolling Red Plains and the Reddish Prairie of Oklahoma, after harvest of wheat crops with heavy residue (2800 to 3360 kg/ha), the initial operation is offset disking with subsequent operations with V-shaped blades. A problem with cheat has developed in the Reddish Prairie of Oklahoma. Inversion of the soil profile with moldboard plows to depths of 20 to 25 cm at 4-year intervals is one method for cheat control. The herbicide metribuzin [4-amino-6-*tert*-butyl-3-(methylthio)-*as*-triazin-5(4*H*)-one] can also be used to control cheat in certain varieties of winter wheat. Currently metribuzin is labeled only for TAM W101 and Newton varieties of wheat (D. M. Fain. 1978. Personal communication).

23-1.1.4 Seeding Rates and Equipment

Seeding rates should be adjusted according to seeding date, water supply, location, and ultimate use of the wheat (grazing, grain or both). The predominant factor is the amount of water available for production. In the drier Southern High Plains Major Land Resource Area, a seeding rate of 33 kg/ha is adequate, whereas in the Reddish Prairie Major Land Resource Area, rates of 44 to 84 kg/ha are recommended. Rates should be increased slightly for late seeding or where the wheat will be grazed.

Two basic types of grain drills are used in seeding wheat—disk and deep-furrow hoe drills. Both types can be found throughout the Southern Great Plains. Disk drills, with single- or double-disk openers, are used to seed smooth, well-prepared seedbeds that have little or no crop residue remaining on the surface. Row spacings range from 20 to 35 cm. The narrow spacings are commonly used in the eastern or high rainfall areas, and wider spacings are used in dry areas.

High-clearance deep-furrow hoe drills, with row spacings of 25 to 35 cm, are adapted to seeding in drier soils and in stubble mulch residues. These drills have a hoe or shovel-type opener that can penetrate dry surface layers and place seed into moist soil. A press wheel in each furrow firms the soil to aid seedling emergence. The resulting furrow provides protection to emerging seedlings from blowing soil.

23-1.2 Cotton

23-1.2.1 General Cropping Practices

Distribution and area of dryland cotton in the Southern Great Plains is shown in Fig. 23-2. The northern limit of production parallels the 210-day frost-free line. Most cotton is grown on coarse-textured, deep soils in the

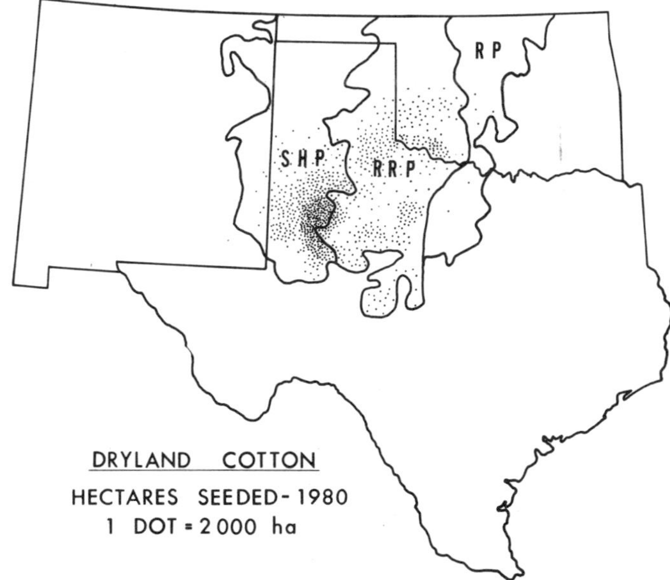

Fig. 23-2. Distribution of dryland cotton in the Southern High Plains (SHP), Rolling Red Plains (RRP), and Reddish Prairie (RP) major land resource areas of the Southern Great Plains. (Drawn from data supplied by Crop and Livestock Reporting Services of Texas, Oklahoma, and New Mexico.)

Table 23-1. Average cotton lint yield for total area, including skip-rows. Cotton was seeded in rows 1 m apart.

Cropping system	Chillicothe, Texas (Mulkey, 1968) 1965-1967	Lubbock, Texas (Newman, 1967) 1963-1965	Big Spring, Texas (Burnett & Welch, 1966) 1958-1962	Spur, Texas (Marion†) 1930-1943
	kg/ha			
All rows planted	268	196	289	299
Skip-row				
1 in-1 out	--	--	292	281
2 in-1 out	291	165	284	285
2 in-2 out	280	178	261	--
4 in-4 out	291	--	226	--

† P. T. Marion. (No date.) Cotton method of planting test A. Texas Agric. Exp. Stn. Mimeo Rep., Spur, Texas.

Southern High Plains and Rolling Red Plains resource areas. Most dryland cotton is upland type grown for once-over stripper harvesting.

Cotton is drought resistant, and its deep root system enables it to produce some lint yield even with adverse soil water conditions. Cotton responds well to increased soil water content at planting (Burnett and Fisher, 1954), but yields are not increased sufficently by fallowing to make the practice profitable (Keating and Mathews, 1957). Thus, cotton is grown continuously or in rotation with sorghum or wheat.

Nearly all cotton grown in areas with less than 56 cm average annual precipitation is grown in a skip-row planting system. Common systems are (i) two rows planted and two rows skipped, (ii) two rows planted and one row skipped, (iii) and four rows planted and two rows skipped. Skip-row planting of cotton is a hold-over practice that developed when acreage limitations were imposed a number of years ago. Highest per hectare yields are usually obtained from planting every row; however, yield decreases attributable to skip row planting are not large (Table 23-1). Skip-row planting has the following advantages over solid planting:

1. Lower production costs because fewer seed are planted and the harvested area is smaller.
2. Weed control is easier.
3. Wheel traffic can be concentrated in skipped rows, resulting in less compaction near planted rows.
4. High-residue crops can be seeded in skipped rows to provide barriers for wind erosion protection. This is not a common practice because yield of cotton, a high-value crop, is often reduced; however, Fryrear (1976) has shown this is an adapted conservation practice on sandy soils in the Southern High Plains.

Weed control in cotton has undergone a quiet revolution since trifluralin (α,α,α-trifluoro-2,6-dinitro-*N,N*-dipropyl-*p*-toluidine) and nitralin [4-(methylsulfonyl)-2,6-dinitro-*N,N*-dipropylaniline] were shown to be effective preplant herbicides (Wiese et al., 1969). Nearly all cotton grown in the Southern Great Plains is treated with preplant or preemergence herbicides for weed control. In addition, johnsongrass in cotton is controlled by postplant treatment with glyphosate [*N*-(phophonomethyl)glycine] using a recirculating sprayer or a rope wick applicator.

23-1.2.2 Seeding Date

Short growing seasons often limit cotton lint yields on the Southern High Plains. It is important to seed as early as practical in the spring to obtain the longest growing period possible. Seeding too early, however, when soil temperatures are low, results in poor emergence, increased incidence of root and seedling diseases, and slowed seedling development. As a guide, Elliot et al. (1968), suggested delaying seeding until the minimum soil temperature at the 20-cm depth in the bed averages 17.7° C for 10 days. On the Southern High Plains (Lubbock, Texas), this normally occurs from 1 to 10 May; yields of cotton seeded after about 1 June begin to decrease, and the latest practical seed date is about 10 June.

In the Rolling Red Plains where the growing season is longer, most cotton is seeded between 15 May and 1 June. Later seeding may result in increased damage to cotton squares and bolls from boll weevil because weevil populations increase as the growing season progresses (Rogers et al., 1976). Cotton often must be reseeded because intense rainfall causes soil crusting, which reduces seedling emergence. Seedlings may also be destroyed by blowing soil or by hail.

23-1.2.3 Seedbed Preparation

Land preparation normally begins in February or March. Residues from the previous crop are shredded, and the land is moldboard plowed or chiseled to reduce compaction and bring soil clods to the surface to prevent soil blowing. Sometime before seeding, most land is listed, which further reduces soil blowing and provides a well-drained seedling environment if the soil is wet at seeding. Preplant herbicides are often incorporated with a tandem disk before listing.

23-1.2.4 Planting Methods and Equipment

Dryland cotton is commonly planted on a bed with unit planters in the loam and clay loam soil areas. Both bed and furrow planting are used in the fine sandy loam areas. Furrow planting with a lister planter is more common on loamy fine sand areas. With the lister planter, a chisel or runner opener behind the lister point opens a slot for placing the seed 4 to 5 cm deep in the bottom of the furrow. A press wheel, operated immediately behind the opener, presses seed into the bottom of the furrow. A drag covers the seed with moist, loose soil.

Cotton emerges better when planted in a broad, shallow lister furrow than when planted in a deep furrow (Colwick, 1957; Elliot et al., 1968).

Deep furrows provide seedlings with protection against soil blowing, which frequently occurs on sandy soils. However, deep furrows are conducive to "washing in," which can cover seedlings when high-intensity rainfall occurs. Furrows must be deep enough so that seed is placed in moist soil; thus, in dry years it is necessary to have deeper furrows. When the soil is wet at planting, listers may be operated at shallow depths. Frequently, herbicides are sprayed in bands behind the planter, with shallow incorporation by vertical fingers or rotary hoes mounted behind the openers.

Double-disk opener unit planters are also used to plant cotton on flat ground or on top of lister beds when the soil is moist near the surface. Sweeps or shovels are frequently operated in front of the planter to kill weeds and loosen the soil.

23–1.2.5 Population and Row Spacing

Plant populations for dryland cotton are not critical. Colwick (1957) reported that 98 800 plants/ha are optimum for Southwest Oklahoma and West Texas, but populations could vary between 74 and 0.124 million plants/ha with little or no effect on yield. N. B. Thomas (1978. Personal communication), at Altus, Oklahoma, and D. W. Fryrear (1979. Personal communication), at Big Spring, Texas, recommended a population of 0.108 million plants/ha for solid planting (1-m row spacing). D. W. Fryrear recommended a plant spacing of 8 to 10 cm within planted rows on sandy loam and fine sandy loam soils at Big Spring, Texas. With this plant spacing, populations would be less for skip-row systems than for solid-planted cotton. Low populations should be avoided when soil water is favorable, because plants grow large and are difficult to harvest.

Cotton is normally seeded in 1-m spaced rows. However, it can be produced successfully with row spacings of 25 to 50 cm. At Lubbock, Texas, Hudspeth and Brashears (1974) reported that cotton seeded in 1-m rows and 25-cm rows produced similar yields when the population was 0.124 million plants/ha or less and when average rainfall occurred. However, yields were significantly higher with narrow rows when precipitation was above average and plant populations were increased to 0.247 million plants/ha. Short-season cotton varieties, adapted to narrow row planting, have been developed, but in dry years the plants are short and harvesting is difficult. Also, herbicides must be 100% effective or hand hoeing will be required because cultivation after planting of narrow row cotton is difficult.

23–1.3 Grain Sorghum

23–1.3.1 General Cropping Practices

Sorghum is well adapted to dryland grain production in the Southern Great Plains. With greater precipitation occurring in the summer months, the peak water use demand for sorghum and high summer rainfall fre-

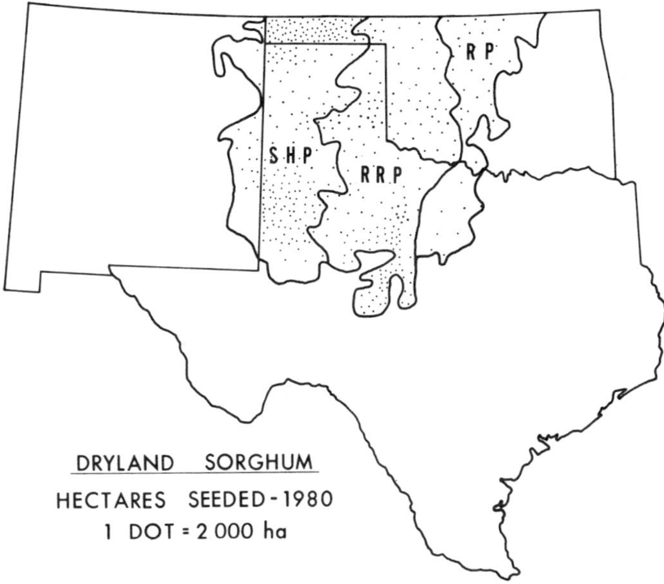

Fig. 23-3. Distribution of dryland sorghum in the Southern High Plains (SHP), Rolling Red Plains (RRP), and Reddish Prairie (RP) major land resource areas of the Southern Great Plains. (Drawn from data supplied by Crop and Livestock Reporting Services of Texas, Oklahoma, and New Mexico.)

quently coincide to produce high sorghum grain yields. When water stress occurs during drought periods, sorghum growth slows and the crop becomes practically dormant. It resumes growth when sufficient soil water is again available; thus, sorghum usually produces some grain, even under adverse moisture conditions (Leonard and Martin, 1963).

The heaviest concentration of sorghum production is on coarse-textured soils in the Southern High Plains, where it is rotated with cotton, and in the Oklahoma panhandle, where sorghum may be seeded if the soil was too dry to seed wheat the preceding fall (Fig. 23-3). Sorghum is the major dryland cash crop only on small areas of sandy soil in the northern panhandle of Texas and the Oklahoma panhandle.

Sorghum uses stored soil water efficiently to produce grain as shown in Fig. 23-4 (Brown and Shrader, 1959; Jones and Hauser, 1974). However, yields are not increased sufficiently by fallowing to justify a sorghum-fallow cropping system, which has a 19-month fallow period (Harvey et al., 1961; Mathews and Barnes, 1940; Fryrear, 1976). In the Rolling Red Plains at Woodward, Oklahoma, sorghum yields from continuous cropping and cropping after fallow were equal (Locke and Mathews, 1953). Yields of continuously cropped sorghum will be lowest in the western areas and increase eastward in response to greater average annual precipitation. Grain yields of continuous sorghum are low on the Southern High Plains (Table 23-2); thus, in drier areas, sorghum is best utilized in a wheat-sorghum-fallow or possibly a cotton-sorghum sequence. Since cotton normally does not leave

Table 23-2. Annual grain sorghum yields per harvested hectare, Southern High Plains.

Cropping sequence	Bushland, Texas (Jones, 1975) 1959–1972	Dalhart, Texas (Mathews & Barnes, 1940) 1909–1938	Lubbock, Texas (Harvey et al., 1961) 1914–1940	Big Spring, Texas (Keating & Mathews, 1957) 1916–1953
	kg/ha			
Wheat-sorghum-fallow	1890	--	--	--
Continuous sorghum	1240	1380	1370	1250
Fallow-sorghum	--	2220	1780	1660
Cotton-sorghum	--	--	1670	730

soil as depleted of moisture and nutrients as sorghum, sorghum yields after cotton usually are as great as yields from continuous sorghum unless influenced by adverse cultural practices. This is demonstrated by the results obtained at Lubbock (Table 23-2). Keating and Mathews (1957) did not explain the extremely low sorghum yields obtained with the cotton-sorghum sequence at Big Spring. These low sorghum yields may be related to the increased wind erosion hazard after the cotton crop was harvested.

23-1.3.2 Seeding Date

Sorghum may be planted from the time the soil warms to 16°C in the spring until the latest date that will allow plants to reach physiological maturity before frost. Major factors that determine seeding date are the ma-

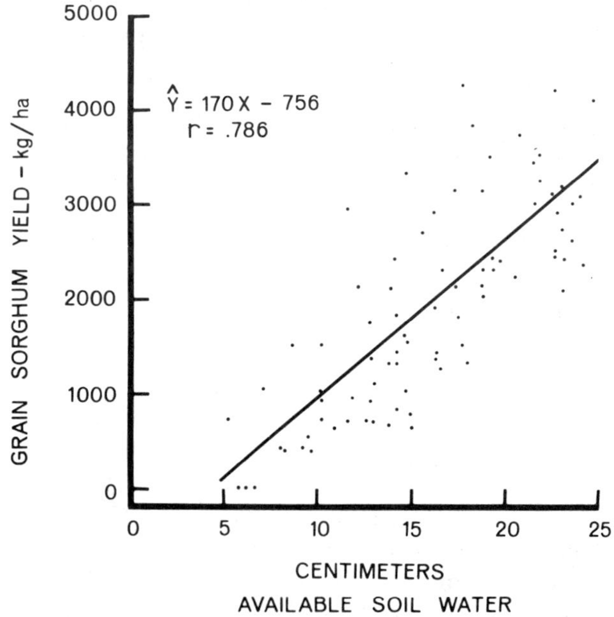

Fig. 23-4. Effect of soil water content at seeding (0- to 180-cm soil depth) on grain yield of dryland sorghum (Jones and Hauser, 1974).

turity group of the sorghum hybrid, soil temperature, seasonal rainfall and temperature patterns, diseases, and insects.

Quinby et al. (1958) reported that sorghum planting can begin on the Southern High Plains about 15 May, but higher yields are produced from plantings made 10 to 25 June. However, to avoid sorghum midge infestations in the southern part of the Southern High Plains, sorghum should be seeded so that blooming occurs before 1 August. For short-maturity hybrids, planting should be before 10 June, and longer-season hybrids should be seeded earlier (Bottrell, 1971).

In the Rolling Red Plains of Texas, similar grain yields are expected from plantings made in April, May, or June. However, during some years, economic infestations of sorghum midge occur in sorghum blooming after 15 July. Sorghum midge can be controlled with insecticides. The insect is seldom observed north of the 35th parallel (E. P. Boring. 1980. Personal communication).

23-1.3.3 Row Spacing and Plant Populations

Sorghum is grown as a row crop. When water is adequate, which is often in the Rolling Red Plains and the Reddish Prairie, sorghum yields are increased by seeding in narrow rows (50- to 75-cm row spacings). Yield increase with narrow rows compared with 1-m row spacing is attributed to more efficient use of water, nutrients, and solar energy (Onken, 1970). With limited water at Bushland, Texas, in the Southern High Plains, Bond et al. (1964) achieved significantly higher production with 1-m spacings than with 50-cm row spacings. They also found that a 2.2-kg/ha seeding rate produced higher yields than a 4.5-kg/ha seeding rate. Brown and Shrader (1959) obtained similar results at Hays, Kansas. They attributed higher yields with wide row spacings to increased competition within the row. They found that the competition reduced forage production but increased grain yield and water use efficiency for grain production.

Sorghum produces the same yield over a wide range of plant populations because of adjustments in tillering and head size. High plant populations with limited water, however, may increase early-season use of soil water with subsequent late-season water stress, causing reduced yield and increased lodging (Onken, 1970). Minimum seeding rate for limited water is about 2.2 kg/ha, or a population of 62 000 plants/ha. Seeding rates can be increased in response to increased water supplies.

23-1.3.4 Seedbed Preparation

The residues of the previous crop should be left on the soil surface to aid in wind erosion control. Residues should be undercut with V-shaped blades or chiseled to loosen the soil and cover seed from the previous crop so that volunteer plants and weeds emerge before seeding. The initial tillage operation should be deep enough to establish a loose, friable seedbed with subsequent operations at shallower depths.

On coarse-textured soils, listing is desirable to prevent or reduce wind erosion. On fine-textured soils, the surface is normally left flat until seeding. Before seeding, shallow tillage with rod weeders or sweeps is usually necessary to kill weeds and volunteer plants. Preemergence herbicides can be applied at planting or any time before seedling emergence. Broadleaf weeds can be controlled by spraying 2,4-D [(2,4-dichlorophenoxy)acetic acid] following emergence in most sorghum hybrids after sorghum is at least 15 cm tall.

23–1.3.5 Planting Methods and Equipment

Sorghum can be seeded with a grain drill, lister planter, or unit planter. Lister planters are commonly used where cotton is grown, and unit planters or grain drills, either disk or deep furrow, are used elsewhere. Spouts on grain drills can be plugged to achieve desired row spacings.

23–1.4 Forages

Forages commonly grown on cultivated drylands in the Southern Great Plains are alfalfa, sorghums, and small grains. Alfalfa is usually grown only in areas of 63 cm or more annual rainfall (Hughes et al., 1962). Alfalfa production in the Southern Great Plains is limited to the eastern half of the Rolling Red Plains and to the Reddish Prairie. On fine-textured soils in North Central Oklahoma, most alfalfa is grown for hay. Yields average about 9 t/ha (D. M. Fain. 1980. Personal communication). Stands are commonly maintained for 5 to 8 years. In Southwestern Oklahoma, on coarser-textured soils, alfalfa is grown primarily for seed. Each year one or two hay crops are harvested, and then the crop is allowed to produce seed. Seed yields are low, ranging from about 170 to 200 kg/ha. Stands are normally maintained for 4 or 5 years, followed by wheat for several years (N. B. Thomas. 1980. Personal communication).

Seedbed preparation for dryland alfalfa is critical, because seed must be sown shallow (1.3 cm) yet must also be placed in moisture. A fine, firm seedbed is obtained by moldboard plowing, disking, and cultipacking. Single- or double-disk grain drills are used to seed 13 to 22 kg/ha of seed inoculated with *Rhizobium meliloti* (Dangeard). Most alfalfa is seeded in August and September, thus annual weeds that emerge after seeding are killed by frost. Seeding of alfalfa is not recommended on land infested with perennial weeds. A 3-year supply of P (about 50 kg/ha) is commonly "plowed down" at establishment.

Maintenance of the alfalfa stand is enhanced by allowing plants to bloom before cutting for hay. Frequent cutting at immature stages increases hay quality but weakens stands, reduces yields, and may allow weeds and grasses to become established. Alfalfa should be cut at a height above the ground so that some leaves remain and basal buds are preserved to promote rapid regrowth (Arnon, 1972).

Sorghum forages are frequently grown for grazing with excess production baled for hay. Sudangrass hybrids and sweet sorghum × sudangrass hybrids are best adapted because they are fine-stemmed and leafy (Doggett, 1970). Sorghum forages are seeded with equipment that is readily available on the farm. Where wheat is the primary crop, grain drills are frequently used, and in other areas, lister or unit planters can be used. Seedbed preparation is similar to that used for grain sorghum or cotton.

Small grains, particularly winter wheat, provide important sources of forage for grazing. Winter wheat is grazed extensively whenever water is sufficient for early growth. Wheat can be grazed until near jointing in the spring without seriously reduced grain yields if adequate nutrients are provided. Yields of 0.70 to 0.90 t/ha of air-dry forage are common on winter wheat in western and central Oklahoma (Hubbard and Harper, 1949; Huffine et al., 1960). With ideal conditions, maximum grazing yield may approach 2.2 t/ha air-dry forage.

Other small grains, frequently seeded for forage are rye, oat, and winter barley. However, the total area seeded is small, amounting to about 7% of the area seeded to wheat.

23–1.5 Sunflower

Commercial sunflower production began in the Southern Great Plains in 1975. Approximately 40 000 ha of oilseed sunflower was grown in Texas in 1977, with most production on the High Plains. Significant production was also recorded in New Mexico and Oklahoma. Although most of the sunflower was irrigated, it was also successfully grown on large dryland areas (Jones, 1978).

The drought tolerance of sunflower is attributed to its extensive root system that can extract water and nutrients to soil depths of 3 m. Thus, on deep soils with a high water-holding capacity, water stored at planting can provide much of the water requirement of sunflower. The effect of soil water content at planting on seed yield of dryland sunflower is shown in Fig. 23–5. High soil water contents at planting can be attained on drylands by fallowing or conservation benching (Jones, 1975).

Sunflower is adapted to a wide range of soil types and can be grown on most of the cultivated soils in the Southern Great Plains. Total area seeded to sunflower has been limited by diseases, insect problems, and economic competition from other summer crops, primarily sorghum and cotton.

Sunflower is grown in row widths of 0.75 to 1 m. Land preparation and planters are the same as for cotton or sorghum. Sunflower can be grown over a wide range of plant populations with little effect on yield. Populations of 25 000 to 35 000 plants/ha are adequate for dryland production. Planting too thickly can cause lodging, and thin stands are difficult to harvest because stalks and heads are large. Standard grain combines with commercially available sunflower heading attachments are used to harvest sunflower (Allen et al., 1979). Row crop headers also work well (Fig. 23–6).

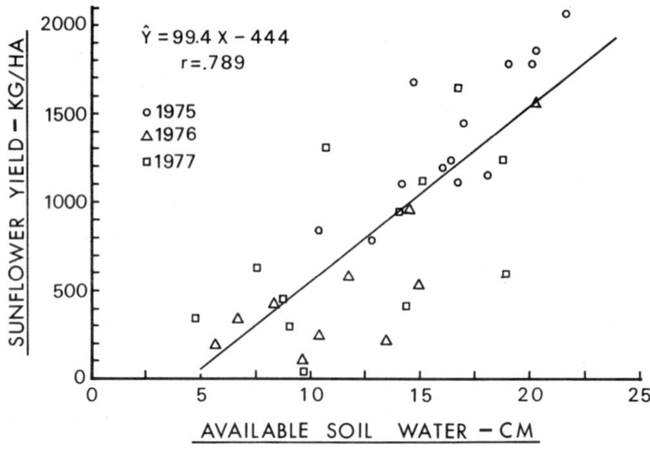

Fig. 23–5. Effect of soil water content at seeding (0- to 180-cm soil depth) on seed yield of dryland sunflower (Jones, 1978).

Fig. 23–6. Sunflower harvest with a row header–equipped combine.

One advantage of sunflower over sorghum or cotton is that a wide range in seeding dates is possible since seedlings are frost resistant and mature heads are seldom damaged by freeze. Thus, sunflower can be seeded from early April until mid-July. Although seed yields and total oil content are reduced by late seeding (June and July), fewer insect and disease problems are encountered. Sunflowers can be successfully grown after hailed-out cotton since herbicides are compatible for the two crops.

Sunflower has the reputation of being hard on soil, and yields of crops following sunflower are frequently reduced because sunflower dried the soil

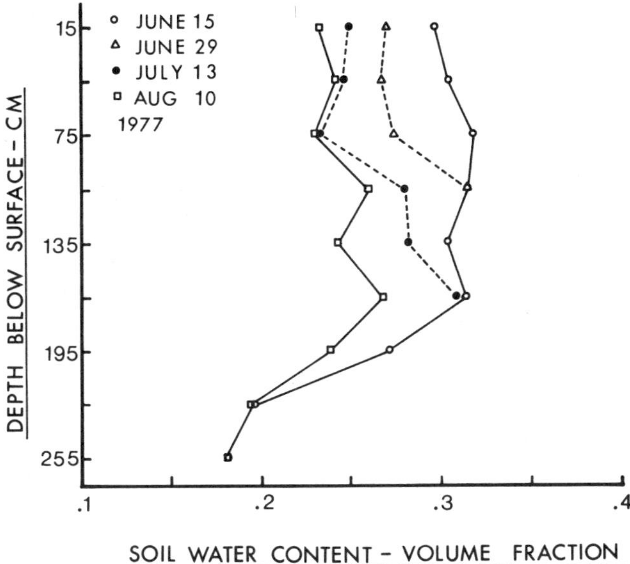

Fig. 23-7. Extraction of soil water by sunflower seeded on 11 April 1977 (Jones, 1978).

profile to much greater depths than most other crops. Soil water depletion by sunflower is shown in Fig. 23-7. The deep extraction of soil water indicates that sunflower can be included in crop sequences to utilize water and nutrients that have accumulated below the normal rooting depth of most crops. Some farm operators on the Southern High Plains that have limited irrigation water supplies are alternating irrigated sorghum with dryland sunflower to increase efficiency of irrigation water use.

23-1.6 Guar

Guar is a summer annual legume grown primarily for its seed (Fig. 23-8). The endosperm of guar seed is high in galactomannan gum, which is used in paper manufacturing, oil well drilling, food, and explosives. Production of guar is centered in the Rolling Red Plains of Texas and Oklahoma, with about 40 000 ha grown annually (Tripp et al., 1977).

Guar is drought resistant. Growth stops or slows when water is limiting and resumes when moisture is again available. It is best adapted to well-drained sandy or sandy loam soils since it cannot tolerate flooding. Guar is usually grown in rotation with cotton because herbicide requirements for the two crops coincide and the lint yield of cotton following guar may be greater than that of cotton following cotton. Mulkey (1971) found that yields of cotton following guar were 22% greater than those of cotton following cotton. Guar can also be interplanted with cotton grown in skip-row cropping systems (Whistler and Hymowitz, 1979).

Seedbed preparation and planting methods are similar to those used for

Fig. 23-8. Guar in the Rolling Red Plains (RRP) near Vernon, Texas. R. E. Stafford, USDA-ARS, Geneticist, compares the branching growth habit common to most guar varieties (*right*) with varieties having a nonbranching growth habit (*left*).

cotton. Guar is usually seeded in 1.0-m rows at a rate of 4.4 to 6.7 kg/ha. The row surface should be above general ground level to facilitate harvest. Guar may be seeded in the Rolling Red Plains between 15 May and 1 July. Earlier seeding is preferred, because with late seeding, fall rains frequently reduce seed quality. Guar is harvested with a grain combine (Tripp et al., 1977).

23-2 CROPPING SYSTEMS

Crops and cropping systems differ as soils, environment, and economic situations change. Useful cropping systems must obtain maximum efficiency of water and nutrient use as well as conserve water and soil resources for future production. Most cropping systems currently employed in the Southern Great Plains result in efficient water and nutrient use and provide wind and water erosion protection when suitable conservation practices are employed. A notable exception occurs, however, with continuous cotton production on sandy lands in the Southern High Plains, where it is much more profitable to grow continuous cotton than to use a cotton-sorghum or other crop sequence.

Continuous winter wheat is the prevalent cropping system in the northern part of the Rolling Red Plains and in the Reddish Prairie. Grain sorghum, alfalfa, and other forages are frequently grown, but areas planted to these crops are small compared with the area planted to wheat. Thus, only a small portion of the area can be rotated from wheat to other crops each year. In the southern part of the Rolling Red Plains, wheat is frequently rotated with cotton. The crops, however, are not seeded in alternate years, because a 10- to 11-month fallow period is required to rotate between these crops, which reduces production. Thus, wheat is grown for 3 or 4 years in succession before rotating to cotton, which is also grown for several years in succession. Guar is sometimes rotated with cotton in this region, but the total area of guar is small compared with the area seeded to cotton and wheat.

Continuous upland cotton is the prevalent dryland cropping system in the southern part of the Southern High Plains. Some areas on coarser soils are seeded only to cotton. This creates problems with wind erosion. However, the high economic return of cotton in relation to sorghum, the other adapted crop, results in cotton being grown exclusively, regardless of wind erosion problems. Tillage is used for emergency wind erosion control on cotton lands. On the drier western edges of the region near the Texas–New Mexico state line, both sorghum and cotton are grown. However, rotations between the crops are seldom fixed. Frequently, sunflower is seeded after cotton has been "blown out" by sand or destroyed by hail. On finer-textured soils, a flexible cropping system that includes winter wheat is used. Wheat is grown in a cotton or sorghum-fallow-wheat sequence. If soil water conditions are favorable in the fall after wheat harvest, wheat will again be seeded. If soil water conditions are poor, the land is fallowed and seeded to cotton or sorghum the following spring. Preferential treatment is given to cotton, the high-value crop, with large areas seeded when soil water conditions are favorable. Wheat is frequently grown as a winter forage crop.

Continuous wheat or wheat-fallow are the prevalent cropping systems in the northern part of the Southern High Plains. Normally, if soil water conditions are favorable at seeding time, wheat is planted. If conditions are poor, the land is fallowed and planted to wheat the following year. Thus, the area fallowed varies from year to year, depending on the amount of rainfall between wheat harvest and seeding. A wheat-sorghum-fallow sequence is also adapted to this region.

The wheat-sorghum-fallow sequence has 11 months of fallow preceding each crop and results in one wheat and one sorghum crop in 3 years. Continuous sorghum is prevalent on small areas of sandy soil in the northwest counties of the Texas Panhandle and in the panhandle of Oklahoma. The total area seeded continuously to sorghum is small compared with the area seeded to wheat.

23-3 PROMISING RESEARCH DEVELOPMENTS

Unger and Phillips (1973) proposed growing irrigated and dryland crops alternately on the same areas, using the high levels of residue remaining after the irrigated crop to increase water storage for dryland crop production. Unger and Wiese (1978) grew wheat and sorghum in an irrigated wheat-fallow-dryland sorghum cropping sequence using different tillage systems. Four-year average fallow efficiency during the 11-month fallow period preceding dryland sorghum was 35% with no-tillage compared with 15% for disk tillage. Subsequent yields of dryland sorghum were 3140 kg/ha with no-tillage compared with 1930 kg/ha with disk tillage. The yield increase was attributed to increased soil water storage during fallow, which resulted from maintaining high levels of crop residue on the surface with the no-tillage system. With large areas on the High Plains currently receiving limited irrigation and with the irrigated area projected to decline, ample opportunity exists for use of alternate irrigated-dryland cropping systems.

No-till and limited tillage (systems that use herbicides to replace tillage) are becoming common on drylands in the Southern Great Plains. Herbicides, primarily atrazine [2-chloro-4-(ethylamino)-6-(isopropylamino)-*s*-triazine] and 2,4-D, can control weeds on fallow after wheat more economically than tillage (A. F. Wiese. 1981. Personal communication). Similarly, weeds in cotton can be controlled more economically with herbicides than by tillage and hand hoeing. Rising fuel and labor costs relative to herbicide costs have improved the economics of weed control with herbicides.

Dryland crop yields can be increased by managing storm runoff either by collecting storm runoff on leveled areas or by adopting conservation practices to retain all precipitation of the field. Jones and Hauser (1974) reported that continuous sorghum yields for 1959 to 1972 were 1240 kg/ha for a 1% slope, 1780 kg/ha for bench terracing, and 2230 kg/ha for conservation benches that received an average runoff contribution of 7 cm/y from adjacent watersheds. Jones and Shipley (1975) reported that construction of bench terraces or conservation bench terrace systems to control runoff could be profitable. Subsequent research (Jones, 1981) showed that any conservation practice that retained runoff in place, e.g., bench terracing or contour listing, increased yields of continuous sorghum from 30 to 98% in comparison to yields of sorghum grown on graded furrows with a 0.25% slope.

Another water conservation practice that has come into widespread use recently on the Southern High Plains is furrow diking (Fig. 23-9 and 23-10), with more than 0.405 million ha furrow diked in 1980 (E. B. Hudspeth. 1980. Personal communication). Bilbro and Hudspeth (1977) reported that furrow diking retained runoff and increased cotton lint yields 15% at Lubbock, Texas. At Bushland, Texas, furrow diking increased average grain sorghum yields by 16% (230 kg/ha) during the period 1975 to 1979 (Clark and Jones, 1981). At Etter, Texas, furrow diking increased sorghum yield 1680 kg/ha in 1980 (O. R. Jones. 1981. Unpublished data). Lyle and Dixon (1977) discuss the history and concepts of furrow diking.

Fig. 23-9. Runoff conservation with furrow dams.

Fig. 23-10. Equipment used to construct furrow dams.

23-4 LITERATURE CITED

Allen, R. R., A. F. Wiese, and E. B. Hudspeth, Jr. 1979. Sunflower plant drying and machine harvest efficiency-Southern Plains. Trans. ASAE 22:992-996.

Arnon, I. 1972. Crop production in dry regions, Vol. II. Systematic treatment of the principal crops. Barnes & Noble Books, New York, p. 580.

Bilbro, J. D. and E. B. Hudspeth, Jr. 1977. Furrow diking to prevent runoff and increase yield of cotton. Texas Agric. Exp. Stn. Progress Report no. 3436.

Bond, J. J., T. J. Army, and O. R. Lehman. 1964. Row spacing, plant populations and moisture supply as factors in dryland grain sorghum production. Agron. J. 56:3-6.

Bottrell, D. G. 1971. Entomological advances in sorghum production. Texas Agric. Exp. Stn. Progress Report no. 2940.

Brown, P. L., and W. D. Shrader. 1959. Grain yields, evapotranspiration and water use efficiency of grain sorghum under different cultural practices. Agron. J. 51:339–343.

Burnett, E., and C. E. Fisher. 1954. Correlation of soil moisture and cotton yields. Soil Sci. Soc. Am. Proc. 18:127–129.

Burnett, E., and N. H. Welch. 1966. Moisture extraction patterns in skip-row cotton systems. p. 24–52. *In* J. R. Runkles et al. (ed.) Skip-row cropping systems on the High and Rolling Plains of Texas. (Mimeo. Report.) Cooperative Texas A&M Univ. and Soil Conserv. Service Conserv. Workshop, College Station, Tex. 15–16 June. Sponsored by USDA-ARS, Soil and Water Conserv. Res. Division, and Texas Agric. Exp. Stn., Texas A&M Univ., College Station.

Clark, R. N., and O. R. Jones. 1981. Furrow dams for conserving rainwater in a semiarid climate. p. 198–206. *In* Proc. Am. Soc. of Agric. Engineers Conf. on Crop Prod. Conserv. 1980's. ASAE Pub. no. 7-81.

Colwick, R. F. 1957. Planting in the mechanization of cotton production. Southern Cooperative Series Bull. no. 49. Mississippi Agric. Exp. Stn., Mississippi State Univ., Mississippi State, Miss.

Daniel, H. A., and H. H. Finnell. 1939. Climatic conditions and suggested cropping systems for northwestern Oklahoma. Oklahoma Exp. Stn. Circ. no. 83.

Doggett, H. 1970. Sorghum. Longmans Green and Co., Ltd., London, p. 268–269.

Elliot, F. C., M. Hoover, and W. K. Porter, Jr. (ed.) 1968. Advances in production and utilization of quality cotton: Principles and practices. Iowa State Univ. Press, Ames, p. 118–149.

Fryrear, D. W. 1976. Continuous cropping as an alternative to fallow in the Southern Great Plains. p. 132–144. *In* Conservation tillage. Great Plains Agric. Coun. Pub. no. 77, Univ. of Nebraska, Lincoln.

Harlan, J. 1962. Hay and pasture seedings for the Southern Great Plains. p. 468–479. *In* H. D. Hughes, M. E. Heath, and D. S. Metcalf (ed.) Forages. Iowa State Univ. Press, Ames.

Harper, H. J. 1959. Sixty-five years of continuous wheat. Oklahoma Agric. Exp. Stn. Bull. no. 531.

Harvey, C., D. L. Jones, and C. E. Fisher. 1961. Dryland crop rotations on the Southern High Plains of Texas. Texas Agric. Exp. Stn. Miscellaneous Report no. 544.

Hubbard, V. C., and H. J. Harper. 1949. Effect of clipping small grains on composition and yield of forage and grain. Agron. J. 41:85–92.

Hudspeth, E. B., and A. D. Brashears. 1974. Dryland narrow row cotton production. Texas Agric. Exp. Stn. Progress Report no. 3280.

Huffine, W. W., N. J. Adams, C. L. Dewald, G. R. Waller, and D. L. Weeks. 1960. Production characteristics of Oklahoma forages: Small grains. Oklahoma Agric. Exp. Stn. Bull. no. 546.

Hughes, H. D., M. E. Heath, and D. S. Metcalf (ed.) 1962. Forages. Iowa State Univ. Press, Ames, p. 468–479.

Johnson, W. C., C. E. Van Doren, and E. Burnett. 1974. Summer fallow in the Southern Great Plains. p. 86–109. *In* Summer fallow in the western United States. USDA-ARS Conserv. Report no. 17.

Jones, O. R. 1975. Yields and water-use efficiencies of dryland winter wheat and grain sorghum production systems in the Southern High Plains. Soil Sci. Soc. Am. J. 39:98–103.

Jones, O. R. 1978. Management practices for dryland sunflower in the U.S. Southern Great Plains. p. 89–98. *In* Proc. Eighth Int. Sunflower Conference, Minneapolis. 23–27 July. Sunflower Assoc. of America, Fargo, N.D.

Jones, O. R. 1981. Land forming effects on dryland sorghum production in the Southern Great Plains. Soil Sci. Soc. Am. J. 45:606–611.

Jones, O. R., and V. L. Hauser. 1974. Runoff utilization for grain production. USDA-AR Series W-22, p. 277–283.

Jones, O. R., and J. L. Shipley. 1975. Economics of land leveling for dryland grain production. J. Soil Water Conserv. 30:177–181.

Keating, F. E., and O. R. Mathews. 1957. Soil and crop studies at Big Spring (Texas) Field Station, 1916–1953. USDA Production Res. Report no. 1.

Leonard, W. H., and J. H. Martin. 1963. Cereal crops. Macmillan, Inc., New York, p. 683, 693.

Locke, L. F., and O. R. Mathews. 1953. Cultural practices for sorghums and miscellaneous field crops at the Southern Great Plains Field Station, Woodward, Oklahoma. USDA Circ. no. 959.

Lyle, W. M., and D. R. Dixon. 1977. Basin tillage for rainfall retention. Trans. ASAE 20: 1013-1017, 1021.

Mathews, O. R., and B. F. Barnes. 1940. Dryland crops at the Dalhart (Texas) Field Station. USDA Circ. no. 564.

Mathews, O. R., and L. A. Brown. 1938. Winter wheat and sorghum production in the Southern Great Plains under limited rainfall. USDA Circ. no. 477.

Mulkey, J. R., Jr. 1968. Skip-row cropping systems with cotton in the Rolling Plains. Texas Agric. Exp. Stn. Progress Report no. 2618.

Mulkey, J. R., Jr. 1971. Cotton-guar rotation. Texas Agric. Exp. Stn. Progress Report no. 2887.

Newman, J. S. 1967. Yields and fiber properties of cotton planted in solid and skip-row systems under minimal soil moisture levels. Texas High Plains, 1963–65. Texas Agric. Exp. Stn. Progress Report no. 843.

Onken, A. B. 1970. Cultural practices for grain sorghum production. Texas Agric. Exp. Stn. Progress Report no. 2930.

Porter, K. B., I. M. Atkins, and C. J. Whitfield. 1952. Wheat production in the panhandle of Texas. Texas Agric. Exp. Stn. Bull. no. 750.

Quinby, J. R., N. W. Kramer, J. C. Stephens, K. A. Lahr, and R. E. Karper. 1958. Grain sorghum production in Texas. Texas Agric. Exp. Stn. Bull. no. 912.

Rogers, C. E., S. N. Oakes, and D. R. Rummel. 1976. Evaluation of infield pheromone traps for boll weevil suppression in the Texas Rolling Plains. Texas A&M Univ. System Monogr. 8, College Station, Tex.

Schlehuber, A. M., and B. B. Tucker. 1967. Culture of wheat. *In* K. S. Quisenberry and L. P. Reitz (ed.) Wheat and wheat improvement. Agronomy 13:117–179. Am. Soc. of Agron., Madison, Wis.

Tripp, L. D., D. A. Lovelace, and E. P. Boring III. 1977. Keys to profitable guar production. Texas Agric. Exp. Stn. Miscellaneous Pub. no. 1321.

Unger, P. W., and R. E. Phillips. 1973. Soil water evaporation and storage. p. 42–54. *In* Conservation tillage. Proc. Natl. Conserv. Tillage Conference. Des Moines, Iowa. March 1973. Soil Conserv. Soc. of Am., Ankeny, Iowa.

Unger, P. W., and A. F. Wiese. 1978. Managing irrigated winter wheat residues for water storage during fallow and subsequent dryland grain sorghum production. Soil Sci. Soc. Am. J. 43:582–588.

Webb, B. B., B. B. Tucker, and R. L. Westerman. 1980. The Magruder plots: Taming the prairies through research. Oklahoma Agric. Exp. Stn. Bull. no. B-750.

Whistler, R. L., and T. Hymowitz. 1979. Guar: Agronomy, production, industrial use, and nutrition. Purdue Univ. Press, West Lafayette, Ind.

Wiese, A. F., E. W. Chenault, and E. B. Hudspeth, Jr. 1969. Incorporation of preplant herbicides for cotton. Weed Sci. 17:481–483.

24

G. O. Hinze
Colorado State University
Akron, Colorado

D. E. Smika
Agricultural Research Service
U.S. Department of Agriculture
Akron, Colorado

Cropping Practices: Central Great Plains

24-1 CROPPING PRACTICES

The majority of the cropping practices in the Central Great Plains region are built around the practice of summer fallow. This dominant practice comprises approximately one third of the cultivated hectarage in each crop year. Individuals from other areas may consider this system wasteful of land resources, but it is necessary, economically sound, and well proved by research (Smika, 1970). In the Central Great Plains, the normal rainfall distribution is such that for winter wheat the fall and spring growth generally must be made from stored soil water. Therefore, no attempt should be made to eliminate the practice of summer fallow immediately before the seeding of winter wheat in this area. As a result of summer fallow, winter wheat production has stabilized, and with proper management the practice offers greater soil and water conservation than any continuous cropping system, thereby utilizing these resources with maximum efficiency. The management objectives of fallow are (i) fallow land should be weed free from harvest to planting 14 months later, (ii) stubble should remain upright during as much of the fallow period as possible to trap snow in the winter and reduce evaporation potential during the rest of the year, (iii) 1680 to 2240 kg/ha of straw mulch should be present at seeding, and (iv) large clods should be maintained during fallow and after seeding. These objectives maximize soil water storage and nitrate supplies, assure better crop seed germination, and reduce wind erosion potential.

Contribution from Colorado State University, Fort Collins, CO 80523, in cooperation with Central Great Plains Research Station, U.S. Department of Agriculture, Agricultural Research Service, Akron, CO 80720.

Dryland Agriculture—Agronomy Monograph no. 23.

24–1.1 Crop Rotation

Crop rotation is not widely practiced in the area, primarily because of the lack of crops other than winter wheat that consistently produce yields that give an economic return compared with what can be obtained with winter wheat. Crops that can be grown in the area include sorghums (both grain and forage types), proso millet and foxtail millet, spring and winter barley, rye, spring oat, sunflower, broomcorn, and corn (Table 20–1). Winter barley frequently suffers significant winter kill, and spring barley and oat usually are injured by hot, drying winds at filling/ripening of the grain. Sunflower appears to have potential as markets develop and as cultural methods are developed. Broomcorn, a specialty crop confined largely to southeastern Colorado, is no longer a significant crop because most of the market is now satisfied by synthetic products.

Beginning in 1971, a practice known as ecofallow was introduced, in which a cropping sequence of winter wheat followed by either grain sorghum, corn, or millet followed by fallow is used. This practice relies largely on the use of herbicides for weed control, with minimal mechanical tillage for weed control. Ecofallow hectarage is increasing each year but still represents less than 15% of the total arable land in the Central Great Plains. Grain sorghum is grown more in the eastern and southern portions of the area, and millet is more important in the drier, cooler portion. Corn hectarage is interspersed throughout the entire area and is showing the greatest increases. Before 1920, dryland corn production exceeded wheat production, but before ecofallow dryland corn production was relatively minor (Greb et al., 1974). The summer crop in the ecofallow system also has the advantage of creating conditions that can disrupt certain insect cycles and does not permit selected weed species to develop in a wheat-fallow system. With ecofallow, four crops are produced every 6 years, whereas with wheat-fallow only three crops are produced every 6 years, with the advantage that the other crop is grown at a time different to wheat and is not subject to the same adverse climatic conditions at the same time.

The wheat-fallow cropping system is presently used by the majority of the dryland farmers. Research has shown this system increases the probability of harvesting a crop that is advantageous for the farmer, and the system is also easy to manage. The fallow period reduces many agronomic problems, e.g., crop volunteer and some weeds, but encourages such weeds as downy brome, cheatgrass, tansy mustard, and jointed goatgrass, all of which are winter annuals and have the same growing habit as winter wheat. The best control for these winter annual weeds is the use of a summer crop in the rotation (Ramig and Smika, 1964) as in ecofallow.

24–1.2 Continuous Cropping

Continuous cropping is not widely practiced in the Central Great Plains but is practiced on occasion in the eastern portion of the area where precipitation is generally greater. In the drier western portions of the area, continuous cropping, if practiced, is largely confined to sandy soils. These soils have low water storage capacity, which negates the benefit of summer fallow, and good water infiltration rates, which maximize water storage from single rainstorms.

Historically, sandy soils are generally not considered to be good for wheat production, but they have proved their value for summer crops. The differentiation between soil and crop types is dramatic, often enabling an observer to note changes in soil type within a short distance by the crop grown.

Corn is grown continuously, but grain and forage sorghums are the most popular crops grown in continuous cropping systems. The availability of newly adapted hybrids, especially of the grain types, has resulted in a proliferation of high-yielding cultivars with a wide range of maturity dates.

Throughout much of the area, proso and foxtail millet are important summer crops. Neither crop is widely grown continuously, because millets do not leave enough residue to protect the soil from blowing. Both millets are grown where the season is too short or too cool for sorghum. Either type is planted from late May through mid-July and can often be harvested 60 days after planting. Briggs and Shantz (1913) showed that proso millet is an especially efficient user of water.

Land devoted to alfalfa and grass for hay should also be considered as continuously cropped. Alfalfa grows especially well on land with a high water table. Taller grasses are usually associated with sandy soils and are preferred for hay. The short types, buffalograss and blue grama, do not grow tall enough for harvesting as hay. Alfalfa is usually harvested only once each year unless a high water table is present, where up to three harvests may be possible. Sweetclover is another potential crop. It grows well, and like alfalfa, it is fairly easy to establish.

Cultural practices for continuous cropping generally involve more soil manipulation though the use of disks or chisels or even moldboard plows to incorporate the crop residue and to destroy growing weeds than those practices used when fallow is part of the crop rotation. Because continuous cropping is usually practiced on the lighter soils more prone to wind erosion, these practices augment the wind erosion problem, but tradition and frequently lack of proper equipment deter farmers from changing their tillage practices. With the advent of herbicides in recent years, some operators are adopting techniques in their farming to reduce erosion potential as well as to increase profits.

Drills and surface planters are widely used, and various forms of listers are extensively used throughout the Central Great Plains. The lister may be used after the soil has been disked or otherwise tilled or without working the soil at all in an attempt to minimize wind erosion, which is a constant threat throughout late spring. When listing is done without prior tillage, the new row is established between the previous year's row (breaking out the middle). This permits last year's stubble to protrude from between the new rows, aiding greatly in reducing wind erosion problems.

Early establishment of weeds between crop rows that are killed by blade cultivation but left standing has been used by some operators to provide wind erosion protection. Although the weeds consume precious soil water, damage from blowing soil may be more severe than that from loss of water and nutrients.

Continuous cropping removes no more nutrients from the soil than the alternate crop–summer fallow system, which is probably attributable to continuous cropping yields being much less than those with a fallow cropping system. Thus, less nutrients are taken off in crop removal. Hobbs and Brown (1965) reported that the loss of total N in the soil was 15% with continuous small grains, 20% for alternate small grain and fallow, 33% for continuous row crop, and 40% for alternate row crop and fallow.

24–2 SEEDBED PREPARATION

A typical time sequence of summer fallow operations, conventional stubble mulch, and minimum tillage systems for the western portion of the Central Great Plains, as developed at both North Platte, Nebraska, and Akron, Colorado, are summarized in Table 24–1 (Greb et al., 1974).

If a farmer follows either of these fallow tillage systems, an optimum seedbed will result with soil covered by crop residue and many large clods, both of which are needed to reduce the hazard of wind erosion and to ensure adequate seedbed soil water for seed germination and plant establishment.

All tillage implements conventionally used in summer fallow tend to loosen the soil surface, with the exception of the mulch treader types. The experienced operator, who begins tillage operations at 10 cm deep or more, tries to reduce depth with each succeeding operation to develop a firm, moist seedbed in which to plant.

The development of herbicides has made no-till cropping possible throughout the Central Great Plains. This practice leaves a firm, moist seedbed and ensures seed germination and plant establishment without destruction of the previous crops' residue necessary for wind erosion protection. The recent introduction of air seeders has made no-till fallow-wheat production a viable option.

Table 24–1. Typical time sequence of summer fallow operations for the western portion of the Central Great Plains (Greb et al., 1974).

Time	Conventional tillage	Minimum tillage
Early July	Harvest wheat	Harvest wheat
25 July–5 Aug.	Blade 10 cm deep to destroy summer and fall weeds in wheat stubble	Apply residual and contact herbicide
1–10 Sept.	Cross blade 10 cm deep to kill volunteer wheat, if needed	--
20 April–10 May	One-way disk if heavy volunteer or excessive stubble, otherwise blade 10 cm deep	--
15–25 June	Blade 8 cm deep	
15–25 July	Rod weed with semichisels 8 cm deep	Sweep 8 cm deep
20 Aug.–1 Sept.	Cross rod weed 8 cm deep	Rod weed 6 to 8 cm deep
10–20 Sept.	Drill wheat with deep furrow hoe drill 25- to 35-cm row spacing, plant seed 5 cm below bottom of 10- to 15-cm furrows	

24-3 SEEDING RATES

Major equipment companies designed and developed implements for the more humid regions that often failed to meet the special needs of semiarid dryland farming. Hargreaves (1957) in her review of dryland farming comments briefly on the special development of machinery for semiarid agriculture.

Where water is limited, regulation of plant population is essential. Too many plants per unit area may induce a drought before the crop attains desired production. For example, W. G. Stewart (1978. Personal communication) concluded that more grain sorghum production failures occurred in eastern Colorado because of overplanting than for any other reason. The recommended 2.3 to 3.4 kg/ha planting rates for dryland grain sorghum are often difficult or impossible to obtain due to design limitations of equipment, grower prejudice, or lack of farmer experience. The above rates will result in a plant stand of 33 000 to 100 000 plants/ha based on a count of 44 100 seeds/kg and 33 to 67% stand establishment (Fenster, 1977; Braunworth et al., 1977). As a general rule, dryland seeding rates are one-third to one-half those rates recommended for humid or irrigated agriculture.

Grain sorghum is frequently grown in 1-m wide row spacings where Greb (1962) showed that 1-m wide rows provide a bank of stored soil water to be used by the crop as grain sorghum root extension develops during the summer season. However, at Temple, Texas, with higher precipitation, Adams et al. (1976) found row widths of 50 to 76 cm yielded better than 101-cm rows if weeds were controlled and plant population was the same.

Summer crops are often seeded with adjustment or minor modification of the same drill used for winter wheat. For example, a certain number of spouts in the grain box may be plugged, thus giving a row effect. This technique reduces equipment costs, but the result is not necessarily the most desirable planting tool. Most farmers producing row crops possess row crop planters, often altered to (i) permit planting at reduced seeding rates and (ii) list so as to reduce wind erosion potential and facilitate seed placement in moist soil.

In the Central Great Plains where tillage has been performed for weed control, the surface soil (6 to 10 cm) is often dry, and planting must be beneath this dry surface into moist soil. Thus, some mechanical means is required to move the dry soil away from the row—generally a lister or deep furrow drill.

24-4 SOIL FERTILITY

Soil fertility studies with winter wheat have been performed since 1946 in the region (Greb et al., 1974). Results of experimental trials have been variable, but generally, response to N application has been more consistent than to other nutrients. Response of wheat to N has varied with time of application, cropping system, and soil type.

Research investigations in Nebraska with fertilizer have been more extensive perhaps than in the other states of the Central Great Plains. Conclusions to date indicate that nitrate release during the fallow season reduces the requirement for applied N (Olson and Rhoades, 1953). However, indications are that wheat on fallow land is increasingly responding to applied N (Nebraska Agric. Exp. Stn., 1952–1970; Smika, 1970). Peterson and Vetter (1971) considered N response to be an effect of both increased time of cultivation since the land was broken from native grass and improved cultural practices, especially those that increase soil water storage. The overall result is higher grain yields with the associated removal of more N from the land.

Generally, early N fertilization has increased grain yield, whereas later application affects grain quality more. Goris and Ludwick (1978) found limited yield increase from fertilizer N applied after seeding, but protein content consistently increased. Recent findings have shown that protein content of wheat grain may be a good guide as to whether yield responses to N fertilizer can be obtained. Indications are that if grain protein is below 11.1%, a yield response will be obtained; from 11.1 to 12.0% protein yields may be limited by N; and when protein is higher than 12%, N fertilizer probably will not increase yields (Goos et al., 1980).

Yield response of wheat to fertilizer N may be uncertain on the medium- to heavy-textured soils because application time may be critical (Smika and Grabouski, 1976). The sandy soils of the region almost always respond to N (Brengle and Greb, 1963).

Wheat may occasionally respond to applied P, especially on sandy soils or on eroded medium-textured soils (Brengle and Greb, 1963).

24–5 FUTURE PRACTICES

Present research indicates many cropping practice changes may be adopted in the near future.

24–5.1 Minimum-till and No-till Fallow

Wittmuss et al. (1973) defined these terms as follows. Minimum tillage is the minimum soil manipulation necessary for producing crops or meeting tillage requirements under existing soil and climatic conditions. Minimum tillage does not define a system but generally refers to a system with fewer operations than a conventional system. Herbicides are usually used for weed control, seedbed preparation, or both.

No-till is a system whereby a crop is planted directly into soil nontilled since harvest of the previous crop.

Research in the Central Great Plains has shown the value of minimum-till and no-till systems. The systems increase water storage during the fallow period, decrease erosion potential, and decrease fuel (and horsepower) re-

Table 24-2. Influence of surface residue on soil water storage of precipitation during fallow (Greb et al., 1967).

Residue level	Precipitation stored
kg/ha	%
0	16
1680	19-26
3360	22-30
6720	28-33
10 080	34

quirements. Wicks and Smika (1973) compared five fallow treatments at North Platte, Nebraska. They found that the no-till treatment (herbicide only) had the least weed growth, most stored soil water, greatest surface mulch, and highest grain yield.

Unger (1974) points out that soil water storage is reduced by poor management practices associated with poor control of weeds and volunteer crop, poor residue management, and excessive and untimely tillage. Under both minimum-till and no-till systems, herbicides replace tillage to obtain weed control. Eliminating or reducing tillage permits crop residues to be maintained, thus protecting the soil surface against wind and water erosion, reducing runoff, reducing surface sealing (and thus maintaining higher infiltration rates), and reducing evaporation. All of these factors increase soil water storage. Crop residue is extremely important in increasing water storage, as Table 24-2 (Greb et al., 1967) shows. Further expansion of the importance of residue and the elimination of tillage for water storage has been shown by Smika (1980), where water storage efficiency of 40% was obtained with minimum tillage fallow, and a 49% storage efficiency was obtained with no-tillage fallow.

Because the main purpose of summer fallow is to store soil water, operators should adopt practices that keep residue on the soil surface and minimize or eliminate tillage. The equipment and techniques necessary to accomplish the operations are now available, and no adverse effects associated with the maintenance of crop residue have been documented in the Central Great Plains.

24-6 RESEARCH NEEDS

No-till and minimum-till operations change the microclimate of the soil where the crop grows. In addition to conserving more precipitation, soils are cooler because of the shading effect of the stubble. Several points concerned with these environmental conditions should be researched:

1. Crop varieties (especially wheat in the Central Great Plains) that can tolerate slightly cooler soils and somewhat more shading should be examined.
2. Plant diseases and insects may become a problem, or potentially there could be a shift in populations of pests.

3. Optimum tillage-herbicide combinations should be developed for the major soil types.
4. Fertilizer requirements and optimum time for their application should be determined.
5. The use of herbicides is essential to a successful no-till or minimum till system. Research may point the way to more effective herbicides than are presently available, but the plant breeder should consider the development of varieties more resistant to present and future chemicals.

24-7 LITERATURE CITED

Adams, J. E., G. F. Arkin,a nd E. Burnett. 1976. Narrow rows increase dryland grain sorghum yields. Texas Agric. Exp. Stn. no. MP-1248.

Braunworth, W. S., V. E. Youngman, and G. O. Hinze. 1977. Grain sorghum date of seeding study at Akron, Colo., 1976. Colorado State Univ. Progress Report no. 3.

Brengle, K. G., and B. W. Greb. 1963. The use of commercial fertilizer with dryland crops in Colorado. Colorado Agric. Exp. Stn. Bull. no. 516-S.

Briggs, L. J., and H. L. Shantz. 1913. The water requirements of plants. I. Investigations in the Great Plains. USDA Plant Industry Bull. no. 285.

Fenster, C. R. 1977. Crop information for western Nebraska. Univ. of Nebraska Extension Service Nebguide no. G77-352.

Goos, R. J., A. E. Ludwick, and D. G. Westfall. 1980. Prediction of nitrogen fertilizer needs for winter wheat by soil analysis and grain protein content. Agron. Abstr. 1980, p. 167.

Goris, J. E., and A. E. Ludwick. 1978. Nitrogen fertilization of dryland winter wheat in eastern Colorado. Colorado State Univ. Progress Report no. 12.

Greb, B. W. 1962. Extra wide row spacing of grain sorghum. Colorado Agric. Exp. Stn. Progress Report no. 39.

Greb, B. W., D. E. Smika, and A. L. Black. 1967. Effect of straw mulch rates on soil water storage during summer fallow in the Great Plains. Soil Sci. Soc. Am. Proc. 31:556–559.

Greb, B. W., D. E. Smika, N. P. Woodruff, and C. J. Whitfield. 1974. Summer fallow in the Central Great Plains. p. 51–85. *In* Summer fallow in the western United States. USDA-ARS Conserv. Res. Report no. 17.

Hargreaves, M. W. M. 1957. Dry farming in the northern Great Plains. Harvard Univ. Press, Cambridge, Mass.

Hobbs, J. A., and P. L. Brown. 1965. Effects of cropping and management on nitrogen and organic carbon contents of a western Kansas soil. Kansas Agric. Exp. Stn. Tech. Bull. no. 144.

Nebraska Agricultural Experiment Station. 1952–1970. Small grain fertilizer experiments in Nebraska, 1951–69. Nebraska Agric. Exp. Stn. Outstate Testing Circ. no. 17, 23, 30, 38, 45, 53, 61, 70, 78, 87, 94, 103, 107, 114, 121, 125, 128, and 133.

Olson, R. A., and H. T. Rhoades. 1953. Commercial fertilizer for winter wheat in relation to the properties of Nebraska soils. Nebraska Agric. Exp. Stn. Res. Bull. no. 172.

Peterson, G. A., and D. J. Vetter. 1971. Soil nitrogen budget of the wheat-fallow area. Univ. Nebr. Q. 17:23–25.

Ramig, R. E., and D. E. Smika. 1964. Fall-wheat-sorghum: An excellent rotation for dryland in central Nebr. Nebraska Agric. Exp. Stn. no. SB-483.

Smika, D. E. 1970. Summer fallow for dryland winter wheat in the semiarid Great Plains. Agron. J. 62:15–17.

Smika, D. E. 1980. Minimum and no-tillage fallow for winter wheat production in the Central Great Plains. p. 101–102. *In* Proc. First Int. Congr. on Dryland Farming, Vol. 1, Adelaide, Australia. 25 Aug.–5 Sept. South Australian Dep. of Agric. and Agric. Technologists of Australia, Adelaide.

Smika, D. E., and P. H. Grabouski. 1976. Anhydrous ammonia applications during fallow for winter wheat production. Agron. J. 68:919–921.

Unger, P. W. 1974. Soil water storage and use. p. 17–26. *In* Soil water storage and use in limited and no-tillage cropping systems. Proc. Symp., Bushland, Tex. USDA Southwestern Great Plains Res. Center, Bushland, Tex.

Wicks, G. A., and D. E. Smika. 1973. Chemical fallow in a winter wheat-fallow rotation. J. Weed Sci. Soc. Am. 21:97–102.

Wittmuss, H. D., G. B. Triplett, and B. W. Greb. 1973. Concepts of conservation tillage systems using surface mulches. p. 5–12. *In* Proc. Natl. Conservation Tillage Conference, Des Moines, Iowa. 28–30 March. Soil Conserv. Soc. of America, Ankeny, Iowa.

25

A. L. Black
Agricultural Research Service
U.S. Department of Agriculture
Mandan, North Dakota

Cropping Practices: Northern Great Plains

The principal factors governing an organized plan of crop management over a number of years on the same land are soil type and climate (quantity and distribution of growing season precipitation). Secondary considerations are the presence or absence of livestock and disease, insect, and weed control management concerns.

Cropping systems involving different crops promote maximum return from the land because of the importance of crop rotations on soil erosion, plant diseases, insect control, and soil nutrient relations. Crop rotation involving winter and spring sown crops is still one of the most effective and practical methods for the control of many winter annual or summer annual weeds. Benefits derived from diversified cropping systems were pointed out in early crop rotation studies (Chilcott, 1931; Conlon et al., 1953; Krall et al., 1965) and compared with the weed problems and other disadvantages of a single-crop system of farming with or without fallow. These studies also demonstrated the importance of the use of summer fallow in sustaining crop yields. Corn in the rotation, when clean-tilled to eliminate weeds, was found to be a good substitute for fallow (Conlon et al., 1953). Because of the dual benefit of summer fallow for increasing soil water supplies and reducing weed problems, the necessity of summer fallow in a given rotation was not questioned.

Conventional use of herbicides for weed control in cereal crops did not become an established practice until the early 1950's. In recent years, a wide range of new and improved herbicides that provide control of most weeds within specific crops have been developed. Therefore, growing different crops in rotation with appropriate herbicide treatment for weed control provides a better opportunity to control a wider spectrum of weeds than a single-crop system having only a limited number of herbicides available for weed control.

Dryland Agriculture—Agronomy Monograph no. 23.

25-1 HISTORY

Cropping systems and the hectarages of spring wheat, winter wheat, spring barley, oat, corn, and flax remained rather stable through the 1950's and 1960's with about 13 million ha of total cropland and 7.4 million ha of summer fallow (USDA, 1955, 1979). In the mid-1970's, cropping systems, crops grown, and total cropland area changed markedly as a result of brighter economic prospects for wheat compared with beef production, the introduction of new oilseed crops, and the identification of summer fallow as the principal cause of the development of saline seeps (Brown et al., 1975; Halvorson and Black, 1974). By 1979, total cropland had increased from 13 to 16.8 million ha, and summer fallow had declined from 7.4 to 6 million ha. The 3.8 million ha of additional land being cropped resulted from a decrease of 0.6 million ha in summer fallow plus an increase of 3.2 million ha of new cropland taken from native rangeland or other types of grassland.

The main crop of the Northern Great Plains is spring wheat (about 6 million ha), followed by alfalfa (2.7 million ha), oilseed crops (2.1 million ha), winter wheat (2 million ha), spring barley and oat (each with 1.5 million ha), and corn (about 0.5 million ha). Oilseed crops (primarily sunflower followed by safflower) and other miscellaneous oilseed crops have increased more than threefold since 1965. The major reason for this increase is the recent availability of preemergence herbicides for the control of annual grass-type weeds. Following an oilseed crop with a cereal grain crop with proper application of herbicides for broadleaf weed control provides a well-balanced weed control program. Therefore, cropping sequences and cultural practices can now be designed to effectively control most common weeds.

25-2 CROPPING PRACTICES

25-2.1 Climatic Influences

The Northern Great Plains climate is distinctly continental with extremes in both temperature and precipitation. Fall and winter (September to March) precipitation across the area is low, averaging about 80 to 100 mm, with 50 to 75 mm received as snow. April through August, precipitation ranges from 150 to more than 280 mm, with about 70% received in May, June, and July.

To facilitate the discussion of cropping systems, I have divided the Northern Great Plains into three general climatic regions based on annual precipitation (Table 25-1). In region I, where annual precipitation is less than 330 mm, cropping systems rely heavily on the use of summer fallow in the cropping system. Winter wheat and spring barley are the principal crops

Table 25-1. Northern Great Plains regions and cropping systems employed as related to precipitation.

Region	Precipitation	Location	Cropping systems
	mm		
I	330	North central and central Montana and Northweastern Wyoming	Winter wheat-fallow Barley-fallow Spring wheat-fallow Winter wheat-barley-fallow
II	330-380	Eastern Montana, western North Dakota, and northwestern South Dakota	Spring wheat-fallow Winter wheat-fallow Spring wheat-winter wheat-fallow Oilseed-fallow-spring wheat Oilseed-barley-winter or spring wheat Barley-corn
III	380	Eastern North Dakota and eastern South Dakota	Spring wheat-barley-fallow Spring wheat-oilseed-fallow Spring wheat-oilseed Spring wheat-oats-oilseed-barley

grown (Table 25-1). Region II, where annual precipitation ranges from 330 and 380 mm, is a transitional area in which fallow is used principally for spring or winter wheat but not necessarily for other crops. In region III, where annual precipitation is more than 380 mm, a wide array of annual cropping systems are interrupted by summer fallow only after two, three, or four consecutive crops.

25-2.2 Fallow

The basic tillage implements for following in the Northern Great Plains are the disk (one-way, offset, or tandem disk), sweep cultivator (sweep width varying from 15 to 45 cm), and the rod weeder (either attached to a sweep cultivator or as a separate implement). Tillage is not usually performed after harvest, because standing stubble holds snow on the field, which is an important aspect of water conservation. Tillage may be needed in some years when late summer and early fall precipitation has caused an abundance of weeds or volunteer cereal grain growth. Undercutting the stubble at a 10- to 12-cm soil depth with wide-blade sweep machines controls weeds in fall while leaving most of the stubble standing to control soil erosion and trap snow.

During the summer of fallow, the choice of implement for the first tillage operation varies depending primarily on the amount of crop residue present. If crop residue levels exceed 4000 kg/ha, a disk-type implement is usually used first. This operation chops the straw into shorter lengths and buries about 50% of it. If crop residue levels are less than 4000 kg/ha, the sweep cultivator without a rod attachment is used for the initial spring operation. Successive operations are then performed three to four times as needed for weed control with the sweep cultivator with or without a rod attachment. Late summer tillage operations are usually performed with the rod weeder alone to facilitate annual grass and weed control and to help establish adequate seed-zone soil water for seeding winter wheat. Fallow land that will be seeded to spring wheat may be tilled with only the sweep cultivator for the last operation in the summer. This practice conserves the most crop residue on the soil surface and provides the best possible soil condition for reducing the wind erosion hazard during the second winter of fallow. Much interest has been shown recently in the potential of partial or complete chemical fallow methods for conserving energy and reducing production costs. However, to date no general acceptable procedures have been established in the Northern Great Plains.

25-2.3 Spring Wheat

25-2.3.1 Seedbed Preparation and Seeding Methods

In regions I and II, spring wheat is planted primarily on clean-till fallow land. In region III, spring wheat is planted on stubble land that is fall- or spring-plowed or is planted on fallow. Early spring tillage with a sweep-cultivator machine (30- to 45-cm sweeps) may precede seeding operations. Usually, the tillage and seeding operations are combined in a single operation. The double-disk press drill with 15-, 18-, or 20-cm row spacing is the principal drill type used for seeding spring wheat. In some instances, a multipurpose semideep-furrow drill with a 25- to 31-cm row spacing is used, because one drill is sufficient for seeding both winter wheat and spring cereal crops, particularly in region I.

25-2.3.2 Seeding Rate and Date

The most common seeding rate for hard red spring wheat is 67 kg/ha; the range is 55 to 80 kg/ha. Durum spring wheat is seeded at slightly higher rates than hard red spring wheat (70 to 90 kg/ha). Land in region I is usually seeded between 20 March and 30 April, and land in regions II and III are seeded between 1 May and 10 June. Numerous studies of seeding dates have shown decreased wheat yields as seeding dates are delayed beyond April 30 in region I and beyond 20 May in regions II and III (Black and Siddoway, 1977a; Olson and Kapusta, 1960; Post, 1966). However, in regions II and III, tillage and seeding operations may be delayed beyond 20 May in an attempt to better control weeds, especially wild oat.

25-2.4 Winter Wheat

25-2.4.1 Seedbed Preparation

Although the major portion of winter wheat is seeded on fallow in regions I and II, an estimated 20% is seeded in standing spring grain stubble in region II. Winterkill is a problem in northeastern Montana and northwestern North Dakota, and seeding winter wheat into standing stubble has minimized the problem (Black and Siddoway, 1977b; Halvorson et al., 1976). To facilitate seeding operations in standing stubble, crop residues passing through the combine should be chopped and uniformly spread on the land or collected and removed.

Tillage before seeding and at the time of seeding winter wheat on fallow is accomplished primarily with a rod weeder operated at 6- to 8-cm soil depth. A sweep machine with narrow sweeps and attached rod weeder also accomplishes the same purpose, that of firming the seedbed to help assure sufficient soil water at a depth of 5 to 10 cm to obtain a satisfactory stand. Deeper tillage operations generally dry the tilled soil layer excessively. September and October precipitation is relatively low and highly variable in the Northern Great Plains, which dictates that rather specific tillage operations are required to maintain adequate seed-zone soil water in the fall for stand establishment.

25-2.4.2 Seeding Methods Date and Rate

Nearly all winter wheat planted on fallow is seeded with a furrow drill in 25-, 31-, or 36-cm row spacings. Winter wheat grown under conditions of high-surface residue or in standing stubble is seeded with the same type of drills modified with narrow shovel, chisel, or spear point openers. Winter wheat is seeded between 1 and 20 September on fallow and between 10 and 30 September in standing stubble. The seeding rate for winter wheat generally increases from 40 to 70 kg/ha from south to north and to some extent from west to east. Winter wheat is seeded at a rather precise depth of 4 to 7 cm, because shallow seeding allows the crown to develop too close to the soil surface and deep seeding delays emergence and weakens nonwinterhardy plants. Both deep and shallow seeding increase the probability of winter injury.

25-2.5 Spring Barley or Oats

25-2.5.1 Seedbed Preparation

Cultural practices for spring barley and oat are similar to those already described for spring wheat. However, a relatively large proportion of these two crops is seeded on nonfallow land following other crops. If crop residues from the previous crop exceed about 4000 kg/ha, a seedbed is prepared in the spring by using either disk-type tillage implements, which chop the stubble, or the plow, which buries enough residue to exclude plugging problems during subsequent seeding operations. In portions of regions II and III, the plow is still used extensively for seedbed preparation for barley and oat following another crop. Preparing a seedbed for barley or oat after low residue–producing crops such as silage corn or oilseed crops is relatively easy, because one tillage operation with a sweep cultivator or a disk is adequate.

25–2.5.2 Seeding Methods, Rate, and Date

Spring barley and oat are planted primarily with a double-disk drill in narrow row spacings of 16-, 18-, or 20-cm, particularly in Regions II and III. In Region I, most of the spring barley is seeded with a semideep or deep-furrow drill in 25-, 31-, or 36-cm rows because the same drills are used for seeding winter wheat. Seeding rates for spring barley range from 60 to 90 kg/ha, with 70 kg/ha being the most common. Oats are planted with the same drills at seeding rates of 55 to 80 kg/ha. Spring barley is usually seeded somewhat earlier than spring wheat and as soon as soil conditions permit. Spring oats are seeded about the same time as spring wheat for grain production or somewhat later for oat-hay production.

25–2.6 Oilseed Crops

25–2.6.1 Sunflower

Sunflower is best adapted to region III, but with continued improvement in varieties, good adaptation has now been shown in region II. Sunflower produces best on summer fallow, but this land is usually reserved for wheat. Therefore, most sunflower is grown on land that was cropped the previous year. Seedbeds need to be prepared only to the extent that previous crop residues do not interfere with herbicide application and subsequent seeding operations. Sunflower is planted in rows 76 to 107 cm with a unit-row planter having special plates. Oilseed varieties of sunflower should be planted 10 to 25 May at a seeding rate of 6 to 7 kg/ha to provide a plant population of 60 000 to 70 000 plants/ha. Sunflower varieties grown for the confectionary market should be seeded at a rate of 4 to 5 kg/ha to obtain a plant population of about 37 000 to 45 000 plants/ha.

25–2.6.2 Safflower

Safflower is grown principally in regions I and II primarily because of less leaf disease potential in the drier regions of the Northern Great Plains. Safflower is a deep-rooted crop that grows well after other crops, particularly after spring wheat. Safflower also produces satisfactorily after winter wheat if the winter wheat stubble remained standing overwinter to trap snow and thus recharge the soil water supply in the upper profile. The seedbed requires two to three tillage operations to reduce previous crop residue levels and to properly incorporate the preemergence herbicide. Safflower is planted with conventional cereal grain drills in rows ranging from 15 to 36 cm apart, at a seeding rate of 16 to 26 kg/ha, and at a seeding depth of 2 to 3 cm. Row spacings greater than 36 cm are not recommended, because the resulting delay in maturity increases the risk of frost damage in the fall. Safflower requires a growing season of about 130 days, which dictates that seeding operations should be completed before 20 May in regions I and II.

25-2.7 Alfalfa

An important aspect of dryland farming in the Northern Great Plains is the continued increase in land devoted to alfalfa. Aside from the importance of alfalfa in providing a rather dependable supply of hay for livestock, alfalfa has become an important crop for two other reasons. First, because alfalfa is a deep-rooted perennial crop that extracts soil water to a depth of 6 m, planting alfalfa on upslope recharge areas can effectively control saline seeps (Brown et al., 1975; Brown and Miller, 1978; Krall and Brown, 1976). Second, alfalfa supplies additional N to the soil for succeeding crops.

Alfalfa is usually planted at a rate of 5.5 kg/ha or less in a firm seedbed at a depth of 1 to 2 cm on land that has been fallowed. Various seeding methods have been successfully used. The seed may be broadcast or metered from a drill box and allowed to drop on the soil surface. The soil is harrowed lightly, or a chain drag is used behind the drill to cover the seed. In many instances, alfalfa is broadcast seeded at the same time a companion crop of spring oat is being seeded at a reduced rate of less than 40 kg/ha.

25-3 ADVANTAGES AND DISADVANTAGES OF CROPPING PRACTICES

Climatic conditions and weed problems are the primary factors governing the potential advantages or disadvantages of any given cropping practice. Strict adherence to a fixed cropping practice that includes fallow will not permit farmers to take full advantage of years when supplies of available water are above average. In addition, a fixed cropping system using the same crop on the same field will promote the propagation of rather specific weed problems. For example, growing winter wheat on the same land with or without fallow causes a buildup of winter annual weeds, e.g., tansy mustard and downy brome. Likewise, growing spring grains continuously with or without fallow causes a buildup of summer annual grass-type weeds, e.g., green foxtail grass and wild oat. Therefore, such crop rotations as winter wheat–barley–fallow in region I, spring wheat–winter wheat–fallow in region II, and spring wheat–sunflower–spring barley–fallow in region III have proved their value in aiding weed control.

The full impact of the recent increase in the hectarage devoted to oilseed crops in the Northern Great Plains is presently unknown. Many farmers and researchers are experimenting to determine where such crops best fit into the overall cropping practices and crop rotations. Because of the good arsenal of preemergence herbicides now available for oilseed crops, the high populations of annual grass-type weeds common to spring grain cropping systems can be reduced substantially. A systematic rotation of an oilseed crop grown on each field every 3 or 4 years would probably suffice. When this procedure is coupled with a good broadleaf weed control program while the land is in cereal grains, a better potential for further improvement will become possible for developing more productive, flexible cropping systems.

25–4 PROMISING RESEARCH DEVELOPMENTS

The advantages and disadvantages of summer fallow in the Northern Great Plains have been documented (Black et al., 1973; Haas et al., 1973). Fallow has been identified as the primary cause of saline seep development (Halvorson and Black, 1974; Krall and Brown, 1976; Doering and Sandoval, 1976), and many farmers have had to change their cropping practices to either reduce or eliminate fallow from the cropping system. Many of these farmers have planted alfalfa on the recharge area or adopted a flexible cropping system based on stored soil water at planting time in the fall or spring (Brown et al., 1981; Brown and Miller, 1978; Krall and Brown, 1976). Dryland cropping strategies for efficient water use to control saline seeps have been developed and discussed in detail (Black et al., 1981).

Snow management has become an important part of the cropping plan through the use of standing stubble (Black and Siddoway, 1977b), the use of wind barriers (e.g., a tall wheatgrass barrier system, Black and Siddoway, 1976), level bench terraces with grass on the dikes (Haas and Willis, 1968), and tree windbreaks (Frank and George, 1975). Seeding winter wheat directly into standing spring grain stubble without tillage has proved successful when combined with proper fertilization practices (Black and Siddoway, 1977b; Halvorson et al., 1976).

Cropping systems are being developed that use available water and nutrient supplies more efficiently than conventional crop-fallow (Black and Siddoway, 1976). For example, spring grains use available water supplies primarily from the upper 1 m of soil; winter wheat uses the upper 2 m of soil; and safflower uses the upper 3 m of soil. Therefore, in any cropping sequence involving these crops, fallowing after either winter wheat or safflower would be more logical than after spring grains. The changes of recharging the upper 1 m of soil through snow management is high in the Northern Great Plains (Black et al., 1973, Haas et al., 1973). A computerized management guide and user's manual have been developed to help farmers in deciding the best cropping and soil management options for different crops based on stored soil water and expected growing season precipitation (Kresge and Halvorson, 1979).

25-5 RESEARCH NEEDED

There is an urgent need to continue the important task of testing and developing new, improved, and economical herbicides for better weed control in all crops. Weeds are probably the number one factor inhibiting many current and future cropping systems from reaching their full potential.

The adaptability of no-till and minimum-till farming concepts need to be fully tested within various cropping systems. Crop residue management techniques that are compatible with the rather strict soil and water conservation needs of the area and that are also suited to minimum-till or no-till farming practices need to be developed. There is an urgent need to develop and select adaptable legume species for use in flexible cropping systems so that a significant portion of the N requirement for other major crops could be supplied. It is extremely important to resource conservation that as improved soil management practices are developed, more efficient cropping systems be evaluated so that water, nutrient, and energy are used with maximum efficiency.

25-6 LITERATURE CITED

Black, A. L., P. L. Brown, A. D. Halvorson, and F. H. Siddoway. 1981. Dryland cropping strategies for efficient water-use to control saline seeps in the Northern Great Plains, USA. Agric. Water Manage. 4:295–311.

Black, A. L., and F. H. Siddoway. 1976. Dryland cropping sequences within a tall wheatgrass barrier system. J. Soil Water Conserv. 31:101–105.

Black, A. L., and F. H. Siddoway. 1977a. Hard red and durum spring wheat responses to seeding date and NP-fertilization. Agron. J. 69:885–888.

Black, A. L., and F. H. Siddoway. 1977b. Winter wheat recropping on dryland as affected by stubble height and nitrogen fertilization. Soil Sci. Soc. Am. J. 41:1186–1190.

Black, A. L., F. H. Siddoway, and P. L. Brown. 1973. Summer fallow in the Northern Great Plains (winter wheat). p. 36–50. *In* Summer fallow in the western United States. USDA-ARS Conserv. Res. Report no. 17.

Brown, P. L., E. C. Cleary, and M. R. Miller. 1975. Water use and rooting depths of crops for saline seep control. p. 125–136. *In* Proc. Regional Saline Seep Control Symp., Bozeman, Mont. 9–11 December. Coop. Extension Service Bull. no. 1132, Montana State Univ., Bozeman.

Brown, P. L., and M. R. Miller. 1978. Soil and crop management practices to control saline seeps in the U.S. Northern Plains. p. 7.9–7.16. *In* Proc. Dryland Saline Seep Control Meeting. Edmington, Alberta. 10 June. 11th Int. Soc. of Soil Sci. Congress, Alberta Agric., Edmonton.

Brown, P. L., A. L. Black, C. M. Smith, J. W. Enz, and J. M. Caprio. 1981. Soil water guidelines and precipitation probabilities for barley and spring wheat in flexible cropping systems. Montana Cooperative Extension Service Bull. no. 356. Montana State Univ., Bozeman.

Chilcott, E. C. 1931. The relationship between crop yields and precipitation in the Great Plains area. USDA Miscellaneous Circ. no. 81.

Conlon, T. J., R. J. Douglas, and L. Moomaw. 1953. Rotation and tillage investigations at the Dickinson Experiment Station, Dickinson, N.D. North Dakota Agric. Exp. Stn. Bull. no. 383.

Doering, E. J., and F. M. Sandoval. 1976. Hydrology of saline seeps in the northern Great Plains. Trans. ASAE 19:856-861, 865.

Frank, A. B., and E. G. George. 1975. Windbreaks for snow management in North Dakota. p. 144-154. *In* Snow management on the Great Plains. Great Plains Agric. Coun. Pub. no. 73, Univ. of Nebraska, Lincoln.

Haas, H. J., and W. O. Willis. 1968. Conservation bench terraces in North Dakota. Trans. ASAE 11:396-398, 402.

Haas, H. J., W. O. Willis, and J. J. Bond. 1973. Summer fallow in the Northern Great Plains (spring wheat). p. 12-35. *In* Summer fallow in the western United States. USDA-ARS Conserv. Res. Report no. 17.

Halvorson, A. D., and A. L. Black. 1974. Saline-seep development in dryland soils of northeastern Montana. J. Soil Water Conserv. 29:77-81.

Halvorson, A. D., A. L. Black, F. Sobolik, and N. Riveland. 1976. Proper management—key to successful winter wheat recropping in northern Great Plains. N.D. Farm Res. 33(4): 3-9.

Krall, J. L., T. J. Army, A. H. Post, and A. E. Seamans. 1965. A summary of dryland rotations and tillage experiments at Havre, Huntley, and Moccasin, Montana. Montana Agric. Exp. Stn. Bull. no. 599.

Krall, J. L., and P. L. Brown. 1976. Cultural practices for the control of saline seep in the Northern Plains. p. 64-77. *In* Conservation tillage. Great Plains Agric. Coun. Pub. no. 77, Univ. of Nebraska, Lincoln.

Kresge, P. O., and A. D. Halvorson. 1979. Flexicrop user's manual: Computer-assisted dryland crop management. USDA Cooperative Extension Service Bull. no. 1214, Montana State Univ., Bozeman.

Olson, H. M., and G. Kapusta. 1960. Wheat and barley yield reduction due to late planting at Williston, 1957-1959. N.D. Farm Res. 21(5):12-15.

Post, A. H. 1966. The effect of rate and date of seeding on yield of spring winter wheat. Montana Agric. Exp. Stn. Bull. no. 609.

U.S. Department of Agriculture. 1955. Agricultural statistics. USDA, U.S. Government Printing Office, Washington, D.C.

U.S. Department of Agriculture. 1979. Agricultural statistics. USDA, U.S. Government Printing Office, Washington, D.C.

26

W. Earl Johnson
University of Saskatchewan
Saskatoon, Saskatchewan, Canada

Cropping Practices: Canadian Prairies

26-1 CROPS AND ENVIRONMENT

Farming systems in the Canadian Prairies must be designed to allow economic production of adapted crops in a region that has a short growing season and is subject to varying degrees of moisture stress. Hard red spring wheat and durum wheat represent a major portion of the seeded acreage. The inclusion of barley or rye in the rotation with wheat frequently results in crop contamination from volunteer growth and quality problems in bread wheat. The acreage devoted to oat is limited. The oilseed crops, flax, mustard, rapeseed, and sunflower do combine well with wheat in cropping programs and also provide for inclusion of barley.

The intensity of summer fallow use increased markedly from about 1942. The benefits of summer fallow, in which the land is fallow for a period of 21 months in the spring wheat system, are improved weed control and larger soil moisture reserves. Research in the last 20 years has shown that a major effect of the summer fallow period is increased mineralization of N. The favorable yields on summer fallow land are frequently attributed to the higher available N supply rather than the difference in moisture reserve.

The dryland area varies in the frequency and intensity of moisture stress. In the Aridic Boroll soil region of southwestern Saskatchewan and southeastern Alberta, the average crop uses 20 to 25 cm more than the average precipitation. The use of summer fallow on medium- to fine-textured soils on an intensive alternate year basis with some shift based on satisfactory moisture reserves is essential. Water use in excess of precipitation in the Typic Boroll region ranges from 15 to 20 cm, and cropping programs can extend to 3 or 4 years with summer fallow considerably reduced. The Udic Boroll soil region average crop use is 12 to 15 cm greater than precipitation, and the use of summer fallow can be significantly reduced to 25% or less of the crop acreage.

Dryland Agriculture—Agronomy Monograph no. 23.

26-2 CONTINUOUS OR EXTENDED CROPPING

The establishment of reasonably precise fertilizer requirements by soil testing and the availability of a wide range of herbicides for weed control allow for extended cropping programs. On the Aridic Borolls, cropping can be based on soil water reserves in the spring. A minimum water reserve of 7.5 cm on medium- to fine-textured soils allows for reasonable yield potential. The result is a 2- to 3-year cropping program. On the Typic Borolls, further extension is possible, and cropping programs on the Udic Borolls can range from 3 to 5 years, with summer fallow used mainly for perennial weed control with emphasis on years in which spring soil water reserves are limited.

Pittman (1977) has reported crop yields for a Typic Boroll loam soil in southern Alberta. Table 26-1 gives the 20-year yields and the total digestible nutrients produced per hectare of rotation.

No commercial fertilizer was used on these plots. The data from these cropping programs show that the main differences in yields per hectare of rotation resulted from the use of livestock manure in rotation D and alfalfa

Table 26-1. Average crop yields and average yields of total digestible nutrients (TDN) from dryland crop rotations, Lethbridge, Alberta, 1954-1973.†

Rotation		Crop yield	TDN yield
		kg/ha	kg/ha of rotation
A	Spring wheat continuously	855.8	718.9
B	Fallow	--	691.8
	Spring wheat	1647.1	
C	Fallow	--	784.8
	Spring wheat	1847.3	
	Spring wheat	955.6	
D	Fallow (manured)	--	840.1
	Spring wheat	1948.5	
	Spring wheat	1050.0	
F	Fallow	--	636.3
	Winter wheat	1843.3	
	Winter wheat‡	715.9	
H	Fallow	--	858.3
	Spring wheat	1770.7	
	Spring wheat	1302.3	
	Crested wheatgrass and alfalfa	--	
	Crested wheatgrass and alfalfa	2580.0	
	Crested wheatgrass and alfalfa	2530.0	
K	Fallow	--	675.1
	Spring wheat	1836.2	
	Spring wheat	1220.7	
	Crested wheatgrass	--	
	Crested wheatgrass	1480.0	
	Crested wheatgrass	1460.0	

† From Pitman, U. J. 1977. Crop yields and soil fertility as affected by dryland rotation in southern Alberta. Commun. Soil Sci. Plant Anal. 8(5):396.

‡ Barley after 1970.

Table 26–2. Grain yields in various rotations on a Udic Boroll clay at Indian Head, Saskatchewan, 1958–1973.

Rotation	Yield		Amount of fertilizer	
	kg/ha of rotation		N	P
	Nonfertilized	Fertilized†		
Continuous wheat	1075	1579	20.6	9.0
Wheat–fallow	1126	1240	3.0	5.9
Wheat–wheat–fallow	1138	1418	7.8	6.9
Wheat–flax–barley	1258	2184	51.8	11.7
Wheat–wheat–sweetclover‡	1378			

† Fertilizer used (kg ha^{-1} y^{-1}).
‡ Partial fallow after incorporation of sweetclover as green manure.

in rotation H. The low yields for winter wheat following winter wheat were caused by unthrifty stands and resulting severe infestations of downy brome.

R. N. McIver (1978. Unpublished data. Research Branch, Agric. Canada, Swift Current, Saskatchewan) has compared production of a number of cropping programs on a Udic Boroll clay soil at Indian Head in the region where crops use 12 to 15 cm more than average precipitation. Yield data for some of the systems are shown in Table 26–2.

If allowance for the seed requirements are used to give net production per hectare of rotation, the yields reported by Pittman and McIver both show that continuous wheat production gives lower yields than a wheat-fallow or wheat-wheat-fallow program without fertilizer.

With moderate applications of N and P, yields for continuous wheat are higher than for the 2- or 3-year wheat summer fallow programs.

The yields of wheat in rotation H at Lethbridge and in the wheat–wheat–sweet clover rotation at Indian Head demonstrate the value of the legume in the rotation. Neither of these rotations was fertilized, and the value of the legume to the N status and physical condition of the soil is shown by the yield data.

The yields in the wheat-flax-barley continuous rotation at Indian Head have been very satisfactory. The wheat yields in this rotation are around 1200 kg/ha higher than those in the continuous wheat rotation or for second crop wheat in the 3-year summer fallow–wheat program. Only about 50% of this yield difference can be attributed to the higher fertilizer rates on the wheat–flax–barley rotation. McIver reports control of Canada thistle as a major problem in the wheat–flax–barley rotation.

Research conducted by the Canada Department of Agriculture at Scott on a Typic Boroll loam and an Aridic Boroll silty clay loam for a 9-year period show that summer fallow is a more essential element of the cropping program on the medium- to fine-textured Aridic Borolls than on the Typic Borolls. Yield data in Table 26–3 show that the yield of second crop wheat is only 61% and continuous wheat only 52% of the yield on summer fallow. The yield of wheat after sunflower again shows that cropping can be extended and total production increased with diversification from major dependence on wheat.

Table 26-3. Spring wheat yields for cropping programs on a Typic (Albic) Boroll and a Aridic Boroll clay, west central Saskatchewan, 1963–1971.

Treatment	Typic boroll loam	Aridic boroll loam
	kg/ha	
Wheat on fallow	1948	1565
Snow-ridged fallow	2016	1646
Wheat after sunflowers (row-seeded)	1760	1081
Wheat after wheat†	1498	960
Continuous wheat	1424	826
Continuous wheat snow-ridged	1370	772

† In wheat–wheat–fallow rotation.

Coarse-textured soils and soils with a salinity hazard are best adapted to perennial forage. Summer fallow is hazardous because of soil drifting susceptibility on coarse-textured soils. Summer fallow generally causes an increase in salinity encroachment. Summer fallow on medium- to fine-textured soils will vary from an intensive crop–summer fallow program on the Aridic Borolls to infrequent or partial fallow on the Udic Borolls.

Fall rye and winter wheat have potential for extending the cropping program. The winter annual crop provides an opportunity to reduce the fallow period from 21 months to 11 to 14 months. The fallow period would extend from harvest in July or August to May or September of the following year.

Extended or continuous cropping requires a shift from monoculture spring wheat to oilseed crops, annual legumes, winter annuals, and biennial or perennial legumes and grasses. Flax–wheat–summer fallow, mustard–wheat–summer fallow, wheat-sunflower, rapeseed–cereal grain–summer fallow, and rapeseed–cereal grain–cereal grain–flax–cereal–cereal–summer fallow (Udic Borolls) rotations have all been used with success. Annual legumes, including lentil on the Typic Borolls and faba bean or field pea on the Udic Borolls soils, offer further opportunity for extended cropping.

The biennial and perennial legumes and grasses in the system will add considerably to maintenance of productivity. A major requirement in successful use of biennial and perennial forages is adjustment for current moisture regimes. Sweetclover can be used in the program for green manure or can be harvested as forage. In a significant moisture deficit regime, sweetclover must be harvested or worked down early to provide some potential for sufficient moisture to grow a crop the following year. Breaking of perennial legume grass sod will require a full-year summer fallow in dry cycles but can be handled as a partial fallow after hay harvest in periods of good moisture supply.

Crop selection on sod breaking will improve results. Flax or oat have been the most successful crops on sod breaking.

Grain-livestock farms have an increased potential for diversified cropping. In addition to utilization of biennial or perennial forage crops, corn silage in regions with sufficient heat energy and continuous oat fodder are added possibilities. Cereal silage seeded and harvested early, with a combination of chemical and cultural weed control after harvest, gives satisfactory continuous cropping total dry matter production and constitutes a good annual weed control program. Production of corn silage is dependent on selection of well-drained, medium-textured, salt-free soils, adequate weed control and fertilization, and early seeding.

26-3 SEEDING MANAGEMENT

26-3.1 Seedbed Preparation and Seeding

Proper seedbed preparation is one of the most important single factors in successful crop establishment and production. The seedbed should allow for uniform, shallow placement of seed with firm seed-soil contact in moist soil. The large-seeded cereal crops should be seeded at a maximum 5 to 6 cm, and small-seeded crops should be seeded at 2.5 cm or less. It is important, however, to attain good contact with moist soil.

The effect of depth of seeding is shown by data obtained by C. H. Anderson (1974. Unpublished data. Research Branch, Agric. Canada, Swift Current, Saskatchewan) for three wheat varieties on an Aridic Boroll loam soil at Swift Current (Table 26–4).

Although there can be rather major differences between seeding machines on a range of seedbed conditions, seedbed condition is the major determinant of good results.

Data for two 5-year periods at Swift Current (Table 26–5) demonstrate that seedbed condition reduces the differences between machines, and improved seedbed conditions improve results with all machines.

In the 1950 to 1954 period, the practice of summer fallow was to leave fields ridged in the fall for additional soil drifting control. For the later period, 1955 to 1960, soil drifting control was achieved by crop residue re-

Table 26–4. Effect of depth of seeding on yield of three varieties of wheat on Aridic Boroll loam soil, Swift Current, Saskatchewan, 1970–1973.

Depth of seeding	Yield	Yield as % of 5.1-cm depth
cm	kg/ha	
2.5	1472	94
5.1	1566	100
7.6	1404	89
10.2	1210	78

Table 26–5. Yields of spring wheat, Swift Current, Saskatchewan, using various machines, 1950–1954 and 1955–1960.

Seeding machine	1950–1954	1955–1960
	kg/ha	
Double-disk press drill	1732	1841
One-way disker	1541	1814
Wide-level disker	1337	1794

Table 26–6. Spring wheat yields on summer fallow at Swift Current, Saskatchewan, using several summer fallow programs.

Fall treatment	Summer fallow program	Yield	Protein in wheat
		kg/ha	%
Nil	Herbicide	1542	14.3
Nil	Herbicide-tillage	1494	14.9
Nil	Tillage-herbicide	1484	14.9
Nil	Tillage	1497	15.1
Herbicide	Tillage	1559	15.0
Tillage	Tillage	1510	15.0

tention, and the surface was left more uniform by rod weeding. In addition, greater care was taken in adjustment and management of the one-way and wide-level diskers. The important element is the increased yield with all machines for the 1955 to 1960 period in spite of the fact that moisture conditions were more favorable during 1950 to 1954.

Research since 1960 has demonstrated that a reduction of tillage on summer fallow using combined chemical and tillage programs or the elimination of tillage through a total chemical control of weeds has a nil to beneficial effect on crop yields. Anderson (1971) reported the results in Table 26–6 for wheat yields on several summer fallow methods.

Anderson (1975) has further shown that preseeding tillage for wheat seeded on summer fallow did not influence the volume weight or yield of spring wheat. There was a trend toward increased protein levels on preseeding tillage, but results were not consistent. The seeding machines used in this study were disker with packer, hoe press drill, high-clearance hoe press drill, double-disk press drill, triple-disk press drill, cultivator–rod weeder drill and the single-disk drill. Summer fallow programs included herbicide-only and tillage-only treatments.

Zero-tillage or no-tillage beyond that required for seeding shows promise on Borolls under optimum management. Lindwall and Anderson (1977) reported results with different seeding machines on no-till summer fallow and stubble. Yields are enhanced on both summer fallow and stubble, but the yield data in Table 26–7 indicates little change in the relative yields on summer fallow and stubble normally obtained. These workers stress that compaction or hard soils may require some form of strip tillage within the drill furrow area to get good results.

Crop residue management is the most important factor in minimum to zero-tillage programs. Lindwall and Anderson note that both the double-disk drill and the triple-disk drill used from 1969 to 1971 did not penetrate residues satisfactorily. Hoe-opener drills do not clear residues that are not chopped and uniformly spread.

Bowren and Dryden (1969, 1971) reported results for fall management of stubble and seeding methods on coarse- to medium-textured Udic Borolls in the region where crop use exceeds precipitation by 12 to 15 cm. The long-term average yields do not show major significant differences for most treatments ranging from no treatment to moldboard plowing. At Melfort, fall-plowing has consistently been the lowest-yielding fall treatment.

Table 26–7. Yields of spring wheat with three machines on no-till stubble in a wheat–wheat–chemical fallow rotation and for the semideep-furrow drill in wheat–chemical fallow and wheat–wheat–chemical fallow rotations on summer fallow.

	Stubble, seeding machine			Summer fallow, Semideep-furrow drill	
	Hoe drill	Double-disk drill	Semideep-furrow drill	3-year	2-year
	kg/ha				
1971	--	1018	1065	1453	1754
1972	1381	1186	1355	2545	2720
1973	600ab*	667a	573b	1308	1710
1974	1152a	964	1018ab	1013	1250
1975	1981	1866	1705	2381	1866
Average					
1971–1975	--	1140	1143		
1972–1975	1278	1171	1163		

* a, b, within years, values followed by the same letter do not differ (P = 0.05).

The direct seeding operation using the disker-seeder and packer has yielded as high or higher than either hoe–press drill or double-disk drill with preseeding tillage. The only exception is the coarse-textured soil at Brandon, where the plow followed by the single-disk press drill gave the highest yields. The result with the plow and single disk or pony press drill on sandy soils has frequently been obtained on coarse-textured soils. At Brandon, fertilizer reduced the difference between methods on sandy soil.

Bowren and Pittman (1975) reported results for barley yields following rapeseed using four treatments. Barley was seeded with the hoe–press drill. The yields in Table 26–8 show the highest yield for no-tillage either fall or spring, provided the rapeseed residues are chopped and uniformly spread.

26–3.2 Seeding Machines

A number of seeding machines have given satisfactory results. The double-disk drill performs well on summer fallow and is probably the best machine for variable soil conditions or where the soil is very loose and friable. The hoe–press drill requires firmness at seeding depth for best operation and depth control. The disker seeder requires a uniform, firm seedbed, careful adjustment and operation, and a packer for best results under most conditions. The rod weeder–seeder combination has performed well,

Table 26–8. Barley yields following rapeseed using various tillage treatments on a Udic Boroll silty clay loam at Melfort, Saskatchewan.

Stubble treatment		Barley yield
Fall	Spring	kg/ha for 4-year average
Disk	Disk	2613
None	None	2934
None	Disk	2779
Disk	None	2642

particularly on coarse- to medium-textured soils. Cultivator-seeders have not been used extensively and have been problematical for depth control.

The rod weeder that develops a uniform, firm condition at seeding depth adds considerably to the quality of seeding for the hoe–press drill and the disker.

The hoe-drills and triple-disk drills have been satisfactory for a number of farmers on zero-tillage programs. Satisfactory soil penetration to allow seed placement well into moist soil is necessary for good results on zero-tillage.

26–3.3 Seeding Rates and Plant Population

Quality seed, proper seed placement, adequate weed control, and fungicide or insecticide where needed, allow for seeding rates below that commonly used. Under moisture stress, high rates of seeding may cause yield reduction. It has been common to increase seeding rates for weed control, to compensate for thinning because of wireworm damage, and, more particularly, to compensate for deep seeding or seeding in loose, dry soil.

Pelton (1969) showed that spring wheat can be seeded at rates as low as 22 kg/ha on the Aridic Borolls. As shown by data in Table 26–9, the yield was significantly reduced on summer fallow at 100 kg/ha and on stubble at 67 kg/ha.

Row spacings of 15 and 18 cm are most common for cereal seeding. The narrower spacings do give slightly better competition with weeds as well as provide somewhat better support for the swath at harvest time.

Results from row spacing and seeding rates for spring wheat on a Typic Boroll loam are shown in Table 26–10. This study by H. Ukrainetz (1978. Unpublished data. Research Branch, Agric. Canada, Saskatoon, Saskatchewan) indicates that row spacing and fertilizer use have some effect on optimum seeding rates.

Higher seeding rates generally result in slightly earlier maturity, which is an important consideration if seeding is delayed. Pelton's results also show a significant effect on grain quality, as indicated by the 1000-kernel weight.

Table 26–9. Yields and 1000-kernel weights for various seeding rates on an Aridic Boroll at Swift Current, Saskatchewan.

	Seeding rate	Grain yield†	1000-Kernel weights†
	kg/ha		g
Summer fallow	22.4	1337a	27.4a
	44.8	1311a	26.6ab
	67.2	1290ab	26.1b
	100.8	1224b	25.1c
Stubble	22.4	859a	24.3a
	44.8	844a	24.0a
	67.2	768b	23.5ab
	100.8	764b	22.6b

† a, b means followed by the same letter are not significantly different (P = 0.05).

Table 26–10. Yields of spring wheat at various seeding rates and row spacing on a Typic Boroll at Scott, Saskatchewan.

Seeding rate	Row spacing	11	23	30
kg/ha			cm	
44.8	Fertilized†	2392	2157	2166
	Check	1989	1606	1935
67	Fertilized	2419	2292	2124
	Check	2157	1908	2083
112	Fertilized	2507	2339	2150
	Check	2325	1882	2076

† Fertilizer applied, N 10.2 kg/ha; P 19.6 kg/ha.

Table 26–11. Seeding rates for crops in the Canadian Prairies.

Crop	Rate
	kg/ha
Spring wheat	50–100
Durum wheat	85–100
Barley	55–100
Oat	55–90
Corn	11–16
Sunflower	5–8
Flax	30
Rye	45–60
Yellow mustard	7–11
Brown and oriental mustard	4–7
Grasses and legumes	11 plants/m^2

Seeding rates recommended for most important crops are given in Table 26–11. The plant population for sunflower is about 62 000 plants/ha and 44 500 plants/ha for corn.

Optimum dates for seeding are important for many crops. Oat and barley give highest average yields when seeded as early as soil conditions allow. Wheat yields are highest when seeded as soon as soil conditions allow or within 10 days of that date.

Corn must be seeded in early to mid-May to allow for sufficient maturity to produce quality silage. It is now established that corn will recover from late frosts if the growing point has not emerged above the soil surface.

Late frost or excessive early heat will damage early-seeded crops that are not seeded properly. Barley is quite susceptible to late frost if seeded too deeply and not supplied with adequate P.

Forage legumes and grasses require shallow seeding. Optimum row spacing ranges from 60 to 70 cm on Aridic Borolls to 30 to 40 cm on the Udic Borolls or in moist locations. Seeding can be done in early or late fall or in early spring.

On the Aridic and Typic Borolls, late fall seeding into standing stubble with adequate weed control is one of the most satisfactory methods of establishment. When companion crops are used for spring seeding of forage, seeding rate for the companion crop should be reduced 25 to 50%.

One practice that has potential for legumes seeded with companion crops is to delay seeding until 7 to 10 days after seeding the companion crop. This practice allows application of postemergent herbicides before germination and emergence of the legumes, which are susceptible to a number of phenoxy and other herbicides.

26-4 CROPPING SYSTEMS

Crop yields in the period 1945 to 1978 show a moderate increase trend. This period has included a major shift to more intensive use of summer fallow. If yields for the 1960 to 1978 period are compared with yields of the 1920 to 1930 period on a per hectare basis (including hectarage in summer fallow), however, yields have been just maintained or slightly lowered.

The potential is good for higher production with a reduction in summer fallow and the proportion of spring wheat. The inclusion of annual grain legumes, oilseeds, silage corn, and forage legumes and grasses can all add to the system.

Selection of adapted varieties or species is important to successful diversification of the cropping system. Pittman (1971) reported yields for two rapeseed varieties (Span and Oro) on a Typic Boroll. The yield for Span averaged 767 kg/ha and averaged 112 kg/ha for Oro over a range of harvesting dates in 1971. The yield of Span rapeseed was 1154 kg/ha seeded 6 May and 401 kg/ha seeded 10 June. The results show the drought-escaping potential of *Brassica campestris* L. rapeseed compared with *B. napus* L. and the benefits of early seeding.

The oilseed and pulse crops provide insufficient residue for summer fallow and must be followed by cereal crops. Durum wheat is one of the more satisfactory annual cereals following these crops. In the higher-moisture deficit areas, fall rye or winter wheat after a partial fallow of 12 to 13 months represents a good program.

Inclusion of biennial and perennial forage legumes will require a flexible program to suit any current moisture regime. On the problem coarse-textured, saline or adverse topography soils, perennial forage should constitute the major component of the system. On the better category soils, legume-grass is best established and broken for grain cropping under favorable moisture conditions.

26-5 RESEARCH—CURRENT AND FUTURE

Satisfactory production and utilization of winter rye and wheat would add a major dimension to crop management in the Canadian Prairies. In addition to development of more winter hardy varieties, there is a need to achieve the best agronomic management. Two elements are of major importance: soil management for best establishment under recurring dry conditions in August-September and protection against winter injury.

Zero-tillage offers major potential for management of both seeded land and summer fallow. The results of research are promising. There is a major need to establish satisfactory fertilizer programs and weed control methods.

Soil incorporated herbicides constitute highly satisfactory weed control. Most of these chemicals, however, require rather excessive soil manipulation. Research should be undertaken on methods to inject these herbicides in liquid form or to place granular materials with minimum tillage by pneumatic or other methods.

Fertilizer practices and nutrient regimes in soil require further research. Studies at the Saskatchewan Institute of Pedology (1981) show larger losses of N from zero-tillage compared with conventional tillage.

Crop residue management requires further work. Using swath widths of 9 m or more causes problems in uniform spreading. The residue concentration is showing not only on the stubble crop but is also beginning to show very strongly on crops seeded on summer fallow. The economics of better harvesting using wide swaths must be balanced with the added costs of handling crop residue from excessively wide swaths.

A major requirement in research is the establishment of site-specific, environment-specific recommendations for cropping practices. The standard deviation of the average yield or other parameter is frequently 50% of the average. Fall treatment of crop land may result in either large increases or large decreases in yield, depending on the moisture situation in the fall. In general, successful crop production in this moisture deficit, short growing season region depends on maximum potential adjustment to the conditions extant in any year or cycle.

26-6 LITERATURE CITED

Anderson, C. H. 1971. Comparison of tillage and chemical summer fallow in a semiarid region. Can. J. Soil Sci. 51:397–404.

Anderson, C. H. 1975. Comparison of preseeding tillage with total and minimal tillage by various seeding machines on spring wheat production in southern Saskatchewan. Can. J. Plant Sci. 55:59–67.

Bowren, K. E., and R. D. Dryden. 1969. Methods of seeding cereal crops on stubble land in Manitoba and Saskatchewan. Can. Agric. Eng. 11:74–77.

Bowren, K. E., and R. D. Dryden. 1971. Effect of fall and spring treatment of stubble land on yield of wheat in the black soil region of Manitoba and Saskatchewan. Can. Agric. Eng. 13:32–35.

Bowren, K. E., and U. J. Pittman. 1975. Production of mustard and rapeseed. p. 229–253. *In* J. T. Harapiak (ed.) Oilseed and pulse crops in western Canada—a symposium, Winnipeg, Manitoba. 21–23 May. Western Co-operative Fertilizers Ltd., Calgary.

Lindwall, C. W., and D. T. Anderson. 1977. Effect of different seeding machines on spring wheat production under various conditions of stubble residue and soil compaction in no-till rotations. Can. J. Soil Sci. 57:81–92.

Pelton, W. L. 1969. Influence of low seeding rates on wheat yield in southwestern Saskatchewan. Can. J. Plant Sci. 49:607–614.

Pittman, U. F. 1971. Papers presented at the 1971 Alberta Soils Advisory Committee meeting, Edmonton, Alberta. 25 November. Alberta Dep. of Agric., Edmonton.

Pittman, U J. 1977. Crop yields and soil fertility as affected by dryland rotations in southern Alberta. Commun. Soil Sci. Plant Anal. 8:391–405.

Saskatchewan Institute of Pedology. 1981. Annual report 1980–81. Saskatchewan Inst. of Pedology, Saskatoon, p. 35–38, 49.

27

Floyd E. Bolton
Oregon State University
Corvallis, Oregon

Cropping Practices: Pacific Northwest

Cropping systems in the Pacific Northwest are limited to crop-fallow rotation, wheat-pea rotations, and annual cropping of cereal grains (wheat, barley, pea) with an occasional crop of alfalfa or fallow (Table 27–1).

A key to any successful cropping system lies in planting the most highly adapted variety available. The combination of soil, crop, and pest management practices that conserves water for crop growth can benefit yields only if a well-adapted crop variety is grown that has adequate yield potential.

Crop varieties should be carefully tested under the particular climatic, soil, and management conditions where they will ultimately be grown to verify their potential yield capacity. Careful consideration should be given to varieties that exhibit the capacity to yield well under a variety of adverse conditions. Dryland crop production involves numerous variable factors. Varieties intended for the drylands should have an inherent capacity to thrive under changing conditions.

27–1 SEEDBED PREPARATION

Soil management practices may vary with climate and soil in different regions, but basic principles remain unchanged. Dryland tillage practices are used (i) to conserve the maximum amount of water in the root zone during the fallow period for use by a succeeding crop and (ii) to prepare a suitable seedbed so that the crop can be established at the optimum time to efficiently use the limited water supply.

In the normal fallow situation, the initial tillage operation, which will ultimately develop a seedbed for the subsequent wheat crop, can be performed any time beginning shortly after harvest until late fall or can be delayed until the following spring. Fall or after-harvest tillage is generally per-

Dryland Agriculture—Agronomy Monograph no. 23.

Table 27–1. Cropping operations sequences for three precipitation ranges in the Pacific Northwest.

Time	Field operations or conditions
1. Fallow-crop (356 mm or less of precipitation)	
July	Harvest grain
Early September	Fall chisel
September to mid-April	Uncultivated wheat stubble
Mid-April	Initial tillage
Mid-April to mid-September	Cultivate fallow
Mid- to late September	Plant grain
Late September to late June	Growth of crop
2. Wheat–peas rotation (356 to 457 mm of precipitation)	
July	Harvest grain
Late July to mid-September	Uncultivated wheat stubble
Mid-September	Plow
Mid-September to early April	Bare
Early April	Plant peas
Late June to early July	Harvest peas (green for canning or freezing)
Early August	Plow
Early August to late September	Bare
Late September	Plant grain
Late September to July	Growth of crop
3. Annual grain (457 mm or more of precipitation)	
July	Harvest grain
Late July to early August	Plow stubble
Early Augst to mid-October	Cultivate fallow and prepare seedbed
Mid-October	Plant grain
Mid-October to July	Growth of crop

formed to control excess weeds and volunteer wheat or to open the soil to prevent runoff from frozen soil. A wide blade sweep can be used for either purpose, but a rotary subsoiler or chisel also is frequently used to open the soil. A moldboard plow should not be used. Disk plows equipped with large wide-spaced disks may be used when stubble is excessively heavy. Such plows must be properly adjusted to avoid burying too much straw and should not operate at depths greater than 15 cm.

Most common problem weeds are Russian thistle, prickly lettuce, tarweed, and downy brome. If the first two listed occur in the wheat crop or germinate immediately after harvest, they should be controlled by either a tillage operation or an application of postemergence herbicide before the ground freezes in the fall. If downy brome is a problem, the field should be tilled lightly after harvest with a disk or similar implement to plant the seed; after emergence a light disking or contact herbicide treatment should be used.

When initial tillage is delayed until spring, the first operation should be performed after spring weed growth has begun and the soil has started to dry.

Whether initial tillage is performed in the fall or spring, the operation should be performed in the depth range of 7 to 13 cm. This depth controls weeds satisfactorily but does not promote soil water loss through drying to greater depths. Uniformity of operation depth is important to assure weed control and ease of implement operation during subsequent tillage and for uniform seeding depth at planting time.

Spring tillage generally requires a second operation immediately after the first to completely kill weeds, particularly grassy and volunteer weeds, and to settle the soil to reduce evaporation, thereby conserving water in the soil for the subsequent wheat crop. Timeliness, choice of implements, and climatic conditions are important factors to consider before performing the second spring operation. The second spring tillage can be performed with a skew treader, narrow sweep, or rod weeder. The choice of implement should be governed by the amount of residue on the soil surface as well as by the type of weeds that are to be killed.

Tillage operations during the summer and just before seeding should be performed only when necessary to control weeds. During the dry summer period, the soil tends to dry out to the depth of tillage. Therefore, summer weed control operations should be at a uniformly shallow 5- to 8-cm depth. All tillages will generally be with a rod weeder. Good fallow practices result in conservation of soil water in both total zone and in the seed zone. If the seed zone is not moist enough, a proper seedbed has not been developed during the fallow period. Subnormal or poorly distributed precipitation, however, may prevent even the most progressive farmer from being able to conserve enough water in the seed zone to ensure plant establishment. Where sufficient precipitation is available for annual cropping, tillage is used only to prepare a suitable seedbed, since generally little or no soil water can be conserved after an annual grain crop. Typical tillage sequence for the three cropping systems of the Pacific Northwest are given in Table 27–1.

27–2 SEEDING METHODS

Dryland cereals planted in fall are subject to a variety of winter climatic conditions that ultimately determine yield. A poorly developed plant facing a cold winter uses more water for recovering during spring than does a healthy, well-developed plant.

Seeding into residual water can be accomplished best by a properly adjusted deep-furrow grain drill, with row spacings ranging from 20 to 35 cm. The predetermined number of seed per given length of row must be placed firmly under 2.5 to 5 cm of moist soil. Seeding equipment should preserve as much residue as possible and securely anchor the residue in the ridges, leaving a rough, cloddy surface.

If there is not enough water close enough to the surface to ensure seedling establishment, the seed should be placed at a shallow depth so that rainfall can trigger germination and emergence. Data clearly demonstrate that as few as 5 to 7 days earlier emergence in fall can significantly improve crop yield.

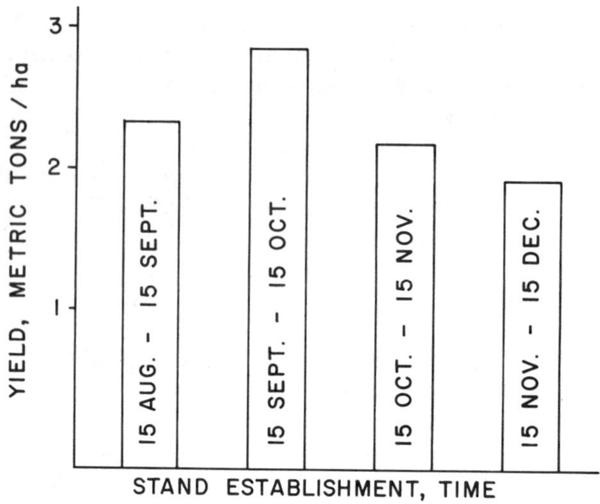

Fig. 27–1. Effect of stand establishment time on grain yield in Oregon (long-term average).

27–3 SEEDING DATE

Stand establishment is the most important single factor affecting crop yields under dryland conditions in the Pacific Northwest. Stand establishment is the preferred term rather than other commonly used terms, such as seeding date, planting date, or date of seeding. Describing the actual time the seeds are placed in the soil does not necessarily ensure time of emergence, crown root development, or the appearance of tillers, all of which must occur before a stand can be defined as established.

In dryland cereal-producing areas with winter precipitation, optimum time of stand establishment may have a considerable effect on economic yield by ensuring that the growth of the crop is adjusted to the available soil water. Data in Fig. 27–1 average many trials over a period of years to show the optimum time range for stand establishment in eastern Oregon. A number of early trials indicated no differences in yield attributable to planting date. Close inspection of the data revealed that often grain seeds were placed in dry soil and several days elapsed before a significant amount of precipitation occurred; the date of seeding could have differed by 2 to 4 weeks, but the time of emergence was the same. Since introduction of deep-furrow drills and the placement of seeds into moist soil as a result of fallow, data have been more consistent and represent truly different dates of emergence.

Long-term observations (F. E. Bolton. 1976. Unpublished data) of precipitation in relation to grain yields at three locations in the Pacific Northwest indicated that grain yields were significantly higher when significant rainfall occurred during the period of optimum stand establishment (15 September to 15 October). Apparently, adequate soil water stimulated emergence and subsequent plant development before low winter temperatures occurred.

As soil water conservation practices improved and deep-furrow drills were introduced, seeding into residual moisture became a standard practice. However, seasons still occur when adequate soil water for seeding is too deep even for the deep-furrow drill. Seeding may be delayed beyond the optimum date while the farmer is waiting for adequate rainfall; this delay often reduces yields because plant development is inadequate before cold weather.

Pahlivanturk (1976) showed that the time from emergence (after adequate water was present) to crown root development and tillering was greatly increased when the mean soil temperature was decreasing. This study indicated that when mean soil temperature at seeding depth dropped below 10°C; elapsed time for seedling development of crown roots, tillers, or both, increased dramatically. Emergence to tillering required 28, 51, and 125 days from 1 October, 15 October, and 1 November, respectively. Mean soil temperature at seeding depth for these planting dates was 16, 9, and 5°C, respectively.

Seeding in dry soil is risky; problems may arise from either surface crusting after rains or (sometime) weeds. However, yield reductions due to late stand establishment are well documented and consistent. The risk in reduced yields from crusting or weeds is not predictable. Even when the soil is dry, timely seeding has the advantage over waiting for adequate precipitation to wet the seed zone.

27-4 SEEDING RATES

The optimum number of plants grown per unit area for maximum economic production varies widely for different environments. When moisture is limited, plant population and distribution can substantially influence water use efficiency and subsequent grain yield if seeding rate is adequate (Harper, 1961).

For cereal crops planted in rows, adjustments in plant populations and distribution are made either within rows or between rows. Increasing distance between individual plants in the row and reducing the distance between rows to achieve a more uniform distribution over a unit area may defeat the purpose under limited water conditions. Brown and Shrader (1959) showed that plants facing minimal competition during the seedling stage tend to produce excessive vegetative growth and deplete soil water that could be reserved for grain production at a later date. Larson and Willis (1957) concluded that, to gain maximum benefit from limited water, the population within rows should be great enough to generate adequate competition during the seedling stage and thereby prevent excessive growth and water consumption. The spaces between rows should be adjusted to fit the water supply.

Seeding rates (within the row) and row spacing traditionally have been used for two purposes, neither actually related to the best plant population for maximum yield. Greater than optimum seeding rates are used to provide ground cover for wind and water erosion control. Close row spacings are often preferred to create increased competition for weeds that may grow between the rows. Although these concepts may have been successful in some areas, they undoubtedly reduce grain yields in many seasons in semiarid areas. Arnon (1972) and Arnon and Blum (1964) suggested that in dryland regions a limited water supply can be used better through greater attention to plant population and distribution.

Several studies conducted in different dryland regions of the world have shown that seeding rates can be reduced and row spacings increased without reducing (and many times increasing) grain yields (Alessi and Power, 1971; Bolton, 1973; Guler, 1975; Wilson, 1969). Reasons for not using more carefully adjusted seeding rates and row spacings are many and varied, but none seems to be related to more efficient use of water. Guler (1975) found that varieties differ in their response to seeding rates and row spacing in different climatic zones. In addition, he reported some varieties had differing optimum plant population numbers at each location, whereas other varieties responded uniformly at a given seeding rate in several locations.

27-5 SOIL FERTILITY

A detailed discussion of soil fertility and plant nutrition in dryland agriculture can be found in Chapter 21. However, the importance of these factors warrants their brief mention here. In areas of limited precipitation, a balanced nutrient supply enables a crop—at all stages of growth—to use the available water supply with greatest efficiency (Arnon, 1975, 1972; Olson et al., 1964). A plant deficient in nutrients, even when not growing or growing slowly, uses water at about the same rate as a nutritionally balanced plant, yet yields considerably less (Asana, 1962; Aspinall et al., 1964; Leggett, 1959).

Highly variable water and temperature typical of the dryland region of the Pacific Northwest tie fertilizer requirements to the prevailing water supply when efficient use of water is to be obtained. Moreover, various soil management methods—bare fallow, stubble mulch, minimum tillage, or no-till (chemical fallow) systems—may require different levels of N to achieve the proper water-fertilizer balance (Brown et al., 1960; Greb et al., 1967; Koehler and Guettinger, 1967; Overson and Appleby, 1971). The quantity of plant residues left on the soil surface or mixed in the upper few centimeters can significantly influence the amount of nitrogen available to the crop plant (Leggett et al., 1974; Smika et al., 1969; Smika, 1970). Therefore, knowledge of the soil's nutrient status as interrelated with cultural practices is essential.

27-6 CULTURAL PRACTICES

Factors that slow or block crop growth and development warrant full attention, and correction where possible. Control of insects and diseases under virtually any condition enhances crop yields. Such control is essential in dryland farming because limited water supplies may preclude recovery of a crop injured by pests. Under high rainfall or irrigated conditions, a damaged or weakened crop may have a chance of recovery. With limited water, yields are irretrievably reduced.

Control of weeds may be more important than other management factors such as choice of crop variety, stand establishment, and water-fertilizer balance. In fact, other improved cultural practices (high-yielding varieties, fertilizers, improved tillage methods or better seedbed preparation) without adequate weed control will generally lead to disappointing yield results. Timely, effective weed control is the first step toward reducing water loss and thereby improving yields in the drylands. Each increment of water used by weeds reduces water available to a crop. Weeds need to be controlled in the early stages of crop growth. Numerous studies document early competition between weeds and crops—for nutrients and light as well as for water—as the most damaging. Generally speaking, the more limited the water supply, the more critical becomes adequate weed control.

Heavy weed infestations in both cropped and fallow lands are common throughout the Pacific Northwest. Even when weed control is practiced, measures often are incorrectly timed, and the substantial yield losses still occur.

27-7 RESEARCH NEEDS

Additional weed control measures, chemical, cultural, and combinations, are needed to control the several species of grassy weeds. Research is needed on minimum tillage practices in heavy residue to include seeding equipment. Insect and disease problems that may be associated with heavy residues need to be determined specifically to eliminate speculated possibilities.

27-8 LITERATURE CITED

Alessi, J., and J. F. Power. 1971. Influences of method of seeding and moisture on winter wheat survival and yield. Agron. J. 63:81–83.

Arnon, I. 1972. Crop production in dry regions, Vol. 1. Leonard Hill Books, London.

Arnon, I. 1975. Physiological principles of dryland crop production. p. 3–145. *In* U. S. Gupta (ed.) Physiological aspects of dryland farming. Oxford & IBH Publishing Co., New Delhi.

Arnon, I., and A. Blum. 1964. Response of hybrid and self pollinated sorghum varieties to moisture regime and intra-row competition. Isr. J. Agric. Res. 14:45–53.

Asana, R. D. 1962. Analysis of drought resistance in wheat. Arid Zone Res. 16:183–190.

Aspinall, D., P. B. Nicholls, and C. H. May. 1964. The effect of soil moisture stress on the growth of barley. I. Vegetative development and grain yield. Aust. J. Agric. Res. 15:729.

Bolton, F. E. 1973. Soil and crop management research at the Wheat Research and Training Center: October 1970 to August 1973, Ankara, Turkey. Termination Report to Rockefeller Foundation and Oregon State Univ. under RF-OSU Grant no. RF70011, Corvallis, Ore.

Brown, B. L., and W. D. Shrader. 1959. Grain yields, evapotranspiration and water use efficiency of grain sorghum under different cultural practices. Agron. J. 51:339–343.

Brown, D. A., G. A. Place, and J. V. Pettiet. 1960. The effect of soil moisture upon cation exchange in soils and nutrient uptake by plants. Int. Congr. Soil Sci., Trans. Seventh (Madison, Wis.) 3:443–449.

Greb, B. W., D. E. Smika, and A. L. Black. 1967. Effect of straw mulch on soil water storage during summer fallow in the Great Plains. Soil Sci. Soc. Am. Proc. 31:556–559.

Guler, M. 1975. Yield and other agronomic characters of winter wheat as affected by five seeding rates and three different environmental conditions. M.S. Thesis. Dep. of Crop Science, Oregon State Univ., Corvallis.

Harper, J. L. 1961. Approaches to the study of plant competition. p. 1–39. *In* F. L. Milthorpe (ed.) Mechanism in biological competition. Symp. 15th Soc. Exp. Biol., Southampton, England. September 1960. Cambridge Univ. Press, Cambridge, England.

Koehler, F. E., and D. L. Guettinger. 1967. Moisture-nitrogen relationships for wheat production in Eastern Washington. p. 93–94. *In* Proc. 18th Annual Pacific Northwest Fertilizer Conference, Twin Falls, Idaho. 11–13 July. Pacific Northwest Plant Food Assoc., Portland, Ore.

Larson, W. E., and W. O. Willis. 1957. Light, soil temperature, soil moisture and alfalfa red clover distribution between corn rows of various spacings and row directions. Agron. J. 49:422–426.

Leggett, G. E. 1959. Relationships between wheat yield, available moisture and available nitrogen in Eastern Washington dryland areas. Washington Agric. Exp. Stn. Bull. no. 609.

Leggett, G. E., R. E. Ramig, L. C. Johnson, and T. W. Masse. 1974. Summer fallow in the Northwest. p. 110–135. *In* Summer fallow in the western United States. USDA-ARS Agric. Service Conserv. Res. Report 17.

Olson, R. A., C. A. Thompson, P. E. Grabouski, K. D. Stukenholtz, K. D. Frank, and A. F. Dreier. 1964. Water requirement of grain crops as modified by fertilizer use. Agron. J. 56:427–432.

Overson, M. M., and A. P. Appleby. 1971. Influence of tillage management in a stubble mulch fallow–winter wheat rotation with herbicide weed control. Agron. J. 63:19–20.

Pehlivanturk, A. 1976. Effect of soil temperature, seeding date, and straw mulch of plant development and grain yield of two winter wheat and two winter barley cultivars. M.S. Thesis. Dep. of Crop Science, Oregon State Univ., Corvallis.

Smika, D. E. 1970. Summer fallow for dryland winter wheat in the semiarid Great Plains. Agron. J. 62:15–17.

Smika, D. E., A. L. Black, and B. W. Greb. 1969. Soil nitrate, soil water, and grain yields in a wheat–fallow rotation in the Great Plains as influenced by straw mulch. Agron. J. 61: 785–787.

Wilson, J. W. 1969. Maximum yield potential. p. 34–56. *In* Transition from extensive to intensive agriculture. Proc. Seventh Colloquium Int. Potash Inst., Bet Dagan, Israel. Int. Potash Inst., Berne, Switzerland.

28

J. L. Hatfield
University of California
Davis, California

Cropping Practices: Pacific Southwest

Areas suitable for dryland agriculture in the Pacific Southwest are scattered throughout the geographic region. However, most of the land is used for rangeland. For example, California contains 4 million ha of land that may be considered dryland agriculture regions, with only 0.8 million ha routinely used for small grains and the balance for rangeland. In California the areas for small grain, wheat, and barley production are located primarily along the fringes of the central valley, with largest concentrations in the western edge of the Sacramento Valley and within San Luis Obispo and Monterey counties.

28-1 ENVIRONMENTAL CONDITIONS

In the dryland regions of California, rainfall ranges from 120 to 450 mm/y. Of primary importance is the seasonal distribution of precipitation relative to potential evaporation. Average values throughout the year are shown in Fig. 28-1 and 28-2 for Davis and San Luis Obispo, respectively. Data from Davis are representative of the dryland areas of the Sacramento Valley. The patterns for San Luis Obispo are similar to those of Davis. Precipitation is concentrated during the winter months from November through April. Precipitation occurs during the time of the lowest potential evaporation. Solar radiation and temperature during the winter months are sufficient to allow for continual growth during the rainy period.

Potential evapotranspiration during the rainy season are below the precipitation amounts, allowing for an increase in the soil water content (Fig. 28-1 and 28-2). During the spring the duration of stored soil water is dependent on the rainfall distribution. The rainfall totals of spring and fall months are quite variable, thus planting and maturity of small grain are de-

Dryland Agriculture—Agronomy Monograph no. 23.

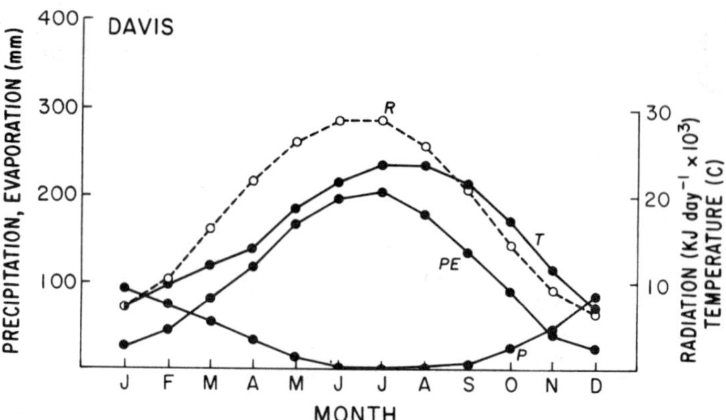

Fig. 28–1. Climatic factors for the northern California region as represented by the data from Davis, California. Data based on 30-year normals.

pendent on the distribution of rainfall (Fig. 28–3). Throughout the rainfall season there is large variation from year to year, with the seasonal totals ranging from 170 to 812 mm at Davis and from 204 to 1130 mm at San Luis Obispo.

The soils in the dryland regions tend to be shallow, with the limited A horizon often overlaying a rocky B and C horizon. More importantly, these soils are located in regions with undulating topography. In the Dunnigan Hills area of the Sacramento Valley the slopes of cultivated land range from 0 to over 40%. Topographic features are similar in the San Luis Obispo area, and the terrain can be described as extremely undulating with cultivation possible on all but the extremely steep slopes.

28–2 CROPPING SYSTEMS

Rainfall in the dryland areas dictates the cropping practices. Irrigation is not practiced because of the sloping terrain and scarcity of water. In addition to annual cropping, two rotations are extensively used: crop-fallow or crop-pasture-fallow (Schaller and Miller, 1960; Miller et al., 1960). Annual cropping of small grains can be practiced where the annual rainfall is above 400 mm in the central valley and above 300 mm in the cooler coastal regions.

The crop-fallow system is common in areas with rainfall below 300 mm. Because of the variability of rainfall, the soil moisture supplies are insufficient to sustain adequate growth without a stored soil moisture supply from a previous season. Even in areas with adequate rainfall, a fallow season is often necessary for the control of wild oat with the fallow year every third to fourth year (Schaller and Miller, 1960). The crop-pasture-fallow rotation has been successfully used in California, with the fallow year preceding the grain crop to ensure adequate soil moisture for the grain crop.

Schaller and Miller (1960) showed that the crop-fallow system could be used to improve soil conditions by incorporating the crop residue. They also

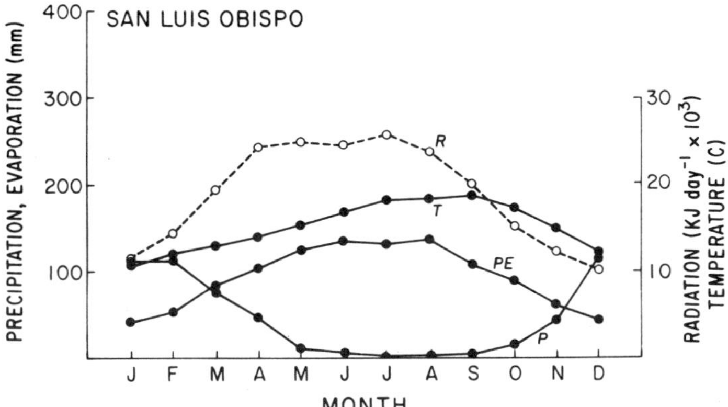

Fig. 28–2. Climatic factors for the coastal area of California as represented by data from San Luis Obispo, California. Data based on 30-year normals.

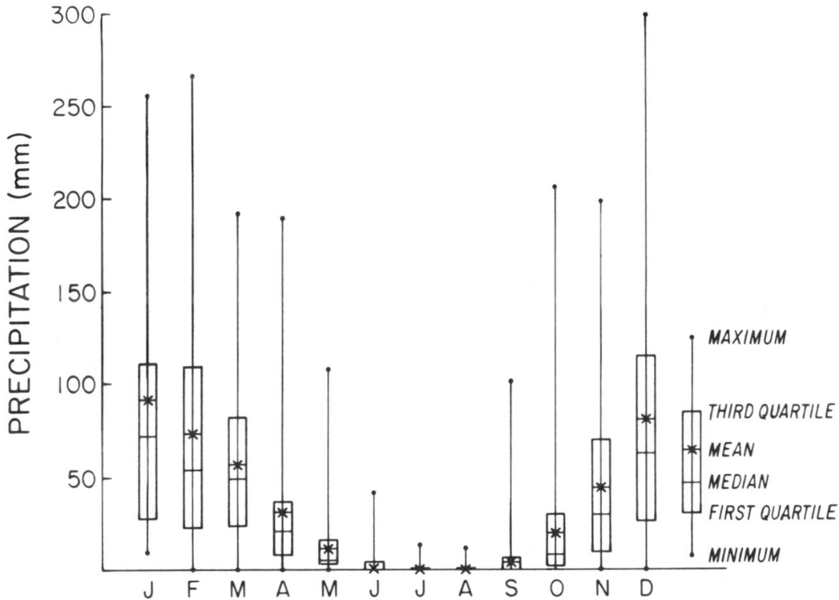

Fig. 28–3. Rainfall variability over a 105-year period from Davis, California.

state that the organic matter in dryland-farmed soils must be managed properly to avoid yield depressions in the fallowing crop. The fallow system also provides more flexibility in planting schedules because the seedbed is prepared in the previous spring or summer, depending on the weed control practice during the summer prior to planting.

All of the cultural operations on the seedbed are necessary to prepare the soil for planting and weed control. Herbicides may reduce the use of mechanical weed control measures. The use of the moldboard plow may be necessary in some locations with heavy soils to properly prepare the seedbed, although it is necessary to vary the depth of plowing so that a plow layer does not develop.

28-3 TILLAGE OPERATION

In the dryland areas of California, the most prevalent planting practice is the crop-fallow rotation system and the tillage operations are used for the control of weeds during the fallow season. After the crop is removed, the first operation generally occurs between October and April, depending on the rainfall conditions; these are the heavy tillage operations and include use of the offset disk, chisel plow, or wheatland plow. As a weed control measure, in late winter a second pass is made over the area with an offset disk and, in some instances, particularly after heavy winter rains, a moldboard plow. In early spring, a spring-tooth harrow or tandem disk may be used to control weeds, and during the early summer a spring-tooth operation may be necessary to destroy any remaining weeds. A chemical spray may also be used at this time to control weeds. Before, seeding operations, another pass is made with a tandem disk or a spring-tooth harrow to prepare the seedbed. These are generally done after the first good fall rains have occurred. In a season when the rains are very late in the fall, the crop may be seeded into dry soil.

The primary objective of tillage operations during the crop-fallow rotation is to conserve soil moisture during the fallow year. Efficiency of the fallow year is realized only if runoff is reduced, evaporation is maintained as low as possible, and any use by unwanted plant growth is minimized. Runoff is minimized by preparing a rough, trashy surface and is done during the early fall before the rains commence. This practice also prepares a surface that allows for good penetration and absorption of the rain yet retains its structure to allow for soil moisture storage. Any volunteer wheat or barley or weed growth must be removed by additional tillage operations. In the years with low rainfall, the removal of volunteer growth is more critical than in the wetter years where moisture conservation is less critical (Schaller and Miller, 1960).

Proper management of the fallow system with proper tillage using the disk-harrow or spring-tooth will leave the seedbed prepared. In the crop-fallow system in the California dryland system, the variability of the fall rains necessitates that the seedbed be prepared to allow for fall planting with a minimum of fieldwork.

28-4 PLANTING METHODS

Three methods are common in the Pacific Southwest for the seeding of the small grain crop: drilling, broadcast, or aerial application of the seed. Drilling the seed has been shown to produce a more uniform stand with less weed problems and less lodging than other methods (Schaller and Miller, 1960). Wheat and barley are generally drilled to a depth of 3 to 5 cm with a

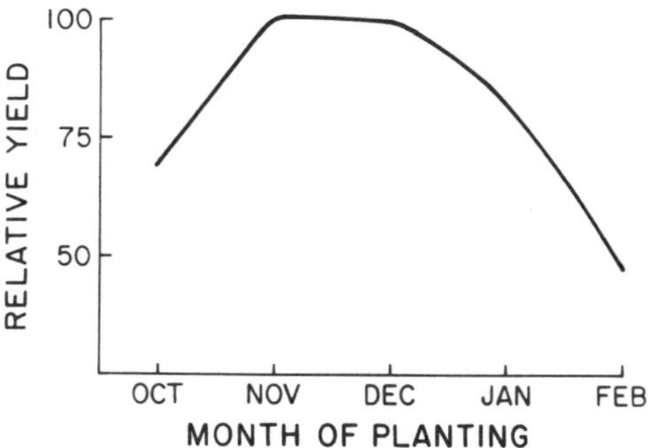

Fig. 28-4. Generalized relationship between small grain yield and time of planting in the Pacific Southwest.

row width of 20 cm. Kearney (1977) suggests that, as a recommended practice, narrower rows would provide a faster ground cover for weed control.

The majority of the wheat and barley seed is broadcast. The seed is placed on the ground surface with a drill box without openers, and a spring-tooth harrow is pulled behind to cover the seed and destroy any remaining weeds. In this planting method the seed operation is generally followed by a harrowing operation to smooth and firm the seedbed. Seed is applied aerially on prepared seedbeds with a final operation of a spike-tooth harrow to cover the seed. Broadcasting of the seed can be done in soil too wet for a drill, however; it should be used only when soil moisture is sufficient for germination and early growth (Schaller and Miller, 1960).

28-5 PLANTING PRACTICES

Small grains are seeded in California in the late fall and early winter, November through early January. The highest yields and most efficient use of the winter rainfall are obtained with planting dates in November and December (Miller et al., 1960; Schaller and Miller, 1960). The advantages of this time of seeding include efficient use of the winter rainfall, maturity before the arrival of the hot summer months, and maturity before the prevalence of the warm-weather diseases. A generalized relationship of yield to planting date is shown in Fig. 28-4. Early planting in October results in reduced yields because of the danger of frost during heading or lack of soil water availability for the grain-filling stage. J. L. Hatfield (1980. Unpublished data) has found that in the Davis area October plantings reach heading at the time when the probability of frost at anthesis is about 50%. Owing to the continual growth during the winter months, early plantings cause excessive vegetative growth because of (i) the warm temperatures and (ii) use of soil moisture resulting in reduced yields from inadequate soil moisture during the grain-filling stage. Wheat is tolerant of the hot weather, and yields begin to decrease with planting dates after mid-January, whereas barley exhibits more of a reduction in yield with late January and early February plantings.

The rate of seeding either wheat or barley is dependent on the location. General recommendations for most areas include 90 to 110 kg/ha (Kearney, 1977). Yield increases are not evident above seeding rates of 90 kg/ha; however, lower rates of seeding often fail to provide adequate competition for weeds early in the season. The recommended seeding rates at row spacings up to 30 cm provide a plant population of approximately 250 to 300 seeds/m^2 (Kearney, 1977). J. L. Hatfield and C. E. Whitman (1978. Unpublished data) have found that dryland wheat will produce the same yield at seeding rates of 70, 110, and 140 kg/ha if the soil moisture supply is adequate for tillering early in the growing season and is adequate throughout the season. If soil moisture become limiting, the plant will reduce the number of tillers reducing the transpiring area.

28-6 FERTILITY PRACTICES

Response to N fertilizer is highly dependent on the rainfall received during the cropping season. Miller et al. (1960) state that N response in wheat or barley may be small following a fallow season unless the rainfall exceeds 250 mm. Where moisture is adequate, yields have shown increases with 22 to 44 kg/ha. Current recommendations for dryland wheat or barley are 20 to 40 kg/ha of N at planting with 18 to 50 kg/ha as a topdress application as needed. Because of the variable rainfall amounts, an additional N application prior to jointing may be applied with aerial applications to enhance the yield of both wheat and barley. Suggested materials for N are ammonium nitrate or ammonium sulfate (Kearney and Garrison, 1977).

Phosphorus application has shown results in increasing yields in some soils, depending on the previous cropping practice and moisture conditions. In soils needing P, it is generally applied with the N at planting. Kearney (1977) has found through field trials in California that 15 to 50 kg/ha of P_2O_5 drilled with the seed will provide the crop with adequate P.

28-7 WEED CONTROL

Weeds of principal importance in California dryland agriculture are field bindweed, wild oat, mustards, fiddleneck, yellow starthistle, and isolated areas of annual ryegrass. Field bindweed, a perennial, is particularly important during the fallow year because of the depletion of soil moisture late in the season. The annual weeds, although important during the fallow, offer more problems from direct competition to the small grain in the year of cropping (J. E. Hill. 1981. Personal communication).

In years of early fall rains, cultural control of weeds is often achieved by delaying cultivation and seeding until after the weed seed has germinated. In years of late fall rains, this practice may result in late planting dates and concomitant yield reductions (Fig. 28-4). After planting, chemical control of broadleaf weeds is often achieved with 2,4-D (2,4-dichlorophenoxyacetic acid) or bromoxynil (3,5-dibromo-4 hydroxybenzonitrile) and wild oat by Avenge (difenzoquat; 1,2-dimethyl-3,5-diphenyl-1*H*-pyrazolium methyl sulfate) to reduce the competition (J. E. Hill. 1981. Personal communication).

Field bindweed is best controlled during the fallow season. Limited soil moisture for field bindweed growth following harvest in the crop year has led to ineffective chemical control. Control is achieved with 2,4-D applied by air or ground or Roundup [glyphosate; *N*-(phosphonomethyl)glycine] applied with ground sprayers. Wild oat and other weeds growing in the fallow season may be controlled by cultural practices during the summer or by grazing. Overall, the summer fallow provides a method of reducing weed problems by reducing the seed population (J. E. Hill. 1981. Personal communication).

28-8 RESEARCH NEEDS

Dryland agriculture is a small part of the California agricultural economy compared with the irrigated areas and, therefore, received much less research attention. Research needs to be conducted on the economics of small grain production to determine where the most benefit can be derived in terms of cultural, fertilizer, and herbicide costs. Research will be needed to determine ways of reducing the production costs either by reducing the number of cultural operations and use of more chemical weed control measures in lieu of mechanical control and the introduction and use of high yielding varieties. The uncertainty of the seasonal rainfall and its effect on harvestable yield is an uncontrollable factor in the Pacific Southwest. Research will have to be directed toward providing a seasonal forecast of the rainfall probability during the winter to determine the effectiveness and economic benefits of fertilizer and chemical weed control.

28-9 LITERATURE CITED

Kearney, T. E. 1977. Wheat and barley trials and production, Yolo County, 1977. Univ. of California Yolo County Farm and Home Advisors Pub., Woodland, Calif.

Kearney, T. E., and K. M. Garrison. 1977. Sample cost of production for dryland wheat and barley. Univ. of California Yolo County Farm and Home Advisors Pub., Woodland, Calif.

Miller, M. D., C. W. Schaller, and P. C. Berryman. 1960. Growing wheat in California. Univ. of California Extension Service Manual no. 29.

Schaller, C. W., and M. D. Miller. 1960. Barley production in California. Univ. of California Extension Service Manual no. 28.

29

J. Ross Wight
Agricultural Research Service
U.S. Department of Agriculture
Boise, Idaho

C. Kerry Gee
Economic Research Service
U.S. Department of Agriculture
Fort Collins, Colorado

R. J. Kartchner
Agricultural Research Service
U.S. Department of Agriculture
Miles City, Montana

Integrated Rangeland and Cropland Management

Rangeland has been described as land on which the native vegetation is predominantly grasses, grasslike plants, forbs, and shrubs that are suitable for grazing or browsing and includes lands revegetated by native or introduced species to provide a cover that is managed like native vegetation (Kothmann, 1974). It is often referred to as residual land—areas left after cultivation has reached the frontiers where inadequate water, difficult topography, and soil instability make cultivation economically and ecologically unfeasible. On a regional basis, the amount of precipitation usually determines rangeland-cropland boundaries. Within dry-farming regions, rangeland occupies those areas that are (i) untillable because of topography, rocks, or water content; (ii) unstable because of water or wind erosion; (iii) unproductive because of inherent low fertility, high salinity, alkalinity, or inadequate soil depth; and (iv) tillable but remain in range through the choice of the farm/ranch operator because of aesthetic or management options. The transition from cropland to rangeland occurs primarily in class IV lands (U.S. Soil Conservation Service land capability classes). Although these rangeland-cropland boundaries are fairly well established, they are not entirely stable; the cropland advances into the rangeland when grain

Dryland Agriculture—Agronomy Monograph no. 23.

Table 29–1. Percent of dryland in farms with beef breeding cows in selected areas (ERS, 1973).

Dryland cropland area	Farms with beef cows
	%
Eastern South Dakota	96
Central North Dakota	53
Montana	77
Western North Dakota and South Dakota	77
Pacific Northwest	87
Southeastern Idaho	43
Eastern Colorado	72
Western Kansas	42
Central Kansas	69
Texas-Oklahoma panhandles	37
Texas High Plains	29
Central Oklahoma	96

prices are high but usually retreats during periods of drought or low grain prices.

An inventory of farms with both rangeland and cropland hectarage has never been made. However, the U.S. Department of Agriculture has estimated the percentage of the total land area in several dryland farming regions that is in farms having beef cow inventories (Table 29–1). These farms commonly use range for part of the annual feed supply. Because some dryland farms with range inventories do not have estimates of beef cows, these percentages are probably conservative. The numbers in Table 29–1 are for 1972 but should be indicative of long-term conditions.

The percentage of the total land in farms having beef cow inventories varies considerably from region to region. The highest percentages are in the Northern Great Plains and the Pacific Northwest. Even in southern Nebraska and western Kansas, where much of the nation's wheat is produced, 40 to 50% of the farmland is associated with beef cow inventories (ERS, 1973). Probably two thirds to three fourths of all dryland farms in the 17 western states combine rangeland hectarage with cropland. The importance of this type of farming operation in western and midwestern feeder cattle production is identified by data from a 1976 USDA survey of farms and ranches with beef cows (C. K. Gee. 1976. Unpublished data). About 86% of these businesses operated dry cropland in conjunction with rangeland, including a few specialized range cattle operations. Regional differences are shown in Fig. 29–1. Only in the mountain, southern Idaho-Utah, and Northern Great Plains range areas were large proportions of cattle businesses operated independently of dryland farming. These data indicate that much rangeland and dry cropland in the Great Plains and western states are on farms and ranches that use both types of operations.

Rangeland forage production is an integral part of most dryland farming operations and appropriately should be included in a discussion of dryland agriculture. This chapter discusses the role and management of rangelands as a forage source in a dryland cropping system. The forage includes both native and introduced species harvested mechanically or by grazing animals.

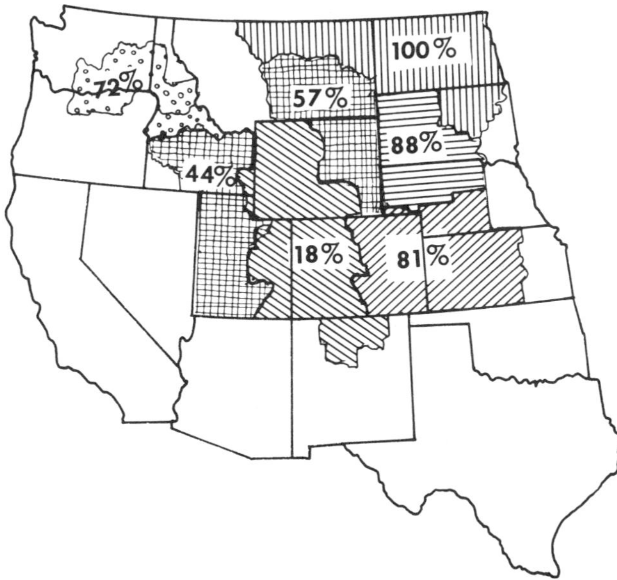

Fig. 29-1. Percentage of cattle operations in feeder cattle production regions that included dry cropland in 1976 (C. K. Gee. 1976. Unpublished data).

29-1 RANGELAND FORAGE CROPS IN DRYLAND AGRICULTURE

29-1.1 Soil Protection and Improvement

In recent geologic time, grass covered most of the world's dryland farming regions. The grass developed stable soils with high mineral and organic reserves for use by the man-seeded crops that followed. Today, grass covers, heals, and gives rest to those portions of earth degraded and damaged by unwise use, natural disasters, and mineral exploration and extraction of fossil fuel. Improved soil structure, stability, and fertility are recognized benefits of grass and legume forage crops (Hanson, 1974). The benefits of rangeland forage crops in dryland agriculture include a wide array of ecologic and economic factors.

In 1972, the U.S. Congress passed Public Law 92-500. Section 208 is directed specifically to the control of nonpoint pollution within the nation's waterways. With implementation of this law, soil stability has become an increasingly important factor in all farming operations. Conversion of marginal croplands to perennial forage crops provides an effective means of reducing soil erosion. For example, the cover and management factor (C) for the universal soil loss equation varies from 0.04 to 0.70 through the various stages of a small grain-fallow rotation; under perennial grass the C value usually ranges between 0.01 and 0.24, depending on type and amount of cover (Wischmeier and Smith, 1978). The C value is the ratio of soil loss from land cropped under specified conditions to the corresponding loss from clean-tilled, continuous fallow at the same location. From data by

Wadleigh et al. (1974), annual soil loss from a grain rotation in Oklahoma was 15.5 t/ha, compared with 0.04 t/ha for continuous grass. Strict enforcement of the law may result in large hectarages of marginal cropland being returned to perennial pastures or range.

In the crop-fallow system common to dryland farming, water often percolates through the soil profile, removes soluble minerals, including essential plant nutrients, and becomes a pollution hazard. Saline seeps, a serious problem in some dryland farming regions, are formed as salt-laden percolate is intercepted by impervious substrata and exposed again to the surface in topographic depressions or at points where impervious substrata intercept a changing topography (Halvorson and Black, 1974). Saline seeps have destroyed crop production on significant portions of dry-farmed land. Vander Pluym (1978) reported that in the Northern Great Plains, 0.8 million ha of cropland was severely affected and another 2.7 million ha was affected to a lesser extent. Seep development is a dynamic process, and without proper remedial action, seep areas continue to develop and expand.

Under perennial crop cover, water percolation through the soil profile is greatly restricted. Thus, perennial forage crops can help in prevention and reclamation of saline seeps. The most common treatment is the planting of perennial grasses, legumes, or both on areas that contribute groundwater to seep development. Under less severe conditions, salt-tolerant grasses can be seeded directly into the seep to remove water and dry the soil sufficiently for tillage. Establishment of perennial cover as a seep-control measure also contributes a significant amount of forage for livestock and wildlife use. However, it may be necessary to remove the bulk of the forage residue from seep areas so as to minimize potential recharge from trapped snow.

The edaphic benefits of grasses and legumes grown in dryland cropping rotations have not been well defined. After reviewing literature for the Northern and Central Great Plains, Brown (1964) concluded that on the basis of increased grain yields, the data did not justify growing grasses and legumes in short rotation on class I, II, and III land. The data, however, did indicate that legumes grown in rotation make significant contributions to soil N reserves, but depletion of soil water by the preceding legume crop may initially offset the beneficial effect of increased soil N. Brown also concluded that in short rotations, grasses and legumes do not measurably affect soil physical properties; but when they were grown for several years, grasses decrease bulk density, increase infiltration, improve aggregation, and reduce erosion. Research has been inadequate to fully assess the value of grasses and legumes in dryland rotations.

29–1.2 Economic Benefits

Native range and perennial grass and legume pastures are the main sources of dryland forage. In some areas, annual crops such as sorghum contribute significantly. The land seeded to grass and legume forages is usually marginal in terms of grain production. Hay crops are produced primarily for on-the-farm livestock. Very little dryland hay enters the market.

Most dryland farms produce large quantities of crop residues from wheat, barley, or grain sorghum. These are potential feeds that may also have limited sales. Livestock is often kept as a means of marketing these commodities that may not be otherwise salable. As a supplementary or complementary enterprise to cereal crops, livestock can consume these low-cost forages, produce meat at a definite cost advantage over livestock produced without these feeds, and contribute significantly to net farm income. By making a dryland farm a diversified business, we can improve economic stability, and returns are no longer dependent on the sale of a single commodity or type of commodity.

29–1.3 Wildlife Benefits

Wildlife is an important economic and aesthetic resource. Its compatibility with cropland can be enhanced when croplands are interspersed with native and seeded rangelands that provide a diversity of habitat and food. Some wildlife species, e.g., pheasant, depend on cultivated crops for food but still need perennial cover to prosper.

In recent decades, wildlife has been increasingly recognized as a valuable resource in terms of its input to local economies through both the range or farm itself and the associated businesses. In South Dakota, for example, where wildlife habitat is interspersed throughout the croplands, hunting makes a significant contribution to the state's economy. A 1975 survey by the U.S. Fish and Wildlife Service revealed that the average resident deer hunter spent $228 on hunting, or a total of almost $6 million (U.S. Fish and Wildlife Service. 1975. Unpublished data). Although most of this expenditure was for license fees, travel, and lodging, some went directly to the ranch or farm operators for hunting rights. The practice of leasing hunting rights has increased during the past 10 years and is a significant source of income for many ranch or farm operations. Ranchers and farmers are encouraged to develop their wildlife resource, and several incentive plans are sponsored by state and federal governments.

29–1.4 Rangeland Forage Production

Forage production on native range within the dryland farming regions is affected not only by the prevailing climate but also by such local site factors as topography and soil texture, depth, and fertility. Thus, production levels vary considerably. For the most part, they average between 100 and 700 kg/ha of grazeable forage annually. In terms of carrying capacity, about 300 kg of grazeable forage is required to support one animal unit (one 454-kg cow plus calf or equivalent) for 1 month. Only a portion of the total forage produced is actually consumed by grazing animals. An old rule of thumb commonly used in grazing native range is to "graze half and leave half" to avoid grazing damage to the perennial species. Grazeable forage is further reduced by the presence of unpalatable species and through trampling and spoilage.

Although native range is the prime forage source in integrated rangeland-cropland farming systems, introduced species are becoming increasingly important. As an indication of this trend, the hectarage in seeded pasture in the provinces of the Canadian Prairies increased from 1.2 million ha in 1956 to 2.0 million ha in 1966, and it is estimated to reach 2.6 million ha by 1980 (Smoliak, 1969). As monocultures or simple mixtures, these seeded pastures usually produce more total forage, a higher percentage of grazeable forage, respond better to favorable water and fertility environments, and are easier to manage than native range. The dryland farming regions contain cropland that, because of erosion hazards, should be seeded to perennial forages and a significant amount of rangeland that could be converted to pastures of introduced or domesticated perennial grasses, legumes, or both.

The production advantage of seeded pastures over native range is significant. In the Canadian Prairies, seeded pastures have demonstrated a threefold to fivefold production advantage over native range (Bratvold, 1969; Kilcher and Cooke [1967, cited by Lodge, 1969]). In one study, Russian wildrye produced six times more beef gain (yearling steers) than did adjacent native range (Smoliak and Slen, 1974). In eastern Montana, experimental plots of alfalfa, Russian wildrye, Altai wildrye, and crested wheatgrass produced two to three times more forage than did adjacent native range during both below- and above-average precipitation years, with the greatest difference occurring during the wet years (J. R. Wight and L. M. White. 1978. Unpublished data). In North Dakota, yearling steer gains on 28-year-old and 6-year-old crested wheatgrass pastures and native range averaged 100, 116, and 47 kg/ha, respectively, over a 6-year period (Rogler et al., 1962). In the same study, an alfalfa–crested wheatgrass mixture produced 25% more beef than crested wheatgrass alone.

Some reported research may overestimate the advantage of seeded pastures over native range because the results cover only the first few years after seeded pasture establishment. During this initial period, indigenous N may be relatively high because of decomposition of the large quantities of organic matter associated with freshly tilled native range. As this indigenous N supply diminishes, yields are reduced, and the yield advantage of the seeded over the native range is less than initially indicated. Lorenz and Rogler (1971) reported a study wherein beef production on a newly established crested wheatgrass pasture decreased from 194 kg/ha to 110 kg/ha 10 years later. In the same study, annual applications of 45 kg of N/ha increased beef yields only 6% the first year, when indigenous N was relatively high, but increased them 143% when applied 10 years later.

The most practical and effective use of seeded pastures is as component parts of range forage systems. Such complementary pastures can have a significant impact on total ranch production. Smoliak (1968) reported that fairway crested wheatgrass and Russian wildrye pastures used in a free-choice grazing system with native range produced 2.0 to 2.2 times more animal gain per hectare that did native range alone. In a later study using the same pastures in a rotation or free-choice system, they increased carrying capacity 87%, extended the grazing season, and produced 55 to 66% more beef per hectare than did native range along (Smoliak and Slen, 1974). In North Dakota, Lorenz (1976) proposed a pasture system composed of 20% crested wheatgrass and 80% native range that would increase carrying capacity 33% over the same amount of native range. In the Central Great Plains, beef production was increased 73% when 10% of the native range was seeded to weeping lovegrass (McIlvain, 1976).

The disproportionately larger production advantage of native range complemented with seeded pastures over seeded or native range alone is attributable in part to increased grazing management flexibility. The fact that most seeded forages are established as monocultures is a distinct management advantage. Different forage species will vary in productivity (Dubbs, 1966), optimum time of use (Westover et al., 1932), palatability (Dubbs, 1966; Gomm, 1969), and resistance to grazing (Wambolt, 1976). These species variations provide a valuable management tool that can be exploited to meet specific needs. As an example, crested wheatgrass and Russian wildrye, which are ready for grazing from one to several weeks earlier than native grasses, are often used for early spring grazing. This advantage has been used to increase rate of gain, reproductive performance, and stocking rate in cattle (Houston and Urick, 1972). The introduced species allow the native range plants to mature enough so that they are more productive and less susceptible to grazing damage than they would be without such deferment. Both crested wheatgrass and Russian wildrye are more resistant to damage from early spring grazing than are most native species. In addition, Russian wildrye remains quite palatable after dormancy, making it particularly suitable for late summer and fall grazing (Smoliak and Slen, 1974). Other species, e.g., intermediate wheatgrass, produce considerable forage but cannot withstand repeated early grazing.

Seeded pastures have the disadvantage of requiring more intensive management in terms of establishment, reestablishment, and sometimes additional fencing costs. They also require management of grazing and fertilization that is different from that of native ecosystems. Another potential disadvantage is that monocultures or simple mixtures may become more susceptible to plant-specific insects and diseases and to drought. To date, this susceptibility has not been an important problem among the more commonly used species, and some introduced species (e.g., crested wheatgrass) have existed in relatively pure and productive stands for over 40 years; they have exhibited a persistence greater than many native species (Hull and Klomp, 1966; Ross et al., 1966).

29-1.5 Forage Species for Seeded Pastures

Numerous plant species and varieties have been used successfully in seeded pastures. The selection and development of plant material is an ongoing process, with some cultivars being abandoned as better cultivars are developed. To date, no artificial species—interspecific crosses such as bluebunch wheatgrass × quackgrass (Dewey, 1976)—have been used other than on an experimental basis, but the potential is great.

Most of the seeded pastures consist of just a few species. Crested wheatgrass, both *Agropyron desertorum* and *A. cristatum,* has been used extensively throughout the Northern Great Plains and intermountain area, and probably accounts for more forage production in these regions than all of the other introduced species combined. This grass was used to revegetate abandoned croplands after the drought of the 1930's. It is one of the more drought tolerant and readily established introduced forage species and has been used almost exclusively in the dry areas of the intermountain region.

In the eastern, more mesic portion of the northern dryland farming region, smooth bromegrass and alfalfa are grown. In other portions of the Northern Great Plains, Russian wildrye, alfalfa, and a few seedings of intermediate, pubescent, and tall wheatgrass are common, along with crested wheatgrass. Russian wildrye is comparable with crested wheatgrass in drought tolerance, but it has not been extensively seeded. Altai wildrye is a recently released species that seems to be well adapted to this area. Weeping lovegrass has been the most frequently seeded pasture species in the Southern Great Plains, along with some King Ranch and Caucasian bluestems. In the Pacific Northwest, crested wheatgrass, Sherman big bluegrass, pubescent and intermediate wheatgrass, and some alfalfas are the most commonly used seeded pasture species within integrated rangeland–cropland farming systems. Within the Mediterranean climatic zone of California, Harding grass and alfalfa are used for perennial pastures and 'Blando' bromegrass with 'Lana' woollypod vetch or a mixture of subterranean and rose clover are common annual pasture components (Hafenrichter et al., 1973).

Occasionally, native grass species are used in seeded pastures. Slender wheatgrass in the Northern Great Plains and Sherman big bluegrass in the Pacific Northwest are examples. In general, the seeded native pastures are not as productive as adapted, introduced species. During a 4-year study in eastern Montana, two introduced species, intermediate wheatgrass and Russian wildrye, each produced about 50% more forage than did the native green needlegrass with or without fertilization (Black and Reitz, 1969).

Although most plant improvement research has been done with introduced species, considerable effort has been expended to improve and domesticate promising native species. A significant portion of this work has been performed under the plant materials program of the Soil Conservation Service. In addition to seeded pastures and range interseeding, these improved species have been used in revegetating strip mine spoils, road cuts, and other problem areas. Schwendiman and Hawk (1973) recently discussed some of these promising native species.

29–1.6 Establishment of Seeded Pastures

Much research has been conducted on the establishment of grass and legume species in dryland areas, and some good reviews are available (Vallentine, 1971). Although seeding techniques may vary with species and climatic region, successful seeding depends on adequate soil water, a firm seedbed, proper seeding depth, proper seeding time, and weed control. The best hedge against vagaries of precipitation common in dryland farming regions is a fallow period before seeding to allow soil water storage. Both mechanical and chemical fallows have been successful. In the Southern Great Plains, plant residues are often used to help maintain sufficient water at the soil surface for seedling emergence and establishment (Herbel and Baltensperger, 1973). A smooth, firm seedbed that retards evaporation is also a good treatment for soil water conservation.

Seedbed preparation is relatively simple on converted croplands; essentially the same procedures can be followed as those required for cultivated crops. Extra firmness can be accomplished by letting the land lie idle before seeding, cultipacking or rolling before planting, and using a press wheel behind the furrow openers during the seeding operation. Converting native range to seeded pastures requires more land preparation than converting cropland to seeded pastures. An effective technique is to grow and harvest a grain crop before seeding the pasture. A common practice in the Southern Great Plains is to grow a forage sorghum crop and seed into the stubble the following spring. The stubble protects the land from wind erosion before and during seedling establishment. Because of the competition for limited soil water, nurse crops are seldom used in dryland pasture seedings.

Because the seeds are small, seeding depth is perhaps more critical than with most other crops. Smoliak et al. (1975) stated that more seedings fail because planting was too deep than for any other reason. They recommended that small seed grasses and legumes should be seeded less than 13 mm deep and the large seed species (e.g., Russian wildrye and tall wheatgrass) should be seeded less than 25 mm deep. Both a firm seedbed and some means of depth control on the planter are needed to maintain good depth control. Depth bands on disk furrowers have proved very effective for controlling seeding depth (Kilcher, 1969; Smoliak et al., 1975).

Seeding time varies with species and climate. In the northern climates, seeding cool season species has been most successful in the fall or early spring. If soil water is adequate, seeding is best in late summer or early fall (usually late August), so that the grasses are well established before freeze-up. Seedlings thus established have a competitive advantage over weeds and can use early spring water supplies effectively. Some recent research at Miles City, Montana, showed that pubescent wheatgrass, crested wheatgrass, and Russian wildrye that were seeded in late August and reached the tillering stage before freeze-up reached full maturity the following year. Plants from later fall seedings developed much more slowly and required up to an additional year of deferment (R. S. White. 1978. Personal communication).

Seeding in the fall just before freeze-up so that the seeds remain dormant until the next spring is often more advantageous than early spring seeding. Wet soils and stormy weather can seriously delay spring seeding. Also, competition for labor and equipment may be greater in the spring than in the fall. Late fall or early spring seeding is recommended in the Canadian Provinces where winterkill is a problem with early fall seeding

(Kilcher, 1969). Early spring seeding generally works well in the Northern Great Plains during normal years. In fields infested with winter annual weeds, e.g., cheatgrass, spring seeding after a tillage treatment is effective. Spring seeding is also recommended for alfalfas and clovers.

For warm-season grasses and seeding in the southern regions, seeding is best in mid- to late spring or when the soil is warm enough to permit rapid germination and growth.

Weed control is necessary both to conserve water and to prevent excessive shading of the new grass seedlings. Fallowing is probably the best weed control tool. Herbicides are effective in some situations, but they are an added expense and may injure the seedlings. In the Southern Great Plains where seeding directly into sorghum stubble is common, undercutting the stubble with shallow sweeps before seeding is often used to control weeds and still leave a protective stubble mulch.

29–1.7 Grazing Management

Most grazing plans fall into four general categories: continuous use, deferred grazing, rotation systems, and deferred or rest rotation systems. Proponents have vigorously espoused the merits of each, but there is probably no one best system. For livestock production, the goal of grazing management is to maximize livestock production efficiency while conserving or improving the soil-water-plant resource. Good management is primarily a matter of applying the grazing system that meets the needs of specific situations.

The basic grazing management objective is to use as much of the annual growth as possible without permanently damaging the important range plants. Regrowth of perennial species after defoliation by grazing animals requires energy from the carbohydrate reserves in the plant's crown and root system until sufficient leaf area is reestablished to support the photosynthetic processes. Continual defoliation under prolonged and intensive grazing gradually depletes carbohydrate reserves, and the weakened plants become susceptible to disease and death. To maintain vigor, perennial plants must have the opportunity to grow sufficient leaf area to rebuild carbohydrate reserves. Unpalatable species and weedy invaders, thus, have a competitive advantage. For annual grasses, the management objective is somewhat different—seed production becomes a critical factor.

Grazing management on native range is especially challenging because the diverse vegetation complex may include many different species within a relatively small area. Grazing season, intensity of use, and animal type all affect plant species composition because various species respond differently to each. Both plant palatability and accessibility affect grazing preference and plant response to grazing. The objective of grazing management on native range is to control season of use, grazing intensity, and other grazing variables to maintain the desired balance of perennial grasses and forbs. Ex-

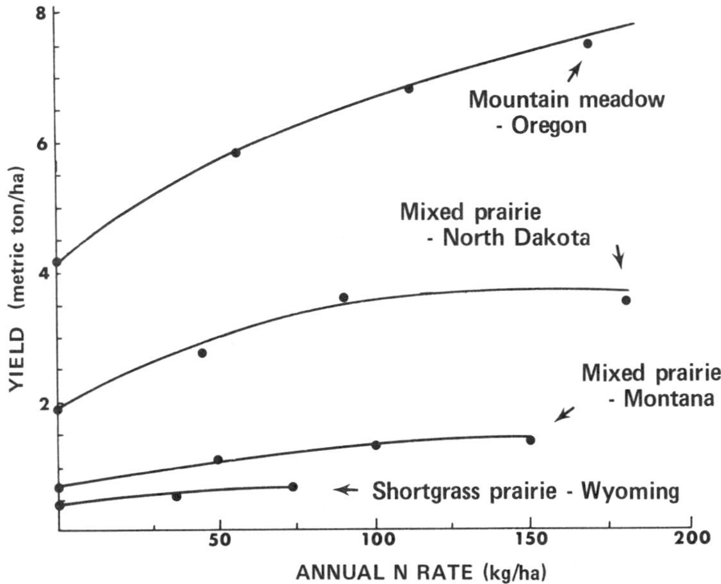

Fig. 29–2. Herbage response to N in four different grassland plant communities (Wight, 1980).

treme variations in annual precipitation and subsequent plant growth make the process even more difficult.

Grazing management on seeded pastures is less complicated than for native range, but the same basic principles apply. Proper management after establishment usually requires 1 and sometimes 2 years of deferment before grazing or haying. With early fall seeding, the pasture may be ready for use by midsummer the next year. Wambolt (1976) suggested that pasture use be deferred until the first seed crop has matured. Although seeded pastures can usually withstand heavier grazing than native range, grazing must still be regulated to avoid damaging the plant's recuperative power. Somewhat unique to seeded pastures is the development of large, coarse, "wolf" plants. These ungrazed plants are relatively unpalatable and can represent a significant forage loss. They can be controlled by periodic heavy grazing, fertilization to improve palatability, or mowing.

29–1.8 Rangeland Fertilization

Most western rangelands are inherently deficient in N especially those within the dryland farming areas of the Northern Great Plains (Rogler and Lorenz, 1974). Fertilizer response varies with climate (primarily precipitation), site characteristics, and N rate (Fig. 29–2). Forage increases of more than 100% are common on favorable sites. Fertilizer response can also be expressed in terms of N use efficiency (Fig. 29–3). In the Northern Great Plains, annual N rates of 30 to 50 kg/ha applied to native range, or higher

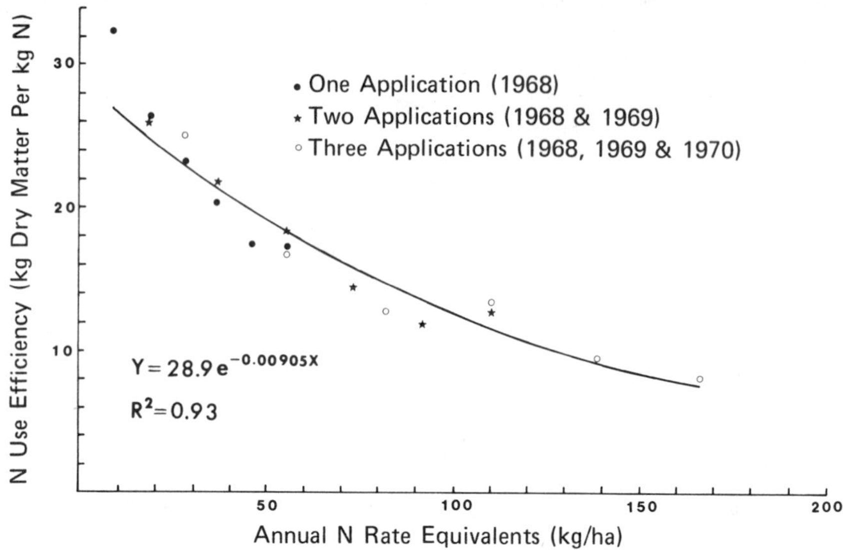

Fig. 29-3. Relationship between N use efficiency and annual rate equivalents of fertilizer N (Data from Houlton, 1975, cited in Wight, 1976).

equivalent rates applied biennially or triennially, optimize N use efficiency at 15 to 20 kg of dry matter/kg of applied N (Wight, 1976). This amounts to about 1 kg of beef/kg of N applied to native range and up to 2 kg of beef/kg of N applied to seeded pastures for beef (Wight, 1976).

Fertilization is proportionally more effective when used as a component part of a forage system. In addition to increased yields, fertilization may improve forage quality and palatability, provide earlier grazing and better livestock distribution, and improve range condition (Lorenz, 1976). These benefits make fertilization a valuable tool in the management of the entire grazing system. Fertilized pastures can also be used to enhance livestock management by providing nutritious forage at critical times in grazing animals' life cycles and by providing areas where livestock can be concentrated for purposes such as breeding.

On rangeland, broadcast application is about the only feasible method of applying fertilizer. For cool-season species, late fall or early spring applications are recommended; for warm-season species, late spring and early summer may be preferable. Perennial plants respond to fertilizer whenever they are in an active growth stage and soil water is adequate. Thus, on native range, fertilizer can be used to stimulate one species in preference to another by varying application time (Rehm et al., 1972). However, residual fertilizer will favor the early growing, cool-season species, which are dominant in the northern climates.

In the Northern Great Plains, and perhaps in other northern climatic regions, a "sink effect" is associated with N fertilization on native range. This effect is defined as the ability of the soil-organic matter complex to immobilize large quantities of inorganic N. Much of the sink effect is attributable to the 20 000 to 40 000 kg/ha of root material found under the native range (Power and Alessi, 1971). At Mandan, North Dakota, the sink is estimated to be as large as 350 kg of N/ha (Power, 1972), and full response to N fertilization is not realized until at least a part of the N sink is satisfied.

The annual rates or annual rate equivalents of 30 to 50 kg of N/ha that are recommended for the Northern Great Plains (Wight, 1976) are probably suitable for most rangelands within the dryland farming region. Because little water moves through the soil profile under perennial cover, there is almost a quantitative carryover of residual fertilizer (Power and Alessi, 1971). Thus, applying higher rates at 2- to 3-year intervals may be more efficient than lower rates annually because it reduces application costs and offsets the N sink effect. In the Northern Great Plains, single, high-rate N applications have been as effective or more effective in increasing forage production over a period of 2 or more years, as have equivalent amounts applied in annual increments (Houlton, 1975; Power and Alessi, 1971; Wight and Black, 1979).

In addition to N, P is sometimes limiting. Significant responses to P occur primarily when N is adequate. In the Northern Great Plains, forage yield increases after P fertilization range from 0 to about 20% (Wight, 1980). Except for S on some annual grasslands, other elements do not generally limit range forage production. They are important primarily in terms of animal nutrition. Fertilizing just to enhance forage quality without a corresponding quantity response is probably uneconomical.

The economic benefits of fertilization are usually greater for seeded pastures than for native range. Seeded pastures occupy the more productive sites; as monocultures or simple mixtures, they produce a much higher proportion of grazeable forage than does native range; and they are more ecologically stable when fertilized than native range. On a 17-year-old crested wheatgrass pasture near Mandan, North Dakota, 67 kg of N/ha applied annually increased the average annual production 274% (from 678 to 2538 kg/ha) from 1949 to 1960 (Lorenz and Rogler, 1971). In other experiments during 1958 to 1963, annual applications of 45 kg of N/ha increased beef production 70 kg/ha (from 98 to 168 kg/ha) on a crested wheatgrass pasture compared with 47 kg/ha (from 54 to 101 kg/ha) on native range (Lorenz and Rogler, 1971).

The use of legumes, alone or in mixtures, can effectively reduce the limiting effect of soil N deficiency. In the study just mentioned (Lorenz and Rogler, 1971), an alfalfa–crested wheatgrass mixture produced an average annual yield of 152 kg of beef/ha compared with 113 kg/ha on unfertilized crested wheatgrass and 189 kg/ha on the crested wheatgrass receiving 45 kg of N/ha annually.

Despite its demonstrated potential to increase production, range fertilization is not a widely accepted cultural practice. For example, a 1974 survey of 20 counties in eastern Montana and western North Dakota indicated that only 1.2% of the native range and 9.5% of the seeded pastures had ever been fertilized (Wight, 1976). Economic constraints and fear of adverse effects of fertilization on native ecosystems have been the main deterrents to range fertilization. Although information is becoming available to make effective economic decisions, the controversy concerning the ecological effects of fertilization have not been resolved. These ecologic concerns, however, apply primarily to native rangelands and are based on the hypothesis that fertilization stimulates disproportionately the weedy components of the vegetation complex. It is our opinion that with proper fertilization techniques and livestock management, such a generalization is unwarranted.

29–1.9 Marketing Forage Produced in Dryland Agriculture

Ruminant animals, e.g., cattle and sheep, play an important role in the productivity of dryland agriculture. Because of their unique digestive systems, they are able to use high fiber crop residues and forages from crops, range, and pasture as well as the more digestible grains when produced in excess of direct human needs. Ruminants not only have the digestive mechanism for using low-quality fibrous plant materials but also supply the means for harvesting these materials from millions of hectares of range, pasture, and cropland that are too rough or otherwise unsuited for mechanical harvesting.

The method of integrating livestock into a dryland crop production unit will vary with the objectives and resources of the producer. For many producers, livestock not only permits diversification of the operation but also greatly increases the flexibility and marketing alternatives for base crops, e.g., cereal grains.

Generally, beef cattle production can be classified as cow–calf, stocker, or some combination of the two. It is variously estimated (Anderson, 1978; Maddox, 1965) that 50 to 75% of the cost of producing a calf is tied up in feeding the cow and calf. Consequently, a cow-calf operation requires a 12-month feed base to support the basic cow herd, and this feed base must cost relatively little to permit a profit to be realized from the sale of the calf. Usually the cow's diet will consist primarily of roughages, including range forage, pasture, crop forage and residues, and hay. During the nonlactation period between weaning and calving, a mature cow's maintenance requirements are relatively low. As a result, she can effectively use low-quality roughages, e.g., crop residues.

Stocker animals increase flexibility because they can be purchased, or the forage is leased at a time and in amounts appropriate to the available feed resource. However, periodic purchasing and selling of stocker animals leaves the operator more subject to the vagaries of the marketplace. Also, many crop residues, e.g., straw, are of such low quality that they are largely unsuitable for feeding to young, growing livestock.

Rental markets for pasture and rangeland are well established in the western states. Although charges have been rising in recent years, rates are still competitive with other sources of range forage (Table 29–2). Increasing real estate prices are an added incentive for livestock producers to acquire rental contracts on pasture and range wherever possible.

Contractual arrangements vary widely. The most common contract requires the landowner to maintain fences and water facilities and the renter

Table 29-2. Average monthly rate per head for pasturing cattle on privately owned land in selected western states, 1972-1978 (ERS, 1978).

States	Monthly rate per head						
	1972	1973	1974	1975	1976	1977	1978
	Dollars						
Montana	4.30	4.80	6.60	7.00	7.40	7.30	7.40
Idaho	4.00	4.40	5.40	5.60	6.10	6.20	6.60
Wyoming	4.40	5.00	5.80	6.30	7.10	7.70	7.50
Colorado	4.40	5.10	5.50	5.80	6.30	6.60	6.80
New Mexico	3.90	4.10	4.40	4.90	5.20	5.80	5.20
Utah	4.30	4.80	5.00	5.50	6.00	6.90	6.40
Washington	3.50	4.40	5.30	5.30	5.80	5.80	5.60
Oregon	3.80	4.20	5.10	5.10	5.20	5.30	5.30
California	4.50	4.70	5.20	5.70	7.00	8.50	7.70
Nine-state avg.	4.30	4.70	5.40	5.80	6.50	7.20	6.90

to provide all inputs for care of the cattle. Because such contracts involve a minimum effort by the landowner, many landowners market their forage through this method.

29-1.10 Economic Competition Between Grain and Forage Production

The earning capacity of forage production has not generally been competitive with that of cereal crops. An analysis of competitive relationships between forage for beef cattle production and dryland wheat or barley indicated that a weaner calf price of well over \$2.20/kg (\$2.20/kg = \$1.00/lb) would be required to net a return to land equal to that of wheat at \$0.07/kg (\$2/bu) or barley at \$0.10/kg (\$2.16/bu) (Gee, 1973). Price and cost relationships generally favor grain production over forage wherever climate, soil, and topography are favorable for growing grain. Objectives other than maximum profit must direct the decision to shift from grain to forage production.

The competitive relationship between forage and grain production on a given hectarage of dry cropland depends on a number of factors, most important of which are:

1. Average grain yields.
2. Expected usable forage yields.
3. Grain prices.
4. Cattle prices, if the forage is to be marketed through cattle.
5. Cost of grain production.
6. Cost of beef production.

A change in any of these factors will affect the competitive relationship between forage and grain. This relation is illustrated in Fig. 29-4 and 29-5, where a shift from wheat to forage is assessed for dry cropland of two productivity levels—1000 and 2000 kg/ha (15 and 30 bu/acre) per crop year in a crop-fallow system. Several forage yields are considered, with yield measured in kilograms of total digestible nutrients consumed per hectare, assum-

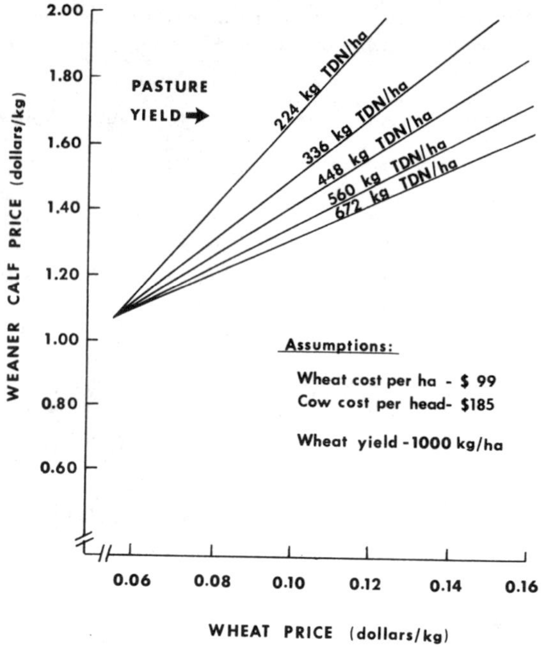

Fig. 29–4. Weaner calf prices needed to make pasture used for beef cows competitive with wheat production.

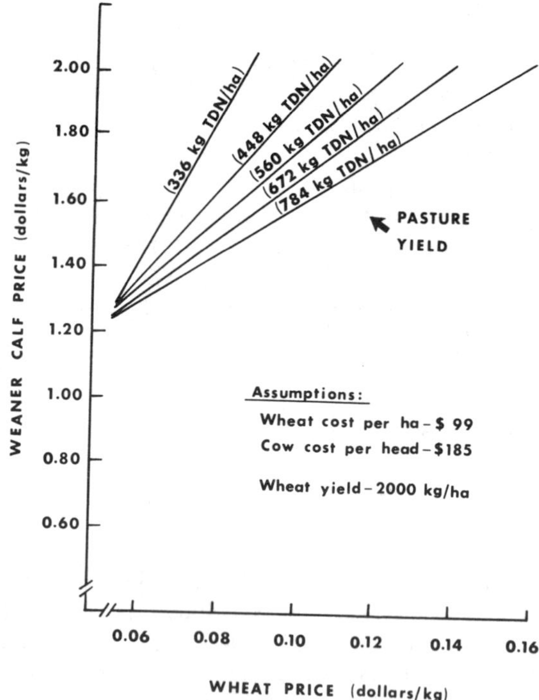

Fig. 29–5. Weaner calf prices needed to make pasture used for beef cows competitive with wheat production.

Table 29-3. Relationship between forage yield and calf prices.

Forage yield	Weaner calf price	
kg of total digestible nutrients/ha	—— $/kg ——	
	A	B
224	1.80	--
336	1.58	>2.20
448	1.47	1.98
560	1.41	1.80
672	1.36	1.72
784	--	1.63

ing that 1 kg of forage is equivalent to 0.5 kg of total digestible nutrients. The forage is consumed by beef cows that produce weaner calves, which are sold. Wheat and beef cow production costs for 1978 were assumed to be constant for the two cropland hectarages and were computed by Gee (1973). With wheat yielding 1000 kg/ha and price at $0.11 kg ($3.00/bu), calf prices per kilogram would have to be at the level shown in Table 29-3 (column A) to make forage competitive with wheat at given forage yields.

It is likely that forage yields on dry cropland that produced 1000 kg of wheat/ha will be rather low. Thus, calf prices must be fairly high—about $1.50/kg—for forage to compete with $0.11/kg of wheat. If wheat prices drop to $0.07/kg ($2/bu), the competitive range of calf prices drops to $1.10 to $1.32/kg. Alternatively, a wheat price of $0.15/kg ($4/bu) would require calf prices over $1.76/kg to make forage competitive.

If the land will produce 2000 kg of wheat/ha (30 bu/acre), calf prices must be even higher to be in the competitive range. To illustrate this relationship, $0.11/kg of wheat and selected forage yields would require the calf prices shown in Table 29-3 (column B).

Wheat prices have been in a $0.07 to $0.15/kg price range in recent years depending on the year, season, and geographic area. Calf prices have ranged all the way from $0.55/kg to above $1.76/kg. Only if calf prices stabilize at near $1.54 to $1.76/kg levels is it likely that forage will be economically competitive with wheat in dry cropland areas.

29-2 CEREAL CROPS AND CEREAL CROP RESIDUES AS FORAGE SOURCES

29-2.1 Cereals for Pasture and Hay

In some areas, immature cereal crops are extensively used as forage for livestock. This includes cereals seeded specifically for pasture, but a much larger area is seeded primarily for grain production and is also grazed. In southern regions, cereals can be grazed throughout the winter under a carefully managed grazing program. The farther north one goes, as winter weather becomes increasingly severe, the more restricted is the period and degree to which these crops can be grazed. In these areas, small grains can provide late fall or early spring grazing, thus extending the grazing season beyond that available from perennial forages alone.

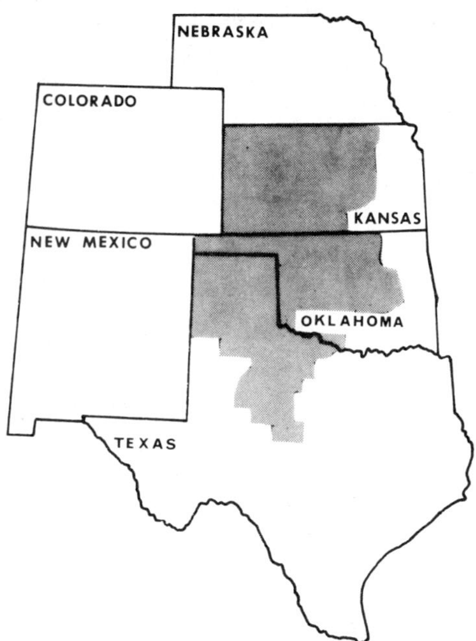

Fig. 29-6. Winter wheat grazing area of the Great Plains (SRS, 1974).

Winter wheat grazing is common in Kansas, Oklahoma, northern Texas, and in a few small areas in eastern Colorado and New Mexico (Fig. 29-6). The area grazed, which may be more than 0.5 million ha, varies from year to year, depending on soil water, plant growth, and winter weather conditions.

Grazing lightweight cattle on wheat fields is common from about October to March. With this growth and conditioning, they are placed in feedlots for fattening to slaughter weights. Winter wheat grazing is an integral part of cattle feeding programs of many Great Plains feedlots.

Sprague (1966) summarized several factors to be considered when using immature cereal crops for grazing: (i) fall seeding is generally practiced, and seeding is done earlier than for grain alone where weather and other conditions permit; (ii) areas to be grazed may be seeded up to 50% more heavily to increase forage production; (iii) overgrazing in the fall increases the danger of winterkill resulting in decreased spring forage yield, and excessive spring or late spring grazing reduces grain yields.

Research and experience have found little or no reduction in wheat yields from winter grazing if cattle are removed before wheat plants reach the jointing state—usually in March (Aase and Siddoway, 1975). Continued pasturing beyond this stage can severely reduce grain yields. If these precautions are taken, there are real financial benefits from grazing wheat. To carry a 182-kg calf through a 5-month grazing period, 2 to 3 ha are needed. Costs for landowners may include fence and water facility maintenance. These costs are usually small. Grazing winter wheat can reduce risks associated with wheat production by providing an additional source of income from the wheat.

There is little difference in chemical composition of the different grains in the young vegetative stage (NRC, 1971). The dry matter of immature cereal forages may contain protein levels of 20 to 30% and high levels of highly digestible carbohydrates. The high quality of these forages makes them suitable for all classes of grazing livestock, and particularly for young growing animals and high-producing stock, e.g., lactating dairy cows. A major disadvantage of grazing immature cereal forage is the possible occurrence of a condition known as grass tetany, a metabolic disorder related to magnesium metabolism (Church, 1971) that is sometimes associated with grazing this type of forage.

Cereal crops make excellent hay for all classes of ruminants. For maximum nutrient availability, forages are generally harvested between the milk and dough stages. Meyer et al. (1957) reported that total digestible nutrients in oat forage decreased slowly from 68% total digestible nutrients in the jointing stage to 60% total digestible nutrients in the flower stage, then decreased rapidly to 50% in the milk stage. Stage of maturity did not adversely affect daily grains until the forage entered the milk stage, when gains decreased drastically. Nutrient yield (total digestible nutrients and lamb production per unit area) was greatest at the 18% flower stage.

Palatability is generally highest for oat hay, intermediate for wheat and beardless barley, and lowest for rye. Bearded barley is not recommended for hay, because the awns are objectionable to livestock and may cause injury.

Growing cereals for hay is not a common practice but may be practical under some dryland farming situations. On a yield/hectare basis, cereal forage crops are competitive with perennial grass–legume mixtures. However, in a crop-fallow system, the total yield over time would be only about half that of a perennial forage crop. During a 9-year study at Swift Current, Saskatchewan, Canada, forage yields of wheat, corn, and oat in a crop-fallow system averaged 2938, 3005, and 2915 kg/ha, respectively, per crop year compared with 2897 and 2414 kg/ha annually on an alfalfa–crested wheatgrass and alfalfa perennial pasture (Kilcher, 1978). In terms of crude protein, the differences were much greater. The alfalfa–crested wheatgrass and alfalfa pastures produced about 50% more crude protein than the cereals on a crop year basis and about 300% more when compared over a time that included the fallow year. Kilcher (1978) concluded that hay for livestock should normally come from perennial forage crops containing alfalfa but that a cereal crop on fallow was a more reliable forage crop than perennial forage and was a good hedge against drought.

Some advantages of cereals as forage sources are (i) they can be readily established on fallow land; (ii) they require no new seeding equipment; and (iii) they produce harvestable forage the same year they are seeded, unlike perennial forages, which are usually not used the first year.

29–2.2 Grazing and/or Feeding Crop Residues

Cereal straw is the principal byproduct on the approximately 31 million ha of cereal crops produced annually on dry cropland in the western states. Livestock feed seems to be the best potential use for excess residues. However, the amount of crop residue that is in excess of the cultural needs of cereal production is probably less than initial inspection indicates. As summarized by Power (1978), crop residues affect soil-water storage, soil fertility, and weed control—all important factors in dryland farming. Residues are an important source of soil nutrients and organic matter and provide water-conserving mulches. Such mulches are also important for wind erosion control. It is not a question of whether or not to leave residue but how much should be left. Livestock's use of residue should be limited primarily to that which is in excess of the amount needed to maintain organic matter and an adequate mulch. Research in the Northern Great Plains has demonstrated that wheat straw residue has a significant effect on soil properties and grain yields (Black, 1973; Black and Siddoway, 1979). Under average conditions in the Northern Great Plains, all of the straw residue in a crop-fallow system is needed to maintain soil organic matter and a protective mulch; in an annual cropping system, about one third of the crop residue would be available for livestock or other uses (A. L. Black. 1978. Personal communication).

Quantities of crop residue consumed by livestock are probably a small percentage of total available. This limited use can be attributed to difficulty and cost of harvesting, a lack of interest by grain producers, and the low nutritional value of the residue.

Grazing or feeding crop residues is generally limited to maintenance of mature animals. Lofgreen and Christensen (1962) found that feeding barley straw instead of alfalfa as a component in a beef cattle fattening ration resulted in a lower rate of gain, lower energy content of the gain, reduced feed intake, lower dressing percent, and increased reticulorumen fill. Anderson (1978) found little justification for use of large quantities of straw in rations designed for high rates of production.

The feeding value of cereal crop residues is relatively low. Lofgreen and Christensen (1962) found the net energy of barley straw to be half that of alfalfa (50.6 vs. 97.7 Mcal/100 kg, respectively). Total digestible nutrient values given by the National Research Council (1971) range from a low of 31% total digestible nutrients for rye straw to 51% for oat straw. Cereal chaff tends to have slightly better feed value than straw, but its potential as a feedstuff is also limited.

All cereal straws are low in protein, generally ranging between 3 and 4.5% crude protein. The apparent digestibility of this protein is also low, ranging from none in rye straw to 30% in oat straw (NRC, 1976). The feeding value of wheat and barley straw are intermediate between rye and oat; wheat straw is generally the more nutritious of the two. The low protein levels in cereal crop residues require that supplemental protein be provided even at maintenance levels in mature livestock.

Efforts to increase the feed value of cereal straws have been concentrated in three areas, namely, supplementation (Andrews et al., 1972; Ball et al., 1971), physical treatments (e.g., chopping and pelleting; Burt, 1966), and chemical treatments (Klopfenstein, 1978). Additional information on feeding cereal residues can be found in an excellent review by Anderson (1978).

Complete utilization of cereal crop residues would require either extensive fencing and water development for livestock grazing or the purchase and operation of residue harvesting equipment, which usually lengthens the time required for harvest. Many farms do not have enough livestock to consume this additional feed. Therefore, livestock would have to be purchased or buyers found for the product. In many areas, the feed would have to be shipped out to find a market. These problems tend to limit use of cereal straw produced in the West.

Where cereal straw is harvested, it can be an economical feed, especially for beef breeding cows. Cost is usually one half to one third of hay cost, and total digestible nutrient content is about 40% compared with 50% for hay. When fed in combination with alfalfa hay, cereal straw can contribute as much as two thirds of the daily feed requirements. If properly supplemented, it can be used as the only roughage. Barley straw is quite palatable, but wheat straw may need to be fed with molasses to induce cattle to eat it. Use of cereal straw in beef cow wintering rations can reduce feed costs 20 to 30%.

29-3 ABANDONED CROPLAND

Rangeland-cropland boundaries are not permanently fixed; they shift in either direction in response to environmental and economic conditions. During the climatic conditions favorable to growth before 1930, cropland boundaries were extended deeply into the rangelands. During the following drought, these marginal croplands were the first to fail and, with loss of cover, were subject to severe wind and water erosion. After the drought, much of this land was returned to a permanent cover of seeded pastures or native vegetation. In Nebraska, for example, an estimated 0.526 million ha of cropland was reconverted to native range (Nebraska Conserv. Needs Comm., 1969). Many fields were simply abandoned and allowed to follow a secondary succession back toward a native climax community. Natural succession is a slow process, often requiring several decades. In many instances, the initial climax community never reestablishes itself because the climax conditions (primarily soils) have been modified or degraded through erosion.

Following the conversion of cropland back to rangeland after the 1930's, the rangeland-cropland boundaries remained relatively stable through 1973. About this time, an expanded wheat market in Russia resulted in high wheat prices. These high wheat prices, plus a drastic drop in beef prices, caused a small surge in the conversion of rangeland to cropland. Some of the rangeland "plow-out" was done by speculators who plow out large sections of rangeland with the intent of obtaining increased wheat hectarage allotments and selling the land as cropland for a quick profit. Good estimates of the total conversion are not available, but some data from Montana and South Dakota may be typical. From July 1974 to June 1978, an estimated 0.406 million ha of seeded pasture and native range in Montana was converted to cropland (Herbert and Stack, 1978). South Dakota reported a

Table 29–4. Summary of estimated grassland-crop conversions from South Dakota, 1 July 1974 through 30 June 1976 (Baumberger, 1977).

Grassland to cropland	ha	Cropland to grassland	ha
Not previously cropped (native grassland)	146 280	Pasture and hayland	104 881
Previously cropped (mostly tame grasses)	120 404	Range seeding (native grasses)	3765
Total	266 684	Total	108 646

net conversion of 141 218 ha (Table 29–4) for the 2-year period of 1974 to 1976 (Baumberger, 1977). Table 29–4 shows that much of the cropland came from previously cropped land that had been seeded to pasture. In most cases, it was probably soil bank land that was being plowed out and put into crop production. Table 29–4 also shows that although the net conversion was toward increased croplands, there was some conversion of cropland to rangeland during this period. In Nebraska, center pivot irrigation in the sandhills region has resulted in the conversion of an estimated 500 175 ha of rangeland to cropland (Bose, 1977).

Limited data are available on the hectarage of different land classes that have been converted. In South Dakota from 1974 to 1976, 26% of the hectarage converted from rangeland to cropland was in land capability class IV, and 16% was in classes VI and VII (Baumberger, 1977). Records kept by an eastern Montana county indicated that of the total coverted land, 25, 50, and 25% were from land classes III, IV, and VI, respectively (M. Saylor. 1978. Personal communication).

If history repeats itself, and it usually does, much of the recently converted rangeland will need to be restored to the permanent cover associated with rangeland uses. In most cases, these lands will be seeded back to tame pastures, providing additional forage resources for the integrated rangeland–cropland farming system. Areas currently under pivot sprinkler irrigation use groundwater, and they are another potential source of lands that may need to be converted to dryland forage crops if groundwater becomes limiting.

Seeding grass and forage species on abandoned cropland is generally easier and more successful than seeding plowed native range. During years of average to above-average precipitation, converting abandoned cropland to rangeland presents no major problems. However, it is during the drought periods, when plant cover is critical, that establishment is most difficult.

29–4 RESEARCH NEEDS

Integrated rangeland–cropland farming systems provide an excellent opportunity to develop forage systems that can supply the year-around nutritional needs of livestock. Research is needed to better determine (i) seasonal nutritional value of various species and pasture systems, (ii) seasonal nutritional needs of grazing animals, and (iii) how to integrate various segments of a forage system to optimize farm income.

Associated with seeded pastures is the need for improving seeding technology and plant materials. Rangeland seeding within the dryland farming region is usually not a critical problem except during times of drought. Because a large hectarage of converted rangeland may have to be abandoned during drought periods, the development of methods for establishing perennial grasses and legumes under water stress conditions is a critical research need. Also, an adequate seed supply may be a problem. Development of palatable and nutritious forage plants that can be established and maintained under water stress conditions is a continuing research need.

The use of annual forage grasses and seeded perennial pastures in crop rotations in integrated rangeland–cropland farming systems is not widely practiced. For weed control, soil stabilization and improvement, and economic returns, such rotations could have significant value. Legume forages in rotation with cereal crops could help alleviate the effects of soil N deficiencies. We need more information about how seeded perennial grass and legume pastures in rotation with cultivated crops affect the soil resource and about the economics involved. With the possibilities of energy and fertilizer shortages and increasing restriction on pesticide use, such information becomes increasingly important.

29–5 LITERATURE CITED

Aase, J. K., and F. H. Siddoway. 1975. Regrowth of spring-clipped winter wheat in the Northern Great Plains of the United States. Can. J. Plant Sci. 55:631–633.

Anderson, D. C. 1978. Use of cereal residues in beef cattle production systems. J. Anim. Sci. 46:849–961.

Andrews, R. P., J. Escuder-Volonte, M. K. Curran, and W. Holmes. 1972. The influence of supplements of energy and protein on the intake and performance of cattle fed on cereal straws. Anim. Prod. 15:167–176.

Ball, C., P. J. Broadbent, and T. L. Dodsworth. 1971. Energy and protein supplements for straw-based winter diets for suckler cows. Anim. Prod. 13:237–243.

Baumberger, R. 1977. South Dakota rangeland resources. Soc. of Range Management, Old West Regional Range Program, Denver, Colo.

Black, A. L. 1973. Soil property changes associated with crop residue management in a wheat-fallow rotation. Soil Sci. Soc. Am. Proc. 37:943–946.

Black, A. L., and L. L. Reitz. 1969. Row spacing and fertilization influences on forage and seed yields of intermediate wheatgrass, Russian wildrye, and green needlegrass on dryland. Agron. J. 61:801–805.

Black, A. L., and F. H. Siddoway. 1979. Influence of tillage and wheat straw residue management on soil properties in the Great Plains. J. Soil Water Conserv. 34:220–223.

Bose, D. R. 1977. Rangeland resources of Nebraska. Soc. of Range Management, Old West Regional Range Program, Univ. of Nebraska, Lincoln.

Bratvold, O. G. 1969. Forage crops in the economy. p. 20–35. *In* K. F. Nielsen (ed.) Canadian Forage Crops Symp. Western Co-operative Fertilizers Ltd., Calgary.

Brown, P. L. 1964. Legumes and grasses in dryland cropping systems in the northern and central Great Plains: A review of literature. USDA Miscellaneous Pub. no. 952.

Burt, A. W. A. 1966. The nutritive value of ground and pelleted barley straw. I. Ground and pelleted vs. long straw for growing Aryshire heifers. J. Agric. Sci. 66:131–137.

Church, D. C. 1971. Digestive physiology and nutrition of ruminants, Vol. 2. Nutrition. Oregon State Univ. Press, Corvallis, p. 704–709.

Dewey, D. R. 1976. Derivation of a new forage grass from *Agropyron repens* × *Agropyron spicatum* hybrids. Crop Sci. 16:175–180.

Dubbs, A. L. 1966. Yield, crude protein, and palatability of dryland grasses in central Montana. Montana Agric. Exp. Stn. Bull. no. 604.

Economic Research Service. 1973. Beef production potential of set-aside land. USDA Report no. ERS-532.

Economic Research Service. 1978. Farm real estate market developments, July. USDA July 1978 Report.

Gee, C. K. 1973. Pasture vs. small grains on dryland farms in the western intermountain area. Colorado Exp. Stn. Tech. Bull. no. 119.

Gomm, F. B. 1969. The palatability of range and pasture grasses in Montana. Montana Agric. Exp. Stn. Bull. no. 619.

Hafenrichter, A. L., D. S. Douglas, and F. L. Brooks, Jr. 1973. Hay and pasture seedings for the Pacific Coast states. p. 510–520. *In* M. E. Heath et al. (ed.) Forages: The science of grassland agriculture. ed. 3. Iowa State Univ. Press, Ames.

Halvorson, A. D., and A. L. Black. 1974. Saline-seep development in dryland soils of northeastern Montana. J. Soil Water Conserv. 29:77–81.

Hanson, A. A. 1974. Importance of forages to agriculture. p. 1–16. *In* D. A. Mays (ed.) Forage fertilization. Am. Soc. of Agron., Inc., Crop Sci. Soc. of Am., and Soil Sci. Soc. of Am., Madison, Wis.

Herbel, C. H., and A. A. Baltensperger. 1973. Ranges and pastures of the southern Great Plains and the Southwest. p. 499–509. *In* M. E. Heath et al. (ed.) Forages: The science of grassland agriculture. ed. 3. Iowa State Univ. Press, Ames.

Herbert, D. L., and R. Stack. 1978. Montana crop and livestock reporting service. Montana Dep. of Agric. and U.S. Dep. of Agric., Economics, Statistics, and Cooperative Services 19 July Report, Helena, Mont.

Houlton, H. A. R. 1975. Range response to nitrogen fertilization in north central Montana. p. 1–15. *In* Proc. Range Fertilization Symp., Havre, Mont. 1 July 1974. Montana Agric. Exp. Stn. Res. Report 70.

Houston, W. R., and J. J. Urick. 1972. Improved spring pastures, cow-calf production, and stocking rate carryover in the northern Great Plains. USDA Tech. Bull. no. 1451.

Hull, A. C., Jr., and g. J. Klomp. 1966. Longevity of crested wheatgrass in the sagebrush-grass type in southern Idaho. J. Range Manage. 19:5–11.

Kilcher, M. R. 1969. Establishment and maintenance of seeded forage crops. p. 90–104. *In* K. F. Nielsen (ed.) Canadian Forage Crops Symp. Western Co-operative Fertilizers Ltd., Calgary.

Kilcher, M. R. 1978. Perennial annuals for winter fodder production in the drier portions of the Canadian Prairies. Can. J. Plant Sci. 58:1081–1085.

Kilcher, M. R., and D. A. Cooke. 1967. Increasing carrying capacity through improvement of existing land. Saskatchewan Beef Industry Conference Report. Canada Dep. of Agric. Res. Stn. Mimeo, Swift Current.

Klopfenstein, T. 1978. Chemical treatment of crop residues. J. Anim. Sci. 46:841–848.

Kothmann, M. M. (chairman.) Range term glossary committee. 1974. A glossary of terms used in range management. ed. 2. Soc. of Range Management.

Lodge, R. W. 1969. Dryland pastures in grassland zone of prairie provinces. p. 282–296. *In* K. F. Nielsen (ed.) Canadian Forage Crops Symp., Edmonton, Alberta. 13–15 May. Western Co-operative Fertilizers Ltd., Calgary.

Lofgreen, G. P., and A. C. Christensen. 1962. Barley straw as an energy source for growing beef steers. J. Anim. Sci. 21:262–265.

Lorenz, R. J. 1976. Resource combination and technological innovations for beef cattle enterprises in the northern Great Plains. p. 37–52. *In* Proc. for Symp. on Integration of Resources for Beef Cattle Production, Omaha, Nebraska. 16–20 Feb. Soc. for Range Management.

Lorenz, R. J., and G. A. Rogler. 1971. Increasing production of native and seeded pastures with nitrogen fertilizer. p. 65–76. *In* Proc. 22nd Annual Fertilizer Conference of the Pacific Northwest, Bozeman, Montana. 13–15 July. Pacific Northwest Plant Food Assoc., Portland, Ore.

Maddox, L. A. 1965. Feeding the cow and calf. Texas Agric. Extension Services Bull. no. B-1043.

McIlvain, E. H. 1976. Seeded grasses and temporary pastures as a complement to native rangeland for beef cattle production. p. 20–31. *In* Proc. for Symp. on Integration of Resources for Beef Cattle Production, Omaha, Nebr. 16–20 Feb. Soc. for Range Management, Univ. of Nebraska, Lincoln.

Meyer, J. H., W. C. Weir, L. G. Jones, and J. L. Hull. 1957. The influence of stage of maturity on the feeding value of oat hay. J. Anim. Sci. 16:623–632.

National Research Council. 1971. Atlas of nutritional data on United States and Canadian feeds. Natl. Res. Coun.-Natl. Academy of Sci., Washington, D.C.

National Research Council. 1976. Nutrient requirements of beef cattle. ed. 5. Natl. Res. Coun.-Natl. Academy of Sci., Washington, D.C.

Nebraska Conservation Needs Committee. 1969. Nebraska conservation needs inventory. U.S. Dep. of Agric., Soil Conserv. Service, Lincoln, Nebr.

Power, J. F. 1972. Fate of fertilizer nitrogen applied to a northern Great Plains rangeland ecosystem. J. Range Manage. 25:367–371.

Power, J. F. 1978. Crop residues—not a waste. Agric. Res. 27(2):11.

Power, J. F., and J. Alessi. 1971. Nitrogen fertilization of semiarid grasslands: Plant growth and soil mineral N levels. Agron. J. 63:277–280.

Rehm, G. W., W. J. Moline, and E. J. Schwartz. 1972. Response of a seeded mixture of warm-season prairie grasses to fertilization. J. Range Manage. 25:452–456.

Rogler, G. A., and R. J. Lorenz. 1974. Fertilization of mid-continent plants. p. 231–254. *In* D. A. Mays (ed.) Forage fertilization. Am. Soc. of Agron., Inc., Crops Sci. Soc. of Am., and Soil Sci. Soc. of Am., Madison, Wis.

Rogler, G. A., R. J. Lorenz, and H. M. Schaaf. 1962. Progress with grass. North Dakota Agric. Exp. Stn. Bull. no. 439.

Ross, J. G., S. S. Bullis, and R. A. Moore. 1966. Grass performance in South Dakota. South Dakota Agric. Exp. Stn. Bull. no. 536.

Schwendiman, J. L., and V. B. Hawks. 1973. Other grasses for the north and west. p. 231–242. *In* M. E. Heath et al. (ed.) Forage: The science of grassland agriculture. ed. 3. Iowa State Univ. Press, Ames.

Smoliak, S. 1968. Grazing studies on native range, crested wheatgrass, and Russian wildrye pastures. J. Range Manage. 21:47–50.

Smoliak, S. 1969. Low cost grazing and forage production. p. 408–420. *In* K. F. Nielsen (ed.) Canadian Forage Crops Symp. Western Co-operative Fertilizers Ltd., Calgary.

Smoliak, S., A. Johnston, and R. A. Wroe. 1975. Seeding pastures in Alberta. Alberta Agric. AGDEX no. 130/22, Lethbridge.

Smoliak, S., and S. B. Slen. 1974. Beef production on native range, crested wheatgrass, and Russian wildrye pastures. J. Range Manage. 27:433–436.

Sprague, M. A. 1966. Cereals as forage. p. 337–345. *In* H. D. Hughes et al. (ed.) Forages: The science of grassland agriculture. ed. 2. Iowa State Univ. Press, Ames.

Statistical Reporting Service. 1974. Wheat pasture report. USDA-SRS July 1974 Report.

Vallentine, J. F. 1971. Range development and improvements. Brigham Young Univ. Press, Provo, Utah, p. 239–298.

Vander Pluym, H. S. A. 1978. Extent, causes and control of dryland saline seepage in the northern Great Plains region of North America, Section 1. p. 48–58. *In* H. S. A. Vander Pluym (ed.) Dryland-saline-seep control. Proc. Meeting of the Subcommission on Salt-Affected Soils, Edmonton. 21-24 June. Int. Soil Sci. Soc. Congr. Alberta, Canada.

Wadleigh, C. H., L. M. Glymph, and H. N. Holtan. 1974. Grasslands in relation to water resources. p. 15–42. *In* H. B. Sprague (ed.) Grasslands of the United States: Their economic and ecologic importance. Iowa State Univ. Press, Ames.

Wambolt, C. 1976. Montana range seeding guide. USDA Cooperative Extension Service Bull. no. 345, Montana State Univ., Bozeman.

Westover, H. L., J. T. Sarvis, L. Moomaw, G. W. Morgan, J. C. Thysell, and M. A. Bell. 1932. Crested wheatgrass as compared with bromegrass, slender wheatgrass, and other hay and pasture crops for the northern Great Plains. USDA Tech. Bull. no. 307.

Wight, J. R. 1976. Range fertilization in the northern Great Plains. J. Range Manage. 29: 180–185.

Wight, J. R. 1980. Range fertilization: The role of phosphorus. Better Crops Plant Food 64: 24–27.

Wight, J. R., and A. L. Black. 1979. Range fertilization: Plant response and water use. J. Range Manage. 32:345–349.

Wischmeier, W. H., and D. D. Smith. 1978. Predicting rainfall-erosion losses from cropland—a guide to conservation planning. USDA Agric. Handbook no. 537.

SECTION E

WEEDS, PESTS, AND DISEASES OF MAJOR CROPS

Allen F. Wiese
Texas Agricultural Research Station
Bushland, Texas

30 Weed Control

Weed control includes mechanical, cultural, chemical, or integrated systems using more than one method. Mechanical and cultural weed control were the only alternatives until 25 years ago. Since then, there has been gradual development of herbicides for chemical weed control.

A variety of tools are used for mechanical and weed control in and between dryland crops. In the wetter eastern part of the Great Plains, moldboard plowing is usually the first tillage operation after harvest. This is followed by disking or cultivation with small sweeps. In the drier parts of the Great Plains, subsurface tillage or one-way disk plowing is used for primary tillage. Subsequent cultivation is done with cultivators, rod weeders, or disk (Onken, 1971).

In the last 10 years, limited tillage, also called chemical fallow, conservation tillage, or ecofallow, has begun to replace conventional tillage between crops. Herbicides, rather than tillage, are used to kill weeds (Wiese and Staniforth, 1973). Advantages for the system are reduced erosion, labor, fuel use, and production cost. Frequently, water is conserved and yields are increased (Wiese et al., 1967; Phillips, 1969; Allen et al., 1977; Fenster, 1977; Unger, 1972).

Cultivation to control weeds in crops is done with small sweeps, rotary hoes, rolling cultivators, and various types of harrows. Thirty years ago, hand hoeing supplemented cultivation, but this practice is no longer economical. The secret of success with mechanical weed control is timeliness. Small weeds are easily uprooted or covered with soil. Before herbicides, when untimely rains interrupted cultivation schedules or when weeds emerged with the crop, it was often necessary to destroy the crop and replant.

Crop rotation is the best cultural method of weed control available. This technique is most effective in the Southern Great Plains because both summer and winter crops are grown. Weeds that infest sorghum and cotton are summer annuals. On the other hand, weeds in winter wheat are primarily winter annuals. When summer annual weeds are troublesome in a sum-

Dryland Agriculture—Agronomy Monograph no. 23.

mer crop, rotating to winter wheat eliminates the problem weeds. Tillage between wheat crops prevents summer annual weeds from going to seed. After a few years of winter crop, seed of summer annuals in the soil will be reduced and a summer crop can be grown again. In the Northern Great Plains, rotating to spring wheat helps but is not as effective as growing winter wheat. In the drier part of the Southern Great Plains, the most common crop rotation is winter wheat-sorghum-fallow. With this cropping sequence, two crops are grown in 3 years and there is an 11-month fallow period between each crop for weed control and soil moisture storage. With this cropping sequence, winter and summer crops are alternated and weed levels do not build up in either crop (Unger, 1972).

Other methods of cultural control are available to growers. Delaying planting until soil temperature is adequate for vigorous crop growth puts weeds at a competitive disadvantage (Onken, 1971).

Chemical weed control will be discussed under a section for each crop. However, successful programs integrate mechanical, cultural, and chemical methods to control weeds in the most economical way. For example, planting sorghum in narrow rows at high seeding rates increases the ability of sorghum to compete with weeds. On the other hand, narrow rows preclude cultivation and favor weeds. With rows closer than 75 cm, herbicides are needed to produce high-yielding crops (Burnside, 1977). In well-managed dryland fields with wide rows in the Western Great Plains, one or two cultivations are usually adequate to control weeds in sorghum, unless rain occurs shortly after planting (Wiese et al., 1964). Then an application of 2,4-D (2,4-dichlorophenoxyacetic acid) may be required. With recent increases in the price of fuel, it is more economical to use one application of propazine [2-chloro-4,6-bis (isopropylamino)-*s*-triazine] to control weeds in sorghum for the season rather than to cultivate and use 2,4-D.

30–1 PERENNIAL WEEDS

Dryland agriculture is plagued with many perennial weeds, primarily because frequent tillage is not necessary to control annual weeds. Infrequent tillage allows perennial weeds to store reserves in roots that penetrate up to 7 m in deep fertile soils (Green et al., 1965; Frazier, 1943).

30–1.1 Problem Weeds

Two perennial grasses affecting dryland farmers in the western USA and Canada are quackgrass in the North and johnsongrass in the South. Fortunately, these two weeds are adapted best in humid rather than semiarid climatic areas (Reed and Hughes, 1970, p. 462).

Field bindweed is the most widespread broadleaved perennial weed in western USA and is a serious problem in both dryland and irrigated crops. It covers 160 000 ha in the panhandle of Texas alone and threatens over 1 million ha of cropland (Wiese and Rea, 1955). Russian knapweed also covers the entire West, particularly in the mountainous areas (Schultz and

Burnside, 1979). Hemp dogbane covers the entire West but does not infest as many hectares as the two previous weeds.

Other perennial weeds tend to be more regionalized. Horsenettle and common milkweed, are most troublesome along the eastern edge of the Great Plains.

Texas blueweed, silverleaf nightshade, woollyleaf bursage, and skeletonleaf bursage infest the Southwest. In the Northern Great Plains and Canadian Prairies, Canada thistle infests almost every field and is the most widespread broadleaved perennial weed in that region. Leafy spurge is particularly troublesome in the Dakotas and central Canada and is found in 458 counties in 26 states from coast to coast in the USA (Dunn, 1979). Perennial sowthistle also plagues northern growers.

30–1.2 Weed Competition

In general, perennial weeds reduce yields about 30 to 50%, but variations are large among weeds and crops. In a 12-year study at Hays, Kansas, wheat averaged 278, barley 458, and oat 385 kg/ha more yield on noninfested land than on land infested with field bindweed. This was an average reduction of 35%. In the same study, sorghum grain yield was reduced 78% by field bindweed (Phillips and Timmons, 1954). In Nebraska, common milkweed reduced sorghum grain yield 21% in a 3-year competition study (Evetts and Burnside, 1973). Yield reductions from hemp dogbane ranged from 0 to 10% in corn, 28 to 41% in soybean, and 37 to 45% in sorghum. Lowest infestations were in alfalfa, pastures, and winter wheat; highest were in oat and soybean (Schultz and Burnside, 1979).

30–1.3 Control of Perennial Weeds

Intensive tillage was the only control method for perennial weeds prior to the development of chemical herbicides after World War II. Best control of field bindweed was obtained when sweep tillage was performed at 2- or 3-week intervals during fallow periods of a wheat-fallow cropping sequence (Phillips and Timmons, 1954; Wiese and Rea, 1959; Zahnley and Pickett, 1934). Similar results were obtained with quackgrass (Lowe and Buchholtz, 1952). Cultivation of johnsongrass growing on fallow land at 2-, 3-, 4-, and 6-week intervals was evaluated in Kansas. Johnsongrass was eliminated with six operations at 3- or 4-week intervals at two locations, and at a third location, eight cultivations at 4-week intervals eliminated the weed (Timmons and Burns, 1951).

Since that time, controlling perennial broadleaf weeds has been made easier with sprays of 2,4-D and other herbicides. Specific research has been reported on field bindweed (Phillips and Timmons, 1954), Woollyleaf bursage (Sand and McCarty, 1963), Canada thistle (Rasmussen, 1956), and Texas blueweed (Wiese, 1956). For 2,4-D to be effective, perennial plants must be growing vigorously and, in some instances, in a particular stage of growth. Wiese and Rea (1962a) showed that good control of field bindweed with 2,4-D occurred in semiarid areas if the soil had at least 4 cm of available water and runners were 15 to 20 cm long. In these studies, 2,4-D was more effective than 2,4,5-T (2,4,5-trichlorophenoxyacetic acid), MCPA {[(4-chloro-*o*-tolyl)oxy]acetic acid} and silvex [2-(2,4,5-trichlorophenoxy)-propionic acid]. Dicamba (3,6-dichloro-*o*-anisic acid) has been useful for controlling Canada thistle and field bindweed (Chang and Vanden Born, 1968; Carson and Bandeen, 1975; Schweizer et al., 1978). Picloram (4-amino-3,5,6-trichloropicolinic acid) has been used alone or in conjunction with other herbicides as a foliar spray (Agbakoba and Goodin, 1970; Sharma et al., 1971). Most recently glyphosate [*N*-(phosphonomethyl)-glycine] has proved toxic to perennial weeds (Banks et al., 1979). Atrazine [2-chloro-4-(ethylamino)-6-(isopropylamino)-*s*-triazine] sprays at 2 to 4 kg/ha have given excellent control of quackgrass in corn (Buchholtz, 1963).

Presently, the most effective method of controlling large areas of broadleaved perennial weeds is with a combination of competitive crops, cultivation, and herbicides. Specific studies have been reported for field bindweed (Phillips and Timmons, 1954; Derscheid et al., 1970; Wiese and Rea, 1959). Usually from 3 to 5 years were required for complete kill of the weed. Using an herbicide increased speed of weed kill and decreased the number of tillage operations compared with intensive tillage alone. Similar control has been obtained for Russian knapweed (Derscheid et al., 1960b; 1963a), leafy spurge (Derscheid et al., 1960a, 1963), as well as Canada thistle and perennial sowthistle (Derscheid et al., 1961).

Greer (1977) says a successful johnsongrass control program integrates tillage, crop rotation, and herbicides. Tillage at 4- to 5-week intervals removes top growth, exposes rhizomes to drying, and prevents new growth. Herbicides useful against johnsongrass are glyphosate, dalapon (2,2-dichloropropionic acid), TCA (trichloroacetic acid), and MSMA (monosodium methanearsonate).

Small areas of perennial weeds are most economically controlled with soil active herbicides. Research by Latshaw and Zahnley (1927) showed that sodium chlorate effectively eliminated field bindweed; however, treated areas stay bare for many years. The chloro-substituted benzoic acids, primarily 2,3,6-TBA (2,3,6-trichlorobenzoic acid), were more effective and persisted less time than sodium chlorate (Phillips, 1960a). Fenac [(2,3,6-trichlorophenyl)acetic acid] was effective against field bindweed, woollyleaf bursage, Texas blueweed, and silverleaf nightshade (Wiese and Rea, 1961; Smith et al., 1972a). Most recently, picloram at 2 to 3 kg/ha was an effective soil sterilant for controlling woollyleaf bursage, field bindweed, Texas blueweed, and silverleaf nightshade (Wiese et al., 1965).

30–2 CORN

Dryland corn is grown from South Dakota south along the eastern edge of the Great Plains through Nebraska and Kansas. The total corn production for these states in 1980 was about 5 million ha. About half of this area is irrigated, so dryland corn production in the area is about 2.5 million ha.

Weed control methods for corn have been primarily developed in Corn Belt states or for irrigation in the Great Plains. However, some research has been done under dryland conditions in Nebraska and Kansas.

30-2.1 History of Weed Control

Some of the basic principles of weeding crops were developed in corn from 1880 to 1930. As early as 1886, Sturdevant showed that weeds seriously reduced yields of corn. In his study, the corn yield was 1200 kg/ha when weeds were allowed to grow. Yields increased to 3760 kg/ha with cultivation and hoeing and 4700 kg/ha when weeds were pulled by hand. He concluded there was no value in cultivation except for removing weeds. In 125 studies conducted from 1906 to 1911, Cates and Cox (1912) compared cultivation with cutting weeds off at the soil surface and found no difference in yield. Similar results were reported by Weimer and Harland (1925, p. 10).

Other research about that time showed that killing existing weeds before planting, followed by shallow cultivation to control weeds in the crop, led to maximum yield (Moiser and Gustafson, 1915). They concluded that under Illinois conditions, weeds primarily competed for soil nutrients and light because irrigating a weedy plot only increased yield 255 kg/ha. Killing weeds without cultivation increased yields 450 kg/ha over cultivation.

30-2.2 Weed Competition

Uncontrolled growth of smooth pigweed in corn reduced grain yield 39% compared with weed-free corn. Total dry matter yield of crop and weed was about equal, regardless of weed population (Moclani et al., 1964). Annual grass weeds in Iowa corn, primarily foxtail, reduced grain yield 940, 670, and 335 kg/ha, respectively, with application of 0, 78, and 156 kg/ha of elemental N. Corn yield was increased two to three times more than yield of foxtail (Staniforth, 1957; Nieto and Staniforth, 1961). Corn hybrids responded differently to competition from yellow foxtail (Staniforth, 1961).

One of the most feared weeds in corn fields is giant foxtail. In a 3-year study, where from 0.5 to 54 giant foxtail plants/0.3 m of row competed with corn, grain yield reduction was 25% for the heaviest population (Knake and Slife, 1962). In another study, giant foxtail was seeded with the crop at 3, 6, 9, and 12 weeks after corn planting. Foxtail seeded with the crop reduced corn yields 13%. Seeding the weed 3 weeks after planting did not decrease corn grain yield, but weeds made seed and produced 560 kg/ha of dry matter (Knake and Slife, 1965).

Quackgrass competition drastically reduced corn height, yield, and delayed maturity (Bandeen and Buchholtz, 1967). Unfertilized corn yields were reduced 84% by quackgrass. Fertilization with both N and K increased corn yields but did not alter the competitive effect of quackgrass.

In the early 1950's, certain authors advanced the idea that clean cultivation of corn fields was neither necessary nor desirable. They speculated that corn grain losses caused by weeds could be offset by extra fertilizer and fertilizer would be cheaper than cultivation. Vengris et al. (1955) exploded this myth and found that high rates of N, P, and K fertilizer did not stop weeds from competing for essential nutrients and suppressing growth and yield of corn.

30–2.3 Herbicides

The use of 2,4-D and other phenoxy herbicides to control broadleaf weeds in corn started before 1950 and still continues. About the same time, ester formulations of 2,4-D were sprayed or granular formulations spread as preemergence herbicides in corn (Frans et al., 1956; Aldrich and Willard, 1952). Atrazine, simazine [2-chloro-4,6-bis(ethylamino-*s*-triazine], and CDAA (*N*-*N*-diallyl-2-chloroacetamide) were evaluated as preemergence herbicides about 1960 (McWhorter and Holstun, 1961; Burnside and Behrens, 1961; Lovely, 1964) and were used widely. Atrazine proved especially useful for controlling quackgrass (Buchholtz, 1963). Propachlor (2-chloro-*N*-isopropylacetanilide) and alachlor [2-chloro-2′-6′-deithyl-*N*-(methoxymethyl)-acetanilide] were two other preemergence herbicides discovered in the mid-1960's. Metolachlor [2-chloro-*N*-(2-ethyl-6-methylphenyl)-*N*-(2-methoxy-1-methylethyl)-acetamide] is an amide that is related chemically to alachlor and propachlor that was tested in the mid-1970's. Cyanazine [2-(4-chloro-6-ethylamino-*s*-triazin-2-ylamino)-2-methylpropionitrile] was discovered about 1970. Aerial application of these herbicides has proved as effective as ground spray (Bovey and Burnside, 1965).

As time progressed, atrazine was used as a preplant incorporated herbicide. The thiocarbamate herbicides EPTC (*S*-ethyl dipropylthiocarbamate) and butylate (*S*-ethyl diisobutylthiocarbamate) were volatile and had to be incorporated to get maximum weed control. These herbicides were toxic to johnsongrass, nutsedge, and shattercane (Roeth, 1973; Burnside, 1970; Hicks and Fletchal, 1967; Chenault and Wiese, 1979). Under high temperature and high soil moisture, EPTC was toxic to corn, but adding the antidote *N,N*-diallyl-2,2-dichloroacetamide to formulations of either butylate or EPTC increased corn tolerance 10 times (Leavitt and Penner, 1978; Burt and Akinsorotan, 1976). Treating corn seed with 1,8-naphthalic anhydride at 0.5% by weight also decreased injury from the two herbicides (Burnside et al., 1971).

An economic evaluation of preemergence herbicides applied in bands over corn rows showed that over a 13-year period the net cost of herbicide was less than the value of corn lost through failure to control weeds in the row (Staniforth and Lovely, 1964).

Postemergence herbicides that can be sprayed over corn are 2,4-D, dicamba, cyanazine, and atrazine (Greer et al., 1978b). Atrazine must be sprayed with a surfactant or phytobland oil in the carrier to increase foliar uptake of atrazine by weeds (Coats and Foy, 1974; Jansen et al., 1961).

30-2.4 Limited Tillage

Because corn is frequently planted in areas subject to erosion, tillage and planting methods that leave crop residues on the soil surface and reduce cost of production are beneficial. The "till-plant system" was developed in Nebraska (Lane and Wittmuss, 1961). With this system, old stalks were shredded before planting with a machine that removed old corn stubble and seeded in a single operation. One cultivation when corn was 25 to 30 cm tall removed weeds and volunteer corn. In weedy fields, a preemergence herbicide at planting reduced early weed growth (Wittmuss and Hurlbut, 1957). Hours of labor to produce a crop was reduced from 11.7 per hectare with conventional methods to 4.8 with the till-plant system. Controlling weeds in conservation tillage systems is different than in conventional tillage. However, adequate systems of weed control are available (Erback and Lovely, 1974; Wiese and Staniforth, 1973).

30-3 WHEAT

30-3.1 Problem Weeds

Weeds troublesome in dryland wheat in the western USA vary with farming regions. In the Southern Great Plains, tansy mustard, pepperweeds, treacle mustard, and smallseed falseflax are troublesome winter annual broadleaf weeds. Also kochia, Russian thistle, and pigweed can emerge in April or May in thin stands of wheat. Cheat, rescue grass, Japanese brome, downy brome, hairy chess, and jointed goatgrass are winter annual grass weeds competing in winter wheat fields in central Nebraska, Kansas, Oklahoma, and northern Texas. Wild oat is a problem in central Oklahoma and the northern part of Texas.

In the northern spring wheat-growing areas of the USA and Canada, wild mustard, lambsquarters, common ragweed, red sorrel, common chickweed, wild garlic, henbit, wild buckwheat, and Tartary buckwheat are problems in spring wheat fields (Nalewaja and Arnold, 1970). Along the West Coast, blue mustard, cow cockle, and fiddleneck are troublesome annual weeds. The most troublesome grass weeds in the Northern Great Plains and West Coast are green foxtail, yellow foxtail, and wild oat (Dunster, 1974; Nalewaja, 1977).

Winter wheat grown in the Pacific Northwest is infested with several members of the mustard family, including blue mustard (Swan, 1971), fiddleneck (Swan and Furtick, 1962), common chickweed, and pineappleweed (Peabody, 1976). Winter annual bromegrasses, especially downy brome, severely compete with the wheat (Rydryeh and Muzik, 1968).

Wild oat originated in Central Asia and have moved to wherever small grains are grown in the world. The major species of wild oat that infest cultivated crops are *Avena fatua* L. and *A. barbata* as well as hybrids of various species and subspecies (Nalewaja, 1977).

30–3.2 Weed Competition and Losses

According to Nalewaja and Arnold (1970), weeds vary in their ability to compete with wheat. Wild oat and wild mustard were much more serious competitors than yellow foxtail and wild buckwheat. When 55 plants/m^2 of wild oat and wild mustard were present, wheat yield was reduced about 20%. On the other hand, 55 plants/m^2 of yellow foxtail and wild buckwheat reduced yield less than 10%. Wheat yield loss continued to increase as weed numbers increased. With 239 plants/m^2 of wild oat and wild mustard, yield was reduced 50%. Wheat yield loss was about 15% with 179 plants/m^2 of wild buckwheat or yellow foxtail. Wild oat became more competitive to wheat with improved fertility. On summer fallow land fertilized with ammonium phosphate, 12 to 48 wild oat plants/m^2 significantly reduced yield of wheat. However, when wheat was not fertilized, 84 to 120 plants/m^2 were needed to reduce wheat yield.

The economic impact of weeds in wheat is large. If wild oat were eliminated from cereal grains and flax in North Dakota, a $150 million to $200 million economic loss would be avoided (Wax et al., 1976). This increase in wheat production would feed 13 million people a diet of 2500 calories/day for 1 year (Nalewaja, 1972).

Nalewaja (1977) states that wild oat reduced grain production more in North America than any other place in the world, and 6.4 million t of wheat and barley are lost because of this pest. A 3-year survey of 142 fields in Manitoba showed that seed yield loss in small grain from green foxtail was 10% (Chow and Dryden, 1973).

In the Southern Great Plains, competitiveness of tansy mustard varied with years (Wiese, 1965). In 1960, 11 mustard plants/m^2 reduced yields 10%. The next year, the same number of weeds reduced yields 6%. In 1962, tansy mustard had no effect on yield. A regression of both 1960 and 1961 data showed that loss was 60 kg/ha, or 9% from 11 weeds/m^2 (Shipley and Wiese, 1969). In 1960 and 1961, soil moisture was adequate and wheat yield above average. In 1962, yield was below the average of 660 kg/ha. Under the dry conditions, tansy mustard germinated only in small depressions where moisture ponded. This additional soil moisture counteracted the competing effect of weeds.

In the Great Plains, competition between weeds and crop is primarily for water. During early fall and winter, neither tansy mustard nor winter wheat used more soil water than what evaporated from bare soil during early fall and winter. However, at full tiller or initiation of early spring growth, both weeds and wheat start using more water than the amount that evaporated from the soil (Wiese, 1960). Allowing weeds to grow beyond this stage caused competition for water and wheat yield was reduced.

A 4-year study in Montana showed that spraying 2,4-D increased wheat yield 13%, or 255 kg/ha. Average yield of unsprayed wheat was 1880 kg/ha. Another study in Canada indicated that uncontrolled wild mustard reduced wheat yield 44% (Alley, 1974).

One fiddleneck/0.09 m^2 markedly reduced grain yield in Oregon (Swan and Furtick, 1962). In California, one fiddleneck plant/0.09 m^2 reduced wheat yield 25% or 660 kg/ha (Dunster, 1974).

30–3.3 Control of Broadleaf Weeds

Shortly after World War II, phenoxy herbicides, including 2,4-D, became available to control weeds in wheat. Most weeds in the mustard family were susceptible. Some, e.g., treacle mustard, were difficult to control if application was made after weed rosettes were over 8 cm in diameter (Wiese et al., 1972). To avoid injury, 2,4-D must be applied when winter wheat is fully tillered. Application made before full tiller usually decreased yields because heads per plant and head weights were reduced (Wiese, 1965; Swan, 1975). Applying 2,4-D before or after frost did not increase risk of injuring wheat (Bradbury and Ashford, 1970). Later research showed that MCPA was less toxic to both winter and spring wheat, and applications could be made from the five-leaf to the boot stage without threat of yield reduction (Hill and Isom, 1978). In Canada, 2,4-D removed wild mustard from spring wheat as safely as removing the weed by hand (Burrows and Olson, 1955). They also found that spraying 2,4-D after jointing frequently decreased yields. After the soft dough stage, large weeds that may interfere with harvest can be controlled with 2,4-D at 1 kg/ha without decreasing yield or quality of wheat (Phillips et al., 1967).

Fiddleneck is a broadleaf weed that is difficult to control with 2,4-D. Fortunately, around the mid-1960's, bromoxynil (3,5-dibromo-4-hydroxybenzonitrile) became available and controlled this weed (Dunster, 1974). An application of 2,4-D ester at 0.56 kg/ha controlled 33% of the weeds, whereas 0.4 kg/ha of bromoxynil eliminated fiddleneck.

Other research showed that bromoxynil could be applied safely at 1.1 kg/ha to winter and spring wheat from the two-leaf until the flag leaf stage of growth. The best time to spray for maximum weed control and crop safety was at the two- to three-leaf stage. Bromoxynil mixed with MCPA increased the spectrum of weed control (Nalewaja and Arnold, 1970; Peabody, 1976). Blue mustard along the West Coast was controlled by diuron [3-(3,4-dichlorophenyl)-1,1-dimethylurea], bromoxynil, and 2,4-D (Swan, 1971).

Tartary and wild buckwheat infest several million hectares of spring wheat in Saskatchewan and Alberta and are not susceptible to 2,4-D. Research prior to 1969 showed that neburon [1-butyl-3-(3,4-dichlorophenyl)-1-methylurea] would selectively control these weeds in wheat and other small grains (Fiesen and Forsbert, 1959). However, this herbicide treatment has never been labeled.

Sometime later, dicamba was proved effective against wild buckwheat in wheat (Quimby and Nalewaja, 1971; Behrens, 1967). Carbon-14 labeled dicamba accumulated in meristem tissue of wild buckwheat but not in young tillers of wheat. The authors concluded that selectivity of dicamba was related to differences in translocation and metabolism. Other research indicated that some wheat varieties were more tolerant to dicamba than others. Yield of 'Chris' wheat treated with 0.56 kg/ha of dicamba equaled the untreated control, whereas yield of Selkirk and Lakota were reduced 38 to 43%, respectively. Applications of higher rates of dicamba, after full tiller, caused malformed plants and severe reduction in yield (Behrens and Johnston, 1965).

30-3.4 Control of Grass Weeds

Wood (1953) lists four characteristics that make wild oat a troublesome weed: (i) early maturity and shedding of seeds, (ii) seed dormancy and after ripening, (iii) low temperature requirements that restrict seed germination, and (iv) long viability of buried seed. Wild oat seed germinate in the spring and fall when soil temperatures range from 4 to 22°C. Higher temperature causes secondary dormancy that prevents germination (Wax et al., 1976). Delayed seeding, fall tillage, preseeding tillage, postseeding tillage, and summer fallow are cultural practices that help reduce wild oat populations (Brown, 1953). In areas where winter wheat is grown, rotating to summer crops is a practical approach to controlling wild oat. Seeding grass and legumes also will reduce wild oat problems. Applying N fertilizers well ahead of seeding broke dormancy, and the plowing operation before planting small grains killed more wild oat than in unfertilized areas (Sexsmith and Pittman, 1963).

Early attempts at controlling wild oat with chemicals began with IPC (propham; isopropyl carbanilate) and CIPC (chloropropham; isopropyl *m*-chlorocarbanilate) (Wiese and Dunham, 1954). Small grains did not have much more tolerance to these herbicides than did wild oat; consequently, the herbicides were not very successful. Another carbamate, barban (4-chloro-2-butynyl *m*-chlorocarbanilate), was the first successful post-emergence herbicide that controlled wild oat in wheat, barley, and rapeseed (Hoffman et al., 1960; Selleck, 1961). Wild oat control with this herbicide decreased with increasing plant age (Friesen, 1961; Friesen, 1967). Best control was obtained when barban was sprayed at 0.55 to 1.1 kg/ha in a low volume of carrier on wild oat in the one- to two-leaf stage of development. At this stage, most wild oat plants were killed and a few surviving plants were suppressed, allowing small wheat to compete and outgrow them. When applied at the three-leaf stage, barban reduced growth and tillering of surviving plants, but fewer plants were killed. Wheat yields increased in proportion to the number of wild oat plants killed. Application of P to a deficient soil increased wild oat control with barban (Friesen, 1967). Sometime later, preplant-incorporated applications of diallate [*S*-(2,3-dichloroallyl) diisopropylthiocarbamate] and triallate [*S*-(2,3,3-trichloroallyl) diisopropylthiocarbamate] prior to wheat planting effectively controlled wild oat (Jacobson and Anderson, 1968). To achieve weed control and crop selectivity, these herbicides were incorporated in the top 2.5 cm of soil and wheat planted deeper than herbicides were incorporated. Hoe and shovel drills placed seed deep enough so that injury did not occur. Using disk drills frequently resulted in injury to wheat. Miller and Nalawaja (1980) found that fall applications of triallate granules gave better wild oat control than spring applications of either liquid or granular formulations. Delaying incorporation of liquid formulations of triallate beyond 8 h reduced wild oat control. When diallate and triallate were evaluated on 214 lines of wild oat, response varied, and many lines were not controlled by recommended rates of the herbicides. Resistance to the herbicides was quantitatively inherited (Jacobsohn and Andersen, 1968).

More recently, dichlofop methyl {2-[4-(2,4-dichlorophenoxy) phenoxy] propanoic acid}, flufenprop-methyl {methyl-2-[benzoyl (3-chloro-4-fluorophenyl) amino] propanoate} (Olson and Nalawaja, 1977), and benzoylprop ethyl [ethyl *N*-benzoyl-*N*-(3,4-dichlorophenyl)-2-amino propionate] (Kirkland and Ashford, 1976) have proved effective against wild oat in wheat. Using an ED_{50}, Todd and Stobbe (1977) found that barley, wild oat, and green foxtail were, respectively, 2, 190, and 1190 times more sensitive to diclofop methyl than wheat. Diclofop methyl has been labeled in the northern USA and Canada, but not in the Southern Great Plains for use on hard red winter wheat.

Green foxtail has always been a problem in spring wheat in the northern USA and Canada. Recently, this problem has been reduced by shallow incorporation of trifluralin (α,α,α-trifluoro-2,6-dinitro-*N,N*-dipropyl-*p*-toludine) into the soil just before or after seeding spring wheat, barley, or flax (Rahman and Ashford, 1972). Selectivity for the treatment was related to the fact that the coleoptile node is the most sensitive area of emerging grass plants. In green foxtail, this node was located less than 1 cm from the soil surface, regardless of seed depth. The coleoptile node of wheat, however, stayed close to the caryopsis and below the depth that trifluralin was incorporated. In addition to the differences in morphology, green foxtail was more susceptible to trifluralin than wheat (Rahman and Ashford, 1970). Unfortunately, 1 to 2 kg/ha of trifluralin took more than one growing season to disappear from the soil under Canadian conditions (Rahman and Ashford, 1973).

In the Southern Great Plains, cheat, downy brome, Japanese brome, rescue grass, and hairy chess are serious competitors in winter wheat (Greer et al., 1980). Before 1979, the only control for these pests was burying seeds temporarily with deep plowing. This practice increased wheat yields over other cultural practices (Davidson and Santleman, 1973). In 1980, spraying metribuzin [4-amino-*tert*-butyl-3-(methylthio)-*as*-triazin-5(4*H*)-one] to fully tillered 'TAM W-101' wheat was labeled for controlling brome species in Oklahoma. The application rate was 0.42 to 0.56 kg/ha depending on the soil type. Rainfall must follow application for this treatment to be effective. Jointed goatgrass is another winter annual grass infesting winter wheat. Crop rotation is the only adequate and economical control method for this weed (Fenster et al., 1980).

30-4 SORGHUM

There are 7.5 million ha of sorghum grown in the USA, primarily in the Great Plains and other dryland areas in the western USA. Almost half of the crop (3.3 million ha) is grown in Texas. Kansas, Nebraska, and Olkahoma grow 1.0, 0.9, and 0.6 million ha, respectively. Arizona, California, Colorado, Missouri, New Mexico, and South Dakota each grow 0.25 million ha or less of the crop. All but 1 million ha are dryland. Most of the irrigated crop is grown in the panhandle of Texas (Phillips, 1970; Wiese, 1976).

30-4.1 Problem Weeds

Pigweed infests almost every field of dryland sorghum in the Great Plains. In addition, kochia, puncture-vine, and lambsquarters are common weeds in the Southern Great Plains. Other broadleaf weeds that occur in the northern part of the sorghum belt are sunflower, velvetleaf, cocklebur, wild buckwheat, smartweed, and mustards (Wrange and Arnold, 1978).

Green foxtail became a serious problem in the eastern part of the sorghum belt from South Dakota through Kansas, when pigweed and other broadleaf weeds were controlled by 2,4-D or low rates of propazine. This species can be controlled by preemergence applications of propachlor mixed with atrazine (Wiese et al., 1977).

Witchgrass took over in many dryland sorghum fields in Kansas when pigweed was controlled. Increasing the rate of propazine from 1.4 to 2.2 kg/ha controlled this grass weed.

Field sandbur (Wiese et al., 1977) threatens many sorghum fields in Kansas now that broadleaf weeds have been brought under control. This weed, once isolated on sandy soil, has invaded other soil types. Sorghum herbicides do not control this weed; thus when other weeds are removed, sandbur increases. Some herbicides used in corn and soybean will control field sandbur, but low rainfall in much of the sorghum area precludes growing these crops without irrigation. Rotation to winter wheat is the only alternative for infested dryland fields, and tillage after wheat harvest prevents seed production of this summer annual weed.

Shattercane, or wild sorghum, is impossible to control in fields of sorghum. Rotating infested fields to winter wheat is the best control in dryland fields (Furrer and Burnside, 1962; Burnside et al., 1977). This weed is difficult to control because seed persist in the soil up to 13 years (Burnside et al., 1977). Growing alfalfa for 3 or more years reduced shattercane stand. Frequent mowings and competition from alfalfa contribute to effective control.

Sorghum seedlings are not vigorous; consequently, they are at the mercy of weeds more than a vigorous crop such as corn (Wiese, 1968). Both root and top growth of sorghum were slower than those of corn, common cocklebur, Palmer amaranth, barnyardgrass, buffalobur, kochia, and large crabgrass. Speed of seedling growth varied among hybrids. Seedlings that grew rapidly had a decided competitive advantage over weeds (Guneyli et al., 1969). This advantage was due to rapid germination and rapid emergence as well as fast root and shoot growth in young plants.

30–4.2 Weed Control

The advent of herbicides markedly improved growers' capabilities of controlling weeds in sorghum. Earliest research with 2,4-D on sorghum was conducted in Oklahoma (Elder, 1949). Research by Phillips (1958, 1960b, p. 11) led to a report outlining the use of 2,4-D for controlling broadleaf weeds in the Midland variety of sorghum. He found that to avoid injury, 2,4-D must be applied to sorghum that was 15 to 25 cm tall. If 2,4-D was applied to smaller sorghum, root development was inhibited, and under certain conditions, plants lodged or were killed. This height restriction was not detrimental to practical weed control because 2,4-D was most effective when weeds were less than 5 cm tall. This stage of weed growth usually occurred before sorghum reached 30 cm. On the other hand, if weeds and sorghum emerge at the same time, there is temptation to apply 2,4-D when sorghum is less than 15 cm tall.

Dicamba is another postemergence herbicide that has been labeled for broadleaf weed control in sorghum. Application of dicamba must be carefully timed between 10 and 25 days after sorghum emergence to avoid injury. Later application may cause varying degrees of sterility in sorghum heads. Dicamba was somewhat safer to use near cotton than most phenoxy herbicides (Smith and Wiese, 1972).

Research started in 1957 (Wiese and Rea, 1962b) showed that propazine was an effective preemergence herbicide. Attempts to locate a preemergence herbicide for sorghum grown on sandy soil led to the use and labeling of terbutryn (2-*tert*-butylamino-4-ethylamino-6-methylthio-*s*-triazine) (Wiese et al., 1975).

Atrazine has proved a practical preemergence herbicide in eastern Nebraska, Kansas, and Oklahoma on soils with over 1.5% organic matter and on clay soils. Atrazine was much more effective than propazine against grass weeds on these soils (Burnside, 1977). In Nebraska, atrazine applied preemergence to sorghum grown in narrow 38-cm rows and high plant populations gave weed control and yield equal to hand weeding (Burnside, 1977).

In a series of studies over 4 years, preplant application of either propazine or terbutryn, followed by incorporation with a rolling cultivator, resulted in effective weed control in sorghum (Wiese et al., 1973). Incorporation of propazine with a disk prior to bedding severely injured sorghum. If application and incorporation with a rolling cultivator was made more than a month before planting, insufficient herbicide remained to give pigweed or barnyardgrass control in the crop. Preplant incorporation of propazine at 3.3 kg/ha shortly before planting improved barnyardgrass control over preemergence treatment with that herbicide (Wiese et al., 1975). Pigweed was eliminated with preplant treatment at 3.3 kg/ha. With lower rates, preemergence application of propazine gave better control of pigweed than preplant application. Results using atrazine at 2.24 kg/ha applied before planting and incorporated at two locations in Nebraska indicated that the herbicide could be applied up to 30 days before planting without reducing weed control (Wicks and Burnside, 1972). However, yield was highest with preemergence applications of atrazine because of better weed control.

Atrazine was applied safely to sorghum in a water carrier with 0.5% surfactant any time after sorghum emerged. Control of pigweed less than 2 cm tall was satisfactory (Wiese, 1966). The treatment was not practical, because most growers attempted to use this combination when weeds were too large. When atrazine was sprayed in an oil-water emulsion carrier, pigweed (up to 18 cm) was controlled (Chenault and Wiese, 1969).

Research over 3 years showed that preemergence and postemergence applications of atrazine and propazine with aircraft were feasible (Chamberlain et al., 1972). One kilogram of 80% wettable powder formulation of either herbicide could be applied in as little as 4.72 L/ha of carrier. Pigweed control in grain sorghum was comparable with that obtained with ground application at 300 L/ha. Liquid formulations of the two herbicides also could be applied in this small amount of carrier.

Crop protectants have been used to reduce toxicity of alachlor to sorghum (Spotanski and Burnside, 1973; Eastin, 1972). The crop protectant 1,8-napthalic anhydride effectively reduce alachlor injury when applied at 0.5% (wt/wt) seed treatment. Anhydride also increased tolerance of sorghum to propachlor. Recently, α-(cyanomethoximino)-benzacetonitrile applied to sorghum seed at 1.25 gm/kg has proved practical for preventing injury to sorghum from metolachlor (Chenault and Wiese, 1980). This practice has the potential of expanding the spectrum of grass weeds that can be controlled in sorghum.

Practical experience and research has shown that best weed control and highest yields occur when a combination of mechanical, cultural, and chemical weed control methods are used. Research in Nebraska at Lincoln, North Platte, and Alliance show that a combination of rotary hoeing or cultivation plus a preemergence application of atrazine gave yields equal to hand weeding (Burnside and Wicks, 1964; Burnside et al., 1964; Burnside et al., 1966).

In another study, preplant soil-incorporated or preemergence applications of herbicides were superior to preemergence application, followed by shallow incorporation with a rotary hoe before crop emergence (Wicks et al., 1970). Postemergence rotary hoeing after the crop was up improved weed control and sorghum yields on all herbicide treatments. The best chemical treatments in conjunction with rotary hoeing were atrazine and atrazine mixed with norea [3-(hexahydro-4,7-methanoindan-5-yl)-1,1-dimethylurea]. These treatments were most effective in eastern and least effective in western Nebraska. The western part of the state has low organic matter, and atrazine caused sufficient injury to reduce yield of sorghum.

In integrated systems involving crop rotation, herbicide residues following or preceding sorghum can be a problem. Burnside and Moomaw (1975) found that atrazine applied at 2.2 kg/ha or less did not persist enough to reduce sorghum yields with repeated annual usage. Atrazine at 3.4 kg/ha reduced sorghum stand; however, the crop recovered and yielded normally. The two authors concluded there was little hazard of reducing sorghum yields from atrazine carryover in clay loam soils in eastern Nebraska. In 11 experiments across Nebraska, neither atrazine nor propazine applied to sorghum at recommended rates reduced wheat yield in a winter wheat–sorghum–fallow rotation (Wicks et al., 1969). Systems of reduced tillage in a winter wheat–sorghum–fallow rotation have been developed in Kansas and Texas (Phillips, 1969; Wiese et al., 1967).

30–5 COTTON

Dryland cotton is produced in the Southern Great Plains in Texas and Oklahoma. The crop is drought tolerant and usually is the most profitable crop grown in this area.

30–5.1 History of Weed Control

Chemical weed control methods for cotton were first developed in the Mississippi Delta (Ennis, 1955), Texas Blacklands (Rea et al., 1953), and under irrigation in California and Arizona (Miller et al., 1962). Salts of dinoseb (DNBP; 2-*sec*-butyl-4,6-dinitrophenol) were the first preemergence herbicides for cotton, and they were used from 1950 through 1954. Starting in 1951, chlorpropham was used to control both annual broadleaf and grass weeds in the crop. However, chlorporpham controlled grass weeds better than broadleaf weeds.

About the same time, diuron and monuron [3-(*p*-chlorophenyl)-1,1-dimethylurea] were being evaluated as preemergence herbicides for cotton (Hill, 1980). These herbicides, particularly diuron, proved to be more dependable than other preemergence herbicides being used for cotton in humid areas at that time.

At the same time that use of preemergence herbicides was coming into practice, researchers were developing postemergence herbicides that had to be directed at the base of cotton plants to reduce injury. Lateral oiling with 10 L/ha of naphtha-type oils controlled most small weeds in cotton that was up to 6 weeks old. Crabgrass, morning-glory, and pigweed that were 2.5 to 5 cm tall were killed. These oils had to have from 18 to 25% aromatic content. The established perennials, johnsongrass and bermudagrass were not killed. However, johnsongrass seedlings were controlled by two to three applications at 5- to 7-day intervals (Elliot and Black, 1961). The practice of spot treating johnsongrass in cotton was developed in the early 1950's. Successful control was achieved with herbicidal oil, sodium salt of dalapon, sodium salt of TCA, and 50:50 mixtures of naphtha and diesel fuel. Four to six applications of these herbicides directed at the base of plants killed established johnsongrass in one season at about the same cost as hoeing (Rea, 1958).

30–5.2 Weed Control

30–5.2.1 Preemergence and Preplant Herbicides

Weed control research on dryland cotton in the Southern Great Plains was not started until 1955. At that time, E. B. Thaxton (1957. Personal communication) evaluated diuron as a preemergence herbicide at Lubbock, Texas. It was not effective because rainfall did not leach the herbicide into the soil before it decomposed. Later research by Weldon and Timmons (1961) and Jordan et al. (1964) showed that phenylurea herbicides (e.g., diuron) were photochemically degraded by either sunlight or ultraviolet light.

In the early 1960's, Wiese and Hudspeth (1968b) determined that shallow incorporation increased weed control with several preemergence herbicides without causing undue injury to cotton on loam and clay loam soil. Incorporation with two helical blades that rolled like a lawn mower gave the best incorporation on loam and clay loam soil. This equipment increased cotton injury from diuron, norea, prometryn [2,4-bis(isopropylamino)-6-(methylthio)-*s*-triazine], and fluometuron [1,1-dimethyl-3-(α,α,α-trifluoro-*m*-tolyl;urea] on sandy loam soil. Metal tines constructed from 0.9-cm rods and dragged over herbicide bands on sandy soil incorporated the herbicides and increased weed control without injuring cotton. However, from a practical standpoint, shallow incorporation was not widely used in the Southern Great Plains.

Other studies showed that if significant rain fell within 1 week after application, optimum depth of incorporation for trifluralin, nitralin (4-methylsulfonyl-2,6-dinitro-*N,N*-dipropylaniline), DCPA (dimethyl tetrachloroterephthalate), and diuron was 2.5 cm with a power tiller (Wiese and Smith, 1970). Under these conditions, fluometuron, alachlor, and prometryn gave the best weed control without incorporation. If it did not rain for 4 weeks after application, percentage weed control dropped but was best with incorporation 7.5 cm deep. Incorporating 22 cm always decreased effectiveness of herbicides.

Studies conducted from 1965 to 1967 showed that trifluralin and nitralin could be incorporated with a disk before planting and frequently eliminated all annual weeds for the season (Wiese et al., 1969; Santelmann et al., 1968). This practice was widely adopted in a few years. Presently over 75% of the cotton is weeded with preplant-incorporated applications of trifluralin and other dinitroaniline herbicides. Wiese and Smith (1971) developed methods of controlling weeds in narrow row cotton with these herbicides.

Trifluralin required more thorough incorporation with a disk, probably to prevent loss by volatility, than did nitralin. Mechanical incorporation of nitralin could be delayed 96 h, whereas trifluralin had to be incorporated within 24 h after application for maximum effectiveness (Smith and Wiese, 1973). Other dinitroaniline herbicides that have given excellent weed control in dryland cotton are fluchloralin [*N*-(2-chloroethyl)-2,6-dinitro-*N*-propyl-4-(trifluoromethyl)aniline], profluralin [*N*-(cyclopropylmethyl)-α,α,α-trifluoro-2,6-dinitro-*N*-propyl-*p*-toluidine], pendimethalin [*N*-(1-ethylpropyl)-3,4-dimethyl-2,6-dinitrobenzenamine], dinitramine (N^4,N^4-diethyl-α,α,α-trifluoro-3,5-dinitrotoluene-2,4-diamine), and butralin [4-(1,1-dimethylethyl-*N*-(1-methylpropyl)-2,6-dinitrobenzenamine] (Greer et al., 1978c). Attempts to incorporate trifluralin and nitralin with either furrow or sprinkler irrigation were not successful (Smith et al., 1973).

30–5.2.2 Postemergence Herbicides

Weed control with postemergence herbicides in cotton was developed in the Southern Great Plains and other cotton-producing areas about the same time. In 1961, diuron and prometryn at 0.55 kg/ha applied as directed sprays containing 0.5% surfactant eliminated 2.5 cm of pigweed in cotton that was 15 cm tall (Wiese et al., 1964). Cotton was produced without hoeing or cultivating using three directed sprays of diuron in 1962. In addition to being more economical, the new practice increased yield of cotton. Similar results on crabgrass were obtained in the Mississippi Delta (McWhorter, 1963).

Directed sprays of DSMA (disodium methanearsonate) and MSMA were toxic to johnsongrass growing in cotton (Elliot, 1963). Other research showed that the herbicides could be used as directed sprays up to flowering without decreasing yield or increasing arsenic content of the seed. Applications over the top could be made only to cotton that was 7.5 to 15 cm tall or crop injury resulted (Wiese, 1968a).

30–5.2.3 Problem Weeds

Pigweed is the most common annual weed in cotton. Before herbicides were used, crabgrass, Texas panicum, volunteer sorghum, and barnyardgrass were common annual grass weeds. Repeated use of dinitroanaline herbicides have changed this picture. Weeds that are resistant to these herbicides are becoming more abundant. Morning-glory, spurred anoda, oakleaf thornapple, and lanceleaf sage have increased but can be controlled with preplant application of a dinitroaniline herbicide followed by preemergence application of prometryn, diuron, or dipropetryn (2-ethylthio-4,6-bis(isopropylamino)-*s*-triazine) (Abernathy et al., 1978).

Fluometuron controlled volunteer castorbean (Smith et al., 1972b) and wild watermelon (Smith and Cooley, 1973). Cocklebur and devilsclaw are two other weeds that escape dinitoraniline herbicides.

30–5.2.4 Herbicide Toxicity to Cotton

Applications of phenoxy and other hormone-type herbicides for control of weeds on rangeland and summer crops other than cotton constitute one of the greatest hazards to dryland cotton production in the Southern Great Plains. As little as 30 g of 2,4-D seriously damaged cotton on 15 ha (Ergle and Dunlap, 1949; Miller et al., 1963). In other studies, herbicide toxicity to cotton in decreasing order came from 2,4-D ester, 2,4-D amine, dicamba, MCPA, picloram, bromoxynil, and 2,3,6-TBA (Smith and Wiese, 1972).

30–6 ALFALFA

Dryland alfalfa grows throughout the Great Plains and northwestern USA. It is grown as a dryland crop, primarily on the eastern edge of the Great Plains. Westward across the Great Plains, dryland alfalfa decreases, and at the western edge it is almost exclusively irrigated. Weed control methods and weed problems are almost the same for either dryland or irrigated alfalfa. However, because of low returns, especially in marginal dryland areas, producers are reluctant to invest much for weed control. In most areas there is not enough price differential between weedy and weed-free alfalfa hay to encourage expenditures for weed control. This is unfortunate, because percentage protein was reduced in proportion to weed content of the forage (Cords, 1973). Cords (1973) arrived at this conclusion by using terbacil (3-*tert*-butyl-5-chloro-6-methyluracil), simazine, diuron, and chlorporpham on dormant alfalfa to control flixweed, downy brome, and shepherdspurse. According to Thomas (1976), if hay was purchased for total digestible nutrient and protein content, weed control would become an integral part of alfalfa production.

30–6.1 Problem Weeds

Winter annual weeds infesting alfalfa in the western part of the Southern Great Plains are tansy mustard and rescue grass (Chenault and Wiese, 1977). Farther east, chickweed and henbit also are troublesome winter annual weeds (Greer et al., 1978a). Summer annual weeds in the Southern Great Plains are barnyardgrass, crabgrass, and yellow foxtail (Chenault et al., 1977). Farther north in Colorado, other weeds infesting alfalfa are blue mustard, cheatgrass, wild salsify, dandelion, foxtail barley, and quackgrass (Heikes and Swink, 1972). Dodder is a parasitic weed common to most alfalfa-producing areas. Canada thistle is a perennial weed that is troublesome in the Northern Great Plains (Schreiber, 1967). In the Pacific Northwest, downy brome, tumble mustard, foxtail millet, prickly lettuce, and white cockle are pests (Swan, 1972; Dawson and Dell'Agostino, 1978; Burr, 1973).

30–6.2 Weed Control

Alfalfa has small seed and must be sown near the soil surface in a well-prepared seedbed to obtain a stand. Seedlings develop slowly and poor stands are especially subject to infestation by weeds. Alfalfa seeded too early in the fall becomes infested with summer annual grasses that reduce stand. Plantings made too late are subject to frost, and winter annual weeds will compete with the crop through the first cutting in the spring (Norris et al., 1977). Chenault et al. (1972) found the most economical method of establishing alfalfa in the Southern Great Plains was planting in September and allowing frost to kill summer annual weeds. Herbicides controlled these weeds but did not increase stand or yield of alfalfa. Herbicides were essential for plantings made in the spring because weeds reduced stand and yield.

Early research in Nebraska showed that TCA and dalapon could be used to control annual grass weeds emerging in young alfalfa (McCarty and Sand, 1958). Shortly thereafter, it was shown that an ester formulation of 2,4-DB [4-(2,4-dichlorophenoxy)butyric acid] controlled annual broadleafed weeds (Sand and McCarty, 1959).

Presently, several preplant-incorporated and postemergence herbicides are labeled for establishing alfalfa (Greer et al., 1978a). Diallate, EPTC (*S*-ethyl dipropylthiocarbamate), benefin (*N*-butyl-*N*-ethyl-α,α,α-trifluoro-2,6-dinitro-*p*-toluidine), and profluralin are preplant herbicides used in the Southern Great Plains. Diallate is effective against henbit and wild oat. EPTC is effective against annual grasses, mustards, and henbit (Greer et al., 1978a; Dawson and Dell'Agostino, 1978). Benefin and profluralin are especially effective against annual grasses and some broadleaf weeds, except mustards.

Postemergence herbicides that can be used on young alfalfa seedlings with at least three trifoliate leaves are pronamide [3,5-dichloro(*N*-1,1-dimethyl-2-propynyl)benzamide] and 2,4-DB. Pronamide is effective against most annual grasses as well as quackgrass but is not toxic to broadleaf weeds that infest alfalfa (Fawcett et al., 1978; Dutt et al., 1979).

Research has shown that chlorpropham, dinoseb (2-*sec*-butyl-4,6-dinitrophenol), and dichlobenil (2,6-dichlorobenzonitrile) will control dodder in established alfalfa fields (Dawson, 1970, 1971b).

Planting alfalfa after 15 August in the state of Washington and applying chlorporpham the following April controlled dodder in new plantings without injuring alfalfa (Dawson, 1971a).

Herbicides useful in dormant stands of alfalfa are dinoseb, trifluralin, diuron, terbacil, and simazine. They control most weeds infesting the crop, including common chickweed (Kapusta, 1973; Greer et al., 1978a).

30-7 LITERATURE CITED

Abernathy, J. R., A. F. Wiese, and J. R. Supak. 1978. Weed control in cotton for West Texas. Texas Agric. Exp. Stn. Tech. Report no. 10.

Agbakoba, C. S. O., and J. R. Goodin. 1970. Pichloram enhances 2,4-D movement in field bindweed. Weed Sci. 18:19–21.

Aldrich, R. J., and C. J. Willard. 1952. Factors affecting the preemergence use of 2,4-D in corn. Weeds 1:338–345.

Allen, R. R., B. A. Stewart, and P. W. Unger. 1977. Conservation tillage and energy. J. Soil Water Conserv. 32:84–87.

Alley, H. P. 1974. Weed control. Bull. Wyo. Agric. Exp. Stn. 603:5–8.

Bandeen, J. D., and K. P. Buchholtz. 1967. Competitive effects of quackgrass upon corn as modified by fertilization. Weeds 15:220–224.

Banks, P. A., L. V. Hill, and P. W. Santelmann. 1979. Control of field bindweed (*Convolvulus arvensis*) in winter wheat (*Triticum aestivum*) with foliar and subsurface layered herbicides. Weed Sci. 27:332–335.

Behrens, R. 1967. The influence of environment on response of wild buckwheat to dicamba. Proc. North Cent. Weed Control Conf. 21:42.

Behrens, R., and D. R. Johnston. 1965. Response of wheat varieties to dicamba. Res. Rep. North Cent. Weed Control Conf. 22:57.

Bovey, R. W., and O. C. Burnside. 1965. Aerial and ground applications of preemergence herbicides incorn, sorghum, and soybeans. Weeds 13:334–336.

Bradbury, I. K., and R. Ashford. 1970. The effect of frost on the response of spring wheat to 2,4-D. Weed Sci. 18:235–237.

Brown, D. A. 1953. Wild oats—progress in cultural control. Weeds 2:293–299.

Buchholtz, K. P. 1963. Use of atrazine and other triazine herbicides in control of quackgrass in corn fields. Weeds 11:202–205.

Burnside, O. C. 1970. Control of wild cane in corn. Weed Sci. 18:272–275.

Burnside, O. C. 1977. Control of weeds in non-cultivated, narrow-row sorghum. Agron. J. 69: 851–854.

Burnside, O. C., and R. Behrens. 1961. Phytotoxicity of simazine. Weeds 9:145–157.

Burnside, O. C., C. R. Fenster, and G. A. Wicks. 1964. Influence of tillage, row spacing, and atrazine on yield components of dryland sorghum in Nebraska. Agron. J. 56:397–400.

Burnside, O. C., and R. S. Moomaw. 1975. Sorghum growth as affected by annual applications of atrazine. Weed Sci. 23:494–498.

Burnside, O. C., and G. A. Wicks. 1964. Cultivation of herbicide treatments for dryland sorghum. Weeds 12:307–310.

Burnside, O. C., G. A. Wicks, and C. R. Fenster. 1971. Protecting corn from herbicide injury by seed treatment. Weed Sci. 19:565–568.

Burnside, O. C., G. A. Wicks, and C. R. Fenster. 1977. Longevity of shattercane seed in soil across Nebraska. Weed Res. 17:139–143.

Burnside, O. C., G. A. Wicks, C. R. Fenster, and P. T. Nordquist. 1966. Weed control in grain sorghum. Nebraska Farm, Ranch Home Quarterly no. QR121, p. 7–10.

Burr, R. J. 1973. Alfalfa weed control. Proc. Annu. Oreg. Weed Conf. 22:9–10.

Burrows, V. D., and P. J. Olson. 1955. Reaction of small grains to various densities of wild mustard and the results obtained after their removal with 2,4-D or by hand. Can. J. Agric. Sci. 35:68–75.

Burt, G. W., and A. O. Akinsorotan. 1976. Factors affecting thiocarbamate injury to corn. I. Temperature and soil moisture. Weed Sci. 24:319–321.

Carson, A. G., and J. D. Bandeen. 1975. Chemical control of Canada thistle. Weed Sci. 23: 116–118.

Cates, J. S., and H. R. Cox. 1912. The weed factor in the cultivation of corn. USDA Bureau of Plant Industry Bull. no. 257.

Chamberlain, E. W., A. F. Wiese, F. F. Owen, A. J. Becton, and W. E. Turner. 1972. Low-volume application of propazine and atrazine to sorghum. Weed Sci. 20:12–16.

Chang, F. Y., and W. H. Vanden Born. 1968. Translocation of dicamba in Canada thistle. Weed Sci. 16:176–181.

Chenault, E. W., and A. F. Wiese. 1969. The effect of oil-water-emulsion carriers on herbicide toxicity. Proc. South. Weed Sci. Soc. 22:165.

Chenault, E. W., and A. F. Wiese. 1977. Annual grass control in irrigated alfalfa. Proc. South. Weed Sci. Soc. 30:147–151.

Chenault, E. W., and A. F. Wiese. 1979. Preplant herbicides for Johnsongrass control in corn. Proc. South Weed Sci. Soc. 32:121.

Chenault, E. W., and A. F. Wiese. 1980. CGA-43089 reduces metolachlor phytotoxicity on sorhgum. Proc. South Weed Sci. Soc. 33:69.

Chenault, E. W., A. F. Wiese, W. Vandiver, D. Lavake, D. Scott, and D. Hollingsworth. 1972. Preplant and preemergence weed control in alfalfa. Texas Agric. Exp. Stn. Miscellaneous Pub. no. 1044.

Chow, P. N. P., and R. D. Dryden. 1973. Wheat tolerance to TCA for green foxtail control. Weed Sci. 21:238–240.

Coats, G. E., and C. L. Foy. 1974. Effect of petroleum oils on the uptake of atrazine-^{14}C by corn. Weed Sci. 22:220–226.

Cords, H. P. 1973. Weeds and alfalfa hay quality. Weed Sci. 21:400–401.

Davidson, J. M., and P. W. Santelmann. 1973. An evaluation of various tillage systems for wheat. Oklahoma Agric. Exp. Stn. Bull. no. 711.

Dawson, J. H. 1970. Dodder control in alfalfa with dichlobenil. Weed Sci. 18:225–230.

Dawson, J. H. 1971a. Establishing alfalfa on dodder-infested soil. Weed Sci. 19:222–225.

Dawson, J. H. 1971b. Dodder control in alfalfa with dinoseb and D(-)(3-chlorophenylcarbamoyloxy)-2N-isopropylpropioamide. Weed Sci. 19:551–554.

Dawson, J. H., and Eduardo Dell'Agostino. 1978. Control of foxtail millet (*Setaria italica*) in new seedlings of alfalfa (*Medicago sativa*) with EPTC applied in surface lines. Weed Sci. 26:637–639.

Derscheid, L. A., R. L. Nash, and G. A. Wicks. 1961. Thistle control with cultivation, cropping, and chemicals. Weeds 9:90–102.

Derscheid, L. A., J. F. Stritzke, and W. G. Wright. 1970. Field bindweed control with cultivation, cropping, and chemicals. Weed Sci. 18:590–596.

Derscheid, L. A., K. E. Wallace, and R. L. Nash. 1960a. Leafy spurge control with cultivation, cropping, and chemicals. Weeds 8:115–127.

Derscheid, L. A., K. E. Wallace, and R. L. Nash. 1960b. Russian knapweed control with cultivation cropping and chemicals. Weeds 8:268–278.

Derscheid, L. A., W. H. Wallace, and L. J. Wrage. 1963a. Cropping, cultivation, and herbicides to eliminate Russian knapweed and prevent reinfestation. Weeds 11:237–243.

Derscheid, L. A., G. A. Wicks, and W. H. Wallace. 1963b. Cropping, cultivation, and herbicides to eliminate leafy spurge and prevent reinfestation. Weeds 11:105–111.

Dunn, P. H. 1979. The distribution of leafy spurge (*Euphorbia esula*) and other weedy *Euphorbia* spp. in the United States. Weed Sci. 27:509–515.

Dunster, K. W. 1974. Small grain weed control justification and California status in barley, wheat, and oats. Proc. Annu. Calif. Weed Conf. 26:53–60.

Dutt, T. E., R. G. Harvey, R. S. Fawcett, N. A. Jorgensen, H. J. Larsen, and D. A. Schlough. 1979. Forage quality and animal performance as influenced by quackgrass (*Agropyron repens*) control in alfalfa (*Medicago sativa*) with pronamide. Weed Sci. 27:127–132.

Eastin, E. F. 1972. Protection of sorghum from alachlor and acetochlor injury by 1,8-naphtalic anhydride seed treatment. Proc. South Weed Sci. Soc. 25:168.

Elder, W. C. 1949. The effect of 2,4-D formulations on one sorghum variety when applied at five stages of growth. Res. Rep. North Cent. Weed Control Conf. 6:62.

Elliott, F. C. 1963. Disodium methyl arsonate finds application as cotton herbicide. Agric. Chem., June issue, p. 59–60, 119.

Elliott, F. C., and G. Black. 1961. Lateral oiling of grass and weeds in cotton. Texas Agric. Extension Service Miscellaneous Pub. no. 506.

Ennis, W. B., Jr. 1955. Weed control in principle crops of the Southern United States. Adv. Agron. 7:251–297.

Erback, D. C., and W. G. Lovely. 1974. Weed control with conservation tillage production of corn and soybeans. J. Soil Water Conserv. 29:44–45.

Ergle, D. R., and A. A. Dunlap. 1949. Responses of cotton to 2,4-D. Texas Agric. Exp. Stn. Bull. no. 713.

Evetts, L. L., and O. C. Burnside. 1973. Competition of common milkweed with sorghum. Agron. J. 65:931–932.

Fawcett, R. S., R. G. Harvey, D. A. Schlough, and I. R. Block. 1978. Quackgrass (*Agropyron repens*) control in established alfalfa (*Medicago sativa*) with pronamide. Weed Sci. 62: 193–198.

Fenster, C. R. 1977. Conservation tillage in the Northern Great Plains. J. Soil Water Conserv. 32:37–42.

Fenster, C. R., H. I. Owens, and F. E. Westbrook. 1980. Jointed goatgrass—how to control it. USDA Fact Sheet AFS 7-6-1.

Frans, R. E., A. M. Davis, and D. A. Hinkle. 1956. Chemical weed control in cotton, soybeans, and corn 1952–1955. Arkansas Agric. Exp. Stn. Report Series no. 55.

Frazier, J. C. 1943. Nature and rate of development of root system of *Convolvulus arvensis.* Bot. Gaz. 104:417–425.

Friesen, G. 1967. The efficiency of barban as influenced by growth stages of wild oats and spring wheat. Weeds 15:160–162.

Friesen, H. A. 1961. Some factors affecting the control of wild oats with barban. Weeds 9: 185–195.

Friesen, H. A., and D. E. Forsbert. 1959. Neburon as a selective herbicide for tartary buckwheat and wild buckwheat. Weeds 7:49–54.

Furrer, J. D., and O. C. Burnside. 1962. Shattercane—once a crop, now a weed. Nebraska Agric. Exp. Stn. Quarterly no. QR-40, p. 4.

Green, J. L., R. G. Davis, and A. F. Wiese. 1965. Root systems of woollyleaf franseria and Texas blueweed. Proc. South. Weed Sci. Soc. 18:649–651.

Greer, H. A. L. 1977. Johnsongrass control in Oklahoma. Oklahoma State Univ. Extension Facts no. 2753.

Greer, H., T. Peeper, and D. Fain. 1980. Cheat control in wheat. Oklahoma State Univ. no. CR-2774.

Greer, H., L. Romman, and W. Bowers. 1978a. Chemical weed control in alfalfa. Oklahoma State Univ. Extension Facts no. 2761.

Greer, H., J. Stiegler, and J. Howell. 1978b. Weed control in corn. Oklahoma State Univ. Extension Facts no. 2757.

Greer, H., N. B. Thomas, and O. H. Williams. 1978c. Chemical weed control in cotton. Oklahoma State Univ. Extension Facts no. 2762.

Guneyli, E., O. C. Burnside, and P. T. Nordquist. 1969. Influence of seedling characteristics on weed competitive ability of sorghum hybrids and inbred lines. Crop Sci. 9:713–716.

Heikes, E., and J. F. Swink. 1972. Weed control in new alfalfa seedlings. Colorado State Univ. Exp. Stn. Progress Report no. 72-6.

Hicks, R. D., and O. H. Fletchal. 1967. Control of Johnsongrass in corn. Weeds 15:16–20.

Hill, G. D. 1980. History and development of the phenylurea herbicides. p. 167–171. *In* Proc. Beltwide Cotton Prod. Res. Conference, St. Louis, Mo. 6–10 Jan. Natl. Cotton Coun. of Am., Memphis, Tenn.

Hill, J. E., and W. H. Isom. 1978. Broadleaf weed control in wheat and barley. Div. of Agric. Sci., Univ. of California Leaflet no. 120;12, Berkeley.

Hoffman, O. L., J. W. Pullen, J. R. Epperly, and T. R. Hopkins. 1960. Factors affecting the activity of 4-chloro-2-butynyl *N*-(3-chlorophenyl) carbamate as a selective herbicide. Weeds 8:198–203.

Jacobsohn, R., and R. N. Andersen. 1968. Differential response of wild oat lines to diallate, triallate, and barban. Weed Sci. 16:491–494.

Jansen, L. L., W. A. Gentner, and W. C. Shaw. 1961. Effects of sufactants on the herbicidal activity of several herbicides in aqueous spray systems. Weeds 9:381–404.

Jordan, L. D., C. W. Coggins, Jr., B. E. Day, and W. A. Clerx. 1964. Photodecomposition of substituted phenylureas. Weeds 12:1–4.

Kapusta, G. 1973. Common chickweed control in established alfalfa. Weed Sci. 21:119–121.

Kirkland, K. J., and R. Ashford. 1976. Benzoylprop ethyl and its analogue for the control of wild oat in wheat. Weed Sci. 24:316–318.

Knake, E. L., and F. W. Slife. 1962. Competition of *Setaria faberii* with corn and soybeans. Weeds 10:26–29.

Knake, E. L., and F. W. Slife. 1965. Giant foxtail seeded at various time in corn and soybeans. Weeds 13:331–334.

Lane, D. E., and H. Wittmuss. 1961. Nebraska till-plant system. Univ. of Nebraska Extension Service no. EC-61-714.

Latshaw, W. L., and J. W. Zahnley. 1927. Experiments with sodium chlorate and other chemicals as herbicides for field bindweed. J. Agric. Res. 35:757–767.

Leavitt, J. R. C., and D. Penner. 1978. Protection of corn (*Zea mays*) from acetanilide herbicidal injury with the antidote R-25788. Weed Sci. 26:653–659.

Lovely, W. G. 1964. Soil incorporation of liquid and granular herbicides for weed control in corn. Proc. North Cent. Weed Control Conf. 20:18–22.

Lowe, H. J., and K. P. Buchholtz. 1952. Cultural methods for control of quackgrass. Weed 1:346–351.

McCarty, J. K., and P. F. Sand. 1958. Chemical weed control in seedling alfalfa. I. Control of weedy grasses. Weeds 6:152–160.

McWhorter, C. G. 1963. Effects of surfactants on the herbicidal activity of foliar sprays of diuron. Weed Sci. 11:265–269.

McWhorter, C. G., and J. T. Holstun, Jr. 1961. Phytotoxicity of *s*-triazine herbicides to corn and weeds as related to structural differences. Weeds 9:592–599.

Miller, J. H., C. L. Foy, H. M. Kempen, L. M. Carter, and M. Hoover. 1962. Weed control in cotton. California Agric. Exp. Stn. Bull. no. 791, Berkeley.

Miller, J. H., H. M. Kempen, J. A. Wilkerson, and C. L. Foy. 1963. Response of cotton to 2,4-D and related phenoxy herbicides. USDA-SEA Bull. no. 1289.

Miller, S. D., and J. D. Nalewaja. 1980. Wild oat (*Avena fatua*) control with fall- and spring-applied triallate. Weed Sci. 28:416–418.

Moclani, M. K., L. Knake, and F. W. Slife. 1964. Competition of smooth pigweed with corn and soybeans. Weeds 12:126–128.

Mosier, J. G., and A. F. Gustafson. 1915. Soil moisture and tillage for corn. Ill. Agric. Exp. Stn. Bull. 12:563–586.

Nalewaja, J. D. 1972. Weeds: Coexistence or control. J. Environ. Qual. 1:344–349.

Nalewaja, J. D. 1977. Wild oats: Global bloom. 1977. Proc. West Sco. Weed Sci. 30:21–32.

Nalewaja, J. D., and W. E. Arnold. 1970. Weed control methods, losses, and costs due to weeds and benefits of weed control in wheat and other small grains. p. 48–61. *In* First Food and Agric. Organization Int. Conference on Weed Control, Davis, Calif. 22 June–1 July. Univ. of Illinois, Urbana.

Nieto, J. H., and D. W. Staniforth. 1961. Corn-foxtail competition under various production conditions. Agron. J. 53:1–5.

Norris, R. F., W. H. Isom, V. L. Marble, and S. R. Radosevich. 1977. Weed control in seedling alfalfa. Univ. of California Leaflet no. 2917, Berkeley.

Olson, W. A., and J. D. Nalewaja. 1977. Wheat and wild oat response to flufenprop-methyl. Weed Sci. 25:355–359.

Onken, A. B. 1971. Cultural practices for grain sorghum production from grain sorghum research in Texas—1970. Texas Agric. Exp. Stn. Progress Report no. 2938, p. 5–15.

Peabody, D. V. 1976. Western Washington weed control guide: Winter wheat. USDA Cooperative Extension Service Em:4007, Washington State Univ., Pullman.

Phillips, W. M. 1958. The effect of 2,4-D on yield of Midland grain sorghum. Weeds 6:271–280.

Phillips, W. M. 1960a. The chloro-substituted benzoic acids for control of field bindweed. Weeds 8:63–70.

Phillips, W. M. 1960b. Weed control in sorghum. Kansas Agric. Exp. Stn. Circ. no. 350, p. 11.

Phillips, W. M. 1969. Dryland sorghum production and weed control with minimum tillage. Weed Sci. 17:451–454.

Phillips, W. M. 1970. Weed control methods, losses, and costs due to weeds and benefits of weed control in grain sorghum. p. 101–108. *In* First Food and Agric. Int. Conference on Weed Control. Weed Sci. Soc. of Am., Davis, Calif. 22 June–1 July. Univ. of Illinois, Urbana.

Phillips, W. M., and F. L. Timmons. 1954. Bindweed—how to control it. Kansas Agric. Exp. Stn. Bull. no. 366.

Phillips, W. M., G. Yip, K. F. Finney, J. H. Hilton, and W. C. Shaw. 1967. Response and analysis of wheat following late preharvest applications of 2,4-D. Weeds 15:107–111.

Quimby, P. C., Jr., and J. D. Nalewaja. 1971. Selectivity of dicamba in wheat and wild buckwheat. Weed Sci. 19:598–600.

Rahman, A., and R. Ashford. 1970. Selective action of trifluralin for control of green foxtail in wheat. Weed Sci. 18:754–759.

Rahman, A., and R. Ashford. 1972. Trifluralin for green foxtail control in barley and flax. Weed Sci. 20:160–163.

Rahman, A., and R. Ashford. 1973. Persistence of trifluralin under field conditions in Saskatchewan. Can. J. Plant Sci. 53:421–423.

Rasmussen, L. W. 1956. The effects of 2,4-D and 2,4,5-T applications on the stand density of Canada thistle (*Cirsium arvense* Scop.). Weeds 4:343–348.

Rea, H. E. 1958. Spot-spraying johnsongrass. Texas Agric. Exp. Stn. Bull. no. 902.

Rea, H. E., F. C. Elliott, and H. P. Smith. 1953. Chemical weed control in cotton. Texas Agric. Extension Service Circ. no. 329.

Reed, C. F., and R. O. Hughes. 1970. Selected weeds of the United States. USDA Agric. Handbook no. 366.

Roeth, F. W. 1973. Johnsongrass control in corn with soil incorporated herbicides. Weed Sci. 21:474–476.

Rydrick, D. J., and T. J. Muzik. 1968. Downbrome competition and control in dryland wheat. Agron. J. 60:279–280.

Sand, P. F., and M. K. McCarty. 1959. Chemical weed control in seedling alfalfa. II. Control of broadleaved weeds. Weeds 7:317–323.

Sand, P. F., and M. K. McCarty. 1963. Control of woollyleaf franseria with 2,4-D. Weeds 11: 25–27.

Santelmann, P. W., H. A. L. Greer, and I. L. Six. 1968. Factors influencing the activity of soil incorporated herbicides. Oklahoma State Univ. Exp. Stn. Bull. no. 658.

Schreiber, M. M. 1967. Effect of density and control of Canada thistle on production and utilization of alfalfa pasture. Weeds 15:138–142.

Schultz, M. E., and O. C. Burnside. 1979. Distribution, competition, and phenology of hemp dogbane (*Apocynum cannabinum*) in Nebraska. Weed Sci. 27:565–570.

Schweizer, E. E., J. F. Swink, and P. E. Heikes. 1978. Field bindweed (*Convolvulus arvensis*) control in corn (*Zea mays*) and sorghum (*Sorghum bicolor*) with dicamba and 2,4-D. Weed Sci. 26:665–668.

Selleck, G. W. 1961. Recent advances in the chemical control of wild oats. Weeds 9:60–71.

Sexsmith, J. J., and U. J. Pittman. 1963. Effect of nitrogen fertilizers on germination and stand of wild oats. Weeds 11:99–101.

Sharma, M. P., F. Y. Chang, and W. H. Van den Born. 1971. Penetration and translocation of picloram in Canada thistle. Weed Sci. 19:349–355.

Shipley, J. L., and A. F. Wiese. 1969. Economics of weed control in sorghum and wheat. Texas Agric. Exp. Stn. Miscellaneous Pub. no. 909.

Smith, D. T., R. C. Berner, and J. P. Walter. 1973. Nitralin and trifluralin incorporation by rainfall and irrigation. Weed Res. 13:359–366.

Smith, D. T., and A. W. Cooley. 1973. Wild watermelon emergence and control. Weed Sci. 21:570–573.

Smith, D. T., A. W. Cooley, D. O. Mooney, and A. F. Wiese. 1972b. Herbicidal control of volunteer castorbean. Weed Sci. 20:338–340.

Smith, D. T., and A. F. Wiese. 1972. Cotton response to low rates of 2,4-D and other herbicides. Texas Agric. Exp. Stn. Bull. no. 1120.

Smith, D. T., and A. F. Wiese. 1973. Delayed incorporation of trifluralin and nitralin. Weed Sci. 21:163–165.

Smith, D. T., A. F. Wiese, and A. W. Cooley. 1972a. Woollyleaf bursage response to selected herbicides. Weed Sci. 20:554–556.

Spotanski, R. F., and O. C. Burnside. 1973. Reducing herbicide injury to sorghum with crop protectants. Weed Sci. 21:531–536.

Staniforth, D. W. 1957. Effects of annual grass weeds on the yields of corn. Agron. J. 49:551–555.

Staniforth, D. W. 1961. Responses of corn hybrids to yellow foxtail competition. Weeds 9: 132–136.

Staniforth, D. W., and W. G. Lovely. 1964. Preemergence herbicides in corn production. Weeds 12:131–133.

Sturtevant, E. L. 1886. Weeds in corn. N.Y. Agric. Exp. Stn. Geneva Annu. Rep. 5:46.

Swan, D. G. 1971. Competition of blue mustard with winter wheat. Weed Sci. 19:340–342.

Swan, D. G. 1972. Effect of herbicides on alfalfa and subsequent crops. Weed Sci. 20:335–337.

Swan, D. G. 1974. Necessity of proper timing of applications of 2,4-D to winter wheat. Down Earth 31:23–25.

Swan, D. G., and W. R. Furtick. 1962. Competition of fiddleneck with wheat. Weeds 10:121–122.

Thomas, R. 1976. Weed control and its effect on quality hay. Proc. Calif. Alfalfa Symp. 6: 59–61.

Timmons, F. L., and V. F. Burns. 1951. Frequency and depth of shoot-cutting in eradication of certain creeping perennial weeds. Agron. J. 43:371–375.

Todd, B. G., and E. H. Stobbe. 1977. Selectivity of dichlofop methyl among wheat, barley, wild oat (*Avena fatua*) and green foxtail (*Setaria viridis*). Weed Sci. 25:382–385.

Unger, P. W. 1972. Dryland winter wheat and grain sorghum cropping systems Northern High Plains of Texas. Texas Agric. Exp. Stn. Bull. no. 1126.

Vengris, J., W. G. Colby, and M. Drake. 1955. Plant nutrient competition between weeds and corn. Agron. J. 47:213–216.

Wax, L. M., W. D. Carpenter, J. H. Dawson, H. A. Friesen, A. Glinz, E. R. Hehn, K. Hempstead, W. Klassen, J. D. Nalewaja, C. I. Seely, and W. C. Shaw. 1976. Wild oats—a situation report. ad hoc Res. Working Group, North Dakota State Univ., Fargo.

Weimer, D. C., and M. B. Harland. 1925. The cultivation of corn. Univ. of Illinois Exp. Stn. Bull. no. 259.

Weldon, L. W., and F. L. Timmons. 1961. Photochemical degradation of diuron and monouron. Weeds 9:111–116.

Wicks, G. A., and O. C. Burnside. 1972. Preplant atrazine applications on sorghum. Weed Sci. 20:49–52.

Wicks, G. A., O. C. Burnside, and C. R. Fenster. 1970. Influence of soil incorporation on herbicide performance in sorghum. Agron. J. 62:252–255.

Wicks, G. A., C. R. Fenster, and O. C. Burnside. 1969. Herbicide residue in soil when applied to sorghum in winter wheat-sorghum-fallow rotation. Agron. J. 61:721–724.

Wiese, A. F. 1956. Chemical control of blueweed. Texas Agric. Exp. Stn. Progress Report no. 1862.

Wiese, A. F. 1960. Effect of tansy mustard on moisture storage during fallow. Weeds 8:683–685.

Wiese, A. F. 1965. Effect of tansy mustard and 2,4-D on winter wheat. Texas Agric. Exp. Stn. Miscellaneous Pub. no. 782.

Wiese, A. F. 1966. Increasing toxicity of propazine, atrazine, and diuron with oils and paraquat. Proc. South. Weed Conf. 19:47–48.

Wiese, A. F. 1968. Rate of root elongation. Weed Sci. 16:11–13.

Wiese, A. F. 1976. Suppressing weeds in sorghum. p. 5–8. *In* The integrated control of the arthropod disease and weed pests of cotton, grain sorghum, and deciduous fruit. Texas Agric. Exp. Stn. Miscellaneous Pub. no. 1276.

Wiese, A. F., E. Burnett, and J. E. Box. 1967. Chemical fallow in dryland cropping sequences. Agron. J. 59:175–177.

Wiese, A. F., O. C. Burnside, W. M. Phillips, and E. F. Eastin. 1977. Controlling tough weeds in sorghum. Weeds Today 8:26–28.

Wiese, A. F., E. W. Chenault, and D. Hollingsworth. 1973. Preplant application of herbicides for weed control in grain sorghum. Agron. J. 65:583–586.

Wiese, A. F., E. W. Chenault, and E. B. Hudspeth, Jr. 1969. Incorporation of preplant herbicides for cotton. Weed Sci. 17:481–483.

Wiese, A. F., E. W. Chenault, D. E. Lavake, and D. L. Scott. 1975. Evaluation of preemergence herbicides for sorghum grown on sandy soil. Texas Agric. Exp. Stn. Progress Report no. 3337C.

Wiese, A. F., J. W. Collier, L. E. Clark, and U. D. Havelka. 1964. Effect of weeds and cultural practices on sorghum yields. Weeds 12:209–211.

Wiese, A. F., R. G. Davis, and D. Degge. 1965. Picloram for controlling small patches of field bindweed and other perennial weeds. Texas Agric. Exp. Stn. Progress Report no. 2347.

Wiese, A. F., and R. S. Dunham. 1954. Preplanting applications of IPC and CIPC for selective control of wild oats (*Avena fatua*). Weeds 3:321–330.

Wiese, A. F., C. Harvey, and E. B. Hudspeth, Jr. 1964. Weed control in cotton with postemergence herbicides. Proc. South Weed Conf. 17:65–66.

Wiese, A. F., and E. B. Hudspeth, Jr. 1968a. Effects of DSMA and MSMA on cotton yield and arsenic content of cottonseed. Texas Agric. Exp. Stn. Miscellaneous pub. no. 877.

Wiese, A. F., and E. B. Hudspeth, Jr. 1968b. Subsurface application and shallow incorporation of herbicides on cotton. Weed Sci. 16:595–598.

Wiese, A.F., D. E. Lavake, E. W.Chenault, and D. Holingsworth. 1975. Herbicide treatments for watergrass control in sorghum. Texas Agric. Exp. Stn. Progress Report no. 3336.

Wiese, A. F., and H. E. Rea. 1955. Bindweed control in the panhandle of Texas. Texas Agric. Exp. Stn. Bull. no. 802.

Wiese, A. F., and H. E. Rea. 1959. Bindweed (*Convolvulus arvensis* L.) control and seedling emergence as affected by tillage, 2,4-D, and competitive crops. Agron. J. 51:672–675.

Wiese, A. F., and H. E. Rea. 1961. Control of field bindweed and other perennial weeds with benzoic and phenylacetic acids. Weeds 9:423–428.

Wiese, A. F., and H. E. Rea. 1962a. Factors affecting the toxicity of phenoxy herbicides to field bindweed. Weeds 10:58–61.

Wiese, A. F., and H. E. Rea. 1962b. Treating irrigated grain sorghum with preemergence herbicides. Crop Sci. 2:29–31.

Wiese, A. F., and D. T. Smith. 1970. Herbicidal activity as affected by soil incorporation and rainfall. Weed Sci. 18:515–517.

Wiese, A. F., and D. T. Smith. 1971. Herbicides in narrow row cotton culture. Crop Sci. 11: 518–520.

Wiese, A. F., and D. W. Staniforth. 1973. Weed control in conservation tillage. p. 108–114. *In* Conservation tillage. Proc. Natl. Conserv. Tillage Conference, Des Moines, Iowa. March 1973. Soil Conserv. Soc. of Am., Ankeny, Iowa.

Wiese, A. F., C. Vandiver, E. W. Chenault, and D. Lavake. 1972. Control of treacle mustard and other weeds in wheat. Texas Agric. Exp. Stn. Progress Report no. 3036.

Wittmus, H. D., and L. W. Hurlbut. 1957. Combining machine functions in corn and sorghum production. Nebraska Agric. Exp. Stn. J. Series, Paper no. 859.

Wood, H. E. 1953. The occurrence and the problem of wild oats in the Great Plains region of North America. Weeds 2:292–294.

Wrange, L. J., and W. E. Arnold. 1978. Chemical weed control in sorghum. South Dakota State Univ. Extension Service no. PS-525D.

Zahnley, J. W., and W. F. Pickett. 1934. Field bindweed and methods of control. Kansas Agric. Exp. Stn. Bull. no. 269.

31

Norris E. Daniels
Texas Agricultural Experiment Station
Bushland, Texas

Insect and Mite (Arthropod) Pests of Major Crops

Before the settlement of a country, native insects live on wild plants. As new plants are introduced and grown in greater abundance, native pests and those accidentally brought in from foreign countries, finding the plants acceptable food, may multiply rapidly to injurious numbers.

Measure for the control of these pests may be divided into several categories, among which are natural, cultural, and chemical. Whatever the method and its effectiveness, the cost of control should be a deciding factor in its practical use. It is evident that no financial profit to an individual will be gained by attempting control when the cost is greater than the probable loss would be without the control measure. Before applying a control method, a farmer should consider crop potential and prospective returns.

Pests described in this chapter attack dryland crops in the Great Plains from Texas north into Canada. They also attack crops in dryland areas on the West Coast.

It is described in the text when pests affect a regional area of economic importance.

31-1 SMALL GRAIN AND GRAIN SORGHUM

Small grains, primarily wheat, are subject to attack from several pests. The most important of these are the greenbug (aphid), mites, and wireworms. Sorghum also shares some of the same insect and mite problems.

31-1.1 Greenbug

The greenbug is a name reserved for the most important of several kinds of aphids that commonly infest small grain. This small, green plant louse has caused periodical crop failure.

Dryland Agriculture—Agronomy Monograph no. 23.

The greenbug is an unusual pest in many ways. It is generally most abundant in winter and spring, long before most other pests appear. It is small, nearly the same green color as the leaf, and to all but the trained observer, its presence goes undetected until yellow or brown spots appear in the field. These spots indicate areas in which the plants have died as a result of greenbug feeding. These aphids are approximately 1.5 mm long and when full grown have a dark green stripe down the back. Both wingless and winged forms occur. All wingless and most winged forms are females and give birth to living young. Most females begin reproduction in 6 to 30 days after birth, depending on the temperature, and continue to produce two or three aphids per day for 20 to 30 days. Many generations may be produced in a year under favorable conditions (Daniels et al., 1956). For the past 24 years, I have found greenbugs every month of each year in the Texas Panhandle.

Optimum temperatures for greenbug reproduction and development is approximately 24°C, but greenbugs can reproduce at temperatures of 13 to 35°C. At 13°C, natural enemies reproduce slowly, which allows a buildup of greenbugs. Temperatures of 5 to 10°C or less reduce greenbug reproduction and activity. Some greenbugs are killed by an overnight temperature of −18°C. Continuous temperatures of −7°C or below for 7 days or longer will reduce populations by 90% or more (Daniels and Chedester, 1980b). Winged forms reproduce less efficiently than wingless forms. However, winged greenbugs can fly and may be blown many miles to establish new colonies. Winged forms are more numerous in dryland wheat than in irrigated wheat (Daniels, 1977).

The greenbug is not new to the country. It was first described in Italy in 1852 and first recorded in the USA in 1882 (Hunter, 1909). A major and very extensive outbreak occurred in 1906 to 1907. This outbreak apparently started in central Texas and spread northeast to within 96 km of Chicago. By the end of the crop season, it had reached eastern Colorado. This was the greatest infestation known from a geographic standpoint, and over 17.62 million hl of small grains were destroyed (Ruggles and Wadley, 1927). In 1926, an infestation that centered in Minnesota was believed to have originated from migrations of greenbugs that overwintered in Oklahoma and Kansas, although some greenbugs were observed near St. Paul, Minnesota, in 1925. This infestation was somewhat unique in that it centered in Minnesota and extended into Wisconsin, North Dakota, and Canada (Wadley, 1931). An outbreak occurred in 1942 on small grain acreages in Texas and southern Oklahoma. Almost total crop losses occurred on oat and barley and on wheat in the central area of the states (Atkins and Dahms, 1945).

Record losses again occurred in 1951 and 1961 in the Texas Panhandle (Daniels et al., 1956; Daniels, 1962a). Damaging infestations of the greenbug have not occurred on small grains every year, at regular intervals, or in any predictable pattern. Greenbugs, with their piercing mouthparts, suck plant juices. Toxins are injected into the plant during the feeding process. Small grains infested with greenbugs first turn yellow to orange. This damage often occurs in a localized spot in a field. As infestations increase, plants die and greenbugs move to uninfested plants, enlarging the affected area. Infestations, however, sometimes appear in much larger areas.

In 1966, greenbugs were found in light numbers on grain sorghum [*Sorghum bicolor* (L.) Moench] in the Texas High Plains (Daniels, 1969). During August and September of 1967, greenbugs again were found in colonies of 10 to 100 on the underside of sorghum leaves. They migrated from the sorghum to early-planted wheat in September and became damaging by November. In 1968, greenbug populations became very heavy on grain sorghum in several states during July and August and caused considerable damage. Wood (1971) stated that the greenbug found on sorghum in 1968 was biotype C. Greenbugs were trapped year round for 28 years 21 km north of Bushland (Daniels, 1981). Sticky traps were used and inspections made once every 2 weeks. From 1953 to 1967, greenbugs were caught during the spring and fall. From 1968 to 1970, they were caught during August. From 1971 to 1975, greenbugs were caught in the spring and summer, and from 1976 to 1980, none was caught. Their migration pattern changed at the time of heavy sorghum infestation and biotype change (to C) in 1968 occurred as it did in 1976. In 1979, biotype E appeared. Greenbugs collected by Daniels and Chedester (1980a) from a wheat field near Bushland in November and December 1979, destroyed the formerly resistant wheat 'Amigo' in a laboratory test (Porter et al., 1981). They were designated biotype E. Specimens collected from the same field and other locations in the Texas Panhandle during the spring and summer of 1980 also proved to be biotype E.

Greenbug infestations in sorghum are detected by reddish spots on the leaves caused by greenbug colonies feeding on the underside of a leaf. The reddened leaf areas enlarge as greenbugs increase. Finally, the leaf begins to die, turning brown from the outer edges toward the center.

31-1.1.1 Control

31-1.1.1.1 Natural Enemies. Greenbugs and other aphids are reduced by several insect enemies, including lady beetles, damsel bugs, syrphid fly larvae, and green lacewing larvae.

31-1.1.1.1.1 Predaceous Natural Enemies. Chief among the predaceous insects are lady beetles, which are small, orange, spotted beetles often found in grain fields. These are well known as adults, but the immature, dark, sluglike grubs from which lady beetles develop are not so well recognized. Clusters of orange-yellow eggs from which these grubs hatch can be seen attached to the wheat leaves when lady beetles are numerous. Both adults and larvae of these predators feed on the greenbugs and in some years aid materially in controlling them. Much publicity has been given to the sale of these beetles for greenbug control during outbreak years. This practice, however, is not recommended. According to Fenton and Dahms (1951), purchasing lady beetles is not an effective control for the greenbug in Oklahoma. "At an average rate of consumption, at favorable temperatures, it would require one gallon of beetles per acre to prevent greenbug increase in a moderately infested field," they reported. Also, if natural conditions were not suitable for lady beetle development at the time of their release, the beetles would, in many cases, disappear from the field. Economically, this type of control would be unreasonable.

Included in the greenbug predators are the damsel bugs, or nabids, which are small, dull gray or brown, elongated insects about 8 mm long. These insects search the wheat plants for greenbugs and other aphids, piercing them with their beaks and sucking the blood. They have been found consistently throughout the winter but always in the adult stage. These predators have been observed moving about and feeding at temperatures below 10°C, at least 6 to 8°C below those at which lady beetles are active. Because they do not begin to multiply until April, they constitute a fixed rather than an expansive biotic force in reducing greenbug populations. When caged with greenbugs, a single nabid has killed as many as 84 of them in a 24-h period (Daniels et al., 1956).

Syrphid flies are classed in the group of greenbug enemies. Medium-sized flies with yellow bands on the body are found sometimes hovering around greenbug-infested plants. They dart from place to place with great speed. The larvae or immature stage of these flies feed on the aphids and are found on the plants in the midst of the greenbug colonies. They are sluglike and vary from green to brown, gray, or mottled. They lack legs and head but possess pointed jaws with which they pierce and suck all the body contents from a greenbug and discard its empty skin. Syrphid larvae are valuable in aphid colonies since they destroy greenbugs rapidly for considerable periods of time.

In this class of feeders is the immature stage of a delicate, green, gauzy-winged insect called the lacewing fly. The larva of this insect is predaceous, crawling about over the plants in search of greenbugs, piercing them with its long jaw, and sucking their blood.

31–1.1.1.1.2 Parasitic Natural Enemies. Greenbug populations are held in check sometimes by a small wasp that usually is present when greenbugs are abundant. This wasp deposits its egg within the aphid's body. The egg hatches into a larva so small that it can feed inside the greenbug (Chedester, 1979). When the larva is fully grown, the skin of the dead aphid turns brown and remains fastened to a leaf as the wasp passes a resting stage inside. A few days later, the adult parasite emerges by cutting a circular lid in the back of the aphid. When the parasite is active, many brown greenbug mummies may be found attached to wheat leaves. These wasps are of less value against the greenbug in the Northern Great Plains than in the Southern Great Plains.

31–1.1.1.2 Cultural Enemies. Most early efforts at control were modifications of cultural practices, e.g., plowing under volunteer grain and adjusting planting dates, variety, and crop selection. Brush drags, plowing, soil-packing devices, and even burning were used without marked success (Webster and Phillips, 1912).

Table 31-1. Guide to determine need for treatment in small grains (Daniels, 1962a).

Plant height, cm	Number greenbugs/linear 0.3 m
7 to 15	100 to 300
10 to 20	200 to 400
15 to 40	300 to 800

Grazing of wheat during greenbug infestations decreased greenbug populations considerably (Daniels, 1975); however, if grazing is continued late in the spring, grain yields will be reduced in proportion to the lateness of grazing and the earliness of the wheat variety. Usually where small grain fields are heavily infested over an area or where a general outbreak is occurring, greenbug populations will increase in a grazed field after removal of livestock. In such cases, chemical control may become necessary a few days after cattle are removed. If the weather is warm and a good predator population is present, however, insecticidal application may be unnecessary.

Cultural measures that will stimulate the growth of small grain and sorghum plants and keep them in a vigorous condition will, in many cases, enable them to withstand more greenbug damage than unhealthy plants. Greenhouse studies conducted at the Bushland station indicated that wheat grown in soil following alfalfa was more tolerant to greenbugs and supported greater populations than wheat grown in soil following wheat (Daniels and Porter, 1956). These results were associated with high N levels in the soil and increased plant vigor.

Large plants frequently survive high infestations of greenbugs, whereas smaller plants are killed. An infestation of 500 greenbugs/linear 0.3 m of drill row in small wheat may be more severe than a population of 1500/linear 0.3 m in large wheat. The smaller wheat will have more insects per gram of foliage.

31-1.1.1.3 Chemical Enemies. Chemical controls are not always practical. In dryland areas where yield are generally low, cost of insecticidal control may be too expensive. It is therefore advisable to contact the local agricultural agent before treatment.

31-1.1.1.3.1 Small Grains. Insecticide need depends on the number of greenbugs present, plant size, vigor, and stage of growth, time of year, moisture conditions, and presence or absence of parasites and predators. Irrigated crops usually can withstand larger greenbug populations than dryland crops before yield loss occurs. Table 31-1 (Daniels, 1962a) may serve as a reliable guide in determining the need for treatment in small grains.

The appearance of small deadened areas caused by greenbug feeding indicates a need for treatment. Occasionally, populations of 25 to 50 greenbugs/linear 0.3 m of row on young small-grain plants (especially oat or barley) may warrant treatment. Oat and barley commonly are more susceptible to greenbug damage than wheat. However, the barley varieties 'Will' and 'Kearney' are resistant to greenbug biotype E. In winter when the temperature is low greenbugs may be down around the crowns of small grains, below the soil surface.

Table 31-2. Guidelines for insecticide applications on sorghum (Texas Agric. Extension Service, 1977).

Plant size	When to treat
Emergence to about 15 cm	Visible damage with colonies of greenbugs on plants
Larger plant to preboot	Before any entire leaves are killed
Boot to hard dough	When greenbug damage is sufficient to cause death of more than two normal-sized leaves

31-1.1.1.3.2 Sorghum. The extent of greenbug damage in grain or forage sorghum depends on greenbug numbers, plant size, vigor, and stage of growth, moisture conditions, and presence or absence of parasites and predators. Producers should closely observe plant conditions, development of greenbug numbers, and damage. Damage from the seedling stage to the early preboot stage may result in plant death and thus stand loss.

Plants from emergence to about 15 cm should be treated if damage is visible and greenbug colonies are present on the lower leaf surface. Research data (Daniels, 1972) indicate that larger plants, up to the preboot stage, will tolerate more greenbugs than smaller plants. Numerous greenbugs on this size plant should be controlled before any entire leaves are killed.

Yield reductions during the preboot, boot, blooming, and grain development stages depend on greenbug numbers, length of time that greenbugs have infested plants, and plant vigor. Based on 4 years of experience and research, heavy populations can cause extensive damage to preboot and older plants. Experience has indicated that yield returns justify treatments only when greenbugs are numerous enough to cause death to more than two normal-sized leaves.

Table 31-2 (Texas Agric. Extension Service, 1977) summarizes guidelines on the need for insecticide applications on sorghum. These general guides assume that beneficial insect numbers are inadequate to prevent economic damage. Plants under stress cannot support many greenbugs without suffering yield reductions.

Parathion (*O,O*-diethyl *O-p*-nitrophenyl phosphorothioate), malathion [*O,O*-dimethyl *S*-(1,2-dicarbethoxyethyl) phosphorodithioate], and dimethoate [*O,O*-dimethyl *S*-(*N*-methylcarbamoylmethyl phosphorodithioate) may be used to control the greenbug on both small grains and grain sorghum. *Do not use* methyl parathion (*O,O*-dimethyl *O-p*-nitrophenyl phosphorothioate) on sorghum because leaves are injured by the chemical.

31-1.1.1.4 Insect Resistance. Sorghums resistant to the sorghum greenbug are available (Harvey and Hackerott, 1971; Johnson et al., 1976; Daniels, 1978a). These hybrids may become the most effective means of controlling the greenbug. Since the appearance of biotype E, greenbug resistance in these sorghum hybrids is uncertain. However, biotype E-resistant sorghums should soon be available. Producers should be aware that the primary type of resistance is tolerance and should not expect plants to be free of greenbugs. However, resistant hybrids generally support lower greenbug populations than do susceptible hybrids. Based on previous research, the damage threshold for resistant sorghums is the same as that for susceptible sorghums.

When tolerance in sorghum was developed, misunderstanding existed relative to the terminology used in describing insect-resistant plants. Therefore, the following terms, which can be applied to insect tolerance or resistance in other plants, were defined:

1. Susceptibility: a susceptible plant is one that displays average or more than average damage by a pest.
2. Immunity: an immune plant is one that a specific pest will never consume or injure under any known condition.
3. Resistance: a resistant plant possesses heritable qualities that enable it to produce a greater yield of better quality than other plants of the same species when exposed to the same pest population or conditions.

Plant resistance is relative and must be determined by comparison with plants that are more severely damaged under the same set of conditions.

Types, bases, or causes of resistance are defined as:

1. Nonpreference: a type of resistance denoting plant characters that cause a pest to desire it less than a susceptible plant for oviposition, food, shelter, or a combination of the three.
2. Tolerance: a type of resistance in which a plant is able to withstand damage in spite of supporting a pest population approximately equal to that which would damage a susceptible host.
3. Antibiosis: a type of resistance denoting some adverse effect of the plant on the pest, e.g., reduced reproduction, decreased size, abnormal length of life, or increased mortality.

31–1.2 Other Types of Aphids

The yellow sugarcane aphid is lemon yellow, covered with setae (or hairs), and has two double rows of dark tubercles down the back. Feeding aphids secrete a plant toxin, and relatively light populations have been known to kill sorghum in the preboot stage. This aphid has been observed causing damage only to sorghum on the Gulf Coast, though its distribution extends into the Texas Panhandle and is found on sorghums and small grains (Bottrell et al., 1972).

The oat bird-cherry aphid spends almost its entire feeding period on various grains and grasses and is particularly abundant on small grains in Texas. Control is seldom required (Bottrell et al., 1972).

The corn leaf aphid, which is greenish or greenish-blue, is found in the whorl or on upper plant parts. Leaves are sometimes entirely covered. Damage most often occurs on small seedling sorghum, causing plant death and stand loss (Daniels and Chedester, 1975b). With heavy populations, plants may show signs of poor head exertion at time of booting. Harvesting problems from heavy honeydew excretion have resulted where heavy populations were present in the heads during the preharvest period. Though extremely common on sorghum, chemical controls are seldom justified.

Small grain infested with the English grain aphid appears about the same as that infested with greenbugs. This species is sometimes mistaken for the greenbug, but its legs and cornicles are longer. After the grain begins to head, numerous aphids may be clustered in the bracts beneath the heads or in the heads. This aphid seldom causes yield loss (Painter et al., 1954).

The western wheat aphid is a slender, pale green, wax-secreting species with cornicles absent or with only rudiments or "pimples." Winged forms are green and black. This aphid usually infests and damages restricted areas within a wheat field, but it was distributed generally over many fields in the Texas Panhandle during the spring of 1959 (Daniels and Thomas, 1961). The appearance of the wheat plant damaged by this aphid is a brilliant yellowing and rolling of the leaves with stunting of growth, giving the plant a yuccalike resemblance. The wheat heads usually are small and club shaped, but severely damaged plants may fail to head. Its distribution is from Texas to Canada. According to Essig (1958), the best control measure is summer fallow.

The crown aphid is found from September to January below ground level about the crowns of small grains on the white parts of the leaf sheath and upper parts of the roots (Painter et al., 1954). The smaller aphids are orange yellow and the large ones olive green. Winged forms are black. There is no evidence of plant injury by this species.

31–1.3 Cutworms

Several species of cutworms injure growing wheat plants in the Great Plains, but only a few are truly destructive. The species involved are grass-inhabiting insects. Their abundance depends to a considerable extent on temperature and rainfall in the fall and late summer preceding outbreaks. Some species, e.g., the army cutworm, are surface feeders; others, e.g., the pale western cutworm, live wholly underground. Others, e.g., the armyworm and the variegated cutworm, are climbing species (Painter et al., 1954).

According to Hagen (1974), sporadic outbreaks of the pale western cutworm have occurred on wheat in the Great Plains region since 1911. Serious infestations were reported in Canada, Montana, and North Dakota in the early 1920's. A serious outbreak occurred in the early 1950's in Nebraska and Wyoming. Painter et al. (1954) reported that this species seldom needs control. Essig (1958) stated that the red-backed cutworm is a destructive pest of cereals in Montana and Canada.

Clean culture, which reduces the number of grassy areas such as fence rows and roadsides, will help to prevent cutworm injury. Toxaphene (chlorinated camphene) or parathion applications will usually control most cutworms.

31-1.4 Corn Earworm and Fall Armyworm

Corn earworm moths deposit eggs on the leaves and heads, and then newly hatched larvae begin to feed almost immediately. On young sorghum plants, corn earworm and the fall armyworm often feed in the whorl. Experimental results indicate that control of corn earworm or fall armyworm in the whorl stage seldom is justified economically (Texas Agric. Extension Service, 1977). Corn earworm larvae are cannibalistic; however, populations exceeding an average of two small larvae per head in maturing grain sorghum are considered a damaging infestation in the absence of effective natural predators.

31-1.5 Sorghum Midge

The sorghum midge is established from Nebraska to the Gulf Coast and can be one of the most damaging insects to grain sorghum in Texas (Thomas, 1969). Females may deposit an average of 150 eggs in blooming florets. The larvae prevent seed development by secreting a chemical, causing "blasted heads" (Petralia et al., 1979). Their damage usually is more severe in late-blooming grain sorghum. Planting hybrids of uniform maturity at the right time to avoid late heading is strongly recommended, especially on the Gulf Coast and in the Rio Grande Valley, where the midge has been a problem. Johnsongrass serves as an alternative host for sorghum midge. Only adults can be controlled with presently approved insecticides, two of which are carbaryl (1-naphthyl *N*-methylcarbamate) and Diazinon (*O,O*-diethyl *O*-2-isopropyl-4-methyl-6 pyrimidyl thiophosphate).

31-1.6 Sorghum Webworm

The sorghum webworm occurs primarily in the Gulf Coast area of the Great Plains but is also found to a lesser degree as far north as Nebraska (Little, 1957). It frequently occurs in large numbers on grain sorghum heads, where it gnaws circular holes in the grain and feeds on the starchy contents. Late-planted crops tend to suffer the heaviest losses. The mature larvae are sluggish caterpillars about 12 mm long. Their basic color is green with four red to blackish stripes on the back. Control consists of destroying crop residues, growing early sorghum, and, if populations are heavy, using carbaryl or toxaphene.

31-1.7 Hessian Fly

The hessian fly was first reported in a few eastern counties of Kansas in 1871, fewer than 100 years after it had been brought to this country by Hessian soldiers during the Revoluntary War (Painter et al., 1954). According to Hagen (1973), it has spread to most wheat-growing areas of the world. The hessian fly overwinters as a full-grown maggot in a brown skin known as a "flaxseed."

Its best-known sign is broken culms, damage that is caused by larval feeding somewhat before harvest time. The flaxseed, or resting state, can be found where the break occurs. Heads on these culms yield 25 to 30% less than heads on uninfested culms. In the fall, the hessian fly kills tillers and young plants. This damage may be identified by finding the flaxseed, or larvae, at the base of dead plants or tillers. Fall damage results in a thinner stand that may be mistaken for winter injury. The white larvae and the brown flaxseed are the stages most frequently found. The adult hessian fly lives for such a short time it is rarely seen.

Control of the hessian fly is prevention. Once flies infest a crop, they cannot be destroyed by any known way until after harvest. Early plowing of wheat stubble is the first step to prevent infestation of the new crop. The destruction of all volunteer wheat as soon as it sprouts prevents infestation of the volunteer and increase of flies during summer. Planting wheat at the time recommended by the state agricultural experiment station aids in avoiding infestation. This date ranges from the last of September to the middle of October. Wheat varieties resistant to the hessian fly are also available (Cartwright et al., 1959; Gallun, 1977).

31–1.8 Soil Insects

Several species of subterranean insects attack planted seed and underground portions of plants. The majority, which originally fed on seeds, roots, and underground stems of native prairie grasses (Say, 1823), have survived cultivation and find wheat and sorghum suitable for survival and reproduction. When growing wheat or sorghum is not available, they find food along fence rows, in ravines, pastures, roadsides, and waste areas. Species that formerly lived on native plants often concentrate on the cultivated crops and cause considerable loss.

The most destructive soil insect pests are false wireworms, white grubs, and true wireworms. These insects have four stages: egg, larva, pupa, and adult.

31–1.8.1 Grubs

31–1.8.1.1 White Grubs. White grubs are the larval stage of May and June beetles. Plant damage results from larvae feeding on the roots. Small plants are often killed, resulting in plant stand loss of both wheat and sorghum. Severely pruned roots of larger plants may cause stunting or plant lodging.

There are 1-, 2-, and 3-year life cycle grubs. The 1-year larvae spend one winter in the soil below the frost line, and the 2-year grubs spend 2 years in the soil. Teetes (1973) found grubs with both 1- and 2-year cycles in

Texas. The 3-year life cycle grub spend three summers feeding in the soil and then two winters below the frost line. The last winter is spent as an adult in a pupal cell below the frost line (Painter et al., 1954). Research has shown that a long-term fallow and crop rotation system is the best control for grubs in dryland agriculture (Daniels, 1971b). Insecticidal soil treatments may be used for quicker results, but their label approval and recommendations change frequently.

31–1.8.1.2 False Wireworms. These soil insects belong to the family Tenebrionidae. Adults are black or grayish black. Their wing covers are marked with longitudinal ridges and are fused together, making the wings useless; therefore, they crawl when they migrate. They vary from about 1 to 2.5 cm long and from 0.5 to 1.5 cm wide. Some species have a reddish brown longitudinal strip on the line where the wing covers are fused.

The larvae are approximately two-thirds grown when the soil becomes cold in the fall. They burrow below the frost line and remain inactive until spring. When the soil temperature rises, the larvae return to the surface and resume feeding on weed seeds, roots, and wheat that did not germinate. Little injury is evident in the spring. An inactive state is then passed by the larvae before they transform to the pupal stage. According to Painter et al. (1954), the larval stages require about 275 days, depending on the species and the conditions under which they develop. Occasionally, a few larvae may change to pupae in the fall. Then adults appear early the next spring.

Considerable injury to fall-sown wheat may be avoided by sowing seed only when there is enough moisture to stimulate rapid germination. Wheat should not be planted too early or dusted in, because the longer the seed lies in the soil without germinating, the greater the opportunity for injury. Clean culture and accumulated soil moisture associated with summer fallowing tend to reduce the amount of injury. Crop rotation and fallowing are cultural controls for these pests. Applying approved chemicals to seed is an effective control (Daniels, 1955, 1957), but chemical control should not be necessary if a good cropping system is followed year after year.

31–1.8.1.3 True Wireworms. Wireworms are the larvae of brownish to black, slender, active beetles of the family Elateridae that have a loose, flexible connection between the prothorax and middle part of the thorax. This loose joint contains a structure that enables the beetle to make a clicking sound and to "jump," or throw itself, some distance when it is disturbed or placed on its back. These pests occur throughout the Great Plains. The importance of *Agriotes* spp. (Eidt, 1953), *Ctenicera destructor* Brown (Burrage, 1964), and wireworms in general as pests of cultivated crops has been well documented in Canada.

Some species may be found around roots of growing wheat or sorghum plants (Daniels, 1971b; Daniels and Chedester, 1976). Some have a 3-year life cycle, but the most common species require only 1 year from egg to adult.

Where control measures are necessary, those effective against white grubs and false wireworms will control wireworms.

31–1.9 Grasshoppers

Locusts have been important pests in the farm economy since agriculture began in the Great Plains. They cause damage over most of the North American continent, but the greatest numbers, and thus the heaviest damage, are found in areas where annual precipitation is from 25 to 75 cm (Campbell and Bell, 1977). Many species inhabit this large area (Daniels and Chedester, 1974; Coppock, 1962; Painter et al., 1954), but about 12 species do 90% of the crop damage. Among these damaging species are *Melanoplus differentialis* (Thos.), *M. femurrubrum* (de G.), *M. bivittatus* (Say), *M. packardii* (Scudder), *M. sanguinipes* (Fabr.), *Aeoloplides turnbulli* (Thos.), and *Camnula pellucida* (Schudder).

Grasshoppers lay eggs in packets or egg pods, each containing from 15 to 125 eggs. Under favorable conditions, a single female may produce several egg pods. Pods are deposited in the soil, on the surface of the soil, or at the base of bunch-forming grasses. Egg laying may begin in early summer and extend into the fall.

Eggs of most crop-damaging grasshoppers hatch from early April and mid-June of the following spring. Hatching time, as with most grasshopper activities, depends on species, temperature, and moisture. When the spring is warm and dry, eggs hatch early, and nymphs develop rapidly with a low death rate; but when the weather is cold and damp during hatching, mortality of nymphs may be high and threatening outbreaks curtailed. Nymphs resemble their parents except in size and lack of functional wings. From 5 to 9 weeks are required for maturity.

One method of control consists of exposing the egg pods to parasites, predators, weather, and desiccation by disking, rod-weeding, sweeping, and by plowing to a depth of 15 cm so that the pods are buried too deep for the hatching nymphs to reach the surface. Other cultural methods include elimination of weedy habitats, planting resistant crops (e.g., sorghum), or using a resistant crop to border a susceptible crop (e.g., wheat).

Grasshoppers may be controlled by applying insecticides as sprays, baits, or dusts. Many insecticides will control grasshoppers. The organophosphates, which include Diazinon, naled (1,2-dibromo-2,2-dichloroethyl dimethylphosphate), malathion, and mevinphos (alpha isomer of 2-carbomethoxyl-1-methylvinyl dimethyl phosphate), are highly effective and if properly used will not leave harmful residues. However, their killing properties do not last as long as the chlorinated hydrocarbon toxaphene. Sevin (carbaryl), a carbamate, may also be used. Sprays are generally preferred, because they give a higher initial kill than dusts and continue to kill over a longer period of time with less total insecticide.

31–1.10 Chinch Bugs

Blissus leucopterus (Say), the true chinch bug, is found primarily in the eastern part of the Great Plains. Adult black and white chinchbugs are about 4 mm long. In early spring, they fly to wheat and other small grains from the bunchgrass, where they pass the winter. The bugs are found in the thinner parts of the fields where they lay eggs behind the wheat leaf sheath near the surface of the ground. The eggs hatch into tiny, active red nymphs that feed on the developing wheat. If extremely abundant, they may stunt wheat or cause it to die in small areas. When wheat is harvested, nymphs and the few adults move on foot to the nearest corn or sorghum field, where severe damage may be done to small sorghum plants. According to Painter et al. (1954), it has not ordinarily been economical to control chinch bugs on wheat chemically.

True bugs of the genus *Nysius* (insects of the false chinch bug group) are general feeders and are widely distributed throughout the Great Plains (Essig, 1958). They feed on grain sorghum, potato, cotton, and many different weeds. *Nysius ericae* (Schilling) has been recognized for many years as a serious pest of crops (Milliken, 1918). An unidentified species of this genus damaged several fields of wheat in the Texas Panhandle during April and May of 1968 (Daniels, 1969). Damaging infestations appeared on both wheat and sugar beet in June. They damaged the plants by sucking the sap from the wheat stems in May and by sucking on the immature heads in June. This caused spots of white or unfilled heads in the fields of early-damaged wheat and shriveled grain in those heads damaged by the June infestation. As these bugs sucked the sap of sugar beet, leaves turned black and dried up. Daniels (1958) and Teetes (1974) reported false chinch bug damage in grain sorghum. The insects caused blasted seed and were most injurious while the grain was in the milk and soft dough stages. Damaging infestations were concentrated in spots 2 to 60 m in diameter.

The nymphs have an oval or pear-shaped body that is widest behind the middle of the abdomen. They become darker with age and are almost black before the last molt. Adults are about 3 mm long and have transparent wings. The newly matured adult is a dull whitish color but in a short time becomes dark. Adults overwinter in grassy areas. Malathion and parathion will control false chinch bugs.

31-1.11 Flea Beetles

Flea beetles occasionally cause serious damage to the edges of winter wheat fields in the fall, particularly near grassy fence rows. These pinhead-sized, shining black beetles jump readily when approached. They feed on wheat, skeletonizing leaves, and occasionally eating small holes in them (Daniels, 1966). The plants become bleached and die when heavily infested. Control can be obtained by applying 1.12 kg of malathion/ha to infested areas.

31-1.12 Wheat Strawworm

The wheat strawworm usually is seen as a small greenish yellow larva inside the wheat stem and just above the top joint. There it feeds on the pith of the inside of the straw, but it rarely reduces yields. It passes the winter in the straw, emerging in February and March as a shiny black, antlike, wingless insect. Most are females that lay eggs in the yound wheat plant. Larvae that hatch feed on the developing head. The wall of the short stem around the larvae enlarges and hardens, forming a gall and eventually killing the tiller that is attached. Plowing under the stubble and summer fallow reduce the populations of this insect (Painter et al., 1954; Fenton, 1952).

31–1.13 Wheat Jointworm

According to Fenton (1952), the wheat jointworm passes the winter in either the larval or pupal stage within infested straws. Adults emerge the following spring usually in April or May. Eggs are inserted into the stems of growing wheat at about the time heads appear. The larvae remain within the cells they have fashioned, and their feeding causes a gall to form that may cause the wheat stem to lodge. They remain as larvae until fall when many pupate. The rest pupate the following spring. Since most of the galls are located on the lower joints, most of these remain in the stubble after the wheat has been harvested. Therefore, they can be controlled by early plowing. There is one generation per year.

31–1.14 Wheat Stem Maggot

The wheat stem maggot attracts attention when the wheat is in bloom, because conspicuous white heads result from infestation by this insect. White heads caused by the maggot always occur singly; the flag leaf is usually green, contrasting strongly with the white head and straw above. The active greenish yellow maggot, about 6 mm long, often may be found above the top joint of the stems with white heads. This is the larva of the first spring generation. Two or three other generations may occur on volunteer wheat or grasses before the insect passes the winter as a larva in the tiller of a volunteer or fall-sown wheat plant. Larvae infesting wheat in the fall may kill the young plant or infested tiller. By the time white heads are visible in wheat, damage already has been done and no control of the insect in wheat is possible (Painter et al., 1954).

31–1.15 Wheat Stem Sawfly

Leaf-feeding sawflies, wheat pests in the Northern Great Plains, are green caterpillar-like larvae without prolegs under the abdomen and with a single black eye on each side of the head.

Winter is passed as a mature larva in the base of a wheat straw at or below the surface of the soil. In the spring, larvae transform to a pupae inside the wheat straw, and adults emerge during June. The females lay eggs by thrusting them into the plant tissues on the upper parts of the stem. The larva bores down through the joints and by late summer reaches the lower parts of the plant close to the surface of the ground. Here it cuts a V-shaped groove entirely around and inside the stem, which causes the stem to break off. The larva plugs itself in the base of the plant with its frass, thus forming a chamber in which it hibernates and later pupates. According to McBride and Colberg (1967), sawfly damage is twofold. First, the insects cause between 10 to 14% reduction in yield by their tunneling activity in the infested stems. Additional loss occurs when sawfly-cut stems fall to the ground and become unharvestable. A combination of deep spring tillage and shallow fall tillage has given as much as 80% control. Early harvest, if possible before infested stems lodge, is also another culture control practice. Wheat varieties resistant to this pest are not available (Spachman and Burkhardt, 1973). Saskatchewan Agriculture Publication 622 (Agric. Canada, 1978), recommended that if 10% of more wheat stems are infested, cultural control should be considered for the following year.

31-1.16 Thrips

Several species of thrips have been collected on wheat. Some of these insects are predaceous, feeding on other thrips or mites, but most are plant feeders. Only one (*Prosopothrips cognatus* Hood) has severely damaged seeded wheat in Kansas (Painter et al., 1954). It causes wheat foliage to turn whitish or rust and, when abundant, injures flowers and grain. Volunteer wheat and various grasses provide alternate hosts during the summer.

Daniels (1974) collected three species from wheat, oat, and sorghum in Texas—western thrips, grass thrips, and onion thrips (Essig, 1958). However, western thrips predominated in wheat and oat and grass thrips predominated in sorghum. These tiny insects migrate from early maturing wheat in June to the young sorghum. Plants less than 7.5 cm tall are very susceptible to their feeding. Their mouth parts are fitted for rasping-sucking, somewhat intermediate between chewing and piercing-sucking. Both nymphs and adults puncture and rasp leaves, suck the contents, and thus destroy the plant cells. Crop rotation and summer fallow will help control these pests, but when wheat and sorghum are side by side, treatments of the insecticides carbaryl, carbofuran (2,3-dihydro-2,2-dimethyl-7-benzofuranyl methylcarbamate), or dimethoate will control thrips long enough for young sorghum plants to grow past the damaging stage.

31-1.17 Mites

31-1.17.1 Brown Wheat Mite

The brown wheat mite is a pest of small grains. Damage by this mite occurs only during dry weather and in some respects resembles that caused by drought (Daniels et al., 1956). A mottling of leaves occurs, and when leaves are observed from a distance, a yellowing or bronzing effect may be noted. Infested leaves first show a silvery cast and later turn brown. When this stage is reached, hundreds of mites can be seen on the leaves and on the ground at the base of the plants. The mite has a rounded, metallic dark brown or blackish body about the size of a period in newsprint, with short hairs on its back. The legs are pale yellow, with the forelegs longer than the other three pairs. It can be identified with a hand lens in the field. The species does not spin webs as do some spider mites.

Brown wheat mites pass the summer as small, shiny, white eggs. The egg is coated with a white waxy material, and one end is flattened in a circular cap somewhat larger than the egg itself. Large numbers of these eggs may be found at the base of the plants in the soil, attached to debris and clods. Hatching begins in the fall as soon as eggs contact a small amount of moisture. Young mites are bright red and have three pairs of legs, but as soon as they feed, most of them become brownish. The next two stages each have four pairs of legs; some of them have greenish bodies, and some resemble the adults in color. After hatching, mites reach the adult stage in 9 to 10 days. Egg laying begins 1 or 2 days later. Winter eggs, laid during the fall, winter, and early spring, are brick red and spherical. They hatch in 6 or 7 days under favorable conditions. Each adult lays 70 to 90 eggs in a 3-week period. According to Baker and Pritchard (1953), males are unknown, and the eggs hatch without fertilization. During the late spring, certain adults begin to lay summer eggs and will lay about 30 during a 3-week period. An adult does not produce both summer and winter eggs.

Heavy rains greatly reduce brown wheat mite populations. Mite numbers usually are much lower on summer-fallowed than on continuously-cropped wheat. If land is fallowed following a wheat crop and summer eggs hatch from stubble or residue in the fall, young mites will die from lack of food.

Controlling this pest with chemicals is difficult. Tests have shown that of several chemicals tested, dimethoate gave best control of the mite but did not increase the yield of wheat (Daniels, 1964). DePew (1962), however, obtained significant yield increases by chemically controlling an extremely heavy population. In general, however, chemical control of the brown mite is not practical (Henderson and Tilton, 1955).

31-1.17.2 Banks Grass Mite

Banks grass mite is a pest of small grains and grain sorghum in parts of the Great Plains. It differs from other mites in that it spins fine webbing on the plants (Harvey, 1954). The mites are whitish or yellowish and about half the size of the brown wheat mite. Feeding symptoms are similar to those of the brown wheat mite (Daniels et al., 1956). During the fall when wheat is small, the mites locate in small colonies on the leaves. During the winter, after the plants have tillered and temperatures are lower, they occur mainly in the crown of the plants near the soil surface. Infested plants can be detected by webbing in the crowns and an abnormal number of dead leaves. The mites generally remain at the plant bases until May and then move to the upper leaves. Sometimes wheat heads will become infested as they emerge from the boot and later will be covered with web. Controlling this pest in wheat with chemicals is not practical.

Although mites are observed earlier in the growing season on sorghum, population increases generally occur after the boot stage of plant growth (Texas Agric. Extension Services, 1977). Initial infestations appear on the lower leaves; however, leaves may be killed as populations become exceedingly heavy and mites move up the plants and into the head of sorghum. Extremely high populations cause extensive webbing of sorghum heads and may be associated with stalk rot and lodging. Although yield reduction can result from severe infestations before the soft-dough stage, the harviest populations have been observed on more mature sorghum. Periods of hot, dry weather favor population increases.

Eggs of the Banks grass mite are laid on the leaf in webbing that the mites produce (Clymer and Daniels, 1975). The eggs are spherical and about one-fourth the size of the adult, which is about half the size of a period in ordinary newsprint. Eggs are transparent to pearly white when first laid bu gradually change to straw yellow. The eggs hatch in 3 to 4 days under optimum conditions, but several weeks may be required when cool moist conditions exist. The small six-legged larvae are light colored when first hatched but become progressively dark green as they feed on sorghum. After several days, they molt (shed outer skin) to the eight-legged nymphal stage and continue feeding. After two additional molts, the mites emerge as adults. The males are smaller than the females. Mating occurs soon after the females emerge. Both males and females mate more than once. Females lay an average of one egg per day for about 48 days, but egg laying may be as long as 80 days. Minimum developmental time from egg to adult is 8 to 10 days.

Controlling this mite on sorghum with chemicals is difficult, although several chemicals have been recommended: carbophenothion {*S*-[(*p*-chlorophenylthio)methyl] *O,O*-diethyl phosphorodithioate}, disulfoton {*O,O*-diethyl *S*-(2-(ethylthio) ethyl] phosphorodithioate}, ethion (*O,O,O,O*-tetraethyl *S,S*-methylene biphosphorodithioate), phorate {*O,O*-diethyl *S*-[(ethylthio)methyl] phosphorodithioate}, Metasystox R {*S*-[2-(ethylsulfinyl)ethyl] *O,O*-dimethyl phosphorothioate}, Supracide [*O,O*-dimethyl phosphorodithioate, *S*-ester with 4-(mercaptomethyl)-2-methoxy-Δ^2-1,3,4-thiadiazolin-5-one], and dimethoate. However, this mite may have developed resistance to the organophosphates in some areas of the Great Plains (Ward and Tan, 1977).

31-1.17.3 Wheat Curl Mite

The wheat curl mite is responsible for transmitting wheat streak mosaic virus. Transmitting mosaic is the only reason for the importance of this mite (Daniels et al., 1956). Other damage is slight and consists of curling and folding of the leaves, causing trapped leaves. Aphids also cause this condition but do not transmit the virus. The wheat curl mite is white, spindle shaped, with only four legs on the front of the body. These mites are so small that they are barely visible when magnified 10 times. They usually are found on the upper leaf surface of the plant in the whorl. Eggs are laid in the grooves of the wheat leaf, with 3 to 25 eggs laid by each female in a 10-day period (del Rosario and Sill, 1964; Somsen, 1966). The mites pass the summer between wheat crops on volunteer wheat and various grasses. They may be present on wheat without transmitting the mosaic. So far, no practical chemical control has been developed.

31–1.17.4 Winter Grain Mite

The winter grain mite causes considerable damage to fall-sown small grains in north central and central Texas (Daniels et al., 1956). Banks (1902) established the first record of the mite in the USA. Both Essig (1939) and Campbell (1941) discussed its damage.

Small grains and grasses are favored hosts of this mite, but it also feeds on legumes, weeds, and vegetables. Heavily infested fields have a grayish or silvery appearance caused by removal of chlorophyll and plant juices by feeding mites. Under heavy infestation, the plants die.

The first-generation mites hatch in the fall, usually around 1 November, from eggs that remained over the summer (oversummer) on grain stubble, straw, or debris in the field. The dark brown to black mites are about 1 mm long and have four pairs of legs. The legs and mouth parts are reddish orange. There usually is a reddish orange spot on the back surrounding the anus. The second generation develops in early January from eggs deposited by first-generation mites. Second-generation mites deposit the eggs that oversummer. Cool and moist conditions are necessary for mite development. When warm weather sets in, usually around 15 April, all winter grain mites disappear.

Mites feed on leaves mainly at night or on cloudy days; on bright days they hide under foliage on the moist soil surface. If the soil is dry, they burrow into it until they reach moisture. Light, loose soils are preferred. Both cultural and chemical control recommendations for the winter grain mite have been developed for small grains (Chada, 1956). Fields planted continuously for 3 or more years usually were heavily infested and damaged. In fields where the previous crop was other than a small grain (e.g., cotton, corn, clover, or sorghum), mites were either absent or very scarce, and there was no damage. Therefore, a change from continuous cropping of small grains to a rotation involving other crops at least every 2 years greatly reduced the damage caused by the winter grain mite. Parathion at 0.29 kg or malathion at 0.5 to 0.84 kg/ha applied in a spray gave adequate control.

31–2 CORN

Great Plains farmers need corn pest management related to water (dryland) management and cultural practices pertinent to their soil type and climatic conditions.

31–2.1 Corn Borer

The European corn borer is one of the limited factors to corn production in some areas of the Great Plains. The European corn borer infests corn from Texas and Oklahoma north into Canada. Corn borer populations vary from year to year as well as from field to field (Kantack et al., 1973). Economically damaging populations in fields cause substantial yield reductions. Damage may be caused by either the first or second generation but seldom by both in the same planting. Early-planted corn is susceptible to first brood (generation) attack, and late-planted corn is susceptible to second brood attack. The first brood burrows in the stalks, whereas the second brood also attacks the ears and ear shanks.

According to Peters and Witkowski (1977), the borer goes through the winter as a full-grown larva in stalks, cobs, and plant debris in corn fields. Most larvae that overwinter will go through the pupal stage and change into moths by the middle of June. Eggs are usually laid near the midrib on the underside of corn leaves in clusters of 15 to 25, overlapping like fish scales. Eggs hatch in about 5 to 7 days in mid- to late June. Those that are about to hatch have a black center, which is the black head of the larva showing through the egg shell. Cool or windy weather tends to prevent moths from flying and laying eggs.

Several factors may reduce European corn borer losses. These include cultural practices and natural controls. Shredding stalks can destroy up to 50% or more of the overwintering larvae. Covering the stalks with several inches of soil will prevent some of the moths from emerging. This is not recommended because of the energy required and potential erosion if plowed in the fall. The longer corn remains in the field after maturity, the greater the amount of stalk lodging, breakage, and ear dropping. Early harvest is usually profitable in late-planted fields where second generation borers are numerous. Most of the commercial hybrids carry some resistance to first-generation European corn borers, but when tassels emerge, this resistance disappears. Varieties with strong stalks tend to break less when entered by borers. Elongated feeding scars are formed on susceptible plants, and small, round feeding scars are formed on highly resistant plants (Dicke and Penny, 1954).

Two disease organisms effectively reduce corn borers some years. One is a fungus disease (*Beauveria bassiana*) that attacks and eventually kills the larvae. The other, a protozoan organism (*Nosema Pyraustea*), often infects the eggs, larvae, pupae, and adults. This infection results in fewer eggs being laid and fewer larvae surviving. Birds may feed on overwintering borers within the stalks. Larvae and adults of lady beetles and lacewing larvae will feed on newly hatched larvae. A parasitic fly (*Lydella thompsoni* Hert.) occasionally parasitizes corn borer larvae (Peters and Witkowski, 1977). Hill et al. (1978) reported that *L. thompsoni* has not been found in Nebraska since 1965 but that *Horogenes punctorius* Roman was widespread in 1976.

With the help of the above cultural practices and natural control factors, insecticides are not needed every year. Insecticidal corn borer control on ensilage corn is not profitable. For first-generation control on corn grown for grain, farmers should observe fields carefully in June if night temperatures have been above 16°C and winds are calm to moderate. The first application should be made when 50% of the plants show leaf feeding scars or "shot holes." A second treatment may be needed 5 to 7 days later if 50% of the plants still have larvae in the whorls. There is no reliable guide to timing insecticide application for control of second-generation larvae. Showers et al. (1978) stated that second-generation European corn borer survival is dependent on feeding sites within the host (i.e., in the sheaths, collars, and ear of the mature corn plant). Therefore, the better the crop, the better the chance of a heavy borer infestation. The summer moths may lay their eggs over several weeks; therefore, more than one insecticide application is usually needed for second-generation corn borer control. Several insecticides will control European corn borers when properly timed and applied. Some recommended insecticides are carbaryl, Diazinon, and fensulfothion {*O,O*-diethyl *O*-[4-(methylsulfinylphenyl] phosphorothioate}. Insecticides may be applied as granules or sprays. Granules usually give better control than do sprays.

The southwestern corn borer, described in 1911 from a specimen collected in Mexico, was found in the USA in 1913 (Davis et al., 1933). These authors also stated that it was probably unknown how long the insect had been in the USA. It spread north and east into the Texas Panhandle and was established in Colorado, Kansas, and Oklahoma by 1931. Wilbur et al. (1943) stated that by 1942 it had reached Nebraska. In 1929 five fields in Lamb and Bailey counties of the Texas Panhandle were 96 to 100% infested (Todd and Thomas, 1930). It was found as far east as Tennessee in 1960 (Pless and Duck, 1969). Several cultural practices, e.g., early planting, stalk destruction, and fall tillage, have resulted in some degree of borer control (Davis et al., 1933; Wilbur et al., 1943; Ralston, 1955; Henderson and Douglas, 1967; Daniels and Chedester, 1974, 1975; Daniels, 1978b).

The southwestern corn borer overwinters as a full-grown larva and usually hibernates at or near the base of the plant crown. The hibernation site is in a tunneled out area within the crown. The upper part of the tunnel is closed with a tough silken plug for protection. In the spring, larvae extend their tunnels to the outer wall of the crown, leaving a thin covering that the moth can easily push aside. Pupation then follows, with the moth soon emerging from the stub of the stalk. Mating occurs within a couple of days, and eggs are glued to the plant, usually on the top side of a leaf about half way up the stalk. Eggs hatch within a few days, depending on the temperature. The average developmental periods are as follows: egg, 5 days; larvae, 25 days; pupal, 10 days; and adult, 5 days. Moths are from 1.5 to 2 cm long and have a wing spread of about 3 cm. They are solid white to pale yellow and fold the wings about the body when at rest. A characteristic snout extends from the head. Daniels and Chedester (1975a) found that spring adults generally emerged from the middle of May into the first week in June in the Texas Panhandle.

Corn yields have also been increased with the use of insecticides in the Texas Panhandle (Daniels and Chedester, 1971, 1975c, 1977; and Owen, 1971). Recommended insecticides for control of this insect are carbaryl, carbofuran, and Diazinon. Workers in Arkansas, Louisiana, and Mississippi reported that great numbers of overwintering southwestern corn borer larvae were destroyed by bird predation (Black et al., 1970; Floyd et al., 1969; Wall and Whitcomb, 1964). Daniels and Chedester (1974) found cold weather in combination with wet weather reduced overwintering borer populations more than dry cold weather, which is in agreement with Chippendale and Reddy (1974). According to Zepp and Keaster (1975), sparse plantings or abundant light may inhibit southwestern corn borer girdling.

31–2.2 Corn Rootworms

Corn rootworms, major economic pests in northern corn-growing areas, are also present as far south as Texas. Three species of rootworms damage corn in the USA: southern corn rootworm, northern corn rootworm, and western corn rootworm. The latter two are of primary importance in the northern corn growing areas of the Great Plains. The spotted cucumber beetle, or southern corn rootworm, is the oblong, pale greenish yellow beetle with six black spots on each wing cover that occurs so commonly on cucumber and melon vines, some weeds, and corn silk. It is about 5 mm long. Western corn rootworm adults resemble the striped cucumber beetle except that the black lines are not as distinct. The rootworm wing color may range from almost all yellow to black. The color underneath the abdomen is yellowish green or brown, whereas the striped cucumber beetle is black. Larvae are small, white worms that destroy roots of corn. The northern corn rootworm is a yellowish green beetle that is slightly smaller than the western and southern corn rootworms. It has no markings on its wings. Eggs are laid in cornfields to overwinter as in the western corn rootworm species (Smith et al., 1943).

Larvae of northern and western rootworms eat on the surface of the roots, especially the crown roots just below the surface, and gouge out holes and tunnel lengthwise. The southern variety feeds on the roots but does not tunnel lengthwise. It also feeds on the growing point of the stalk while it is still near the soil line and may cause the plant to wilt and die. Root systems are weakened by all three, and lodging is common. The lodged plants later grow upward but have a crook at the base. The adults feed on corn silks. According to Aldrich and Leng (1965), the capacity of corn to tolerate rootworm attacks is determined by the strength of the root system, the plant's ability to regenerate roots after they are eaten, the stage of growth when attacked, and favorability of growing conditions after the attack. According to Kantack et al. (1977), several insecticides—carbofuran, Dyfonate (*O*-ethyl-*S*-phenylethylphosphonodithioate), and Lorsban [*O,O*-diethyl *O*-(3,5,6-trichloro-2-pyridyl)-phosphorothioate]—will control rootworms, but corn growers should check each year for the best ones to use.

31-2.3 Corn Earworm

The corn earworm is well known to corn growers when found in the ear of corn. However, when found on other parts of the plant, they may be mistaken for other larvae because of color variations. According to Wilbur et al. (1943), the corn earworm will tunnel into the stalk. In Texas, Wene and Blanchard (1955) found that as high as 32% of the ear shoots were infested before silking had begun. Dicke (1939) stated that when the average mean temperature for December, January, and February fell to −10°C or below, a reduction in earworm abundance followed. According to Luckmann (1963), the incubation period for corn earworm eggs was 2 to 4 days when temperatures were between 32 and 24°C. Gross and Young (1977) stated that the mean time from oviposition to pupation of the corn earworm ranged from 21.8 days when they were reared on a corn-soy blend diet to 30.6 days when they were reared on corn. Oatman (1966) found the average parasitization of corn earworm eggs by *Trichogramma pretiosum* Riley to be 38% over a 3-year period. In studying plant resistance to the corn earworm, McMillan et al. (1970) showed a relationship between the amount of earworm feeding and the presence of certain chemical constituents of the plant.

According to Hare and Harrell (1977), malathion and carbaryl were each effective in controlling the corn earworm. Methods of determining corn earworm injury have been devised and used through the years (Collins and Kempton, 1917; Poole, 1934; Yarnell, 1952; Widstrom, 1967). It is doubtful, however, that chemical control of the corn earworm would be feasible in dryland corn.

31-2.4 Fall Armyworm, Cutworms, and Maggots

The fall armyworm is often mistaken for the corn earworm. Painter (1955) stated that the fall armyworm was by far the most conspicuous pest of corn in Guatemala, and Wiseman et al. (1966) declared it to be one of the most conspicuous major pests of corn in the Americas. The worms feed on leaves and in the whorl. They also enter the ear and the upper parts of the stalk. The worms later move to the ground to pupate. Grain and forage yields were not increased with the use of insecticides against the fall armyworm in corn (Daniels, 1963). However, in 1971, I observed effective control of heavy fall armyworm infestations in large corn acreages in Dallam County, Texas, with carbaryl and parathion. Daniels (1970) also reported that early corn planting reduced corn earworm and fall armyworm infestations.

The western bean cutworm, originally a pest of beans, has become a pest of corn in some of the Great Plains. About 1 July, moths begin to emerge from the soil. Eggs are laid in the corn plants shortly after emergence. The majority of cutworms feed until the middle or end of August, with a few feeding to mid-September. When mature, cutworms enter the soil and change to the prepupal stage to overwinter. To control this cutworm, at least 95% of the field should be tasseled. According to Hagen (1976), very few ears should be formed before a control is applied. Once the cutworms are inside the husk, they seldom expose themselves until they are mature and therefore cannot be controlled. At present, the only chemicals registered for western bean cutworm on corn are carbaryl and trichlorfon [dimethyl (2,2,2-trichloro-1-hydroxyethyl) phosphonate].

The black cutworm, a larva with a greasy gray appearance, occurs from southern Canada throughout the USA and Mexico (Walkden, 1950). It feeds on a wide variety of plants, but in the Great Plains states it is principally a pest of young corn. Many times, however, it will destroy wheat during the winter and is often mistaken for the army cutworm. They destroy plants and quickly move to other plants and repeat the process. The larvae make burrows into which they drag their food. They are quite vicious and cannibalistic. They can be controlled with parathion.

Insects often eat the entire corn kernel, leaving only the seed coat (Aldrich and Leng, 1965). Partly eaten kernels may germinate, but in most cases they soon die. The damage is scattered over the field and is often unnoticed. It becomes worse when any other condition delays germination. The seed-corn maggot is one of the worst seed damaging pests. It is a small black, or gray to black, fly whose body is 2 to 4 mm long. The thorax varies from yellowish through gray to black and has no stripes. The abdomen is gray with a median black line and slender, and each segment is marked off with a narrow black band. The pale or yellowish white maggots of this flay attack germinating seeds in cold springs when germination is slow.

The corn seed beetle, a yellowish brown beetle with a broad black strip on each wing cover, measures 6 mm or a little less. The head is black, with shallow furrows on the wing covers. According to Smith et al. (1943), this pest injures corn seed planted in wet soils. The maggots and beetles can be controlled by insecticidal seed or soil treatments, but unless the pests occur every year, it would probably not be feasible to treat dryland corn.

31–2.5 Mites

The Banks grass mite is the principal mite involved in corn injury. Infestations usually begin on lower leaves of corn plants. As the number of mites increases, they migrate to the upper leaves. They suck plant juices and, when populations are large, kill the leaves. Loss of leaves reduces the food manufacturing ability of the plants. If corn is to be used for silage, the number of succulent leaves is reduced and the quality of silage will be poor. Since the lower leaves seldom contribute to grain yields, infestations late in the season on lower leaves are not serious. Under these conditions, control is not profitable. However, infestations that have reached the area of the ear by 1 August on silage corn or before dent stage on grain corn can cause serious yield losses. These infestations should be controlled.

31-2.6 Wireworms and Grasshoppers

Wireworms often attack kernels and young seedlings. Grasshoppers are sometimes serious pests of dryland corn.

31-3 ALFALFA

Dryland alfalfa, grown in the Great Plains since 1870 in areas with adequate rainfall, is of major importance. Several insects that attack the legume include grasshoppers, blister beetles, lygus and other plant bugs, webworms, armyworms, alfalfa seed chalcid, pea aphid, spotted alfalfa aphid, and alfalfa weevil.

31-3.1 Alfalfa Weevil

According to Kantack et al. (1976), the alfalfa weevil was first found in the USA in Utah in 1904. It was detected in South Dakota in 1936 and in Texas in 1968 (Texas Agric. Extension Service, 1973). They produce one generation during each growing season in most areas of the Great Plains. However, a partial second generation is produced in south Texas. When cold weather sets in, adult weevils crawl down into the crown of the alfalfa or seek other sheltered places in alfalfa fields, e.g., nearby ditch banks or field borders, where they overwinter. The weevils become active in the spring. Each female lays several hundred eggs in clusters of 2 to 25 eggs each. Weevils lay most of their eggs in green alfalfa. Eggs hatch in 1 to 2 weeks in warm weather, but in cool weather, they may accumulate in the field until a favorable hatching temperature occurs. It is not uncommon that some eggs are laid in the fall before cold weather begins. Most of these eggs do not hatch until the following spring in the Central Great Plains.

The larvae, on hatching, are almost white but soon become green with a white strip down the center of the back. The small larvae feed in the interior of the stems for 3 to 4 days and then migrate to the opening leaf buds at the tips of the plants. The larval feeding period usually requires 30 to 60 days. When full grown, the larvae are about 9 mm long. They migrate to the soil surface to pupate. After about 10 days in the pupal stage, they emerge as adults. These adults feed on alfalfa for the remainder of the summer and go into their winter quarters early in the fall for semihibernation.

Both the larvae and adults take the nutritional value out of the alfalfa by feeding on the plant tips, leaves, and buds. Larvae do the greatest damage by feeding within plant tips, on the upper leaves as they open, and then on the lower foliage, skeletonizing the leaves. Damaged leaves dry rapidly, and injured plants take on a grayish to whitish color. Young alfalfa stands can be destroyed.

Early cutting helps control alfalfa weevil by killing eggs, young leaves, and pupae. Where fields are heavily infested, this procedure is recommended. This reduces injury to the first crop and helps the second crop get an early start.

A hymenopterous parasite (*Bathyplectes curculionis* Thomson) is distributed throughout alfalfa-producing areas of Oklahoma and has steadily increased its parasitism of the alfalfa weevil (Berberet and Gibson, 1976). This species is becoming an important agent for control of the weevil in other states (Wilson and Armbrust, 1970; Pike and Burkhardt, 1974; Eklund and Simpson, 1977).

Such insecticides as malathion, Guthion [*O,O*-dimethyl *S*-{[4-oxo-1,2,3-benzotriazin-3(4*H*)-yl]methyl}phosphorodithioate], carbofuran, carbaryl, methoxychlor [1,1,1-trichloro-2,2-bis (*p*-methoxyphenyl)ethanol], and parathion can effectively control the weevil. Close field inspection is a necessity in determining the need for an insecticide application. Insecticidal application should begin when 25 to 30% of the alfalfa tips show feeding damage or when 25 or more larvae/0.09 m^2 are present. Additional applications may be necessary for severe infestations. Testing the stubble as soon as the first crop is removed gives the best control.

31-3.2 Aphids

The spotted alfalfa aphid is small and grayish yellow with four to six rows of black spots on its back. Infestations begin on lower leaves; as the population increases, however, the aphid can be found on all parts of the plant. This aphid copiously produces honeydew, and extremely heavy infestations in established alfalfa stands make leaves and stems sticky.

The pea aphid is a bright green, long-legged aphid about twice the size of the spotted alfalfa aphid. Pea aphids generally cause the heaviest damage in the spring and early summer and then again in the fall. They feed on the more succulent stems, buds, and leaves near the stem terminals. Heavy infestations cause tops of the plant to wilt. Planting of aphid-resistant alfalfa varieties has been the most effective means of controlling both the spotted alfalfa aphid and pea aphid. Growers should determine the cultural adaptability of any resistant variety selected for hay and seed production under local conditions. 'Cody,' 'Lahontan,' and 'Zia' are well-adapted, varieties resistant to spotted alfalfa aphid (Daniels, 1971a). 'Kanza,' 'Washo,' and 'Dawson' are resistant to both the spotted alfalfa aphid and pea aphid (Kindler and Schalk, 1975).

Infestations can increase rapidly under favorable conditions. However, rain and high humidity often wipe out an infestation or create conditions unfavorable for the pest. Sometimes after cutting a stand, aphid populations will decrease or even disappear. Occasionally large numbers of winged aphids can be found, but they fail to produce colonies. Insecticides, diazinon, parathion, or malathion should be applied only when large numbers of wingless aphids and honeydew can be found.

31-3.3 Lygus Bugs

Lygus bugs and plant bugs are often serious pests of alfalfa grown for seed but are not considered major pests of alfalfa grown for hay. Lygus bugs are present throughout the Great Plains. Plant bugs are found primarily in the northern states and Canada (Craig, 1973). Heavy infestations cause blasted buds, blossom drop, and shriveled seed. Lygus counts should be made when the wind velocity is less than 16 km/h preferably in early morning or late afternoon. This sampling period is suggested because lygus bugs are difficult to sample during periods of high temperature or high wind velocity. Counts are based on two-sweep counts taken in 10 to 20 randomly selected sites scattered over the field using a standard sweep net. Chemical control measures should be based on the average population. Rarely is it practical to apply insecticides to only portions of a field. Each lygus nymph should be counted as two adults. When insects reach an average of 75 per 100 sweeps, control measures should be applied. Plant bugs should be controlled when they reach four per sweep. Malathion, parathion, carbaryl, or trichlorfon may be used.

31-3.4 Stink Bugs

Stink bugs may cause extensive damage to seed crops (Texas Agric. Extension Service, 1973). They often move into fields in large numbers soon after pods begin to form. The most serious damage is caused when they suck juices from the immature seed. A sweep net should be used to determine whether stink bugs are numerous enough (12/100 sweeps) to warrant insecticidal control. Carbaryl and trichlorfon have been used with success.

31-3.5 Chalcids

The clover seed chalcid and the alfalfa seed chalcid are small, jet black wasps that usually deposit one egg per immature alfalfa seed. The developing larva hollows out the seed. Populations of chalcids increase from spring through late summer. Consequently, late-season seed is often more severely attacked. Chemical control is not recommended for seed chalcids. The following cultural practices are effective in reducing chalcid populations and damage:

1. Manage the crop so that it matures uniformly as early as possible.
2. Keep field margins, borders, and ditchbanks free of volunteer alfalfa that may produce seed. According to Metcalf et al. (1951), chalcids overwinter as mature larvae in seed.
3. Destroy chaff and screenings from previous year before spring.

31-3.6 Webworms

The garden, alfalfa, and beet webworms are general feeders of alfalfa. They also feed on several weeds, especially pigweed.

Cultivated crops are frequently attacked by these worms migrating from weeds that they have devoured. The larvae feed primarily on the underside of leaves, more or less skeletonizing them. They spin webs and pull other leaves within their webbing as additional food is needed. Flimsy webs near the plant terminals are noticeable in alfalfa infested with these insects.

On forage alfalfa, mowing is recommended if the infested crop is near the cutting stage. The webworms will usually disappear before the new crop comes in (Brooks and Gates, 1979). If the crop is more than 2 weeks from cutting and heavily infested, one of the following insecticides should be used: methoxychlor, carbaryl, or methyl parathion.

31-3.7 Alfalfa Caterpillar

The alfalfa caterpillar is a common insect pest of alfalfa. This larva feeds on foliage, and populations usually increase when not held in check by disease or beneficial insects. Damage can sometimes be avoided by early cutting or, when necessary, by using the insecticides carbaryl or methoxychlor.

31-3.8 Armyworm, Cutworm, and Grasshopper

The fall armyworm, true armyworm, beet armyworm, and army cutworm are sometimes found on alfalfa and may develop damaging populations requiring insecticidal control (Walkden, 1950). The larvae are foliage feeders but sometimes feed on flowers and green seed pods. Descriptions and control measures are given in section 31-1.9.

Grasshopper damage may occur any time between May and frost.

31-4 COTTON

Dryland producers should select a cotton variety that produces well under local conditions. Short-season cottons fruit and mature more rapidly than Delta-type varieties; consequently, the short-season varieties are subject to insect damage for a shorter time period.

Overfertilization of cotton can delay maturity, produce excessive stalk growth, and result in a higher level of pest damage. Fertilization of cotton should be based on a soil test.

Dryland cotton should be planted so that there are 67 000 to 131 000 plants/ha (two to four plants per 30 cm on 102-cm rows). Higher plant populations lead to greater competition between plants for moisture and nutrients and can lead to increased pest damage (Texas Agric. Extension Service, 1979).

31-4.1 Bollworm

The bollworm, commonly known as corn earworm, tomato fruitworm, and cotton bollworm, is selfom a problem in dryland fields. However, it may become a problem following insecticidal application that destroys beneficial insects. Insecticides should be applied only when necessary as determined by frequent field inspection. Azodrin [monochrotophos; *O,O*-dimethyl-*O*-(2-methylcarbamoyl-1-methyl-vinyl)-phosphate] and methyl parathion are effective in controlling the bollworm.

31-4.2 Boll Weevil

The boll weevil, found near Brownsville, Texas, in 1892, had spread northward from cotton-producing areas in Mexico. From this point, it spread rapidly over the Cotton Belt (Fenton, 1952). Its spread was more rapid northeastward than in other directions because of more favorable climatic conditions for its development in that part of the Cotton Belt. It did not occur in west Texas until the 1960's (J. G. Thomas. 1979. Personal communication).

The most distinctive characteristic of the weevil is the presence of two spurs on the inner surface of the front femur. The larva is a white, wrinkled, fat, legless grub that is found only in the squares or bolls. It attains a length of 5 to 7 mm.

It overwinters as an adult in all types of shelter, e.g., leaves, gin trash, seed houses, and cornstalks. Weevils that are best protected have the best chances of survival. According to Little (1957), winter mortality is quite high; the average survival is only about 6%. Emergence begins in the spring when the mean air temperaure reaches 17.8°C and may extend over a period of several months. The first weevils to emerge from hibernation feed on the terminal portions of young cotton plants. When squares become available, they puncture them and feed on the highly nutritious pollen sacs. Eggs are laid in punctures made by the females, one egg being deposited in each square as long as there is sufficient supply. When the squares are punctured, they usually flare and later drop or hang on the stalks, where they wither and turn brown. Bolls are heavily attacked when squares are not available. Small bolls may drop, but large ones do not, although one or more locules may be destroyed. During the summer, weevils live an average of 50 days, and those that hibernate in the fall may live into the next summer. Female weevils oviposit an average of about 100 eggs, which are laid over a period of about 1 month.

In areas where the boll weevil is present, producers should plant cotton uniformly after mid-May (Texas Agric. Extension Service, 1979). Earlier planting may result in the production of a complete weevil generation before late planted cotton begins to square. Early weevils will greatly increase the damage caused during the entire growing season.

The destruction of cotton stalks as soon as the crop is harvested eliminates late-season increases of weevils and forces the adults into a starvation period before the time of hibernation. This practice, when conducted on a communitywide basis, materially reduces the number of weevils to emerge the next season.

The boll weevil is difficult to control with insecticides, because the adult is largely an internal feeder and the immature stages are found only in the squares and bolls. However, a number of insecticides will kill the adult if they are correctly applied. Toxaphene, carbaryl, methyl parathion, malathion, and Guthion are recommended.

31-4.3 Pink Bollworm

The first specimens of the pink bollworms were found near Hearne, Texas, in 1917 (Fenton, 1952). Early attempts at eradication were made in which the establishment of noncotton zones played an important part. Success rewarded the efforts in east Texas and Louisiana, but in the more western districts, it was impossible to eradicate the pest. Evidence points to the reinfestation of that region by moths blown northward by prevailing winds. However, in the dryland cotton area of the Great Plains, climatic conditions are unfavorable for its survival and development.

The egg is oval, more or less flattened, and averages 0.5×0.25 mm. The egg is pearly white when first laid but turns reddish before hatching. The shell is minutely reticulated. Eggs are laid singly or in small groups. The younger larvae are white, but this color is replaced by pink or salmon red in the last instar. The mature larva is about 1.5 cm long with a reddish brown head. The pupa is brown and about 9 mm long.

Winter is passed in the larval stage in a cocoon in various situations, e.g., the surface soil in a cotton field, on the tap root of the cotton, in shed squares, in blooms and small bolls in the field, in gin trash, in cotton picking sacks, and in stored cotton seed. Overwintering worms are known as long-cycle larvae. The larvae pupate in the spring, and moths soon emerge. The period of spring emergence lasts for several weeks, and some moths may emerge as late as August. The life cycle is from 26 to 32 days. The early destruction of stalks and bolls stops reproduction of the pink bollworm.

Cotton and cotton products, harvesting and ginning machinery, and other regulated articles may move without restriction within Texas or from Texas to Oklahoma, New Mexico, and Arizona and back to Texas. However, movement of such cotton products or regulated articles to other cotton-producing areas may be made only under permit, except for the following products: samples of lint and linters, cottonseed meal or cottonseed cake, and compressed baled lint, if moved by common carrier. Movement of these materials is unrestricted (Texas Agric. Extension Service, 1974).

31–4.4 Cabbage Looper and Armyworms

Rarely do foliage feeders such as the cabbage looper, fall armyworm, yellow-striped armyworm, or beet armyworm cause sufficient damage to warrant control late in the season. Since well-defined economic thresholds do not exist for these pests, control in dryland cotton is a matter of judgment. Azodrin is most effective if applied when worms are small (J. G. Thomas. 1979. Personal communication).

31–4.5 Cotton Fleahopper

The cotton fleahopper damages small squares. Both adults and nymphs suck sap from tender portions of the plant, including the small squares. Squares are subject to attack from pinhead size until they are one-third grown. According to Fenton (1952), the adult is a pale, greenish, oval insect about 3 mm long with fine black specks uniformly distributed over the body. The duration of the egg stage averages 8 days. Upon hatching, nymphs progress through five instars and become adults in about 11 days. Three days after maturity, the females begin laying eggs. An entire life cycle may be completed in 22 days. Little (1957) stated that seven to eight generations may be completed in one season.

Winter is passed in the egg stage in various weed hosts. The eggs hatch in the early spring, and the nymphs feed on any succulent weed nearby. When a scarcity of host plants occur, winged forms migrate to cotton fields. They may also drift with air currents from large weedy areas and infest cotton fields some distance away. When cotton has reached the fruiting stage, it should be examined weekly for fleahoppers. When 25 to 35 fleahoppers/100 plant terminals are found, control measures should be employed. Carbaryl, dicrotophos (dimethyl *cis*-2-dimethyl-carbamoyl-1-methylvinyl phosphate), or methyl parathion will give adequate control. Spray applications should be made at 7- to 10-day intervals until the insect is controlled.

31–4.6 Cotton Aphids

The cotton aphid also called the cotton louse, occurs wherever cotton is grown. Aphids are soft-bodied, sucking insects that range from light yellow to green or almost black. Natural controls in the form of predators, parasites, and unfavorable weather are almost always effective in holding aphid populations below damaging levels.

Heavy infestations of aphids can cause young leaves to cup downward, older leaves to turn yellow and shed, and squares and small bolls to shed. Severe infestations early in the boll maturation period can cause defoliation and premature boll opening, resulting in incomplete development of fiber. Honeydew excreted by the aphids can drop on fiber of open bolls. A black, sooty fungus sometimes develops on honeydew deposits during wet periods. Fiber from such bolls is stained, sticky, and of lower quality. According to S. Coppock (1979. Personal communication) and J. G. Thomas (1979. Personal communication), the cotton aphid is seldom a problem in dryland cotton. Insecticide treatment with dicrotophos or parathion may be necessary in rare instances when infestations are extremely heavy in large areas of the field and natural control has not been sufficient to limit aphid buildup.

31-4.7 Grasshopper

Grasshoppers sometimes damage dryland cotton. Description and control measures are given in section 31-1.9. However, one species, the lubber grasshopper, was not mentioned. This species cannot fly but has strong hopping legs. If nothing bothers its eggs in the soil (weather, disease, or predators) and conditions are favorable for several years, this species may build up, and with its voracious appetite and large size, it may become a damaging local pest to dryland cotton (E. P. Boring III. 1979. Personal communication; Little, 1957).

31-5 LITERATURE CITED

Aldrich, S. R., and E. R. Leng. 1965. Modern corn production. A. and L. Pub., Champaign, Ill.

Agriculture Canada. 1978. Insect control on field crops. Saskatchewan Agric. Pub. no. 622.

Atkins, I. M., and R. G. Dahms. 1945. Reaction of small grain varieties to greenbug attack. USDA Tech. Bull. no. 901.

Baker, E. W., and A. E. Pritchard. 1953. A guide to the spider mites of cotton. Hilgardia 22: 203-234.

Banks, N. 1902. New genera and species of acarians. Can. Entomol. 34:172.

Berberet, R. C., and W. P. Gibson. 1976. *Bathyplectes curculionis* in Oklahoma: Distribution and effective parasitism of the alfalfa weevil. Ann. Entomol. Soc. Am. 69:205-208.

Black, E. R., F. M. Davis, C. A. Henderson, and W. A. Douglas. 1970. The role of birds in reducing overwintering populations of the southwestern corn borer, *Diatraea grandiosella* (Lepidoptera:Crambidae) in Mississippi. Ann. Entomol. Soc. Am. 53:701-706.

Bottrell, D. G., N. E. Daniels, and J. G. Thomas. 1972. Greenbugs and other aphids on sorghum and small grains. Texas Agric. Extension Service Pub. no. L-819.

Brooks, L., and D. E. Gates. 1979. Kansas field crop insect control recommendations. USDA Coop. Extension Service, Pub. no. C-431, Kansas State Univ., Manhattan.

Burrage, R. H. 1964. Trends in damage by wireworms (Coleoptera:Elateridae) in grain crops in Saskatchewan, 1954-1961. Can. J. Plant Sci. 44:515-519.

Campbell, J. B., and J. H. L. Bell. 1977. Grasshoppers. Univ. of Nebraska Extension Service Nebguide no. G74-106.

Campbell, R. E. 1941. Further notes on the blue oat or pea mite, *Penthaleus major* (Duges) in Californai. Bull. California Dep. Agric. 30:312–314.

Cartwright, W. B., R. M. Caldwell, and L. E. Compton. 1959. Response of resistant and susceptible wheat to Hessian fly attack. Agron. J. 51:529–531.

Chada, H. L. 1956. Biology of the winter grain mite and its control in small grains. J. Econ. Entomol. 49:515–520.

Chedester, L. D. 1979. Development of a braconid wasp, *Lysiphlebus testaceipes* (Cresson). texas Agric. Exp. Stn. Progress Report no. 3456.

Chippendale, G. M., and A. S. Reddy. 1974. The effects of tillage and weather on southwestern corn borer populations. Environ. Entomol. 3:431.

Clymer, B. C., and N. E. Daniels. 1975. Insect and mite pests of field corn. Texas Agric. Exp. Stn. and Extension Service.

Collins, G. N., and J. H. Kempton. 1917. Breeding sweet corn resistant to the corn earworm. J. Agric. Res. 11:549–572.

Coppock, S., Jr. 1962. The grasshopper of Oklahoma (Orthoptera: Acrididae). Oklahoma State Univ. Exp. Stn. (Bull.) Processed Series P-399:1–143.

Craig, C. H. 1973. Insect pests of legume and grass crops in Western Canada. Canada Dep. of Agric. Pub. no. 1435.

Daniels, N. E. 1955. False wireworm control in sorghum. Texas Agric. Exp. Stn. Progress Report no. 1793.

Daniels, N. E. 1957. False wireworm control in wheat. Texas Agric. Exp. Stn. Progress Report no. 1978.

Daniels, N. E. 1958. Chemical control of a species of the false chinch bug group in grain sorghum. Texas Agric. Exp. Stn. Progress Report no. 2063.

Daniels, N. E. 1962a. Economic greenbug control. Tex. Agric. Prog. 8:27–28.

Daniels, N. E. 1962b. Insecticidal control of the greenbug. Texas Agric. Exp. Stn. Progress Report no. 2247.

Daniels, N. E. 1963. Chemical control of the fall armyworm. Texas Agric. Exp. Stn. Progress Report no. 2280.

Daniels, N. E. 1964. The effect of brown wheat mite control on wheat yield. Texas Agric. Exp. Stn. Progress Report no. 2328.

Daniels, N. E. 1966. Flea beetle control in wheat. Texas Agric. Exp. Stn., Tech. Report no. 1.

Daniels, N. E. 1969. Alternate hosts of the greenbug. J. Econ. Entomol. 62:521.

Daniels, N. E. 1970. Corn insect control. Texas Agric. Exp. Stn. Progress Report no. 2730.

Daniels, N. E. 1971a. Area adaptation studies of alfalfa varieties and their resistance to the spotted alfalfa aphid. Texas Agric. Exp. Stn. Progress Report no. 2896.

Daniels, N. E. 1971b. Detection of insecticidal residue and control of soil insects. J. Econ. Entomol. 64:175–177.

Daniels, N. E. 1972. Insecticidal control of greenbugs in grain sorghum. J. Econ. Entomol. 65:25–40.

Daniels, N. E. 1974. Control of thrips in grain sorghum. Texas Agric. Exp. Stn. Progress Report no. 3250.

Daniels, N. E. 1975. Factors influencing greenbug infestations in irrigated winter wheat. Texas Agric. Exp. Stn. Miscellaneous Pub. no. 1187.

Daniels, N. E. 1977. Spring migration of the greenbug in the Texas Panhandle. Southwestern Entomol. 2:191–196.

Daniels, N. E. 1978a. Greenbug control in grain sorghum. Texas Agric. Exp. Stn. Progress Report no. 3527.

Daniels, N. E. 1978b. Insecticidal and cultural control of the southwestern corn borer. Southwest. Entomol. 3:308–314.

Daniels, N. E. 1981. Migration of greenbugs in the Texas Panhandle in relation to their biotypes. Texas Agric. Exp. Stn. Miscellaneous Pub. no. 1487.

Daniels, N. E., H. L. Chada, D. Ashdown, and E. A. Cleveland. 1956. Greenbugs and some other pests of small grains. Texas Agric. Exp. Stn. Bull. no. 845.

Daniels, N. E., and L. D. Chedester. 1971. Insect control in corn. Texas Agric. Exp. Stn. Progress Report no. 2903.

Daniels, N. E., and L. D. Chedester. 1974. The effects of tillage and weather on southwestern corn borer populations. Texas Agric. Exp. Stn. Progress Report no. 3431.

Daniels, N. E., and L. D. Chedester. 1975a. Cultural control and biological studies of the southwestern corn borer. Texas Agric. Exp. Stn. Progress Report no. 3375C.

Daniels, N. E., and L. D. Chedester. 1975b. Experimental aphid control in grain sorghum. Texas Agric. Exp. Stn. Progress Report no. 3335C.

Daniels, N. E., and L. D. Chedester. 1975c. Insect and mite control in corn. Texas Agric. Exp. Stn. Progress Report no. 3328C.

Daniels, N. E., and L. D. Chedester. 1976. Grub and wireworm populations after manure and nitrogen applications. Texas Agric. Exp. Stn. Miscellaneous Pub. no. 1308.

Daniels, N. E., and L. D. Chedester. 1977. Insecticidal control and damage evaluation of the southwestern corn borer. Texas Agric. Exp. Stn. Progress Report no. 3431.

Daniels, N. E., and L. D. Chedester. 1980a. Greenbug control in dryland wheat. Texas Agric. Exp. Stn. Progress Report no. 3732.

Daniels, N. E., and L. D. Chedester. 1980b. Winter survival of greenbugs in the Texas Panhandle. Texas Agric. Exp. Stn. Miscellaneous Pub. no. 1447.

Daniels, N. E., and K. B. Porter. 1956. Greenbug damage to winter wheat as affected by proceeding crop. J. Econ. Entomol. 49:600–602.

Daniels, N. E., and J. D. Thomas. 1961. Pathological studies with the western wheat aphid. Texas Agric. Exp. Stn. Progress Report no. 2215.

Davis, E. G., J. R. Horton, C. H. Gable, E. V. Walters, and R. A. Blanchard. 1933. The southwestern corn borer. USDA Tech. Bull. no. 388.

del Rosario, M. E., and W. H. Sill. 1964. Additional biological and ecological characteristics of *Aceria tulipae* (Acrina:Eriophyedae). J. Econ. Entomol. 57:893–896.

DePew, L. J. 1962. Evaluation of brown wheat mite control on yield of winter wheat in Kansas. J. Econ. Entomol. 55:10–11.

Dicke, F. F. 1939. Seasonal abundance of the corn earworm. J. Agric. Res. 59:237–257.

Dicke, F. F., and L. H. Penny. 1954. Breeding for resistance to European corn borer. Proc. Hybrid Corn Industry-Res. Conference 9:44–53.

Eidt, D. C. 1953. European wireworms in Canada with particular reference to Nova Scotian infestations. Can. Entomol. 85:408–414.

Eklund, L. R., and R. G. Simpson. 1977. Correlation of activities of the alfalfa weevil and *Bathyplectes curculionis* with alfalfa height and degree-day accumulation in Colorado. Environ. Entomol. 6:69–71.

Essig, E. O. 1939. The blue oat or pea mite, *Penthaleus major* (Duges), in California. Calif. Dep. Agric. Bull. 28:507–508.

Essig, E. O. 1958. Insects and mites of western North America. Macmillan, Inc., New York.

Fenton, F. A. 1952. Field corn insects. Macmillan, Inc., New York.

Fenton, F. A., and R. G. Dahms. 1951. Attempts at controlling the greenbug by the importation and release of lady beetles in Oklahoma. Proc. Okla. Acad. Sci. 32:1–2.

Floyd, E. H., L. Mason, and S. Phillips. 1969. Survival of overwintering southwestern corn borer in corn stalks in Louisiana. J. Econ. Entomol. 62:1016–1019.

Gallun, R. L. 1977. Genetic basis of Hessian fly epidemics. Ann. N.Y. Acad. Sci. 287:223–229.

Gross, H. R., Jr., and J. R. Young. 1977. Comparative development and fecundity of corn earworm reared on selected wild and early season hosts common to the Southeastern U.S. Ann. Entomol. Soc. Am. 70:63–65.

Hagen, A. F. 1973. Hessian fly on wheat. Univ. of Nebraska Extension Service Nebguide no. G73-46.

Hagen, A. F. 1974. Pale western cutworms in wheat. Univ. of Nebraska Extension Service Nebguide no. G74-130.

Hagen, A. F. 1976. Western bean cutworm in corn. Univ. of Nebraska Extension Service Nebguide no. G76-290.

Hare, W. W., and E. A. Harrell. 1977. Field tests to evaluate experimental spray equipment and techniques for protection of sweet corn from insects. Soil Sci. Soc. Am. Proc. 70: 185–186.

Harvey, T. L. 1954. Chemical control of a spider mite on wheat in Eastern New Mexico. J. Econ. Entomol. 47:593–597.

Harvey, T. L., and H. L. Hackerott. 1971. Research on greenbugs and resistance in sorghum. p. 84–86. *In* Proc. Seventh Biennial Grain Sorghum Res. and Utilization Conference, Lubbock, Tex. Texas A&M Univ., College Station.

Henderson, C. A., and W. A. Douglas. 1967. Fall cultivation for control of the southwestern corn borer in Mississippi. J. Econ. Entomol. 60:602–603.

Henderson, C. F., and E. W. Tilton. 1955. Tests with acaricides against the brown wheat mite. J. Econ. Entomol. 48:157–161.

Hill, R. E., D. P. Corpino, and Z. B. Mayo. 1978. Insect parasites of the European corn borer *Ostrinia nubilalis* in Nebraska from 1948–1976. Environ. Entomol. 7:249–253.

Hunter, S. J. 1909. The greenbug and its enemies. Univ. of Kansas Dep. of Entomology Bull. no. 8, Lawrence.

Johnson, J. W., G. L. Teetes, and C. A. Schaefer. 1976. Greenhouse and field techniques for evaluating resistance to sorghum cultivars to the greenbug. Southwest. Entomol. 1:150–154.

Kantack, B. H., W. L. Berndt, and R. J. Walstrom. 1976. The alfalfa weevil and its control. South Dakota Agric. Extension Service Pub. no. FS276.

Kantack, B. H., P. A. Jones, and W. L. Berndt. 1973. European corn borer control in South Dakota. USDA Cooperative Extension Service Pub. no. FS527, South Dakota State Univ., Brookings.

Kantack, B. H., P. A. Jones, and W. L. Walstrom. 1976. The alfalfa weevil and its control. South Dakota Agric. Exp. Service Pub. no. FS276.

Kantack, B. H., D. Walgenbach, and W. L. Berndt. 1977. Corn rootworm control in South Dakota. USDA Cooperative Extension Service Pub. no. FS491. South Dakota State Univ., Brookings.

Kindler, S. D., and J. M. Schalk. 1975. Frequency of alfalfa plants with combined resistance to the pea aphid and spotted alfalfa aphid in aphid-resistant cultivars. J. Econ. Entomol. 68:716–718.

Little, V. A. 1957. General and applied entomology. Harper & Row, Publishers, Inc., New York.

Luckmann, W. H. 1963. Measurements of the incubation period of corn earworm eggs. J. Econ. Entomol. 56:60–62.

McBride, D. K., and W. J. Colberg. 1967. The wheat stem sawfly in North Dakota. North Dakota State Univ. Agric. Extension Service Circ. no. A-513.

McMillian, W. W., B. R. Wiseman, and A. A. Sekul. 1970. Further studies on responses of corn earworm larvae to extracts of corn silks and kernels. Ann. Entomol. Soc. Am. 63: 371–378.

Metcalf, C. L., W. P. Flint, and R. L. Metcalf. 1951. Destructive and useful insects. McGraw-Hill Publications, New York.

Milliken, I. B. 1918. Nysius eracae, the false chinch bug group. J. Agric. Res. 13:571–578.

Oatman, E. R. 1966. Parasitization of corn earworm eggs on sweet corn silk in Southern California, with notes on larval infestations and predators. J. Econ. Entomol. 59:830–835.

Owen, D. 1971. Revenge seen for corn borer. Irrig. Age 5:40.

Painter, R. H. 1955. Insects of corn and teosinte in Guatemala. J. Econ. Entomol. 48:36–42.

Painter, R. H., H. R. Bryson, and D. A. Wilbur. 1954. Insects that attack wheat in Kansas. Kansas Agric. Exp. Stn. Bull. no. 367.

Peters, L. L., and J. F. Witkowski. 1977. European corn borer, 1977. Univ. of Nebraska Extension Service Nebguide no. G75-217.

Petralia, R. S., A. L. Wuensche, G. L. Teetes, and A. Sorensen. 1979. External morphology of the mouthparts of larvae of sorghum midge, *Contarinia sorghicola.* Ann. Entomol. Soc. Am. 72:850–855.

Pike, K. S., and C. C. Burkhardt. 1974. *Bathyplectes curculionis* on the western strain of the alfalfa weevil. J. Kans. Entomol. Soc. 47:405–411.

Pless, C. O., and B. N. Duck. 1969. Control of the southwestern corn borer with furadan insecticide. Tennessee Farm and Home Sci. Progress Report no. 70, p. 27–28.

Poole, C. F. 1934. Corn earworm resistance in maize varieties at Davis, California in 1934. Proc. Am. Soc. Hortic Sci. 32:453–457.

Porter, K. B., G. L. Peterso, and O. Vise. 1981. A new greenbug biotype. p. 1–11. *In* Crop Production Utilization Symp., Section H, Amarillo, Tex. 12 February. Texas A&M Univ., College Station.

Ralston, L. H. 1955. The southwestern corn borer in Arkansas. Arkansas Agric. Experiment Bull. no. 553.

Ruggles, A. G., and F. M. Wadley. 1927. The greenbug in Minnesota. J. Econ. Entomol. 20: 34–27.

Say, T. 1823. Descriptions of coleopterous insects collected in the late expedition to the Rocky Mountains; performed by order of Mr. Calhoun, Secretary of War, under the command of Major Long. J. Natl. Sci. Phila. 3:139–216.

Showers, W. B., M. B. DeRozori, G. L. Reed, and R. H. Shaw. 1978. Temperature-related climatic effects on survivorship of the European corn borer. Environ. Entomol. 7:717–723.

Smith, R.C., E. G. Kelly, G. A. Dean, H. R. Bryson, and R. L. Parker. 1943. Common insects in Kansas. Kansas State Board of Agric. Report, Topeka.

Somsen, H. W. 1966. Development of migratory form of wheat curl mite. J. Econ. Entomol. 59:1283–1284.

Spachman, E., and C. C. Burkhardt. 1973. Control of field crop insects. Univ. of Wyoming Agric. Extension Service Bull. no. 543R.

Teetes, G. L. 1973. *Phyllophaga crinita*: Damage assessment and control in grain sorghum and wheat. J. Econ. Entomol. 66:773–776.

Teetes, G. L. 1974. Damage assessment of a false chinch bug, *Nysius raphanus* (Howard) in grain sorghum. Texas Agric. Exp. Stn. Program Report no. 3260.

Texas Agriculture Extension Service. 1973. Texas guide for controlling insects on legumes and grasses. Texas Agric. Extension Service Miscellaneous Pub. no. 1014.

Texas Agriculture Extension Service. 1974. Ways to fight the pink bollworm in Texas. Texas Agric. Extension Service Pub. no. L-219.

Texas Agriculture Extension Service. 1977. Suggestions for controlling insects and mites on corn, sorghum and small grains. Texas Agric. Extension Service Miscellaneous Pub. no. 339.

Texas Agriculture Extension Service. 1979. Management of cotton insects on the High Plains, Rolling Plains and Trans-Pecos areas of Texas. Texas Agric. Extension Service Bull. no. 1209.

Thomas, J. G. 1969. The sorghum midge and its control. Texas Agric. Extension Service no. L-842.

Todd, C. J., and F. L. Thomas. 1930. Notes on the southwestern corn borer, *Diatraea grandiosella* Dyar. J. Econ. Entomol. 23:118–121.

Wadley, F. M. 1931. Ecology of *Toxoptera graminum,* especially as to factors importance in the Northern United States. Ann. Entomol. Soc. Am. 24:325–395.

Walkden, H. H. 1950. Cutworms, armyworms and related species attacking cereal and forage crops in the Central Great Plains. USDA Circ. no. 849.

Wall, A. D., and W. H. Whitcomb. 1964. Effect of bird predators on winter survival of the southwestern and European corn borers in Arkansas. J. Kans. Entomol. Soc. 37:265–266.

Ward, C. R., and F. M. Tan. 1977. Organophosphate resistance in the Banks grass mite. J. Econ. Entomol. 70:250–252.

Webster, F. M., and W. J. Phillips. 1912. The spring grain aphid or "greenbugs." USDA Bureau of Entomology Bull. no. 110.

Wene, G. P., and R. A. Blanchard. 1955. Relation between corn silking and earworm infestation in the lower Rio Grande Valley of Texas. Fla. Entomol. 38:13–16.

Widstrom, N. W. 1967. A valuation of methods for measuring corn earworm injury. J. Econ. Entomol. 60:791–794.

Wilbur, D. A., H. R. Bryson, and R. H. Painter. 1943. Southwestern corn borer in Kansas. Kansas State Univ. Bull. no. 317.

Wilson, M. C., and E. J. Armburst. 1970. Approach to integrated control oif the alfalfa weevil. J. Econ. Entomol. 63:554–557.

Wiseman, B. R., R. H. Painter, and C. E. Wossom. 1966. Detecting corn seedling differences in the greenhouse by visual classification of damage by the fall armyworm. J. Econ. Entomol. 59:1211–1214.

Wood, E. A., Jr. 1971. Designation and reaction of three biotypes of the greenbug cultured on resistant and susceptible species of sorghum. J. Econ. Entomol. 64:183–185.

Yarnell, S. H. 1952. Breeding sweet corn for resistance to the corn earworm. Am. Soc. Soc. Proc. 60:379–386.

Zepp. D. B., and A. J. Keaster. 1975. Southwestern corn borer girdline activity: Relationship to physiological stress of host plants. J. Kans. Entomol. Soc. 48:505.

32

Ben Doupnik, Jr.
University of Nebraska
Clay Center, Nebraska

Richard A. Frederiksen
Texas A&M University
College Station, Texas

Diseases of Major Crops

Plant diseases are one of the limiting factors in the production of the important crops (wheat, corn, grain sorghum, alfalfa, and cotton) grown under dryland agriculture in the Great Plains and other areas of North America. In recent years, production losses caused by plant diseases have been confined to localized epiphytotics rather than being widespread. In years past, however, widespread disease outbreaks did occur. Widespread wheat stem rust epiphytotics caused by *Puccinia graminis* Pers. f. sp. *tritici* Eriks. & Henn. were common because of the lack of host resistance or the ability of the pathogen to produce new races, strains, (or both) (Wiese, 1977). A more recent example of a widespread epiphytotic was the 1970 outbreak of southern corn leaf blight caused by *Helminthosporium maydis* Nisik. & Miy. (Shurtleff et al., 1980) and brought about by the almost total use of a single source of male sterility (Texas male-sterile cytoplasm) to produce hybrid seed corn. All corn containing this source of cytoplasm was susceptible to a new race of the southern corn leaf blight fungus, race T.

Genetic vulnerability and narrow genetic bases of resistance are of particular significance to the Great Plains because of the sequential planting and maturing of many of the crops grown there. Those pathogens that can be disseminated over great distances may thus find a susceptible host each step of the way. If movement of the pathogen coincides with favorable environmental conditions, an epiphytotic can become widespread. Present and future concern about conserving our natural resources has and will continue to affect agricultural production systems. Some of the land that is now being irrigated may be returning to dryland production in the not too distant future. Innovative cropping systems will need to be developed to ensure that such land remains productive. In addition, much of the current dryland acreage is being and will be put under new cropping systems. Re-

Dryland Agriculture—Agronomy Monograph no. 23.

duced tillage (conservation tillage) is rapidly increasing as a new crop production system throughout the Great Plains area (Fenster et al., 1977). The short- and long-term disease consequences of reduced tillage will thus have to be taken into consideration. Unfortunately, we already have some early warning signs that reduced tillage and monoculture can lead to increased disease problems (Boosalis and Doupnik, 1976; Boosalis et al., 1981; Cook et al., 1978; Sumner et al., 1981). On the other hand, an innovative reduced tillage cropping system known as ecofallow, which rotates wheat with corn or grain sorghum in a 3-year rotation system, has been shown to reduce Fusarium stalk rot caused by *Fusarium moniliforme* Sheld. (Doupnik and Boosalis, 1980).

In this chapter we will not attempt to cover every disease that occurs on the major crops grown in the Great Plains and other areas but will discuss those diseases that are presently important as well as those that may become important. The common name of the crop disease is designated with the causative pathogen in parentheses. For more detailed information on these diseases, see the Disease Compendia Series published by the American Phytopathological Society, St. Paul, MN 55121, and the Disease Atlas Series published by the Texas Agricultural Extension Service, Texas A&M University, College Station, TX 77843. Possible control measures and pitfalls will be discussed where appropriate. Several review articles on the effects of reduced tillage on plant diseases (Boosalis and Doupnik, 1976; Boosalis et al., 1981; Cook et al., 1978; Sumner et al., 1981), and several books dealing with disease epidemiology, management, and control (Frederiksen et al., 1978; Horsfall and Cowling, 1977; Zodoks and Schein, 1979), are cited as useful references.

32-1 WHEAT DISEASES

The disease complexes commonly referred to as "crown and root rots" and "seedling blights" occur every year to some extent wherever wheat is grown. These diseases are caused primarily by fungi. Examples of fungi involved are species of *Helminthosporium, Fusarium, Pythium,* and *Rhizoctonia* (Wiese, 1977). Losses are brought about by stand loss and by weakened plants that are more susceptible to winter injury. Control is obtained primarily by employing the following practices: (i) using clean, certified seed, (ii) applying a fungicidal seed treatment, (iii) planting at recommended dates, (iv) and planting into a firm, mellow seedbed using adapted varieties.

Another severe fungal root rot disease of wheat is take-all, which is caused by *Gaeumannomyces graminis* var. *tritici* Walker. Losses are brought about by stunting, reduced tillering, and premature ripening, which results in sterile heads. Control is very difficult. High disease incidence is associated with continuous monoculturing of wheat and unbalanced fertility. Resistance is not available, thus crop rotation out of wheat for as long as 3 to 4 years is recommended. Grasses, soybean, or alfalfa should not be used in the rotation, because take-all disease is sometimes severe following these crops (Wiese, 1977).

Two relatively old and serious diseases of wheat that have increased in recent years are loose smut caused by *Ustilago tritici* (Pers.) Rostr. and common bunt (stinking smut) caused by *Tilletia foetida* (Wallr.) Liro and *T. caries* (DC.) Tul. (Wiese, 1977). Both of these diseases cause grain to be replaced by dark spore masses, resulting in severe yield and quality losses. Smut diseases were once controlled by appropriate seed treatments and the use of certified seed. In recent years, however, growers have become lax in the use of seed treatments and clean certified seed, which has resulted in an increase in these two diseases. Resistant varieties have also been developed; thus, if adapted varieties are available, they should be planted.

Several foliar diseases affect wheat production throughout North America. These include stem rust (*Puccinia graminis* f. sp. *tritici*), leaf rust (*P. recondita* Rob. ex. Desm. f. sp. *tritici*), stripe rust (*P. striiformis* West.), tan spot [*Pyrenophora trichostoma* (Fr.) Fckl.], Septoria leaf and glume blotch (*Septoria* spp.), and to a lesser extent, bacterial leaf blight (*Pseudomonas syringae* van Hall), and Cephalosporium stripe (*Cephalosporium gramineum* Nis. & Ika.) (Wiese, 1977). Historically, the rust diseases have played an important role in wheat production and have changed the course of history by destroying this major food source. Stem rust is now effectively controlled through the use of resistant varieties. Occasionally, however, yield losses are still significant from leaf rust, particularly in the Central and Northern Great Plains. These losses usually occur when abundant spore showers coincide with extended periods of moist, moderate temperature conditions. Often when these conditions exist, other foliar diseases, e.g., tan spot and Septoria leaf and glume blotch, will also be occurring. These multiple infections are able to overpower the wheat plant's resistance, and yield losses result. Tan spot, Septoria leaf and glume blotch, and leaf rust can be effectively controlled with foliar fungicides; however, unless the leaf spotting caused by these three diseases is severe, it is probably not always economical. Resistance is not readily available for either tan spot or Septoria leaf and glume blotch. Rotation to nonhosts is a recommended control for both of these diseases. Stripe rust occurs commonly in the western mountains and valleys from Canada to Mexico but is rare in the Great Plains. Resistant varieties and fungicides help control this disease.

The long-term effects of reduced tillage practices in wheat production, as well as other crops, on disease incidence and severity remains to be seen. Early warnings may already be appearing with regard to the foliar disease consequences on wheat. With the exceptions of stem rust and leaf rust, several diseases seem to be increasing in those areas where reduced tillage is being extensively practiced. The undisturbed residue is providing an overwintering reservoir for many of these diseases. Tan spot and Septoria leaf and glume blotch, in particular, appear to be increasing in incidence and severity under these situations. In addition, diseases that have not been of importance in the past are now showing up. Recent problems with Cephalosporium stripe and bacterial leaf blight in some areas of the Great Plains, for example, appear to be associated with reduced tillage practices. The net gains in the conservation of our natural resources may eventually have to be weighted against the net changes in disease consequences unless resistant varieties can be developed or other cultural methods adopted to lessen the impact of these diseases (Boosalis and Doupnik, 1976; Boosalis et al., 1981; Cook et al., 1978; Sumner, 1981).

An example of another foliar disease that has increased in recent years in the Central Great Plains is powdery mildew, caused by *Erysiphe graminis* DC. f. sp. *tritici* Marchal. This disease has, in part, been brought on by reduced tillage, but probably of more importance is the increased acreage of the semidwarf cultivars, many of which are highly susceptible. Control can be achieved with fungicides but may not be economical. Control is best obtained by using resistant or highly tolerant cultivars (Wiese, 1977).

At present two major virus diseases are affecting wheat in the Great Plains—wheat streak mosaic and wheat soilborne mosaic. Wheat streak mosaic also occurs in the Western Great Plains. Wheat streak mosaic is transmitted by the wheat curl mite (*Aceria tulipae* Keif.), and thus control of the vector will effectively control the disease. Destruction of volunteer wheat at least 2 weeks before planting time, and delaying the planting dates, are effective cultural control practices. Although highly resistant varieties are not yet available, some varieties are moderately resistant and should be planted when adapted varieties are available and wheat streak mosaic is a problem (Wiese, 1977).

Wheat soilborne mosaic is often the number one disease problem on wheat in the Central Great Plains. This disease is transmitted by a soilborne fungus (*Polymyxa graminis* Led.), which invades the wheat roots. Control of the vector is not economical. Fortunately, resistant varieties are now available and should be planted in those areas where wheat soilborne mosaic is a problem (Wiese, 1977).

Until recently, barley yellow dwarf has not been considered an important virus disease of wheat for most of the Great Plains and western dryland areas. Increasing incidences of this disease, however, may redirect our thinking, especially in view of the fact that the barley yellow dwarf virus has a wide grass host range and that several strains of the virus are known to exist. Little is known about the disease reactions of many of our current varieties, and other effective control measures are not available.

The importance of nematode diseases on wheat production in the dryland areas of North America has not been adequately determined; however, they may be more important than is presently realized. Of particular concern is the cereal cyst nematode (*Heterodera avenae* Woll.), which has been found in Oregon. The spring wheats appear to be more susceptible than the winter wheats.

32-2 CORN AND GRAIN SORGHUM DISEASES

Corn and grain sorghum have several diseases in common, and most of these occur throughout the Great Plains and western areas of North America, thus these crops will be covered together. Fortunately, however, relatively few of them are serious problems. Only the more important diseases will be discussed (Edmunds and Zummo, 1975; Horne and Berry, 1980; Shurtleff et al., 1980; Ullstrup, 1974).

Seed and root rots and seedling blights are often associated with poor stand establishment of both corn and grain sorghum, which usually occurs during extended periods of cool weather and moist soil conditions. Control is difficult, but the use of adapted hybrids and fungicidal seed treatments will help reduce seed decay and stand loss.

Several mildew diseases occur on corn and sorghum. The most important are sorghum downy mildew [*Peronosclerospora sorghi* (Weston & Uppal) Shaw] and crazy top [*Sclerophthora macrospora* (Sacc.) Thir., Shaw, & Naras.]. Sorghum downy mildew has historically been a problem on grain sorghum on the Texas Gulf Coast, but it is now being found with increasing incidence and severity on both corn and grain sorghum throughout the Southern and Central Great Plains. Control is primarily through resistant hybrids and crop rotation with nonhost crops. Obviously, a grain sorghum–corn rotation should be avoided where this disease has been a problem.

Crazy top commonly occurs throughout North America on both corn and grain sorghum under conditions of high soil moisture and flooding. Resistance is not commonly available in either crop, but the disease can be prevented by providing adequate soil drainage and by not planting in low, wet spots in the field. Fortunately, substantial losses only occur in very localized areas.

Several smut diseases occur on corn and grain sorghum. The most prevalent on corn is common smut caused by *Ustilago maydis* (DC.) Cda. Hot, dry weather in association with mechanical damage caused by hail, cultivation, or insects are often associated with high disease incidences. Unbalanced soil fertility may also be associated with this disease. Severe losses caused by common smut, however, are seldom seen because of the planting of more resistant hybrids. A balanced soil fertility program will also aid in control. The most important smut on grain sorghum is sorghum head smut caused by *Sphacelotheca reiliana* (Kuhn) Clint. Sorghum head smut can be controlled effectively by growing resistant varieties and by crop rotation.

The most common foliar fungal diseases of corn include northern corn leaf blight (*Helminthosporium turcicum* Pass.), southern corn leaf blight (*H. maydis*), common corn rust (*Puccinia sorghi* Schwabe), and Physoderma brown spot (*Physoderma maydis* Miyabe). These diseases are usually of little economic importance in the dryland agricultural areas; however, increasing incidences are often associated with reduced tillage cropping systems. Control is most effective by using resistant varieties and/or crop rotation.

The most common fungal leaf spotting diseases of grain sorghum are leaf blight (*H. turcicum* Pass.), target leaf spot (*H. sorghicola* Lefebvre & Sherwin), grey leaf spot (*Cercospora sorghi* Ell. & Ev.), zonate leaf spot (*Gloeocercospora sorghi* Bain & Edg.), rough leaf (*Ascochyta sorghina* Sacc.), sooty stripe [*Ramulispora sorghi* (Ell. & Ev.) Olive & Lefebvre], rust (*Ruccinia purpurea* Cke.), and anthracnose [*Colletotrichum graminicolum* (Ces.) G. W. Wils.]. Of these, anthracnose can be the most damaging, especially in the Southern Great Plains. As with the foliar fungal diseases of corn, reduced tillage is often associated with the increased incidence of the foliar fungal diseases of sorghum. Control measures for these leaf-spotting diseases include using resistant varieties and/or crop rotation.

Bacterial diseases of importance on corn include holcus spot (*Pseudomonas syringae* van Hall), Stewart's bacterial wilt [*Erwinia stewartii* (Smith) Dye], and Goss's wilt and blight (*Corynebacterium nebraskense* Vidaver & Mandel). Of these, Goss's wilt and blight is of most concern, although it is not widespread in the Great Plains at this time (Kansas, Nebraska, Colorado, and South Dakota) and has not been found in Canada or the western areas. Reduced tillage is often associated with increased incidences of these bacterial diseases; however, all three can be controlled by using resistant varieties and/or crop rotation.

Bacterial diseases occurring on grain sorghum are bacterial stripe [*Pseudomonas andropogoni* (E. F. Smith) Stapp], bacterial streak [*Xanthomonas holcicola* (Elliott) Starr & Burkh.], and bacterial spot (*Pseudomonas syringae* van Hall). These are seldom damaging but, if necessary, can be controlled by using crop rotation, clean tillage, resistant varieties, or a combination of these practices.

Stalk rots are the most prevalent and serious diseases of corn and grain sorghum throughout the Great Plains. The principal stalk rot diseases are charcoal rot [*Macrophomina phaseolina* (Tassi) Goid] and Fusarium stalk rot. Losses are incurred by premature death of plants and/or lodging. Stalk rot diseases are very complex and difficult to control. An early-season environment favors the development of stalk rots. In addition, the host plant becomes more susceptible to invasion by stalk-rotting pathogens at senescence; thus, the "stalk rot complex" is often referred to as a disease of stress and old age. Stalk rot losses can be reduced by: (i) selecting tolerant adapted hybrids, (ii) reducing plant populations, (iii) using soil sample analyses to determine fertilizer requirements, (iv) planting longer-season hybrids, (v) rotating to nonhosts, (vi) conserving soil moisture and reducing soil temperature through reduced tillage, and (vii) harvesting problem fields as soon as practical to reduce losses from lodging and ear droppage. In summary, practices that reduce stress—particularly during pollination, grain development, and maturation —will help reduce stalk rot problems.

An innovative reduced tillage cropping system called ecofarming (ecofallow) that has been developed for the semiarid Central Great Plains reduces stalk rot. This system involves a 3-year winter wheat–corn or grain sorghum–fallow rotation that has resulted in increased yields and decreased incidences of stalk rot inboth corn and grain sorghum. The reduction in stalk rot is related to the increased soil moisture storage and lower, more constant soil temperature brought about by the wheat straw residue left on the soil surface. In addition, the planting of one crop species directly into the residue of a different crop species rather than the same species avoids the disease pitfalls often encountered when monoculture is practiced under reduced tillage. The growing of two different crop species breaks the overwintering reservoir link in the buildup of diseases that often occurs under monoculture (Doupnik and Boosalis, 1980).

Fusarium head blight of sorghum caused by *F. moniliforme* is often the cause of poor grain quality and can occur wherever high moisture conditions delay harvest. Control is difficult, but some varieties are less susceptible than others to head invasion by this fungus.

Virus diseases are probably second only to the stalk rots as causes of production losses to corn and grain sorghum. Maize dwarf mosaic virus is prevalent throughout the Great Plains and occurs on both crops. This virus is carried primarily by aphids, but control through vector elimination is not feasible. Two strains of maize dwarf mosaic virus exist. Strain A will infect johnsongrass, whereas strain B will not. In certain areas, elimination of johnsongrass and other grass hosts may provide some control of this virus but the migratory flights of aphid vectors reduces the effectiveness of controlling alternate hosts. Although no corn or grain sorghum varieties appear to be totally resistant to maize dwarf mosaic virus infection, highly tolerant varieties are available and should be grown where the virus is a problem.

A new virus disease recently found on corn in Kansas and Nebraska is corn lethal necrosis. This disease is caused by a combination of two viruses, maize dwarf mosaic virus and maize chlorotic mottle virus. Southern and western corn rootworm beetles are the primary vectors of maize chlorotic mottle virus. Control of the vectors is not effective, but the use of varieties that are tolerant to corn lethal necrosis will greatly reduce losses. Whether corn lethal necrosis will spread to other areas of North America remains to be seen, but its potential for spreading is obvious because of the migratory habits of the vectors of the component viruses (Shurtleff et al., 1980).

Other virus diseases that are not present or widespread in the dryland areas of North America but are potential threats to corn producers include white line mosaic, American wheat striate mosaic, maize chlorotic dwarf, and maize rayado fino. Control for most virus diseases will be through the use of resistant or highly tolerant varieties and crop rotation since control of vectors and alternate hosts is not feasible in most cases. Virus diseases as a group are not greatly affected by reduced tillage practices, but wheat streak mosaic is an exception. Wheat streak mosaic virus can infect both corn and wheat as well as a number of other grasses and can use volunteer wheat and/or corn as an oversummering reservoir to infect the new crop.

The importance of nematode diseases of corn and grain sorghum has not been well established. The recent discovery of several species affecting these crops, however, suggests that nematodes may be much more important than previously thought. They appear to cause more severe problems in sandy soils. The root lesion (*Pratylenchus* spp.), needle (*Longidorus* spp.), sting (*Belonolaimus* spp.), dagger (*Xiphinema* spp.), and root knot (*Meloidogyne* spp.) nematodes appear to be the most important species. Control is primarily through crop rotation to nonhosts and the use of nematicides.

32-3 ALFALFA DISEASES

A number of plant diseases affect alfalfa production throughout North America, and excessive moisture is often associated with the development of many of them. Although relatively few of the more serious diseases are controlled by genetic resistance, losses can be greatly reduced by proper cultural practices (Graham et al., 1979).

During stand establishment, the newly germinated seed and small seedlings are subject to invasion by a number of soilborne fungi, especially under extended periods of excessive soil moisture. These pathogens include species of *Pythium, Phytophthora, Fusarium,* and *Rhizoctonia* (Graham et al., 1979). Control is primarily through seed treatment with fungicides and properly prepared seedbeds.

Plants in established stands are also susceptible to several root and crown rots. These include anthracnose (*Colletotrichum trifolii* Bain & Essary) and bacterial wilt [(*Corynebacterium insidiosum* (McCull.) Jens.], in addition to those caused by species of *Rhizoctonia, Pythium, Phytophthora,* and *Fusarium* (Graham et al., 1979). Resistant varieties are continually being developed, with good resistance now available to bacterial wilt, anthracnose, and Phytophthora root rot, but resistance is not readily available against the others. A combination of good management practices that encourage vigorous growing conditions and the planting of resistant varieties will greatly reduce losses from root and crown rots.

Of the vascular wilt diseases, Fusarium wilt caused by *Fusarium oxysporum* Schlecht. f. sp. *medicaginis* (Weimer) Snyd. & Hans. has historically been serious in alfalfa and is often associated with the decline of older stands, as has bacterial wilt and anthracnose. Varieties are now available with resistance to Fusarium wilt and should be grown where this disease is a problem. Verticillium wilt, caused by *Verticillium albo-atrum* Reinke & Berth., occurs primarily in the western states and is a potential disease problem in the Great Plains (Graham et al., 1979).

Many foliar diseases occur on alfalfa, and under ideal environmental conditions for disease development, they can cause loss of yield as well as quality (Graham et al., 1979). The more important include common leaf spot [*Pseudopeziza medicaginis* (Lib.) Sacc.], lepto leaf spot [*Leptosphaerulina briosiana* (Poll.) Graham & Luttrell], Stemphylium leaf spot (*Stemphylium botryosum* Wallr.), spring black stem (*Phoma medicaginis* Malbr. & Roum.), summer black stem (*Cerospora medicaginis* Ell. & Ev.), bacterial leaf spot [*Xanthomonas alfalfae* (Riker, Jones, & Davis) Dows.], downy mildew (*Peronospora trifoliorum* d By.), and rust (*Uromyces striatus* Schroet.). Control is primarily through the use of resistant varieties, where available, and cutting early to reduce losses from premature defoliation.

The importance of virus diseases of alfalfa in the North American dryland areas has not been adequately determined. Infected plants may be symptomless and undetectable. Diagnostic procedures have not been developed for identifying some of the viruses capable of infecting alfalfa.

Alfalfa mosaic is widespread and probably occurs wherever alfalfa is grown. Control is primarily through the use of virus-free certified seed and elimination of the insect vectors to help reduce spread, but most established stands have infected plants.

The effect of plant parasitic nematodes on alfalfa production in North America has also not been adequately determined. Of the species known to attack alfalfa, the stem [*Ditylenchus dipsaci* (Kühn) Filip.] root knot (*Meloidogyne* spp.), and root lesion (*Pratylenchus* spp.) are probably most important. Control is difficult because resistant varieties are not readily available. Chemical control with nematicides is often effective but not economical. Crop rotation is also not effective because of the wide host ranges of the nematodes but may be useful in reducing populations below threshold levels that cause damage.

32-4 COTTON DISEASES

In the dryland agricultural areas of North America, cotton is grown only in the southern portion of the Great Plains and in the Southwest and, for the most part, has few damaging disease problems (Watkins et al., 1981). During the past 10 years, tremendous progress in developing cottons with better emergence in cooler soils along with improved seed quality has greatly reduced losses from seed and seedling diseases. Other serious diseases include Verticillium wilt (caused by *Verticillium dahliae* Kleb.) and bacterial blight [caused by *Santhomonas campestris* pr malvacearum (E.F. Smith) Dye] (Watkins et al., 1981). Such cultural practices as reduced N application, skip planting, furrow diking, and the use of higher levels of disease resistance have reduced the severity of Verticillium wilt. High levels of host resistance in combination with reduced N levels has also reduced the severity of bacterial blight on dryland cotton. Phymatotrichum or Texas root rot [caused by *Phymatotrichum omnivorum* (Shear) Dug.] is important in soils with high pH. Control is primarily through proper cultural practices, including rotation to cereals or grasses and early planting with adequate fertility. A disease that has increased in recent years is Alternaria leaf spot caused by *Alternaria macrospora* Zimm. This is attributable in part to the development and wide use of new cultivars that are susceptible. Cotton root knot caused by *Meloidogyne incognita* (Kofoid & White) Chitwood is the major nematode problem. This disease can be effectively controlled with the use of nematocides, crop rotation, and resistant varieties.

32-5 LITERATURE CITED

Boosalis, M., and B. Doupnik. 1976. Management of crop diseases in reduced tillage systems. Bull. Entomol. Soc. Am. 22:300–302.

Boosalis, M., B. Doupnik, and G. Odvody. 1981. Conservation tillage in relation to plant diseases. p. 445–474. *In* CRC handbook of pest management, Vol. 1. CRC Press, Inc., Boca Raton, Fla.

Cook, R., M. Boosalis, and B. Doupnik. 1978. Influence of crop residues on plant diseases. p. 147–163. *In* W. R. Oschwald (ed.) Crop residue management systems. ASA Spec. Pub. no. 31, Am. Soc. of Agron., Crop Sci. Soc. of Am., and Soil Sci. Soc. of Am., Madison, Wis.

Doupnik, B., and M. B. Boosalis. 1980. Ecofallow—a reduced tillage system and plant disease. Plant Dis. 64:31–35.

Edmunds, L. K., and N. Zummo. 1975. Sorghum diseases in the United States and their control. Agric. Handbook no. 468.

Fenster, C. R., H. I. Owens, and R. H. Follett. 1977. Conservation tillage for wheat in the Great Plains. USDA Extension Series no. PA-1190.

Frederiksen, R. A., G. L. Teetes, D. T. Rosenow, and J. W. Johnson. 1978. Utilizing disease and insect resistance in North America crop plants. p. 77–89. *In* R. Teranishi (ed.) Agricultural and food chemistry: Past, present, future. AVI Publishing Co., Westport, Conn.

Graham, J. H., F. Frosheiser, D. Stuteville, and D. Erwin. 1979. A compendium of alfalfa diseases. Am. Phytopathological Soc., St. Paul, Minn.

Horne, C. W., and R. W. Berry. 1980. Sorghum disease atlas. Texas A&M Univ. Agric. Extention Series no. B-1323.

Horsfall, J. G., and E. B. Cowling. 1977. Plant disease—an advanced treatise, Vol. 1. How disease is managed. Academic Press, Inc., New York.

Shurtleff, M. C., T. L. Whatley, and T. Kommedahl. 1980. Compendium of corn diseases. ed. 2. Am. Phytopathological Soc., St. Paul, Minn.

Sumner, D., B. Doupnik, and M. Boosalis. 1981. Effects of reduced tillage and multiple cropping on plant diseases. Annu. Rev. Phytopathol. 19:167–187.

Ullstrup, A. J. 1974. Corn diseases in the United States and their control. USDA Agric. Handbook no. 199.

Watkins, G. M. (ed.) 1981. Compendium of cotton diseases. Am. Phytopathological Soc., St. Paul, Minn.

Wiese, M. V. 1977. Compendium of wheat diseases. Am. Phytopathological Soc., St. Paul, Minn.

Zodoks, J. C., and R. D. Schein. 1979. Epidemiology and plant disease management. Oxford Univ. Press, New York.

SECTION F

RELATED TOPICS

33

Odell L. Walker
Oklahoma State University
Stillwater, Oklahoma

Ronald D. Krenz
Economic Research Service
U.S. Department of Agriculture
Stillwater, Oklahoma

Economics

The economic attractions and pitfalls of dryland farming are depicted well by the master storyteller James Michener in a chapter called "Drylands" of the book *Centennial* (Michener, 1974). The attractiveness of the broad, level expanses of free land in western Kansas, Oklahoma, and Colorado is shown through the eyes of his character in 1911. The settlers were enticed by a story character called Dr. Thomas Doles Creeney, who enunciated 10 farming principles for dryland areas:

1. Catch, store, and protect from evaporation all the rain that falls.
2. Fallow at least 60% of the land.
3. Choose your soil carefully, and keep a close check on the status of the soil by means of a soil auger.
4. Keep a mulch on the field to avoid evaporation.
5. Disk immediately after a rain to hold the moisture.
6. Plow in the fall.
7. Plow 25.4 cm deep, disk, and harrow.
8. Plant wheat in the fall.
9. Grow a green manure crop occasionally.
10. Farm every day as though each year is the drought.

The settlers prospered following these rules although there were good and bad crop years because of rainfall variability. A few voices were raised suggesting that the system of straight rows and clean tilling without regard to wind erosion was dangerous. The drought and winds of the "dust bowl" proved them right, and many farmers were forced to move on, both in the book and in real life.

Despite the experiences of farmers with the uncertain financial results and hard work of dryland farming, it retains an attractiveness today. To many eyes, the land vista is flat, bare, and beautiful. If farmers get a series

Dryland Agriculture—Agronomy Monograph no. 23.

of good years early in their tenure and utilize appropriate management strategies for their uncertain environment, e.g., maintaining financial reserves, they are likely to be successful dryland farmers. However, they must achieve sufficient farm growth to keep pace with technological changes, tendencies for product prices to fall relative to input costs, and inflation. If they get a series of bad years and have only average financial reserves, they will almost certainly have economic difficulty.

The key economic characteristics of dryland farming that are discussed in this chapter are partly implicit in Michener's book. The number and diversity of production alternatives are small. Small grains, mostly wheat grown on fallow, and livestock, predominate in all areas. Variability of yields is great both among years and between production areas within a year. Drought and hail account for most of this variability. The flat, big fields are well suited to mechanization with large machines. Substantial economies of size are obtained when machinery, equipment, and operator labor, and management are used to capacity. Energy requirements per hectare are low in the dryland areas but are comparable with other production areas on a per unit of production basis. Because of the large size of farming units, population in dryland farming areas is sparse and distances between neighbors and shopping centers are great. Thus, unique marketing and social service problems are created.

Wheat and feed grains are subject to booms and busts in prices as well as yields. Exports, the livestock market, and governmental programs all affect prices. Many federal farm programs have centered on wheat and feed grains, and the producer seldom knows what programs will be in effect in the future. Thus, investments in land and other long-term assets are based on uncertain payoffs. Farm programs have included price supports, crop insurance, and payments for conservation practices and diverting land. Programs have also aided the construction of grain and water storage facilities and provided low-cost credit when disaster strikes.

The purpose of this chapter is to provide the reader with a basis for assessing economic opportunities and difficulties in dryland farming. These opportunities and difficulties arise because of the characteristics outlined above. A bibliography is provided to allow the reader to dig deeper into areas for which space does not allow full treatment here.

33–1 DRYLAND AREAS AND CROPS

Table 33–1 shows selected characteristics of major U.S. dryland farming regions—the Northern and Southern Great Plains, the mountain region, and the Pacific region based on the 1974 agricultural census. Figure 33–1 outlines the areas from which data were taken. The regions shown in Fig. 33–1 have a small percentage of pasture because the selection of areas in the regions focused on dryland farming and omitted most of the ranching areas. The ranching aspects of dryland resource use are not treated in this chapter. Data on irrigated and dryland farms and hectares are provided in

Table 33-1. Dryland and irrigated land and crops in four major farming regions, USA.†

	Northern Great Plains‡		Southern Great Plains§		Mountain¶		Pacific#	
	Irrigated	Dryland	Irrigated	Dryland	Irrigated	Dryland	Irrigated	Dryland
Total land, ha	1 685 548	38 130 838	1 926 959	18 604 698	1 611 579	37 403 727	2 827 852	9 480 559
Number of farms	18 335	63 585	13 438	38 278	17 741	20 409	33 481	9 164
Area in farms, ha	9 969 860	29 846 526	6 426 748	14 104 909	13 809 360	25 211 048	6 354 872	5 953 418
Average farm size, ha	543	469	478	368	778	1 235	190	650
Land utilization, ha								
Corn	1 156 876	1 407 604	186 365	38 230	295 469	24 072	125 482	2 104
Sorghum	146 894	1 342 314	487 559	823 494	70 351	134 066	46 713	1 138
Wheat	149 062	6 855 681	250 337	2 035 351	167 140	2 624 067	243 828	1 294 398
Hay	140 909	2 679 024	81 274	426 897	559 568	908 455	478 777	105 721
Oat, barley, rye	--	745 091	--	73 901	123 753	184 393	69 069	197 055
Soybean	24 866	413 781	18 108	38 041	455	--	--	--
Sugar beet	32 043	--	10 021	--	67 102	--	96 346	--
Cotton, peanut	--	--	456 193	277 534	--	--	--	--
Potato	--	--	--	--	82 589	--	50 277	--
Pasture	7 347 072		14 482 695		799 872		373 771	
Summer fallow	4 237 519		299 415		2 596 613		773 339	
Livestock summary, head								
Cattle calves	9 206 712		5 838 618		4 824 853		2 932 142	
Hogs, pigs	3 235 373		302 615		82 214		100 153	
Sheep, lambs	812 219		771 728		1 619 269		889 669	

† Fig. 1.3 delineates the major geographic areas containing dryland farming in the USA. The figure was used to define the counties included in the region in each state. Major range areas are excluded, but the counties included have substantial areas that are not cropland or pasture, e.g., wooded pasture and wasteland. Data are from the Bureau of Census (1974).

‡ Includes dryland areas of Kansas, Nebraska, North Dakota, and South Dakota.

§ Includes areas of Texas and Oklahoma.

¶ Includes areas of Colorado, Idaho, Wyoming, Montana, and New Mexico.

Includes areas of California, Oregon, and Washington.

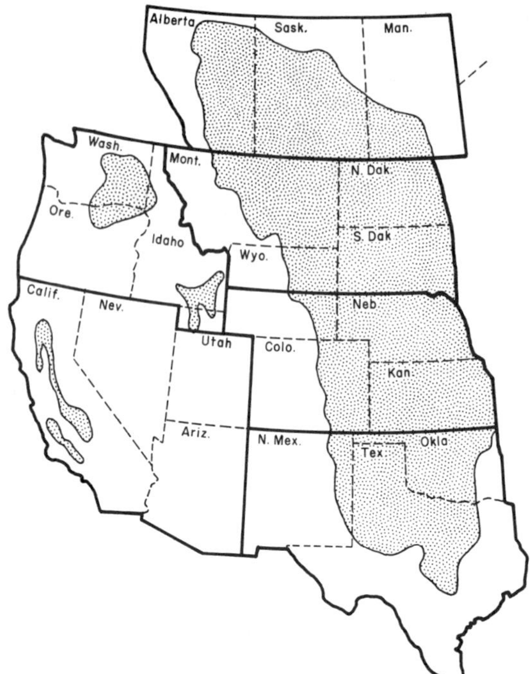

Fig. 33–1. Major dryland farming areas of the USA and Canada.

Table 33–1. Many of the acres in irrigated farms are actually used for dryland cropland or pasture.

The irrigated farms in the Northern and Southern Great Plains appear to be larger than dryland farms. The reason is probably that the dryland farms include more small, part-time, or other noncommercial farm units. Irrigated farming is most predominate in the Pacific region, particularly California, as indicated by total farms and total hectares.

Wheat has approximately 12% of the total land hectares and ranks a strong first in each of the four regions. Fallow uses 7% of the total land. Other crops in order of economic importance are sorghum, hay, corn, other small grains, soybean, cotton, peanut and sugar beet.

No attempt was made to classify farms by type, e.g., cash crop or livestock. However, each region has large numbers of cattle and sheep and part of the Northern Plains is an important producer of hogs. In the Southern Plains, many of the cattle are grazed on wheat pasture and sorghum stubble. Because of the way the counties in the regions were chosen, some range cattle probably are included in all areas.

33–2 TYPICAL DRYLAND FARMS

A USDA typical farm study provides a basis for analyzing the cost and returns from dryland farms in selected areas for one crop year, 1976 (Strickland and Fawcett, 1978). The farms in the USDA study are located in major commodity production areas. Summary data for typical farms in dryland areas are presented in Tables 33–2, to 33–4. An irrigated farm in Nebraska and a Corn Belt farm in Illinois are included for comparison. The analysis focuses on (i) the size, capital investment, and organization of farms in different areas; (ii) the economic status of each farm as reflected for the 1976 crop year; and (iii) the economic future of dryland farming.

33–2.1 Size, Investment, and Organization

Table 33–2 provides information on acres and capital investments on each farm and Table 33–3 provides labor data. Table 33–4 compares the farms' economic performance. In terms of hectares, the dryland farms of the West are clearly larger than the more intensive irrigated and Corn Belt farms. Total average capital investment per farm (an impressive $0.75 million or more) is also greater on three of the dryland farms. The Oklahoma and Kansas farms probably need to be bigger to meet income needs of a farm family. In all dryland areas, wheat is the major crop, with pea and barley added in the Northwest and cotton, sorghum, and cattle in the Southwest. Corn is the major crop on the irrigated farm and the Illinois farm.

Table 33–3 shows the labor distributions for each of the farms and should be studied in combination with the land uses in Table 33–2. The large Washington farm requires the majority of labor in the late summer. Even though only 2002 h of labor are used, 345 h of labor must be hired because of the distribution of labor needed. The operator has excess labor for the year but income problems are not apparent. The same can be observed for the Corn Belt farm. The more diversified farms—Colorado with cattle and wheat, Oklahoma with wheat-cotton and cattle, and the irrigated farm in Nebraska—spread labor across the year and use more total labor. But increased income depends on more than just selling more labor.

33–2.2 Economic Status

Because the data are for only 1 year, some caution must be exercised in developing economic implications. One way to evaluate the economic position of the farmer is to estimate net returns to the resources contributed. Three estimates are provided in Table 33–2: (i) net returns to land, other capital and family, and operator labor and management (e.g., $81 841 in the Palouse, Washington area); (ii) net return to nonland capital, family and operator labor and management ($49 600 in the Palouse area); and (iii) net returns to owned resources and family, and operator labor and management for the modal tenure arrangement in the production area ($69 209 in the Palouse area).

A typical farmer with the modal financial and tenure situation in any of the areas but Colorado appears to have had a viable income position in 1976. The income on the Kansas and Oklahoma farms was substantially lower than in other areas. The 1976 year was a below average year but not a disaster crop year in the Southwest. For example, wheat yields (reported in

Table 33–2. Economic characteristics of selected typical dryland farms, an irrigated Nebraska farm, and an Illinois Corn Belt farm, 1976b (Strickland and Fawcett, 1978).

	Palouse area, Washington	North central Montana	East central Colorado	Southwest Oklahoma	South central Kansas	South central Nebraska	East central Illinois
	Winter wheat	Wheat fallow	Winter wheat, fallow	Cotton, wheat, beef cow	Winter wheat	Irrigated	Corn Belt
Total area, ha†	520	1230	1295	390	260	260	160
Cropland, ha	505	1100	620	250	185	195	155
Pasture, ha	--	--	655	130	65	60	--
Enterprises							
Winter wheat, ha	225	380	290	170	135	30	--
Fallow, ha	120	550	290	--	--	--	--
Hay, ha	--	--	40	20	10	30	--
Miscellaneous, ha	80 (Peas)	80 (Spring wheat)	--	60 (Cotton)	40 (Sorghum)	--	75 (Soybean)
Corn, ha	--	--	--	--	--	135	80
Barley, ha	80	90	--	--	--	--	--
Beef cows, head	--	--	107	30	30	30	--
Stocker cattle, head	--	--	--	--	100	--	--
Capital investment, $	805 985	865 888	845 073	619 476	420 101	768 637	682 203
Land	595 200	681 120	(621 160)	(471 360)	(256 640)	(586 240)	(539 200)
Improvements, equipment, and machinery	210 785	184 768	(190 342)	(137 817)	(154 271)	(173 207)	(143 003)
Breeding livestock			(33 571)	(10 300)	(9190)	(9190)	

(continued on next page)

Table 33-2. Continued.

	Palouse area, Washington	North central Montana	East central Colorado	Southwest Oklahoma	South central Kansas	South central Nebraska	East central Illinois
	Winter wheat	Wheat fallow	Winter wheat, fallow	Cotton, wheat, beef cow	Winter wheat	Irrigated	Corn Belt
Gross farm receipts‡	146 115	111 768	46 914	63 870	42 523	137 130	107 463
Total cash costs§	45 846	40 978	31 673	30 633	19 259	42 783	35 247
Depreciation	18 428	14 528	13 489	9143	9479	12 610	10 885
Total cash costs plus depreciation	64 274	53 506	45 163	39 776	28 738	55 393	46 132
Net returns to land, other capital, family and operator labor and managment¶	31 841	56 262	1751	24 094	13 785	81 737	61 331
Net returns (after full land rent) to other capital, family and operator labor and management#	49 600	34 302	8824	9478	8707	35 447	28 097
Net returns under the modal tenure arrangement in production area††	69 209	45 437	−1862	13 657	11 225	55 536	36 808
Discretionary income for the modal tenure situation‡‡	49 967	33 614	3141	11 300	9364	36 071	28 000
Land value adjustment for modal tenure $/y§§	56 149	47 947	26 399	18 854	15 398	38 106	58 234

† Other land, e.g., farmstead, roads, and wasteland, made up the difference between total area and the sum of cropland and pasture.
‡ Gross farm receipts are the value of production sold and do not exclude share rent.
§ Total cash costs include cash production expenses, property taxes, hired labor, insurance, and other general farm cash expenses.
¶ No charge has been made for land, capital in improvements, equipment and machinery, and family and operator labor and management.
# A charge has been made for land as though it is all rented. Thus, the residual return is to other capital and family and operator labor and management.
†† The study estimated the modal financial and tenure position for farmers in the production area. This return reflects the residual after cash costs and depreciation are paid. It is the residual to owned resources and family and operator labor and management.
‡‡ Discretionary income is the residual cash remaining after cash costs, principal payments, and all taxes, including self-employment taxes. The value given is for the model tenure and financial situation found in the production area in 1976.
§§ The land value adjustment estimates the change in the market value of all real estate owned by the operator during the production year 1976.

Table 33-3. Labor distributions for selected typical dryland and nondryland farms in the USA for 1976.

	Palouse, Washington	North central Montana	East central Colorado	Southwest Oklahoma	South central Kansas	South central Nebraska	East central Illinois
	Winter wheat	Wheat, fallow	Wheat, fallow	Cotton, wheat, cattle	Wheat	Irrigated	Corn-soybean
	hours						
January	30	82	170	114	73	64	23
February	82	82	166	110	91	47	23
March	182	82	208	208	75	208	23
April	204	82	240	136	170	342	167
May	41	277	195	306	143	426	353
June	80	252	138	243	297	467	274
July	451	239	642	331	352	518	155
August	371	786	289	261	177	326	23
September	471	609	567	325	77	242	23
October	30	82	153	214	311	64	343
November	30	82	162	255	89	544	58
December	30	82	170	183	73	64	23
Yearly total	2002	2734	3099	2685	1928	3312	1488

Table 33–4. A comparison of economic performance of selected typical dryland and nondryland farms in the USA for 1976.

	Palouse, Washington	North central Montana	East central Colorado	Southwest Oklahoma	South central Kansas	South central Nebraska	East central Illinois
	Winter wheat	Wheat, fallow	Wheat, fallow	Cotton, wheat, cattle	Wheat	Irrigated	Corn-soybean
				$			
Returns to all land, other capital, family and operator labor and management	$81 841	$56 262	$ 1751	$24 094	$13 785	$81 737	$61 331
Accounting charge for family and operator labor and management	9035	12 366	11 585	13 431	10 747	13 277	6869
Returns per hour of family and operator labor	5.45	5.35	5.90	5.15	5.76	5.43	5.14
Returns to all capital	72 745	43 896	−9834	10 664	3038	68 470	54 471
Returns/$ to capital, %	9.0	5.1	−1.2	1.7	0.7	8.9	8.0
Land value adjustment	89 280	68 112	52 799	37 709	38 496	93 798	194 112
Returns/$ to capital, including a land value adjustment, %	20.0	12.9	5.1	7.8	9.9	21.1	36.4

bu/acre) were 17.3 in the Colorado area, 25.5 in Kansas, and 23.0 in Oklahoma. However, cattle prices were reported as less than $40/cwt and wheat less than $3/bu. Also, as mentioned earlier, the Kansas and Oklahoma farms may not be comparable in size with the Northwest farms.

The cash income position of each farm in 1976 is estimated by the item labeled discretionary income, which is cash remaining after cash costs, income tax, self-employment tax, and principle and interest payments. It reflects the residual return to the farmer's resources after all obligations are met except those associated with family living, education, vacation, new investments, etc. It can be used to judge ability of the farmer to survive a given year without further borrowing or other adjustments.

33–2.3 Economic Future

Over the longer run, owners of resources (capital, labor, and management) tend to use resources where they earn the highest return. One way to evaluate the current use of a resource is to compare returns with potential returns in other uses. Table 33–4 was used to partition returns to land and other capital, family, and operator labor and management. The analysis used the net returns to land, other capital, and family and operator labor. An accounting charge was made for labor and management, and the residual was designated as "returns to all capital." The before-tax returns to capital, without considering land appreciation, are marginally acceptable to an investor in three areas and low in the other four—all dryland. When returns are adjusted for increases in land values during the 1976 production year, the before-tax rate of return is attractive in three areas, marginally acceptable in two and low in two. The inflation in land value is speculative from year to year and the land must be sold to capture that return, unless a lender will allow a loan on the increment in land value.

33–3 COST AND RETURN RELATIONSHIPS FOR WHEAT

The budgets on which the USA typical farm study was based are for a specific year (ERS, 1978). However, they are useful for comparing returns and costs for the major enterprise, wheat, in dryland areas. Table 33–5 contains wheat enterprise budget summaries for five areas. Under the production section, yields and prices reflect market and production conditions in the area in 1976. Except for Washington, all budgets are for hard red winter wheat. Oklahoma and Kansas receipts include a value for wheat pasture.

Production costs in section 2 of Table 33–5 are summarized into four groups plus the total. The major difference between areas in preharvest variable costs is for fertilizer and pesticides. All labor is charged at the rate for hired labor. Variable harvest and ownership costs vary depending on the amount of custom combining. The land charge is based on the crop share rent prevailing in the area after subtracting the landlord's share of expenses.

Table 33–5. Cost and return relationships for winter wheat in selected dryland areas, 1976.

	Palouse area, Washington	North central Montana	East central Colorado	Southwest Oklahoma	South central Kansas
	Winter wheat on fallow	Winter wheat on fallow	Winter wheat	Winter wheat	Winter wheat
Production					
Yield, bu/ha	88.7	67.7	47.2	46.2	68.2
Price, $/bu	2.85	2.45	2.45	2.95	2.70
Value of pasture, $/ha	--	--	--	7.00	5.00
Total receipts, $/ha	252.71	165.81	115.57	153.54	196.41
Production costs, $/ha					
Variable, preharvest†	128.44	57.06	44.46	70.32	81.68
Variable, harvest‡	11.78	11.73	13.31	17.71	11.31
Ownership costs§	54.61	38.93	36.26	25.19	42.58
Land charge (share rent)	73.51	53.40	36.36	48.63	56.51
Total costs	268.34	161.12	130.39	161.85	192.08
Residual to management, overhead, and risk, $/ha	15.63	4.69	−14.82	−8.32	4.33
Estimated normal yield, bu/ha	124	74	57	64	84
Coefficient of variation, %¶	18	32	51	48	42
Central two-thirds yield range, bu/ha#	101–146	49–98	27–86	35–94	49–119
Residual return to management, overhead, and risk for the central two-thirds yield range, $/ha					
Low	10.05	−24.40	−46.44	−46.51	−41.45
High	95.07	54.02	47.25	77.51	94.50
Normal	51.94	22.70	0.44	24.16	67.63

† These costs include purchased inputs, machinery, fuel, lubrication and repairs, and all preharvest labor. Costs of fallow are included (ERS, 1978).
‡ Harvest costs include machinery and labor used by the operator and costs for the portion of the crop custom harvested.
§ Ownership costs are depreciation, taxes, interest, and insurance on the machinery used per hectare of wheat.
¶ Coefficient of variation is (standard deviation of the average yield/average yield) × 100.
# For illustrative purposes, a normal distribution is assumed in the one standard deviation range.

Total receipts minus total costs is the residual to management, overhead, and risk. In 1976, receipts covered all costs in only two of these five areas. A negative return means that owned capital or operator labor and management did not get paid a full opportunity cost.

The approximate, recent average or normal yield for each area is also given in Table 33–5. The 1976 crop yield was below normal in all five areas. To further describe the nature of yields in the production areas, an estimate of the coefficient of yield variation was obtained from previous research (Bostwick, 1963). The coefficient of variation is based on 1940 to 1962 yields and the assumption is that variation increased with yields so that the coefficient of variation remained the same.

The coefficient of variation can be used to make probability statements about the yield distributions. Given a coefficient of variation of X%, two thirds of the yields will be within 100 ± X% of the average yield, if we assume normally distributed yields. That range is given in Table 33–5 for each area. To reflect the income variations faced by dryland farmers, estimates of the residual return to management, overhead, and risk with the low and high of the range and the normal yield are provided in section 7 of Table 33–5. Of course, yield and income extremes exist on each end of the distribution beyond the range given.

33–4 ECONOMICS OF FARM SIZE

Many factors affect the size of farms. Technology of farming and economic conditions determine the size of farm needed to provide an adequate family income. Economies of size, or decreasing per-unit costs with increasing size, tend to prevail because of changing proportions of the inputs as size changes. Typically, the operator's management is a fixed quantity for the farm, and thus the management input per unit of output decreases as farm size increases. Also in a highly mechanized agriculture, such as that found in the dryland areas, adequate farm size must be attained to fully utilize the machinery. Hence, opportunities exist for cost savings through expansion (Madden, 1967; Heid et al., 1977).

The number of empirical studies of the economics of size in the dryland cropping acres of the Great Plains has been limited. A study by Krenz et al. (1974) is the most recent available. It reports results from a survey of 80 large-scale wheat farms in Montana, North Dakota, South Dakota, Colorado, and Kansas. The authors reported that 4860-ha wheat farms were using essentially the same machinery technology and machine size as 1215-ha farms, hence little or no economies were found in use of machines in this size range. The larger farms in some areas used more fertilizer and obtained higher yields.

Per unit production costs were similar from a size of 607.5 to 4860 ha. Below 607.5 ha, per-unit production costs are higher because of the use of smaller machines and more labor per hectare. If larger machines were used on the smaller farms, the investment costs were greater per hectare. Also, on the smaller farms the overhead and management costs were greater per hectare.

The major differences in costs were attributable to volume discounts on purchased inputs. The 1215-ha farms received average discounts of 6.6% on machinery purchases, 8.9% on machinery repairs, 3% on diesel fuel, and only 1% on fertilizer and herbicide. The 4860-ha farms averaged a discount of 9.5% on machine purchases and repairs, 14.5% on diesel fuel, 7.1% on fertilizer, and 4.3% on herbicides. These discounts reduced costs by approximately $2.47/ha at 1970 prices.

Premiums on prices received for wheat, due to volume, were reported as high as 10¢/bu with an average of 4.4¢/bu for the 4860-ha farms. This premium would amount to over $4000/y in added income.

Interactions between size and management incentives are demonstrated in regard to pecuniary economics. As size increases, chances improve to obtain volume discounts and the economic value of such discounts increases. The same is true for sales of wheat. For example, if a farmer has 5000 bu of wheat to sell he or she is not likely to find a buyer who will give a price premium for volume, and a 4¢/bu premium would earn only $200. On the other hand, with 150 000 bu, a farmer may get a 4¢ premium, which would be worth $6000. Hence as size increases, farmers find it profitable to spend more effort in marketing their product and buying their inputs.

Bostwick (1963) concluded that a relationship exists between size of farm and both average yield and yield variability. Average yield appeared to increase with farm size, and variability decreased. Farm unit dispersion necessitated by the larger farm is a major reason for lower variability.

The questions may be raised, "If economies of size exist, why aren't all farms big?" or "Why do smaller farms persist?" Krenz et al. (1974) suggested that some cost economies are exhausted as a certain size is reached. However, periodically technological developments shift this size to a higher level. Pecuniary economies may continue over a wider range as size is increased, or they may actually start only above sizes at which production economies have stoppped. Lack of uniformity of resources to be added to an existing unit, higher costs caused by spreading the farm over a larger area, and uncertainty about future conditions are deterrents to expansion according to Madden (1967). Some farmers have chosen part-time, off-farm, and custom farm work as an alternative to a larger farm.

Acquiring resources to enter farming or expand the farm unit to an economic size is a challenge in a period when resources may total $0.25 million to $0.75 million per modest size of farm (Minnick and Walker, 1976). Assumption of family interests through partnerships and eventual inheritances are the most common way of entering and growing. Careful business organization and estate planning are needed to assure that the older farmer's and the younger farmer's objectives are considered.

Direct acquisition of resources by a prospective farmer may include use of contract purchases of land, share or cash leasing of land, and custom hiring or renting of some machinery to conserve scarce capital resources. Share leases also reduce risks. Individual lenders are a major source of long-term capital, particularly loans associated with transferring land holdings to a different farmer. The Farmers Home Administration (FHA), in cooperation with the Federal Land Bank, is an important source of capital for low resource, beginning farmers.

33-5 RISKS AND UNCERTAINTY IN DRYLAND FARMING

Variability of production and income is a major characteristic of dryland farming. Managers must be ready to adjust practices and farming plans for weather conditions in a given production year. Drought periods within and across years are the major problem, along with hail and windstorms. Strong winds bringing dust or snow may occur in any season, particularly in the Great Plains.

It is also possible for dryland farmers producing wheat, feed grains, and cattle to get caught in wide price swings. These swings are attributable to many factors, including the ability of the dryland areas to oversupply those commodities in a period of good years and attractive prices. Governmental actions, world supply and demand for the commodities, international trade conditions, and technological developments contribute to uncertainties facing dryland farmers and the agricultural businesses and communities serving them.

Dryland farming in most areas has several critical time periods within which tilling, pest control, seeding, and harvesting need to be accomplished. Problems with machines, inputs, or labor in those periods may have serious effects on income. Farmers must live with and manage around uncertainties of weather disruptions, machinery breakdowns, input shortages, and labor vagaries.

Management under uncertainty has been the subject of several research studies by agricultural economists working in the Great Plains and the West (Walker and Nelson, 1977). Part of the research attention has been on developing data for use in decision making, e.g., estimates of yield distributions (means, variances, and skewness) (Larson and Thompson, 1966). Other studies have concentrated on the decision makers in attempts to understand their goals and attitudes toward risks (Harmon et al., 1972). Much effort has been devoted to developing tools for decision making under uncertainty, including specifying alternative actions and possible outcomes and choosing actions that are suited to one or more farmer goals (Halter and Dean, 1971). In the following sections, decision data, farmer's risk attitudes, and guides for decision making under uncertainty are reviewed.

33-5.1 Improved Knowledge Environments for Decisions under Uncertainty

One way to assist decision makers in an uncertain environment is to provide technical alternatives that reduce uncertainty and control or shift variability. Research and extension programs that increase understanding of interrelationships among key variables affecting production are proving useful. For example, the goal of work reported by Mapp et al. (1975) is to model the plant growth process under a wide range of conditions and to design a system of production practices that maximizes economic benefits. Farmers are using results from these studies to monitor plant and soil conditions and make irrigation decisions in the Southern Great Plains.

Irrigation developed in dryland areas because the expansive flat areas and available water were ideal and profitable, particularly when aided by governmental water projects, and the practice removed effects of rainfall variability. Shelterbelts, contour, and terraced farming; stubble mulch equipment; big, fast machinery; and drought and pest tolerant varieties are other examples of technological developments that reduced farm income variation.

Federal farm programs that attempt to lessen the burden of risk in crop production, such as federal crop insurance, disaster loans, disaster payments, and conservation programs, are well known. In Manitoba, Canada, a rancher may contract to be part of a governmental beef income assurance plan designed to "stabilize the producer's income and encourage production of more finished beef" (Agric. Faculty, 1977).

33-5.2 Farmers' Risk Attitudes

Logically, managers' feelings about taking a chance in business affect their choice of a vocation, place to farm, farm organization, investment decisions, and production practices. Harmon et al. (1972) evaluated the relative importance of the eight goals in Table 33-6. The three goals relating closest to risk avoidance—"avoid being forced out of business," "avoid years of low profits or losses," and "reduce borrowing needs"—rated high with farmers in the Great Plains study area. Other studies indicate that farmers individually and in aggregate tend to act as risk averters (Just, 1978; Lin et al., 1974). However, individual risk takers are found. Age, financial position, and family obligations, among other variables, affect the degree of risk aversion.

The implication of multiple goals, including risk avoidance, is that farmers need information about the whole distribution of returns rather

Table 33-6. Respondents' ranking of selected goals; first and last (Harmon et al., 1972).

Goal	Respondents preferring goal the least	Respondents preferring goal the most
	Number	
Control more hectares	38	12
Avoid being forced out of business	9	21
Maintain or increase family living	3	32
Avoid years of low profits or losses	4	26
Increase leisure time	67	6
Increase net worth	5	16
Reduce borrowing needs	11	18
Make most annual profits	3	38
Total	140	169

than just average or expected returns. They also need decision rules and analytical procedures that are useful for their diverse goals, risk attitudes and uncertain environment.

33–5.3 Guides for Decision Making under Uncertainty

Many strategic principles and tools are part of everyday use in dryland farming. Choice of sure (less variable) enterprises, diversification of enterprises, use of financial and physical reserves, maintenance of enterprise and practice flexibility, farm unit dispersion, forward pricing, and insurance have applicability in dryland farming.

33–5.3.1 Diversification

The admonition to not "put all your eggs in one basket" is familiar to all. Producing a combination of enterprises rather than one may be motivated by technical and economic reasons or concern for risk. If diversification is achieved by dividing the farm resources among more enterprises, income variance can decrease even if correlation is positive. If variance of the new enterprise is less than that of the old, income variance will always decrease. However, if the new crop is less profitable, income must be sacrificed to reduce income variance. A farmer may diversify by adding a new enterprise along with the necessary new resources to an existing farm operation. Total variance of income will increase if correlation between the two enterprises is positive, and variance will decrease if correlation is negative.

33–5.3.2 Reserves

Financial reserves allow a farmer to borrow in one or a series of bad years to survive. The most economical way to maintain reserves is as a credit reserve rather than cash. The manager keeps the debt/net worth ratio (leverage) low enough that the creditor is willing to make additional loans if they are needed. The opposite would be to borrow as much as lenders will allow. If a bad year occurs, the farmer would have a difficult time getting funds for expenses.

For example, Farmer Jones has assets of $400 000 and has borrowed $200 000 for operating expenses and equipment. His banker insists on a minimum net worth to asset ratio of 1:4 (25% equity). Thus, he could borrow an additional $100 000 if necessary to take him through a series of bad years. This $100 000 is his credit reserve. Jones's neighbor, Smith, has the same amount of assets but has borrowed $300 000 for expansion in land and equipment. If a series of adverse events occur, Smith has no credit reserve and may become insolvent. Jones may have to give up income to maintain a financial reserve. The manager must decide how he wants to trade—higher leverage (debt/equity), potential returns and risks vs. lower leverage, returns, and risk of sacrificing liquidity.

Physical reserves also have a place in dryland farming. Hay and other feeds, water supplies, soil moisture (e.g., from fallowing), excess labor and machinery capacity, and input supplies are examples. The economic problem concerns whether it is cheaper to maintain the physical reserve or go into the market in a crisis to buy what is needed. The farmer would be competing with neighbors for possibly limited supplies in those times.

Unsold grain in the bin may also serve as a reserve. Usually unsold grain is kept as a marketing strategy in the anticipation that grain prices may be higher at some future date; however, some farmers hold actual grain as a reserve against drought. If the operator is on a cash basis for income tax purposes, no tax has to be paid on the grain until the grain is actually sold. In a drought year, enough grain can be sold to meet expenses and hence, income taxes are minimized and the grain has served as a reserve. If grain prices were steady over time, because of the tax treatment, grain would be a better reserve than cash.

33–5.3.3 Flexibility

Flexibility is the ability to modify enterprises, production practices, and production resources to offset effects of short-term economic and production conditions. Dryland farmers may desire to plant a "catch" crop, find a substitute for drought-stricken pastures, adjust cattle numbers quickly, reduce farm expenses when a crop fails, and store wheat on the farm rather than sell. Farming plans featuring good flexibility may be less profitable. Again, the manager must analyze trade-offs and make decisions that suit the operation.

For example, a common practice of range-livestock producers is to adjust their strategy between selling calves and selling yearlings. This adjustment can be used for changes in price and changes in feed supply. However, sometimes feed conditions indicate that one should sell calves, and price conditions indicate that one should sell yearlings.

33–5.3.4 Dispersion

A Montana study gave modest empirical support to the idea that farm unit dispersion in a dryland area of the Great Plains would reduce yield variability for the farm (Bostwick, 1963). The logic is that storms with intermittent rain and hail have a fairly narrow directional pattern across a given area. Separate storms move in parallel across the farming area, creating a spotty pattern of rainfall and hail effects. Dispersion of the farm unit in a north-south direction in an area with east-moving storms gives a better chance of rain somewhere on the farm. At the same time, there is less chance the whole farm will have a hailstorm (or rainstorm). The farmer must decide whether the risk reduction from dispersion is worth added costs of travel and coordination for the dispersed farm. Those familiar with large-scale dryland farming will understand that many farms are widely dispersed because farmers take land where they find it, and travel in the wide open spaces is not too costly. Perhaps they are also taking advantage of the reduced risk.

33–5.3.5 Forward Pricing

Future prices of products or inputs can be made virtually certain by contracting or hedging. For example, the buyer of the forward contract or futures contract of a product accepts the risk of price variation. The seller retains risks of yield and quality variation and, in a hedge, "basis" variation. The basis is the difference between the cash and futures price and is particularly critical at selling time. A farmer using forward pricing gives up opportunities for favorable price movements while avoiding unfavorable ones.

The major drylands commodity, wheat, is traded on three different exchanges: soft red winter wheat on the Board of Trade of Chicago; hard red winter wheat on the Board of Trade of Kansas City; and spring wheats on the Minneapolis Grain Exchange. Both fat and feeder cattle are traded on the Chicago Mercantile Exchange. The volume of trade in some markets has increased nearly twofold since 1970. Grain merchandizers, processors, and speculators trade much more heavily than producers.

33–5.3.6 Insurance

Insurance is another means of shifting risk with the cost being the insurance premium and its opportunity cost. A farmer may buy life, liability, fire and windstorm, property damage, disability, medical, hospitalization, and crop insurance. Federal all-risk crop insurance is available in most dryland counties for some of the following important dryland crops: wheat, barley, cotton, grain sorghum, and oat. Guaranteed yields and insurance costs per hectare are established for the insurable areas in the counties. The farmer can determine his dollar level of coverage by the crop yield and price level he selects. The intent of the federal crop insurance program is to cover enough of production costs so that a farmer can survive a low yield.

Crop hail insurance is sold by commercial companies with rates established for townships using actuarial data. For example, in Oklahoma, rates range up to $15/$100 of coverage. Farmers may choose a deductible policy so that they are protected for losses above some minimum level. Insurance can be canceled by a preset date in case of a crop failure.

The decision to insure or not depends on the chance of a loss, consequences of a loss, and cost of insurance protection, all as perceived by the decision maker. Individuals view risk differently. In general, the farmer should (i) insure against losses that may result in financial ruin, (ii) establish most critical insurance needs and fill them first, and (iii) not insure risks that are damaging but not major financial disasters.

Table 33-7. A hypothetical payoff matrix for alternative dryland grain farm organizations.

Events	Actions: Diversified farm plan A_1	Specialized farm plan A_2	Flexible farm plan A_3	P_i
	$ return to equity, operator, labor, and management			
s_1 Winter rainfall most favorable	19 500	5000	21 000	0.2
s_2 Normal rainfall	24 000	32 000	21 500	0.6
s_3 Summer rainfall most favorable	10 000	15 000	14 000	0.2

33-5.4 Decision Theory

A useful framework for risky decision making suggests that the decision maker evaluate possible *outcomes* of alternative *actions* under different *events* that might occur during the production and marketing period (Nelson et al., 1978). The events are not under the control of the decision maker, and their occurrence is uncertain. The outcome of each action for each event is termed a payoff, and a table showing the actions, events, and payoffs is called a *payoff matrix.* Table 33-7 provides a hypothetical example. The events are rainfall conditions that may prevail. For simplicity, price variation is not included in the events. The actions reflect three alternative organizations of a dryland grain farm. One plan is designed to be as diversified as practical; one specializes in a single crop (e.g., grain sorghum); and the third may be diversified or specialized but maintains input, product, and cost flexibility. In the example, the payoffs are chosen to reflect logical results. A farmer would need to carefully budget payoffs for each event and action combination. Subjective probabilities in Table 33-7 reflect the decision maker's information and judgment about the likelihood of each event.

Table 33-8 describes possible action choices the farmer might make, depending on the decision rule used. If the farmer is not concerned about

Table 33-8. Alternative decision criteria and actions for the dryland grain farm decision problem.

	Action	Outcomes: Maximum	Expected	Minimum
Choose the action that will:				
A. Maximize the expected return	A_2	32 000	23 000	5000
B. Maximize the worst return (Maximum)	A_3	21 500	19 900	14 000
C. Maximize the maximum possible return	A_2	32 000	23 200	5000
D. 1. "Assure" $10 000 or more and 2. give the highest expected return	A_1	24 000	20 300	10 000
E. Maximize the probability that income is $12 000 or more	A_3	21 500	19 900	14 000

occasional $5000 income years, he or she could choose criterion A and maximize expected return. The very conservative farmer might choose A_3 to "assure" at least a $14 000 income (criterion B). Action 2 might be chosen to provide a chance for the $32 000 return (criterion C). A multiple criterion could be used as in criterion D. The decision maker first assures a minimum acceptable return by eliminating A_2 and then chooses the remaining action that gives the highest expected return (A_1). This is a type of safety-first decision rule as is E (Roumasset, 1976). Criterion E chooses the action that gives the best chance of the $12 000 minimum income desired, A_3. Neighboring farmers may use different farming plans and practices because their feelings about risk and criteria for decisions differ.

Decision theory provides a procedure that farmers can use on important risky decisions. If farmers are to use such tools, agricultural research and extension institutions need to supply data concerning outcomes of production practices over the entire range of technical and economic conditions faced by dryland farmers.

33-6 GOVERNMENT PROGRAMS

Several factors work together to make the dryland agricultural areas of the Great Plains and West unique in regard to government programs. The fact that one crop, wheat, is so important to income gives the area a particular dependence on the price and yield of this crop. Since 1960, approximately 55% of the total U.S. wheat production has been exported, with exports ranging from 35% of the annual crop to 74% (ESCS, 1978). This dependence on the export market contributes substantially to wheat price variability. Thus, U.S. agricultural programs and policies affecting grain production, storage, and export are of keen interest to the dryland farmers of the Great Plains and West. Arguments exist as to which group benefits from price stability, farmers or consumers, but there is little doubt that government programs are of great concern to the wheat farmer (Miller, 1975).

33-6.1 Supply and Price Programs

The common reoccurring problem has been a tendency for wheat supply to exceed the domestic demand, with resulting surpluses and low prices. This problem has been "dealt with" through programs to reduce supply through various hectarage controls or allotments and inducements to divert or set aside land from production. For a brief history of government commodity programs, see Tweeten (1970). Efforts to increase demand include food stamp programs, school lunch programs, and increasing foreign demand through Public Law 480 special trade agreements.

Management of wheat stocks has been an important aspect of the government programs. Stocks of wheat can serve as a buffer against both low and high prices and against food shortages in the world economy (Walker and Sharples, 1975).

33-6.2 Yield and Crop Disasters

Two major government programs have been used to reduce the impacts of yield variability (Miller and Walter, 1977). The Federal Crop Insurance Corporation (FCIC) was created in 1938 to insure crops against losses from a wide variety of causes. Its programs were updated by Congressional action in 1980. The FCIC is partially supported by the federal government through appropriations limited to $12 million/y for administration and through a premium subsidy to farmers.

The Agricultural and Consumer Protection Act of 1973 contained provisions for a Disaster Payments Program, which permits increased income deficiency payments if production falls below two thirds of a farm's normal production. Under this program approximately $540 million was paid out in 1974, with 62% going to farms in 10 Great Plains states (Miller, 1975). Changes in FCIC in 1980 were designed to replace the disaster program, with crop insurance being available for all crops in all farming areas.

33-6.3 Conservation Programs

In addition to cost-sharing programs available in all areas of the country, a special Great Plains Conservation Program has been in existence in 10 Great Plains states since 1956. The Great Plains Conservation Program differs from the other programs in that it enters into long-term (3 to 10 years) contracts with cooperating farmers and ranchers. The program is available in about 400 counties in the 10 Great Plains states in an area from about the 98th meridian on the east to the foothills of the Rocky Mountains on the west. Under this program, major expenditures have been made for improving vegetation on existing range land, for water management, installing irrigation practices, and regrassing cropland (ERS, 1970).

36-7 ENERGY

With rising energy prices it is appropriate to consider the energy requirements of dryland agricultural production as it may affect future production and comparative advantage. The dryland areas of the Great Plains use comparatively low energy per hectare of crop because of the extensive agriculture involved. As a result of low rainfall, low levels of fertilization are practiced on crops. Partially because of low rainfall, low levels of chemical pesticides are needed.

Table 33-9 presents some energy estimates for the production of wheat in two dryland areas, under irrigation, and in two more humid areas. The main energy use in dryland wheat production is power for tillage. Some small quantities of pesticides are used in all of these wheat producing areas but available data only describe costs and not physical quantities. The costs

Table 33-9. Energy and cost comparisons for wheat, 1976, assuming average production practices in each area.†

Item	Units	Winter wheat on fallow, western Kansas	Spring wheat on fallow, northeastern Montana	Irrigated wheat, west Texas High Plains	Winter wheat, eastern Kansas	Winter wheat, southern Illinois
Yield	bu/ha	49.9	70.6	82.5	62.5	90.2
Fertilizer use						
N	kg/ha	13	--	88	67	55
P_2O_5	kg/ha	--	1.8	11	47	54
K_2O	kg/ha	--	--	22	15	44
Ancillary energy in fertilizer‡	BTU × 10^3/ha	740	15	5049	4342	3915
Fuel use						
Gasoline	gallon/ha	9.26	9.68	7.98	9.41	7.90
Diesel	gallon/ha	11.39	11.09	18.92	9.85	8.40
Natural gas	MCF/ha	--	--	42.56	--	--
BUT's in fuel§	BTU × 10^3/ha	2752	2764	48 338	2556	2164
BTU's	BTU × 10^3/bu					
In fertilizer		14.9	0.2	61.2	69.5	43.4
In fuel		55.2	39.1	585.9	40.9	24.0
Variable and machinery ownership cost	$/bu	1.86	1.60	4.10	2.23	1.78

† Wheat budgets are file no. 227, 405, 441, 1015, and 1220, USDA-FEDS system (ERS, 1978).
‡ Davis and Blouin (1977).
§ Sloggett (1977).

of pesticides are slightly lower per bushel in the dryland areas. Calculations were not made of the energy embodied in machinery. Water pumping is the primary energy user on irrigated wheat. In the two dryland wheat areas described in Table 33–9, the common wheat production system involves use of summer fallow, and the costs of tilling summer fallow have been included.

Energy requirements for fertilizer are much lower, both per hectare and per bushel, in the two dryland areas here examined. Energy requirements per acre for tillage are comparable for all areas, but on a per bushel basis, energy requirements for tillage are higher in the two dryland areas.

Without question, rising energy prices place irrigated wheat at a distinct disadvantage. The difference in per-bushel energy requirements between the dryland and the humid areas is too small to indicate that either the dryland or the humid areas have any distinct advantage. Costs of all inputs will affect future location of wheat production. Currently the irrigated wheat area is at a definite cost disadvantage. Costs per bushel change considerably from year to year in the dryland areas because of varying yields. Hence 1 year of results is inconclusive in determining comparative advantage.

Table 33–10 presents a similar comparison for two dryland and two other grain sorghum–producing areas. In the two dryland areas, grain sorghum is produced in extensive systems. Fertilizer usage is low on a per hectare basis, but the energy requirement in fertilizer on a bushel basis is similar in western Kansas to the other areas examined.

Table 33–10. Energy and cost comparisons for grain sorghum, 1976, assuming average production practices in each area.†

Item	Units	Dryland western Kansas	Dryland Texas High Plains	Irrigated Texas High Plains	Missouri
Yield	bu/ha	56.3	54.8	173.6	144.5
Fertilizer use					
N	kg/ha	44	11	139	73
P_2O_5	kg/ha	44	6.6	5.5	29
K_2O	kg/ha	1.1	--	--	31
Ancillary energy in fertilizer‡	BTU × 10^3/ha	2514	677	7830	4471
Fuel use					
Gasoline	gallon/ha	5.80	12.84	6.84	19.71
Liquid propane	gallon/ha	0.54	--	--	6.27
Diesel	gallon/ha	11.46	17.04	18.40	8.13
Natural gas	MCF/ha	--	--	58.51	--
BTU's in fuel§	BTU × 10^3/ha	2381	3992	64 870	4197
BTU's					
In fertilizer	BTU × 10^3/bu	44.6	12.4	45.1	30.9
In fuel		42.3	72.8	373.6	29.0
Variable and machinery ownership cost	\$/bu	1.84	2.05	2.42	1.36

† Sorghum budgets are from the USDA-FEDS system (ERS, 1978).
‡ Sloggett (1977).
§ Davis and Blouin (1977).

The energy use for tillage is similar in all areas on a per-hectare basis; however, on a per-bushel basis the dryland areas show high energy requirements compared to the more humid area. The energy requirement for irrigation again places irrigation at a distinct disadvantage with rising energy prices, and irrigation farmers face adjustments (Skold, 1977).

Dryland farmers and agricultural researchers have always had a strong interest in farming systems that conserve water and energy (the number of times they must move machinery over the field) and give acceptable economic results. Such systems have been named "stubble mulch," "minimum tillage," and simply "fallow."

Technology has provided chemicals that aid in weed control, and a current system of considerable interest is called "ecofallow." The latter leaves crop residue on the field after harvest, controls weeds by chemical means if necessary, and allows the land to be idle for a year to add moisture. Its potential economic benefits center on yield gains from moisture and cost savings derived from reduced mechanical tillage. Clearly, the system must be used in a production area where fallow pays as a practice. If one-half of the land is fallowed, yields per hectare must double or decreased receipts must be offset by decreased costs. Also, new systems require development of management skills sufficient to successfully carry out the practices required.

33-8 SHIFTING BACK TO DRYLAND FARMING FROM IRRIGATION

Table 33-1 shows that irrigation is important in terms of total hectares and numbers of farms in many of the dryland farming regions. Because of the substantially higher yields, the magnitude of total production is especially significant in the irrigated hectares. The irrigated land through much of the Plains uses high quality irrigation water from the Ogallala formation. Recharge of the aquifer is negligible and the stock of water is "mined" by the users. The economic life of the water stock will be exhausted at some time, depending on the price of agriculture products, the depth to water, and cost of energy.

Studies by Hughes and Harmon (1969) estimate that the adjustment period back to dryland will range from 5 to 50 years, depending on water in a particular region, the economy and technology. They estimated that, in the Texas High Plains, irrigated hectares would decrease from 1.4 million ha in 1966 to 50 625 ha in 2015. Substantial declines in the volume of agricultural production and farm income would result, and the structure of farms and agricultural business would change. Hughes and Harmon (1969) wrote

"Thus the adjustments have important individual, area, state, and national implications. The projected decline in net farm income poses some difficult adjustment problems for farm operators. They must adopt a more remunerative crop or expand the size of farm operations. The prospects for adjusting through an increase in farm size are more promising, but even they have uncertainties. Current land values, high interest rates, and the scarcity of farm labor are deterrents to an increase in farm size. These deterrents, along with declining farm income, strongly suggest that current land values are likely to be among the early casualties in the process of adjusting to declining water supplies.

"The economic strategy for a farmer in the situation described is to (i) use existing water in as economically efficient manner as possible, (ii) decide the period of time over which the use of water resources is profitable, and (iii) make plans to move into the realities of dryland farming (or out of farming in this area). Because water rights belong to the landowner who claims and uses them first, the farmer has an incentive to take all the water he can profitably use every year."

The visitor to intensively irrigated areas that have a declining water table must be impressed with the bustling trading centers, prosperous business serving agriculture, high employment, and the high income. The prospects for the future are fewer and larger farms, less total area income and production, and fewer businesses and governmental services. Dryland farming will increase in these areas in the future. That is a certainty if the areas are farmed at all.

33-9 THE FUTURE OF DRYLAND FARMING

Vast, tillable spaces that offer abundance in some years and the "good life" in most years assure the future of dryland farming. Several important variables are on the positive side of dryland agriculture's future. The drylands are highly adapted to mechanization and extensive production of the food and feed grains. Large farm units offer size economies in production and in buying and selling. These advantages fit well with the kinds of technological advances that have occurred in agriculture and might be expected in the future.

The energy needs in dryland farming compare favorably on a per unit of production basis with those in the higher yielding, humid areas. Dryland farming has distinct advantages over irrigation farming with respect to energy requirements. Energy inputs should not be emphasized alone, because costs to all inputs are the major determinants of production advantages. If energy costs increase substantially, even transportation costs of moving inputs and products to and from the dryland regions would increase rapidly. However, it appears that in a relative sense the dryland areas may compete well with the humid areas.

The possibility of one or several good crop years and a high payoff keeps many in dryland farming. In some cases, the farmer may not have to risk a large fortune in quest of the big payoff, if the individual has a highly leveraged financial position. If the farmer has bad years, the individual moves elsewhere.

The long-term outlook is for expanding international markets. These markets arise from economic development in other parts of the world, population increases, and growing needs for food and fiber. Such growth provides an underpinning for dryland agriculture's food and feed grain production. However, there are no assurances of stable dollar demand for products and high economic returns.

The rural areas of dryland farming regions may be aided by the tendency in the USA for an urban to rural movement. Increased population and some new industry can help support social capital investments that are now either lacking or must be borne by the sparse populations. Roads, medical services, schools, and other forms of social services could be improved and made available at a lower cost as a result of growth in the population. However, careless and overly optimistic planning for growth could increase costs.

Technology of agricultural production has always had an important influence on the productivity of dryland agriculture. Private and public agricultural research is important in the development of technology. Prospects for the future include improvements in water use either by mechanical means associated with tilling practices or on the biological side by improved understanding of growth processes and increased genetic adaptation to dryland conditions. Production costs may be lowered as a result of technology, e.g., through the use of new chemicals, tillage techniques, and decreases in labor. Some refinements in product uses may also arise from technology. Research may lead to improvements in managerial skills, particularly ability to manage under uncertainty. All of these factors are positive for dryland farming.

On the troublesome side of dryland farming are significant variables. Price and yield variation are perennial problems. Periodically many farmers will have rough economic times. Many kinds of farm programs will continue to be designed to reduce the inefficiency caused by uncertainty and to assist people who are affected by that uncertainty.

The capital requirements for dryland farming are large. It is currently almost impossible for an individual to put together enough resources to enter farming in a scale sufficient for economic survival. Research is needed on ways that entry can be made easier and talented managerial skills can be brought into agriculture. Large capital requirements may hasten the entry of corporate forms of business with absentee ownership of resources by stockholders. However, considerable sentiment exists for farmily farms, family farm corporations, and maintenance of an atomistic flavor in agriculture.

In summary, the prospective producer must focus on the variability of the economic returns in dryland farming and the large capital requirements necessary to operate a viable dryland farm. The research worker's focus will most certainly be on increasing the level and stability of production from given resources by improving technology, plant materials, and understanding of the production process. Finding ways that scarce water can be used more efficiently is a priority. In economics research, attention should be on managerial procedures and the market and institutional arrangements necessary to acquire the resources for large-scale dryland farming, achieve efficient resource use through time, effect intergeneration resource transfers, and plan under uncertainty.

33-10 LITERATURE CITED

Agriculture Faculty. 1977. Principles and practices of commercial farming. Univ. of Manitoba, Winnipeg.

Bostwick, D. 1963. Studies in yield variability. Montana Agric. Exp. Stn. Bull. no. 374.

Bureau of Census. 1974. Census of agriculture. U.S. Dep. of Commerce, Bureau of Census, U.S. Government Printing Office, Washington, D.C.

Davis, C. H., and G. M. Blouin. 1977. Energy consumption in the U.S. chemical fertilizer system from the ground to the ground. p. 315–333. *In* W. Lockeretz (ed.) Agriculture and energy. Academic Press, Inc., New York.

Economic Research Service. 1970. An economic evaluation of the Great Plains conservation program. USDA-ERS. Agric. Economic Report no. 440.

Economic Research Service. 1978. Costs of producing selected crops in the United States—1976, 1977 and projections for 1978. USDA-ERS, Senate Committee on Agriculture, Nutrition and Forestry Committee. U.S. Government Printing Service, Washington, D.C.

Economics, Statistics, and Cooperative Services. 1978. Wheat situation. USDA-ESCS Report no. WS-244.

Halter, A. N., and G. W. Dean. 1971. Decisions under uncertainty with research applications. South-Western Publishing Co., Cincinnati.

Harmon, W. L., R. E. Hatch, V. R. Eidman, and P. L. Claypool. 1972. An evaluation of factors affecting the hierarchy of multiple goals. Oklahoma Agric. Exp. Stn. Tech. Bull. no. T-134.

Heid, W., H. M. Neghassi, and F. Walters (ed.) 1977. The U.S. food industry: Description of structural changes. Colorado State Univ. Tech. Bull. no. 129.

Hughes, W. F., and W. L. Harmon. 1969. Projected economic life of water resources, subdivision number 1 High Plains underground water reservoir. Texas Agric. Exp. Stn. Tech. Monogr. no. 6.

Just, R. E. 1978. The welfare economics of agricultural risk. p. 1–19. *In* Market risks in agriculture: Concepts, methods and policy issues. Texas A&M Univ. Tech. Report no. 78-1.

Krenz, R. D., W. G. Heid, and H. Sitler. 1974. Economics of large wheat farms in the Great Plains. USDA-ERS Agric. Economic Report no. 264.

Larson, D. K., and L. S. Thompson. 1966. Variability of wheat yields in the Great Plains. USDA-ERS Economic Report no. 287.

Lin, W., G. W. Dean, and C. V. Moore. 1974. An emperical test of utility versus profit maximization in agricultural production. Am. J. Agric. Econ. 56:497–508.

Madden, J. P. 1967. Economies of size in farming. USDA-ERS Agric. Economic Report no. 107.

Mapp, H. P., Jr., V. R. Eidman, J. E. Stone, and J. M. Davidson. 1975. Simulating soil water and atmospheric stress-crop yield relationships for economic analysis. Oklahoma Agric. Exp. Stn. Tech. Bull. no. T-140.

Michener, J. A. 1974. Centennial. Random House, Inc., New York.

Miller, T. A. 1975. Testing the Plains uniqueness hypotheses relating to agricultural instability and policy needs. p. 47–56. *In* Stability: The continuing quest. Great Plains Agric. Coun. Pub. no. 74, Univ. of Nebraska, Lincoln.

Miller, T. A., and A. S. Walter. 1977. Options for improving government programs that cover crop losses caused by national hazards. USDA-ERS Agric. Economic Report no. 659.

Minnick, D. L., and O. L. Walker. 1976. An assessment of alternative credit sources for low resource, beginning farmers. Oklahoma Agric. Exp. Stn. Bull. no. B-728.

Nelson, G. A., G. L. Casler, and O. L. Walker. 1978. Making farm decisions in a risky world: A guidebook. Oregon State Univ. Extension Service.

Roumasset, J. A. 1976. Rice and risk. Elsevier North-Holland, Inc., New York.

Skold, M. D. 1977. Farmer adjustments to higher energy prices. USDA-ERS no. 663.

Sloggett, G. 1977. Energy used for pumping irrigation water in the United States, 1974. p. 113–130. *In* W. Lockeretz (ed.) Agriculture and energy. Academic Press, Inc., New York.

Strickland, P. L., and D. Fawcett. 1978. Selected typical farming operations in the United States. Unpublished manuscript. U.S. Dep. of Agric., Economic, Statistics, and Cooperative Services.

Tweeten, L. G. 1970. Foundations of farm policy. Univ. of Nebraska Press, Lincoln.

Walker, O. L., and A. G. Nelson. 1977. Agricultural research and education related to decision-making under uncertainty: An interpretative review of literature. Oklahoma Agric. Exp. Stn. Res. Report no. P-747.

Walker, R. L., and J. A. Sharples. 1975. Reserve stocks of grain: A review of research. USDA-ERS Agric. Economic Report no. 304.

34

J. Allen Williams, Jr., Lynn K. White, and David R. Johnson
University of Nebraska
Lincoln, Nebraska

Sociology

34-1 PURPOSE

The physical environment of the Great Plains region in the USA is unique to the nation. It presents a set of climate and land conditions so extreme that for many years it was known as the "great American desert." As late as 1823, Major Long of the Army Engineers reported that most of the land between the Missouri River and the Rocky Mountains "is almost wholly unfit for cultivation, and of course uninhabitable by a people depending on agriculture for their subsistence." Today, by adapting techniques to fit a semiarid location and through the development of new technologies, the Great Plains has become one of the most productive agricultural areas in the world.

This chapter will examine important social characteristics of the people living on the Great Plains. The guiding question in the preparation has been whether the physical nature of the region and subsequent pattern of settlement has led to any significant differences between the inhabitants of the area and those living elsewhere in American society. How, if at all, has living in a semiarid or dryland region affected the quality of life of its people? Do they have any special problems? Do they have any special advantages?

These questions are addressed by looking at the demographic characteristics of the Great Plains, the quality of life on the Great Plains in comparison with other regions in the nation, the characteristics, quality of life, and problems of the Great Plains farming population. Comparisons are made between farmers living in the Great Plains with farmers living elsewhere in the nation, between those using dryland methods with those who use irrigation, and between farmers and urban dwellers. General conclusions and suggestions for further study will be discussed in the final section.

Dryland Agriculture—Agronomy Monograph no. 23.

34-2 DEFINITION OF THE GREAT PLAINS REGION

The major dryland agricultural area in North America (the Great Plains) lies approximately between the eastern slope of the Rockies and the 100th meridian. Its southern border is in the Texas Panhandle, and it extends north into the Canadian provinces of Alberta and Saskatchewan. Smaller dryland areas outside the Great Plains are in eastern Washington, north central Oregon, the central part of California, southeastern Idaho, and northern and southeastern Utah. Discussion will focus on the states of Montana, Wyoming, North Dakota, South Dakota, Nebraska, Kansas, and Oklahoma, and to simplify discussion, these seven states are referred to as the Great Plains region. The other three Great Plains states are Colorado, New Mexico, and Texas. Because social and demographic data are most often collected within designated political boundaries, e.g., nations, states, counties, it is not practical to include the dryland portions of Canada and the areas outside the seven Great Plains states in the USA. The demographic and social characteristics of people living in dryland areas outside the seven states here defined as the Great Plains region may be similar, but this is an assumption that may or may not prove to be true through additional research. Other regions of the nation, exclusive of the Great Plains states, that will be used for comparative purposes are the U.S. Census Bureau's East, South, Midwest, and West regions (U.S. Bureau of the Census, 1970).

34-3 POPULATION CHARACTERISTICS[1]

34-3.1 Settlement

Initial westward expansion of agriculturists in the 19th century stopped abruptly at the Missouri River, held back by the belief that the climate of the Great Plains rendered it unsatisfactory for farming. In 1860 no state or territory in the Great Plains had a population density of even one person per square mile.

The Homestead Act of 1862 opened millions of hectares of land to settlement by making it possible for a person to obtain 61 ha at little cost. All that was required for ownership was to live on a claim for at least 5 years and to improve it through cultivation. Another important source of low-cost land was made available by railroad companies. The federal government had granted the companies alternate blocks of land in a strip 17 km wide on either side of their rights of way to be used to finance construction of railroads across the Great Plains. These companies and newly formed boards of immigration set up by state and territorial governments made great efforts to attract settlers. Often cooperating with ethnic and religious

[1]All findings presented in this section are based on data provided by U.S. Bureau of the Census (1900, 1970, 1975, 1977).

societies, they advertised the Great Plains as an agricultural Eden, a region of fertile and rockless soil.

The availability of inexpensive land on the Great Plains coincided with a time of extensive immigration into the USA. Between 1870 and 1900, the nation received nearly 13 million people from European nations. Along with Americans from the eastern part of the country, many immigrants moved onto the Great Plains. For example, in 1900, immigrants and persons of foreign parentage composed 47% of Nebraska's white population and made up 78% of the white population of North Dakota. Excluding Indians, the population of the Great Plains grew from only 0.531 million in 1870 to 4.383 million in 1900, an increase of more than 700%.

Development of the Great Plains region in the early years of settlement owes a great deal to the foreign born and their children. This is especially true in agriculture. Although they tended to form somewhat different kinds of settlement patterns, depending on their national origin, these ethnically diverse people usually arrived on the Great Plains as family units. Unlike many native-born Americans who viewed land ownership as a means of wealth, they saw land as necessary for the style of life they preferred. Luebke (1977) points out that the immigrants were either farmers or persons who intended to work in some form of agriculture. Thus, it is not surprising that such groups as the Germans, Norwegians, and Swedes were more likely than the native born to locate in the rural, farming areas of the region. And, as suggested above, these settlers played an important role in transforming the Great American Desert into one of the nation's most agriculturally productive regions. It was the Russian Germans, for example, who introduced the hard red winter wheat (Turkey Red) that revolutionized wheat production in the Southern Great Plains, and it was this group that had a most instrumental part in the success of the sugar beet culture in Nebraska and Colorado.

34–3.2 Population Growth

Except for the depression years from 1930 to 1940, the population of the Great Plains has continued to grow (Table 34–1). However, this increase has been due almost entirely to an excess of births over deaths. The region continued to receive newcomers in substantial numbers until 1910, but even as early as 1890, Kansas and Nebraska were losing large numbers of people through out-migration. Harsh winters, drought, grasshopper plagues, and other disturbing conditions took their toll. Inspection of the net migration figures in Table 34–1 shows that after the initial settling of the area, the Great Plains encountered a 60-year migration drain. This extended period of population loss was felt throughout the region, and except for Wyoming, all of the states experienced one or more decades of actual population decline. Small towns and rural areas were the most severely affected. Cities, in fact, gained in population but generally at a much slower rate than cities in other sections of the nation.

Table 34–1. Great Plains population size, growth, and net migration, 1880–1970 for seven states (U.S. Bureau of the Census, 1970).

Year	Population size in 1000's	Ten-year net growth	Ten-year net migration†
		%	%
1880	1643	209.4	127.6
1890	3496	112.8	52.6
1900	4384	25.4	9.7
1910	6223	42.0	18.9
1920	7120	14.4	−0.2
1930	7793	9.4	−5.7
1940	7548	−3.1	−10.3
1950	7619	0.9	−10.5
1960	8236	8.1	−8.2
1970	8599	4.4	−5.8

† Positive numbers: in-migration exceeds out-migration; negative numbers: out-migration exceeds in-migration.

Between 1970 and 1975, the Great Plains showed a net gain from migration (0.9%) for the first time since 1910. This gain has not occurred evenly throughout the region, however. Kansas and the Dakotas have continued to lose population through emigration, although the net loss appears to be declining. Montana and Wyoming have showed the largest gains, 3.7 and 7.5%, respectively.

Despite losses from out-migration, the Great Plains population increased throughout most of this century because of the growth of urban areas. The rate of growth, however, has been much less than for the USA as a whole. For example, from 1960 to 1970, the USA had an average annual growth of 1.34% compared with 0.45% for the Great Plains. Two kinds of changes are presently occurring that may change this picture in the future.

First, the U.S. growth rate is declining while the Great Plains rate is increasing. Average annual growth for the USA between 1970 and 1975 was 0.96, a 28.4% decline from the rate in the 1960's. From 1970 to 1975 the average annual growth rate for the Great Plains was 0.84, an increase of 86.7% over the average rate from 1960 to 1970. Thus, although the nation continued to grow at a faster rate than the Great Plains during the first half of the last decade, simple projection of trends suggests that the growth rate for the Great Plains region will surpass the national rate in the near future.

Second, a reversal of the rural to urban migration has characterized American society for more than a century. We can see this change by comparing growth in metropolitan and nonmetropolitan counties. A metropolitan county is one that lies within a standard metropolitan statistical area. This is an area made up of one or more entire counties that surround and are socially and economically tied to a city of 50 000 or more population. From 1960 to 1970, metropolitan counties grew by 17.1% compared with a 4.1% increase in nonmetropolitan counties. From 1970 to 1975, the U.S. population residing in nonmetropolitan counties increased by 6.5% while that in metropolitan counties grew by only 4.2%. The Great Plains

Table 34-2. Total population and population density of seven Great Plains states, 1970 (U.S. Bureau of the Census, 1970).

State	Population in 1000's	Population density per square kilometer
Oklahoma	2559	14.3
Kansas	2247	10.6
Nebraska	1483	7.5
Montana	694	1.8
South Dakota	666	3.4
North Dakota	618	3.4
Wyoming	332	1.3

has a relatively small proportion of its population living in metropolitan counties, 39% compared with 73% for the nation's population in 1970. However, during the 1960's, the metropolitan counties in the Great Plains increased at a rate nearly comparable with the rate of growth for the U.S. metropolitan population, an average yearly rate of 1.59%. And, just as the overall metropolitan growth rate declined in the first half of the 1960's, the average rate in the Great Plains declined to 0.96%. The nonmetropolitan counties in the Great Plains actually lost population during the 1960's. The average rate of loss was 0.17%. This pattern reversed beginning in 1970, and the average rate of increase through 1975 was 0.75%. Montana and Wyoming experienced the highest nonmetropolitan growth during this time, an annual average increase of 1.92%. This, of course, represents nearly a 10% increase in nonmetropolitan population in the span of 5 years! It seems possible that the extensive population loss from rural areas and small towns in the Great Plains has come to an end.

34-3.3 Population Density

The Great Plains population has not increased a great deal since 1910, and population density has remained fairly constant. The density was 4.5 persons/km^2 in 1920 and 5.4 persons/km^2 in 1970. This is by far the lowest density of any region in the country. In comparison, density in the USA in 1920 was 11.5 persons/km^2 and increased by 1970 to 22.2 persons/km^2.

Although no state in the Great Plains region has a density comparable with that of the USA as a whole, the density varies considerably among the states within the region. Because population density is one of the most important demographic characteristics of the Great Plains, state-level data on population size and density are provided in Table 34-2.

34-3.4 Other Demographic Characteristics

As might be expected from the large percentage of people living in nonmetropolitan counties, the Great Plains has a higher proportion of its labor force employed in agriculture—10.8% in 1970 compared with 3.7% of the total U.S. civilian labor force. In 1970, the labor force in the Great Plains employed in agriculture varied from a low of 5.3% in Oklahoma to a high of 20.9% in North Dakota. Other Great Plains–USA comparisons show a slightly larger proportion over 65 years of age (12% compared with 10%), a somewhat higher sex ratio (97 males per 100 females vs. 95 per 100), and only 6% of the people in the Great Plains regions are considered nonwhite compared with 13% in the USA. However, these and other small demographic differences evidenced through analysis appear to be of little consequence. Aside from size, density, growth, and rural-urban composition, the Great Plains population is very similar to that of the USA as a whole.

Some evidence, based on research in Nebraska, suggests that ethinic identity may be greater among Great Plains residents than in other regions (Williams et al., 1980). Such identity could cause problems if associated with ethnic discrimination or value conflict, but the extent of the problem is currently unknown.

34–4 QUALITY OF LIFE

34–4.1 Definition and Measurement

People in all societies must engage in various individual- and system-maintaining activities. For example, the physical environment chosen for habitation must not only be able to support human life but must also be protected so that its ecological system continues to function properly after settlement. Necessary goods and services, e.g., food and shelter, must be produced and distributed in sufficient quantity and quality. Because people are guided more by culture than by instincts, they must be educated to perform as members of the society. Mortality requires population replacement, and given the long years of early dependency, the group needs to maintain social organizations (e.g., the family) to protect and nurture children. Activities within the society's division of labor need to be coordinated and some degree of order maintained. Adequate physical and mental health are necessary, and this requires providing appropriate health services. The more successful a society is in performing activities related to meeting basic individual and social objectives, the higher the quality of life for its members. Failure to cope effectively with these social functions gives rise to social problems.

The Great Plains area was settled in a relatively short period of time by people from a variety of different cultural heritages. After initial settlement, the people experienced at least 60 years of extensive out-migration. Although the number of births exceeded deaths, net loss through emigration kept the population size relatively stable. Urban growth has occurred more slowly and to a lesser extent than in other regions of the nation, and population density has remained low. The proportion of people employed in agriculture has declined from more than 50% of the labor force at the turn of the century to only 10.8% in 1970, but this figure is still nearly three times greater than the average for the USA. Furthermore, many nonagricul-

Table 34-3. Satisfaction with climate by region of the USA (Campbell et al., 1971).

	Great Plains	West	Midwest	East	South
	%				
Very good	26.7	51.0	26.4	26.3	44.7
Fairly good	50.5	38.7	43.4	41.2	42.7
Neither good nor bad	16.1	5.2	16.6	15.4	6.0
Not very good	6.6	2.6	10.4	12.0	5.3
Not good at all	0.0	2.5	5.0	5.1	1.6
Number of cases	85	359	535	464	706

tural occupations are linked to agriculture in one way or another, e.g., railroads, meat packing, fertilizer production, flour milling, and manufacture of farm machinery. Have these and related characteristics affected the quality of life of Great Plains residents?

Quality of life can be measured with respect to both objective and subjective components. That is, the quality of various aspects of a population can be assessed through objective indicators, e.g., per capita income or median education, and through the subjective attitudes of people about their lives and surrounding social conditions.

34-4.2 Physical Environment

Many facets of the physical environment affect the quality of life. Of particular importance are the climate, air, water quality, and recreation opportunities.

34-4.2.1 Climate

Blizzards, heat, drought, dust storms, and tornadoes represent the kinds of weather that originally persuaded people to either remain east of the Missouri River or to cross over the Great Plains as quickly as possible toward what they believed to be a more hospitable climate on the other side of the Rockies. Apart from the perception of economic advantages to be gained by moving west, climate also may have contributed to the long years of emigration from the Great Plains after initial settlement. Nevertheless, the settlers proved that the region can support human life, and the tremendous agricultural production of the area is ample testimony to the value of the land. Thus, perhaps the most salient point regarding an association between climate and the quality of life on the Great Plains is how do people feel about it? Do they see the climate as having a negative impact on their lives?

In a national survey, Campbell et al. (1971) asked people, "Another way people judge a place to live is what the weather throughout the year is like—As far as you are concerned, how good is the climate here?" Residents of the West and South are the most likely to say that their climates are very good (Table 34-3). However, responses of Great Plains residents are not different from people living in the East and Midwest, and, in fact, only the West has a smaller percentage stating that their climate is not very good or not good at all. The majority (77.2%) of the Great Plains population think the climate is fairly good or very good.

34–4.2.2 Air and Water Pollution

Primarily because of its smaller, less densely settled population and limited industrialization, the Great Plains region has the lowest levels of air and water pollution in the USA. The U.S. Environmental Protection Agency (USEPA. 1973. Personal communication) evaluations show that only Kansas and Oklahoma have a serious air pollution problem among the Great Plains states. Among the metropolitan counties in the Northern Great Plains, only Omaha has an air pollution problem. The region has water pollution, especially along the eastern frings of the Plains (e.g., the Missouri and Arkansas rivers), but even this problem tends to involve fewer stream kilometers and to be of less duration and intensity than in most other areas in the nation (USEPA, 1972).

34–4.2.3 Recreation

To the extent that life is enhanced by providing a portion of the physical environment for recreational use, the Great Plains region offers a comparatively high quality. For example, in addition to such traditional vacation spots as Yellowstone and the Black Hills, more hectares per 100 000 population have been set aside by local and state governments for parks and recreational use than in any other region. The national average is 17.4 ha, and land areas per 1000 000 population for regions are Great Plains 45.1; East, 39.5; West, 19.8; Midwest, 10.6; and South, 6.3 (U.S. Bureau of the Census, 1971). Among the Plains states only North Dakota and Kansas fall below the national average. In addition, the Great Plains states added recreational areas at a faster rate than other regions throughout the 1970's.

34–4.3 Economy

Compared with the world average of $1650, the per capita gross national product of $7890 for the USA is very high (Population Reference Bureau, 1978). Comprising about 6% of the world's population, the USA uses more than 25% of the earth's nonrenewable resources (Meadows et al., 1972) and over 50% of all basic resources (Allen, 1969). Despite these high levels of production and consumption, the nation is not free of economic difficulties (e.g., inflation, unemployment, and poverty). Regional variations in economic problems, and hence in the economic quality of life, can stem from differences in such factors as availability of natural resources,

Table 34–4. Selected indicators of economic quality by region of the USA.

	Great Plains	West	Midwest	East	South
Cost-adjusted personal income per capita†	$3634	$3973	$3895	$3974	$3427
Percent of families with incomes below poverty level‡	12.2	9.9	8.2	7.7	16.4
Percent unemployed§	5.6	8.5	8.2	10.4	8.6
Percent of workers dissatisfied with main job¶	6.5	11.9	7.4	9.4	6.4

† Source: Liu (1973).
‡ Source: U.S. Bureau of the Census (1970).
§ Source: U.S. Bureau of the Census (1976).
¶ Source: Campbell et al. (1971).

type and amount of production, technological development, and population composition.

According to a national assessment of the quality of life by the Midwest Research Institute (MRI), Kansas City, Missouri, the Great Plains region is approximately equal to the national average in economic productivity and only slightly lower in technological development (Liu, 1973). Exceptions to this are Nebraska's high standing in the rate of economic productivity and North Dakota's relatively low ranking in technology. The Midwest Research Institute's general evaluation tends to be supported by indicators that relate to an individual's chances of economic success (Table 34–4). With the exception of the South, cost-adjusted personal income is somewhat lower in the Great Plains states, and the proportion of families with incomes below the poverty level is larger. On the other hand, the Great Plains has the lowest level of unemployment among the five regions, and the likelihood of being dissatisfied with one's job is lower than in most other regions. More will be said about the economic quality of life in the next section dealing with the farming population in the Great Plains.

There is at least one newly developing source of economic production in the Great Plains states that could have a major impact on the quality of life for the majority, if not all, of the residents. The rapidly dwindling supply of crude petroleum, natural gas, and uranium in the nation and the world makes the deposits of coal and shale oil in the Northern Great Plains an especially important natural resource. In 1967, none of the Great Plains states was among the top 10 coal-producing states. However, in estimated coal reserves, North Dakota, Montana, and Wyoming ranked first, second, and fourth, respectively (Schwartz, 1967). Current estimates indicate that the Great Plains has about 37% of the nation's minable coal reserve by weight and 60% of the surface minable coal (Northern Great Plains Resources Program, 1975). Research suggests that the presence of these mineral resources in the Great Plains region may represent a mixed blessing for the quality of life in the region. As mentioned, population and urban development are increasing in the region at the present time. Rapid development of mining and auxiliary industries could add substantially to that growth. High growth in a short period of time has been found to be associated with a variety of negative consequences, e.g., alienation, mental illness, crime, family disruption, and a drain on social, medical, educational, and recreation services. Some evidence suggests these consequences are now occurring in Montana and Wyoming (Gold, 1974). In addition, the extraction and distribution of coal and other minerals can have unfortunate consequences for the natural ecology of an area, including deterioration of the quality of air, land, and water and destruction of animal habitat (Northern Great Plains Resources Program, 1975). To this listing can be added the possibility that mining and related industries will compete with agriculture for labor, water, and land (Dalsted et al., 1974; White, 1974). The research groups for the Northern Great Plains Resources Program have indicated that some of the available approaches to using the region's mineral reserves will have less harmful effects than others. Thus, the extent to which negative results occur depends on how rapidly development takes place and the types of choices made by the decision makers. Furthermore, the mineral resources in the Great Plains could offer benefits for the quality of life. In addition to providing needed sources of energy for agriculture and industry, development may increase employment opportunities, raise the general standard of living, further halt the emigration of young people from the region, and increase the availability of services. Much depends on the willingness of various interest groups to work together for the common welfare of the region.

34-4.4 Education

Although there are a number of learning sources (e.g., that provided by the family, the church, on the job training, and the mass media), the best indicators of the quality of life in relation to education in an industrial society are those associated with the effectiveness of the public school system.

The survey conducted by MRI led to the construction of several indices, each based on a number of different variables known to be associated with the quality of life in that area. After combining measures for an index, MRI standardizes raw scores by setting the national average at 1.00 and computing the standard deviation. States having an index score more than one standard deviation above the mean are rated excellent, those with more than one standard deviation below the mean are judged substandard, and the rest are rated average. These ratings are arbitrary, but they do provide evidence of rather clear superiority or inferiority on a comparative basis.

The MRI index of education is made up of 10 variables, including public school pupil/teacher ratio, percent of median school years completed among persons 25 years old and over, ratio of higher education enrollment to population 18 to 34 years old, and the ratio of cost-adjusted public school expenditures to personal income per capita. By their rating procedure, 10 states received excellent scores, and half of these are in the Great Plains. Kansas and Oklahoma are defined as average, but Kansas does have a score

Table 34–5. Satisfaction with schools by region of the USA (Campbell et al., 1971).

	Great Plains	West	Midwest	East	South
			%		
Very good	43.3	33.2	42.2	37.4	35.9
Fairly good	46.5	44.2	39.9	37.8	44.4
Neither good nor bad	6.5	5.5	6.6	7.7	6.6
Not very good	0.0	11.2	6.8	10.5	7.3
Not good at all	3.6	6.0	4.5	6.6	5.8
Number of cases	62	273	435	351	560

above the national mean. By combining state scores and computing regional averages, MRI rated the Great Plains region as excellent, the South as substandard, and all other regions as average. Thus, according to the MRI index, the Great Plains has the highest quality of education in the USA.

Table 34–5 shows that MRI's high rating of the quality of education in the Great Plains region is consistent with public opinion. Campbell et al. (1971) asked people, "How do you feel about the quality of the public schools that the children from around here go to?" The Great Plains has the largest proportions saying "very good" and "fairly good" and has the smallest proportion of those who are dissatisfied.

34–4.5 Population and Family

Population size and growth per se need not affect the quality of life. However, problems can arise when a population is either too large or too small in relation to the base of economic and social support, when the age structure creates a high dependency ratio, or when the population is distributed geographically in a manner that makes it difficult to provide access to support services.

The Great Plains states have managed to tread the line between too many and too few people. Births exceeded deaths in every decade of this century, but extensive out-migration kept the overall growth to manageable proportions. The resulting population, generally, has provided a labor force well matched in size and quality to the economic needs of the region. Out-migration of young people has created a proportionally higher number of elderly persons in the Great Plains than in the U.S. average, but demographic trends from the 1970's may have reduced this problem. Potential problems associated with low density in rural areas are examined in section 34–4.6.1.

Even though the family appears to be undergoing some fundamental changes in American society, this institution continues to have primary responsibility for nurturing, protecting, and caring for children and for meeting many of the psychological and social needs of adults. Thus, an analysis of the quality of life in relation to population would not be complete without some assessment of the quality of marriage and the family. A

Table 34–6. Selected indicators of marital and family quality by region of the USA (Campbell et al., 1971).

	Great Plains	West	Midwest	East	South
Percent divorced or separated	3.2	10.7	7.3	5.7	9.9
Mean martial adjustment	22.9	23.5	23.9	23.8	23.5
Mean martial satisfaction	20.9	22.5	21.2	21.7	21.1
Percent never wished to be free from parental responsibility	91.2	79.2	79.1	78.8	84.7
Percent always or nearly always enjoys being a parent	82.0	80.4	77.6	75.2	80.1
Percent obtaining satisfaction from family life	91.3	83.0	86.1	85.5	88.6

thorough examination of family life is far beyond the scope of this chapter, but data that may be used to obtain a fairly reasonable estimate of marital stability and how people feel about their marriages and their role as parents are available.

Table 34–6 shows that the proportion of people who are divorced or separated is lower in the Great Plains states than in other regions of the USA. The next two measures in the table give some idea of the degree of adjustment and satisfaction with marriage. The marital adjustment index is a combination of several items dealing with such subjects as perceived agreement on spending, how well the respondent feels that he or she understands the spouse and is understood by him or her, and the extent of companionship in the marriage. The range of this index is 12 to 53; the lower the score, the better the marital adjustment. Differences among regions are not large, but the average for the Great Plains population indicates better adjustment than in other regions. It is possible for people to be well adjusted to each other in a marriage and still not derive a great deal of satisfaction from the relationship. The marital satisfaction index taps this dimension. The range is from 10 to 70; the lower the score, the higher the satisfaction. Once again, regional differences are not large, but the Great Plains population shows a somewhat higher degree of marital satisfaction.

Data in Table 34–6 also indicates the amount of satisfaction obtained from parenthood and family life. Asking parents whether they ever wished to be free from parental duties and how much they enjoy being a parent is not a direct measure of the quality of child care, but it seems plausible that there is a positive relationship between quality and enjoying being a parent and not wishing to be free from parental respon ibility. Furthermore, responses to these questions provide an idea of whether people see their own needs as being thwarted or fulfilled through parenthood. The final item, how much satisfaction people find from family life, is a good measure of perceptions of how well their families are assisting in meeting general social and psychological needs. All of these measures suggest that the great

Table 34-7. Attitudes toward selected community services by region of the USA (Campbell et al., 1971).

	Great Plains	West	Midwest	East	South
Percent saying residence not very or not at all conveniently located	23.8	10.7	14.1	16.6	19.9
Percent dissatisfied with upkeep of streets and roads	18.5	14.4	18.5	24.8	23.0
Percent dissatisfied with parks and playgrounds	9.4	22.0	24.1	18.5	31.3
Percent dissatisfied with public transportation	20.8	30.0	23.1	20.7	19.6
Percent dissatisfied with garbage collection	7.9	7.7	5.4	6.1	9.9
Percent saying local taxes are very high	23.4	24.7	16.2	24.9	15.6
Percent neutral or dissatisfied with community	8.2	21.9	28.2	28.5	22.6

majority of parents accept and enjoy their responsibilities as parents and are obtaining satisfaction from their families. To the extent that answers to these questions measure the quality of family life, people of the Great Plains enjoy a higher quality, on the average, than do people in other regions.

34-4.6 Government

For the most part, all Americans share in essentially the same form of government. Hence, regional differences in quality stem largely, but certainly not exclusively, from the capacity of local and state governments to provide public services and secure the safety of the population.

34-4.6.1 Public Services

Measures of the quality of public services by region are not available. However, the information presented in Table 34-7 gives an indication of whether people have access to certain services and how they feel about others. The first item in the table shows responses to the question, "First, thinking about the kinds of things you would like to have near where you live—places you go fairly often—how convenient would you say this location is?" The Great Plains has the largest proportion saying they are not conveniently located. This would appear to be a function of the lower population density of the region. Half of the Great Plains sample live in nonmetropolitan counties, and among these respondents only 8.4% say they are conveniently located. A much smaller proportion (13.4%) of the sample in other regions live in nonmetropolitan counties, but 35.0% of these people

Table 34–8. Rates per 0.1 million for selected crimes and attitudes toward police protection and safety of neighborhood by region of the USA.†

	Great Plains	West	Midwest	East	South
Murder and nonnegligency manslaughter	5.6	8.4	7.3	4.8	12.0
Forcible rape	16.7	34.4	22.3	15.6	22.8
Robbery	58.5	153.5	185.2	160.3	142.5
Aggrevated assault burglary —breaking or entering	946.7	1804.6	1284.3	1350.7	1338.3
Larceny-theft	2363.0	3686.7	2881.5	2340.3	2369.5
Motor vehicle theft	243.0	479.6	411.7	594.5	305.5
Percent saying police protection not very good or not good at all	8.1	15.1	13.1	17.6	18.1
Percent saying not safe to walk in neighborhood at night	7.8	23.2	22.6	26.5	29.1

† Sources: Crime rates computed from U.S. Bureau of the Census (1976) and attitudes toward police protection and safety of neighborhood from Campbell et al. (1971).

indicate that they are conveniently located to services. There is no difference in perceived convenience between metropolitan residents in the Great Plains states and those living in metropolitan counties in other regions. It can be seen from the attitudes toward specific services that people in the Great Plains region seem to be no more or even less dissatisfied with their quality. Furthermore, the lack of convenient location does not appear to interfere with community satisfaction. The last item in Table 34–7 shows that a much smaller proportion of people in the Great Plains states than in any other region are either neutral to or dissatisfied with their community "as a place to live in."

34–4.6.2 Safety

Table 34–8 shows the regional rates per 100 000 population for major crimes and for the attitudes of people regarding police protection and the safety of their neighborhoods. The East has the lowest rates of murder, rape, and larceny, with the Great Plains next to lowest. The Great Plains has the lowest rates of robbery, assault, burglary, and motor vehicle theft. Perhaps the most important information shown by these data in terms of regional differences is that the Great Plains is consistently low on all forms of crime. This relatively low crime rate is reflected in the attitudes of people. As indicated, a smaller proportion of people in the Great Plains are dissatisfied with the quality of police protection, and a much smaller proportion than in other regions believe that their neighborhoods are not safe to walk in at night. It could be argued that these attitudes are more related to the low incidence of crime in a sparsely populated area than they are to the quality of a governmental service. Undoubtedly there is some truth to this idea, but two points should be made. First, among people living in metro-

Table 34-9. Attitudes toward personal health by region of the USA (Campbell et al., 1971).

	Great Plains	West	Midwest	East	South
Percent saying they have health problems	20.4	31.1	26.3	25.3	31.5
Percent indicating dissatisfaction with health	6.8	12.4	9.4	7.8	11.8
Among those with health problems, percent saying this prevents them from "doing lots of things"	20.8	39.8	37.7	33.5	42.1

politan counties, Great Plains residents are less likely to be dissatisfied with police protection (5.0 compared with 15.7%) and much less likely to feel that their neighborhoods are not safe (13.1 compared with 31.2%). Second, regardless of the reasons, relatively low incidence of crime in the Great Plains contributes to the quality of life in the region.

34-4.7 Health

The Great Plains has fewer physicians per 100 000 population than any other region: Great Plains, 113; West, 154; Midwest, 137; East, 183; and South, 134 (U.S. Bureau of the Census, 1976). This is at least partly due to the higher proportion of Great Plains people living in rural or nonmetropolitan counties and the reluctance of physicians to locate in rural areas. Thus, for example, 15 counties in Nebraska have no physicians, and all or part of 46 other counties are defined by the U.S. Department of Health, Education, and Welfare as areas of "primary medical shortage" (Terrill, 1978). Living in a sparsely settled area represents an inconvenience for some and perhaps a serious problem for others in obtaining health services (Kraenzel and Macdonald, 1971). On the other hand, the situation is not as grave as it might appear. With modern methods of transportation, patients usually are within reasonable driving distance of a physician. Regardless of the adequacy of health delivery, perhaps in spite of it, the general health of the Great Plains population appears to be at least as good, and probably better, than in other regions.

Life expectancy, especially for women, is the most sensitive and least biased measure of the health of a population. In 1970, life expectancy for women averaged 76.4 years for the Great Plains states compared with 74.8 years for the USA (USDHEW, 1970). Each state in the Great Plains has a life expectancy for women exceeding the national figures and no significant difference in either fertility or mortality between metropolitan and nonmetropolitan counties. Furthermore, the conclusion based on these objective measures is supported by the attitudes of Great Plains residents toward their own health. Compared with other regions, smaller proportions say they have health problems and are dissatisfied with their health (Table 34-9). Among those who do have problems, the Great Plains has the smallest proportion saying this prevents them from "doing lots of things."

34-5 FARMERS

34-5.1 Objectives

This section focuses on the characteristics, quality of life, and problems of the farming population. Farming is more likely to impose structural constraints on life-style than most other occupations. Farming the land normally means living in a low-density area. This reduces the convenience of access to necessary services and to the wide choices among other services typically found in higher-density urban areas. Economic uncertainties related to the exigencies of the weather and fluctuations in the agricultural market are very real to the farmer, who must combine the tasks of running a business that varies considerably in demand during different times of the year with supplying the continuous social and economic needs of the family. Life on the farm has a strong tradition in American society as a unique life-style that affects all aspects of how a person thinks and behaves as a member of the society.

The broad question of how the quality of life of farmers differs from that of other Americans has not been addressed, although some findings are suggestive. We shall discuss how farming in a dryland region may affect the opportunities and outlook of farmers and their families in comparisons with several other groups. The first question is whether the farming population in the Great Plains differs in any important respects from the farming population in other regions of the nation. Second, because some farmers in the Great Plains use irrigation, a comparison is made between the irrigation and those using dryland methods to see if the practice has any important consequences for the quality of life. Finally, the effect of living in a sparsely populated area can be examined by comparing farmers with persons living in urban areas.

34-5.2 Sample

Although several national studies that include variables relevant to assessing quality of life are available, none was found that contained an adequate number of farmers representative of the Great Plains region. The 1978 *Nebraska Annual Social Indicators Survey* (NASIS) (Bureau of Soc. Res., 1978) includes a representative sample of 1940 persons. A sufficient number of farmers and their spouses could be identified to provide a good assessment of the farming population in this Great Plains state.[2] Nebraska farm families may not be representative of the farming population for all

[2] Data were collected by the Bureau of Sociological Research, Univ. of Nebraska, Lincoln. The Bureau of Sociological Research assumes no responsibility for the views or interpretations contained in the present study report. A detailed description of the research design and sampling method can be found in White (1978).

the Great Plains region, but they should not differ in any important respects. Among seven Great Plains states, Nebraska has a higher population density than Montana, the Dakotas, and Wyoming but a lower density than Kansas and Oklahoma (Table 34–2).

34–5.3 Farming in Dryland Areas

The conditions that make the farming occupation unique are likely to be accentuated for farmers in dryland areas. Weather fluctuations can be extreme, and the annual variability can be large. Attempts to reduce some of these weather effects can require large capital investments in equipment, e.g., center-pivot irrigation systems. The nature of the occupation requires most farmers to live in a low-density area, but farming in a dryland area typically means even greater isolation than elsewhere. For example, in 1969 the average size of farms in the USA was 156 ha, whereas the farms of the Great Plains states averaged 577 ha (U.S. Bureau of the Census, 1976). Generally, the lower the population density, the smaller the town or urban place that can be supported by those in the surrounding countryside. Thus, many farmers in the Great Plains are within commuting distance only to very small communities that do not have the large variety of services found in a city. Although the records are far from adequate, they suggest that in the early period of settlement (the era of the sod house) and later during the period of the "dust bowl" and the Great Depression, the climate, barren terrain, social isolation, and difficulty in obtaining necessary goods and services were accompanied by mental illness, suicide, homicide, and a variety of other psychological and social problems. But, to what extent is this pattern true today with vastly improved transportation, mass communication, and other technological and social developments of modern America? Are the rural residents of the Great Plains still affected in negative ways—ways that lower the quality of their lives? One method for approaching an answer to these questions is to compare farm families living on the Great Plains with farm families in other parts of the nation.

The National Opinion Research Center (1977) asked people how satisfied they are with the place they live and how satisfied they are with their lives in general. Responses from the national sample, which includes 270 farmers and their spouses, can be compared with responses given by 350 Nebraska farm families. If conditions in the rural areas of the Great Plains are indeed worse than in other parts of rural America, the Nebraskans would be less satisfied. However, as shown in Table 34–10, even though neither group can be characterized as very dissatisfied, the Nebraska farmers are no more likely to be unhappy than others. In fact, the Nebraska farmers are much less likely to say they are not too happy with their lives.

Table 34–10 also gives a partial profile of the social characteristics of Nebraska farm families, and the comparisons with farmers in the nation provide a possible explanation of why the Nebraskans tend to be happier.

Table 34-10. Selected demographic and socioeconomic characteristics and attitudes of Nebraska and U.S. farm families.

	Nebraska†	USA‡
Percent with some, little, or no satisfaction with where they live	5.7	8.7
Percent saying they are not too happy with their lives	2.1	11.8
Average age	49.4	53.0
Percent nonwhite	1.4	12.6
Percent Catholic	20.0	17.0
Average number of school years completed	11.4	9.3
Percent with total family income less than $10 000	26.3	43.2
Percent married	86.1	83.7
Average household size	3.3	3.2
Average number of children born	2.6	2.8
Percent saying their family gives them less than a great deal of satisfaction	17.9	21.8
Percent saying their health is only fair or poor	21.6	39.8

† Source: Bureau of Sociological Research (1978).

‡ Source: National Opinion Research Center (1977). All findings are based on survey conducted from 1972 through 1977, with the exception of family income, which is based only on the 1977 survey.

The more important differences bearing on this issue appear to be that Nebraska farmers are better educated, less likely to perceive their health as only fair or poor, and much more likely to be earning in excess of $10 000. In most other respects. Nebraska farm families appear to be similar to their counterparts throughout the USA.

34-5.4 Dryland Farmers

Dryland agriculture means the production of crops under natural precipitation in semiarid regions where water is a major limiting factor. The Nebraska data were analyzed to see if there are any important differences between farmers using dryland methods as opposed to those who use various systems of irrigation. To make the comparisons as precise as possible, farm laborers, ranchers, and retired farmers were deleted from the sample. Thus, the present analysis is based on 131 farm families—70 on dryland and 61 who use irrigation.

The two groups were compared on more than 100 characteristics covering a wide range of demographic, ecological, cultural, and social variables. The picture that emerges is one of striking similarity. For example, the two groups are almost identical in the proportion owning their farms, age, educational attainment, marital status, family size, and even political affiliation and recent voting behavior, but the dryland farmer is somewhat more likely to have always lived in the same place, and the farm tends to be larger and further from an urban area.

Income does represent an important difference between persons using these two methods of farming. As can be seen from Table 34-11, the two groups have approximately the same median income, with half of each hav-

Table 34–11. Financial situation and attitudes of dryland and irrigation farmers in Nebraska, 1977 (Bureau of Soc. Res., 1978).

	Dryland	Irrigation
Median family income	$20 079	$20 034
Mean family income	$20 183	$28 884
Percent below federal poverty level	16.9	7.9
Percent not at all satisfied with financial situation	34.0	26.8
Percent saying family income is below average	40.2	26.6
Median percent total income from farm	95.2	72.5

ing earned in excess of $20 000 in 1977. Thus, either method can yield a substantial income. On the other hand, 16.9% of dryland farmers had incomes below the federal poverty level, making them more than twice as likely to be in this category as farmers using irrigation. Furthermore, in a comparison of mean incomes, farmers using irrigation average $8701 more than dryland farmers. These differences are reflected in attitudes: a somewhat larger proportion of dryland farmers is dissatisfied with their financial situation, and a much larger proportion believe their family income is below average.

Several factors appear to at least partly explain the observed income differences between dryland and irrigation farmers. With respect to the mean difference, farmers with very high earnings, e.g., $50 000 or more, are more likely to be using irrigation. It is not known whether this result occurs because farmers with financially very successful operations could more easily afford to invest in irrigation systems or because irrigation helped to produce this high income. Perhaps it is a combination of both. In any case, the earnings of these farmers substantially affect the mean income of the total group. The substantively more important finding is the difference at the lower income range. Although the median income for the two groups is about the same, the dryland farmer is more likely to be at the low end of the income scale. The median percent total income from the farm suggests an explanation. For 50% of the dryland farmers, 95.7% or more of their total income is derived from the farm. The corresponding figure for irrigation farmers is only 72.5%. In other words, for 50% of the irrigation farmers 27.5% or more of their income is obtained from sources other than the farm. This difference could reflect the lack of available employment opportunities in the less densely populated areas in which dryland farmers are more likely to reside. Greater dependency on one income source increases the vulnerability of the dryland farmer to the variations in weather and market and thus accentuates the financial uncertainty of the family.

Another difference between irrigation and dryland farmers observed in the analysis appears to be directly related to differences in farming technology. Respondents were asked if the state government is doing enough, too much, or not enough in several potential problem areas. Among farmers using irrigation, 68.3 compared with 55.5% of dryland farmers said the state needs to do more about energy problems. Irrigation farmers were even more likely to differ from the dryland group in feeling that the state is not doing enough about water resources (58.1 compared with 36.5%). The in-

Table 34–12. Selected attitudes of dryland and irrigation farmers in Nebraska, 1978 (Bureau of Soc. Res., 1978).

	Dryland	Irrigation
Percent with some, little, or no satisfaction with where they live	4.3	7.1
Percent saying they dislike neighborhood somewhat or very much	0.0	4.0
Percent saying a little or very dissatisfied with job	3.9	8.4
Percent saying neighborhood is not very or not conveniently located	10.0	14.2
Percent saying their health is only fair or poor	10.8	17.7
Percent saying they are not too happy with their lives	1.5	0.0

creasing cost of energy and the declining water table in the area raise serious questions about irrigation as a long-run solution to farming problems in an arid area unless appropriate conservation measures are taken. These figures indicate that many irrigation farmers are aware and apprehensive about potential shortages.

As previously mentioned, compared with farmers who use irrigation, dryland farmers tend to live in more sparsely populated areas and are more likely to be poor. It is possible that these conditions are interrelated and that income differences are reflected in satisfaction with the family's financial situation and perceptions of having a below-average income. On the other hand, even though these conditions can affect the quality of a person's life, the differences between the two groups do not appear to be large enough to produce any important divergences in satisfaction with general aspects of living. Table 34–12 shows that as a group, dryland farmers are no more likely than irrigation farmers to be dissatisfied with their place of residence, neighborhood, or occupation. They are no more likely to say their neighborhood is inconveniently located, their health is only fair or poor, or they are not too happy with their lives. In fact, the small, but generally consistent, differences tend to favor the dryland farmer.

34–5.5 Farmers and Urban Dwellers

Farm families in the dryland part of Nebraska do not appear to be disadvantaged compared with farmers elsewhere in the USA, and those using dryland methods tend to fare as well in most aspects of their lives as irrigation farmers. Are there important advantages or disadvantages of farming and living in a low-density area compared with living in a city? To answer this question, working farmers, exclusive of laborers and ranchers, were compared with people living in Nebraska's two complete standard metropolitan statistical areas, Omaha and Lincoln. Initial analysis indicated that the average farm population is older, less educated, and more likely to be married, and that a higher proportion of the respondents are men. Because these characteristics may affect quality of life independent of whether a person lives on a farm or in a city, the findings reported below are statistically

Table 34–13. Quality of life differences between farmers and standard metropolitan statistical area residents in Nebraska, 1978 (Bureau of Soc. Res., 1978).

	Farm	Standard metropolitan statistical area
Percent with some, little, or no satisfaction with the place where they live	6.1	13.3
Percent saying they are not too happy with their lives	0.0	4.8
Percent saying their family gives them some, little, or no satisfaction	0.0	7.8
Percent saying they have thought about a divorce in the last 2 years	4.1	10.7
Percent saying they were victimized by a crime in 1977	13.0	29.0

adjusted for these differences in population distribution (Andrews et al., 1973). By controlling for age, education, marital status, and gender, one can more easily determine if differences found between these groups are caused by occupation and residence.

34–5.5.1 Residential Location

Access to services is one of the greatest potential problems of living in a low-density area. It was shown earlier (Table 34–7) that people living in the Great Plains states are somewhat more likely than others to say their residence is either not very or not at all conveniently located, and those living in nonmetropolitan counties are much more apt to give one of these same responses. Not surprisingly, the Nebraska study shows much the same, with 59.4% of urban dwellers saying their residence is very conveniently located compared with only 18.1% of the farmers. However, although distance from services may present an inconvenience, examination of the data suggests that it does not create a significant social problem. For example, among the various necessary services, close and rapid access to medical care may be the most important. Respondents were asked how long it normally takes them to get to a physician. Farmers reported an average of 21.7 min and urban residents reported an average of 19.0 min. Of course, it is a considerable problem for some. Among the farmers, 4.0% said that it takes 1 hour or more. But for the great majority, whether urban or rural, the time is about the same.

Comparisons on a number of other indicators that theoretically might be associated with place of residence fail to indicate much difference between farmers and city people. The few differences that were found tend to show a possible advantage to living in a rural area. As can be seen from Table 34–13, smaller percentages of farmers than urban residents express dissatisfaction with the place where they live and say that they are not too happy with their lives, that they derive only some, little, or no satisfaction from their families, and that they have contemplated a divorce. The largest

Table 34–14. Economic characteristics and attitudes of farmers and standard metropolitan statistical area residents in Nebraska, 1978 (Bureau of Soc. Res., 1978).

	Farm	Standard metropolitan statistical area
Percent saying they are dissatisfied with the work they do	6.1	11.8
Mean family income	$22 336	$19 200
Percent with family income less than $15 000	55.9	66.1
Percent with family income less than $25 000	67.2	73.8
Percent below federal poverty level	14.4	4.5
Percent not at all satisfied with financial situation	44.2	11.3
Percent saying family income is below average	34.3	18.0
Percent saying financial situation is getting worse	61.3	7.6

difference among the various quality of life measures, apart from convenience of location, is the much smaller proportion of farmers than urban dwellers who say they were victimized by a crime in the past year.

34–5.5.2 Occupation

Table 34–14 shows that a smaller proportion of farmers than metropolitan residents say they are dissatisfied with their occupation, but farm families had a higher average income and a smaller proportion earning less than $15 000 and $25 000 in 1977. On the other hand, compared with their urban counterparts, farmers tend to be considerably less satisfied with their financial situation, more apt to perceive the family's income as below average, and far more likely to think their financial situation is getting worse. Part of the explanation for the apparent inconsistency between income level and satisfaction can be found in the disproportionate representation of farmers at both ends of the income range. For example, whereas a higher proportion of farmers than urban people earned more than $25 000 in 1977, a higher proportion of farm families also had incomes below the federal poverty level. However, this is only part of the answer; the proportion of farmers with very low incomes is not nearly large enough to account for the large difference in financial dissatisfaction between the urban and farm groups. Nevertheless, the high incomes of some farmers and the verv low incomes of others do suggest that farming may be perceived as involving a somewhat larger degree of economic uncertainty or risk than many other occupations. If this statement is true, it is possible that this risk could lead to feeling financially less secure and hence to a higher rate of economic dissatisfaction. Another likely possibility is that given the size of monetary investment, farmers may have higher financial expectations. That is, despite having a higher average income than their urban counterparts, farmers generally receive a much lower return for the amount they have invested. Thus, farmers feel deprived relative to what they expect or think they should earn, and they feel deprived relative to what others with comparable capital investment earn. These explanations may or may not be correct, but it is clear that there is a substantial difference in financial satisfaction between farmers and urban residents and that this difference may have important implications for behavior.

34-6 SUMMARY, CONCLUSIONS, AND SUGGESTIONS FOR FURTHER RESEARCH

34-6.1 Summary

Perhaps because of its nearly unique environmental characteristics, the Great Plains has been the subject of considerable discussion and fascination. From an examination of the literature, both fictional and factual, it seems fair to say that its disagreeable nature has tended to receive the major emphasis. Among novelists, for example, Rolvaag (1927) describes the hardships of Norwegian settlers in South Dakota who had to contend with the overwhelming boredom of the terrain, social isolation, grasshopper plagues, lack of access to services, and blizzards. The final chapter of Book II is entitled, "The Great Plain Drinks the Blood of Christian Men and Is Satisfied." If early settlement was difficult, the "dust bowl" years in the 1930's may have been equally as harsh. No one has described the problems in this era more forcefully than Steinbeck (1942) in *The Grapes of Wrath.* More recently, Michener (1974) discusses the tremendous problem of obtaining water and the conflict over right of access. The Platte, he says, has been the subject of more jokes than any other river on earth, "too thick to drink, too thin to plow." Social scientists have scarcely been more complimentary. Kraenzel and Macdonald (1971) report that in eastern Montana, local employment is low, per capita income is low, and health and social services are few. They suggest that these conditions are associated with feelings of helplessness, anomie, social disorganization, and mental illness. Loftsgard and Voelker's (1978) examination of North Dakota farmers takes a more optimistic view indicating that the socioeconomic gap between Great Plains farmers and urban residents is closing and that farmers have increasingly been able to secure goods and services once considered luxuries. Immediately following this report, Parsons (1978) takes these authors to task by saying that they should have analyzed the relative economic position of farmers and urban residents. His own research indicates that over two thirds of upper Midwest farmers have inadequate incomes and are underemployed. Many of the "luxuries" that farmers have are frequently of ancient vintage. Powers, at a symposium on the cultural heritage of the Great Plains held at the University of Nebraska, Lincoln, in 1977, addressed the subject of quality of life on the Great Plains and concluded that the available data are inadequate for the assessment. He offered the view, however, that objective indicators suggest the Great Plains has a significant problem with access to services but that subjective data show a high level of satisfaction.

This chapter began by asking whether the physical characteristics of the Great Plains and the nature of its settlement have led to any important differences between the region's inhabitants and those living elsewhere in the USA. The primary objective was to determine if the quality of life of Great Plains people, especially the farming population, was superior, equal to, or inferior to other Americans. In the light of previous literature, one might expect to find serious social problems in the Great Plains states. Much of what has been said about life on the Great Plains in past years is true, but past difficulties have been overly emphasized or too broadly generalized. More important, the Great Plains region, along with all other sections of the nation, has experienced extensive social and technological changes. How have these changes affected the lives of Great Plains residents?

Demographic analyses revealed at least three major differences between the Great Plains and other regions. The Great Plains states were settled very rapidly by people from a variety of national and cultural backgrounds. After initial settlement, despite an excess of births over deaths, the population size remained comparatively small and grew at a slow pace because of extensive out-migration. Although preceded by Great Plains Indians, cattlemen, and sheepherders, homesteading farmers were largely responsible for the great influx of population in the settlement period. Because of the nature of ranching and farming and the land, population density was low. Analysis of data from 1970 to 1975 indicates a cessation of out-migration for the first time in 60 years and an increase in the growth rate of both rural and urban areas. Nevertheless, compared with the entire USA, the Great Plains region has a small population in relation to its land area. Furthermore, some evidence indicates that ethnic identity remains stronger in the Great Plains than in the rest of the nation. All three of these population characteristics could be, and in the research literature have been found or hypothesized to be, associated with social problems.

Comparative analysis with other regions showed no evidence of significant social problems associated with living in the Great Plains region. With reference to the physical environment, Great Plains residents were no more likely to express dissatisfaction with the climate than people in other regions, and the region was found to have less of a problem of air and water pollution. More area is set aside in the Great Plains for recreational use than in any other section of the nation. The region does have the next to lowest per capita income and next to highest proportion of its population below the federal poverty level, but the differences among regions are small, and the Great Plains has the lowest unemployment rate and little job dissatisfaction. A variety of indicators suggest that the quality of education is high. Marriage and family adjustment and satisfaction appear to be as strong as, or stronger than, that of other regions. A relatively high proportion of people say they are not conveniently located to services, but satisfaction with services appears to be high, and the region has the lowest proportion dissatisfied with their communities. Compared with other regions, crime rates are low and people are more likely to be positive about the quality of police protection and to feel safe in their neighborhoods. The Great Plains has the smallest number of physicians per capita but the longest life expectancy and the smallest proportion of people thinking they have health problems or indicating dissatisfaction with their health.

If the semiarid characteristics of the region and the comparative isolation arising from the pattern of settlement were associated with problems, it seemed likely that farmers and their families would encounter the most difficulty. They, more than most others, are dependent on the weather and the land, and the nature of the occupation requires living in comparative isolation. Data are not available on the entire farming population in the Great Plains region, but a recent Nebraska study includes a large enough sample of farm families for analysis. Generalizations from this study to the Great Plains region must be made with caution. Nevertheless, the findings provide useful information about whether Great Plains farmers encounter unusual problems in coping with their environment.

A comparison of Nebraska farm families with those in other parts of the nation showed little difference with respect to a number of social and demographic characteristics. However, the differences found suggest that the Great Plains farmer may enjoy a somewhat higher quality of life. The Nebraskans have a higher median level of education, are less likely to be dissatisfied with their health, have a much larger proportion earning $10 000 or more, and are less apt to say they are not too happy with their lives.

A second comparison was made between Nebraska farmers using irrigation and dryland methods. Analysis over a wide range of variables showed the two groups to be similar in most respects. Among the few differences found, those using irrigation are more likely to be concerned about energy and water resources; however, the most important difference appears to be economic. As a whole, both groups have relatively high incomes (a median slightly above $20 000), but dryland farmers have a higher proportion with earnings below the poverty level. Analysis of possible reasons for this led to the discovery that irrigation farmers were obtaining a larger proportion of their income from nonfarm sources. One possible reason is that irrigation farmers tend to live closer to urban places and find it easier to obtain off-farm employment. There may be additional reasons, but the income difference cannot be accounted for by age, education, or number of gainfully employed people in the family. And, it is doubtful that irrigation farmers have more free time from farming responsibilities to carry a second job, although this remains an open question. Perhaps it should be reemphasized that although the proportional difference in numbers below the poverty level is important, the proportions are relatively small (16.9 vs. 7.9%), and the great majority of farmers, both dryland and irrigation, express high levels of satisfaction with nearly all aspects of their lives.

After controlling for distributional differences in age, education, martial status, and gender, the study comparing Nebraska farmers with residents of Lincoln and Omaha allowed comparison of possible problems associated with residential location and farming in a dryland area. It is generally recognized that farmers tend to be less conveniently located relative to services than urban dwellers, and farmers' comments support this observation. However, the data suggest that location does not create a significant problem of access. For example, on the average it takes farmers only 2.7 min longer to reach a physician. On the other hand, location does present a real problem for a minority of farm families, with 4% saying it takes 1 hour or more to travel to a physician's office, and it is possible that location away from services is a more sizable problem in those Great Plains states that have a lower population density than Nebraska. An examination of other problems potentially related to place of residence revealed little difference between farm and urban families. If anything, farmers are more likely to be satisfied with where they live, their families, and their lives in general. Furthermore, they are less likely to be victimized by crime.

Few farmers (6.1%) are dissatisfied with their occupations, and the average farmer has a higher income than a city resident. On the other hand, farmers tend to be overly represented at both ends of the income scale such that a larger proportion earned more than $25 000 in 1977 and a larger proportion had incomes below the poverty level. Economic dissatisfaction is much greater among farmers than urban workers, and this difference is much too great to be accounted for solely by farmers earning low incomes. This element of dissatisfaction may be associated with the economic uncertainty of farming, dissatisfaction with income relative to capital investment, and, perhaps, dissatisfaction with being deprived in relation to the perceived income earned by the average urban resident.

34–6.2 Conclusions

The assessment of the social characteristics of people living on the Great Plains indicates that they are much like other Americans. They enjoy as high, and insome cases a higher, quality of life as people in the other regions of the nation, by a wide variety of indicators, including measurements concerned with the physical environment, family, education, economics, government, and health. This is not to say that Great Plains people do not have problems. This study has provided much evidence to the contrary. Rather, the point here is that the Great Plains region has no more, and in some respects fewer, problems than the rest of the country. Apparently, any problems that once may have arisen from the nature of the region, the capricious environment, and the small, isolated, ethnically diverse population have been largely overcome.

Examination of current trends suggests that the residents of the Great Plains are now facing some important challenges to preserving and improving the quality of life they presently have. Many of these they share with all Americans. However, some tend to be endemic to the region. One of the advantages of living on the Great Plains may stem from its social and demographic characteristics. For example, relatively clean air and water, a low crime rate, good health, and satisfaction with one's community seem to be associated with low population density. The Great Plains population is growing at a faster rate than the national average. Nationally, and no less true of the Great Plains, people have begun to leave the cities, perhaps seeking a higher quality of life. A 1971 survey found that more than one third of the American people would prefer to live in a rural area and an additional third in a small town (Fuguitt and Zuiches, 1975). These are the environments in greatest abundance in the Great Plains. Even the metropolitan counties in the region are relatively small and provide the amenities of close-by rural or small town living. Harris, Louis and Associates in a 1978 national survey for the U.S. Department of Housing and Urban Development reports that 35% of urban residents say they plan to move in the next 2 to 3 years, and over half anticipate moving away from the city, with rural areas being the most popular intended destination. We doubt that all of these people will actually move, but their plans are consistent with population trends. The highest growth rates in the Great Plains have been in the non-metropolitan counties of Wyoming and Montana. These counties are not the ones designated as principal impact areas for coal development. Rather, this growth appears to have resulted from people seeking "the good life." Nevertheless, the Great Plains is experiencing rural and small town growth as part of a national trend, but it also can be expected to grow from development of its energy-producing resources: coal and shale oil. Unless carefully managed, exploitation of these resources can have disruptive consequences: rapid population expension, effects on the physical environment, and deleterious competition with agriculture for water, land, and labor. Water supply may become a particularly pressing problem, especially because competition is arising not only from the mining industry but also from farmers with the increasing use of irrigation.

Offering possible solutions for social problems is beyond the scope of this chapter. We have endeavored to describe some of the more basic social and demographic characteristics of the Great Plains population and advantages and disadvantages of living in the region, especially for the farming population. This approach has allowed us to uncover some of the problems of the area. As more research is conducted, increasingly better means will be available to solve the problems that now exist and to prevent others from occurring.

34-6.3 Suggestions for Further Research

Difficulties encountered in trying to answer the questions posed delineate some needs for future research. First, because most data were collected within politically defined boundaries, it was not possible to examine the entire population living in the Great Plains portion of North America's dryland region. Second, findings reported in this chapter are based on secondary analysis. That is, we had to rely on data that were not obtained specifically to answer our research questions. This problem could have been more severe than it was since much of the data used was intended to assess the quality of life. However, data are not available to assess those problems potentially associated with living in a rural, dryland area and with farming. Third, because farmers compose a very small segment of the American population, representative samples do not include a sufficient number of them for analysis. A Nebraska study was used that included a number of farm families, but the extent to which findings based on these respondents may be generalized and extrapolated to farm families living in other Great Plains states is unknown Furthermore, although the Nebraska study contained enough farmers for examination, even this sample was not large enough for the kind of detailed statistical analysis desired.

Without going into particulars, we believe that a very valuable contribution to knowledge could be made through a study design that would avoid the problems faced by using limited data. Several techniques are available to gain the necessary data base, but the preferred method would use disproportional stratified sampling. This method would ensure that the research included enough people living in dryland areas and enough farm families for proper statistical analysis. For comparative purposes, the study should ask certain questions of all respondents, but it should also obtain in-depth information from farm families and others living in low-density environments. This study should be repeated on a regular basis to record any changes that may be occurring, e.g., from coal development, in-migration, availability of water, and need for new or additional services. Knowing what the problems are and why they exist could result in substantial improvements in social policy, thus helping to solve existing problems and preventing new ones.

34-7 LITERATURE CITED

Allen, D. L. 1969. Population, resources, and the great complexity. p. 450–461. *In* Transactions of the 34th North American Wildlife and Natl. Resources Conference. Wildlife Management Inst., Washington, D.C.

Andrews, F., J. Morgan, and J. Sonquist. 1973. Multiple classification analysis. Survey Res. Center, Inst. of Soc. Res., Univ. of Michigan, Ann Arbor.

Bureau of Sociological Research. 1978. Nebraska annual social indicators survey. Univ. of Nebraska, Lincoln.

Campbell, A., P. E. Converse, and W. L. Rodgers. 1971. The quality of American life. Inst. for Social Res., Ann Arbor, Mich.

Dalsted, N. L., F. L. Leistritz, and T. A. Hertsgaard. 1974. Energy resources development in the northern great Plains. Northern Great Plains Resources Program, Denver.

Fuguitt, G. V., and J. J. Zuiches. 1975. Residential preferences and population distribution. Demography 12:491–504.

Gold, R. L. 1974. A comparative case study of the impact of coal development on the way of life of people in the coals areas of eastern Montana and northeastern Wyoming. Northern Great Plains Resources Program, Denver.

Kraenzel, C. F., and F. H. Macdonald. 1971. Social forces in rural communities of sparsely populated areas. Montana Agric. Exp. Stn. Bull. no. 647, Montana State Univ., Bozeman.

Leubke, F. C. 1977. Ethnic group settlement on the great plains. West. Hist. Q. 8:405–430.

Liu, B. C. 1973. The quality of life in the U.S., 1970. Midwest Res. Inst., Kansas City, Mo.

Loftsgard, L. D., and S. W. Voelker. 1978. Changing rural life in the Great Plains. p. 59–62. *In* R. D. Rodefeld et al. (ed.) Change in rural America. The C. V. Mosby Co., St. Louis.

Meadows, D. H., D. L. Meadows, J. Randers, and W. Behrens. 1972. The limits to growth. Universe Books, New York.

Michener, J. A. 1974. Centennial. Random House, Inc., New York.

National Opinion Research Center. 1977. National data program for the social sciences; cumulative, 1972–77, general social survey. Univ. of Chicago Press, Chicago.

Northern Great Plains Resources Program. 1975. Effects of coal development in the northern great plains—work group report. Northern Great Plains Resources Program, Denver.

Parsons, F. L. 1978. Discussion: Changing rural life in the Great Plains. p. 63. *In* R. D. Rodefeld et al. (ed.) Change in rural America. The C. V. Mosby Co., St. Louis.

Population Reference Bureau. 1978. World population data sheet. Population Reference Bureau, Washington, D.C.

Powers, R. C. 1977. Quality of life on the plains. *In* Nebraska Symp. on the Cultural Heritage of the Plains, Univ. of Nebraska, Lincoln. (Unpublished.)

Rolvaag, O. E. 1927. Giants in the earth. Harper & Row, Inc., New York.

Schwartz, D. 1967. Lignite in the economy. p. 125–128. *In* C. C. Zimmerman and S. Russell (ed.) Symp. on the Great Plains of North America, Fargo, N.D. 1965. The North Dakota Inst. for Regional Studies, Fargo.

Steinbeck, J. 1942. The grapes of wrath. The Viking Press, New York.

Terrill, D. 1978. Counties need larger dose of doctors. *Sunday Journal and Star,* Lincoln, Nebr. 1 Oct. 1978, Section B, p. 1, 3.

U.S. Bureau of the Census. 1900. Census of the U.S. U.S. Bureau of the Census, U.S. Government Printing Office, Washington, D.C.

U.S. Bureau of the Census. 1970. U.S. census of the population, 1970, Vol. 1. U.S. Bureau of the Census, U.S. Government Printing Office, Washington, D.C.

U.S. Bureau of the Census. 1971. Statistical abstract of the U.S. U.S. Bureau of the Census, U.S. Government Printing Office, Washington, D.C.

U.S. Bureau of the Census. 1975. Historical statistics, colonial times to 1970. U.S. Bureau of the Census, U.S. Government Printing Office, Washington, D.C.

U.S. Bureau of the Census. 1976. Statistical abstract of the U.S. U.S. Bureau of the Census, U.S. Government Printing Office, Washington, D.C.

U.S. Bureau of the Census. 1977. Estimate of the population of counties and metropolitan areas: July 1, 1974 and 1975. U.S. Bureau of the Census, U.S. Government Printing Office, Washington, D.C.

U.S. Department of Health, Education, and Welfare. 1970. Vital statistics of the U.S. Natality, Vol. 1 and Mortality, Vol. 2. USDHEW, U.S. Government Printing Office, Washington, D.C.

U.S. Environmental Protection Agency. 1972. The economics of clean water, Vol. 1. USEPA, U.S. Government Printing Office, Washington, D.C.

White, L. K. 1978. Designs, procedures, instruments and forms for the 1978 NASIS. Bureau of Sociological Research, Lincoln, Nebr.

White, W. G. 1974. Socio-economic impacts of coal development in Nebraska. Northern Great Plains Resources Program, Denver.

Williams, J. A., D. R. Johnson, and M. A. Carranza. 1980. Ethnic assimilation and pluralism in Nebraska. p. 210–229. *In* F. C. Leubke (ed.) Ethnicity on the Great Plains. Univ. of Nebraska Press, Lincoln.

SECTION G

GLOSSARY

35

Common and Scientific Names

These glossaries were compiled from common names listed in the individual chapters. Several sources were used to verify the names or to resolve questions about concurrence of scientific and common names. These sources include:

1. R. E. Buchanan and N. E. Gibbons (ed.) 1974. *Bergey's Manual of Determinative Bacteriology.* ed. 8. The Williams & Wilkins Co., Baltimore.
2. R. L. Metcalf's revision of C. L. Metcalf and W. P. Flint. 1962. *Destructive and Useful Insects: Their Habits and Control.* McGraw-Hill Book Co., New York.
3. E. E. Terrell (ed.) 1977. *A Checklist of Names for 3,000 Vascular Plants of Economic Importance.* USDA-ARS Agriculture Handbook no. 505, U.S. Government Printing Office, Washington, D.C.
4. U.S. Department of Agriculture. 1960. *Index of Plant Diseases.* USDA-ARS Agriculture Handbook no. 165, U.S. Government Printing Office, Washington, D.C.
5. J. C. Walker. 1969. *Plant Pathology.* ed. 3. McGraw-Hill Book Co., New York.
6. A. C. Zeven and P. M. Zhukovsky. 1975. *Dictionary of Cultivated Plants and Their Centres of Diversity: Excluding Ornamentals, Forest Trees, and Lower Plants.* Centre for Agricultural Publishing and Documentation, Wageningen, The Netherlands.

The glossaries should not be considered authorities to the use of common and scientific names.

Table 35-1. Names of plants.

Common name	Scientific name
Alfalfa	*Medicago sativa* L.
Yellow-flowered	*Medicago falcata* L.
Amaranth, Palmer	*Amaranthus palmeri* (S.) Wats.
Anoda, spurred	*Anoda cristata* (L.) Schlecht.

(continued on next page)

Table 35-1. Continued.

Common name	Scientific name
Barley	*Hordeum vulgare* L. and other species
Foxtail	*Hordeum jubatum* L.
Barnyardgrass	*Echinochloa crusgalli* (L.) Beauv.
Bean	*Phaseolus vulgaris* L.
Dry	*Phaseolus vulgaris* L.
Faba	*Vicia faba* L.
Mung	*Vigna radiata* (L.) Wilczek
Beet, sugar	*Beta vulgaris* L.
Bermudagrass	*Cynodon dactylon* (L.) Pers.
Bindweed, field	*Convolvulus arvensis* L.
Bluegrass	*Poa* species
Sandberg	*Poa secunda* Presl
Sherman big	*Poa ampla* Merr.
Bluestem	*Andropogon* species
Caucasian	*Bothriochloa caucasica* (Trin.) C.E. Hubb.
King Ranch	*Bothriochloa ischaemum* var. *songarica* (Rupr. ex Fisch. & Mey) Celar. & Harlan
Blueweed, Texas	*Helianthus ciliaris* DC.
Brome	
Downy	*Bromus tectorum* L.
Japanese	*Bromus japonicus* Thunb. ex Murr.
Bromegrass	*Bromus mollis* L.
Smooth	*Bromus inermis* Leyss.
Broomcorn	*Sorghum bicolor* (L.) Moench
Buckwheat	
Tartary	*Fagopyrum tataricum* (L.) Gaertn.
Wild	*Polygonum convolvulus* L.
Buffalobur	*Solanum rostratum* Dunal
Buffalograss	*Buchloe dactyloides* (Nutt.) Engelm.
Bunchgrass	*Stipa viridula* Trin.
Bursage	
Skeletonleaf	*Franseria discolor* Nutt.
Woolyleaf	*Franseria tomentosa* A. Gray
Caragana	*Caragana arborescens* L.
Castorbean	*Ricinus communis* L.
Cheat or cheatgrass	*Bromus secalinus* L. and *B. tectorum* L.
Chess, hairy	*Bromus commutatus* Schrad.
Chickweed, common	*Stellaria media* (L.) Cyrill.
Clover	*Trifolium* species
Rose	*Trifolium hirtum* All.
Subterranean	*Trifolium subterranean* L.
Clubmoss	*Selaginella densa* Rybd.
Cockle	
Cow	*Vaccaria segetalis* (Neck.) Garcke
White	*Lychnis alba* Mill.
Cocklebur	*Xanthium pensylvanicum* Wallr., *X. strumarium* L., and other species
Common	*Xanthium pensylvanicum* Wallr.
Corn	*Zea mays* L.
Cotton	*Gossypium hirsutum* L. and other species

(continued on next page)

Table 35-1. Continued.

Common name	Scientific name
Cowpea	*Vigna unguiculata* (L.) Walp.
Crabgrass and large crabgrass	*Digitaria sanguinalis* (L.) Scop.
Dandelion	*Taraxacum officinale* Weber
Devilsclaw	*Proboscidea louisianica* (Mill.) Thell.
Dodder	*Cuscuta* species
Dogbane, hemp	*Apocynum cannabinum* L.
Falseflax, smallseed	*Camelina microcarpa* Andrz. ex DC.
Fiddleneck	*Amsinckia intermedia* and other species
Flax	*Linum usitatissimum* L.
Flixweed	*Descurainia sophia* (L.) Webb ex Prantl
Foxtail	*Setaria* species
Giant	*Setariva faberi* Herrm.
Green	*Setaria viridis* (L.) Beauv.
Yellow	*Setaria lutescens* (Weigel) F. T. Hubb.
Garlic, wild	*Allium vineale* L.
Goatgrass, jointed	*Aegilops cylindrica* Host
Grama	
Black	*Bouteloua eriopoda*
Blue	*Bouteloua gracilis* (H.B.K.) Lag. ex Steud.
Rothrock	*Bouteloua rothrockii*
Grass	
Barnyard	*Echinochloa crusgalli* (L.) Beauv.
Galleta	*Hilaria jamesii*
Green foxtail	*Setaria viridis* L. Beauv.
Harding	*Phalaris stenoptera* Hack
Needle-and-thread	*Stipa comata*
Rescue	*Bromus willdenowii* Kunth
Western needle	*Stipa sparta* Trin.
Wire	*Cynodon dactylon*
Guar	*Cyamopsis tetragonoloba* (L.) Taub.
Henbit	*Lamium amplexicaule* L.
Horsenettle	*Solanum carolinense* L.
Johnsongrass	*Sorghum halepense* (L.) Pers.
Knapweed, Russian	*Centaurea repens* L.
Knotweed, prostrate	*Poygonum aviculare* L.
Kochia	*Kochia scoparia* (L.) Schrad.
Lambsquarter	*Chenopodium album* L.
Lentil	*Lens culinaris* Medik.
Lettuce, prickly	*Lactuca serriola* L.
Lovegrass, weeping	*Eragrostis curvula* (Schrad.) Nees
Maize	*Zea mays* L.
Mesquite, curly	*Hilaria belangeri*

(continued on next page)

Table 35-1. Continued.

Common name	Scientific name
Milkweed, common	*Asclepias syriaca* L.
Millet	
Foxtail	*Setaria italica* (L.) Beauv.
Hay	*Panicum miliaceum* L.
Pearl	*Pennisetum americanum* (L.) Leeke
Proso	*Panicum miliacem* L.
Morning-glory	*Ipomoea* species
Mustard	*Brassica hirta* Moench, *B. juncea* L., and other species
Blue	*Chorispora tenella* DC.
Tansy	*Descurainia pinnata* (Walt.) Britt.
Treacle	*Erysimum repandum* L.
Tumble	*Sisymbrium altissimum* L.
Wild	*Brassica kaber* (DC.) L.C. Wheeler var.
Needlegrass, green	*Stipa viridula* Trin.
Nightshade, silverleaf	*Solanum elaeagnifolium* Cav.
Nutsedge	*Cyperus* species
Oat	*Avena sativa* L.
Wild	*Avena barbata* and *A. fatua* L.
Panicum, Texas	*Panicum texanum* Buckl.
Pea	*Pisum sativum* L.
Austrian winter	*Pisum* species
Field	*Pisum sativum* L.
Peanut	*Arachis hypogaea* L.
Pepperweed	*Lepidium* species
Pigweed	*Amaranthus retroflexus* L. and other species
Smooth	*Amaranthus hybridus* L.
Pineapple-weed	*Matricaria matricarioides* (Less.) Porter
Potato	*Solanum tuberosum* L.
Puncture-vine	*Tribulus terrestris* L.
Quackgrass	*Agropyron repens* (L.) Beauv.
Ragweed, common	*Ambrosia artemisiifolia* L.
Rape	*Brassica* species
Rapeseed	*Brassica campestris* L. and *B. napus* L.
Rye	*Secale cereale* L.
Ryegrass	*Lolium multiflorum* Lam.
Safflower	*Carthamus tinctorius* L.
Sage, lanceleaf	*Salvia reflexa* Hornem.
Salsify, wild	*Tragopogon porrifolius* L.
Sandbur, field	*Cenchrus incertus* M.A. Curtis
Sedges	*Carex* species
Shattercane	*Sorghum bicolor* (L.) Moench
Shepherdspurse	*Capsella bursa-pastoris* (L.) Medik.

(continued on next page)

Table 35–1. Continued.

Common name	Scientific name
Smartweed	*Polygonum* species
Sorghum	*Sorghum bicolor* (L.) Moench and *S. vulgare* (L.) Pers.
Forage	*Sorghum bicolor* (L.) Moench and other species
Grain	*Sorghum bicolor* (L.) Moench and *S. vulgare* (L.) Pers.
Sweet	*Sorghum bicolor* (L.) Moench
Wild	*Sorghum bicolor* (L.) Moench
Sorrel, red	*Rumex acetosella* L.
Sowthistle, perennial	*Sonchus arvensis* L.
Soybean	*Glycine max* (L.) Merr.
Spurge, leafy	*Euphorbia esula* L.
Starthistle, yellow	*Centaurea solstitialis* L.
Sudangrass	*Sorghum sudanense* (Piper) Stapf and other species
Sunflower	*Helianthus annuus* L. and other species
Sweetclover	*Melilotus alba, M. indica* (L.) All., *M. officinalis* Lam., and other species
Tarweed	*Amsinckia intermedia*
Thistle	
Canada	*Cirsium arvense* (L.) Scop.
Russian	*Salsola kali* L. var. *tenuifolia* Tausch and *S. postifer*
Thornapple, oakleaf	*Datura quercifolia* H.B.K.
Tumblegrass	*Schedonnardus paniculatus* (Nutt.) Trel.
Velvetleaf	*Abutilon theophrasti* Medik.
Vetch, 'Lana' woollypod	*Vicia dasycarpa* Ten.
Watermelon, wild	*Citrullus vulgaris* Schrad. var. *citroides* Bailey
Wheat	*Triticum aestivum* L. em. Thell.
Durum	*Triticum durum* Desf.
Hard red spring	*Triticum aestivum* L. em. Thell.
Hard red winter	*Triticum aestivum* L. em. Thell.
Soft red winter	*Triticum aestivum* L. em. Thell.
Spring	*Triticum aestivum* L. em. Thell.
Winter	*Triticum aestivum* (L.) em. Thell.
Wheatgrass	*Agropyron elongatum* (Host) Beauv.
Bluebunch	*Agyropyron spicatum* (Pursh) Scribn. & Smith
Crested	*Agropyron desertorum* (Fisch. ex Link) Schult.
Fairway crested	*Agyropyron cristatum* (L.) Gaertn.
Intermediate	*Agyropyron intermedium* (Host.) Beauv. var. *intermedium*
Pubescent	*Agyropyron intermedium* var. *trichophorum*
Streambank	*Agropyron riparium* Scribn. & Smith
Slender	*Agropyron trichycaulum* (Link) Malte ex H. F. Lewis
Thickspike	*Agropyron dasystachym* (Hook.) Scribn.
Tall	*Agropyron elongatum* (Host.) Beauv.
Western	*Agropyron smithii* Rydb.
Wildrye	
Altai	*Elymus angustus* Trin.
Russian	*Psathyrostachys juncea* (Fisch.) Nevski
Witchgrass	*Panicum capillare* L.

Table 35-2. Names of diseases.

Common name	Pathogen
Anthracnose	*Colletotrichum graminicolum* (Ces.) G. W. Wils. and *C. trifolii* Bain & Essary
Black stem	
Spring	*Phoma medicaginis* Malbr. & Roum.
Summer	*Cercospora medicaginis* Ell. & Ev.
Blight	
Bacterial	*Xanthomonas campestris* pr *malvacearum* (E. F. Smith) Dye
Bacterial leaf	*Pseudomonas syringae* van Hall
Goss's wilt and	*Corynebacterium nebraskense* Vidaver & Mandel
Leaf	*Helminthosporium turcicum* Pass.
Northern corn leaf	*Helminthosporium turcicum* Pass.
Southern corn leaf	*Helminthosporium maydis* Nisik. & Miy.
Blotch	
Glume	*Septoria* species
Septoria leaf	*Septoria* species
Bunt, common	*Tilletia caries* (DC.) Tul. and *T. foetida* (Wallr.) Liro
Crazy top	*Sclerophthora macrospora* (Sacc.) Thir., Shaw, & Naras.
Dwarf, barley yellow	*Heterodera avenae* Woll.
Eyespot	*Cercosporella herpotrichoides* Fron and *Pseudocercosporella herpotrichoides* (Fron) Dei.
Leaf, rough	*Ascochyta sorghina* Sacc.
Mildew	
Downy	*Peronospora trifoliorum* d By.
Powdery	*Erysiphe graminis* DC. f. sp. *tritici* Marchal.
Sorghum downy	*Peronosclerospora sorghi* (Weston & Uppal) Shaw
Mosaic	
Wheat streak	*Aceria tulipae* Keif. and *Eriophyes tulipae* Keif.
Wheat soilborne	*Polymyxa graminis* Led.
Nematode	
Cereal cyst	*Heterodera avenae* Woll.
Cotton root knot	*Meloidogyne incognita* (Kofoid & White) Chitwood
Dagger	*Xiphinema* species
Needle	*Longidorus* species
Root knot	*Meloidogyne* species
Root lesion	*Pratylenchus* species
Stem	*Ditylenchus dipsaci* (Kühn) Filip.
Sting	*Belonolaimus* species
Rot	
Charcoal	*Macrophomina phaseolina* (Tassi) Goid
Dryland foot	*Fusarium culmorum* W. G. Smith, *F. graminearum*

(continued on next page)

Table 35-2. Continued.

Common name	Pathogen
	Schwabe, *F. roseum* 'Graminearum,' and *F. roseum* Link. emend. Snyd. & Hans. 'Culmorum'
Phymatotrichum root	*Phymatotrichum omnivorum* (Shear) Dug.
Fusarium stalk	*Fusarium moniliforme* Sheld.
Root	*Gaeumannomyces graminis* (Sacc.) Arx & Oliv.
Texas root	*Phymatotrichum omnivorum* (Shear) Dug.
Wheat root	*Helminthosporium sativum* P. K. B.
Rust	*Puccinia purpurea* Cke. and *Uromyces striatus* Schroet.
Common corn	*Puccinia sorghi* Schwabe
Leaf	*Puccinia recondita* Rob. ex Desm. f. sp. *tritici*
Stripe	*Puccinia striiformis* West.
Wheat stem	*Puccinia graminis* Pers. f. sp. *tritici* Eriks. & Henn. and *P. graminis* f. sp. *tritici*
Smut	
Common	*Ustilago maydis* (DC.) Cda.
Loose	*Ustilago tritici* (Pers.) Rostr.
Sorghum head	*Sphacelotheca reiliana* (Kühn) Clint.
Stinking	*Tilletia caries* (DC.) Tul. and *T. foetida* (Wallr.) Liro
Spot	
Alternaria leaf	*Alternaria macrospora* Zimm.
Bacterial	*Pseudomonas syringae* van Hall
Bacterial leaf	*Xanthomonas alfalfae* (Riker, Jones, & Davis) Dows.
Common leaf	*Pseudopeziza medicaginis* (Lib.) Sacc.
Grey leaf	*Cercospora sorghi* Ell. & Ev.
Holcus	*Pseudomonas syringae* van Hall
Lepto leaf	*Leptosphaerulina briosiana* (Poll.) Graham & Luttrell
Physoderma brown	*Physoderma maydis* Miyabe
Stemphylium	*Stemphylium botryosum* Wallr.
Tan	*Pyrenophora trichostoma* (Fr.) Fckl.
Target leaf	*Helminthosporium sorghicola* Lefebvre & Sherwin
Zonate leaf	*Gloeocercospora sorghi* Bain & Edg.
Straw breaker	*Cercosporella herpotrichoides* Fron and *Pseudocercosporella herpotrichoides* (Fron) Dei.
Streak, bacterial	*Xanthomonas holcicola* (Elliott) Starr & Burkh.
Stripe	
Bacterial	*Pseudomonas andropogoni* (E. F. Smith) Stapp
Cephalosporium	*Cephalosporium gramineum* Nis. & Ika.
Sooty	*Ramulispora sorghi* (Ell. & Ev.) Olive & Lefebvre
Take-all	*Gaeumannomyces graminis* var. *tritici* Walker
Wilt	
Bacterial	*Corynebacterium insidiosum* (McCull.) Jens.
Fusarium	*Fusarium oxysporum* Schlecht. f. sp. *medicaginis* (Weimer) Snyd. & Hans.
Goss's, and blight	*Corynebacterium nebraskense* Vidaver & Mandel
Stewart's bacterial	*Erwinia stewartii* (Smith) Dye
Verticillium	*Verticillium albo-atrum* Reinke & Berth. and *V. dahliae* Kleb.

Table 35-3. Names of insects.

Common name	Scientific name
Aphid	*Schizaphis graminum* Rond.
Corn leaf	*Rhopalosiphum maidis* Fitch
Cotton	*Aphis gossypii* Glover
Crown	*Rhopalosiphum subterraneum* Mason
English grain	*Macrosiphum granarium* Kirby
Oat bird-cherry	*Rhopalosiphum padi* L.
Pea	*Macrosiphum pisum* Harris
Spotted alfalfa	*Therioaphis maculata* Buckton
Western wheat	*Brachycolus tritici* Gillette
Yellow sugarcane	*Sipha flava* Forbes
Armyworm	*Pseudaletia unipuncta* Haworth
Beet	*Spodoptera exigua* Hübner
Fall	*Spodoptera frugiperda* Smith
Yellow-striped	*Prodenia ornithogalli* Guér.
Beetle	
Corn seed	*Agonoderus lecontei* Chaud.
Flea	*Chaetocnema* species
June	*Cyclosephala* species, *Dynastes* species, and *Phyllophaga* species
Lady	*Hippodamia convergens* Guér.
May	*Cyclosephala* species, *Dynastes* species, and *Phyllophaga* species
Bollworm	
Cotton	*Heliothis zea* Boddie
Pink	*Pectinophora gossypiella* Saund.
Borer, southwestern corn	*Diatraea grandiosella* Dyar
False chinch	*Nysius* species and *N. ericae* Schilling
Bug	
Damsel	*Nabis ferus* L.
Lygus	*Lygus* species
Plant	*Plagiognathus* species
Stink	Pentatomidae
True chinch	*Blissus leucopterus* Say
Caterpillar, alfalfa	*Colias eurytheme* Boisduval
Chalcid	
Alfalfa seed	*Bruchophagus roddi* Gussakovsky
Clover seed	*Bruchophagus gibbus* Boh.
Corn borer, European	*Ostrinia nubilalis* Hübner
Cutworm	
Army	*Euxoa auxiliaris* Grote
Black	*Agrotis ipsilon* Rott.
Pale western	*Agrotis orthogonia* Morrison
Red-backed	*Euxoa ochrogaster* Guenée
Variegated	*Peridroma margaritosa* Haworth
Western bean	*Loxagrotis albicosta* Smith
Earworm, corn	*Heliothis zea* Boddie

(continued on next page)

Table 35-3. Continued.

Common name	Scientific name
Fleahopper, cotton	*Psallus seriatus* Reuter
Fly	
Hessian	*Phytophaga destructor* Say
Lacewing	*Chrysopa* species
Fruitworm, tomato	*Heliothis zea* Boddie
Grasshopper	*Aeoloplides turnbulli* Thos., *Camnula pellucida* Schudder, *Melanoplus bivittatus* Say, *M. differentialis* Thos., *M. femur-rubrum* DeG., *M. packardii* Scudder, and *M. sanguinipes* Fabr.
Lubber	*Brachystola magna* Girard
Greenbug	*Schizaphis graminum* Rond.
Grub, white	*Cyclosephala* species, *Dynastes* species, and *Phyllophaga* species
Jointworm, wheat	*Harmolita tritici* Fitch
Locust	(See Grasshopper)
Looper, cabbage	*Trichoplusia ni* Hübner
Louse, cotton	*Aphis gossypii* Glover
Maggot	
Seed-corn	*Hylemya cilicura* Rond.
Wheat stem	*Meromyza americana* Fitch
Midge, sorghum	*Contarinia sorghicola* Coq.
Mite	
Banks grass	*Oligonychus pratensis* Banks
Brown wheat	*Petrobia latens* Müller
Wheat curl	*Eriophyes tulipae* Keif.
Winter grain	*Penthaleus major* Dugès
Napid	*Nabis ferus* L.
Rootworm	
Northern corn	*Diabrotica longicornis* Say
Southern corn	*Diabrotica undecimpuctata howardi* Barber
Western corn	*Diabrotica virgifera* LeC.
Sawfly, wheat stem	*Pachynematus* species
Strawworm, wheat	*Harmolita grandis* Riley
Syrphid	*Syrphidus* species
Thrip	*Prosopothrips cognatus* Hood
Grass	*Anaphothrips obscurus* Müller
Onion	*Trips tabaci* Lindeman
Western	*Frankliniella occidentalis* Pergande
Wasp	*Lysiphlebus testaceipes* Cress.

(continued on next page)

Table 35–3. Continued.

Common name	Scientific name
Webworm	*Loxostege* species
Alfalfa	*Loxostege commixtalis* Walker
Beet	*Loxostege sticticalis* Linné
Garden	*Loxostege similalis* Guenée
Sorghum	*Celama sorghiella* Riley
Weevil	
Alfalfa	*Hyper postica* Gyllenhal
Boll	*Anthonomus grondis* Boh.
Wireworm	
False	*Eleodes opaca* Say, *E. suturalis* Say, *E. tricostatus* Say, and Tenebrionidae
True	*Aeolus mellilus* Say, *Agriotes* species, *Conoderus auritus* Herbst, *C. vespertinus* Fabr., *Ctenicera destructor* Brown, and Elateridae

Subject Index